T0396564

Handbook of Research Methods in Health Social Sciences

Pranee Liamputtong
Editor

Handbook of Research Methods in Health Social Sciences

Volume 1

With 192 Figures and 81 Tables

 Springer

Editor
Pranee Liamputtong
School of Science and Health
Western Sydney University
Penrith, NSW, Australia

ISBN 978-981-10-5250-7   ISBN 978-981-10-5251-4 (eBook)
ISBN 978-981-10-5252-1 (print and electronic bundle)
https://doi.org/10.1007/978-981-10-5251-4

Library of Congress Control Number: 2018960888

This Springer imprint is published by the registered company Springer Nature Singapore Pte Ltd.
The registered company address is: 152 Beach Road, #21-01/04 Gateway East, Singapore 189721, Singapore

To my mother:
Yindee Liamputtong
and
To my children:
Zoe Sanipreeya Rice and Emma Inturatana
Rice

Preface

Research is defined by the Australian Research Council as "the creation of new knowledge and/or the use of existing knowledge in a new and creative way so as to generate new concepts, methodologies, inventions and understandings." Research is thus the foundation for knowledge. It produces evidence and informs actions that can provide wider benefit to a society. The knowledge that researchers cultivate from a piece of research can be adopted for social and health programs that can improve the health and well-being of the individuals, their communities, and the societies in which they live. As we have witnessed, in all corners of the globe, research has become an endeavor that most of us in the health and social sciences cannot avoid. This Handbook is conceived to provide the foundation to readers who wish to embark on a research project in order to form knowledge that they need. The Handbook comprises four main parts: Traditional Research Methods in Health and Social Sciences, Innovative Research Methods in Health Social Sciences, Doing Cross-Cultural Research in Health Social Sciences, and Sensitive Research Methodology and Approach. This Handbook attests to the diversity and richness of research methods in the health and social sciences. It will benefit many readers, particularly students and researchers who undertake research in health and social science areas. It is also valuable for the training needs of postgraduate students who wish to undertake research in cross-cultural settings, with special groups of people, as it provides essential knowledge not only on the methods of data collection but also salient issues that they need to know if they wish to succeed in their research endeavors.

Traditionally, there are several research approaches and practices that researchers in the health social sciences have adopted. These include qualitative, quantitative, and mixed-methods approaches. Each approach has its own philosophical foundations, and the ways researchers go about to form knowledge can be different. But all approaches do share the same goal: to acquire knowledge that can benefit the world. This Handbook includes many chapters that dedicate to the traditional ways of conducting research. These chapters provide the "traditional ways of knowing" that many readers will need.

As health and social science researchers, we are now living in a moment that needs our imagination and creativity when we carry out our research. Indeed, we are now living "in the new age" where we will see more and more "new experimental works" being invented by researchers. And in this new age, we have witnessed many innovative and creative forms of research in the health and social sciences. In this Handbook, I also bring together a unique group of health and social science researchers to present their innovative and creative research methods that readers can adopt in their own research. The Handbook introduces many new ways of doing research. It embraces "methodological diversity," and this methodological diversity will bring "new ways of knowing" in the health and social sciences. Chapters in this Handbook will help to open up our ideas about doing research differently from the orthodox research methods that we have been using or have been taught to do.

Despite the increased demands on cross-cultural research, discussions on "culturally sensitive methodologies" are still largely neglected in the literature on research methods. As a result, researchers who are working in cross-cultural settings often confront many challenges with very little information on how to deal with these difficulties. Performing cross-cultural research is exciting, but it is also full of ethical and methodological challenges. This Handbook includes a number of chapters written by researchers who have undertaken their research in cross-cultural settings. They are valuable to many readers who wish to embark on doing a cross-cultural research in the future.

Globally too, we have witnessed many people become vulnerable to health and social issues. It will be difficult, or even impossible, for health and social science researchers to avoid carrying out research regarding vulnerable and marginalized populations within the "moral discourse" of the postmodern world, as it is likely that these population groups will be confronted with more and more problems in their private and public lives as well as in their health and well-being. Similar to undertaking cross-cultural research, the task of conducting research with the "vulnerable" and/or the "marginalized" presents researchers with unique opportunities and yet dilemmas. The Handbook also includes chapters that discuss research that involves sensitive and vulnerable/marginalized people.

This Handbook cannot be born without the help of others. I would like to express my gratitude to many people who helped to make this book possible. I am grateful to all the contributors who worked very hard to get their chapter done timely and comprehensively. I hope that the process of writing your chapter has been a rewarding endeavor to you as well. My sincere appreciation is given to Mokshika Gaur who believes in the value of the volume on research methods in health social sciences and has given me a contract to edit the Handbook. I also thank Tina Shelton, Vasowati Shome, and Ilaria Walker of Springer who helped to bring this book to life.

I dedicate this book to my mother Yindee Liamputtong, who has been a key person in my life. It was my mother who made it possible for me to continue my education amidst poverty. Without her, I would not have been where I am now. I also dedicate the book to my two daughters, Zoe Sanipreeya Rice and Emma Inturatana Rice, who have formed an important part of my personal and professional lives in Australia.

Sydney, Australia   Pranee Liamputtong

Contents

Volume 2

About the Editor

Pranee Liamputtong is a medical anthropologist and Professor of Public Health at the School of Science and Health, Western Sydney University, Australia. Previously, Pranee held a Personal Chair in Public Health at the School of Psychology and Public Health, College of Science, Health and Engineering, La Trobe University, Melbourne, Australia, until January 2016. She has also previously taught in the School of Sociology and Anthropology and worked as a Public Health Research Fellow at the Centre for the Study of Mothers' and Children's Health, La Trobe University. Pranee has a particular interest in issues related to cultural and social influences on childbearing, childrearing, and women's reproductive and sexual health. She works mainly with refugee and migrant women in Melbourne and with women in Asia (mainly in Thailand, Malaysia, and Vietnam). She has published several books and a large number of papers in these areas.

Some of her books in the health and social sciences include *The Journey of Becoming a Mother Among Women in Northern Thailand* (Lexington Books, 2007); *Population Health, Communities and Health Promotion* (with Sansnee Jirojwong, Oxford University Press, 2009); *Infant Feeding Practices: A Cross-Cultural Perspective* (Springer, 2011); *Motherhood and Postnatal Depression: Narratives of Women and Their Partners* (with Carolyn Westall, Springer, 2011); *Health, Illness and Wellbeing: Perspectives and Social Determinants* (with Rebecca Fanany and Glenda Verrinder, Oxford University Press, 2012); *Women, Motherhood and Living with HIV/AIDS: A Cross-Cultural Perspective* (Springer, 2013); *Stigma, Discrimination and Living with HIV/AIDS: A Cross-Cultural*

Perspective (Springer, 2013); *Contemporary Socio-Cultural and Political Perspectives in Thailand* (Springer, 2014; *Children and Young People Living with HIV/AIDS: A Cross-Cultural Perspective* (Springer, 2016); *Public Health: Local and Global Perspectives* (Cambridge University Press, 2016, 2019); and *Social Determinants of Health: Individuals, Communities and Healthcare* (Oxford University Press, 2019).

Pranee is a Qualitative Researcher and has also published several method books. Her most recent method books include *Researching the Vulnerable: A Guide to Sensitive Research Methods* (Sage, 2007); *Performing Qualitative Cross-Cultural Research* (Cambridge University Press, 2010); *Focus Group Methodology: Principle and Practice* (Sage, 2011); *Qualitative Research Methods, 4th Edition* (Oxford University Press, 2013); *Participatory Qualitative Research Methodologies in Health* (with Gina Higginbottom, Sage, 2015); and *Research Methods in Health: Foundations for Evidence-Based Practice*, which is now in its third edition (Oxford University Press, 2017).

About the Contributors

Mauri K. Ahlberg was Professor of Biology and Sustainability Education at the University of Helsinki (2004–2013). On his 69th birthday, January 01, 2014, he had to retire. He has since the 1980s studied research methodology. He is interested in theory building and its continual testing as a core of scientific research. He has developed an integrating approach to research. He is an expert in improved concept mapping.

Farah Ahmad is an Associate Professor at the School of Health Policy and Management, Faculty of Health, York University. Applying equity perspective, she conducts mixed-method research to examine and improve the healthcare system for psychosocial health, vulnerable communities, access to primary care, and integration of eHealth innovations. The key foci in her research are mental health, partner violence, and cancer screening.

Jo Aldridge is a Professor of Social Policy and Criminology at Loughborough University, UK. She specializes in developing and using participatory research methods with "vulnerable" or marginalized people, including children (young carers), people with mental health problems and learning difficulties and women victims/survivors of domestic violence.

Julaine Allan is a Substance Use Researcher and Practitioner with over 30 years' experience in social work. Julaine's research is grounded in a human rights approach to working with stigmatized and marginalized groups. Julaine is also Senior Research Fellow and Deputy CEO at Lyndon, a substance treatment, research, and training organization. She holds conjoint positions at the National Drug and Alcohol Research Centre at UNSW and Charles Sturt University.

Jordan M. Alpert received his Ph.D. in Communication from George Mason University and is currently an Assistant Professor in the Department of Advertising at the University of Florida.

Andrea Montes Alvarado is an M.A. candidate in Sociology at California State University, Northridge. Her main research interests are in education, immigration, demography, and social inequality. Her current research explores the interpersonal relations and lived experiences of drag queens, specifically how drag queens construct and maintain families of choice within the drag community.

Stewart Anderson is a Professor of Biostatistics and Clinical and Translational Medicine at the University of Pittsburgh, Graduate School of Public Health.

His areas of current methodological research interests include (1) general methodology in longitudinal and survival data analysis, (2) modern regression techniques, and (3) methods in the design and analysis of clinical trials. He has over 25 years of experience in cancer clinical trial research in the treatment and prevention of breast cancer and in mental health in mid- to late-life adults.

Mike Armour is a Postdoctoral Research Fellow at NICM, Western Sydney University. Mike's research focus is on implementing experimental designs that can replicate complex clinical interventions that are often seen in the community. Mike has extensive experience in the design and conduct of clinical trials using a mixed-methods approach to help shape trial design, often including clinicians and community groups.

Amit Arora is a Senior Lecturer and National Health and Medical Research Council (NHMRC) Research Fellow in Public Health at Western Sydney University, Australia, where he teaches public health to undergraduate and postgraduate students. His research focuses on developing interventions to improve maternal and child health and oral health. Amit's research expertise includes mixed-methods research, health promotion, and life course approach in health research.

Kathy Arthurson is the Director of Neighborhoods, Housing and Health at Flinders Research Unit, Flinders University of South Australia. Her past experiences as a Senior Policy Analyst in a range of positions including public health, housing, and urban policy are reflected in the nature of her research, which is applied research grounded in broader concepts concerning social inclusion, inequality, and social justice.

Anna Bagnoli is an Associate Researcher in the Sociology Department of the University of Cambridge. Anna has a distinctive interest in methodological innovation and in visual, arts-based, and other creative and participatory approaches. She teaches postgraduates on qualitative analysis and CAQDAS and supervises postgraduates engaged with qualitative research projects. Her current work looks at the identity processes of migrants, with a focus on the internal migrations of Europeans, particularly young Italians.

Amy Baker is an academic in the Occupational Therapy Program at the University of South Australia. Her teaching and research focuses on mental health and suicide prevention, particularly working with people of culturally and linguistically diverse backgrounds and research approaches that are qualitative and participatory in nature.

Ruth Bartlett is an Associate Professor based in the Faculty of Health Sciences, University of Southampton, with a special interest in people with dementia and participatory research methods. Ruth has designed and conducted several funded research projects using innovative qualitative techniques, including diary and visual methods. Ruth has published widely in the health and social sciences and teaches and supervises postgraduate students.

Pat Bazeley has 25 years' experience in exploring, teaching, and writing about the use of software for qualitative and mixed-methods analysis for social and health research. Having previously provided training, consulting, and retreat facilities to researchers through Research Support P/L, she is now focusing on writing and researching in association with an Adjunct Professorial appointment at Western Sydney University.

Linda Liska Belgrave is an Associate Professor of Sociology in the Department of Sociology at the University of Miami. Her scholarly interests include grounded theory, medical sociology, social psychology, and social justice. She has pursued research on the daily lives of African-American caregivers of family with Alzheimer's disease, the conceptualization of successful aging, and academic freedom. She is currently working on research into the social meanings of infectious disease.

Robert F. Belli is a Professor of Psychology at the University of Nebraska–Lincoln. He received his Ph.D. in Experimental Psychology from the University of New Hampshire. Robert's research interests focus on the role of memory in applied settings, and his published work includes research on autobiographical memory, eyewitness memory, and the role of memory processes in survey response. The content of his work in surveys focuses on methods that can improve retrospective reporting accuracy.

Jyoti Belur is a Lecturer in the UCL Department of Security and Crime Science. She worked as a Lecturer before joining the Indian Police Service as a Senior Police Officer. She has undertaken research for the UK Home Office, College of Policing, ESRC, and the Met. Police. Her research interests include countering terrorism, violence against women and children, crime prevention, and police-related topics such as ethics and misconduct, police deviance, use of force, and investigations.

Denise Benatuil obtained her Ph.D. in Psychology from Universidad de Palermo, Argentina. She has a Licenciatura in Psychology from UBA, Argentina. She is the Director of the Psychology Program and Professor of Professional Practices in Psychology at Universidad de Palermo, Argentina. She is member of the Research Center in Social Sciences (CICS) at Universidad de Palermo, Argentina.

Sameer Bhole is the Clinical Director of Sydney Dental Hospital and Sydney Local Health District Oral Health Services and is also attached to the Faculty of Dentistry, the University of Sydney, as the Clinical Associate Professor. He has dedicated his career to address improvement of oral health for the disadvantaged populations with specific focus on health inequities, access barriers, and social determinants of health.

Mia Bierbaum is a Research Assistant at Macquarie University's Centre for Healthcare Resilience and Implementation Science, within the Australian Institute of Health Innovation. Mia had worked on the Healthy Living after Cancer project, providing a coaching service for cancer survivors, and conducted population monitoring research to examine the perceptions of cancer risk factors and behavioral change for patients. Her interests include health systems enhancement, behavioral research, and the prevention of chronic disease.

Simon Bishop is an Associate Professor of Organizational Behavior at Nottingham University Business School and a founding member of the Centre for Health Innovation and Learning. His research focuses around the relationship between public policy, organizational arrangements, and frontline practice in healthcare. His work has been published in a number of leading organization and policy and health sociology journals including *Human Relations, Journal of Public Administration Research and Theory,* and *Social Science & Medicine.*

Karen Block is a Research Fellow in the Jack Brockhoff Child Health and Wellbeing Program in the Centre for Health Equity at the University of Melbourne. Karen's research interests are social inclusion and health inequalities with a focus on children, young people, and families and working in collaborative partnerships with the community.

Chelsea Bond is a Munanjali and South Sea Islander Australian woman and Health Researcher with extensive experience in Indigenous primary healthcare, health promotion, and community development. Chelsea is interested in the emancipatory possibilities of health research for Indigenous peoples by examining the capabilities rather than deficiencies of Indigenous peoples, cultures, and communities.

Virginia Braun is a Professor in the School of Psychology at the University of Auckland, Aotearoa/New Zealand. She is a feminist and Critical Psychologist whose research has explored the intersecting areas of gender, bodies, sex/sexuality, and health/well-being across multiple topics and the possibilities, politics, and ethics of lives lived in neoliberal times.

Wendy Brodribb is an Honorary Associate Professor in the Primary Care Clinical Unit at the University of Queensland in Brisbane, Australia. She has a background as a medical practitioner focusing on women's health. Her Ph.D. investigated the breastfeeding knowledge of health professionals. Wendy's research interests are focused on infant feeding, especially breastfeeding, and postpartum care of mother and infant in the community following hospital discharge.

Mario Brondani is an Associate Professor and Director of Dental Public Health, University of British Columbia. Mario has developed a graduated program in dental public health (as a combined M.P.H. with a Diploma) in 2014 in which he currently directs and teaches. His areas of research include dental public health, access to care, dental geriatrics, psychometric measures, policy development, and dental education.

Jessica Broome has designed and applied qualitative and quantitative research programs for clients including top academic institutions, Fortune 500 companies, and grassroots community groups, since 2000. Jessica holds a Ph.D. in Survey Methodology from the University of Michigan.

Mark Brough is a Social Anthropologist with extensive experience in social research and teaching. Mark specializes in the application of qualitative methodologies to a wide range of health issues in diverse social contexts. He has a particular focus on strength-based approaches to community health.

Katerina Bryant completed a Bachelor of Laws and Bachelor of Arts degree in 2016. She is currently completing an honors degree in the Department of English and Creative Writing at Flinders University. Her nonfiction has been published widely, including in the *Griffith Review, Overland, Southerly,* and *The Lifted Brow.*

Lia Bryant is an Associate Professor in the School of Psychology, Social Work and Social Policy, University of South Australia, and is Director of the Centre for Social Change. She has vast experiences in working with qualitative methods and has published several books and refereed journal articles. She has authored two books and edited books. One of these relates specifically to research methodologies and methods. Lia also teaches innovation in research to social science students.

Anne Bunde-Birouste is Director of Yunus Social Business for Health Hub and Convener of the Health Promotion Program at the Graduate School of Public Health and Community Medicine, University of New South Wales in Sydney Australia. She specializes in innovative health promotion approaches for working with disadvantaged groups, particularly community-based research in sport for development and social change, working with vulnerable populations particularly youth.

Vivien Burr is a Professor of Critical Psychology at the University of Huddersfield. Her publications include *Invitation to Personal Construct Psychology* (2nd edition 2004, with Trevor Butt) and *Social Constructionism* (3rd edition 2015). Her research is predominantly qualitative, and she has a particular interest in innovative qualitative research methods arising from Personal Construct Psychology.

Valentina Buscemi is an Italian Physiotherapist with a strong interest in understanding and managing chronic pain conditions. She completed her M.Sc. in Neurorehabilitation at Brunel University London, UK, and is now a Ph.D. student at the Brain Rehabilitation and Neuroplasticity Unit, Western Sydney University. Her research aims to investigate the role of psychological stress in the development of chronic low back pain.

Rosalind Bye is the Director of Academic Program for Occupational Therapy in the School of Science and Health at Western Sydney University. Rosalind's research interests include adaptation to acquired disability, family caregiving experiences, and palliative care. Rosalind employs qualitative research methods that allow people to tell their story in an in-depth and meaningful way.

Breanne Byiers is a Researcher in the Department of Educational Psychology at the University of Minnesota. Her research interests include the development of measurement strategies to document changes in cognitive, health, and behavioral function among individuals with severe disabilities, assessment and treatment of challenging behavior, and single-subject experimental design methodology.

Fiona Byrne is a Research Officer in the Translational Research and Social Innovation (TReSI) Group, part of the School of Nursing and Midwifery at Western Sydney University. She works in the international support team for the Maternal Early Childhood Sustained Home-visiting (MECSH) program.

Robert Campbell is a Reader in Information Systems and Research Coordinator for the School of Creative Technologies at the University of Bolton. He holds a Ph.D. in this area and is on the board of the UK Academy for Information Systems. He has been professionally recognized as a Fellow of the British Computer Society and as a Senior Fellow of the Higher Education Academy. For many years, he was an Information Technology Specialist in the UK banking sector.

Michael J. Carter is an Associate Professor of Sociology at California State University, Northridge. His main research interests are in social psychology and microsociological theory, specifically the areas of self and identity. His current work examines how identities motivate behavior and emotions in face-to-face versus virtual environments. His research has appeared in a variety of academic journals, including *Social Psychology Quarterly* and *American Sociological Review*.

Rocco Cavaleri is a Ph.D. candidate and Associate Lecturer at Western Sydney University. He has been the recipient of the Australian Physiotherapy Association (APA) Board of Director's Student Prize and the New South Wales APA Award for academic excellence. Rocco has been conducting his research with the Brain Rehabilitation and Neuroplasticity Unit at Western Sydney since 2014. His research interests include chronic illnesses, continuing education, and exploring the nervous system using noninvasive technologies.

Jane Chalmers is a Lecturer in Physiotherapy at Western Sydney University and is also undertaking her Ph.D. through the University of South Australia. Jane's research focus has been on the assessment, management, and pathology of pelvic pain and in particular vulvodynia. She has extensive research experience using a range of methodologies and has recently used the Delphi technique in a novel way to create a new tool for assessing pelvic pain in women by using a patient-as-expert approach.

Nancy Cheak-Zamora is an Assistant Professor in the Department of Health Sciences at the University of Missouri. She conducts innovative research to inform policy-making, advocacy, service delivery, and research for youth with special healthcare needs and adults with chronic medical conditions.

Betty P.M. Chung, a graduate from the University of Sydney, Australia, has worked in the field of palliative and end-of-life care as practicing nurse and researcher. Her focus is particularly in psychosocial care for the dying and their families, family involvement to care delivery, good "quality of death," and well-being at old age. She has methodological expertise in qualitative interpretation and close-to-practice approaches in health sciences research.

Victoria Clarke is an Associate Professor in Sexuality Studies at the University of the West of England, Bristol, UK. She has published three prize-winning books, including most recently *Successful Qualitative Research: A Practical Guide for Beginners* with Virginia Braun (Sage), and over 70 papers in the areas of LGBTQ and feminist psychology and relationships, appearance psychology, human sexuality, and qualitative methods.

Bronwyne Coetzee is a Lecturer in the Department of Psychology at Stellenbosch University. She teaches introductory psychology and statistics to undergraduate students and neuropsychology to postgraduate students. Bronwyne's main interests are in health psychology, specifically in behavioral aspects of pediatric HIV and strengthening caregiver and child relationships to improve adherence to chronic medications.

Ana Lucía Córdova-Cazar holds a Ph.D. in Survey Research and Methodology from the University of Nebraska–Lincoln and is a Professor of Quantitative Methods and Measurement at the School of Social and Human Sciences of Universidad San Francisco de Quito, Ecuador. She has particular interests in the use of calendar and time diary data collection methods and the statistical analysis of complex survey data through multi-level modeling and structural equation modeling.

Jonathan C. Craig is a Professor of Clinical Epidemiology at the Sydney School of Public Health, the University of Sydney. His research aims to improve healthcare and clinical outcomes particularly in the areas of chronic kidney disease (CKD) and more broadly in child health through rigorous analysis of the evidence for commonly used and novel interventions in CKD, identifying gaps/inconsistency in the evidence, conducting methodologically sound clinical trials, and application of the research findings to clinical practice and policy.

Fiona Cram has tribal affiliations with Ngāti Pahauwera (Indigenous Māori, Aotearoa New Zealand) and is the mother of one son. She has over 20 years of Kaupapa Māori (by Māori, for Māori) research and evaluation experience with Māori (Indigenous peoples of Aotearoa New Zealand) tribes, organizations, and communities, as well as with government agencies, district health boards, and philanthropic organizations.

Anne Cusick is Professor Emeritus at Western Sydney University, Australia, Professor and Chair of Occupational Therapy at the University of Sydney, Editor in Chief of the *Australian Occupational Therapy Journal*, and inaugural Fellow of the Occupational Therapy Australia Research Academy. She has widely published in social, medical, health, and rehabilitation sciences.

Angela Dawson is a Public Health Social Scientist with expertise in maternal and reproductive health service delivery to priority populations in Australian and international settings. Angela has also undertaken research into innovative approaches to deliver drug and alcohol services to Aboriginal people. Angela is nationally recognized by the association for women in science, technology, engineering, mathematics, and medicine and is a recipient of the 2016 Sax Institute's Research Action Awards.

Kevin Deane is a Senior Lecturer in International Development at the University of Northampton, UK. His educational background is in development economics, but his research draws on a range of disciplines including political economy, development studies, economics, public health, and epidemiology. His research interests continue to focus on mobility and HIV risk, local value chains, transactional sex, and female economic empowerment in relation to HIV prevention.

Virginia Dickson-Swift was a Senior Lecturer in the La Trobe Rural Health School, La Trobe University, Bendigo, Australia. She has a wealth of experience in teaching research methods to undergraduate and postgraduate students throughout Australia. Her research interests lie in the practical, ethical, and methodological challenges of undertaking qualitative research on sensitive topics, and she has published widely in this area.

Rebecca Dimond is a Medical Sociologist at Cardiff University. Her research interests are patient experiences, clinical work, the classification of genetic syndromes and their consequences, and reproductive technologies. Her work is currently funded by an ESRC Future Research Leaders Award.

Agnes Dodds is an Educator and Evaluator and Associate Professor in the Department of Medical Education, the University of Melbourne. Agnes' current research interests are the developmental experiences of young people. Her current projects include studies of the development of medical students, selection into medical school, and school-related experiences of refugee children.

Suzanne D'Souza is a literacy tutor with the School of Nursing and Midwifery at Western Sydney University. Suzanne is keenly interested in the academic writing process and is actively involved with designing literacy strategies and resources to strengthen students' writing. Suzanne is also a doctoral candidate researching the hybrid writing practices of nursing students and their implications for written communicative competence.

Joan L. Duda is a Professor of Sport and Exercise Psychology in the School of Sport, Exercise and Rehabilitation Sciences at the University of Birmingham, UK. Joan is one of the most cited researchers in her discipline and is internationally known for her expertise on motivational processes and determinants of adherence and optimal functioning within physical and performance-related activities such as sport, exercise, and dance.

Clemence Due is a Postdoctoral Research Fellow in the Southgate Institute for Health, Society and Equity at Flinders University. She is the author of more than 50 peer-reviewed academic articles or book chapters, with a primary focus on research with adults and children with refugee backgrounds. Her research has focused on trauma, child well-being, housing and health, access to primary healthcare, and oral health.

Tinashe Dune is a Clinical Psychologist and Health Sociologist with significant expertise in sexualities and sexual and reproductive health. Her work focuses on the experiences of the marginalized and hidden populations using mixed methods and participatory action research frameworks. She is a recipient of a Freilich Foundation Award (2015) and one of Western's *Women Who Inspire* (2016) for her work on improving diversity and inclusivity in health and education.

Jane Edwards is a Creative Arts Therapy Researcher and Practitioner and is currently Associate Professor for Mental Health in the Faculty of Health at Deakin University. She has conducted research into the uses of the arts in healthcare in a range of areas. She is the Editor in Chief for *The Arts in Psychotherapy* and edited *The Oxford Handbook of Music Therapy* (2016). She is the inaugural President of the International Association for Music & Medicine.

Carolyn Ee is a GP and Research Fellow at NICM, Western Sydney University. Carolyn has significant experience in randomized controlled trials having conducted a large NHMRC-funded trial on acupuncture for menopausal hot flushes for her Ph. D. Carolyn combines clinical practice as a GP and acupuncturist with a broad range of research skills including health services and translational research as well as mixed-methods approaches to evaluating the effectiveness of interventions in the field of women's health.

Ulrike Felt is a Professor in the Department of Science and Technology Studies and Dean of the Faculty of Social Sciences at the University of Vienna. Her research focuses on public engagement and science communication; science, democracy, and governance; changing knowledge cultures and their institutional dimensions; as well as development in qualitative methods. Her work is often comparative between national context and technoscientific fields (especially life sciences, biomedicine, sustainability research, and nanotechnologies).

Monika Ferguson is a Research Associate in the Mental Health and Substance Use Research Group at the University of South Australia. Her current program of research focuses on educational interventions to reduce stigma and improve support for people at risk of suicide, both in the health sector and in the community.

Pota Forrest-Lawrence is a University Lecturer in the School of Social Science and Psychology, Western Sydney University. Pota holds a Ph.D. in Criminology from the University of Sydney Law School. Her research interests include drug law and policy, criminological and social theory, criminal law, and people and risk. She has published in the areas of cyber security, illicit drugs, and media.

Bronwyn Fredericks is an Indigenous Australian woman from Southeast Queensland (Ipswich/Brisbane) and is Professor and Pro Vice-Chancellor (Indigenous Engagement) and BHP Billiton Mitsubishi Alliance Chair in Indigenous Engagement at Central Queensland University, Australia. Bronwyn is a member of the National Indigenous Research and Knowledges Network and the Australian Institute of Aboriginal and Torres Strait Islander Studies.

Caroline Fryer is a Physiotherapist and a Lecturer in the School of Health Sciences at the University of South Australia. Her interest in equity in healthcare for people from culturally diverse backgrounds began as a clinician and translated to a doctoral study investigating the experience of healthcare after stroke for older people with limited English proficiency.

Emma George is a Lecturer in Health and Physical Education at Western Sydney University. Emma teaches across a range of health science subjects with a focus on physical activity, nutrition, health promotion, and evidence-based research methodology. Her research aims to promote lifelong physical activity and improve health outcomes. Emma's research expertise includes men's health, intervention design, implementation and evaluation, mixed-methods research, and community engagement.

Lilian A. Ghandour is an Associate Professor at the Faculty of Health Sciences, American University of Beirut (AUB). She holds a Ph.D. from the Johns Hopkins Bloomberg School of Public Health and a Master of Public Health (M.P.H.) from the American University of Beirut. Her research focuses on youth mental health and substance use, and Dr. Ghandour has been involved in the implementation and analyses of several local and international surveys. She has over 40 publications in high-tier peer-reviewed international journals.

Lisa Gibbs is an Associate Professor and Director of the Jack Brockhoff Child Health and Wellbeing Program at the University of Melbourne. She leads a range of complex community-based public health studies exploring sociocultural and environmental influences on health and well-being.

Claudia Gillberg is a Research Associate with the Swedish National Centre for Lifelong Learning (ENCELL) at Jonkoping University, Sweden. Her research interests include feminist philosophy, social citizenship, lifelong learning, participation, methodology, and ethics. Claudia divides her time between the UK and Sweden, closely following developments in health politics in both countries and worldwide. In her work, she has expressed concerns about the rise in human rights violations of the sick and disabled.

Brenda Gladstone is Associate Director of the Centre for Critical Qualitative Health Research and Assistant Professor at Dalla Lana School of Public Health,

University of Toronto. Brenda's research uniquely examines intergenerational experiences and effects of mental health and illness, focusing on parental mental illnesses. She teaches graduate-level courses on qualitative methodologies and uses innovative visual and participatory research methods to bring young people's voices into debates about their mental/health and social care needs.

Namrata Goyal is a Postdoctoral Fellow at the New School for Social Research. She received her Ph.D. in Psychology from the New School for Social Research and completed her B.A. at York University. Her research interests include cultural influences on reciprocity, gratitude, and social support norms, as well as developmental perspectives on social expectations and motivation.

Adyya Gupta is a Research Scholar in the School of Public Health at the University of Adelaide, Australia. She has an expertise in applying mixed methods. Her interests are in social determinants of health, oral health, and behavioral epidemiology.

Talia Gutman is a Research Officer at the Sydney School of Public Health, the University of Sydney. She has an interest in patient and caregiver involvement in research in chronic kidney disease and predominantly uses qualitative methods to elicit stakeholder perspectives with the goal of informing patient-centered programs and interventions. She has conducted focus groups with nominal group technique both around Australia and overseas as part of global studies.

Elizabeth Halcomb is a Professor of Primary Healthcare Nursing at the University of Wollongong. Elizabeth has taught research methods at an undergraduate and postgraduate level and has been an active supervisor of doctoral candidates. Her research interests include nursing workforce in primary care, chronic disease management, and lifestyle risk factor reduction, as well as mixed-methods research.

Marit Haldar is a Professor of Sociology. Her studies are predominantly on childhood, gender, and families. Her most recent focus is on the vulnerable subjects of the welfare state and inequalities in healthcare. Her perspectives are to understand the culturally and socially acceptable in order to understand what is unique, different, or deviant. She has a special interest in the development of new methodologies and analytical strategies.

Camilla S. Hanson is a Research Officer at the Sydney School of Public Health, University of Sydney. Her research interest is in psychosocial outcomes for patients with chronic disease. She has extensive experience in conducting systematic reviews and qualitative and mixed-methods studies involving pediatric and adult patients, caregivers, and health professionals worldwide.

John D. Hathcoat is an Assistant Professor of Graduate Psychology and Associate Director of University Learning Outcomes Assessment in the Center for Assessment and Research Studies at James Madison University. John has taught graduate-level courses in educational statistics, research methods, measurement theory, and performance assessment. His research focuses on instrument development, validity theory, and measurement issues related to "authentic" assessment practices in higher education.

Nikki Hayfield is a Senior Lecturer in Social Psychology in the Department of Health and Social Sciences at the University of the West of England, Bristol,

UK. Her Ph.D. was an exploration of bisexual women's visual identities and bisexual marginalization. In her research she uses qualitative methodologies to explore LGB and heterosexual sexualities, relationships, and (alternative) families.

Lisa Hodge is a Lecturer in Social Work in the College of Health and Biomedicine, Victoria University, and an Adjunct Research Fellow in the School of Nursing and Midwifery, University of South Australia. Her primary research interests include eating disorders and sexual trauma in particular and mental health more broadly, as well as self-harm, the sociology of emotions, and the use of creative expression in research methodologies.

Anne Hogden is a Research Fellow at Macquarie University's Centre for Healthcare Resilience and Implementation Science, within the Australian Institute of Health Innovation, and a Visiting Fellow at the Centre for Health Stewardship, Australian National University (ANU). Her expertise is in healthcare practice and research. Her research uses qualitative methodology, and she is currently focusing on patient-centered care, patient decision-making, and healthcare communication in Motor Neurone Disease.

Nicholas Hookway is a Lecturer in Sociology in the School of Social Sciences at the University of Tasmania, Australia. Nick's principle research interests are morality and social change, social theory, and online research methods. He has published recently in *Sociology* and *The British Journal of Sociology*, and his book *Everyday Moralities: Doing It Ourselves in an Age of Uncertainty* (Routledge) is forthcoming in 2017. Nick is current Co-convener of the Australian Sociological Association Cultural Sociology group.

Kirsten Howard is a Professor of Health Economics in the Sydney School of Public Health at the University of Sydney. Her research focuses on methodological and applied health economics research predominantly in the areas of the assessment of patient and consumer preferences using discrete choice experiment (DCE) methods as well as in economic evaluation and modeling. She has conducted many discrete choice experiments of patient and consumer preferences in diverse areas.

Martin Howell is a Research Fellow in Health Economics in the Sydney School of Public Health at the University of Sydney. His research focuses on applied health economics predominantly in the areas of assessment of preferences using discrete choice experiment (DCE) methods applied to complex health questions including kidney transplant and the trade-offs to avoid adverse outcomes of immunosuppression, preferences, and priorities for the allocation of deceased donor organs and relative importance of outcomes in nephrology.

Erin Jessee is a Lord Kelvin Adam Smith Research Fellow on Armed Conflict and Trauma in the Department of Modern History at the University of Glasgow. She works primarily in the fields of oral history and genocide studies and has extensive experience conducting fieldwork in conflict-affected settings, including Rwanda, Bosnia-Hercegovina, and Uganda. Her research interests also include the ethical and methodological challenges that surround conducting qualitative fieldwork in highly politicized research settings.

James Rufus John is a Research Assistant and Tutor in Epidemiology at Western Sydney University. He has published quantitative research articles on community perception and attitude toward oral health, utilization of dental health services, and public health perspectives on epidemic diseases and occupational health.

Craig Johnson is from the Ngiyampaa tribe in western NSW. Craig did over 30 years of outdoor and trade-based work; he moved to Dubbo in 1996 and became a trainee health worker in 2010. Craig is now a registered Aboriginal Health Practitioner and qualified diabetes educator and valued member of the Dubbo Diabetes Unit at Dubbo Base Hospital. Craig is dedicated to lifelong learning and improving healthcare for Aboriginal people.

Monica Johnson is from the Ngiyampaa tribe in western NSW. Monica has always been interested in health and caring for people and has worked in everything from childhood immunization to dementia. Monica works for Marathon Health (previously Western NSW Medicare Local) in an audiology screening program in western NSW. Monica has qualifications in nursing and has been an Aboriginal Health Worker for the past 8 years.

Stephanie Jong is a Ph.D. candidate and a sessional academic staff member in the School of Education at Flinders University, South Australia. Her primary research interests are in social media, online culture, and health. Stephanie's current work adopts a sociocultural perspective using netnography to understand how online interactions influence health beliefs and practices.

June Mabel Joseph is a Ph.D. candidate in the University of Queensland. She has a keen interest in researching vulnerable groups of populations – with a keen interest on mothers from refugee background in relation to their experiences of displacement and identity. She is also passionate in the field of breastfeeding, maternal and infant health, sociology, and Christian theology.

Angela Ju is a Research Officer at the Sydney School of Public Health, the University of Sydney. Her research interest is in the development and validation of patient-reported outcome measures. She has experience in conducting primary qualitative studies with patients and health professionals and in systematic review of qualitative health research in the areas of chronic disease including chronic kidney disease and cardiovascular disease.

Yong Moon Jung, with a long engagement in quantitative research, has strong expertise and extensive experience in statistical analysis and modeling. He has employed quantitative research skills for the development and evaluation purposes at either a policy or program level. He has also been teaching quantitative methods and data analysis for the undergraduate and postgraduate courses. Currently, he is working as part of the Quality and Analytics Group of the University of Sydney.

Ashraf Kagee is a Distinguished Professor of Psychology at Stellenbosch University. His main interests are in health psychology, especially HIV and mental health, health behavior theory, and stress and trauma. He has an interest in global mental health and lectures on research methods, cognitive psychotherapy, and psychopathology.

Ida Kaplan is a Clinical Psychologist and Director of Direct Services at the Victorian Foundation for Survivors of Torture. Ida is a specialist on research and practice in the area of trauma, especially for refugees and asylum seekers.

Bernie Kemp was born in Wilcannia and is a member of the Barkindji people. Bernie has spent most of the past 25 years working in Aboriginal health in far western NSW providing health checks and preventing and treating chronic disease. Bernie has qualifications in nursing and diabetes education and is a registered Aboriginal Health Practitioner. Bernie recently relocated to Dubbo to be closer to family and works at the Dubbo Regional Aboriginal Health Service.

Lynn Kemp is a Professor and Director of the Translational Research and Social Innovation (TReSI), School of Nursing and Midwifery, Western Sydney University. Originally trained as a Registered Nurse, Lynn has developed a significant program of community-based children and young people's research that includes world and Australian-first intervention trials. She is now leading an international program of translational research into the implementation of effective interventions at population scale.

Priya Khanna holds a Ph.D. in Science Education and Master's degrees in Education and Zoology. She has been working as a Researcher in medical education for more than 10 years. Presently, she is working as an Associate Lecturer, Assessment at the Sydney Medical School, University of Sydney, New South Wales.

Alexandra King graduated from the University of Tasmania in December 2014 with a Ph.D. in Rural Health. Her doctoral thesis is entitled "Food Security and Insecurity in Older Adults: A Phenomenological Ethnographic Study." Alexandra's research interests include social gerontology, social determinants of health, ethnographic phenomenology, and qualitative research methods.

Kat Kolar is a Ph.D. candidate, Teaching Assistant, and Research Analyst in the Department of Sociology, University of Toronto. She is also completing the Collaborative Program in Addiction Studies at the University of Toronto, a program held in collaboration with the Centre for Addiction and Mental Health, the Canadian Centre on Substance Abuse, and the Ontario Tobacco Research Unit.

Gary L. Kreps received his Ph.D. in Communication from the University of Southern California. He currently serves as a University Distinguished Professor and Director of the Center for Health and Risk Communication at George Mason University.

Tahu Kukutai belongs to the Waikato, Ngāti Maniapoto, and Te Aupouri tribes and is an Associate Professor at the National Institute of Demographic and Economic Analysis, University of Waikato. Tahu specializes in Māori and Indigenous demographic research and has written extensively on issues of Māori and tribal population change, identity, and inequality. She also has an ongoing interest in how governments around the world count and classify populations by ethnic-racial and citizenship criteria.

Ian Lacey comes from a professional sporting background and was contracted to the Brisbane Broncos from 2002 to 2007. Ian has expertise in smoking cessation, completing the University of Sydney Nicotine Addiction and Smoking Cessation

Training Course and the Smoke Check Brief Intervention Training. He has recently completed a Certificate IV in Frontline Management and Diploma in Management.

Michelle LaFrance teaches graduate and undergraduate courses in writing course pedagogy, ethnography, cultural materialist, and qualitative research methodologies. She has published on peer review, preparing students to write across the curriculum, e-portfolios, e-research, writing center and WAC pedagogy, and institutional ethnography. She has written several texts and her upcoming book, *Institutional Ethnography: A Theory of Practice for Writing Studies Researchers*, will be published by Utah State Press in 2018.

Kari Lancaster is a Scientia Fellow at the Centre for Social Research in Health, UNSW, Sydney. Kari is a Qualitative Researcher who uses critical policy study approaches to contribute to contemporary discussions about issues of political and policy significance in the fields of drugs and viral hepatitis. She has examined how policy problems and policy knowledges are constituted and the dynamics of "evidence-based" policy.

Jeanette Lawrence is a Developmental Psychologist and Honorary Associate Professor of the Melbourne School of Psychological Sciences. Jeanette's current research specializes in developmental applications to cultural and refugee studies. With Ida Kaplan, she is developing age- and culture-appropriate computer-assisted interviews to assist refugee and disadvantaged children to express their thoughts, feelings, and activities.

Eric Leake is an Assistant Professor of English at Texas State University. His areas of research include the intersection of rhetoric and psychology as well as civic literacies and writing pedagogy.

Raymond M. Lee is Emeritus Professor of Social Research Methods at Royal Holloway University of London. He has written extensively about a range of methodological topics including the problems and issues involved in research on "sensitive" topics, research in physically dangerous environments, and the use of unobtrusive measures. He also provides support and advice to the in-house researchers at Missing People, a UK charity that works with young runaways, missing and unidentified people, and their families.

Caroline Lenette is a Lecturer of Social Research and Policy at the University of New South Wales, Australia. Caroline's research focuses on refugee and asylum seeker mental health and well-being, forced migration and resettlement, and arts-based research in health, particularly visual ethnography and community music.

Doris Leung graduated from the University of Toronto and worked there as a Mental Health Nurse and Researcher focused on palliative end-of-life care. Since moving to Hong Kong, she works at the School of Nursing, the Hong Kong Polytechnic University. Her expertise is in interpretive and post-positivistic qualitative research approaches.

Nicole Loehr is a Ph.D. candidate in the School of Social and Policy Studies at Flinders University. Her doctoral work examines the integration of the not-for-profit, governmental, and commercial sectors in meeting the long-term housing needs of resettled refugees and asylum seekers in the private rental market. She has worked as a Child and Family Therapist and in community development and youth work roles.

Janet Long is a Health Services Researcher at the Australian Institute of Health Innovation, Macquarie University, Australia. She has a background in nursing and biological science. She has published a number of studies using social network analysis of various health and medical research settings to demonstrate silos, key players, brokerage and leadership, and strategic network building. Her other research interests are in behavior change, complexity, and implementation science.

Colin MacDougall is concerned with equity, ecology, and healthy public policy. He is currently Professor of Public Health at Flinders University and Executive member of the Southgate Institute of Health, Society and Equity. He is a Principal Fellow (Honorary) at the Jack Brockhoff Child Health and Wellbeing Program at the University of Melbourne. He studies how children experience/act on their worlds.

Catrina A. MacKenzie is a Mixed-Methods Researcher investigating how incentives, socioeconomic conditions, and household well-being influence conservation attitudes and behaviors, with a particular focus on tourism revenue sharing, loss compensation, and resource extraction. For the last 8 years, she has worked in Uganda studying the spatial distribution of perceived and realized benefits and losses accrued by communities as a result of the existence of protected areas.

Freya MacMillan is a Lecturer in Health Science at Western Sydney University, where she teaches in health promotion and interprofessional health science. Her research focuses on the development and evaluation of lifestyle interventions for the prevention and management of diabetes in those most at risk. Particularly relevant to this chapter, she has expertise in working with community to develop appropriate and appealing community-based interventions.

Wilson Majee is an Assistant Professor in the Master of Public Health program and Health Sciences Department at the University of Missouri. He is a Community Development Practitioner and Researcher whose primary research focuses on community leadership development in the context of engaging and empowering local residents in resource-limited communities in improving their health and well-being.

Karine E. Manera is a Research Officer at the Sydney School of Public Health, the University of Sydney. She uses qualitative and quantitative research methods to generate evidence for improving shared decision-making in the area of chronic kidney disease. She has experience in conducting focus groups with nominal group technique and has applied this approach in global and multi-language studies.

Narendar Manohar is a National Health and Medical Research Council Postgraduate Research candidate and a Research Assistant at Western Sydney University, Australia. He teaches public health and evidence-based practice to health science students and has published several quantitative research studies on several areas of public health. His interests include evidence-based practice, quantitative research, and health promotion.

Oana Marcu is a Researcher at the Faculty of Political and Social Sciences of the Catholic University in Milan. Her interests include qualitative methods (the ethnographic method, visual and participatory action research). She has extensively worked with Roma in Europe on migration, transnationalism, intersectionality, and health. She teaches social research methodology and sociology of migration.

Anastasia Marinopoulou is teaching political philosophy, political theory, and epistemology at the Hellenic Open University and is the coeditor of the international edition *Philosophical Inquiry*. Her publications include the recent monograph *Critical Theory and Epistemology: The Politics of Modern Thought and Science* (Manchester University Press, 2017). Her latest research award was the Research Fellowship at the University of Texas at Austin (2017).

Kate McBride is a Lecturer in Population Health in the School of Medicine, Western Sydney University. Kate teaches population health, basic and intermediate epidemiology, and evidence-based medicine to undergraduate and postgraduate students and has also taught at Sydney University. Kate's research expertise is in epidemiology, public health, and the use of mixed methods to improve health at a population level through the prevention of and reduction of chronic and non-communicable disease prevalence.

Angela McGrane is Director of Review and Approvals at Newcastle Business School, Northumbria University. She is currently investigating the effect of placement experience on student perceptions of themselves in a work role through a longitudinal study following participants from first year to graduation.

Jane McKeown is a Mental Health Nurse specializing in dementia research, practice, and education, working for the University of Sheffield and Sheffield Health and Social Care NHS Foundation Trust. Jane has a special interest in developing and implementing approaches that enable people with dementia to have their views and experiences heard. She also has an interest in the ethical aspects of research with people with dementia.

Karen McPhail-Bell is a Qualitative Researcher whose interest lies in the operation of power in relation to people's health. Karen facilitates strength-based and reciprocal processes in support of community-controlled and Aboriginal and Torres Strait Islander-led agendas. She has significant policy and program experience across academic, government and non-government roles in health equity, health promotion, and international development.

Joan Miller is Professor of Psychology at the New School for Social Research and Director of Undergraduate Studies in Psychology at Lang College. Her research interests center on culture and basic psychological theory, with a focus on interpersonal motivation, theory of mind, social support, moral development, as well as family and friend relationships.

Rodrigo Mariño is a Public Health Dentist and a Principal Research Fellow at the Oral Health Cooperative Research Centre (OH-CRC), the University of Melbourne. Rodrigo has an excellent publication profile and research expertise in social epidemiology, dental workforce issues, public health, migrant health, information and communication technology, gerontology, and population oral health. Rodrigo has also been a consultant to the Pan American Health Organization/World Health Organization in Washington, DC.

Anna Maurer-Batjer is a Master of Social Work and Master of Public Health student at the University of Missouri. She has a special interest in maternal and child health, particularly in terms of neurocognitive development. Through her graduate research assistantship, she has explored the use of photovoice and photo-stories among youth with Autism Spectrum Disorder.

Cara Meixner is an Associate Professor of Psychology at James Madison University, where she also directs the Center for Faculty Innovation. A scholar in brain injury advocacy, Cara has published mixed-methods studies and contributed to methodological research on this genre of inquiry. Also, Cara teaches a doctoral-level course on mixed-methods research at JMU.

Zelalem Mengesha has Bachelor's and Master's qualifications in Public Health with a special interest in sexual and reproductive health. He is passionate about researching the sexual and reproductive health of culturally and linguistically diverse communities in Australia using qualitative and mixed-methods approaches.

Dafna Merom is a Professor in the School of Science and Health, Western Sydney University, Australia. She is an expert in the area of physical activity measurement, epidemiology, and promotion. She has recently led several clinical trials comparing the incidence of falls, heart health, and executive functions of older adults participating in complex motor skills such as dancing and swimming versus simple functional physical activity such as walking.

Magen Mhaka-Mutepfa is a Senior Lecturer in the Department of Psychology, Faculty of Social Sciences, the University of Botswana. Previously, she was a Teaching Assistant in the School of Public Health at the University of Sydney (NSW) and was a recipient of Australia International Postgraduate Research Scholarship. Her main research interests are on health and well-being, HIV and AIDS, and ageing.

Christine Milligan is a Professor of Health and Social Geography and Director of the Centre for Ageing Research at Lancaster University, UK. With a keen interest in innovative qualitative techniques, Christine is an active researcher who has led on both national and international research projects. She has published over a hundred books, journal articles, and book chapters – including a recently published book (2015) with Ruth Bartlett on *What Is Diary Method?*

Naomi Moller is a Chartered Psychologist and Lecturer in Psychology at the Open University. Trained as a Counseling Psychologist, she recently coedited with Andreas Vossler *The Counselling and Psychotherapy Research Handbook* (Sage). Her research interests include perceptions and understandings of counselors and counseling, relationship, and family research including infidelity.

Marta Moskal is a Sociologist and Human Geographer based at the University of Glasgow, UK. Her research lies within the interdisciplinary area of migration and mobility studies, with a particular interest in family, children, and young people's experiences and in using visual research methods.

Elias Mpofu is a Rehabilitation Counseling Professional with a primary research focus in community health services intervention design, implementation, and evaluation applying mixed-methods approaches. His specific qualitative inquiry orientation is interpretive phenomenological analysis to understand meanings around and actions toward health-related quality of community living with chronic illness or disability.

Sally Nathan is a Senior Lecturer at the School of Public Health and Community Medicine, UNSW, Sydney. She has undertaken research into consumer and community participation in health as well as research approaches which engage and partner directly with vulnerable and marginalized communities and the organizations that represent and advocate for them, including a focus on adolescent drug and alcohol treatment and Aboriginal health and well-being.

Alison Nelson is an occupational therapist with extensive research, teaching, and practice experience working alongside urban Aboriginal and Torres Strait Islander people. Alison has a particular interest in developing practical strategies which enable non-Indigenous students, researchers, and practitioners to understand effective ways of working alongside Aboriginal and Torres Strait Islander Australians.

Ingrid A. Nelson is an Assistant Professor of Sociology at Bowdoin College, in Brunswick, Maine, USA. Her research examines the ways that families, schools, communities, and community-based organizations support educational attainment among marginalized youth. Her current work examines the educational experiences of rural college graduates, through the lens of social capital theory.

Christy E. Newman is an Associate Professor and Social Researcher of health, sexuality, and relationships at the Centre for Social Research in Health, UNSW, Sydney. Her research investigates social aspects of sexual health, infectious disease, and chronic illness across diverse contexts and communities. She has a particular passion for promoting sociological and qualitative approaches to understanding these often culturally and politically sensitive areas of health and social policy.

Mark C. Nicholas is Executive Director of Institutional Assessment at Framingham State University. He has published qualitative research on program review, classroom applications of critical thinking, and its assessment.

Kayi Ntinda is a Lecturer of Educational Counseling and Mixed-Methods Inquiry Approaches at the University of Swaziland. Her research interests are in assessment, learner support services service learning, and social justice research among vulnerable populations. She has authored and coauthored several research articles and book chapters that address assessment, inclusive education, and counseling practices in the diverse contexts.

Chijioke Obasi is a Principal Lecturer in Equality and Diversity at the University of Wolverhampton, England. She is a qualified British Sign Language/English Interpreter of Nigerian Origin and was brought up in a bilingual and bicultural household.

Felix Ogbo is a Lecturer in the School of Medicine, Western Sydney University. Felix studied at the University of Benin (M.B.B.S.), Benue State University (M.H.M.), and Western Sydney University – M.P.H. (Hons.) and Ph.D. He is an active collaborator in the landmark Global Burden of Disease Study – the world's largest scientific effort to quantify the magnitude of health loss from all major diseases, injuries, and risk factors by year-age-sex-geography and cause, which are essential to health surveillance and policy decision-making.

Andrew Page is a Professor of Epidemiology in the School of Medicine, Western Sydney University. He has been teaching basic and intermediate epidemiology and population health courses to health sciences students for 10 years and has published over 150 research articles and reports across a diverse range of population health topics. He has been a Research Associate at the University of Bristol and has also worked at the University of Queensland and University of Sydney in Australia.

Bonnie Pang is a Lecturer at the Western Sydney University and a school-based member of the Institute for Culture and Society. As a Sociocultural Researcher, she specializes in ethnographic methods, social theories, and youth health and physical activity. She has over 10 years of research experiences in exploring Chinese youth communities in school, familial, and neighborhood environments in Australia and Hong Kong.

Gemma Pearce is a Chartered Psychologist currently working in the Health Behaviour and Interventions Research (BIR) group at Coventry University, UK. Her research focuses on women's health, public health, and self-management support of people with long-term conditions. She specializes in systematic reviews and pragmatic research methods (including qualitative, quantitative, and mixed-methods research). She developed a synchronous method of online interviewing when conducting her Ph.D.

Karen Peres is an Associate Professor of Population Oral Health and Director of the Dental Practice Education Research Unit (DPERU) at the University of Adelaide. Her main research interests are in the epidemiology of oral diseases, particularly in social inequality in oral health, mother and child oral health-related issues, and life course and with over 100 publications in peer-reviewed journals.

Janette Perz is a Professor of Health Psychology and Director of the Translational Health Research Institute, Western Sydney University. She researches in the field of reproductive and sexual health with a particular focus on gendered experiences, subjectivity, and identity. She has demonstrated expertise in research design and analysis and mixed-methods research.

Rona Pillay (nee' Tranberg) is a Lecturer and Academic Course Advisor BN Undergraduate. She has unit coordinated research units and teaches primarily in research. Her research interests are in health communication, teams and teamwork, decision-making, and patient safety based on ethnomethodology and conversational analysis. Her additional interests lie in communication of cancer diagnosis in Aboriginal people and treatment decision-making choices.

Jayne Pitard is a Researcher in education at Victoria University, Melbourne, Australia. She has completed a thesis focused on her teaching of a group of students from Timor-Leste. She has held various teaching and research positions within Victoria University for the last 27 years.

Jennifer Prattley is a Research Associate at the Cathie Marsh Institute for Social Research, University of Manchester, UK. She works across a variety of ageing and life course studies, with research questions relating to women's retirement, frailty, and social exclusion. Jennifer has expertise in the analysis of longitudinal data and an interest in applying and appraising new and innovative quantitative methods. She also maintains an interest in statistics and mathematics education and the teaching of advanced methods to nonspecialists.

Nicholas Procter is a Professor and Chair of Mental Health Nursing and Convener of the Mental Health and Substance Use Research Group at the University of

South Australia. Nicholas has over 30 years' experience as a mental health clinician and academic. He works with various organizations within the mental health sector providing education, consultation, and research services in the suicide prevention and trauma-informed practice areas.

Marsha B. Quinlan is a Medical Anthropologist focusing on the intersection of health with ethnobiology and an Associate Professor in the Department of Anthropology and affiliate of the School for Global Animal Health at Washington State University, Pullman, WA, USA. Her research concentrates on ethnomedical concepts including ethnophysiology, ethnobotany, and ethnozoology, along with health behavior in families, and psychological anthropology.

Frances Rapport is a Professor of Health Implementation Science at Macquarie University's Centre for Healthcare Resilience and Implementation Science within the Australian Institute of Health Innovation. She is a social scientist with a background in the arts. Frances has won grants within Australia and the United Kingdom, including the Welsh Assembly Government, the National Institute for Health Research in England, and Cochlear Ltd. to examine the role of qualitative and mixed methods in medical and health services research.

Liz Rix is a Researcher and Academic working with Gnibi College of Indigenous Australian Peoples at Southern Cross University, Lismore. She is also a practicing Registered Nurse, with clinical expertise in renal and aged care nursing. Liz currently teaches Indigenous health to undergraduate and postgraduate health and social work professionals. Her research interests include Indigenous health and chronic disease, reflexive practice, and improving the cultural competence of non-Indigenous clinicians.

Norma R. A. Romm is a Research Professor in the Department of Adult Education and Youth Development at the University of South Africa. She is author of *The Methodologies of Positivism and Marxism: A Sociological Debate* (1991); *Accountability in Social Research: Issues and Debates* (2001); *New Racism* (2010); *People's Education in Theoretical Perspective: Towards the Development* (with V. McKay 1992); *Diversity Management: Triple Loop Learning* (with R. Flood 1996); and *Assessment of the Impact of HIV and AIDS in the Informal Economy of Zambia* (with V. McKay 2006).

Rizwana Roomaney is a Research Psychologist, Registered Counselor, and Lecturer in the Psychology Department at Stellenbosch University. She teaches research methods and quantitative data analysis to undergraduate and postgraduate students. Rizwana's main interests are health psychology and research methodology. She is particularly interested in women's health and reproductive health.

Lauren Rosewarne is a Senior Lecturer at the University of Melbourne, Australia, specializing in gender, sexuality, media, and popular culture. Her most recent books include *American Taboo: The Forbidden Words, Unspoken Rules, and Secret Morality of Popular Culture* (2013); *Masturbation in Pop Culture: Screen, Society, Self* (2014); *Cyberbullies, Cyberactivists, Cyberpredators: Film, TV, and Internet Stereotypes* (2016); and *Intimacy on the Internet: Media Representations of Online Connections* (2016).

Simone Schumann is a doctoral candidate and former Lecturer and Researcher in the Department of Science and Technology Studies, University of Vienna. Her main research interest is in public engagement with emerging technologies, especially on questions of collective sense-making and the construction of expertise/power relations within group settings. In her dissertation project, she focuses on the case of nano-food to understand how citizens encounter and negotiate an emerging food technology in Austria.

Claudia G. Schwarz is a Postdoc Researcher at the University of Vienna and Lecturer and former Researcher in the Department of Science and Technology Studies. Her current research interests lie mainly in the public understanding of and engagement with science and technology. In her dissertation project, she examined how laypeople use analogies as discursive devices when talking about nanotechnology.

Lizzie Seal is a Senior Lecturer in Criminology at the University of Sussex. Her research interests are in the areas of feminist, historical, and cultural criminology. She is the author of *Women, Murder and Femininity: Gender Representations of Women Who Kill* (Palgrave, 2010); *Transgressive Imaginations: Crime, Deviance and Culture* (with Maggie O'Neill, Palgrave, 2012); and *Capital Punishment in the Twentieth-Century Britain: Audience, Justice, Memory* (Routledge, 2014).

Kapriskie Seide is a doctoral student in the Department of Sociology at the University of Miami with concentrations in Medical Sociology and Race, Ethnicity, and Immigration. Her current research explores the impacts of infectious disease epidemics on everyday life and human health and their overlap with lay epidemiology, social justice, and grounded theory.

Nicole Sharp is a Lecturer in the School of Science and Health at Western Sydney University. Nicole's research interests include the in-depth experiences of people with disability, particularly at times of significant life transition. Nicole is focused on using inclusive research methods that give voice to people who may otherwise be excluded from participation in research.

Norm Sheehan is a Wiradjuri man and a Professor. He is currently Director of Gnibi College SCU. Basing on his expertise in Indigenous Knowledge and Education that employs visual and narrative principles to activate existing strengths within Indigenous education contexts for all individuals and learning communities, he has led the development of two new degrees: the Bachelor of Indigenous Knowledge and the Bachelor of Aboriginal Health and Wellbeing.

Patti Shih is a Postdoctoral Research Fellow at Macquarie University's Centre for Healthcare Resilience and Implementation Science, within the Australian Institute of Health Innovation. She teaches qualitative research methods in public health and has conducted qualitative and mixed-methods projects across a number of healthcare settings including HIV prevention in Papua New Guinea, e-learning among medical students, and professional training in the aged care sector.

Abla Mehio Sibai is a Professor of Epidemiology at the Faculty of Health Sciences, American University of Beirut (AUB). She has led a number of population-based national surveys, including the WHO National Burden of Disease

Study and the Nutrition and Non-communicable Disease Risk Factor (NNCD-BRF) STEPWise study. She is the author of over 150 scholarly articles and book chapters. Abla holds a degree in Pharmacy from AUB and a Ph.D. in Epidemiology from the London School of Hygiene and Tropical Medicine.

Ankur Singh is a Research Fellow in Social Epidemiology at the University of Melbourne, Australia. Ankur has published in different areas of public health including oral epidemiology, tobacco control, and health promotion. He is an invited reviewer for multiple peer-reviewed journals.

Valerie Smith was born in Dubbo and is a Wiradjuri woman. Val started in the healthcare industry in 2012, becoming an Aboriginal Health Worker with the Dubbo Regional Aboriginal Health Service in 2014. Val has recently moved to Port Macquarie and is currently taking a career break to spend time with her family.

Jennifer Smith-Merry is an Associate Professor of Qualitative Health Research in the Faculty of Health Sciences at the University of Sydney. She has methodological expertise in a range of qualitative methods including inclusive research, process evaluation of policy and services, narrative analysis, and critical discourse analysis. Her research focuses on mental health service and policy and consumer experiences of health and healthcare.

Helene Snee is a Senior Lecturer in Sociology at Manchester Metropolitan University, UK. Her research explores stratification with a particular focus on youth and class. Helene is the author of *A Cosmopolitan Journey? Difference, Distinction and Identity Work in Gap Year Travel* (Ashgate, 2014), which was short-listed for the BSA's Philip Abrams Memorial Prize for the best first and sole-authored book within the discipline of Sociology.

Andi Spark led the animation program at the Griffith Film School for 10 years following two decades in various animation industry roles. Working as an animation director and producer, she has been involved with projects ranging from large-scale outdoor projection installations to micro-looping animations for gallery exhibitions and small-screen devices and to children's television series and feature-length films and supervised the production of more than 200 short animated projects.

Elaine Stasiulis is a community-based Research Project Manager and Research Fellow at the Hospital for Sick Children and a doctoral candidate at the University of Toronto. Her work has involved an extensive range of qualitative and participatory arts-based health research projects with children and young people experiencing mental health difficulties and other health challenges.

Genevieve Steiner is a multi-award-winning NHMRC-ARC Dementia Research Development Fellow at NICM, Western Sydney University. Gen uses functional neuroimaging and physiological research methods to investigate the biological bases of learning and memory processes that will inform the prevention, diagnosis, and treatment of cognitive decline in older age. She also conducts many rigorous clinical trials including herbal and lifestyle medicines that may reduce the risk of dementia.

Neil Stephens is a sociologist and Science and Technology Studies (STS) scholar based at Brunel University London. He uses qualitative methods, including ethnography, to research innovation in biomedical contexts and explore the cultural and political aspects of the setting. He has researched topics including human embryonic stem cell research, biobanking, robotic surgery, and cultured meat.

Sara Stevano is a Postdoctoral Fellow in Economics at SOAS. She has a background in development economics, with interdisciplinary skills in the field of political economy and anthropology. Her research interests include the political economy of food and nutrition, agrarian change, and labor markets, with a focus on sub-Saharan Africa (Mozambique and Ghana in particular).

James Sutton is an Associate Professor of Anthropology and Sociology at Hobart and William Smith Colleges in Geneva, New York, USA, where he serves as a member of the Institutional Review Board. His research is focused in the areas of criminology and criminal justice, and his substantive research emphases include prisons, gangs, sexual assault, white-collar crime, and criminological research methods.

Roy Tapera is an Epidemiologist and Medical Informatician with more than 10 years of experience in the field of public health. He is currently a Lecturer of Epidemiology and Health Informatics at the University of Botswana, School of Public Health.

Gail Teachman is a Researcher and an Occupational Therapist. She is currently completing postdoctoral studies at McGill University, Montreal, Canada, and was a trainee with the Critical Disability and Rehabilitation Studies Lab at Bloorview Research Institute in Toronto, Canada. Gail's interdisciplinary research draws on critical qualitative inquiry, rehabilitation science, social theory, and social studies of childhood.

Gareth Terry is a Senior Research Fellow at the Centre for Person Centred Research at the Auckland University of Technology. He comes from a background in critical health and critical social psychologies and currently works in critical rehabilitation studies. His research interests are in men's health, gendered bodies, chronic health conditions, disability and accessibility, and reproductive decision-making.

Cecilie Thøgersen-Ntoumani is an Associate Professor in the School of Psychology and Speech Pathology at Curtin University, Western Australia. She conducts mixed-methods research with a particular emphasis on the development, implementation, and evaluation of theory-based health behavior interventions.

Natalie Thorburn is a Ph.D. candidate at the University of Auckland, New Zealand. Natalie works as Policy Advisor in the community sector, and her background is in sexual violence and intimate partner violence. Her research interests focus on sexual violence, sexual exploitation, and trafficking, and she sits on the board of Child Alert, an organization committed to ending sexual exploitation of children through child trafficking, child prostitution, and child pornography.

Irmgard Tischner is a Senior Lecturer in Social Psychology at the University of the West of England, Bristol, UK, and member of the Centre for Appearance Research. Focusing on poststructuralist, feminist, and critical psychological approaches, her research interests include issues around embodiment and subjectivity, particularly in relation to (gendered) discourses of body size, health, and physical activity in contemporary western industrialized societies.

Allison Tong is an Associate Professor at the Sydney School of Public Health, the University of Sydney. She developed the consolidated criteria for reporting qualitative health research (COREQ) and the enhancing transparency in reporting the synthesis of qualitative health research [ENTREQ], which are both endorsed as key reporting guidelines by leading journals and by the EQUATOR Network for promoting the transparency of health research.

Graciela Tonon is the Director of the Master Program in Social Sciences and the Research Center in Social Sciences (CICS) at Universidad de Palermo, as well as the Director of UNICOM, Faculty of Social Sciences, Universidad Nacional de Lomas de Zamora, Argentina. She is Vice-President of External Affairs of the International Society for Quality of Life Studies (ISQOLS) and Secretary of the Human Development and Capability Association (HDCA).

Billie Townsend is a non-Aboriginal woman born in Dubbo with family connections to the local area. Billie graduated with a double degree in International Studies and History in 2015 from the University of Wollongong. Billie worked on the Aboriginal people's stories of diabetes care study as a Volunteer Research Assistant for the University of Sydney, School of Rural Health, prior to returning to Wollongong to complete her honors year.

Nicole Tujague is a descendant of the Gubbi Gubbi nation from Mt Bauple, Queensland, and the South Sea Islander people from Vanuatu and the Loyalty Islands. Nicole has extensive experience interpreting intercultural issues for both Indigenous and non-Indigenous stakeholders. She has the ability to engage with Aboriginal and Torres Strait Islanders effectively to enable and support changes on an individual and family level.

Jane M. Ussher is Professor of Women's Health Psychology, in the Centre for Health Research at the Western Sydney University. Her research focuses on examining subjectivity in relation to the reproductive body and sexuality and the gendered experience of cancer and cancer care. Her current research include sexual health in CALD refugee and migrant women, sexuality and fertility in the context of cancer, young women's experiences of smoking, and LGBTI experiences of cancer.

Bram Vanhoutte is Simon Research Fellow in the Department of Sociology of the University of Manchester, UK. His research covers a wide field both in terms of topics and methods used. He has investigated aspects of the political socialization of youth and network influences on community-level social cohesion to more recently life course perspectives on ageing. Methodologically, he is interested in how to improve measurement and modeling using survey and administrative data and has recently developed an interest in life history data.

Lía Rodriguez de la Vega is a Professor and Researcher of the University of Palermo and Lomas de Zamora National University (Argentina). She obtained her doctoral degree in International Relations from El Salvador University, Argentina, in 2006. She is also a postdoctoral stay at the Psychology Faculty, Universidade Federal de Rio Grande do Sul (UFRGS, Porto Alegre, Brazil, in the context of the Bilateral Program Scholarship MINCYT-CAPES, 2009).

Randi Wærdahl is an Associate Professor of Sociology. Her work has an everyday perspective on transitions and trajectories in childhood and for families, in consumer societies, in times of social and economic change, and in changing contexts due to migration. She combines traditional methods of interviews and surveys with participatory methods such as photography or diary writing.

Morten Wahrendorf is a Senior Researcher at the Centre for Health and Society, University of Duesseldorf, Germany, with substantial expertise in sociology and research methodology. His main research interest are health inequalities in ageing populations and underlying pathways, with a particular focus on psychosocial working conditions, patterns of participation in paid employment and social activities in later life, and life course influences.

Maggie Walter is a member of the Palawa Briggs/Johnson Tasmanian Aboriginal family. She is Professor of Sociology and Pro Vice-Chancellor of Aboriginal Research and Leadership at the University of Tasmania. She has published extensively in the field of race relations and inequality and is passionate about Indigenous statistical engagement.

Emma Webster is a non-Aboriginal woman who has lived in Dubbo over 20 years and has family connections to the area. Emma joined the University of Sydney, School of Rural Health, in 2015 after working in the NSW public health system for 21 years. She has extensive experience in guiding novice researchers through their first research study having worked in the NSW Health Education and Training Institute's Rural Research Capacity Building Program.

Johannes Wheeldon has more than 15 years' experience managing evaluation and juvenile justice projects. He has worked with the American Bar Association, the Open Society Foundations, and the World Bank. He has published 4 books and more than 25 peer-reviewed papers on aspects of criminal justice, restorative justice, organizational change, and evaluation. He is an Adjunct Professor at Norwich University.

Sarah J. White is a Qualitative Health Researcher and Linguist with a particular interest in using conversation analysis to understand communication in surgical practice. She is a Senior Lecturer at the Faculty of Medicine and Health Sciences at Macquarie University, Sydney.

Matthew Wice is a doctoral student at the New School for Social Research where he studies Developmental and Social Psychology. His research interests include the influence of culture on moral reasoning, motivation, and the development of social cognition.

Nicola Willis is a Pediatric HIV Nurse Specialist. She is the Founder and Director of AfricAid, Zvandiri, and has spent 14 years developing and scaling up differentiated HIV service delivery models for children and adolescents in Zimbabwe. Nicola's work has been extensively influenced by the engagement of HIV-positive young people in creative, participatory approaches as a means of identifying and informing their policy and service delivery needs.

Noreen Willows uses anthropological, qualitative, and quantitative methodologies for exploring the relationships between food and health, cultural meanings of food and health, how food beliefs and dietary practices affect the well-being of communities, and how sociocultural factors influence food intake and food selection.

Noreen aims to foster an understanding among academics and nonacademics of the value of community-based participatory research.

Denise Wilson is a Professor in Māori Health and the Director of Taupua Waiora Centre for Māori Health Research at Auckland University of Technology. Her research and publications focus on Māori health, health services access and use, family violence, cultural responsiveness, and workforce development.

Elena Wilson is a Ph.D. candidate in the rural health research program, "Improving the Health of Communities through Participation," in the Rural Health School at La Trobe University, Bendigo. Elena's research interests are research methods, research ethics, and community participation with a focus on rural health and well-being.

Leigh Wilson is a Senior Lecturer in the Ageing Work and Health Research Group at the University of Sydney. Leigh comes from a public health and behavioral science background and has worked as an Epidemiologist and Health Service Manager in the NSW Health system. Her key research interests are the epidemiology of ageing, the impact of climate change (particularly heatwaves) on the aged, and the impact of behavior and perceptions of age on health.

Shawn Wilson has worked with Indigenous people internationally to apply Indigenist philosophy within the contexts of Indigenous education, health, and counselor education. His research focuses on the interrelated concepts of identity, health and healing, culture, and well-being. His book, *Research is Ceremony: Indigenous Research Methods*, has been cited as bridging understanding between mainstream and Indigenist research and is used as a text in many universities.

Jess Woodroffe works in research, academic supervision, teaching, and community engagement. Her research interests include community partnerships and engagement, social inclusion, health promotion, inter-professional learning and education, health sociology, qualitative research, and evaluation.

Kevin B. Wright is a Professor of Health Communication in the Department of Communication at George Mason University. His research focuses on social support processes and health outcomes in both face-to-face and online support groups/communities, online health information seeking, and the use of technology in provider-patient relationships.

Anna Ziersch is an Australian Research Council Future Fellow based at the Southgate Institute for Health, Society and Equity at Flinders University. She has an overarching interest in health inequities, in particular multidisciplinary and multi-method approaches to understanding the social determinants of health, and in particular for refugees and asylum seekers.

Contributors

Mauri Ahlberg Department of Teacher Education, University of Helsinki, Helsinki, Finland

Farah Ahmad School of Health Policy and Management, York University, Toronto, ON, Canada

Jo Aldridge Department of Social Sciences, Loughborough University, Leicestershire, UK

Julaine Allan Lyndon, Orange, NSW, Australia

Jordan Alpert Department of Advertising, University of Florida, Gainesville, FL, USA

Stewart J. Anderson Department of Biostatistics, University of Pittsburgh Graduate School of Public Health, Pittsburgh, PA, USA

E. Anne Marshall Educational Psychology and Leadership Studies, Centre for Youth and Society, University of Victoria, Victoria, BC, Canada

Mike Armour NICM, Western Sydney University (Campbelltown Campus), Penrith, NSW, Australia

Amit Arora School of Science and Health, Western Sydney University, Sydney, NSW, Australia

Discipline of Paediatrics and Child Health, Sydney Medical School, Sydney, NSW, Australia

Oral Health Services, Sydney Local Health District and Sydney Dental Hospital, NSW Health, Sydney, NSW, Australia

COHORTE Research Group, Ingham Institute of Applied Medical Research, Liverpool, NSW, Australia

Kathy Arthurson Southgate Institute for Health, Society and Equity, Flinders University, Adelaide, SA, Australia

Anna Bagnoli Department of Sociology, Wolfson College, University of Cambridge, Cambridge, UK

Amy Baker School of Health Sciences, University of South Australia, Adelaide, SA, Australia

Ruth Bartlett Centre for Innovation and Leadership in Health Sciences, Faculty of Health Sciences, University of Southampton, Southampton, UK

Pat Bazeley Translational Research and Social Innovation Group, Western Sydney University, Liverpool, NSW, Australia

Linda Liska Belgrave Department of Sociology, University of Miami, Coral Gables, FL, USA

Robert F. Belli Department of Psychology, University of Nebraska-Lincoln, Lincoln, NE, USA

Jyoti Belur Department of Security and Crime Science, University College London, London, UK

Denise Benatuil Master Program in Social Sciences and CICS, Universidad de Palermo, Buenos Aires, Argentina

Sameer Bhole Sydney Dental School, Faculty of Medicine and Health, The University of Sydney, Surry Hills, NSW, Australia
Oral Health Services, Sydney Local Health District and Sydney Dental Hospital, NSW Health, Surry Hills, NSW, Australia

Mia Bierbaum Centre for Healthcare Resilience and Implementation Science, Australian Institute of Health Innovation (AIHI), Macquarie University, Sydney, NSW, Australia

Simon Bishop Centre for Health Innovation, Leadership and Learning, Nottingham University Business School, Nottingham, UK

Karen Block Melbourne School of Population and Global Health, The University of Melbourne, Melbourne, VIC, Australia

Chelsea Bond Aboriginal and Torres Strait Islander Studies Unit, The University of Queensland, St Lucia, QLD, Australia

Virginia Braun School of Psychology, The University of Auckland, Auckland, New Zealand

Wendy Brodribb Primary Care Clinical Unit, Faculty of Medicine, The University of Queensland, Herston, QLD, Australia

Mario Brondani University of British Columbia, Vancouver, BC, Canada

Jessica Broome University of Michigan, Ann Arbor, MI, USA
Sanford, NC, USA

Mark Brough School of Public Health and Social Work, Queensland University of Technology, Kelvin Grove, Australia

Katerina Bryant Department of English and Creative Writing, Flinders University, Bedford Park, SA, Australia

Lia Bryant School of Psychology, Social Work and Social Policy, Centre for Social Change, University of South Australia, Magill, Australia

Anne Bunde-Birouste School of Public Health and Community Medicine, UNSW, Sydney, NSW, Australia

Viv Burr Department of Psychology, School of Human and Health Sciences, University of Huddersfield, Huddersfield, UK

Valentina Buscemi Western Sydney University, Sydney, NSW, Australia

Rosalind A. Bye School of Science and Health, Western Sydney University, Campbelltown, NSW, Australia

Breanne Byiers Department of Educational Psychology, University of Minnesota, Minneapolis, MN, USA

Fiona Byrne Translational Research and Social Innovation (TReSI) Group, School of Nursing and Midwifery, Ingham Institute for Applied Medical Research, Western Sydney University, Liverpool, NSW, Australia

Robert H. Campbell The University of Bolton, Bolton, UK

Michael J. Carter Sociology Department, California State University, Northridge, Northridge, CA, USA

Rocco Cavaleri School of Science and Health, Western Sydney University, Campbelltown, NSW, Australia

Jane Chalmers Western Sydney University (Campbelltown Campus), Penrith, NSW, Australia

Nancy Cheak-Zamora Department of Health Sciences, University of Missouri, Columbia, MO, USA

Betty P. M. Chung School of Nursing, The Hong Kong Polytechnic University, Hong Kong, SAR, China

Victoria Clarke Department of Health and Social Sciences, Faculty of Health and Applied Sciences, University of the West of England (UWE), Bristol, UK

Bronwyne Coetzee Department of Psychology, Stellenbosch University, Matieland, South Africa

Ana Lucía Córdova-Cazar Colegio de Ciencias Sociales y Humanidades, Universidad San Francisco de Quito, Diego de Robles y Vía Interoceánica, Quito, Cumbayá, Ecuador

Jonathan C. Craig Sydney School of Public Health, The University of Sydney, Sydney, NSW, Australia

Centre for Kidney Research, The Children's Hospital at Westmead, Sydney, NSW, Australia

Fiona Cram Katoa Ltd, Auckland, Aotearoa, New Zealand

Anne Cusick Faculty of Health Sciences, Sydney University, Sydney, Australia

Angela J. Dawson Australian Centre for Public and Population Health Research, University of Technology Sydney, Sydney, NSW, Australia

Kevin Deane Department of Economics, International Development and International Relations, The University of Northampton, Northampton, UK

Lía Rodriguez de la Vega Ciudad Autónoma de Buenos Aires, University of Palermo, Buenos Aires, Argentina

Virginia Dickson-Swift LaTrobe Rural Health School, College of Science, Health and Engineering, LaTrobe University, Bendigo, VIC, Australia

Rebecca Dimond School of Social Sciences, Cardiff University, Cardiff, UK

Agnes E. Dodds Melbourne Medical School, The University of Melbourne, Melbourne, VIC, Australia

Suzanne D'Souza School of Nursing and Midwifery, Western Sydney University, Sydney, NSW, Australia

Joan L. Duda School of Sport and Exercise Sciences, University of Birmingham, Birmingham, West Midlands, UK

Clemence Due Southgate Institute for Health, Society and Equity, Flinders University, Adelaide, SA, Australia

Tinashe Dune Western Sydney University, Sydney, NSW, Australia

Jane Marie Edwards Deakin University, School of Health and Social Development, Geelong, VIC, Australia

Carolyn Ee NICM, Western Sydney University (Campbelltown Campus), Penrith, NSW, Australia

Ghinwa Y. El Hayek Department of Epidemiology and Population Health, Faculty of Health Sciences, American University of Beirut, Beirut, Lebanon

Ulrike Felt Department of Science and Technology Studies, Research Platform Responsible Research and Innovation in Academic Practice, University of Vienna, Vienna, Austria

Monika Ferguson School of Nursing and Midwifery, University of South Australia, Adelaide, SA, Australia

Pota Forrest-Lawrence School of Social Sciences and Psychology, Western Sydney University, Milperra, NSW, Australia

Bronwyn Fredericks Office of Indigenous Engagement, Central Queensland University, Rockhampton, Australia

Caroline Elizabeth Fryer Sansom Institute for Health Research, University of South Australia, Adelaide, SA, Australia

Emma S. George School of Science and Health, Western Sydney University, Sydney, NSW, Australia

Lilian A. Ghandour Department of Epidemiology and Population Health, Faculty of Health Sciences, American University of Beirut, Beirut, Lebanon

Lisa Gibbs Melbourne School of Population and Global Health, The University of Melbourne, Melbourne, VIC, Australia

Claudia Gillberg Swedish National Centre for Lifelong Learning (ENCELL), Jonkoping University, Jönköping, Sweden

Geoffrey Jones Centre for Welfare Reform, Sheffield, UK

Brenda M. Gladstone Dalla Lana School of Public Health, Centre for Critical Qualitative Health Research, University of Toronto, Toronto, ON, Canada

Namrata Goyal Department of Psychology, New School for Social Research, New York, NY, USA

Adyya Gupta School of Public Health, The University of Adelaide, Adelaide, SA, Australia

Talia Gutman Sydney School of Public Health, The University of Sydney, Sydney, NSW, Australia

Centre for Kidney Research, The Children's Hospital at Westmead, Westmead, NSW, Australia

Elizabeth J. Halcomb School of Nursing, University of Wollongong, Wollongong, NSW, Australia

Marit Haldar Department of Social Work, Child Welfare and Social Policy, Oslo and Akershus University College of Applied Sciences, Oslo, Norway

Camilla S. Hanson Sydney School of Public Health, The University of Sydney, Sydney, NSW, Australia

Centre for Kidney Research, The Children's Hospital at Westmead, Westmead, NSW, Australia

John D. Hathcoat Department of Graduate Psychology, Center for Assessment and Research Studies, James Madison University, Harrisonburg, VA, USA

Nikki Hayfield Department of Health and Social Sciences, Faculty of Health and Applied Sciences, University of the West of England (UWE), Bristol, UK

Lisa Hodge College of Health and Biomedicine, Victoria University, Melbourne, VIC, Australia

Anne Hogden Centre for Healthcare Resilience and Implementation Science, Australian Institute of Health Innovation (AIHI), Macquarie University, Sydney, NSW, Australia

Nicholas Hookway University of Tasmania, Launceston, Tasmania, Australia

Kirsten Howard School of Public Health, University of Sydney, Sydney, NSW, Australia

Martin Howell School of Public Health, University of Sydney, Sydney, NSW, Australia

Erin Jessee Modern History, University of Glasgow, Glasgow, UK

James Rufus John Translational Health Research Institute, School of Medicine, Western Sydney University, Penrith, NSW, Australia
Capital Markets Cooperative Research Centre, Sydney, NSW, Australia

Craig Johnson Dubbo Diabetes Unit, Dubbo, NSW, Australia

Monica Johnson Marathon Health, Dubbo, NSW, Australia

Stephanie T. Jong School of Education, Flinders University, Adelaide, SA, Australia

June Joseph Primary Care Clinical Unit, Faculty of Medicine, The University of Queensland, Herston, QLD, Australia

Angela Ju Sydney School of Public Health, The University of Sydney, Sydney, NSW, Australia
Centre for Kidney Research, The Children's Hospital at Westmead, Westmead, NSW, Australia

Yong Moon Jung Centre for Business and Social Innovation, University of Technology Sydney, Ultimo, NSW, Australia

Ashraf Kagee Department of Psychology, Stellenbosch University, Matieland, South Africa

Ida Kaplan The Victorian Foundation for Survivors of Torture Inc, Brunswick, VIC, Australia

Bernie Kemp Dubbo Regional Aboriginal Health Service, Dubbo, NSW, Australia

Lynn Kemp Translational Research and Social Innovation (TReSI) Group, School of Nursing and Midwifery, Ingham Institute for Applied Medical Research, Western Sydney University, Liverpool, NSW, Australia

Priya Khanna Sydney Medical Program, University of Sydney, Camperdown, NSW, Australia

Alexandra C. King Rural Clinical School, Faculty of Health, University of Tasmania, Burnie, Tasmania, Australia

School of Pharmacy, Faculty of Health, University of Tasmania, Hobart, Tasmania, Australia

Nigel King Department of Psychology, University of Huddersfield, Huddersfield, UK

Kat Kolar Department of Sociology, University of Toronto, Toronto, ON, Canada

Gary L. Kreps Department of Communication, George Mason University, Fairfax, VA, USA

Tahu Kukutai University of Waikato, Hamilton, New Zealand

Ian Lacey Deadly Choices, Institute for Urban Indigenous Health, Brisbane, Australia

Michelle LaFrance George Mason University, Fairfax, VA, USA

Kari Lancaster Centre for Social Research in Health, Faculty of Arts and Social Sciences, UNSW, Sydney, NSW, Australia

Jeanette A. Lawrence Melbourne School of Psychological Science, The University of Melbourne, Melbourne, VIC, Australia

Eric Leake Department of English, Texas State University, San Marcos, TX, USA

Raymond M. Lee Royal Holloway University of London, Egham, UK

Caroline Lenette Forced Migration Research Network, School of Social Sciences, University of New South Wales, Kensington, NSW, Australia

Doris Y. Leung School of Nursing, The Hong Kong Polytechnic University, Hong Kong, SAR, China

The Lawrence S. Bloomberg Faculty of Nursing, University of Toronto, Toronto, ON, Canada

Pranee Liamputtong School of Science and Health, Western Sydney University, Penrith, NSW, Australia

Nicole Loehr School of Social and Policy Studies, Flinders University, Adelaide, SA, Australia

Janet C. Long Australian Institute of Health Innovation, Macquarie University, Sydney, NSW, Australia

Colin MacDougall Health Sciences Building, Flinders University, Bedford Park, SA, Australia

Catrina A. MacKenzie Department of Geography, McGill University, Montreal, QC, Canada

Department of Geography, University of Vermont, Burlington, VT, USA

Freya MacMillan School of Science and Health and Translational Health Research Institute (THRI), Western Sydney University, Penrith, NSW, Australia

Wilson Majee Department of Health Sciences, University of Missouri, Columbia, MO, USA

Karine Manera Sydney School of Public Health, The University of Sydney, Sydney, NSW, Australia

Centre for Kidney Research, The Children's Hospital at Westmead, Westmead, NSW, Australia

Narendar Manohar School of Science and Health, Western Sydney University, Sydney, NSW, Australia

Oana Marcu Università Cattolica del Sacro Cuore, Milan, Italy

Rodrigo Mariño University of Melbourne, Melbourne, VIC, Australia

Anastasia Marinopoulou Department of European Studies, Hellenic Open University, Patra, Greece

Anna Maurer-Batjer Department of Health Sciences, University of Missouri, Columbia, MO, USA

Kate A. McBride School of Medicine and Translational Health Research Institute, Western Sydney University, Sydney, NSW, Australia

Angela McGrane Newcastle Business School, Northumbria University, Newcastle-upon-Tyne, UK

Jane McKeown School of Nursing and Midwifery, The University of Sheffield, Sheffield, UK

Karen McPhail-Bell University Centre for Rural Health, University of Sydney, Camperdown, NSW, Australia

Poche Centre for Indigenous Health, Sydney Medical School, The University of Sydney, Camperdown, NSW, Australia

Abla Mehio Sibai Department of Epidemiology and Population Health, Faculty of Health Sciences, American University of Beirut, Beirut, Lebanon

Cara Meixner Department of Graduate Psychology, Center for Faculty Innovation, James Madison University, Harrisonburg, VA, USA

Zelalem Mengesha Western Sydney University, Sydney, NSW, Australia

Dafna Merom School of Science and Health, Western Sydney University, Penrith, Sydney, NSW, Australia

Translational Health Research Institute, School of Medicine, Western Sydney University, Penrith, NSW, Australia

Joan G. Miller Department of Psychology, New School for Social Research, New York, NY, USA

Christine Milligan Division of Health Research, Lancaster University, Lancaster, UK

Naomi Moller School of Psychology, Faculty of Social Sciences, The Open University, Milton Keynes, UK

Andrea Montes Alvarado Sociology Department, California State University, Northridge, Northridge, CA, USA

Marta Moskal Durham Univeristy, Durham, UK

Elias Mpofu University of Sydney, Lidcombe, NSW, Australia
Educational Psychology and Inclusive Education, University of Johannesburg, Johannesburg, South Africa

Magen Mhaka Mutepfa Department of Psychology, University of Botswana, Gaborone, Botswana

Sally Nathan School of Public Health and Community Medicine, Faculty of Medicine, UNSW, Sydney, NSW, Australia

Alison Nelson Allied Health and Workforce Development, Institute for Urban Indigenous Health, Brisbane, Australia

Ingrid A. Nelson Sociology and Anthropology Department, Bowdoin College, Brunswick, ME, USA

Christy Newman Centre for Social Research in Health, Faculty of Arts and Social Sciences, UNSW, Sydney, NSW, Australia

Mark C. Nicholas Framingham State University, Framingham, MA, USA

Kayi Ntinda Discipline of Educational Counselling and Mixed-Methods Inquiry Approaches, Faculty of Education, Office C.3.5, University of Swaziland, Kwaluseni Campus, Manzini, Swaziland

Chijioke Obasi University of Wolverhampton, Wolverhampton, UK

Felix Ogbo School of Medicine and Translational Health Research Institute, Western Sydney University, Campbelltown, NSW, Australia

Andrew Page School of Medicine and Translational Health Research Institute, Western Sydney University, Campbelltown, NSW, Australia

Bonnie Pang School of Science and Health and Institute for Culture and Society, University of Western Sydney, Penrith, NSW, Australia

Gemma Pearce Centre for Advances in Behavioural Science, Coventry University, Coventry, West Midlands, UK

Karen G. Peres Australian Research Centre for Population Oral Health (ARCPOH), Adelaide Dental School, The University of Adelaide, Adelaide, SA, Australia

Janette Perz Translational Health Research Institute, School of Medicine, Western Sydney University, Sydney, NSW, Australia

Rona Pillay School of Nursing and Midwifery, Western Sydney University, Sydney, NSW, Australia

Jayne Pitard College of Arts and Education, Victoria University, Melbourne, VIC, Australia

Jennifer Prattley Department of Social Statistics, University of Manchester, Manchester, UK

Nicholas Procter School of Nursing and Midwifery, University of South Australia, Adelaide, SA, Australia

Marsha B. Quinlan Department of Anthropology, Washington State University, Pullman, WA, USA

Frances Rapport Centre for Healthcare Resilience and Implementation Science, Australian Institute of Health Innovation (AIHI), Macquarie University, Sydney, NSW, Australia

Elizabeth F. Rix Gnibi Wandarahn School of Indigenous Knowledge, Southern Cross University, Lismore, NSW, Australia

Lia Rodriguez de la Vega Ciudad Autónoma de Buenos Aires, Buenos Aires, Argentina

Norma R. A. Romm Department of Adult Education and Youth Development, University of South Africa, Pretoria, South Africa

Rizwana Roomaney Department of Psychology, Stellenbosch University, Matieland, South Africa

Lauren Rosewarne School of Social and Political Sciences, University of Melbourne, Melbourne, VIC, Australia

Simone Schumann University of Vienna, Vienna, Austria

Claudia G. Schwarz-Plaschg Research Platform Nano-Norms-Nature, University of Vienna, Vienna, Austria

Lizzie Seal University of Sussex, Brighton, UK

Kapriskie Seide Department of Sociology, University of Miami, Coral Gables, FL, USA

Nicole L. Sharp School of Science and Health, Western Sydney University, Campbelltown, NSW, Australia

Norm Sheehan Gnibi Wandarahn School of Indigenous Knowledge, Southern Cross University, Lismore, NSW, Australia

Patti Shih Centre for Healthcare Resilience and Implementation Science, Australian Institute of Health and Innovation (AIHI), Macquarie University, Sydney, NSW, Australia

Ankur Singh Centre for Health Equity, Melbourne School of Population and Global Health, The University of Melbourne, Melbourne, VIC, Australia

Valerie Smith Formerly with Dubbo Regional Aboriginal Health Service, Dubbo, NSW, Australia

Jennifer Smith-Merry Faculty of Health Sciences, The University of Sydney, Sydney, Australia

Helene Snee Manchester Metropolitan University, Manchester, UK

Andi Spark Griffith Film School, Queensland College of Art, Griffith University, Brisbane, QLD, Australia

Elaine Stasiulis Child and Youth Mental Health Research Unit, SickKids, Toronto, ON, Canada
Institute of Medical Science, University of Toronto, Toronto, ON, Canada

Genevieve Z. Steiner NICM and Translational Health Research Institute (THRI), Western Sydney University, Penrith, NSW, Australia

Neil Stephens Social Science, Media and Communication, Brunel University London, Uxbridge, UK

Sara Stevano Department of Economics, University of the West of England (UWE) Bristol, Bristol, UK

James E. Sutton Department of Anthropology and Sociology, Hobart and William Smith Colleges, Geneva, NY, USA

Roy Tapera Department of Environmental Health, University of Botswana, Gaborone, Botswana

Gail Teachman McGill University, Montreal, QC, Canada

Gareth Terry Centre for Person Centred Research, School of Clinical Sciences, Auckland University of Technology, Auckland, New Zealand

Michelle Teti Department of Health Sciences, University of Missouri, Columbia, MO, USA

Cecilie Thøgersen-Ntoumani Health Psychology and Behavioural Medicine Research Group, School of Psychology and Speech Pathology, Curtin University, Perth, WA, Australia

Natalie Thorburn The University of Auckland, Auckland, New Zealand

Irmgard Tischner Faculty of Sport and Health Sciences, Technische Universität München, Lehrstuhl Diversitätssoziologie, Munich, Germany

Allison Tong Sydney School of Public Health, The University of Sydney, Sydney, NSW, Australia

Centre for Kidney Research, The Children's Hospital at Westmead, Westmead, Australia

Graciela Tonon Master Program in Social Sciences and CICS-UP, Universidad de Palermo, Buenos Aires, Argentina

UNICOM- Universidad Nacional de Lomas de Zamora, Buenos Aires, Argentina

Billie Townsend School of Rural Health, Sydney Medical School, University of Sydney, Dubbo, NSW, Australia

Nicole Tujague Gnibi Wandarahn School of Indigenous Knowledge, Southern Cross University, Lismore, NSW, Australia

Jane M. Ussher Translational Health Research Institute, School of Medicine, Western Sydney University, Sydney, NSW, Australia

Bram Vanhoutte Department of Sociology, University of Manchester, Manchester, UK

Randi Wærdahl Department of Social Work, Child Welfare and Social Policy, Oslo and Akershus University College of Applied Sciences, Oslo, Norway

Morten Wahrendorf Institute of Medical Sociology, Centre of Health and Society (CHS), Heinrich-Heine-University Düsseldorf, Medical Faculty, Düsseldorf, Germany

Maggie Walter University of Tasmania, Hobart, Tasmania, Australia

Emma Webster School of Rural Health, Sydney Medical School, University of Sydney, Dubbo, NSW, Australia

Johannes Wheeldon School of Sociology and Justice Studies, Norwich University, Northfield, VT, USA

Sarah J. White Macquarie University, Sydney, NSW, Australia

Matthew Wice Department of Psychology, New School for Social Research, New York, NY, USA

Nicola Willis Zvandiri House, Harare, Zimbabwe

Noreen D. Willows Faculty of Agricultural, Life and Environmental Sciences, University of Alberta, Edmonton, AB, Canada

Denise Wilson Auckland University of Technology, Auckland, New Zealand

Elena Wilson Rural Health School, College of Science, Health and Engineering, La Trobe University, Melbourne, VIC, Australia

Leigh A. Wilson School of Science and Health, Western Sydney University, Penrith, NSW, Australia

Faculty of Health Science, Discipline of Behavioural and Social Sciences in Health, University of Sydney, Lidcombe, NSW, Australia

Shawn Wilson Gnibi Wandarahn School of Indigenous Knowledge, Southern Cross University, Lismore, NSW, Australia

Jessica Woodroffe Access, Participation, and Partnerships, Academic Division, University of Tasmania, Launceston, Australia

Kevin B. Wright Department of Communication, George Mason University, Fairfax, VA, USA

Anna Ziersch Southgate Institute for Health, Society and Equity, Flinders University, Adelaide, SA, Australia

Section I

Traditional Research Methods in Health and Social Sciences

Traditional Research Methods in Health and Social Sciences: An Introduction

1

Pranee Liamputtong

Contents

Keywords

Research · Research approaches · Knowledge · Evidence · Qualitative research · Quantitative research · Mixed methods research

---

1   Introduction

> Acquiring knowledge through the use of research findings that were derived from the scientific method is the most objective way of knowing something. (Grinnell et al. 2014, p. 12)

Researchers conduct research in order to obtain knowledge that health and social care practitioners can use in the provision of health and social care programs to those in need. Knowledge can be cultivated in many ways. Cohen and Manion (2000) suggest that there are three primary forms of knowing: experience, reasoning, and research. However, for researchers, they generate new knowledge through research (Fawcett and Pockett 2015). In health and social care in particular, scientific research is seen to "yield the best source of evidence" (Schmidt and Brown 2015, p. 7). In order to provide evidence-based health care, research is conducted so that knowledge that addresses the practice concern can be obtained (Grove et al. 2013).

P. Liamputtong (✉)
School of Science and Health, Western Sydney University, Penrith, NSW, Australia
e-mail: p.liamputtong@westernsydney.edu.au

© Springer Nature Singapore Pte Ltd. 2019
P. Liamputtong (ed.), *Handbook of Research Methods in Health Social Sciences*,
https://doi.org/10.1007/978-981-10-5251-4_52

Research has been referred to as a "planned and systematic activity" (Schmidt and Brown 2015, p. 14) that results in the construction of new knowledge that can be used to provide answers to some health problems or as evidence for health care practice (Polit and Beck 2011; Liamputtong and Schmied 2017). Traditionally, there are three main research approaches that health and social science researchers have adopted in their research. They are known as a qualitative, a quantitative, and a mixed methods research approach (Babbie 2016; Bryman 2016; Liamputtong 2017; Creswell and Plano Clark 2018). To put it simply, qualitative research approaches yield "qualitative data in the form of text" whereas quantitative research methods offer "quantitative data in the form of numbers" (Grinnell et al. 2014, p. 20). For the mixed methods research, both forms of data are included (Creswell and Plano Clark 2018).

Researchers need to make a decision what type of knowledge they need to generate, which will allow them to decide about the research method that is appropriate to their research questions (Liamputtong and Schmied 2017).

The word "research," according to Walter (2013, p. 4), "evokes a popular imagery of a scholarly endeavour pursue using complicated formulas, and uninterpretable language and techniques." Due to this perception, research can be seen as "far removed from our everyday lives and our social world." But as readers will see in the section, not all research is like that. This is particularly research in the health and social science areas. Research, as Walter (2013, p. 4) points out, "is everywhere, and it touches many aspects of our social lives." Readers can witness this in all chapters in the section (as well as those in other sections of the Handbook).

2   About the Section

Research, as defined by the Australian Research Council (2015, p. 9), is "the creation of new knowledge" as well as "the use of existing knowledge in a new and creative way" in order to obtain novel understandings, concepts, methodologies, and inventions. Research can also encompass "synthesis and analysis of previous research to the extent that it leads to new and creative outcomes." This definition is clearly reflected in the section of the Handbook.

The section on traditional research methods in health and social sciences comprises four parts. It is the largest section within the Handbook. Section 1 is an introduction to research approaches and it comprises four chapters. Pranee Liamputtong introduces the nature of the qualitative inquiry that health and social science researchers can use (▶ Chap. 2, "Qualitative Inquiry"). Leigh Wilson provides discussions regarding the quantitative approach in health research (▶ Chap. 3, "Quantitative Research"). Cara Meixner and John Hathcoat write about the nature of mixed methods research (▶ Chap. 4, "The Nature of Mixed Methods Research"). The last chapter in this part is about the recruitment of research participants and was written by Narendar Manohar, Freya MacMillan, Genevieve Steiner and Amit Arora (▶ Chap. 5, "Recruitment of Research Participants").

Section 2 contains a number of chapters that discuss the philosophical foundations and methodological frameworks that have been adopted in the health and social sciences. It contains 17 chapters. John Hathcoat, Cara Meixner, and Marck Nicholas provide in-depth discussion about ontology and epistemology that underpins the research approaches in health social science areas (▶ Chap. 6, "Ontology and Epistemology"). This chapter is followed by several chapters that dedicate to different epistemological approaches in the health social sciences. Viv Burr writes about the social constructionism (▶ Chap. 7, "Social Constructionism"), while Anastasia Marinopoulou provides discourses about the critical theory (▶ Chap. 8, "Critical Theory: Epistemological Content and Method"). Priya Khanna offers readers ideas about positivism and realism that underpins quantitative approaches (▶ Chap. 9, "Positivism and Realism"). In the next four chapters, the authors provide great insights into the methodological frameworks that most qualitative researchers have adopted: symbolic interactionism by Michael Carter and Andrea Alvarado (▶ Chap. 10, "Symbolic Interactionism as a Methodological Framework"); hermeneutics by Suzanne D'Souza (▶ Chap. 11, "Hermeneutics: A Boon for Cross-Disciplinary Research"); a feminist pragmatic model by Claudia Gillberg (▶ Chap. 12, "Feminism and Healthcare: Toward a Feminist Pragmatist Model of Healthcare Provision"); and critical ethnography by Patti Shih (▶ Chap. 13, "Critical Ethnography in Public Health: Politicizing Culture and Politicizing Methodology").

Eric Leake offers readers a novel methodological framework in social research, what he refers to as empathy (▶ Chap. 14, "Empathy as Research Methodology"). The Indigenist and decolonizing research methodology is written by Elizabeth Rix and colleagues (▶ Chap. 15, "Indigenist and Decolonizing Research Methodology"). Rona Pillay confers ideas about ethnomethodology (▶ Chap. 16, "Ethnomethodology") and Elena Wilson writes about the community-based participatory action research (▶ Chap. 17, "Community-Based Participatory Action Research"). Linda Belgrave and Kapriskie Seide deliberate in great details about the grounded theory methodology (▶ Chap. 18, "Grounded Theory Methodology: Principles and Practices"). Case study research is written by Pota Forrest-Lawrence (▶ Chap. 19, "Case Study Research"). Two chapters on evaluation are conferred by Angela Dawson (▶ Chap. 20, "Evaluation Research in Public Health") and Gary Kreps and Jordan Alpert (▶ Chap. 21, "Methods for Evaluating Online Health Information Systems"). In the last chapter in this part, it is about translational research that has become an increasingly important in the health sciences and is written by Lynn Kemp (▶ Chap. 22, "Translational Research: Bridging the Chasm Between New Knowledge and Useful Knowledge").

Section 3 of this section dedicates to the ways researchers in the health social sciences collect their empirical materials. It is divided into three sub-parts. In the first sub-part, chapters involved data collection methods in qualitative research. Sally Nathan and colleagues examine the qualitative interviewing method that has been popularly adopted in qualitative research projects (▶ Chap. 23, "Qualitative Interviewing"). Narrative research is offered by Kay Ntinda (▶ Chap. 24, "Narrative Research"), and the life history interview by Erin Jessee (▶ Chap. 25, "The Life History Interview"). Two chapters concerning ethnography are discussed by two authors: Bonnie Pang on the ethnographic method (▶ Chap. 26, "Ethnographic Method") and Michelle LaFrance on the institutional ethnography (▶ Chap. 27, "Institutional Ethnography"). Sarah White provides insights into the conversation analysis (▶ Chap. 28, "Conversation Analysis: An Introduction to Methodology, Data Collection, and Analysis") whereas Raymond Lee discusses issues relating to the unobtrusive methods (▶ Chap. 29, "Unobtrusive Methods"). A great insight about autoethnography is written by Anne Bunde-Birouste and colleagues (▶ Chap. 30, "Autoethnography"). The last chapter in this sub-part is the feminist-based method of memory work discussed by Lia Bryant and Katerina Bryant (▶ Chap. 31, "Memory Work").

In the quantitative approach and practice sub-part, authors offer knowledge about different types of quantitative methods. Megan Mhaka and Roy Tapera discuss issues surrounding the traditional survey and questionnaire platforms (▶ Chap. 32, "Traditional Survey and Questionnaire Platforms"). Kate McBride and colleagues provide good insights into epistemology in health research (▶ Chap. 33, "Epidemiology"). Single-case designs are written by Breanne Byiers (▶ Chap. 34, "Single-Case Designs") and longitudinal study designs by Stewart Anderson (▶ Longitudinal Study Designs). Martin Howell and Kirsten Howard confer the discrete choice experiments (▶ Chap. 36, "Eliciting Preferences from Choices: Discrete Choice Experiments") whereas Mike Armour and colleagues discuss the randomized controlled trials (▶ Chap. 37, "Randomized Controlled Trials"). The last chapter in his sub-part is about measurement issues in quantitative research written by Dafna Merom and James Rufus John (▶ Chap. 38, "Measurement Issues in Quantitative Research").

The last sub-part on research methods includes chapters that discuss the multi-methods, mixed methods, collaborative research, and systematic reviews. There are eight chapters in this sub-part. Graciela Tonon reviews the use of integrated methods in research (▶ Chap. 39, "Integrated Methods in Research"), while Kate McBride and colleagues address the use of mixed methods in research (▶ Chap. 40, "The Use of Mixed Methods in Research"). The Delphi technique is conferred by Jane Chalmers and Mike Armour (▶ Chap. 41, "The Delphi Technique"). Karine Manera and colleagues discuss the nominal group technique (▶ Chap. 42, "Consensus Methods: Nominal Group Technique"), and Tinashe Dune and colleagues tell us about the Q methodology that researchers use to examine human subjectivity (▶ Chap. 43, "Jumping the Methodological Fence: Q Methodology"). Social network analysis is written by Janet Long and Simon Bishop (▶ Chap. 44, "Social Network Research"). The last two chapters in this sub-part contain a systematic review of literature as a research method. Angela Dawson provides great insights into the meta-synthesis of qualitative literature (▶ Chap. 45, "Meta-Synthesis of Qualitative Research") while Freya MacMillan and colleagues do so for the systematic review method (▶ Chap. 46, "Conducting a Systematic Review: A Practical Guide").

The last part of this section comprises chapters that discuss how to make sense of the research data, research representation, and evaluation of research. Doris Leung and Betty Chung confer content analysis (▶ Chap. 47, "Content Analysis: Using Critical Realism to Extend Its Utility") whereas Virginia Braun and colleagues provide great insights into the thematic analysis method (▶ Chap. 48, "Thematic Analysis"). Nicole Sharp and Rosaline Bye write about how to analyze data from the narrative research (▶ Chap. 49, "Narrative Analysis"), and Jane Ussher and Janette Perz suggest ways that researchers can adopt the critical discourse analysis in research (▶ Chap. 50, "Critical Discourse/Discourse Analysis"). The schema analysis of qualitative data, a team-based approach, is written by Frances Rapport and colleagues (▶ Chap. 51, "Schema Analysis of Qualitative Data: A Team-Based Approach"). Patricia Bazeley provides great discussions and details about using qualitative data analysis software (QDAS) to assist data analyses (▶ Chap. 52, "Using Qualitative Data Analysis Software (QDAS) to Assist Data Analyses"). The chapter sequence analysis of life history data is written by Bram Vanhoutte, Morten Wahrendorf, and Jennifer Prattley (▶ Chap. 53, "Sequence Analysis of Life History Data"). Yong Moon Jung explains how researchers can analyze quantitative data (▶ Chap. 54, "Data Analysis in Quantitative Research"). The next few chapters are dedicated to the reporting of research findings. Allison Tong and Jonathan Craig write about reporting of qualitative health research (▶ Chap. 55, "Reporting of Qualitative Health Research"), while Ankur Singh and colleagues confer the way that quantitative research studies can be written up (▶ Chap. 56, "Writing Quantitative Research Studies"). Graciela Tonon tells us about a traditional academic presentation of research findings and public policies (▶ "Traditional Academic Presentation of Research Findings and Public Policies"). The last three chapters in this part are about the appraisal of a research study. Camilla Hanson and colleagues discuss critical appraisal of qualitative studies (▶ Chap. 58, "Appraisal of Qualitative Studies") while Rocco Cavaleri and colleagues do so for the appraisal of quantitative research (▶ Chap. 59, "Critical Appraisal of Quantitative Research"). The very last chapter in this part is written by Elizabeth Halcomb, who provides explanations about appraising mixed methods research (▶ Chap. 60, "Appraising Mixed Methods Research").

References

Australian Research Council. Excellent in research for Australia: ERA 2015 submission guidelines. 2015. Retrieved from http://archive.arc.gov.au/archive_files/ERA/2015/Key%20Documents/ERA_2015_Submission_Guidelines.pdf.

Babbie E. The practice of social research. 14th ed. Boston: Cengage Learning; 2016.

Bryman A. Social research methods. 5th ed. Oxford: Oxford University Press; 2016.

Cohen L, Manion L. Research methods in education. 2nd ed. New York: Routledge; 2000.

Creswell JW, Plano Clark VL. Designing and conducting mixed methods research. 3rd ed. Thousand Oaks: Sage; 2018.

Fawcett B, Pockett R. Turning ideas into research: theory, design & practice. London: Sage; 2015.

Grinnell RM, Unrau YA, Williams M. Introduction. In: Grinnell RM, Unrau YA, editors. Social work research and evaluation: foundations of evidence-based practice. 10th ed. Oxford: Oxford University Press; 2014. p. 3–29.

Grove SK, Burns N, Gray JR. The practice of nursing research: appraisal, synthesis, and generation of evidence. 7th ed. St. Louis: Elsevier; 2013.

Liamputtong P. The science of words and the science of numbers. In: Liamputtong P, editor. Research methods in health: foundations for evidence-based practice. 3rd ed. Melbourne: Oxford University Press; 2017. p. 3–28.

Liamputtong P, Schmied V. Getting started: designing and planning a research project. In: Liamputtong P, editor. Research methods in health: foundations for evidence-based practice. 3rd ed. Melbourne: Oxford University Press; 2017. p. 29–48.

Polit DF, Beck CT. Nursing research: principles and methods. 9th ed. Philadelphia: Lippincott Williams & Wilkins; 2011.

Schmidt NA, Brown JM. What is evidence-based practice? In: Schmidt NA, Brown JM, editors. Evidence-based practice for nurses: appraisal and application of research. 3rd ed. Burlington: Jones & Bartlett Learning; 2015. p. 3–41.

Walter M. The nature of social science research. In: Walter M, editor. Social research methods. 3rd ed. Melbourne: Oxford University Press; 2013. p. 3–24.

Qualitative Inquiry

2

Pranee Liamputtong

Contents

Abstract

This chapter discusses the nature of the qualitative inquiry. Qualitative inquiry refers to "a broad approach" that qualitative researchers adopt as a means to examine social circumstances. The inquiry is based on an assumption which posits that people utilize "what they see, hear, and feel" to make sense of social experiences. There are many features that differentiate qualitative inquiry from the quantitative approach. Fundamentally, it is interpretive. The meanings and interpretations of the participants are the essence of qualitative inquiry. Qualitative researchers can be perceived as constructivists who attempt to find answers in the real world. Fundamentally, qualitative researchers look for meanings that people have constructed. Qualitative research is valuable in many ways. It offers

P. Liamputtong (✉)
School of Science and Health, Western Sydney University, Penrith, NSW, Australia
e-mail: p.liamputtong@westernsydney.edu.au

© Springer Nature Singapore Pte Ltd. 2019
P. Liamputtong (ed.), *Handbook of Research Methods in Health Social Sciences*,
https://doi.org/10.1007/978-981-10-5251-4_53

9

researchers to hear silenced voices, to work with marginalized and vulnerable people, to address social justice issues, and to contribute to the person-centered healthcare and the design of clinical trials. The chapter discusses in great depth the distinctive features of the qualitative inquiry. In particular, it includes the inductive nature of qualitative research, methodological frameworks, purposive sampling technique, saturation concept, qualitative data analysis, and the trustworthiness of a qualitative study.

Keywords

Qualitative inquiry · Meaning and interpretation · Qualitative researcher · Methodological framework · Saturation · Trustworthiness

1   Introduction

Qualitative inquiry seeks to discover and to describe narratively what particular people do in their everyday lives and what their actions mean to them. It identifies meaning-relevant *kinds* of things in the world—kinds of people, kinds of actions, kinds of beliefs and interests—focusing on differences in forms of things that make a difference for meaning. (Erikson 2018, p. 36, original emphasis)

Qualitative inquiry refers to "a broad approach" that qualitative researchers adopt as a means to examine social circumstances. The inquiry is based on an assumption which posits that people utilize "what they see, hear, and feel" to make sense of social experiences (Rossman and Rallis 2017, p. 5). Fundamentally, qualitative research contributes to the social inquiry which aims to interpret "the meanings of human actions" (Bradbury-Jones et al. 2017, p. 627). It is a type of research that embodies individuals as the "whole person living in dynamic, complex social arrangements" (Rogers 2000, p. 51).

The word "qualitative" derives from the Latin word "*qualitas*", which pertains to "a primary focus on the qualities, the features, of entities" (Erikson 2018, p. 36). In contrast, the word "quantitative" is from the word "*quantitas*" which relates to "a primary focus on differences in amount" (Erikson 2018, p. 36). The term qualitative, according to Denzin and Lincoln (2011, p. 8), has its focus on "the qualities of entities" as well as on "processes and meanings that are not experimentally examined or measured in terms of quantity, amount, intensity, or frequency."

Qualitative inquiry, according to Rossman and Rallis (2017, p. 5), is grounded in "empiricism," the philosophical tradition which theorizes that "knowledge is obtained by direct experience through the physical senses." Aristotle, one of the first qualitative researchers, attempted to interpret things in the world by listening and watching. He theorized that ideas that we have are concepts that are derived from our experiences with actual beings, objects, and events. This is what qualitative researchers do. Often, qualitative researchers work closely with individuals, and they listen attentively on what the participants say, probe further, and try to make sense of what the participants tell them.

Qualitative inquiry has its focus on the social world. The qualitative research approach, according to Hesse-Biber (2017, p. 4), offers "a unique grounding position" for researchers to undertake research that "fosters particular ways of asking questions" and "provides a point of view onto the social world," which in turn will help researchers "to obtain understanding of a social issue or problem that privileges subjective and multiple understandings." Qualitative research offers "explanations for objects or social actions" (Rossman and Rallis 2017, p. 8). In the social world, we deal with the subjective experiences of individuals. In different social situations and over time, people's "understanding of reality" can change (Dew 2007, p. 434). This makes qualitative research different from researching the natural world, which can be treated as "objects or things" (Taylor et al. 2016). In order to capture and understand the perspectives of individuals, qualitative inquiry relies heavily on words or stories that these individuals tell researchers (Liamputtong 2013; Patton 2015; Creswell and Poth 2018). Thus, qualitative research has also been recognized as "the word science" (Denzin 2008; Liamputtong 2017).

According to Yin (2016), Marshall and Rossman (2016), Taylor et al. (2016), Tolley (2016), Rossman and Rallis (2017), and Creswell and Poth (2018), there are common features of qualitative inquiry. These are presented in Table 1.

Qualitative research traverses many fields and disciplines (Denzin and Lincoln 2018). Qualitative research has been adopted extensively in the social sciences, particularly in anthropology and sociology. We have also witnessed the wider adoption of a qualitative approach in criminology, social work, education, nursing, and even psychology and medicine. In the past decade or so, qualitative data or interpretive information has been gradually accepted as a crucial component in our understanding of health (Green and Thorogood 2014; Liamputtong 2016, 2017). In public health, particularly the new public health that recognizes the need to "understand" people, Baum (2016, p. 201) suggests that qualitative research is needed since

Table 1 Common features of qualitative inquiry

Qualitative inquiry: Common features
• It is fundamentally interpretive
• The meanings and interpretation of the participants is the essence of qualitative inquiry
• It asks why, how, and under what circumstance things arise
• It explicitly attends to and account for the contextual conditions of the participants
• It takes place in the natural settings of human life
• It emphasizes holistic accounts and multiple realities
• It is situated within some methodological frameworks
• It makes use of multiple methods
• It is emergent rather than rigidly predetermined
• Participants are treated as an active respondents rather than as subject
• The researcher is the means through which the research is undertaken.

it offers "considerable strength in understanding and interpreting complexities" of people's behavior and their health issues (see Liamputtong 2016).

Readers may notice that qualitative research is often used interchangeably with the term "qualitative inquiry." Dimitriadis (2016) suggests that the word "research" should be replaced with the word "inquiry." He argues that the word "research" is "tainted by a lingering positivism." This is the consequence of the paradigm wars operating within the influence of evidence-based practices in medicine that has treated qualitative research badly (Liamputtong 2016, 2017; Torrance 2017; Denzin 2018). Instead, inquiry "implies an open-endedness, uncertainty, ambiguity, praxis, pedagogies of liberation, freedom, [and] resistance" (Denzin and Lincoln 2018, p. 11). I will mainly use the term qualitative inquiry in this chapter, but occasionally, I will also use the term qualitative research.

2   Why Qualitative Inquiry

Readers might ask why should researchers use qualitative inquiry in their research. Creswell and Poth (2018, p. 45) contend that qualitative research is adopted because there is a problem or issue that needs to be explored. This is particularly when "silenced voices" need to be heard or problems or issues cannot be "easily measured." Qualitative inquiry permits researchers to ask questions, and to find answers, that can be difficult or impossible with the quantitative approach (Hesse-Biber 2017). For example, What makes many working-class men continue to smoke? Why do some health and social care programs do not work? What makes some of these programs succeed? Why are screening programs underused? What contributes to stigma and discrimination in the local areas despite extensive media and educational campaigns in the country? Why do women with low incomes put motherhood before marriage? How do single mothers confront their social and economic challenges? How do young refugee people deal with social isolation? How do migrant mothers deal with infant feeding practices in their new social environment? These are some examples of what qualitative research can find answers for health and social care policy-makers and professionals.

According to Yin (2016, p. 3), every real-world issue can be examined by the qualitative inquiry. These real-world issues are what Patton (2015) refers to as "stuff." Yin writes:

> Stuff happens everywhere. Qualitative inquiry documents the stuff that happens among real people in the real world in their own words, from their own perspectives, and within their own contexts; it then makes sense of the stuff that happens by finding patterns and themes among the seeming chaos and idiosyncrasies of lots of stuff.

Qualitative research is essentially crucial for research involving marginalized, vulnerable, or hard-to-reach individuals and communities (Liamputtong 2007, 2010; Taylor et al. 2016; Flick 2018). This is particularly so when they are "too small to become visible" in quantitative research (Flick 2018, p. 452; Greenhalgh et al. 2016;

Yin 2016). More importantly, due to their marginalized, vulnerable status, and distrust in research, most of these individuals tend to decline to participate in research. The nature of qualitative inquiry will permit qualitative researchers to be able to engage with these individuals. In their qualitative research (using life history narratives and intersubjective dialogue) with previously incarcerated HIV-positive women in Alabama, Sprague et al. (2017, p. 725) tell us that:

> The qualitative design and accompanying methods applied in this study enabled us to explore the timing, sequence, and import of life events and behaviors driving risk for HIV and incarceration—while grounding these realities firmly in their unique social context to foster a greater understanding of women's lived experiences. These methods allowed for exploring the interaction and intersection between different variables, such as abuse and drug use, across the life course. Such interaction effects are poorly controlled for in more quantitative or modeling approaches... Qualitative research methods are uniquely able to wrestle with multiple social structural factors and influences in ways that remain unmatched by other research methods.

We have also witnessed the rise of qualitative inquiry in research relating to the social justice issues (see Denzin and Lincoln 2011; Taylor et al. 2016; Daniels et al. 2017). However, this trend has also been practiced by many social scientists in the past. Becker (1967), in his piece *"Whose Side Are We On?,"* discussed the role of social scientists in presenting the voices of some marginalized groups of people. Becker contended that "since powerful people have many means at their disposal to present their versions of reality, we should side with society's underdogs, the powerless" (Taylor et al. 2016, p. 26). C Wright Mills (1959) believed that it is the political responsibility of social scientists to help individuals to understand their "personal troubles," which are also "social issues" that confronted others in the society as well.

Qualitative research has increasingly received interest in the patient-centered care and evidence-based practice in healthcare (Olson et al. 2016). This is mainly because qualitative inquiry can cultivate great understanding of the beliefs, attitudes, and behaviors of the patients and consumers (Liamputtong 2013, 2016, 2017; Olson et al. 2016; Thirsk and Clark 2017). These in-depth understanding will permit health professionals to better accommodate health interventions to suit the needs of the consumers (Yardley et al. 2015). In the areas of HIV/AIDS, for example, HIV/AIDS is a global public health concern, and antiretroviral therapy (ART) has saved many lives worldwide. However, nonadherence to ART is still a huge issue in most parts of the world (Liamputtong et al. 2015; Fields et al. 2017). Sankar et al. (2006, p. S54) contend that qualitative research can contribute greatly to our understanding of the nonadherence of combination antiretroviral therapy (ART). Qualitative inquiry provides a unique means for understanding the complex factors that influence adherence (see Sankar et al. 2006; Liamputtong et al. 2015; Fields et al. 2017).

According to Sankar et al. (2006), qualitative inquiry offers persuasive means for exploring adherence to ART. Often, there are many factors that influence adherence among individuals including culture, stigma, medication beliefs, and access. Sankar et al. (2006, p. S54) argue that "a disregard for the social and cultural context of adherence or the imposition of adherence models inconsistent with local values and practices is likely to produce irrelevant or ineffective interventions." Sankar et al. (2006, p. S65) also contend that thus far qualitative research has been essential in studies concerning ART adherence as it has unfolded "key features and behaviors that could not have been identified using quantitative methods alone" (see Liamputtong et al. 2015).

Qualitative research can contribute effectively to the design of clinical trials in healthcare (Duggleby and Williams 2016; Russell et al. 2016; Toye et al. 2016). Qualitative inquiry helps trial researchers to know their trial participants better. It can also help the researchers to comprehend the successes and failure of trial interventions, which can save money on trials that might not work (Toye et al. 2016). Essentially, the qualitative inquiry can "show the human faces behind the numbers, providing critical context when interpreting statistical outcomes," as well as make sure that "the numbers can be understood as constituting meaningful changes in the lives of real people" (Patton 2015, p. 179).

3   Qualitative Researchers

Fundamentally, qualitative researchers look for meanings that people have constructed (Taylor et al. 2016; Hesse-Biber 2017; Denzin and Lincoln 2018). They are interested in learning about "how people make sense of their world and the experiences they have in the world" (Merriam and Tisdell 2016, p. 15). Qualitative researchers can be perceived as "constructivists" who "seek answers to their questions in the real world" (Rossman and Rallis 2017, p. 4) and then "interpret what they see, hear, and read in the worlds around them" (p. 5).

Qualitative researchers are individuals who commit themselves to several issues. Figure 1 presents the characteristics of qualitative researchers.

Qualitative researchers are also seen as craftspersons (Taylor et al. 2016, p. 11). They are malleable in how they carry out their research. They are social scientists who are inspired to be a research methodologist who "deploy a wide range of interconnected interpretive practices, hoping always to get a better understanding of the subject matter at hand" (Denzin and Lincoln 2018, p. 10).

Qualitative inquiry is personal and sensitize to personal biography as the researcher acts as "the instrument of inquiry" (Patton 2015, p. 3; Rossman and Rallis 2017, p. 9). What brings qualitative researchers to their research matters. Thus, qualitative researchers tend to acknowledge who they are and how their personal biography frames their research. They value their "unique perspective as a source of understanding rather than something to be cleansed from the study" (Rossman and Rallis 2017, p. 9). Qualitative research is an exquisite sensitivity to personal biography.

4   Qualitative Inquiry: The Distinctiveness

I have pointed out to general features of qualitative inquiry at the beginning. In the following sections, I will discuss some salient characteristics in more details.

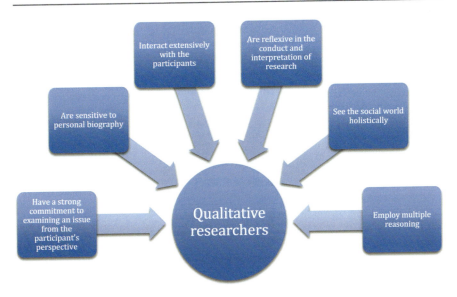

Fig. 1 Salient characteristics of qualitative researchers

4.1   Theory Generation: Inductive Approach

Typically, qualitative research is inductive. The qualitative research adopts "a logic of 'theory generation' rather than 'theory testing'" as practiced in quantitative research (Hesse-Biber 2017, p. 11; see also ▶ Chap. 3, "Quantitative Research"). Inductive reasoning will allow researchers to adopt particular understandings and develop a general conceptual understanding of the issue they examine (Babbie 2016; Taylor et al. 2016; Rossman and Rallis 2017).

Qualitative research projects typically commence with the collection of particular data and follow with the analysis that will allow the researchers to form a more general understanding of the issue they examine (Green and Thorogood 2014; Berger and Lorenz 2016; Merriam and Tisdell 2016; Hesse-Biber 2017). Thus, guiding questions that they use to collect their data are open-ended that permit allow multiple meanings to surface. Usually, the questions that researchers ask in qualitative research start with words like "how," "why," or "what" (Hesse-Biber 2017, p. 4), rather than questions about "how many" or "how much" (Green and Thorogood 2014, p. 5). For instance, How do women living with breast cancer cope with their health condition during chemotherapy treatment?, Why do many working mothers experience struggles to continue breastfeeding?, and What prejudice and discrimination that transwomen experience in their everyday life?

Nevertheless, despite being inductive, often qualitative researchers commence their study with some conceptual frameworks that shape their decisions in undertaking their research. However, this framework is adjustable. It can change as the study is being conducted (Rossman and Rallis 2017).

4.2   Framing the Study: Methodological Frameworks

Qualitative research tends to be situated within some methodological frameworks. Methodology determines a method for researchers to produce data for analysis (Taylor et al. 2016; Hesse-Biber 2017; Silverman 2017; Gaudet and Robert 2018). Avis (2003, p. 1003) contends that qualitative researchers need to provide their "methodological justification" by discussing the reason why they select a particular method in their research. The methods that researchers select and what they expect to get out of those methods is strongly formed by their "methodological position" (Dew 2007, p. 433).

A methodological framework provides "ways of seeing" (Morgan 1986, p. 12) in the conduct of qualitative research. Qualitative researchers must defend the adoption of their methods based on an appropriate methodological framework (Avis 2003; Taylor et al. 2016). It is not enough to say that we will use an in-depth interview in our research to examine the experience of living with chronic illness among our research participants. We must provide some framework to justify our method. In this case, we may say that we are interested in the lived experience of participants, and hence phenomenology will be our methodological framework, and with this type of research, an in-depth interview is appropriate because it will allow participants to tell their stories in great depth (Liamputtong 2013).

There are many methodological frameworks and research approaches practiced by qualitative researchers (Bradbury-Jones et al. 2017; Denzin and Lincoln 2018; Gaudet and Robert 2018). These include ethnography, symbolic interactionism, phenomenology, feminism, postmodernism, ethnomethodology, empathy, participatory action research, grounded theory, case study, and Indigenist methodology.

According to Hesse-Biber (2017, p.10), "the diversity of the methods with which qualitative researchers work is one of the distinguishing features of the qualitative landscape, which makes for a vast range of possible research topics and questions" (see also Bradbury-Jones et al. 2017; Denzin and Lincoln 2018). Bradbury-Jones et al. (2017, p. 628) suggest that the diversity of approaches that are available to qualitative researchers provides "a significant benefit" as it "provides a rich pool of methodological and technical options" that researchers can employ in their research.

In my own writing, I have advocated several methodological frameworks that qualitative researchers can draw on. These include ethnography, phenomenology, symbolic interactionism, hermeneutics, feminism, and postmodernism. Taylor et al. (2016) recommend several frameworks. These include phenomenology, symbolic interactionism, ethnomethodology, feminism, institutional ethnography, postmodernism, narrative analysis, and multi-sited, global research. In her recent book, Hesse-Biber (2017) suggests different methodological lenses including symbolic interactionism, dramaturgy, phenomenology, ethnomethodology, postmodernism, feminisms, critical race theory, and queer theory. Gaudet and Robert (2018) propose five frameworks: phenomenology, grounded theory, discourse analysis, narrative analysis, and ethnography (see also ▶ Chaps. 8, "Critical Theory: Epistemological Content and Method," ▶ 10, "Symbolic Interactionism as a Methodological Framework," ▶ 11, "Hermeneutics: A Boon for Cross-Disciplinary Research,"

▶ 13, "Critical Ethnography in Public Health: Politicizing Culture and Politicizing Methodology," ▶ 14, "Empathy as Research Methodology," ▶ 16, "Ethnomethodology," ▶ 15, "Indigenist and Decolonizing Research Methodology," ▶ 17, "Community-Based Participatory Action Research," and ▶ 27, "Institutional Ethnography").

Different methodological frameworks offer different perspectives in the conduct of qualitative research (Silverman 2017). Each methodological framework may also be suitable for some research questions than others. Qualitative researchers must carefully consider which methodological framework will be more appropriate for their research project.

4.3   "Data Enhancers": Qualitative Data Collection Methods

Qualitative research methods have been coined as "data enhancers" as they disclose "elements of empirical reality" which cannot be revealed by numbers (Neuman 2011; Berger and Lorenz 2016). All qualitative research methods attempt to obtain "rich, highly detailed accounts," what Clifford Geertz (1973) refers to as "thick description" of a small number of people (Berger and Lorenz 2016).

Traditionally, there are a number of methods that qualitative researchers use to collect empirical materials in their research. These include in-depth interviewing method, focus group, life or oral history, and unobtrusive methods (see also ▶ Chaps. 23, "Qualitative Interviewing," ▶ 25, "The Life History Interview," and ▶ 29, "Unobtrusive Methods"). Among these methods, the in-depth interviewing method is the most commonly known technique and is widely employed by qualitative researchers (Minichiello et al. 2008; King and Horrocks 2010; Brinkmann and Kvale 2015; Serry and Liamputtong 2017). As Fontana and Frey (2000, p. 646) suggest, we reside in "an interview society." Thus, interviews have become essential for individuals to make sense of people's lives (Brinkmann and Kvale 2015).

In-depth interviewing method refers to "face-to-face encounter between the researcher and informants directed toward understanding informants' perspectives on their experiences, or situations as expressed in their own words" (Taylor et al. 2016, p. 102). Gubrium et al. (2012) suggest that interviewing is a way of collecting empirical data about the social world of individuals by inviting them to talk about their lives in great depth. In an interview conversation, the researcher asks questions and then listens to what individuals say about their experiences such as their dreams, fears, and hopes. The researcher will then hear about their perspectives in their own words and learn about their family and social life and work (Kvale 2007; Brinkmann and Kvale 2015). Rubin and Rubin (2012, p. 3) say that in-depth interviewing "helps reconstruct events the researchers have never experienced." It is particularly valuable when researchers wish to learn about the life of a wide range of individuals, "from illegal border crossing to becoming a paid assassin" (see also ▶ Chap. 23, "Qualitative Interviewing").

There are other qualitative approaches that embody some forms of qualitative method. However, they are often referred to as the method (although some may refer them to as research traditions). These include the ethnographic method, narrative research, grounded theory, memory work, autoethnography, case study, and participatory action research. It is not feasible to discuss all of these methods in this chapter. Readers can read more about them in the Qualitative Approaches and Practices in the handbook (see also ▶ Chaps. 24, "Narrative Research," ▶ 26, "Ethnographic Method," ▶ 27, "Institutional Ethnography," ▶ 18, "Grounded Theory Methodology: Principles and Practices," ▶ 30, "Autoethnography," and ▶ 31, "Memory Work").

4.4   The Source of Data: Research Participants, Sampling Strategy, and Saturation Concept

Qualitative research is concerned with the in-depth understanding of the issue or issues under examination; it thus relies heavily on individuals who are able to provide information-rich accounts of their experiences (Patton 2015; Yin 2016; Creswell and Poth 2018). This is referred to as the purposive sampling strategy (Liamputtong 2013; Houser 2015; Patton 2015; Bryman 2016; Yin 2016; Hesse-Biber 2017; Creswell and Poth 2018). Purposive sampling is a deliberate selection of specific individuals, events, or settings because of the crucial information they can provide, which cannot be obtained as adequately through other channels (Patton 2015; White 2015; Babbie 2016; Yin 2016; Creswell and Poth 2018). For example, in research that is concerned with how women with breast cancer deal with the side effects of chemotherapy treatment, purposive sampling will require the researcher to find women who have experienced the side effects of chemotherapy, instead of randomly selecting any cancer patients from an oncologist's patient list. The purposive sample is also referred to as a "judgment sample" (Hesse-Biber 2017, p. 62). Research participants are selected to be included in a study due to their particular characteristics as determined by the particular objective of the research.

The powers of purposive sampling techniques, Patton (2015, p. 264, original emphasis) suggests, "lie in selecting *information-rich cases* for study in depth." These information-rich cases are individuals or events or settings from which researchers can learn extensively about issues they wish to examine (Houser 2015; Liamputtong 2017; Creswell and Poth 2018).

The focus of decisions about sample size in qualitative research is on "flexibility and depth." As I have suggested earlier, a fundamental concern of qualitative research is quality, not quantity. A qualitative research usually involves a small number of individuals. Qualitative research aims to examine the "meanings" or a "process" that individuals give to their own social situations. It does not require a generalization of the findings as in positivist science (Houser 2015; Hesse-Biber 2017). Qualitative research trades "breadth for depth" (Hesse-Biber 2017, p. 63). The important question to ask is whether the sample provides data that will allow the research questions or aims to be thoroughly addressed (Mason 2002; Green and Thorogood 2014; Houser 2015).

In qualitative research, there is no set formula that can be adopted to determine the sample size rigidly (Malterud et al. 2016). Often, at the commencement of the research project, the number of participants to be recruited is not definitely known. However, data saturation, a concept associated with grounded theory, tends to be adopted by qualitative researchers to determine the number of research participants (see Hennink et al. 2017). Saturation is considered to occur when little or no new data is being generated (Padgett 2008; Liamputtong 2013; Green and Thorogood 2014; White 2015). O'Reilly and Parker (2012, p. 192) suggest that saturation is reached at "the point at which there are fewer surprises and there are no more emergent patterns in the data." The sample is adequate when "the emerging themes have been efficiently and effectively saturated with optimal quality data" (Carpenter and Suto 2008, p. 152) and when "sufficient data to account for all aspects of the phenomenon have been obtained" (Morse et al. 2002, p. 12).

Saturation is usually established during the data collection process (Houser 2015; Malterud et al. 2016). This means that in most qualitative research, data collection and data analysis tend to occur concurrently. This is the only way that researchers can know if they have reached saturation or not (Liamputtong 2013; Liamputtong and Serry 2017). Despite its usefulness, saturation has received some criticism, and many qualitative researchers have also questioned its use (see Morse 2015a). There are also different levels and types of saturation (see code saturation and meaning saturation discussed by Hennink et al. 2017). Due to space of the chapter, I encourage readers to refer to some of the references I have included here.

4.5   Qualitative Research and Trustworthiness

Qualitative research has criteria that can be used to evaluate the quality of the research. Within the qualitative approach, we use the term "rigor" or "trustworthiness," which refers to the quality of qualitative inquiry (Liamputtong 2013; Coleman and Unrau 2014; Morse 2015b; Bryman 2016; Marshall and Rossman 2016; Yin 2016; Rossman and Rallis 2017). A trustworthy research, according to Houser (2015, p. 146), is a research that researchers have "drawn the correct conclusions about the meaning of an event or phenomenon." In health research and practice in particular, trustworthiness means that "the findings must be authentic enough to allow practitioners to act upon them with confidence" (Raines 2011, p. 497).

Lincoln and Guba (1989) propose four criteria that many qualitative researchers have adopted to evaluate the trustworthiness of their qualitative research. The criteria were invented in responses to the influence of quantitative research that relies heavily on the issues of validity and reliability (Morse 2015b, 2018; see ▶ Chaps. 3, "Quantitative Research," and ▶ 38, "Measurement Issues in Quantitative Research"). These criteria include credibility, dependability, confirmability, and transferability.

Credibility relates to the question "how believable are the findings?" (Bryman 2016, p. 44). A credible study, according to Yin (2016, p. 85), refers to a study that "provides assurance that you have properly collected and interpreted the data, so the findings and conclusions accurately reflect and represent the world that was studied." Dependability focuses on "the consistency or congruency of the results" (Raines

2011, p. 456). It asks whether the research findings fit the data that have been collected (Carpenter and Suto 2008). Confirmability attempts to show that the findings and the interpretations of the findings do not derive from the imagination of the researchers but are clearly linked to the data.

Transferability (also referred to as applicability) begs the question of "to what degree can the study findings be *generalised* or applied to other individuals or groups, contexts, or settings?" or "do the findings apply to other contexts?" (Bryman 2016, p. 44). It attempts to establish the "generalisability of inquiry" (Tobin and Begley 2004, p. 392). Transferability in qualitative research emphasizes the theoretical or analytical generalizability of research findings. Transferability suggests that the theoretical knowledge that researchers obtained from qualitative research can be applied to other similar individuals, groups, or settings (Sandelowski 2004; Carpenter and Suto 2008; Padgett 2008). This is particularly so for research employing the ethnographic method and grounded theory research (see Gobo and Marciniak 2016; Wong et al. 2017; Creswell and Poth 2018; see also ▶ Chaps. 26, "Ethnographic Method," and ▶ 18, "Grounded Theory Methodology: Principles and Practices").

There are several strategies that qualitative researchers employ to ensure the rigor of their study. Often, these include prolonged engagement, persistent observation, thick description, peer review, member checking, external audits, triangulation, and reflexivity (Liamputtong 2013; Coleman and Unrau 2014; Morse 2015b; Rossman and Rallis 2017; Creswell and Poth 2018). Each of these strategies may not be suitable for all types of qualitative research, and some strategies may create difficulties for some qualitative projects than others (see Morse 2015b). Qualitative researchers need to take into account the type of qualitative method they use and social context on which their research is situated.

4.6 Making Sense of Qualitative Data: Analytic Strategies

As I have suggested, qualitative inquiry relies heavily on words and stories, how qualitative researchers make sense of their data also reflect this. Data analysis and interpretation of data are an exciting process that qualitative researchers bring meaning to the piles of data that they have gathered (Rossman and Rallis 2017, p. 227). It is a complex process and involves the following: "organizing the data, familiarizing with the data, identifying categories, coding the data, generating themes, interpreting, and searching for alternative understanding" (Rossman and Rallis 2017, p. 237).

There are several analytic strategies that qualitative researchers analyze and organize their data. These include content analysis, thematic analysis, discourse analysis, narrative analysis, and semiotic analysis method (Liamputtong 2013; see ▶ Chaps. 47, "Content Analysis: Using Critical Realism to Extend Its Utility," ▶ 48, "Thematic Analysis," ▶ 49, "Narrative Analysis," and ▶ 50, "Critical Discourse/Discourse Analysis"). Thematic analysis method seems to be the most commonly adopted in qualitative research. It is seen as "a foundational method for qualitative analysis" (Braun and Clarke 2006, p. 78). Thematic analysis is "a method for identifying, analysing and reporting patterns (themes) within the data" (Braun and Clarke 2006, p. 79; see also ▶ Chap. 48, "Thematic Analysis").

Nowadays, there are some software programs, such as NVivo, ATLAS.ti, and MAXQDA, that can assist qualitative researchers with the analysis. This is referred to as "computer assisted qualitative data analysis" (CAQDAS) (Yin 2016; Serry and Liamputtong 2017; see also ▶ Chap. 52, "Using Qualitative Data Analysis Software (QDAS) to Assist Data Analyses"). However, the software programs do not analyze the data per se; they help to manage the qualitative data more efficiently (Creswell and Poth 2018). The researchers still have to perform "all the analytic thinking" (Yin 2016, p. 188). Another word, "the real analytical work" still "takes place in your head" (Patton 2015, p. 531).

It must be noted that qualitative data analysis is a time-consuming process. However, it is also "creative and fascinating." Sufficient time needs to be dedicated to data analysis in qualitative research. Taylor et al. (2016) contend that data analysis is the most difficult thing for qualitative researchers to teach or communicate to others. Most novice qualitative researchers tend to "get stuck" when they start to analyze their data. Taylor et al. (2016, p. 168) contend that this is because qualitative analysis "is not fundamentally a mechanical or technical process; it is a process of inductive reasoning, thinking and theorizing." For most qualitative researchers, their ability to analyze their data arises out of their experience. The best way that researchers can learn more about inductive analysis is to work with "a mentor who helps them learn to see patterns or themes in data by pointing these out" and by reading qualitative papers and studies to see how others have come up with their data. This is my recommendation as well.

5 Conclusion and Future Directions

Qualitative research is now a well-established and important mode of inquiry for the social and health science fields (Marshall and Rossman 2016). However, the field of qualitative research is ever advancing (Hesse-Biber 2017). Gaudet and Robert (2018, p. 2) suggest that "qualitative research is a never-ending journey." This is because "there are always new phenomena to learn about, new methods to invent and new forms of knowledge to create." We have witnessed this, and chapters in the Innovative Research Methods in Health Social Sciences attest to this.

We have also witnessed that qualitative inquiry is increasingly used as part of a mixed methods research (Hesse-Biber 2017). This is clearly articulated in ▶ Chaps. 60, "Appraising Mixed Methods Research," ▶ 39, "Integrated Methods in Research," and ▶ 40, "The Use of Mixed Methods in Research". Creswell and Poth (2018, p. 47) suggest that qualitative research "keeps good company" with quantitative research. However, it is not "an easy substitute for a 'statistical' or quantitative study."

Most importantly, we are now living in a fractured world, where we have and continue to be confronted with social inequalities and injustices in all corners of the world (Flick 2014). We need qualitative research that can help us to find better answers that better suit people, particularly those who are marginalized and vulnerable (Mertens 2009; Denzin 2015; Flick 2018). Qualitative inquiry can lead to a positive change in the lives of many people. This is what Denzin (2010, 2015, 2017) has advocated. Denzin (2017, p. 8) puts this clearly when he calls for qualitative research that "matters in the lives of those who daily experience social injustice." We have witnessed critical qualitative research in the last decades or so, and this is also reflected in several chapters in this handbook (see ▶ Chaps. 12, "Feminism and Healthcare: Toward a Feminist Pragmatist Model of Healthcare Provision," ▶ 15, "Indigenist and Decolonizing Research Methodology," and ▶ 13, "Critical Ethnography in Public Health: Politicizing Culture and Politicizing Methodology"). I contend that critical qualitative research will continue to play a crucial role in qualitative inquiry in the years to come.

I would like to end this chapter with the Chinese proverb: "The journey is the reward." The proverb indeed, as Maschi and Youdin (2012, p. 206) contend, underscores "the importance of the process of using a qualitative approach to understanding human experience." I hope readers who adopt qualitative inquiry will "find it exhilarating and deeply moving." And indeed, it "can change your worldview" (Rossman and Rallis 2017, p. 10).

References

Avis M. Do we need methodological theory to do qualitative research? Qual Health Res. 2003;13(7):995–1004.

Babbie E. The practice of social research. 14th ed. Boston: Cengage Learning; 2016.

Baum F. The new public health. 5th ed. Melbourne: Oxford University Press; 2016.

Becker HS. Whose side are we on? Soc Probl. 1967;14(3):239–47.

Berger RJ, Lorenz LS. Disability and qualitative research. In: Berger RJ, Lorenz LS, editors. Disability and qualitative inquiry: methods for rethinking an ableist world (Chapter 1). Hoboken: Taylor & Francis; 2016.

Bradbury-Jones C, Breckenridge J, Clark MT, Herber OR, Wagstaff C, Taylor J. The state of qualitative research in health and social science literature: a focused mapping review and synthesis. Int J Soc Res Methodol. 2017;20(6):627–45.

Braun V, Clarke V. Using thematic analysis in psychology. Qual Res Psychol. 2006;3:77–101.

Brinkmann S, Kvale S. InterViews: learning the craft of qualitative research interviewing. 3rd ed. Thousand Oaks: Sage; 2015.

Bryman A. Social research methods. 5th ed. Oxford: Oxford University Press; 2016.

Carpenter C, Suto M. Qualitative research for occupational and physical therapists: a practical guide. Oxford: Blackwell; 2008.

Coleman H, Unrau YA. Qualitative data analysis – a step-by-step approach. In: Grinnell RM, Unrau YA, editors. Social work research and evaluation: foundations of evidence-based practice. 19th ed. Oxford: Oxford University Press; 2014. p. 554–72.

Creswell JW, Poth CN. Qualitative inquiry and research design: choosing among five approaches. 5th ed. Thousand Oaks: Sage; 2018.

Daniels K, Hanefeld J, Marchal B. Social sciences: vital to improving our understanding of health equity, policy and systems. Int J Equity Health. 2017;16:57. https://doi.org/10.1186/s12939-017-0546-6.

Denzin NK. The new paradigm dialogs and qualitative inquiry. Int J Qual Stud Educ. 2008;21(4):315–25.

Denzin NK. The qualitative manifesto: a call to arms. Walnut Creek: Left Coast Press; 2010.

Denzin NK. What is critical qualitative inquiry? In: Cannella GS, Perez MS, Pasque PA, editors. Critical qualitative inquiry: foundations and futures. Walnut Creek: Left Coast Press; 2015. p. 31–50.

Denzin NK. Critical qualitative inquiry. Qual Inq. 2017;23(1):8–16.

Denzin NK. The elephant in the living room, or extending the conversation about the politics of evidence. In: Denzin NK, Loncoln YS, editors. The Sage handbook of qualitative research. 5th ed. Thousand Oaks: Sage; 2018. p. 839–53.

Denzin NK, Lincoln YS, editors. The Sage handbook of qualitative research. 4th ed. Thousand Oaks: Sage; 2011.

Denzin NK, Lincoln YS. Introduction: the discipline and practice of qualitative research. In: Denzin NK, Loncoln YS, editors. The Sage handbook of qualitative research. 5th ed. Thousand Oaks: Sage; 2018. p. 1–26.

Dew K. A health researcher's guide to qualitative methodologies. Aust N Z J Public Health. 2007;31(5:433–7.

Dimitriadis G. Reading qualitative inquiry through critical pedagogy: some reflections. Int Rev Qualit Res. 2016;9:140–6.

Duggleby W, Williams A. Methodological and epistemological considerations in utilizing qualitative inquiry to develop interventions. Qual Health Res. 2016;26(2):147–53.

Erikson F. A history of qualitative inquiry in social and educational research. In: Denzin NK, Lincoln YS, editors. The Sage handbook of qualitative research. 5th ed. Thousand Oaks: Sage; 2018. p. 36–65.

Fields EL, Bogart LM, Thurston IB, Hu CH, Skeer MR, Safren SA, Mimiaga MJ. Qualitative comparison of barriers to antiretroviral medication adherence among perinatally and behaviorally HIV-infected youth. Qual Health Res. 2017;27(8):1177–89.

Flick U. An introduction to qualitative research. 5th ed. Los Angeles: Sage; 2014.

Flick U. Triangulation. In: Denzin NK, Lincoln YS, editors. The Sage handbook of qualitative research. 5th ed. Thousand Oaks: Sage; 2018. p. 444–61.

Fontana A, Frey JH. The interview: from structured questions to negotiated text. In: Denzin NK, Lincoln YS, editors. Handbook of qualitative research. 2nd ed. Thousand Oaks: Sage; 2000. p. 645–72.

Gaudet S, Robert D. A journey through qualitative research: from design to reporting. London: Sage; 2018.

Geertz C. The interpretation of cultures. New York: Basic Books; 1973.

Gobo G, Marciniak LT. What is ethnography? In: Silverman D, editor. Qualitative research. 4th ed. London: Sage; 2016. p. 103–19.

Green J, Thorogood N. Qualitative methods for health research. 3rd ed. Los Angeles: Sage; 2014.

Greenhalgh T, Annandale E, Ashcroft R, Barlow J, Black N, Bleakly A, … Ziebland S. An open letter to the BMJ editors on qualitative research. Br Med J. 2016;352:i563.

Gubrium JF, Holstein JA, Marvasti AB, McKinney KD. Introduction: the complexity of the craft. In: Gubrium JF, Holstein JA, Marvasti AB, McKinney KD, editors. The Sage handbook of interview research: the complexity of the craft, Introduction section. 2nd ed. Thousand Oaks: Sage; 2012.

Hennink MM, Kaiser BN, Marconi VC. Code saturation versus meaning saturation: how many interviews are enough? Qual Health Res. 2017;27(4):591–608.

Hesse-Biber SN. The practice of qualitative research. 3rd ed. Thousand Oaks: Sage; 2017.

Houser J. Nursing research: reading, using, and crating evidence. 3rd ed. Sudbury: Jones & Bartlett Learning; 2015.

King N, Horrocks C. Interviews in qualitative research. London: Sage; 2010.

Kvale S. Doing interviews. London: Sage; 2007.

Liamputtong P. Researching the vulnerable: a guide to sensitive research methods. London: Sage; 2007.

Liamputtong P. Performing qualitative cross-cultural research. Cambridge: Cambridge University Press; 2010.

Liamputtong P. Qualitative research methods. 4th ed. Melbourne: Oxford University Press; 2013.

Liamputtong P. Qualitative research methodology and evidence-based practice in public health. In: Liamputtong P, editor. Public health: local and global perspectives. Melbourne: Cambridge University Press; 2016. p. 169–87.

Liamputtong P. The science of words and the science of numbers. In: Liamputtong P, editor. Research methods in health: foundations for evidence-based practice. 3rd ed. Melbourne: Oxford University Press; 2017. p. 3–28.

Liamputtong P, Serry T. Making sense of qualitative data. In: Liamputtong P, editor. Research methods in health: foundations for evidence-based practice. 3rd ed. Melbourne: Oxford University Press; 2017. p. 421–36.

Liamputtong P, Haritavorn N, Kiatying-Angsulee N. Local discourse on antiretrovirals and the lived experience of women living with HIV/AIDS in Thailand. Qual Health Res. 2015;25(2):253–63.

Lincoln YS, Guba EG. Fourth generation evaluation. Newbury Park: Sage; 1989.

Malterud K, Siersma VD, Guassora AD. Sample size in qualitative interview studies: guided by information power. Qual Health Res. 2016;26(13):1753–60.

Marshall C, Rossman GB. Designing qualitative research. 6th ed. Thousand Osks: Sage; 2016.

Maschi T, Youdin R. Social worker as researcher: integrating research with advocacy. Boston: Pearson; 2012.

Mason J. Qualitative researching. 2nd ed. London: Sage; 2002.

Merriam SB, Tisdell EJ. Qualitative research: a guide to design and implementation. 4th ed. San Francisco: Jossey-Bass; 2016.

Mertens D. Transformative research and evaluation. New York: Guildford Press; 2009.

Mills CW. The sociological imagination. Harmondsworth: Penguin; 1959.

Minichiello V, Aroni R, Hays T. In-depth interviewing: principles, techniques, analysis. 3rd ed. Frenchs Forest: Pearson Education Australia; 2008.

Morgan G. Images of organizations. Beverhill: Sage; 1986.

Morse JM. Data are saturated. … Qual Health Res. 2015a;25(5):587–8.

Morse JM. Critical analysis of strategies for determining rigor in qualitative inquiry. Qual Health Res. 2015b;25(9):1212–22.

Morse JM. Reframing rigor in qualitative inquiry. In: Denzin NK, Lincoln YS, editors. The Sage handbook of qualitative research. 5th ed. Thousand Oaks: Sage; 2018. p. 796–817.

Morse, J.M., Barrett, M., Mayan, M., Olson, K., & Spiers, J.. Verification strategies for establishing reliability and validity in qualitative research. Int J Qual Methods. 2002;1(2), Article 2. www.ualberta.ca/~iiqm/backissues/1_2Final/pdf/morseetal.pdf.

Neuman WL. Social research methods: qualitative and quantitative approaches. 7th ed. Boston: Allyn & Bacon; 2011.

O'Reilly M, Parker N. 'Unsatisfactory saturation': a critical exploration of the notion of saturated sample sizes in qualitative research. Qual Res. 2012;13(2):190–7.

Olson K, Young RA, Schultz IZ. Handbook of qualitative health research for evidence-based practice. New York: Springer; 2016.

Padgett DK. Qualitative methods in social work research. 2nd ed. Los Angeles: Sage; 2008.

Patton MQ. Qualitative research and evaluation methods. 4th ed. Thousand Oaks: Sage; 2015.

Raines JC. Evaluating qualitative research studies. In: Grinnell RM, Unrau YA, editors. Social work research and evaluation: foundations of evidence-based practice. 9th ed. New York: Oxford University Press; 2011. p. 488–503.

Rogers AG. When methods matter: qualitative research issues in psychology. In: Brizuela BM, Stewart JP, Carrillo RG, Berger JG, editors. Acts of inquiry in qualitative research. Cambridge, MA: Harvard Educational Review; 2000. p. 51–60.

Rossman GB, Rallis SF. Learning in the field: an introduction to qualitative research. 4th ed. Thousand Oaks: Sage; 2017.

Rubin HJ, Rubin IS. Qualitative interviewing: the art of hearing data. 3rd ed. Thousand Oaks: Sage; 2012.

Russell J, Berney L, Stanfeld S, Lanz D, Kerry S, Chandola T, Bhui K. The role of qualitative research in adding value to a randomized controlled trial: lessons from a pilot study of a guided e-learning intervention for managers to improve employee wellbeing and reduce sickness absence. Trials. 2016;17:396. https://doi.org/10.1186/s13063-016-1497-8.

Sandelowski M. Using qualitative research. Qual Health Res. 2004;14(10):1366–86.

Sankar A, Golin C, Simoni JM, Luborsky M, Pearson C. How qualitative methods contribute to understanding combination antiretroviral therapy adherence. J Acquir Immune Defic Syndr. 2006;43(Supplement 1):S54–68.

Serry T, Liamputtong P. The in-depth interviewing method in health. In: Liamputtong P, editor. Research methods in health: Foundations for evidence-based practice. 3rd ed. Melbourne: Oxford University Press; 2017. p. 67–83.

Silverman D. Doing qualitative research. 5th ed. London: Sage; 2017.

Sprague C, Scanlon ML, Pantalone DW. Qualitative research methods to advance research on health inequities among previously incarcerated women living with HIV in Alabama. Health Educ Behav. 2017;44(5):716–27.

Taylor SJ, Bogdan R, DeVault M. Introduction to qualitative research methods: a guidebook and resource. Hoboken: Wiley; 2016.

Thirsk LM, Clark AM. Using qualitative research for complex interventions: the contributions of hermeneutics. Int J Qual Methods. 2017;16:1–10. https://doi.org/10.1177/1609406917721068.

Tobin GA, Begley CM. Methodological rigour within a qualitative framework. J Adv Nurs. 2004;48(4):388–96.

Tolley EE. Qualitative methods in public health: a field guide for applied research. Hoboken: Wiley; 2016.

Torrance H. Experimenting with qualitative inquiry. Qual Inq. 2017;23(1):69–76.

Toye F, Williamson E, Williams MA, Fairbank J, Lamb SE. What value can qualitative research add to quantitative research design?: an example from an Adolescent Idiopathic Scoliosis Trial Feasibility Study. Qual Health Res. 2016;26(13):1838–50.

White AH. Using samples to provide evidence. In: Schmidt NA, Brown JM, editors. Evidence-based practice for nurses: appraisal and application of research. 3rd ed. Burlington: Jones & Bartlett Learning; 2015. p. 294–319.

Wong P, Liamputtong P, Rawson H. Grounded theory in health research. In: Liamputtong P, editor. Research methods in health: foundations for evidence-based practice. 3rd ed. Melbourne: Oxford University Press; 2017. p. 138–56.

Yardley L, Morrison L, Bradbury K, Muller I. The person-based approach to intervention development: application to digital health-related behavior change interventions. J Med Internet Res. 2015;17(1):e30. https://doi.org/10.2196/jmir.4055.

Yin RK. Qualitative research: from start to finish. 2nd ed. New York: Guildford Press; 2016.

Quantitative Research

Leigh A. Wilson

Contents

L. A. Wilson (✉)
School of Science and Health, Western Sydney University, Penrith, NSW, Australia

Faculty of Health Science, Discipline of Behavioural and Social Sciences in Health, University of Sydney, Lidcombe, NSW, Australia
e-mail: l.wilson@westernsydney.edu.au; leigh.wilson@sydney.edu.au

© Springer Nature Singapore Pte Ltd. 2019
P. Liamputtong (ed.), *Handbook of Research Methods in Health Social Sciences*,
https://doi.org/10.1007/978-981-10-5251-4_54

Abstract

Quantitative research methods are concerned with the planning, design, and implementation of strategies to collect and analyze data. Descartes, the seventeenth-century philosopher, suggested that *how* the results are achieved is often more important than the results themselves, as the journey taken along the research path is a journey of discovery. High-quality quantitative research is characterized by the attention given to the methods and the reliability of the tools used to collect the data. The ability to critique research in a systematic way is an essential component of a health professional's role in order to deliver high quality, evidence-based healthcare. This chapter is intended to provide a simple overview of the way new researchers and health practitioners can understand and employ quantitative methods. The chapter offers practical, realistic guidance in a learner-friendly way and uses a logical sequence to understand the process of hypothesis development, study design, data collection and handling, and finally data analysis and interpretation.

Keywords
Quantitative · Research · Epidemiology · Data analysis · Methodology · Interpretation

1   Introduction

Many health professionals consider the conduct of research beyond the realms of their capabilities and, therefore, leave quantitative research to those who are experienced in the field. Others focus on the rich narrative data provided by qualitative research rather than focus on numbers. Quantitative and qualitative research methods are complementary in that they each provide unique and specific information on which to inform and/or develop new clinical or professional practice, change policy, or merely add to the body of academic literature in a given topic area (see also ▶ Chaps. 2, "Qualitative Inquiry," and ▶ 4, "The Nature of Mixed Methods Research").

Many healthcare professionals do not wish to undertake research. However, health decisions are increasingly based on the best available evidence, and, therefore, health professionals need to know how to read and interpret the best available evidence published in their field (Liamputtong 2017). This chapter is intended to take the anxiety out of conducting and/or interpreting quantitative methods and to provide a simple overview of the way new researchers and practitioners can understand and employ quantitative research. The chapter offers practical, realistic guidance in a learner-friendly way and uses a logical sequence to understand the process of hypothesis development, study design, data collection and handling, and finally data analysis and interpretation. This is not a chapter designed to give researchers a detailed lesson in statistical analysis or sophisticated techniques; these are best left to experienced statisticians.

Research is fun – it is merely detective work! It is about investigating something that is not fully understood, collecting all the evidence, reviewing the information, and coming up with an explanation. If more people thought this way, research would be far more popular!

2   What Is Quantitative Research?

Quantitative research focuses on the objective measurement of data that are collected through questionnaires, surveys, clinical measurement, or polls and that are analyzed numerically using statistical techniques. The value of quantitative research is that results gained from numerical data in a sample population can be used to generalize or explain a particular phenomenon in the general population (Babbie 2016).

Quantitative research can be either experimental or descriptive (nonexperimental, i.e., describes a population in specific terms). Experimental research is used to identify whether there is any relationship between two things (called variables) and, if so, whether that relationship is positive or negative. For example, it is known that the greater the amount of exercise one does in a given day is positively associated with levels of fitness (i.e., the more exercise you do the fitter you get) (types of variables will be discussed later in this chapter).

Researchers who use experimental methods seek to identify and isolate specific variables that sit within the framework of the study. Once these are identified, the researchers review the data to determine whether there is any correlation or relationship between the variables or whether there may be a likelihood of causality between variables (i.e., one variable could cause the other). It is critical that researchers attempt to control the environment in which the study is conducted to avoid the risk of variables other than those under study accounting for any relationship identified (McNabb 2007; Babbie 2016; see also ► Chaps. 34, "Single-Case Designs," ► 36, "Eliciting Preferences from Choices: Discrete Choice Experiments," and ► 37, "Randomized Controlled Trials").

There are a number of strengths and benefits of using quantitative research methods. They include:

- An objective viewpoint – thereby minimizing bias and distancing the researcher from the participants of the study
- Greater accuracy of results – gained by employing pre-established statistical techniques
- Larger study sizes with greater generalizability across populations
- The ability to compare data over long time periods to assess changes over time
- The ability to replicate research and compare between studies or similar populations

In order to provide robust and accurate results, researchers assume that quantitative methods are employed consistently and objectively. The results will provide a series of facts (as results), and, while these may be meaningful in statistical terms, they may not be of clinical use or application.

However, there are some limitations to quantitative research which need to be considered:

- Quantitative data may not provide enough context to explain results (unlike qualitative research) (see ► Chap. 2, "Qualitative Inquiry").
- The use of standardized questionnaires and surveys developed by researchers may lead to bias and reflect the researchers' ideas or beliefs (see ► Chaps. 32, "Traditional Survey and Questionnaire Platforms," and ► 76, "Web-Based Survey Methodology").
- The results may not provide behavioral aspects of research, such as opinions, motivation, attitudes, or feelings (unlike qualitative research).

- Preset answers (such as are contained in surveys) may not be reflective of actual results but are the closest response.

3   Describing Data

In order to get a "feel" for the data under investigation, it needs to be summarized in a meaningful way. Diagrams are a useful way of presenting data visually so that the location and distribution (or spread) of the data can be evaluated. There are standard terms to describe the location and distribution of data within a dataset. The most frequently used are mean, median, mode, range, variance, and standard deviation (see also ▶ Chap. 54, "Data Analysis in Quantitative Research"). Within population data, terms such as incidence and prevalence are also used (see also ▶ Chap. 33, "Epidemiology").

3.1   The Mean

The mean (or average) of a set of values is calculated by adding up all the values and dividing the total by the number of values in the dataset. In formulaic terms this calculation is:

$$x^1 + x^2 + x^3 \ldots ../n$$

where

x = each individual value and
n = the number of values

3.2   The Median

To calculate the median value of a dataset, the data is arranged in order of magnitude, starting with the smallest value and ending with the largest value. The median value is the middle value of the ordered set. Where there is an odd number of values in a dataset, the middle value is the median. However, where there is an even number of values, the two middle values are added together and divided by 2 to obtain the median. The formula for this is:

$$(n + 1)/2$$

where

n = the number of values

3.3   The Mode

The mode is the value in the dataset that occurs most frequently. Some datasets may not have a mode if the values occur only once. Similarly, a dataset can have multiple modes if there is the same frequency of differing values.

3.4   The Range

The range of a dataset is the difference between the largest and the smallest value. These two values are often represented in parentheses rather than the difference between them (x to y). If the dataset contains a number of outliers (data at extreme ends of the dataset), the range is not a good measure of spread. In this situation, percentiles are used to calculate the range.

3.5   Percentiles, Deciles, and Quartiles

Ordering the data from the smallest to largest value will enable percentiles to be calculated. The value of x that has 1% of the observations lying below it is the first percentile. The value with 2% of the values lying below it is the second percentile and so on. The value that has 10% of the values lying below it is the first decile. The values that divide the dataset into four equally sized groups are called quartiles (25%, 50%, 75%). The 50th percentile is the mean.

The interquartile range is the difference between the first and the third quartiles. This range contains the central 50% of the observations in the dataset with each 25% above and below this limit.

3.6   Variance

Another method of determining the spread of the data is to measure the variance or the extent to which each observation (value) deviates from the mean. The larger the deviation from the mean, the greater the variability of the observations. It is not possible, however, to use the mean of these deviations as a measure of spread as the positive differences cancel out the negative differences. To overcome this, each deviation is squared with the mean of the squared deviation described as the variance.

3.7   Standard Deviation

The standard deviation is the square root of the variance. The standard deviation can be considered an "average" of the deviations of the observations from the mean. Dividing the standard deviation by the mean and expressing this as a percentage gives the coefficient of variation. This is a measure of spread that is independent of the units of measurement.

3.8   Prevalence

The prevalence (commonality) of a disease or given health indicator within a given population is useful in identifying the current health status of a population. Prevalence can be calculated as follows:

> Prevalence
> = the number of people in a population with a given health indicator during a given time period divided by the population in the same time period × 100.

3.9   Incidence

Incidence can be described as the number of new cases of a given disease or health indicator that occurs over a given time period in a given population. It can be calculated as follows:

Incidence
= the total number of new cases of a given health indicator that occur over a given time period divided by the number of the population at risk during the same time period.

4   Probability

Probability measures the chance that a particular event will occur. A probability value is a positive number that lies between zero and one. If the probability of an event is zero, then the event cannot occur; however, if the probability is one, the event will occur. The rules of probability are the addition rule and the multiplication rule.

4.1   The Addition Rule

If the probability of two events X and Y are mutually exclusive (i.e., each event precludes the other), then the probability of either X or Y occurring is equal to the sum of their probabilities. This can be described as:

$$\text{Probability } (X \text{ or } Y) = \text{Probability } (X) + \text{Probability } (Y)$$

4.2   The Multiplication Rule

If the probability of two events is completely independent from each other, the probability of both events will occur is equal to the product of the probability of each event occurring.

$$\text{Probability } (X \text{ and } Y) = \text{Probability of } (X) \times \text{Probability } (Y)$$

5   The Normal Distribution

In any given population, it can be assumed that the population is normally distributed. That is, there are some population values that lie at the lower end of a curve, with similar numbers of values at the higher end of the curve and the majority somewhere in the middle. This is called the normal distribution, the "bell-shaped curve," or the Gaussian curve (Wilson and Black 2013) (see Fig. 1).

The normal distribution has a number of distinct features:

- It is symmetrical about its mean.
- It can be described by two parameters: the mean (μ) and the variance (σ^2).
- It is bell shaped.
- It becomes flatter as the variance increases.

The normal distribution has the probability that 68% of the values in a normally distributed dataset will lie within one standard deviation of the mean, 95% within two standard deviations, and 99.7% within three standard deviations of the mean (see Fig. 2).

6   The Skewed Distribution

Skew is a measure of the asymmetry of the probability distribution of a real-valued random variable about its mean. Not all data is normally distributed and may be skewed either negatively or positively. The position of the "tail" in skewed data is the way the skew is described (see Fig. 3).

Fig. 1 The normal distribution (Gaussian curve)

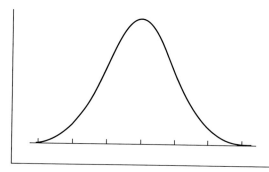

Fig. 2 Percentages of probability under the normal distribution

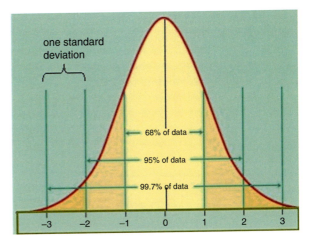

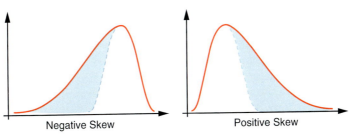

Fig. 3 Negatively and positively skewed distributions

7 Sampling

Because it is costly, time consuming, and impractical to collect data on an entire population of interest, a sample of the population is used. Samples need to be representative of the population of interest so researchers can make inferences about the wider population. Population samples may not fully reflect the normal population, and if this occurs, a researcher is said to have introduced sampling error. There are ways of choosing a population sample that minimize the likelihood of sampling error.

7.1 Representative Sampling

Representative sampling is important so that results are accurate and generalizable to the wider population (see also ▶ Chap. 54, "Data Analysis in Quantitative Research"). The best way to ensure sampling is representative is to take a random sample of the population. To do this, all people in the data frame of interest are listed and a sample is chosen at random. In this way, everyone in the data frame has an equal chance of inclusion in the study. Where this is cost prohibitive or impossible, convenience sampling (i.e., choosing a sample that is convenient to the researcher but still representative) may be used. Other nonrandom methods are sometimes used; however, it is critical to remember that the sample should be as representative of the entire population of interest as possible.

Because a sample population is a subset of the normal population, a normally distributed sample population curve will be flatter than that of a normal population (see Fig. 4).

7.2   Sample Size and Statistical Power

When sample populations are recruited, it is important to ensure the sample is large enough to detect any effect of an intervention. The larger the study, the more statistical power the study has. If the statistical power is high, the chances of accepting the null hypothesis when in fact the null hypothesis is incorrect (there actually is an effect) are less. Power calculations are best conducted by statisticians who can determine the optimal sample size to detect an effect without oversampling which can be costly and time consuming.

8   Developing a Research Hypothesis

Quantitative research is used when a researcher seeks to answer a specific research question or has a problem (hypothesis) that they want to investigate (or test). Hypothesis testing (sometimes called significance testing) in quantitative research is conducted on either the "null" or the "alternative" hypothesis.

8.1   The Null Hypothesis

The null hypothesis is a statement that assumes the problem being investigated is absent or has no effect (i.e., the difference between means equals zero). For example,

Fig. 4 Population distribution (solid) compared to sample distribution (dotted)

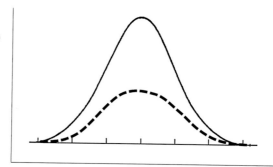

if a researcher were interested in comparing smoking rates between men and women in the population, the null hypothesis (H_0) would be:

H_0: There is no difference in smoking rates between men and women in the population.

8.2 The Alternative Hypothesis

The alternative hypothesis is defined when the null hypothesis is not true. The alternative hypothesis relates more closely with researcher's beliefs about the research and is expressed as H_1. Using the previous example, the alternative hypothesis would be:

H_1: Smoking rates are different between men and women in the population.

9 Study Designs

Once a researcher has developed a hypothesis or a research question, the appropriate study design for that question needs to be chosen. Study designs are chosen by researchers based on the hypothesis and the type of intervention or group under study and are guided by ethical considerations (see ▶ Chap. 106, "Ethics and Research with Indigenous Peoples"). Study designs can be classified as either nonexperimental designs or experimental designs (see also ▶ Chap. 33, "Epidemiology"). These study designs provide varying levels of evidence. It is not always possible to conduct a study that provides high-level evidence for ethical or logistical reasons.

9.1 Experimental Study Designs

Experimental studies are those where the researcher determines the exposure (or conducts an experiment).

- The randomized controlled trial (RCT) (see ▶ Chap. 37, "Randomized Controlled Trials").

9.2 Nonexperimental (Observational) Study Designs

Nonexperimental studies (sometimes called observational studies) are those where the exposure to a particular event or condition is predetermined. Studies in this category are:

- Cohort studies
- Case-control studies

- Cross-sectional studies
- Other study types (case studies, ecologic studies, case series studies)

10   Levels of Evidence

The evidence (or result) that research provides is based on the strength of the study design. The evidence pyramid visually depicts the strength of evidence for differing study designs (see Fig. 5).

10.1   Systematic Reviews

The highest level of evidence is systematic reviews and meta-analyses. Systematic reviews are research papers which have reviewed a number of high-quality studies on a given topic and consolidated (or pooled) the evidence. Where data from a number of studies has been pooled and reanalyzed (meta-analyses), the resulting data provides much stronger evidence than a single study alone (see ► Chap. 46, "Conducting a Systematic Review: A Practical Guide").

10.2   The Randomized Controlled Trial (RCT)

The randomized controlled study is the next highest level of study design. In a randomized controlled trial, a sample of the population of interest is chosen. Each should have similar characteristics to ensure they are representative of the wider population. Subjects are then randomly allocated to either the "intervention" group

Fig. 5 Evidence pyramid for quantitative research

or the "control" group. Data on a range of factors are collected prior to the study commencement and then at the end of the study. Results are then analyzed to determine whether there is any change between these measurements and whether there is any relationship to the intervention. Generally, researchers and participants are "blinded" to the intervention. This is known as a "double blind" RCT. Where either the researcher or the participant are not blinded to the intervention, this is known as a "single blind" RCT. Blinding minimizes bias as subjects cannot choose the group they are in, and it also minimizes the placebo effect (discussed further in this chapter). The RCT is the "gold standard" in study design; however in some cases, the RCT cannot be conducted for ethical or logistical reasons (Wilson and Black 2013; see also ▶ Chap. 37, "Randomized Controlled Trials").

10.3 The Cohort Study

Cohort studies are usually large studies conducted over a long period of time. Researchers follow the cohort of participants in the study and observe the differences between those who have been exposed to a particular phenomenon and those who were not exposed. A good example of a cohort study is the British Doctors Study (Doll and Hill 1954). In this study, doctors were followed over 20 years, and the health of those who smoked was compared to those who did not. This was one of the earliest studies linking smoking to cancer. In a cohort study, all participants must be healthy at the commencement of the study and not have the outcome of interest. Prospective cohort studies follow people over time; however, retrospective cohort studies can also be conducted. In these studies, data collected retrospectively (previously) can be reviewed to ascertain exposure to a particular phenomenon. Databases and medical records are often used to collect retrospective data. There are both strengths and limitations to cohort studies. The strengths are that they provide an unbiased assessment of exposure in a large group of people, it is easy to establish incidence of disease, and rare exposures can be studied. However, cohort studies are expensive and time consuming to conduct and because of their longevity may be prone to bias as researchers change over time or may even lose their relevance by the time the data collection provides results.

10.4 The Cross-Sectional Study

Cross-sectional studies measure exposure and outcome at the same time. They are generally a "snapshot" of a given population at a given time. Cross-sectional studies are especially useful for determining prevalence of a disease within a community and are useful in that they can be conducted relatively, quickly, easily, and cheaply. They are also able to provide prevalence estimates, and they can inform the development of a larger cohort study. Cross-sectional studies are not useful for examining rare conditions and may be prone to bias as they may not be totally representative of the broader population.

10.5   The Case-Control Study

Case-control studies are used to study rare diseases or conditions. They are generally retrospective and start with a "case" who is then matched retrospectively to a "control" (someone without the condition). Differences in exposure to particular factors are then assessed to try and ascertain whether what is observed is different to that expected. Case-control studies are often used in cancer studies to examine exposure to particular substances or conditions. Case-control studies are retrospective and cases are selected based on disease. There are methodological challenges with case-control studies. There is often bias in the selection of cases and controls, it is difficult to match controls to cases, and there may be information bias in what is reported by participants (see ▶ Chap. 38, "Measurement Issues in Quantitative Research"). However, case-control studies are often quick, require a smaller sample size, are relatively inexpensive to conduct, and are useful in diseases that are rare or have long latency periods.

10.6   The Case Series

This is a series of cases that are unusual or novel. They may appear in a cluster or over a short period of time. Cases may be highly relevant and in some circumstances are the best evidence available. The cases may or may not be related in time and space, such as an infectious disease outbreak. Case-series studies are not generalizable to the wider population due to their small size; however, they may highlight a particular phenomenon or emerging new disease.

10.7   Expert Opinion and Reports

Expert opinion and editorials are the lowest form of evidence on the evidence hierarchy. This is because opinion is only one person's perspective and may be heavily influenced by personal values, beliefs, or attitudes rather than quantitative evidence.

11   The Placebo Effect

The placebo effect is defined as "a perceived improvement or change in condition that is due to a psychological response rather than to any active intervention" (Wilson and Black 2013, p. 142). A placebo is any medical treatment that is inert (inactive), such as a sugar pill or "sham treatment." Around one third of people who take placebos (believing them to be medication) will experience an end to their symptoms. This is called the placebo effect.

Because the chance of any intervention having a placebo effect is equal across the population, the likelihood of the placebo effect occurring in either intervention or control groups is also equal. Thus, the chance of bias from the placebo effect is minimized in RCTs as each group is equally at risk of developing the placebo effect (see ▶ Chap. 37, "Randomized Controlled Trials").

12 Bias in Quantitative Research

Bias is defined as any influence that prejudices an outcome. The Webster dictionary defines research bias as that when "systematic error [is] introduced into sampling or testing by selecting or encouraging one outcome or answer over others" (Merriam-Webster.com 2017). Researchers often have a "hunch" about what they are likely to find in their research; however, it is critical that the influence of the researcher is eliminated from the research study; otherwise biased results will eventuate.

Bias can occur at any phase of research, including study design or data collection, as well as in the process of data analysis and publication. The most common types of bias are measurement bias, selection bias, researcher bias, and publication bias (see ▶ Chap. 38, "Measurement Issues in Quantitative Research").

12.1 Selection Bias

According to Panucci and Wilkins (2010, p. 3), "selection bias may occur during identification of the study population." In order to ensure a suitable study population, the sample should be clearly defined, accessible, reliable, and at increased risk to develop the outcome of interest. Selection bias occurs when the criteria used to recruit and enrol patients are inherently different (Portney and Watkins 2009). This can be a particular problem with case-control and retrospective cohort studies where exposure and outcome have already occurred at the time individuals are selected for study inclusion (Pannucci and Wilkins 2010).

12.2 Measurement Bias

Measurement bias occurs when there is a systematic error in the way data is measured. This may include an incorrectly calibrated instrument, incorrect measurement tools, recall inconsistencies (sometimes called recall bias), or incorrect measurement between two researchers.

12.3 Researcher Bias

This can occur when a researcher selectively "chooses" particular patients to be included in a study because they suspect they will benefit from the intervention. Although the researcher may be blinded as to the group in which the patient is included, the researcher may treat the patient differently to others or make comments to the patient that may influence the way any intervention or control medication works.

12.4   Publication Bias

Publication bias may occur when researchers and/or funding organizations are unwilling to publish negative research results, because they believe that this may impact on their reputation or credibility. As a result, positive results are more likely to be submitted for publication than negative results.

13   Collecting and Handling Data

The purpose of all quantitative research is to collect data to obtain information about a particular topic or given area of research. Data are observations or measurement of a particular variable. Variables are characteristics of interest in a given population. For example, researchers may collect data on the following variables: age, sex, height, weight, and smoking status. Variables can be classified into either categorical (qualitative) data or numerical (quantitative) data.

13.1   Categorical Data

Categorical data are those which can be put into categories as the data are not numerical. Categorical data can be further classified as nominal, ordinal, or dichotomous.

13.1.1   Nominal Data
Nominal data are data in categories that have names. Examples include eye color (blue, green, brown, gray), blood group (A, B, AB, O), and marital status (married, widowed, single, divorced).

13.1.2   Ordinal Data
Ordinal data are data in categories that are ordered in some way. The data may be ordered on a "scale" series of levels, for example, pain scores (none, mild, severe, extreme) or disease progression (Stage 1, Stage 2, Stage 3, Stage 4).

13.1.3   Dichotomous Data
Dichotomous data (sometimes called binary data) has only two categorical options. This may include Yes/No, Diseased/Non-diseased, Dead/Alive, and Male/Female.

13.2   Numerical Data

Numerical data occur when the variable has a numerical value. These can be divided into two types: discrete data and continuous data.

13.2.1 Discrete Data

Discrete data are those where the variable can only take certain numerical values. Examples include number of GP visits in 1 year, number of falls in the last month, or number of episodes of asthma in 12 months.

13.2.2 Continuous Data

Continuous data are those that have an infinite number of values. Examples are blood pressure, height, weight, and length.

14   Types of Variables

Experiments cannot exist without variables as these are the data points that are held constant or manipulated in an experiment. A variable is a characteristic that can be varied or changed, for example, height or weight. There are six types of variables which are usually identified in experiments: dependent, independent, extraneous, confounding, demographic, and environmental. Each type of variable should be clearly defined at the commencement of a study, with the last four variables discussed as limitations of the study.

14.1   The Independent Variable

The independent variable is the cause of the outcome. It is manipulated so it is independent of any effect. The independent variable is always the intervention that is applied (Wilson and Black 2013).

14.2   The Dependent Variable

The dependent variable is the effect or the consequence of the independent variable, so it depends on it for any change to occur. It is observed and measured and, therefore, must have the ability to vary or change. Multiple independent and dependent variables can be used in a study, and they should be explicitly stated.

14.3   The Extraneous Variable

Extraneous variables have the potential to impact upon the reliability of a study. It is possible to control for extraneous variables at the commencement of a study by considering what external factors may impact upon the outcome of interest. For example, if researchers were measuring temperature in a sample of participants but half of them were sitting next to a hot radiator, this would impact upon results. The extraneous variable would be "location when temperature was taken."

14.4   The Confounding Variable

Confounding variables are those that occur outside the study framework but that can impact upon the reliability of the results. Unlike extraneous variables, it is impossible to control for confounding variables. Such factors include the weather, genetic disposition, past experiences, and age.

14.5   The Demographic Variable

Demographic variables are those that are collected to describe the characteristics of the sample. These include such characteristics as age gender, height, weight, socioeconomic status, work history, education level, and so on. Depending on the study hypothesis and design, the demographic variables may be constrained (e.g., only females) or very diverse.

14.6   The Environmental Variable

Environmental variables are similar to extraneous variables, in that they can be controlled for in a laboratory. These variables include characteristics such as temperature, light, wind speed, humidity, and so on.

15   Hypothesis Testing

Quantitative researchers gather data in order to test a hypothesis. This enables the researcher to quantify a belief against a particular hypothesis. After collecting the data, the resulting values can then be used in a formula to undertake a specific test and obtain a test statistic. The test statistic will provide the amount of evidence against the null hypothesis. Common test statistics include the P-value, confidence intervals (CI), z statistic and χ^2 statistic, and t-value (see also ► Chap. 54, "Data Analysis in Quantitative Research").

15.1   The P-Value

All test statistics follow known theoretical probability distributions. The test statistic from the sample is then related to the known population distribution to determine the P-value (the area in both (sometimes one) tails of the distribution. The P-value is the probability of obtaining these results or something more extreme, if the null hypothesis is true. Put another way, the P-value is the likelihood that the result we have obtained occurred by chance. The smaller the *P*-value, the less likely the result occurred by chance. Generally, researchers are satisfied that if the *P*-value is less than 0.05 there is a statistically significant result. This is described as being significant at the 5% level.

The choice of 5% is an arbitrary figure and means that on 5% of occasions we may reject the null hypothesis when it is true. Where this has a high level of clinical significance, researchers often use 0.01 or 0.001 as the chosen cutoff for the *P*-value so that the chance of error is less.

15.2   The Confidence Interval (CI)

According to Petrie and Sabin (2005, p. 28), "the confidence interval quantifies the difference in means and enables researchers to interpret the clinical implications of the results." Because the CI provides a range of values for the true effect, it can be used to make a decision, even in the absence of a *P*-value.

16   Analyzing Data

Most data analysis is now done using sophisticated computer programs that can calculate test statistics and perform complex data analysis in the blink of an eye. The following section is not written to enable the reader to calculate statistics; this is best undertaken by statisticians that have studied the theories of statistical analysis for years. It is written to assist the reader to understand what types of tests are used and to familiarize them with the terminology related to statistical analysis.

16.1   Descriptive Data Analysis

Demographic variables and descriptive data can be presented using mean, median, range, and percentages to outline the characteristics of the sample population.

16.1.1   Continuous, Ordinal, and Numeric Normally Distributed Data

The One Sample T-Test

Researchers generally assume (if all care has been taken with sampling) that the sample population is normally distributed with a given (usually unknown) variance. Where there is one sample dataset to be analyzed, researchers use a one sample *t*-test to compare the sample to the normal distribution. This will give a *P*-value and confidence intervals.

The Paired T-Test

Where there are two samples who are related to each other (such as in a case-control study or where data are being measured on the same participants over two occasions), a paired *t*-test is used. Normal distribution is assumed, and because of the paired nature of the data, the samples must be of the same size. Applying the *t*-test will identify any differences in the means between the two populations. The test statistic *(t)* is compared to values from a known population distribution. This will then give the *P*-value so the CI can be calculated.

The Unpaired T-Test

Where the sample sizes are large (i.e., have enough power), it can be assumed that the data are normally distributed. In two unrelated groups (such as an intervention group and a control group), an ordinal or numerical variable can be investigated to determine whether the means or distribution of the groups is the same. Assuming the data is normally distributed, the null hypothesis suggests that the difference between the means will be zero (that is, an intervention will have no effect). Where there is a difference in the means, the *t*-statistic will be greater than 1.96 (at the 5% level). Where this is the case, the researchers can reject the null hypothesis and report the *P*-value as statistically significant (i.e., the likelihood that the result occurred by chance is less than <5%). Assuming variances in the groups are equal, CIs can be calculated for the difference in the two means.

Analysis of Variance (ANOVA)

Where there are more than two groups in a sample dataset, the statistical test applied is known as an "analysis of variance" (ANOVA). This test is a more complex statistical analysis that compares the "between group" variation and the "within group" variation. The components of variation are measured using variances (hence the term ANOVA). Based on the null hypothesis that the group means are the same, the between group variance will be similar to the within group variance. However, if there are differences between groups, then the between group variance will be larger than the within group variance. The test is based on the ratio of these variances.

16.1.2 Continuous, Ordinal, and Numeric Non-normally Distributed Data

Where continuous, ordinal, or numeric data is not normally distributed, there are a number of ways to analyze the data, either nonparametric tests are used to investigate the data depending on the sample (single, paired, or unpaired) or the data can be log transformed. Log transformation is a process whereby a formula is applied to the data to make the distribution more normal.

16.2 Analyzing Categorical Data

Analytical methods on categorical data are different to continuous and ordinal data. The test statistics used are the χ^2 (Chi-square statistic), z statistic. These test statistics give us the final P-value and CI for the sample.

16.2.1   The Test of a Single Proportion

Where there is a single sample with a given number of participants, each participant either possesses a characteristic of interest (male, pregnant, alive) or does not possess those characteristics (female, not pregnant or dead). It is useful to determine how many individuals in the sample display the characteristic, and this is defined as the proportion.

16.2.2   The Test of Two Proportions

A test of two proportions can be conducted when there are two independent groups of individuals or when there are two related groups. Investigating the proportion of individuals in a group with a particular characteristic may identify whether there is any significant difference between one group and the other. In related samples (the same group measured twice or where individuals are matched), the differences between results can be measured using proportions. In this situation, a Chi-square (χ^2) test is applied to the data. This will provide a test statistic which can be compared to a normal distribution to get a P-value, and the confidence interval can then be determined.

16.2.3   The Test of Three or More Proportions

Where there are three or more proportions (e.g., blood group A, B, AB, or O), we can investigate the frequencies of each category. The expected frequency and the observed frequency are calculated and the χ^2 test is applied. The test statistic that results focuses on the discrepancy between the observed and expected frequency of each result. The test statistic can then be compared to a normal distribution to give a *P*-value.

16.3   Correlation

Correlational analysis is concerned with measuring the relationship between two numerical variables (x and y). These variables can be plotted on a scatter graph with the values for x on the horizontal axis and the values for y on the vertical axis. The resulting scatter of points may indicate a relationship between the variables (see Fig. 6). Where there is a linear relationship between variables (a line can be drawn in a particular direction – either positive or negative), a correlation between the variables exists. The correlation coefficient describes the strength of the correlation and is always between zero and 1 or zero and -1. A strong positive correlation would approximate 1, and a strong negative association would approximate -1. A dataset with no correlation has a correlation coefficient of zero.

17   Interpreting Quantitative Research

Health professionals are bombarded with new research information everyday. It is estimated that around 2.5 million peer-reviewed papers are published each year (Olesen Larsen and von Ins 2010). One of the most difficult aspects for health

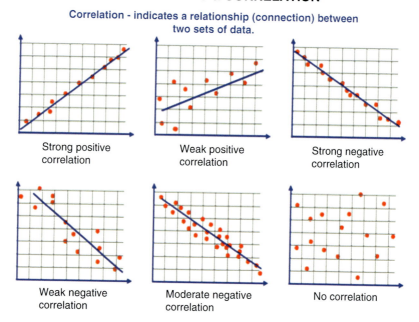

Fig. 6 Scatterplots and correlation

professionals is the ability to interpret research articles and to distinguish "high" quality research from "poor" quality research. The majority of journals that publish medical and health research have a peer review process that is thorough and robust. The peer review process includes a detailed review by experts in the field, who will assess the methods used, sample size, and limitations of the study. A study may be rejected for publication if it does not meet the rigorous standards of the journal or is deemed to be poorly conducted (see also ▶ Chap. 59, "Critical Appraisal of Quantitative Research").

Interpreting results published in journals is best undertaken by thoroughly investigating the methods used by the researchers. Health professionals should question whether the methods used are appropriate for the type of study (study design) and whether the statistical analysis is thorough and robust (was the right test used for the type of data collected). Interpreting P-values and confidence intervals enables the reader to determine whether the researchers obtained statistically significant results in a study. On occasion, statistically significant results are not obtained due to limitations of the study (there may not be enough participants, or there was a confounding factor that was not considered). Where this type of study is published, readers are able to determine the methods used by researchers and familiarize themselves with some of the challenges and pitfalls of research.

18   Conclusion and Future Directions

Quantitative research methods are concerned with the planning, design, and implementation of strategies to collect and analyze data (Sheehan 1986). Descartes (1637) suggests that *how* the results that are achieved is often more important than the results themselves, as the journey taken along the research path is a journey of discovery. High-quality quantitative research is characterized by the attention given to the methods and the reliability of the tools used to collect the data. The ability to critique research in a systematic way is an essential component of a health professional's role in order to deliver high quality, evidence-based healthcare. The rise in technology and the ability to easily calculate complex statistics will impact on the type and complexity of data available using quantitative methods. Although these data provide a numerical overview of the population, it is important to remember that quantitative methods are only one way of analyzing data. Triangulation with qualitative methods in mixed methods research provides an in-depth overview of the population.

References

Babbie ER. The practice of social research. 14th ed. Belmont: Wadsworth Cengage; 2016.

Descartes. Cited in Halverston, W. (1976). In: A concise introduction to philosophy, 3rd ed. New York: Random House; 1637.

Doll R, Hill AB. The mortality of doctors in relation to their smoking habits. BMJ. 1954;328 (7455):1529–33. https://doi.org/10.1136/bmj.328.7455.1529.

Liamputtong P. Research methods in health: foundations for evidence-based practice. 3rd ed. Melbourne: Oxford University Press; 2017.

McNabb DE. Research methods in public administration and nonprofit management: quantitative and qualitative approaches. 2nd ed. New York: Armonk; 2007.

Merriam-Webster. Dictionary. http://www.merriam-webster.com. Accessed 20th December 2017.

Olesen Larsen P, von Ins M. The rate of growth in scientific publication and the decline in coverage provided by Science Citation Index. Scientometrics. 2010;84(3):575–603.

Pannucci CJ, Wilkins EG. Identifying and avoiding bias in research. Plast Reconstr Surg. 2010;126 (2):619–25. https://doi.org/10.1097/PRS.0b013e3181de24bc.

Petrie A, Sabin C. Medical statistics at a glance. 2nd ed. London: Blackwell Publishing; 2005.

Portney LG, Watkins MP. Foundations of clinical research: applications to practice. 3rd ed. New Jersey: Pearson Publishing; 2009.

Sheehan J. Aspects of research methodology. Nurse Educ Today. 1986;6:193–203.

Wilson LA, Black DA. Health, science research and research methods. Sydney: McGraw Hill; 2013.

The Nature of Mixed Methods Research

4

Cara Meixner and John D. Hathcoat

Contents

Abstract

Mixed methods research (MMR) has gained traction in the social sciences, evolving as a genre of inquiry that intentionally and systematically connects qualitative and quantitative methods in order to address substantive questions.

C. Meixner (✉)
Department of Graduate Psychology, Center for Faculty Innovation, James Madison University, Harrisonburg, VA, USA
e-mail: meixnecx@jmu.edu

J. D. Hathcoat
Department of Graduate Psychology, Center for Assessment and Research Studies, James Madison University, Harrisonburg, VA, USA
e-mail: hathcojd@jmu.edu

© Springer Nature Singapore Pte Ltd. 2019
P. Liamputtong (ed.), *Handbook of Research Methods in Health Social Sciences*,
https://doi.org/10.1007/978-981-10-5251-4_76

Mixed methods projects are often preplanned, resulting in a fixed design. MMR can also be emergent; a researcher may craft a follow-up qualitative study, for instance, to make meaning of elusive quantitative findings. We unveil the nature of mixed methods research in the context of philosophical positions (e.g., constructivism, postpositivism, advocacy, and pragmatism) with critical attention to successes and challenges (e.g., incompatibility, conditional incompatibility, integration). Drawing from examples in the health social sciences, we showcase major mixed methods approaches – such as convergent parallel, exploratory, explanatory, intervention, and hybrid designs – while attending to notions of mixing, timing, and weighting data. Criticisms and accolades regarding MMR are thread throughout the chapter, given their parallels to phases in research design. Pivotally, the chapter also addresses ways that readers can "get started" on their own mixed methods projects.

Keywords

Mixed methods research · Research design · Philosophical positions · Convergent parallel design · Exploratory design · Explanatory design · Intervention design · Hybrid design · Constructivism · Postpositivism · Advocacy · Pragmatism

1 Introduction

This chapter orients students and novice researchers in the health social sciences to mixed methods research (MMR), a genre of inquiry that intentionally and systematically connects qualitative and quantitative methods in order to address substantive research questions. MMR is particularly well suited to the health social sciences, wherein researchers use social science tools to investigate and make meaning of human health. Inquiry into the topic of health is as deep as it is vast, necessitating a diverse set of methods. To understand the layered experience of caregiving for cancer survivors, for instance, one would employ a fundamentally different set of methodological tools than those we would use to predict the effect of the rural opioid epidemic on income security.

To fathom MMR as a vibrant, integrative approach to health social sciences research, we first turn to its evolution. From there, we will gaze into MMR as it stands currently, with attention given to the contemporary landscape of inquiry in the healthcare field. Parcel to understanding MMR is exploring related philosophical positions, implicit and explicit, that frame methodological work. Then, we orient the reader to major mixed methods designs (i.e., convergent parallel, exploratory, explanatory, and intervention) and their affiliated assumptions. Each of these facets is intended, ultimately, to guide our audience in contemplation of the chapter's final intent: to invite readers to "get started" on their own MMR endeavors.

1.1 Defining Mixed Methods

Meanings, classifications, and implications of MMR have varied across each of the periods described in the following subsection. Across the disciplines, among the more known definitions arises from Creswell (2015, p. 2), who views MMR as:

An approach to research in the social, behavioral, and health sciences in which the investigator gathers both quantitative (closed-ended) and qualitative (open-ended) data, integrates the two, and then draws interpretations based on the combined strengths of both sets of data to understand research problems.

Within this chapter, Creswell's definition will be endorsed, with attention given to its implications for the health social sciences. However, we would be remiss not to add the rich, thoughtful conjectures on MMR drawn from Johnson and Onwuegbuzie (2004, p. 17):

Mixed methods research also is an attempt to legitimate the use of multiple approaches in answering research questions... It is an expansive and creative form of research, not a limiting form of research. It is inclusive, pluralistic, and complementary, and it suggests that researchers take an eclectic approach to method selection and the thinking about and conduct of research.

1.2   The Evolution of Mixed Methods Research

From the Latin verb *evolvere*, the term *evolution* points to steady development, from simple to complex, and to a pattern of movements or turns. Both definitions aptly describe the growth of MMR and the various factors – philosophical, sociological, and political, among others – that have influenced its development. While Creswell and Plano Clark (2018) trace MMR's formal beginning to the late 1980s, during which there was a confluence of seminal publications across varying disciplines, credit can be drawn to the early work of Campbell and Fiske (1959). They pioneered a multi-method approach, suggesting the inclusion of multiple quantitative measures to study psychological traits. As the decades turned, researchers like Sieber (1973), Denzin (1978), and Jick (1979) combined quantitative and qualitative methods in an effort to better understand research enigmas and phenomena.

This formative period in the evolution of MMR was interjected by paradigmatic debates (see Rossman and Wilson 1985; Bryman 1988; Reichardt and Rallis 1994; Greene and Caracelli 1997), wherein scholars took varying stances regarding the philosophical legitimacy of mixing qualitative and quantitative data. At the center of the argument was the issue of incompatibility between the overarching paradigms under which qualitative and quantitative inquiry respectively reside. Just because data *can* be mixed to formulate a holistic understanding of a research problem, *should* they be? As the reader might imagine, this historical stage had its drawbacks, particularly in advancing MMR. Johnson and Onwuegbuzie (2004, p. 14) opined:

The quantitative versus qualitative debate has been so divisive that some graduate students who graduate from educational institutions with an aspiration to gain employment in the world of academia or research are left with the impression that they have to pledge allegiance to one research school of thought or the other... A disturbing feature of the paradigm wars has been the relentless focus on the differences between the two orientations.

In spite of the schism between qualitative and quantitative researchers, what emerged during this timeframe, and is explored later in this chapter, is the notion of understanding and critically interrogating research methods within their broader philosophical assumptions (e.g., ontology, epistemology), mental models (Greene 2007), and associated claims. Though the paradigmatic debates linger, scholars' contributions to issues like transparency (see Greene and Caracelli 1997) and reflexivity have grown increasingly significant.

In the 1980s, attention "began to shift toward the early procedural development period in the history of mixed methods in which writers focused on methods of data collection, data analysis, research designs, and the purposes for conducting a mixed methods study" (Creswell and Plano Clark 2018, p. 27). These developments established ground for the advocacy and expansion of mixed methods, wherein authors such as Johnson and Onwuegbuzie (2004, p. 17) pronounced MMR as "the 'third wave' of the research movement, a movement that moves past the paradigm wars by offering a logical and practical alternative." Further, mixed methods have gained credence within publications, conferences, text and reference books, and funding agencies (e.g., National Science Foundation, National Institutes of Health).

Creswell and Plano Clark (2018) suggested that a reflective and refinement period for MMR commenced in 2003, with two prominent and related features: (1) discussions assessing the MMR field and projecting its future and (2) constructive critiques on the field's emergence and evolution. Therein, the authors cited criticisms from nursing research. For instance, varied voices (e.g., Giddings 2006; Holmes 2006; Freshwater 2007) challenged issues such as the false polarization of quantitative and qualitative methods, a risk on the part of MMR to marginalize qualitative inquiry, and the flat presentation of MMR (i.e., rhetoric, discourse). Scholars also sought clarity on how to *evaluate* mixed methods contributions (Morse and Niehaus 2009), a question that continues to pervade discourse (see ▶ Chap. 60, "Appraising Mixed Methods Research").

1.3   Mixed Methods: A Contemporary Context

Halcomb, Andrew, and Brannen (2009) connected current trends in nursing and health sciences to a need for vigorous mixed methods studies. Advances and changes in healthcare, rising costs, an aging population, and the complexities related to disease, among other factors, co-create a need to review provider roles and assess the viability of best practices. Of note, Halcomb et al. (2009) purported that MMR is not necessarily a new approach in the health sciences; it has been parcel to many scholars' approaches – often covertly. The authors further stated that:

> What are new, are the emerging impetuses that are leading researchers to methodological change and advancement. Where previously people were reluctant to disclose this combination of approaches, researchers are now discussing frank and meaningful information regarding methodological issues leading to innovation and the greater potential to have a repertoire of skills appropriate to a range of research questions. (p. 7)

In consideration of MMR in the health social sciences, Halcomb et al. (2009) invited novice and skilled scholars to consider five points of impetus that drive a continued need for methodological pluralism. These points are summarized in

Table 1. We have added reflection questions to guide the reader's personal consideration of these premises.

Also germane to those reading this chapter is a contemporary context etched in the landscape of funding, which is especially relevant given costs associated with studies of human health. To this end, the Office of Behavioral and Social Sciences Research (OBSSR) of the NIH published *Best Practices for Mixed Methods Research in the Health Sciences* (Creswell et al. 2011). Freely available, the document guides the reader through an abbreviated introduction to MMR, leading into how a researcher might prepare applications for NIH-funded research projects, small grants, and exploratory/developmental research grants. Among the core

Table 1 Impetus for use and range of mixed methods in the health social sciences

Impetus	Meaning	Reader reflection
Increased reflexivity in relationships	Researcher sensitivity and reflection drive not only what is studied but also how and with whom the study takes place	*What is my relationship to the phenomena I study? Why is that so? What drives this interest? With whom do I forge research relationships? Where are my beliefs, biases, and values present? Am I transparent about these facets?*
Increased political awareness of research	Recognition that research has a political bent, influencing policymaking, practice, and change	*For what purposes or goals (micro, meso, and macro), am I engaged in research in the health social sciences? How might my research be interpreted and used by various stakeholders? To what end do the methods I intend to use reflect valued aims of inquiry?*
Growing formalization of ethics and governance	Awareness that research into human health must be situated in ethical frameworks that require approval and consent	*What are the ethical implications of my purported study? Have I sought all requisite approvals (e.g., human research ethics committees)? In what ways could the methods serve to oppress or liberate participant perspectives?*
Availability and ease of technologies	Technologies present ever-changing opportunities for data collection, analysis, and integration	*What technologies make it plausible to collect, analyze, compare, and integrate data? How might these technologies also help ensure a diverse array of participants – Or consumers of research, for that matter?*
International research collaboration	Globalization makes collaboration possible, but this is also a heightened call from funding bodies	*What can we learn through international partnerships into the investigation of human health? How do international partnerships inform the ethical, methodological, and productive features of health social science research?*

contributions is an acknowledgment that in writing the innovation section of a research grant, "the use of mixed methods researchers may be an innovation in and of itself" (p. 19). This positions MMR not only as a tool to addressing complicated phenomena but also as a methodology worthy of recognition for its own advance. That the NIH and other prominent foundations and agencies (e.g., NSF) forwardly invite MMR submissions is another indication of the "third wave's" merits.

1.4   Example of MMR in Health Social Sciences

We offer an example, drawn from the first author's own research (see Meixner et al. 2013), to provide insight into MMR in the health social sciences. As with many research questions, ours arose from practice and is situated in an advocacy agenda for survivors of brain injury. Our initial inquiry, which focused on obstacles to accessing crisis intervention services in rural loci, surfaced from multiple vantage points. The intersected identities on our research team included that of clinical expert, director of a nonprofit organization serving persons with brain injury, caregiver, and research methodologist. Together, we had observed, over time and firsthand, what appeared to be substantive gulfs in services, particularly when survivors experienced acute crisis. Gaps in the literature further sensitized us to the need for additional inquiry. Our overarching question thus became: What are the barriers to accessing crisis intervention services for persons with brain injuries?

Practically speaking, MMR provided us with the most integrative, comprehensive toolkit to understand a complex question. Yet, our choice to pursue MMR was multifold. One, the canon of literature helped us see a complex connection between psychiatric comorbidities, cognitive and behavioral changes, funding limitations, systemic obstacles, and more. These factors had been explored quantitatively, and often discretely, leaving scholars with unresolved questions. Specifically, the literature precluded an interdependent analysis of barriers, a focus on rural loci, and a research design principled on rigorous mixed methods inquiry. We also knew that qualitative inquiry would invite novel perspectives given the deep, nuanced explorations of social processes that robust narrative and phenomenological methods make possible.

As scholar-practitioners, we also acknowledged that our intentions were action-oriented, with a desire to inform plans to improve intervention and mental health services while building relationships across providers. Relevant to action research is the intent to disrupt the status quo, challenge norms, and transform systems (Argyris et al. 1985; Anderson and Herr 2005). To this end, we noted: "Equally vital in action research is the empirical analysis of diverse perspectives, which can neither be ascertained by quantitative nor qualitative data alone. [MMR arrives]... at a more complex understanding of the research question" (Meixner et al. 2013, p. 379). Finally, we surmised our stakeholders (i.e., academics, providers, policymakers, survivors, etc.) would value the juxtaposition of qualitative and quantitative methods. Though the resident literature was dominated by quantitative and quasi- and non-experimental designs, we observed the field of practice to be one that valued the narrative, non-experimental tradition.

The study involved two phases of data collection and analysis; we commenced with a web-based survey, comprised primarily of quantitative questions. Data from

110 participants (49% response rate) were analyzed to inform the development of a second, qualitative strand of inquiry. Herein, we reflected on the tremendous advantage MMR provided – such as the opportunity to utilize qualitative inquiry to explore, identify, and unpack the barriers elicited in the quantitative strand. We conducted seven focus groups (n = 25), analyzing the data for resonance and divergence with its quantitative counterpart. A rigorous analysis yielded themes and minor inconsistencies across both strands of inquiry:

> A multipart analysis of the focus group data revealed. . .the most critical barriers to accessing services. . . [T]hese obstacles are connected, affecting each other and deepening fragmentation in the system. The nature of these findings as symbiotic adds significant value to the literature, expanding providers' awareness of the layered complexity of brain injury and access to services. Of note, the qualitative phase supported quantitative findings with the exception of [one subtheme]. (p. 383)

In spite of the obvious and surprising benefits of our approach, we were not without obstacles and encumbrances. MMR is labor intensive, requiring scholars both literate and skilled in rigorous statistical and qualitative methodologies. As discussed in the next section, such proficiency is not just methodological; researchers must also attend to paradigmatic assumptions, explicit and covert, governing inquiry. Within our postpositivist neurotrauma research community, for instance, some of our peers discounted the value of the qualitative strand, yet their counterparts in the advocacy community upheld the rich, bountiful perspectives as evidence of a flawed system of care. Not all of the publications we considered were welcoming of MMR so we had to be vigilant in locating venues of mutual advantage. These challenges, while present, did not undermine the vigor of the process we upheld. In all, the MMR approach resulted in a thorough understanding of a complex phenomenon through the collection, analysis, and interpretation of quantitative and qualitative data.

2   Philosophical Positions

Philosophical stances, what some refer to as paradigms, have been a topic of much debate in the MMR literature. This is partly due to the historical evolution of MMR that was previously discussed. Can we integrate quantitative and qualitative data within a single study if they are informed by seemingly contradictory philosophical positions? If not, is the notion of MMR simply nonsensical? To contextualize these issues, we provide an overview of four philosophical stances that are discussed by Creswell (2014). This is followed by an examination of some of the paradigmatic successes and challenges manifest within MMR.

2.1   Frameworks

As professors, we find ourselves engaged in conversations with budding scholars about the intersection between ontology, epistemology, and axiology in social science research. "What's *philosophy* got to do with my research project?" inquired one precocious student keen on pursuing a mixed methods thesis. Our answer is, well, *"Everything!"* Whether overt or implicit, scholarly inquiry exists within and around philosophical perspectives – also referred to as paradigms or worldviews. To abscond consideration of philosophy in research methods would be akin to traveling to a new country without having considered its cultural manifestations, like collective rites, language, laws, history, and so on. For individuals new to the concept of philosophical positions, Creswell's (2014) primer is a useful starting place for understanding their role in research. Though numerous philosophical stances exist, he elaborated on four views: postpositivist, constructivist, transformative, and pragmatist. Each view is laden with assumptions regarding *ontology* (i.e., the nature of knowledge), *epistemology* (i.e., how we come to know something), *axiology* (i.e., the role of values in inquiry), *rhetoric* (i.e., how we write or project our findings), and *methodology* (i.e., the processes we utilize to understand our question) (see also ▶ Chap. 6, "Ontology and Epistemology").

Our students tend to find postpositivism the easiest stance to grasp; inclusive of the procedures traditionally adopted within science, this philosophical position contends that there is an independent reality that can, at least approximately, be known (Crotty 1998). In other words, there are objective aspects of human health that researchers aim to discover using rigorous methodological techniques (see also ▶ Chap. 9, "Positivism and Realism"). Though we may know never know with certainty that our best scientific theories are true, they serve as our closest approximations to an objective reality since they have withheld empirical scrutiny. Generally speaking, the standards of rigor include what is taught in many research methods textbooks (e.g., sampling, reliability, validity, generalizability, and so on). Researchers are, therefore, expected to remain distant from the topic of investigation since they could introduce various forms of bias – as illustrated by the need for researchers to be unaware of whether a participant is in an experimental or control group in randomized control trials (see ▶ Chap. 37, "Randomized Controlled Trials").

In contrast to postpositivism, social constructivism assumes that meanings are not only co-constructed but also multiplistic, varied, and highly contextual (see also ▶ Chap. 7, "Social Constructionism"). Constructivism guides deep inquiry into phenomena that may be impervious to the reductive techniques and standards that dominate the milieu of postpositivism. To compare these two, let us consider the musings of two scientists, Lanza and Berman (2009, p. 4), intent on understanding an ineffable construct:

Consciousness is not just an issue for biologists; it's a problem for physics. Nothing in modern physics explains how a group of molecules in your brain create consciousness. The beauty of a sunset, the miracle of falling in love, the taste of a delicious meal—these are all mysteries to modern science. Nothing in science can explain how consciousness arose from matter. Our current model simply does not allow for [it].

The essence of consciousness, thus, is fodder for constructivist inquiry and would likely draw from a qualitative method, such as phenomenology, for exploration.

Herein, the role of researcher diverges from that explored above; the scholar is, herself, a tool in the elicitation and construction of meaning.

The third position that Creswell (2014, p. 9) reviews is the transformative worldview, which "arose during the 1980s and 1990s from individuals who felt that the postpositivist assumptions imposed structural laws and theories that did not fit marginalized individuals in our society or issues of power and social justice, discrimination, and oppression that needed to be addressed." Thus, studies under the umbrella of advocacy – focused on change, emancipation, and/or justice – are often integrated with critical theoretical perspectives (e.g., feminist, queer, disability) and draw from both quantitative and qualitative methods to advance desired aims. Under this approach, researchers are often participant-observers who work with other participants to develop questions, make meaning of results, and evoke changes to promote social justice (see also ▶ Chap. 17, "Community-Based Participatory Action Research"). An advocacy approach is technically consistent with a range of philosophical stances (Hathcoat and Nicholas 2014; Nicholas and Hathcoat 2014). However, at the center of an advocacy approach resides the recognition that inquiry is inherently value-laden, collaborative, and transformative. Described above, our illustrative study (Meixner et al. 2013) is emblematic of this claim.

The pragmatist knowledge claim, fourth in Creswell's (2014) heuristic, may abscond commitment to any one ontological or epistemological underpinning, focusing instead on researchers' freedom "to choose the methods, techniques, and procedures of research that best meet their needs and purposes" (p. 12). Primary importance is given to aligning the question asked with the MMR approach and assuring focus is paid to the *consequences* of research (Creswell and Plano Clark 2018; see also ▶ Chap. 40, "The Use of Mixed Methods in Research"). Such liberalism leads many mixed methodologists to situate their studies under the auspices of this claim, touting pragmatism as the "what works" approach. This can be problematic, as we believe pragmatism can be as misunderstood as it is haphazardly applied (Hathcoat and Meixner 2017). Bryman (2007) would concur; he found that many MMR scholars cast aside philosophical issues in order to procure funding and publish their results. This issue is further attended to below.

2.2   Paradigmatic Challenges and Successes

Several years ago, we were working with a doctoral student, to whom we will refer as Nadia, whose mixed methods study was dominantly quantitative and supported by a qualitative strand. Well trained in advanced statistical methods, Nadia was likewise aware of her postpositivist inclinations. Still, she held immense respect for qualitative research, recognizing that subjecting her qualitative method, techniques, and data to the "standards" of the postpositivist paradigm would be deleterious. But, this did not assuage her worries about how the data would be integrated and understood. Various questions came to mind: How might Nadia situate her study, philosophically? Why does doing so matter, to begin with? Is the study pragmatist in philosophical orientation? Or, should Nadia consider a shift among paradigms, allowing postpositivism to guide the quantitative phase, with constructivism orienting the qualitative phase? Further, what are the repercussions of conjoining paradigms, and their methods, in the interpretation process?

These are the very puzzles that continue to vex mixed methods scholars. As we touched on above, issues of incompatibility emerged in the 1960s and 1970s, a time during which qualitative research became more widely advanced in the social sciences (Hanson et al. 2005). The incompatibility thesis states that quantitative and qualitative research, informed by opposing philosophical assumptions, cannot be integrated. Scholars offer diverse opinions on this matter. The purist argues for philosophical consistency, upholding the notion of paradigmatic and methodological incompatibility. The dialectical position, on the other hand, acknowledges the "assumptive sets of different paradigms are different in important ways, but paradigms themselves are historical and social constructions and so are not inviolate or sacrosanct" (Greene 2007, p. 69). This perspective maintains a position wherein the philosophies underlying quantitative or qualitative data collection, however contradictory, are respected, valued, honored, and made transparent. Doing so requires flexibility and self-awareness on the part of the researcher.

Tashakkori and Teddlie (2003), reasoning that the research question is the force driving the methodological approach, found pragmatism to be *the* best philosophical position to anchor MMR. Howe (1988) rejected postpositivism and constructivism as dichotomous constructs, promoting pragmatism as an alternative position wherein researchers move from "whether" to combine to "*how* this combination can be accomplished" (p. 14). Despite thoughtful, informed efforts by scholars to apprise the MMR community of pragmatism's nuances (see Howe 1988; Tashakkori and Teddlie 2003; Denscombe 2008), what has emerged is a "what works" and "anything-goes" attitude toward pragmatism that fails to adequately inform methodological decisions, techniques, and questions. While beyond the scope of this chapter, our own *conditional incompatibility thesis* (Hathcoat and Meixner 2017) is an attempt to reconcile these tensions and position both seasoned and novice researchers, like Nadia, for success.

Returning to the dialectical ideal, there is one more thing for the reader to ponder in light of the tensions articulated above: *the mental model* – a "set of assumptions, understandings, predispositions, and values and beliefs with which a social inquirer approaches his or her work" (Greene 2007, p. 53). We perceive the mental model to prescribe philosophical stances, meaning that particular aspects of research are unjustifiable absent philosophical position taking. Does research happen without philosophical consideration? Yes, it does. However, this assumes philosophical agnosticism (Hathcoat and Meixner 2017), which can endanger various facets of the research design, most especially the interpretation phase. Let us imagine, for instance, that Nadia is unwittingly steeped in a postpositivist perspective with little awareness of a constructivist epistemology. What are the repercussions, covert and explicit, of embedding interviews into an experimental design? Might she attempt to ascribe deterministic, empirical dispositions to make meaning of her data despite it being inconsistent with an espoused philosophical stance?

We recognize that these are tedious issues to consider, especially for a scholar new to MMR. However, we would argue that the cost of approaching MMR without considering philosophical positions could outweigh the benefits. The reader, thus, is advised to think critically about her own mental model, contemplating how beliefs, experiences, and biases may color her ability to take a dialectical stance, interweave both paradigms and methods, and interpret results in the spirit of the question asked. If approaching MMR through a pragmatist gaze, it is incumbent on the reader to move beyond a "what works" maxim, assuring she understand the philosophy's rich origins and emerging insights. The issue is not that pragmatism lacks substance, or fails as a paradigm within which MMR can be situated, but that it is often misunderstood. As Cherryholmes (1999, p. 1) cleverly opined, "pragmatism looks simple at first glance. It is ferociously complex." For a reader inclined to dig into this more, we advise a review of Biesta (2010).

3   Major Approaches

Across the arch of mixed methods research – and during the procedural development period, in particular – scholars have advanced an array of taxonomies and classifications that guide researchers in the design of MMR studies. In his concise, practitioner-oriented primer on MMR, Creswell (2015) agilely distilled the major approaches into three basic designs and three advanced designs. This chapter will reflect on the chief basic approaches (i.e., convergent, explanatory, and exploratory) and the most common of the advanced designs (i.e., intervention). Before doing so, we want to clarify that mixed methods designs can be *fixed* or *emergent* (Creswell and Plano Clark 2018). A medical anthropologist studying the culture of teen pregnancy within a rural locus may design, from the start, a *fixed* study that draws from both quantitative and qualitative methods. Herein, the deficits associated with any one method of data may pull strength from the other method – and vice versa. On the other hand, it is entirely plausible that the same scholar may have begun the study with a focus on addressing a quantitative hypothesis. Imagine, however, that the results contradict extant theory, as well as the researcher's hypothesis. The researcher may build in a second phase wherein qualitative interviews could help explain the first phase's curious results. This can be thought of as an *emergent* mixed methods study. The study (Meixner et al. 2013) reviewed above was intentionally *fixed*, prioritizing the use of second-phase focus groups to explore and triangulate survey results.

Before we peer into each of the four major approaches, it is also important to elucidate three methodological decisions that often confuse researchers who are new to MMR: timing, weighting, and mixing. Adapted from the work of Creswell and Plano Clark (2018), each practice is described with relevant examples in Table 2. We stress that the researcher's choice regarding these practices must (1) align with the research question and (2) cohere with the *fixed* or *emergent* design of the study.

Morse and Niehaus (2009) utilized the term *core component* in reference to the strand that has more weight in any given study; this core is to be thought of as "the backbone...onto which all other components, methods, or strategies will be attached" (p. 23). If a mixed methods study commences with a dominant quantitative strand, that particularly strand should, hypothetically, stand on its own. The logic

Table 2 Common practices within MMR approaches

Practice	Meaning	Example
Timing	This refers to the temporal and ordinal placement of quantitative and qualitative strands. Does one precede the other, or do they happen concurrently?	Jeremiah develops a grounded theory that explores the experiences of persons coping with lupus. If he develops an instrument based on the theory, the timing of the quantitative strand follows the qualitative one
Weighting	This refers to relative priority of strands. Are the qualitative and quantitative strands equally important, or does one support the other? Are both equally or differentially represented in analysis and interpretation?	Maris designs a quantitative survey to gauge perceptions of a mental health intervention. In support of her items, she adds several open-ended qualitative questions allowing participants to explain the rationale for the scores they have indicated. Herein, the quantitative strand has greater priority or emphasis
Mixing	This regards the point where the researcher chooses to integrate the strands. Does mixing happen at the level of design, collection, analysis, and/or interpretation?	In a study on crisis intervention services for sexual assault survivors, Fletcher's team collects survey data and focus group data separately. Each is analyzed according to its research tradition, later merged into a comparison heuristic. Here, the data are mixed at the level of interpretation

follows that the *supplement component* exists to support the *core*; it may not, therefore, be independent. Herein, we can see that MMR is much more than the knitting together of a qualitative study and a quantitative study, as is often the presumption of novice researchers. MMR, rather, is a thoughtful and integrative methodological endeavor.

The practices of mixing, weighting, and timing are sometimes depicted in the literature with symbols and nomenclature. For instance, one's choice to prioritize a qualitative strand over its quantitative counterpart would result in a truncated capitalization of qualitative (i.e., "QUAL") and lowercase, italicized denotation of quantitative (i.e., "*quan*"). Timing decisions can be represented in research articles with arrows (e.g., "QUAL ➜ *quan*" means that a prioritized qualitative strand was followed by a quantitative strand) or plus marks (e.g., "*qual* + QUAN" means that data were timed concurrently, with priority given to the quantitative strand) (see also ▶ Chap. 40, "The Use of Mixed Methods in Research").

In reference to our research study on barriers to crisis intervention services, the appropriate classification is as follows: [QUAN + *qual*] ➜ QUAL. Data were mixed at the levels of design, collection, and analysis. The first phase (i.e., QUAN + *qual*] refers to the web-based survey, which prioritized the quantitative items. Within the survey, participants were given the opportunity to offer open-ended comments to contextualize their item responses. While the qualitative data were reviewed, our team prioritized a statistical analysis premised upon frequentist assumptions. In this regard, the timing of the second strand was deliberate; we utilized the survey results to inform the development of our focus group protocol. The second QUAL strand was weighted equally to the first phase (i.e., QUAN + *qual*), given the extent to which it addressed the research question and aligned with our goals for the study. To this end, we also employed an intensive qualitative coding, analysis, and interpretation drawn from respected grounded theory techniques (Bowers 1988; Charmaz 2006).

3.1   Convergent Design

The convergent design, also referred to as *convergent parallel*, is as elegant as it is simple, composed of one phase. Timing is concurrent with equally weighted qualitative and quantitative strands that are ordinarily mixed at the level of interpretation (Creswell 2015; Creswell and Plano Clark 2018). Of note, the idea of concurrent timing does not mean that data from the two strands are collected at exactly the same time. It is possible, for instance, for a researcher to deploy a survey instrument and conduct interviews on different days, or weeks, for that matter. Concurrent timing is not an exact measure; it relates more to the intentionality of the design. The strands are regarded as independent, merged after each set of data has been collected and analyzed separately. Then, the data sets are often compared for convergence and divergence.

Kukla, Bonfils, and Salyers (2015) utilized the convergent approach to explore factors impacting work success in veterans with post-traumatic stress disorder (PTSD) or other severe mental illness (SMI). The authors chose this design for "the strength of offering a more comprehensive understanding of the phenomenon...in that different, yet complementary, information was sought through qualitative and quantitative methods" (p. 54). Therein, qualitative data (i.e., a survey of factors impacting vocational functioning) and quantitative data (i.e., narrative accounts of work experiences) were collected within one phase and analyzed separately, mixed at the level of interpretation. We might understand the nomenclature of this design as "QUAN+QUAL." Kukla et al. (2015) reported that the qualitative data corroborated survey findings, with additional insight produced. As the first-published study of this topic, MMR was especially important to the authors, as it evidenced a rich, triangulated picture of the complex factors affecting work successes for persons with PTSD and SMI.

3.2   Explanatory Sequential Design

The explanatory sequential design consists of two phases, always starting with the quantitative, the results from which drive the development of a second qualitative phase (Creswell 2015; Creswell and Plano Clark 2018). This design can be *fixed* or *emergent;* the latter is often precipitated by curious or unexpected quantitative findings that bear further exploration. In the explanatory design, researchers analyze and interpret quantitative data, which informs the design of a follow-up qualitative protocol. Qualitative findings exist to explain results from the first phase, rendering this strand subordinate to its quantitative counterpart. Though "QUAN→*qual*" is standard in this design, there are known deviations. Creswell and Plano Clark (2018) explore a variant of interest to health social scientists: the participant selection approach (i.e., *quan*→ QUAL). In this approach, let us imagine that researcher plans to explore a phenomenon that requires a diverse, purposive sample. She has access to a database that allows for her to select, and later invite, participants with different scores into her primary study. Thus, the quantitative data exist, in this case, to select participants into the qualitative phase.

Restall and Borton (2009) employed an explanatory sequential design (e.g., "QUAN→*qual*") to apprehend the prevalence of parental concern about their children's development at the point of school entry. Broadly, this study reflected a desire to support child development in Canada, within which, at the time of study, there were "no guidelines for the systematic screening of children so little is known about the extent of developmental problems in young children or parental experiences in addressing developmental concerns" (p. 208). To address their quandary, the researchers first surveyed parents utilizing the Parents' Evaluation of Developmental Status (PEDS; Glascoe 2004). Data (n = 290) were analyzed, and a select cadre of parents was identified (i.e., of children in the moderate- or high-risk categories with variable health services) for phenomenological interviewing (n = 9). Of the value of this MMR design, the study underscored the importance of juxtaposing data on prevalence of parental concerns across varied domains (e.g., expressive language, social-emotional, school and health) with a narrative elicitation of parental concern about their children's development. Further, the study allowed the researchers, in the qualitative strand, to employ a heterogeneous sampling technique.

3.3 Exploratory Sequential Design

The exploratory sequential design, comprised of two or three phases, commences with qualitative data collection and proceeds to an instrument or intervention development process that is then tested quantitatively (Creswell 2015). Remarking on the nuanced challenges of this approach, Creswell (2015, p. 40) stated:

> With three phases, it also becomes the most difficult of the three basic designs. Like the explanatory sequential design, this design takes times, but these phases are extended out in time much more than the other basic designs. This design is also challenging to conduct because of the difficulty in taking qualitative results and turning them into a new variable, a new instrument, or a new set of intervention activities.

Historically, the design was referred to as two-phased (Creswell and Plano Clark 2011), which lingers in some MMR circles. Therein, the core intent was to generalize first-phase qualitative findings, such as a grounded theory, through a follow-up quantitative study. Weighting and prioritizing of strands depends on the goal of the study. If the researcher's intent is to develop and refine a theory or taxonomy, for instance, it is plausible to consider the study as follows: "QUAL→*quan*." If developing an instrument or assessment, on the other hand, it is reasonable to consider the first phase as subordinate to the second (e.g., "*qual*→QUAL").

Asante, Meyer-Weitz, and Peterson (2016) used a two-phase exploratory design (e.g., "QUAL➔*quan*") to investigate mental health and health risk behaviors of Ghanaian homeless youth. As the topic had not been explored in an African context, the authors began with a qualitative inquiry (n = 16) into reasons for engagement in risky behaviors. These data informed a quantitative survey (n = 227) wherein the investigators looked at clustering effects and predictors relative to the identified behaviors. As the authors incisively reflected, "the qualitative data and results enabled participants to provide views regarding health and wellbeing, while the quantitative data explores the extent of such problems and their determinants" (p. 436). In this study, neither a new instrument nor intervention was developed; rather, the qualitative phase appeared to have been influenced researchers' selection of existing, internationally validated measures.

A study by Hansen, Okolonko, Ogynbajo, North, and Niccolai (2017) used an exploratory sequential design to "examine acceptability of and perceived facilitators and barriers to HPV vaccination at [school-based health centers] among parents and adolescents" (p. 706). The study bears similarities and differences to that of Asante et al. (2016). Similarly, Hansen et al. (2017) sought to explore a phenomenon around which extant literature offered insufficient depth or explanation. Differently, the authors developed a survey based on the results of the qualitative strand and utilized the results from both strands to address the core question. First, they conducted in-depth interviews with parents (n = 20) and adolescents (n = 20) until they reached thematic saturation. Then, they developed a web-based survey based on key themes from the qualitative segment. The survey was piloted and administered to parents only (n = 131). Data appear to have been weighted equally (e.g., "QUAL➔QUAN"). While the authors describe the study as two-phased, Creswell (2015) would likely consider this a three-phased design. Hansen et al. (2017) collected qualitative data, used those data to develop an instrument, and gathered quantitative data using the newly designed survey.

3.4   Intervention Design

Creswell (2015) introduced the intervention design, also referred to as the experimental design (Creswell and Plano Clark 2018), as an addendum to a previously conceptualized embedded design (Creswell and Plano Clark 2011). Within this design, one conducts a quantitative experiment – or an intervention – that includes a qualitative strand. Let us imagine that we are testing a new therapy designed to help individuals diagnosed with panic disorder. Participants are randomly assigned to three groups: *new* intervention, *status quo* therapeutic intervention, and *no* intervention. Pre- and post-measures consist of a battery of items drawn from psychometrically robust instruments. As the reader might imagine, a design of this nature has its limitations, among them the rival hypotheses about causal processes leading to the results. Qualitative inquiry may yield valuable insights, allowing for a clearer picture of participants' therapeutic experiences.

The intervention design, adding to one or more of the basic designs explored above, allows for the collection of qualitative data before an intervention (i.e., exploratory sequential), during the intervention (i.e., convergent), or after the intervention (i.e., explanatory) (Creswell 2015). Throughout the intervention period, for instance, the researchers may invite participants to engage in brief interviews or even journal about their experiences. In another variant, researchers may conduct a quantitative-only intervention that yields results into which they hope to elicit depth. This could point to an emergent explanatory sequential design.

Curry et al. (2017) reported upon a 2-year intervention designed to improve five facets of performance in ten hospitals: learning environment, senior management support, psychological safety, commitment to the organization, and time for improvement. Using two surveys and a morality measure, each mapping to different outcomes (i.e., changes in organizational culture, uptake of evidence-based strategies, and mortality rates), measures were taken at baseline, 12 months, and 24 months. What made this a mixed methods study was the collection of qualitative interview data at baseline, 6 months, and 18 months; this was intended "to enhance the assessment of culture change with greater validity than possible with only quantitative data" (p. 3). In addition, the researchers used ethnographic observations at baseline and at 18 months. Of note, all forms of data appeared to have been prioritized in the integration process; furthermore, the qualitative analyses were conducted with adherence to strong standards, including the use of iterative and multi-rater coding. Among the contributions of this study is its attention to how mixed methods – most especially the embedded qualitative strands – enhanced the researchers' understanding of a complex issue. Curry et al. (2017) opined, "...although the quantitative magnitude of changes in culture was relatively modest, the qualitative experiences were compelling, a finding we attribute in part to the difficulty of quantitative measurement of nuanced concepts such as culture" (p. 8).

3.5   Hybrid Design Choices

As the reader might have already conceived, not every mixed methods study fits squarely into one or more of the models denoted above. Two studies serve as examples. The first is a four-phase study that examined counselor and client perceptions of the authors' emergent approach to treatment, neuroscience-informed cognitive behavior therapy (nCBT) (Field et al. 2016, 2017). To test the approach, the authors merged qualitative (*qual*) with quantitative (QUAN) data at four time periods (i.e., baseline, 3, 6, and 12 months). Peering into the fourth phase of data collection (12-month interval) was of intrigue, as the authors "followed an explanatory sequential process whereby the qualitative data were connected to earlier merged quantitative data to better understand initial quantitative findings" (Field et al. 2017, p. 354). That is, quantitative data – with supporting qualitative items – were collected during phases one, two, and three. At the fourth phase, only qualitative data were collected, via open-ended survey questions, to help the researchers better understand the effectiveness of nCBT. These data allowed for development of a pathway that depicts client and counselor expectancy development for nCBT treatment.

Kovacs Burns et al. (2014) utilized a hybrid approach to determine the resources and supports needed for constituents (e.g., patients, families, providers) to engage in patient-centered decisions with an overarching intent to "improve access, quality, safety, and sustainability" (p. 234) of a Canadian healthcare model. Four objectives were mapped to distinct phases, independent though amalgamated across the arc of the study. Phase one entailed a need assessment, wherein constituents were interviewed individually or in groups. Results were used to guide the second and third phases. Phase two was comprised of a systematic literature review (qualitative), used also to inform phase three's development of a draft resource kit that was tested, through qualitative and quantitative means, in phase four with some of the same individuals who participated in phase one. The approach, deemed "practical and significant for this type of study" (p. 244), was intentionally integrative. While the authors did not discuss relative weighting among components, a hypothetical diagram for this study might be: ["QUAL+*quan*" *needs assessment*]→["QUAL+*quan*" *systematic review*] →["QUAL+QUAN" *intervention*]→["QUAL+QUAN" *evaluation*].

4   Getting Started

Getting started on a mixed methods project entails multiple considerations. One of the best ways to begin is by reviewing mixed methods studies published within one's own discipline or field. Therein, one can appraise research studies of interest, contemplating some of the seminal issues that we vet in this chapter, such as: How does a mixed methods approach address the research question in a more robust way than a qualitative or quantitative study could do on its own? What are the philosophical underpinnings of the study – and in what ways were these rendered transparent or considered in the integration of qualitative and quantitative data? How were the data mixed, weighted, and timed? Does the design reflect one of the better-known classifications, or did the authors utilize a novel approach? Were the data integrated and discussed in a coherent, systematic manner? As Morse and Niehaus (2009) summarily stated, "the more you know about research methods, the easier mixed methods will be!" (p. 77).

We would argue that reflection into one's own skills, biases, and mental models is an integral precursor to all social science research. MMR is no exception. An individual whose worldview is inherently constructivist should not be dissuaded from engaging in MMR; instead, that researcher is advised to think critically about how her epistemological, ontological, and methodological stances might flavor the collection, analysis, and integration of qualitative and quantitative data. Correspondingly, one would not unwittingly situate an MMR study under a pragmatist guise without having considered how philosophy guides method, data, and design. Furthermore, doing MMR well requires expertise in quantitative and qualitative approaches, but this does not imply that the reader has to be *the* specialist in both areas. Indeed, it is rare to discover solo-authored mixed methods studies, in part because so many researchers partner with colleagues whose skill sets complement their own.

Expanding on the importance of reflection, Morse and Niehaus (2009) underscored thinking and planning in the context of what they call "the armchair walkthrough" (p. 78) – which takes place well before one starts a study. Herein, a researcher reflects on hypothetical variations of the envisioned study. Start with "if" thinking (e.g., "*If* I pursue this research question, what approach might I take? *If* this is the best approach, who comprises my sample and what methods do I consider?). Walking through a hypothetical approach to the research project provides an opportunity to envision multiple pathways, navigate plausible dilemmas or obstacles, identify research partners, contemplate a timeline, and interrogate the extent to which MMR is (or is not) the best approach.

Creswell et al. (2011), writing for the National Institutes of Health (NIH), acknowledged that while there "is no rigid formula for designing a mixed methods study" (p. 6), researchers are advised to consider several steps. Once the above-noted preliminary considerations (e.g., philosophy) are resolved, and the goals of the study are clear, the researcher should determine methods; select the design best aligned to the goals and question (e.g., exploratory); collect, analyze, and integrate qualitative and quantitative data in accordance with the design chosen; (4) make meaning of how mixed methods provides the most robust understanding of the phenomenon under investigation; and write the report in such a way that not only the research question is better understood but that the reader walks away with a greater appreciation for MMR (Creswell et al. 2011) (see also ▶ Chap. 40, "The Use of Mixed Methods in Research").

5   Conclusion and Future Directions

As with quantitative or qualitative methods alone, MMR is not without its limitations. As noted above, lack of methodological expertise can jeopardize data collection, analysis, and integration; thus, we recommend the cultivation of a research team. Given the transdisciplinary and interprofessional nature of the health social sciences, such a team may not only be diverse with respect to methodological prowess but also with regard to disciplinary and clinical experience. Researchers are also advised to be mindful of time and resources; since MMR entails multiple phases or strands, a study may take longer than one might ordinarily anticipate. Finally, we recommend attending as much to interpretive issues as one would to the design itself. In our experience, novice mixed methods researchers often "lose steam" at the time during which analytical power is most important – the integration of data sets in order to arrive at a full-bodied interpretation.

In spite of these limitations, MMR offers a strategic, innovative, and methodologically diverse approach to investigating health phenomena. Specifically, MMR "reflects the nature of the problems facing public health, such as disparities among populations, age groups, ethnicities, and cultures; poor adherence to treatment thought to be effective; behavioral factors contributing to disability and health; and translational needs for health research" (Creswell et al. 2011, p. 2). In other words, the very methods used, like the thoughtful juxtaposing of critical ethnography with survey research, may allow for inquiry into a third-world health epidemic that holistically, ethically, and carefully considers the character of the research itself.

References

Anderson G, Herr K. The action research dissertation: a guide for students and faculty. Thousand Oaks: Sage; 2005.

Argyris C, Putnam R, Smith DM. Action science. San Francisco: Jossey-Bass; 1985.

Asante KO, Meyer-Weitz A, Petersen I. Mental health and health risk behaviours of homeless adolescents and youth: a mixed methods study. Child Youth Care Forum. 2016;45:433–49.

Biesta G. Pragmatism and the philosophical foundations of mixed methods research. In: Tashakkori A, Teddlie C, editors. Handbook of mixed methods in social & behavioral research. 2nd ed. Thousand Oaks: Sage; 2010. p. 95–118.

Bowers BJ. Grounded theory. In: Sarter B, editor. Paths to knowledge: innovative research methods for nursing. New York: National League for Nursing; 1988. p. 33–59.

Bryman A. Quantity and quality in social research. London: Routledge; 1988.

Bryman A. Barriers to integrating quantitative and qualitative research. J Mix Method Res. 2007;1 (1):8–22.

Campbell DT, Fiske DW. Convergent and discriminant validation by the multitrait-multimethod matrix. Psychol Bull. 1959;56:81–105.

Charmaz K. Constructing grounded theory: a practical guide through qualitative analysis. London: Sage; 2006.

Cherryholmes CH. Reading pragmatism. New York: Teachers College Press; 1999.

Creswell J. Research design: qualitative, quantitative and mixed methods approaches. 4th ed. - London: Sage; 2014.

Creswell JW. A concise introduction to mixed methods research. Thousand Oaks: Sage; 2015.

Creswell JW, Klassen AC, Plano Clark VL, Smith KC. Best practices for mixed methods research in the health sciences. Bethesda: National Institutes of Health; 2011.

Creswell JW, Plano Clark VL. Designing and conducting mixed methods research. 2nd ed. Thousand Oaks: Sage; 2011.

Creswell JW, Plano Clark VL. Designing and conducting mixed methods research. 3rd ed. Thousand Oaks: Sage; 2018.

Crotty M. The foundations of social research: meaning and perspective in the research process. Thousand Oaks: Sage; 1998.

Curry LA, Brault MA, Linnander EL, McNatt Z, Brewster AL, Cherlin E, Peterson Fleiger S, Ting HH, Bradley EH. Influencing organisational culture to improve hospital performance in care of patients with acute myocardial infarction: a mixed-methods intervention study. BMJ Quality & Safety. 2017;0:1–11. Retrieved from http://qualitysafety.bmj.com

Denscombe M. Communities of practice: a research paradigm for the mixed methods approach. J Mix Method Res. 2008;2(3):270–83.

Denzin NK. The research act: a theoretical introduction to sociological methods. New York: McGraw-Hill; 1978.

Freshwater D. Reading mixed methods research: contexts for criticism. J Mix Method Res. 2007;1 (2):134–45.

Field TA, Beeson ET, Jones LK. Neuroscience-informed cognitive-behavior therapy in clinical practice: A preliminary study. Journal of Mental Health Counseling. 2016;38:139–154. https://doi.org/10.17744/mehc.28.2.05

Field TA, Beeson ET, Jones LK, Miller R. Counselor allegiance and client expectancy in neuroscience-informed cognitive-behavior therapy: A 12-month qualitative follow-up. Journal of Mental Health Counseling. 2017;39:351–365. https://doi.org/10.17744/mehc.39.4.06

Glascoe FP. Parents' evaluation of developmental status. Nashville: Ellsworth and Vandermeer Press 2004.

Giddings LS. Mixed-methods research: positivism dressed in drag? J Res Nurs. 2006;11 (3):195–203.

Greene JC. Mixed methods in social inquiry. San Francisco: Jossey-Bass; 2007.

Greene JC, Caracelli VJ, editors. Advances in mixed-method evaluation: the challenges and benefits of integrating diverse paradigms: new directions for evaluation, 74. San Francisco: Jossey-Bass; 1997.

Halcomb EJ, Andrew S, Brannen J. Introduction to mixed methods research for nursing and the health sciences. In: Andrew S, Halcomb EJ, editors. Mixed methods research for nursing and the health sciences. Sussex: Wiley-Blackwell; 2009. p. 3–12.

Hansen CE, Okoloko E, Ogynbajo A, North A, Niccolai LM. Acceptability of school-based health centers for human papillomavirus vaccination visits: a mixed-methods study. J Sch Health. 2017;87:705–14.

Hanson WE, Creswell JW, Plano Clark VL, Petska KS, Creswell JD. Mixed methods research designs in counseling psychology. J Couns Psychol. 2005;52(2):224–35.

Hathcoat JD, Meixner C. Pragmatism, factor analysis, and the conditional incompatibility thesis in mixed methods research. J Mix Method Res. 2017;11(4):433–49.

Hathcoat JD, Nicholas M. Epistemology. In: Coghlan D, Brydon-Miller M, editors. The Sage encyclopedia of action research ("E" entries). Washington, DC: Sage; 2014. p. 285–332.

Holmes CA. Mixed (up) methods, methodology and interpretive frameworks. Paper presented at the Mixed Methods Conference, Cambridge, UK; 2006.

Howe KR. Against the quantitative-qualitative incompatibility thesis or dogmas die hard. Educ Res. 1988;17:10–6.

Jick TD. Mixing qualitative and quantitative methods: triangulation in action. Adm Sci Q. 1979;24:602–11.

Johnson RB, Onwuegbuzie AJ. Mixed methods research: a research paradigm whose time has come. Educ Res. 2004;33(7):14–26.

Kovacs Burns K, Bellows M, Eigenseher C, Jackson K, Gallivan J, Rees J. Exploring patient engagement practices and resources within a health care system: applying a multi-phased mixed methods knowledge mobilization approach. Int J Mult Res Appro. 2014;8(2):233–47.

Kukla M, Bonfils KA, Salyers MP. Factors impacting work success in veterans with mental health disorders: a veteran-focused mixed methods pilot study. J Vocat Rehabil. 2015;43(2015):51–66.

Lanza R, Berman B. Biocentrism: how life and consciousness are the keys to understanding the true nature of the universe. Dallas: BenBella Books, Inc.; 2009.

Meixner C, O'Donoghue CR, Witt M. Accessing crisis intervention services after brain injury: a mixed methods study. Rehabil Psychol. 2013;58(4):377–85.

Morse JM, Niehaus L. Mixed method design: principles and procedures. New York: Taylor & Francis; 2009.

Nicholas M, Hathcoat JD. Ontology. In: Coghlan D, Brydon-Miller M, editors. The sage Encyclopedia of action research ("O" entries). Washington, DC: Sage; 2014. p. 285–332.

Reichardt CS, Rallis SF, editors. The qualitative-quantitative debate: new perspectives. San Francisco: Jossey-Bass; 1994.

Restall G, Borton B. Parents' concerns about their children's development at school entry. Child Care Health Dev. 2009;36(2):208–15.

Rossman GB, Wilson BL. Numbers and words: combining quantitative and qualitative methods in a single large-scale evaluation study. Eval Rev. 1985;9(5):627–43.

Sieber SD. The integration of fieldwork and survey methods. Am J Sociol. 1973;78:1335–59.

Tashakkori A, Teddlie C, editors. Handbook of mixed methods in social and behavioral research. Thousand Oaks: Sage; 2003.

Recruitment of Research Participants

Narendar Manohar, Freya MacMillan, Genevieve Z. Steiner, and Amit Arora

Contents

N. Manohar (✉)
School of Science and Health, Western Sydney University, Sydney, NSW, Australia
e-mail: narendar.manohar@hotmail.com

F. MacMillan
School of Science and Health and Translational Health Research Institute (THRI), Western Sydney University, Penrith, NSW, Australia
e-mail: f.macmillan@westernsydney.edu.au

G. Z. Steiner
NICM and Translational Health Research Institute (THRI), Western Sydney University, Penrith, NSW, Australia
e-mail: g.steiner@westernsydney.edu.au

A. Arora
School of Science and Health, Western Sydney University, Sydney, NSW, Australia

Discipline of Paediatrics and Child Health, Sydney Medical School, Sydney, NSW, Australia

Oral Health Services, Sydney Local Health District and Sydney Dental Hospital, NSW Health, Sydney, NSW, Australia

COHORTE Research Group, Ingham Institute of Applied Medical Research, Liverpool, NSW, Australia
e-mail: a.arora@westernsydney.edu.au

© Springer Nature Singapore Pte Ltd. 2019
P. Liamputtong (ed.), *Handbook of Research Methods in Health Social Sciences*,
https://doi.org/10.1007/978-981-10-5251-4_75

Abstract

Successful recruitment and retention of study participants are essential for the overall success of a research study. The recruitment process involves identifying potential research participants and providing them with the information to establish their interest to join a proposed research study. Research studies are often time and labor intensive, and inappropriate recruitment of research participants can significantly impact the study findings. This chapter will introduce readers to a range of associated issues and offer possible solutions and mitigation strategies to enhance research participant engagement. First, this chapter describes the issues surrounding investigators' and potential participants' expectations related to their involvement in research. Next, the chapter will identify the facilitators, barriers, and challenges associated with recruitment and retention of participant. Then, the chapter will highlight some traditional and modern recruitment and retention techniques, for participation in health research across the life course: children and adolescents, adults, and seniors. Last, the chapter will detail the specific attention, resources, and sensitivity required to maximize recruitment and retention when conducting research with specific population groups such as minority populations and medically compromised people.

Keywords

Recruitment strategies · Study participants · Participant engagement ·
Participation · Target population · Minority recruitment · Opt in–opt out

1   Introduction

Recruitment of participants in health research can be challenging. An appropriate selection of participants is essential for accurate representation of the population of interest. However, poor recruitment is still a significant drawback for many studies (McDonald et al. 2006; Glasser 2014). Recruitment can be defined as a "dialogue between an investigator and a potential participant prior to initiation of the consent process" (Patel et al. 2003, p. 229). The recruitment process involves identifying, targeting, and enlisting potential participants, followed by provision of information to potential participants and establishing their interest in the proposed study (Patel et al. 2003). To ensure appropriate recruitment in a quantitative research project, for example, it is important to identify participants that closely represent the target population and meet the sample size and power requirements of the study (Hulley et al. 2013). Thus, it is important to identify appropriate environments to gain access to the intended population. Overall, the recruitment process depends on the type of research, collaboration between researchers and the recruitment/referral pathway (e.g., clinicians), characteristics and preferences of participants, and the strategies employed for recruiting (Patel et al. 2003). This chapter describes the issues surrounding investigators' and potential participants' expectations related to involvement in research, the facilitators, barriers, and challenges associated with recruitment and retention of participants and explains the process of recruitment across the life-span and in various population groups.

2 Investigator: Participant Interface in Recruitment

2.1 Researcher's Expectation and Personal Perspective

Prior to initiating any study, it is essential to consider the relevant roles, responsibilities, and interactions of the primary parties involved, that is, the investigator, who asks scientific questions, and the participant, who consents to answer scientific questions put forward by the investigator (Rodriguez et al. 2003). Both the investigator and the participant have preconceived expectations from the relationship that they will be establishing and how it will progress. The investigator goes through an extensive process before a study begins that shapes expectations about recruitment. This includes developing a research protocol, participant information sheets and consent forms, and review and approval by relevant scientific and ethics committees. The investigator expects to recruit participants that closely match the target population and aims to achieve the defined sample size. In the co-author's own words regarding a researcher's expectations toward participant recruitment:

> I have over 1,000 participants to recruit; in that case, I would love to have easy access to my potential participants so that I am closer to my estimated sample size without delaying my study timelines and budget. (Arora)

2.2 Expectations of Research Participants

Participants also have expectations from the study investigator(s) and from the research study they join. It is expected that a research participant will be assertive and protect their own rights, which will actually improve the quality of research that they are enrolled in (Rodriguez et al. 2003). Potential research participants should have access to detailed information about what they are consenting to (see also ▶ Chap. 106, "Ethics and Research with Indigenous Peoples"). This will inform their expectations:

– To be informed about the potential benefits and harms associated with the planned research; and if they are ill, then how will this research help them
– To ensure that their interests are protected. That is, a clear explanation of what is involved and what they will have to do, who will be in charge of their health interests, any safety standards in place to protect them from harm, whether the research has been reviewed and approved by relevant committees, whether they have to spend money out of their own pocket or will be compensated for taking part, whether they have the right to withdraw from the study without prejudice, and whether the results will be shared with them upon study completion
– To have some understanding about the study design and leadership. That is, who designed the protocol and whether it is well designed, whether the investigator is competent to undertake the research, why the research is important, who else is involved, and whether any community advocate was involved in the design and review of the research
– To be informed if there is any conflict of interest or controversy associated with the study, if a similar study has been done before, any financial beneficiaries, and the self-interest of the investigator

3 Barriers and Facilitators to Participant Recruitment and Retention

Problems with recruitment can adversely affect the timeline of a study, increase workload, compromise the study findings, and eventually lead to study abandonment (Patel et al. 2003). Despite the extensive recommendations for improving participants' recruitment in research (Caldwell et al. 2010), poor recruitment remains a fairly common and costly issue.

3.1 Barriers and Challenges in Recruitment

The following list illustrates barriers or challenges for recruitment in any study (Liljas et al. 2015; Stein et al. 2015; Quay et al. 2017) and has been categorized into "researcher- and trial-related" and "participant- and treatment-related" factors.

3.1.1 Researcher- and Trial-/Protocol-Related Factors
– Risk of test therapy/uncertainties inherent in investigational drugs
– Perceived risk associated with randomization
– Clinical visits required
– Systemic therapy involved
– Added stress from trials
– Additional tests required

- Fatalism
- Variations in Institutional Review Board (IRB)/Ethics Committee regulations across partnering institutions
- Culture or language related: language (lack of study materials and communication in target populations' language) or cultural issues (e.g., lack of respect for gender or segregation, religious practices)
- Logistics: under-presentation of ethnic population at recruitment sites, costs associated with elevated recruitment requirements, limited time to recruit or requirement for repeated recruitment efforts, the absence of a care coordinator, ineffective informed consent processes, and unfeasible study and recruitment goals
- Study design: lack of appropriate (racial/ethnic/language specific or validated) assessment tools, narrow entry criteria, and recruitment based on convenience
- Awareness: stereotypes about difficulties of engaging with different racial/ethnic population groups, researcher attitudes (e.g., apathy), ineffective guidance to study staff, and limited knowledge about methods to promote a study, recruit, and retain participants.

3.1.2 Participant- or Treatment-Related Factors

- Disinterest or lack of feeling of belonging: immigrant perceptions of not belonging to society meant to benefit from the proposed research, lack of interest, misgivings about scientific importance or benefit, prior treatment for disease (trial participation perceived as unnecessary), and utilization of disease-specific services (e.g., diabetes services)
- Conflict: decisional hierarchy and gender, substance abuse or mental health issues, and religion or cultural conflicts
- Education or training related: poor understanding of research intentions among community or religious leaders and lack of understanding about the consent process
- Logistic or opportunity costs: potential costs associated with participation, time spent away from work, travel time, family and other commitments, and logistical issues related to transportation or location
- Fear or inhibitions: fear of being reported to immigration, stigma of being labeled with a health condition, concerns about adverse effects (i.e., related to interventions), fear of finding out health issues, mistrust of research, previous poor experiences participating in research, conservative attitude toward risk-taking, unique health situations, and no or limited health insurance.

3.2 Facilitators of Recruitment

A range of recruitment interventions can be adopted to improve participation in research (Mapstone et al. 2007), for example, changing aspects of the design (e.g., not including a placebo group) (Welton et al. 1999), making the participant's experience in data collection easy and simple, providing monetary incentives for participants (must be balanced with the ethical considerations of economically disadvantaged participants), and engaging collaborators and developing referral pathways via newsletters to provide updates on study progress. The major facilitators to health research participation can be categorized into (i) design and logistics, (ii) benefits and low risk in participation, (iii) altruism, and (iv) family approval and community involvement; these will now be described in detail below.

3.2.1 Design and Logistics

During the planning and designing phase of any study, it is advisable to hire staff to whom participants can relate (e.g., same racial/ethnic background as the target population, insider approach) (Hayfield and Huxley 2015; Stein et al. 2015; Manohar et al. 2017). This is because most racial/ethnic or marginalized population participants can communicate effectively and comfortably in their own language and rhythm of expression (Liamputtong 2010; George et al. 2014), thereby improving recruitment by establishing a level of trust. For instance, in a study involving immigrant Filipino women, the investigators noticed that the potential participants were more likely to provide consent if they personally knew the recruiter (Maxwell et al. 2005). Similarly, in a study involving recruitment of transgender participants (Owen-Smith et al. 2016), participants felt that research staff who are not a transgender himself/herself would not be able to effectively connect with them.

However, there is also always a risk of exploitation which needs to be considered (Brugge et al. 2005). Furthermore, language-appropriate material should be available so that potential participants can understand their involvement (Arora et al. 2017b). Also, it is important that the recruiting staff are well informed about the study and appropriately answer potential participants' queries; otherwise, low recruitment is a potential outcome.

In terms of logistics, it is advisable to facilitate participants' involvement by providing free parking or by utilizing a study site close to public transport. It is also important to provide flexibility in scheduling participation on weekdays, weekends, and afterhours where possible, as this will facilitate active participation without compromising participants' daily routines. Furthermore, participants should be provided different options for obtaining recruitment approval (e.g., in person, phone, postal mail, and/or email) (Owen-Smith et al. 2016). Last, the research staff should acquire employer support where applicable (for the potential participants) to take time off from their work for participation (Wyatt et al. 2003).

3.2.2 Benefits and Low Risk in Participation

Incentives play an essential role in improving recruitment rates and can include money, transport cover, complementary refreshments, gift vouchers, health checks/examinations, or access to healthcare facilities (Liamputtong 2007, 2010; George et al., 2014). Incentive choice has been reported to be desirable by many potential participants (Owen-Smith et al. 2016). Certain marginalized population groups, such as transgender individuals, are willing and motivated to participate in research (relevant to their population group) in order to receive information on the latest scientific developments on their health, interact with fellow community members, and receive free medical services (if part of a study) (Owen-Smith et al. 2016).

Furthermore, it has been observed that a high level of detail on the risks and safeguards associated with the intended research can serve as a facilitator for recruitment (Farmer et al. 2007). Additionally, studies which are noninvasive, or have the least risk of discomfort, such as questionnaire-based surveys, educational interventions, or minimal intervention-related health risks, tend to have better recruitment rates (Sengupta et al. 2000). Those individuals who are terminally ill (such as stage 4 cancer patients) and have no other treatment options available except the proposed "test therapy" are often willing to participate as they feel it is the last resort (Lee et al. 2016).

3.2.3 Altruism

Altruism is defined as the "willingness to do things that bring advantages to others, even if it results in disadvantage for yourself" (Satar et al. 2015, p. 271). Such ambition is quite evident in studies targeting particular racial/ethnic or marginalized population groups. For example, Owen-Smith et al. (2016) reported that many of the recruited transgender participants felt motivated to join the study to contribute to research which might benefit their respective community and possibly provide essential data to fill knowledge gaps and to share their information with the scientific community. Lee et al. (2016, p. 4) identified altruism as one of the important facilitators for recruitment in their clinical trial and stated that the participants saw trials as a "way to advance treatment for others."

3.2.4 Family Approval and Community Involvement

Opinion and approval of family (sibling, spouse, child) and friends serve as a significant facilitator for recruitment in research (Giarelli et al. 2011). Such an influence was evident in a recent clinical trial conducted in Asian women suffering from breast cancer. In that study, a significant number of participants cited that if the opinions of family members and friends were positive, they will certainly join the trial; however, if the opinions were negative, they would not be a part of it (Lee et al. 2016). Likewise, in certain indigenous population groups, community mediation is integral for research participation, and its decision (either approval or refusal) will be taken as a facilitator or barrier to recruitment (George et al. 2014). Hence, in such cases, it is very important to seek community elder's endorsement in the intended research.

4   Recruitment Strategies

A wide range of recruitment strategies can be adopted to facilitate participant engagement and enrolment in research. The strategies employed will depend on factors such as the study's intended target population, the study size (number of participants and recruitment sites), the study and intervention design, ethics approval processes, and the recruitment budget. Below, we detail several online and offline recruitment strategies that can be adopted in research.

More traditional recruitment means can be effective if designed, coordinated, and distributed strategically. For example, clear, simple, and well-designed letters and mail-outs (either via post, e.g., the electoral roll, or email, e.g., listserv), flyers, posters, and pamphlets can increase awareness of a study. Newsletters can also help to maximize retention in studies or foster engagement in future studies (Bower et al. 2014).

Moderate- to large-sized research studies benefit from a comprehensive and well-coordinated marketing campaign. This can often be organized and planned with the assistance of media units at the university, institute, or hospital that the study is being coordinated at. Marketing strategies will often include a media release and targeted calls to local print media (i.e., newspapers), radio, and television for further study promotion. Paid advertisements can be placed in classified sections of local papers (such strategies are particularly effective for the recruitment of older adults) (Martinson and Hindman 2005).

Grassroots and community outreach approaches are popular and successful methods of engaging potential study participants. Well-promoted public lectures, seminars, or stalls at supermarkets or local fetes can be an effective way of increasing public awareness of research that can also enable news of a study to spread via word of mouth. Stalls at special interest exhibitions, with recruitment and enrolment of people onsite, are particularly effective for recruiting populations suffering from uncommon disease conditions (e.g., people with coeliac disease have been successfully recruited from gluten-free expos) (Chiu et al. 2016).

Clinical cohorts are often recruited via healthcare referral pathways. A comprehensive and engaged network of clinicians is essential for these strategies to work. Clinical trials may have a number of participating hospital sites, and a range of specialist clinicians involved in the trial may refer potential participants to a participating hospital for screening from their private rooms or other hospitals. General practitioners (GPs) or family doctors can be engaged to refer patients. GP referrals are best facilitated by simple eligibility criteria, a strategy whereby referrals reduce practitioner workload, patient incentives, and involvement of a discipline champion (Ngune et al. 2012). Similarly, community-based healthcare workers (e.g., social workers or community nurse consultants) may identify and refer potential participants into trials.

Online methods for recruiting research participants have become increasingly popular. A comprehensive online marketing strategy for a new study can include targeted social media campaigns (e.g., via Facebook or Twitter), search engine marketing (e.g., Google AdWords), display advertising (e.g., web banners), and mobile advertising (e.g., via apps). Advertising through existing online communities is also an effective means of recruitment, for example, Craigslist (2017) and Facebook groups and pages (particularly for a disease focus area, e.g., Younger Onset Dementia Support Group 2017; Endometriosis Australia 2017). Some countries have online portals where potential participants can browse through and sign up for studies that are open for recruitment (e.g., Join Dementia Research, UK (Join

Dementia Research 2017), Clinical Connection, USA (Clinical Connection 2017), ResearchMatch, USA (ResearchMatch 2017)). Having an online point of engagement for people to find further information about the study (e.g., study website, Facebook page) can enhance recruitment.

Clinical trial recruitment agencies are emerging as a one-stop shop for online recruitment of participants and are becoming increasingly popular. Agencies, such as Trialfacts (2017), offer complete recruitment packages combining a mixture of online media strategies (e.g., social media campaigns, search engine marketing) together with a media release and online newscaster approach (i.e., feature pieces in online news articles) that have a call to action which funnels people into an online screener. Potential participants can learn more about the study and complete a web form to assess their eligibility for the trial. Their information is then forwarded to the research team for assessment and follow-up contact.

Thus far, we have discussed participant and researcher expectations in recruitment of study participants, facilitators and barriers in recruitment of research participants, and recruitment strategies. The rest of the chapter will discuss recruitments of potential research participants from different groups.

5   Recruitment of Children and Adolescents

Accessing and recruiting children and young people in any research is challenging and consists of a number of anticipated and unanticipated ethical complexities, dilemmas, and issues. Any research involving children and adolescents should ensure that their opinions and assistance are sought prior to project commencement since their perspectives may impact the research design (Hibberd 2016). Ethics is critical in every research study; however, if a project involves children and young people, it needs to justify certain prerequisites before being considered for research ethics committee approval (UCL Research Ethics Committee 2016):

- Is the planned research study important for the health and well-being of children and adolescents? Does it really need to be done?
- Is the participation of children and adolescents important to answer the research question or can the intended information be obtained in other ways?
- Is the study method appropriate for children and young people?
- Do the circumstances in which research is conducted provide for the physical, emotional, and psychological safety of the child or adolescent?

In any research, informed consent is a vital part of the recruitment process (see also ▶ Chap. 106, "Ethics and Research with Indigenous Peoples"). Informed consent is based on the Declaration of Helsinki, which applies to all human subjects, adults and children. In relation to children, consent of the child and adult guardian should both be sought: "When the subject is a minor, permission from the responsible relative replaces that of the participant in accordance with national legislation. Whenever the minor child is in face able to give a consent, the minor's consent must be obtained in addition to the consent of the minor's legal guardian" (World Medical Association 2001, p. 374).

5.1   Utilization of Midwives, Child and Family Health Nurses, and Childcare for Recruiting Newborn and Infants

Depending on the study design, if the aim of research is to recruit newborn children, an innovative recruitment approach is to utilize the services of midwives or child and family health nurses (CFHN) (Arora et al. 2012). Midwives are probably the first health professional that new mothers come in contact with, and they provide maternity care to the majority of women worldwide (Johnson et al. 2015). This contact allows them not only to deliver key health messages but also to recruit potential mother-infant dyads. Furthermore, the CFHN serve as a first point of contact between postpartum mothers and health professionals (Arora et al. 2012) and can connect effectively, efficiently, and with trust, respect, and safety to families (Ramsden 2002). In regard to recruiting infants, it is estimated that every second child in Australia attends formal childcare for an average of 28 h per week (Australian Government Productivity Commission 2015). The childcare staff share a close association with the families and work effectively to develop a healthy lifestyle for infants. Hence, childcare serves as an ideal venue to successfully recruit infant-parent dyads.

5.2   Utilization of Schools for Recruiting Older Children and Young People

Schools can serve as a valuable venue for recruiting research participants such as children, adolescents, as well as their parents. Childhood and adolescence are very receptive stages of life; hence, effective health promotion or intervention initiated at these development stages will have great benefits. Furthermore, primary and secondary schools have significant power to influence child/adolescent decision-making capabilities at their respective ages (Walsh 2011). The advantages of school-based research are:

(i)   The researcher can obtain a diverse range of participants from a target population (Mishna et al. 2012).
(ii) It provides better retention rates compared to other recruitment methods, since school attendance is mandatory.
(iii) It improves participation by easing the suspicions among student-family populations since potential participants trust the school administration which may transfer onto the study also (Bruzzese et al. 2009).

Research participation can be further improved by including a school staff member in the recruitment process, especially among underserved populations

(Alibali and Nathan 2010). If research is conducted on the school premises, students are already at the site, saving the allocated research budget. Schools may be incentivized to participate through some remuneration or small fund (Mishna et al. 2012).

A number of strategies can be employed for recruiting children or adolescents in a school-based research study. Research team members should thoroughly explain the proposed study and its benefits, address participants' concerns and queries, and obtain contact information through sign-up forms to facilitate follow-up. This process alongside consent form completion may facilitate interaction between researchers and parents/guardians and improve recruitment rates (Berry et al. 2013; Daley 2013). To make recruitment more attractive, refreshments may be arranged, and research team members can answer questions and guide the participants, parents, and school personnel. Furthermore, recruitment can be enhanced by promoting or advertising the study at parents-teacher meetings; the presence of the research team at these meetings reflects the school's support for the project (Bartlett et al. 2017). Additionally, hiring a parent who has participated in an earlier phase of the same research can greatly assist in building trust in potential participants. Involving a school staff as a faculty sponsor or navigator (Bartlett et al. 2017) also improves recruitment rates. Such school-based navigator can assist in identifying and enabling contact with potential participants, demonstrate endorsement of the project, build trust of students and parents toward the research team, and promote participation (Bartlett et al. 2017). If the target population consists of adolescents from vulnerable populations, it is a viable strategy to employ recruiters and interventionists who serve as "insiders" in view of the target population (Craig et al. 2014).

6   Recruiting Adults into Health Research

There is a lack of evidence to guide the best strategies for recruitment of adults in health research, as few studies explore or fully report on the effectiveness of their recruitment strategies. In quantitative research, when designing a recruitment plan for adults, as for any target group, it is important to identify strategies that are most likely to result in a representative sample. If this is not done, then external validity is reduced and study results are less generalizable to the wider population (Liamputtong 2017).

Strategies to recruit adults into health research should, therefore, be tailored to the characteristics of the adult sample being sought. In order to capture the target groups' attention effectively, it is important to ask yourself:

- Who are your target adult population?
- What are their characteristics (e.g., gender, age, employment status, ethnicity, etc.)?
- Where are they likely to congregate?

Then think about methods of contact that may appeal to this group. For example, selected recruitment methods may need to be different for young adults versus middle-aged or older adults, unemployed adults versus employed, women versus men, English- versus non-English-speaking adults, adults from low versus higher socioeconomic status backgrounds, and those from different ethnic backgrounds.

You will also need to tailor your recruitment strategies as per the design of your study. It is important to consider how much of a burden you are imposing on your participants and how you might be able to encourage their participation, without coercing them into participating. It may be particularly challenging to recruit certain subgroups of adults such as working or caring adults who have busy schedules or adults lacking basic needs, as other priorities are often above participating in your study. Flexible and creative ways to attract and incentivize such adults will likely be necessary.

Joseph and colleagues (2016) identify five elements of successful recruitment:

1. Leveraging pre-existing social networks and personal contacts
2. Identification of community gatekeepers and fostering collaborations with them
3. Creating maps/lists of recruitment platforms and settings/locations
4. Developing concise recruitment materials
5. Developing trusting and respectful relationships with potential participants

6.1 Approaching Adults

6.1.1 Consulting the Target Adult Group

If there is a specific group of adults that you are trying to reach or a specific location that you are attempting to recruit adults from, consultation early on with the potential target group is worthwhile. This could be done by identifying an organization that individuals are commonly associated with or an individual who is well connected in the network of adults you wish to recruit (a "champion" or "gatekeeper"). During consultation, the following should be explored in relation to the target group, to help design an appealing and relevant recruitment plan:

- The best location of recruitment and study visits
- The best time to approach (and what the factors might be that could affect this)
- The best methods of contact for study invitation
- Who should be delivering the invitation
- Appropriateness and understanding of wording of any invitations (face to face, written or visual materials)
- When a follow-up on an initial invite should be made, how often and through what methods

Recruitment of adults into health research may be directly via the research team or via a third person (e.g., a healthcare professional or community leader). Interaction with potential participants and effectiveness in negotiating participation are essential in ensuring effective recruitment, whether this is done by the researcher or a third person.

6.1.2 Recruitment Location

Notoriously, researchers recruit adult samples through tertiary education settings by drawing on their students and colleagues, as often students and staff are an easier-to-reach pool of potential participants than recruiting through external locations to education. The issue with this recruitment setting, however, is that skewed samples (e.g., younger and healthier student cohorts) can be recruited, in comparison with the wider population outside of education. Although such convenient approaches can result in large numbers of adhering participants, the applicability of results to the wider population are, therefore, questionable.

Thinking about the target group of adults, it is important to identify locations that are easily accessible and highly visited by your target population. For example, GPs and other health services are regularly used by adults, and so recruitment through healthcare providers may be an effective recruitment avenue. Workplaces often have newsletters and internal mail systems that can be utilized to promote your research. You may need to think creatively about where it is possible to cross paths with your target group of adults, such as places of worship, shopping centers and supermarkets, public transport centers, bookmakers, soccer clubs, and Returned and Services League (RSL) clubs and hotels. If you are delivering an intervention as part of your health research, then recruitment via the location of intervention delivery may be an option, particularly if your intervention is community based. For example, a male-only weight management intervention study in Scotland successfully recruited their required sample ($N = 747$ men) through a combination of professional football club-based activities, media coverage, workplace advertisements, and word of mouth (Wyke et al. 2015). The intervention was delivered via each of 13 professional football clubs by community coaches and included participants from a range of socioeconomic backgrounds. Participating men spoke of being drawn into the intervention because of its setting – the football club that they support and associate with Wyke et al. (2015).

6.1.3 Recruitment Methods

Methods of recruitment will also play a part in the adults that subsequently participate in your research (e.g., recruiting through only passive methods where the interested individual has to make first contact with you, such as newspaper advertisements and through social media, will likely draw a different sample to recruiting through active face-to-face strategies where you make first contact with individuals). Selected methods need to be flexible but also feasible. There is limited evidence that the number of strategies used results in greater recruitment rates of adults into health research, but one would assume that the more strategies used, the more likelihood that the target group will be reached in some way.

The financial budget is often a deciding factor on the methods to be used for recruitment. Recruitment strategies do not have to be expensive to be effective.

For example, research with young Australian adults found that utilizing unpaid online recruitment methods resulted in a greater response rate than paid methods (Musiat et al. 2016). Interestingly, in this particular study, the unpaid methods generally resulted in a sample that were younger and had lower emotional well-being, compared to those recruited via paid methods, highlighting the slight differences in characteristics of samples recruited through different means. Table 1 details different recruitment methods and pros and cons to consider in regard to their use in recruiting individuals into health research.

6.1.4 Recruiting Adults via Healthcare Professionals

If you are recruiting via healthcare, understanding the particular health system and the roles of team members in that system is essential in ensuring that your recruitment strategies can be introduced with as minimal disruption as possible and to effectively capture the target adult audience. Building a strong relationship early on with the healthcare team is essential, particularly if they will be inviting individuals into your study. This might be done by presenting a summary of your research, meeting with team members, and being present on site for informal conversations with the team. Administrative staff may be particularly vital to engage in the process of developing and rolling out of your recruitment strategy, especially reception staff who are the first point of contact with a patient during visits in a health setting. Providing a one-page flow chart, short information sheet, or recruitment card about your study to administrative staff and healthcare professionals may be useful in reminding them about your study and the role they have in engaging patients in your research.

Often, it is a healthcare professional that decides if an individual is eligible to be invited to participate in a study or not, which depending on their investment, understanding and interest in the project can be detrimental or assistive to recruitment rates. Healthcare professionals are busy and have many responsibilities. Thus, promoting your research study may not always be at the top of their priorities. Identifying ways of recruitment to assist healthcare professionals in the screening eligibility process can be extremely useful. For example, consider all team members that may be able to assist in screening for eligibility. Clinicians are often the most time poor. Perhaps nurses or other team members can assist. It might be possible for the research team to actually make the invite once eligible individuals are identified to relieve the expectation of the healthcare team from doing this (e.g., can the research team recruit via face-to-face interaction in the clinic waiting room or draft and send out invitation letters on behalf of the healthcare team).

If you are working with a particular subgroup of adults, then working collaboratively with clinicians and appropriate community organizations (as mentioned earlier, "champions" or "gatekeepers") can assist in your recruitment. Often, the researcher will still need to drive the recruitment, but working collaboratively with champions expands the researcher's reach into the potential pool of participants.

Table 1 The pros and cons of different recruitment methods for health research

Recruitment method	Pros	Cons
"Snowball" sampling (initial participants invite others to participate)	Tapping into existing networks so likely to speed up the recruitment process	Samples are likely to be similar in their characteristics and not a representative sample of the wider population
	Utilizing participants to complete recruitment rather than the researcher	Difficult to evaluate success of this method
	Useful to reach individuals that you may not be able to reach through other methods (e.g., homeless individuals, illegal immigrants)	It may take time to establish a strong relationship with initial participants before they commence snowball recruitment
	Useful in qualitative research	Difficult to evaluate success of this method (requires individuals to keep a record of who they have invited and when)
Letters or postcards (either direct from the research team or via third person)	Can be addressed from someone that the individual already trusts (e.g., a healthcare professional)	Costly (print costs and person costs if hand delivering)
	Quick to administer in large numbers if delivering via postal services	Several mail-outs or visits to houses may be required to reach the target sample size
		May be discarded as "junk" mail
		Addresses are needed for mail-out
		Several personnel may be required to deliver invites door to door
Referrals from healthcare providers	Utilize established trusted relationships	A lack of control as to who is invited – the healthcare provider's role and interest in the research will determine the effort put in to actively promote recruitment
		Difficult to evaluate success of this method (requires the healthcare provider to keep a record of who they have invited and when)

(continued)

Table 1 (continued)

Recruitment method	Pros	Cons
Flyers, pamphlets, brochures, posters	Can be left at convenient locations	Cost (printing and personal time if being distributed by hand)
		Need to be placed in a prime position to capture attention
		May need to be replaced often
		Difficult to evaluate success of this method (e.g., how do you capture number of people that have read a poster)
Videos	Can be played at times when the research team are not present to a wide audience	Require expertise in creating appealing and high-quality videos
	Can be utilized on website, social media, public locations (e.g., general practice surgeries)	May be costly if hiring individuals to participate in the video
		Difficult to evaluate success of this method (e.g., how do you capture number of people that have read a poster)
Face-to-face recruitment at healthcare provider setting (can go through clinic lists, registers)	Builds rapport with the individual	Costly (research time)
	Validation of the research through recruitment at a trusted site (the individual already has a relationship with the healthcare provider)	Time intensive (e.g., may require screening from a healthcare professional and face-to-face interactions can take longer than other approaches)
	Clinic lists and registers can be used to identify large numbers of eligible individuals relatively quickly allowing a more targeted recruitment approach	

Face-to-face recruitment at community outreach activity	Reach potentially large numbers Builds rapport with the individual	Can be difficult to evaluate success of this method Time intensive May require incentives to attract individuals to your stall if competing with others at a community event
Phone calls, emails, internal mail in workplaces, etc.	Direct access to a large number of individuals via a known third party	A lack of control as to who is invited – reliance on the third party to make initial contact with individuals through their lists (consent will need to be sought prior to researcher involvement) Difficult to evaluate success of this method (e.g., relying on a third person that has access to individual's contact details to record who has been contacted and when)
Newsletters, local newspapers, radio, television	Potentially can reach large numbers	Prime time and location can be costly Difficult to evaluate success of this method (e.g., how do you capture number of people that have watched a TV advertisement)
Social media and the web	Highly used by adults of all background Cost-effective Website analytics are useful to guide and manage online strategies Particularly useful strategy for people of low SES, as a large proportion of the population have access to a phone and have social media accounts	Advertisement needs to be well thought out to grasp attention of the target group (may require payment to be located in a prime position) May require management by research team (e.g., Facebook where posts can be inappropriate)

(continued)

Table 1 (continued)

Recruitment method	Pros	Cons
Registries	Access to large number of potential participants	Bias in sample (the most motivated individuals and often those that are well informed and educated in regard to research tend to join registries)
	Immediate contact with the researcher may be possible to build rapport	
	Useful for building community research registries at locations where large numbers of minority groups converge, such as during community health and cultural events	
Word of mouth	Can be successful particularly if participants through "champions" and "gatekeepers" that are enthusiastic about the research	Relying on others to recruit
	Individuals already have a relationship with those inviting them – rapport already exists	Difficult to evaluate success of this method (e.g., relying on others to record who has been invited and when)
Recruitment through existing research studies	Individuals already have a research-focused relationship with a team known to you (rapport has already been developed)	Reliance on other research team members inviting participants into your study
		Participant overburden needs considered
		More motivated individuals tend to provide their time again and again

6.1.5 Recruiting Adults Through Existing Databases and Organization Contacts

Research organizations often have websites that can be useful avenues for recruitment. Detailed information on studies can be made available for potential individuals to read at their leisure on such websites prior to making a decision on whether to participate or not. Such organizations often have databases for adults to register their interest in participating in a particular research area or type of research. Other organizations, including workplaces, community organizations, and sports and recreation clubs, may also provide access to the target group. The first point of contact may be someone within the organization or you may be asked to provide a presentation or similar to introduce your research to the target audience.

6.1.6 Flexibility in Location and Timing

For working or caring adults, time can be a constraining factor for participation in research. Flexibility in where visits can occur can ensure better recruitment and retention. For example, if home, work, or other public space visits are possible, which allow the researcher to go to the participant, recruitment and retention rates may increase. Allowing for flexibility in timing of visits will also be important (e.g., outside of work hours and at a convenient time for those with caring responsibilities). For safety reasons, this may mean having two researchers or recruiters available, particularly if it is likely that the recruiter will need to enter the potential participant's home. When working with dyads or family members, extra time in recruitment may be necessary as additional juggling to find a time that all can attend may be required.

6.1.7 Incentives, Compensation, and Acknowledgments

Where incentives and compensation (e.g., for travel, childcare, and so on) are utilized, the timing and size are an important consideration. Providing incentives at the point of recruitment, particularly if they are very appealing, may boost initial motivation to participate in research. However, participants may become less motivated as time progresses, and drop-out rates could be high. On the flip side, if incentives are not provided until the end of study participation, then retention may be poor as participants may drop out early as they have not received an incentive. Compensation should be provided as timely as possible to the study visit, so that the individual is not out of pocket (Robinson et al. 2015).

Incentives used for adults need to be specifically tailored for the adult audience being targeted and within the approvals of ethics. If designing an intervention, then the intervention itself should be designed to be appealing as possible for adults, sparking interest in their own health, to entice their participation. Feeding back on health data collected can be of interest to some participants, such as providing a summary of health information collected as part of the research. The timing of this feedback will need to be carefully considered so as to not affect your results (e.g., in an intervention study, providing this data after baseline data collection may impact participant's behavior in a control group). Acknowledgments, such as thank-you notes and birthday greetings, as well as small gifts such as water bottles and drawstring bags with the study name and logo, can also be used to keep rapport and encourage retention once participants have commenced in your research (Arora et al. 2017b).

6.1.8 Reminders and Retention

Depending on the recruitment method, it may be possible to alert potential participants to a recruitment event. For example, if you plan to recruit through a workplace, then reminders on the run-up and day before the introduction of a recruitment stall may be appropriate. It is unlikely that you will have direct access to individual's contact details prior to recruitment for ethical reasons, but a third party might, who may be able to provide such reminders. With most adults owning mobile phones, a text or phone call reminder the day before launch of a recruitment strategy may be useful.

Just as for recruitment plans, plans for retaining participants also need to be tailored to the participant group. Retention can be helped by:

- Continued contact with participants (the various methods used needs recorded, as does the number of attempts and when). Using the same data collector can help build relationships with the participant.
- Planning a follow-up well in advance (alerting the participant to future visits or study involvement and thanking them after each involvement is key).
- Study branding provides participants with the opportunity to self-identify with the study (merchandise or study ID cards).
- Incentives and acknowledgments.
- Research team training (such as scripts to help with retention).

---

7 Recruitment of Older Adults

Recruitment and retention of older adults into research studies can be challenging, particularly for researchers engaging with clinical groups such as people with dementia (see also ▶ Chap. 114, "Researching with People with Dementia"). Accordingly, there are a number of recommendations and guidelines to facilitate working with older adults in a research context. For example, the NIH National Institute on Aging has developed the Recruiting Older Adults into Research (ROAR) Toolkit (National Institute of Ageing 2015), which contains a user guide, frequently asked questions, sample social media messages, and exemplar PowerPoint presentations and flyers to assist in helping seniors and carers learn why research participation is important, the benefits of research, and how to find opportunities.

As with other cohorts and age groups, the study design will play a large role in determining optimal recruitment strategies. One systematic review detailed that older adults' response rates to postal surveys could be increased by around half when participants were prenotified about the survey; using colored ink and offering an incentive also increased the response rate (Auster and Janda 2009). Another study testing a physical activity program for older adults effectively utilized a similar approach by randomly identifying potential participants from the Australian Federal Electoral Roll, sending them postcards with information about the study to establish program credibility, and then calling them to invite them to join the project (Jancey et al. 2006). This strategy was found to be particularly cost-effective, with an estimated cost of $30 AUD per recruit. Another Australian study found that direct mail-outs cost $21 AUD per recruit, and this was more cost-effective and yielded a higher percentage of enrolments than oral presentations to community groups ($21 AUD vs. $144 AUD, 51.6% vs. 10%, respectively) (Clemson et al. 2004). However, other studies have found flyers to be the least effective method of recruitment when compared with referrals and face-to-face presentations (Adams et al. 1997). This suggests that fliers can be helpful if they are utilized as a priming mechanism to build trust and awareness in older adults before secondary contact is made by another means.

Face-to-face community-based recruitment strategies can be effective in facilitating grassroots engagement with research, particularly in tightly networked communities of older adults. One study found that the highest rate of face-to-face contact with older adults produced the greatest number of successful recruits (Ford et al. 2004). In our experience, we have found that involving organizations who have frequent touch-points with older adults including local councils (e.g., libraries, community centers, and leisure centers), seniors' clubs (e.g., RSL clubs, Probus, Lions Clubs, Rotary), and residential aged care facilities can dramatically facilitate recruitment, particularly if they have a dedicated liaison for seniors' outreach.

This level of grassroots engagement can produce referrals from friends and relatives, which have been shown to be one of the most effective methods of recruiting older adults into clinical trials (Adams et al. 1997). When recruiting for our mild cognitive impairment and dementia studies, our team asks our study participants to take a flyer and pass it on to a friend or relative. In addition, we ask carers (who attend study appointments with enrolled research participants) if they would like to be involved as a participant in control groups; they are often very willing to participate. Other referral strategies can come through networks of clinicians (e.g., primary care, staff specialists in hospitals or their private rooms, and community healthcare workers). This is a particularly effective method for recruiting clinical groups of older adults with specific eligibility criteria. A simple referral pathway that includes clear recruitment criteria, a site coordinator or recruitment champion, and incentives for referrals (e.g., access to clinical investigations, such as genetic testing or imaging, which would not otherwise be available, authorships on manuscripts, decreased workload) can facilitate recruitment of older adults via clinicians.

Other recruitment strategies for older adults can increase awareness of research studies (effectively priming) but may not yield recruits directly. This includes outreach such as newsletters or information on websites of advocacy groups (e.g., Dementia Australia, Australian Men's Shed Association, Older Women's Network NSW) and health governance organizations (e.g., Primary Health Networks, local health districts). Media outreach strategies (as detailed above) can also be an effective means of increasing engagement and awareness about research participation opportunities for seniors. Other direct recruitment strategies include running classified advertisements in local newspapers, online campaigns (e.g., through social media; however, there are varying degrees of success here due to mixed levels of information literacy in older adults (Chesser et al. 2016)), and campaigns coordinated through clinical trial recruitment agencies (as detailed above).

8 Recruitment of Difficult-to-Reach Communities and Medically Compromised People

Hard-to-reach subgroups of the population will require additional recruitment planning. Such subgroups include vulnerable population groups such as children, pregnant women and newborn babies, prisoners, the homeless, terminally ill patients or those with life-threatening conditions, and cognitively impaired individuals (Liamputong 2007, 2013). Indigenous, culturally and linguistically diverse (CALD), refugee population groups are all disadvantaged communities that will also require extra consideration when planning recruitment. Many of these groups will be from low-socioeconomic status (SES) backgrounds, where often basic needs are of a higher priority than research involvement (Liamputtong 2007, 2010).

In those with terminal disease or life-threatening conditions, there may be unforeseen circumstances (e.g., unplanned medical treatments) that arise unexpectedly and that will require extra flexibility from the research team (such as offering home visits) to ensure successful recruitment and retention. Eligibility criteria of this group may also change over time and will require consideration when calculating sample size estimates for studies. Timing of recruitment is a particularly important consideration for this population – although diagnosis may allow for longer follow-up from a research point of view, this can be a highly stressful time for the patient and therefore may not be an appropriate time for an invitation to participate in research. If recruitment at diagnosis is particularly important, then invitation from a healthcare professional may be more appropriate than researcher.

8.1 Community Ownership

When researching with hard-to-reach groups, working with the community from the inception of your research planning onward provides the opportunity to build community ownership of the research. Planning recruitment strategies with the feedback of the community is more likely to result in effective approaches. Empowering communities to undertake the research being conducted allows for trust to be developed; sustainability of the research relationship; better uptake, adherence, and involvement in research; and often research capacity building within community members. Providing community ownership is particularly important for indigenous groups where devolution of power has often led to distrust (Voyle and Simmons 1999). Guidance for undertaking research with indigenous populations exists (Henderson et al. 2002). In terms of recruitment, it is suggested that partnership committees are developed and that consultation with the community is key around the research being proposed and the most appropriate methods and avenues for recruitment. Community committees can also supervise researchers throughout the recruitment process to ensure cultural appropriateness of implemented approaches. Where trust can be an issue initially, having a community champion or gatekeeper on board either from that community, whom already has an established and trusting relationship with the community, is vital for first introductions to a project and gaining access to that community. Employing or involving champions from the community to actually drive recruitment can also be beneficial or at least have cultural training of researchers who are involved in recruitment.

8.2   Incentives and Acknowledgments

Incentives may be particularly appealing for people of low socioeconomic status but should not be too big that they coerce individuals into participating. Learning about health or having access to health screening can in itself be an incentive for disadvantaged communities rather than the need for any monetary gifts. For example, diabetes health education and screening, prior to recruitment into a community-based diabetes risk reduction program, overcame barriers to participating in low socioeconomic communities from a range of minority backgrounds (Santoyo-Olsson et al. 2011).

8.3   Setting

Often, recruiting hard-to-reach groups through healthcare providers will not work, as visitation rates can be lower than for the general population. This may be due to a range of factors, including socioeconomic factors affecting access to healthcare; a lack of availability of health services; religious beliefs that "god dictates health" and, therefore, health is outside of the individual's control; and a lack of trust or connection with healthcare providers (Simmons and Voyle 2003; Arora et al. 2018).

Extra work may be necessary to study the hard-to-reach target group to ensure an effective culturally appropriate recruitment strategy. This may mean adjusting intervention design and intervention and/or data collection settings, to take these out of conventional locations (such as universities and healthcare settings) and into communities. Recruitment of Maori people was effective in a study conducted in New Zealand where a diabetes prevention program was delivered via marae (the local community meeting grounds) (Simmons and Voyle 2003).

8.4   Several Means of Contact

Multiple recruitment strategies may be important for hard-to-reach groups to ensure that they actually receive invitation to a study, as often hard-to-reach groups are highly mobile. It is recommended that multiple contact details are collected from such individuals, and this may need to include next of kin and friend's contact details too, in order to keep track of participants effectively. Rolling recruitment strategies may be necessary to allow enough time to sufficiently reach required sample sizes. Any materials used for recruitment will need to be culturally and linguistically tailored to the target group, to increase the chance of invitations being read.

In those with life-threatening illnesses, terminal disease attrition rates will be high, and it may be that others closely connected to the individual of interest are targeted to collect data from in this instance (e.g., proxy measures).

8.5   Recruitment of Medically Compromised Groups

Involving medically compromised (and often vulnerable) groups in research can be ethically and logistically challenging (Liamputtong 2007; McMurdo et al. 2011). There are guidelines available to assist with the ethical design and conduct of such research including the National Health and Medical Research Council (NHMRC) National Statement on Ethical Conduct in Human Research (National Health and Medical Research Council 2015), which details specific ethical considerations that need to be considered for certain groups of participants, for example, people highly dependent on medical care who may be unable to give consent (e.g., neonatal intensive care, terminal care, emergency care, intensive care, and the care of unconscious people) and people with a cognitive impairment, an intellectual disability, or mental illness. The consent process when recruiting medically compromised individuals for research involves careful consideration. The NHMRC guidelines recommend seeking consent from the individual wherever they are capable of providing consent, and it is practicable to do so. For people with a cognitive impairment or mental illness, their capacity to consent may fluctuate over time, and this needs to be taken into consideration. If it is not possible to obtain consent from an individual, it may need to be obtained from the participant's guardian (see also ▶ Chap. 106, "Ethics and Research with Indigenous Peoples").

In addition to some of the ethical issues outlined above, medically compromised populations can also be difficult to reach, making recruitment logistically challenging. Barriers to recruiting these populations include rare/low-prevalence populations; participants' willingness to participate and attitudes toward research, clinician characteristics, attitudes, and practices; perceived harms of studies; mistrust of research; low health literacy; and lack of time (Bonevski et al. 2014). There are a range of strategies that can be adopted to overcome these potential barriers to recruiting vulnerable and medically compromised groups (see Bonevski et al. 2014 for a comprehensive review). For example, to overcome the barrier of perceived harm of research, softer wording when describing the project can be utilized (e.g., "study" rather than "investigation," community-driven research in disadvantaged communities to overcome mistrust, using peer recruiters to overcome fear of authority).

9   Conclusion and Future Directions

To ensure successful recruitment and retention of suitable study participants, it is essential to consider the expectations and interests of both investigator and potential participants. It is vital that participants are best facilitated to ensure their utmost commitment toward the research and/or clinical trial, thereby lowering the risk of participant attrition, which can negatively impact research budgets and timelines. This chapter highlighted broad recruitment strategies to aid in effective design of the recruitment phase of research studies. Furthermore, detailed explanations of recruitment strategies across the life-span and in particular population groups have been provided.

Recruitment is very crucial for any research, and since research is continuously evolving and so are the health and social issues which tend to increase in future. Therefore, there is a strong need to carefully think about the effective recruitment methods of research participants in the planning stage of any research study. Furthermore, there might be additional innovative recruitment methods that researchers will invent in order to ensure that they can effectively recruit and retain research participants.

References

Adams J, Silverman M, Musa D, Peele P. Recruiting older adults for clinical trials. Control Clin Trials. 1997;18(1):14–26.

Alibali MW, Nathan MJ. Conducting research in schools: a practical guide. J Cogn Dev. 2010;11 (4):397–407.

Arora A, Bedros D, Bhole S, Do LG, Scott J, Blinkhorn A, Schwarz E. Child and family health nurses' experiences of oral health of preschool children: a qualitative approach. J Public Health Dent. 2012;72(2):149–55.

Arora A, Manohar N, Liamputtong P, Do LG, Eastwood J, Bhole S. Researching the perceptions of Vietnamese migrant caregivers for an oral health literacy study in Australia. SAGE research methods cases. London: Sage; 2017b. https://doi.org/10.4135/9781526423320.

Arora A, Manohar N, Bedros D, Hua APD, You SYH, Blight V, Ajwani S, Eastwood J, Bhole S. Lessons learnt in recruiting disadvantaged families to a birth cohort study. BMC Nursing. 2018;17(7):1–9. https://doi.org/10.1186/s12912-018-0276-0.

Auster J, Janda M. Recruiting older adults to health research studies: a systematic review. Australas J Ageing. 2009;28(3):149–51.

Australian Government Productivity Commission. Early childhood education and care. 2015. Retrieved from http://www.pc.gov.au/research/ongoing/report-on-government-services/2015/childcare-education-and-training/early-childhood-education-and-care/rogs-2015-volumeb-chapter3.pdf

Bartlett R, Wright T, Olarinde T, Holmes T, Beamon ER, Wallace D. Schools as sites for recruiting participants and implementing research. J Community Health Nurs. 2017;34(2):80–8.

Berry DC, Neal M, Hall EG, McMurray RG, Schwartz TA, Skelly AH, Smith-Miller C. Recruitment and retention strategies for a community-based weight management study for multi-ethnic elementary school children and their parents. Public Health Nurs. 2013;30(1):80–6.

Bonevski B, Randell M, Paul C, Chapman K, Twyman L, Bryant J,... Hughes C. Reaching the hard-to-reach: a systematic review of strategies for improving health and medical research with socially disadvantaged groups. BMC Med Res Methodol. 2014;14(1):42. https://doi.org/10.1186/1471-2288-14-42.

Bower P, Brueton V, Gamble C, Treweek S, Smith CT, Young B, Williamson P. Interventions to improve recruitment and retention in clinical trials: a survey and workshop to assess current practice and future priorities. Trials. 2014;15(1):399. https://doi.org/10.1186/1745-6215-15-399.

Brugge D, Kole A, Lu W, Must A. Susceptibility of elderly Asian immigrants to persuasion with respect to participation in research. J Immigr Minor Health. 2005;2(7):93–101.

Bruzzese JM, Gallagher R, McCann-Doyle S, Reiss PT, Wijetunga NA. Effective methods to improve recruitment and retention in school-based substance use prevention studies. J Sch Health. 2009;79(9):400–7.

Caldwell PH, Hamilton S, Tan A, Craig JC. Strategies for increasing recruitment to randomised controlled trials: systematic review. PLoS Med. 2010;7(11):e1000368.

Chesser AK, Woods NK, Smothers K, Rogers N. Health literacy and older adults a systematic review. Gerontol Geriatr Med. 2016;2:2333721416630492.

Chiu CL, Hearn NL, Lind JM. Development of a risk score for extraintestinal manifestations of coeliac disease. Medicine (United States). 2016;95(15):1–6. [e3286]. https://doi.org/10.1097/MD.0000000000003286.

Clemson L, Cumming RG, Kendig H, Swann M, Heard R, Taylor K. The effectiveness of a community-based program for reducing the incidence of falls in the elderly: a randomized trial. J Am Geriatr Soc. 2004;52(9):1487–94.

Clinical Connection. 2017. Retrieved 23 Nov 2017, from https://www.clinicalconnection.com/

Craig SL, Austin A, McInroy LB. School-based groups to support multiethnic sexual minority youth resiliency: preliminary effectiveness. Child Adolesc Soc Work J. 2014;31(1):87–106.

CraigsList. 2017. Retrieved 23 Nov 2017, from https://newyork.craigslist.org/search/vol

Daley AM. Adolescent-friendly remedies for the challenges of focus group research. West J Nurs Res. 2013;35(8):1043–59.

Endometriosis Australia. 2017. Retrieved 23 Nov 2017, from https://www.facebook.com/EndometriosisAustralia.

Farmer D, Jackson S, Camacho F, Hall M. Attitudes of African American and low socioeconomic status white women toward medical research. J Health Care Poor Underserved. 2007;18(1):85–99.

Ford ME, Havstad SL, Davis SD. A randomized trial of recruitment methods for older African American men in the prostate, lung, colorectal and ovarian (PLCO) cancer screening trial. Clin Trials. 2004;1(4):343–51.

George S, Duran N, Norris K. A systematic review of barriers and facilitators to minority research participation among African Americans, Latinos, Asian Americans, and Pacific Islanders. Am J Public Health. 2014;104(2):e16–31.

Giarelli E, Bruner D, Nguyen E, Basham S, Marathe P, Dao D, . . . Nguyen G. Research participation among Asian American women at risk for cervical cancer: exploratory pilot of barriers and enhancers. J Immigr Minor Health. 2011;13(6):1055–68.

Glasser SP. Recruitment and retention in clinical research. In: Glasser SP, editor. Essentials of clinical research. Switzerland: Springer International; 2014. p. 177–92.

Hayfield N, Huxley C. Insider and outsider perspectives: reflections on researcher identities in research with lesbian and bisexual women. Qual Res Psychol. 2015;12(2):91–106.

Henderson R, Simmons DS, Bourke L, Muir J. Development of guidelines for non-indigenous people undertaking research among the indigenous population of north-east Victoria. Med J Aust. 2002;176(10):482–5.

Hibberd S. Involving children and their parents in research design. Arch Dis Child. 2016;101(9):e2–e2.

Hulley SB, Cummings SR, Browner WS, Grady DG, Newman TB. Designing clinical research. An epidemiological approach. 2nd ed. London: Lippincott Williams & Wilkins; 2013.

Jancey J, Howat P, Lee A, Clarke A, Shilton T, Fisher J, Iredell H. Effective recruitment and retention of older adults in physical activity research: PALS study. Am J Health Behav. 2006;30(6):626–35.

Johnson M, George A, Dahlen H, Ajwani S, Bhole S, Blinkhorn A, . . . Yeo A. The midwifery initiated oral health-dental service protocol: an intervention to improve oral health outcomes for pregnant women. BMC Oral Health. 2015;15(2):1–9. https://doi.org/10.1186/1472-6831-15-2.

Join Dementia Research. 2017. Retrieved 23 Nov 2017, from https://www.joindementiaresearch. nihr.ac.uk/

Joseph R, Keller C, Ainsworth B. Recruiting participants into pilot trials: techniques for researchers with shoestring budgets. Californian J Health Promot. 2016;14(2):81–9.

Lee GE, Ow M, Lie D, Dent R. Barriers and facilitators for clinical trial participation among diverse Asian patients with breast cancer: a qualitative study. BMC Womens Health. 2016;16(43):1–8. https://doi.org/10.1186/s12905-016-0319-1.

Liamputtong P. Researching the vulnerable: a guide to sensitive research methods. London: SAGE Publications; 2007.

Liamputtong P. Performing qualitative cross-cultural research. Cambridge: Cambridge University Press; 2010.

Liamputtong P. Qualitative research methods. 4th ed. Melbourne: Oxford University Press; 2013.

Liamputtong P. The science of words and the science of number. In: Liamputtong P, editor. Research methods in health: foundations of evidence-based practice. 3rd ed. South Melbourne: Oxford University Press; 2017. p. 3–27.

Liljas AE, Jovicic A, Kharicha K, Iliffe S, Manthorpe J, Goodman C, Walters K. Facilitators and barriers for recruiting and engaging hard-to-reach older people to health promotion interventions and related research: a systematic review. Lancet. 2015;386:S51.

Manohar N, Liamputtong P, Bhole S, Arora A. Researcher positionality in cross-cultural and sensitive research. In: Liamputtong P, editor. Handbook of research methods in health social sciences. Singapore: Springer; 2017. p. 1–15. https://doi.org/10.1007/978-981-10-2779-6_35-1.

Mapstone J, Elbourne D, Roberts I. Strategies to improve recruitment to research studies. Cochrane Database Syst Rev. 2007;2:MR000013.

Martinson BE, Hindman DB. Building a health promotion agenda in local newspapers. Health Educ Res. 2005;20(1):51–60.

Maxwell AE, Bastani R, Vida P, Warda US. Strategies to recruit and retain older Filipino–american immigrants for a cancer screening study. J Community Health. 2005;30(3):167–79.

McDonald AM, Knight RC, Campbell MK, Entwistle VA, Grant AM, Cook JA, . . . Roberts I. What influences recruitment to randomised controlled trials? A review of trials funded by two UK funding agencies. Trials. 2006;7(9):1–8. https://doi.org/10.1186/1745-6215-7-9.

McMurdo ME, Roberts H, Parker S, Wyatt N, May H, Goodman C, . . . Ali K. Improving recruitment of older people to research through good practice. Age Ageing. 2011;40(6):659–65.

Mishna F, Muskat B, Cook C. Anticipating challenges: school-based social work intervention research. Child Sch. 2012;34(3):135–44.

Musiat P, Winsall M, Orlowski S, Antezana G, Schrader G, Battersby M, Bidargaddi N. Paid and unpaid online recruitment for health interventions in young adults. J Adolesc Health: Off Publ Soc Adolesc Med. 2016;59(6):662–7.

National Health and Medical Research Council. National statement on ethical conduct in human research. Canberra: NHMRC; 2015.

National Institute of Ageing. Recruiting older adults into research (ROAR) toolkit overview and user guide. 2015. Retrieved 23 Nov 2017, from https://www.nia.nih.gov/sites/default/files/d7/roar_user_guide.pdf

Ngune I, Jiwa M, Dadich A, Lotriet J, Sriram D. Effective recruitment strategies in primary care research: a systematic review. Qual Prim Care. 2012;20(2):115–23.

Owen-Smith AA, Woodyatt C, Sineath RC, Hunkeler EM, Barnwell LT, Graham A, . . . Goodman M. Perceptions of barriers to and facilitators of participation in health research among transgender people. Transgender Health. 2016;1(1):187–96.

Patel MX, Doku V, Tennakoon L. Challenges in recruitment of research participants. Adv Psychiatr Treat. 2003;9(3):229–38. https://doi.org/10.1192/apt.9.3.229.

Quay T, Frimer L, Janssen P, Lamers Y. Barriers and facilitators to recruitment of South Asians to health research: a scoping review. BMJ Open. 2017;7(5):e014889.

Ramsden I. Cultural safety and nursing education in Aotearoa and Te Waipounamu. Wellington: Victoria University of Wellington; 2002.

ResearchMatch. 2017. Retrieved 23 Nov 2017, from https://www.researchmatch.org/

Robinson KA, Dinglas VD, Sukrithan V, Yalamanchilli R, Mendez-Tellez PA, Dennison-Himmelfarb C, et al. Updated systematic review identifies substantial number of retention strategies: using more strategies retains more study participants. J Clin Epidemiol. 2015;68 (12):1481–7.

Rodriguez LL, Hanna KE, Federman DD. The participant-investigator interface. In: Rodriguez LL, Hanna KE, Federman DD, editors. Responsible research: a systems approach to protecting research participants. Washington, DC: National Academies Press; 2003. p. 108–35.

Santoyo-Olsson J, Cabrera J, Freyre R, Grossman M, Alvarez N, Mathur D, … Stewart AL. An innovative multiphased strategy to recruit underserved adults into a randomized trial of a community-based diabetes risk reduction program. Gerontologist. 2011;51(suppl_1):S82–93.

Satar MMA, Abdel-Raouf O, Baset MA, El-henawy I. Altruism as a tool for optimization: literature review. Int J Eng Trends Technol. 2015;22(6):270–5.

Sengupta S, Strauss R, DeVellis R, Quinn S, DeVellis B, Ware W. Factors affecting African-American participation in AIDS research. J Acquir Immune Defic Syndr. 2000;24(3):275–84.

Simmons D, Voyle JA. Reaching hard-to-reach, high-risk populations: piloting a health promotion and diabetes disease prevention programme on an urban marae in New Zealand. Health Promot Int. 2003;18(1):41–50.

Stein MA, Shaffer M, Echo-Hawk A, Smith J, Stapleton A, Melvin A. Research START: a multimethod study of barriers and accelerators of recruiting research participants. Clin Transl Sci. 2015;8(6):647–54.

Trialfacts. 2017. Retrieved 23 Nov 2017, from https://trialfacts.com/

UCL Research Ethics Committee. Guidance note 1: research involving children. 2016. Retrieved 23 Nov 2017, from https://ethics.grad.ucl.ac.uk/forms/guidance1.pdf

Voyle J, Simmons D. Community development through partnership: promoting health in an urban indigenous community in New Zealand. Soc Sci Med. 1999;49(8):1035–50.

Walsh R. Helping or hurting: are adolescent intervention programs minimizing racial inequality? Educ Urban Soc. 2011;43(3):370–95.

Welton A, Vickers M, Cooper J, Meade T, Marteau T. Is recruitment more difficult with a placebo arm in randomised controlled trials? A quasirandomised, interview based study. Br Med J. 1999;318(7191):1114–7.

World Medical Association. World Medical Association Declaration of Helsinki. Ethical principles for medical research involving human subjects. Bull World Health Organ. 2001;79(4):373–4.

Wyatt SB, Diekelmann N, Henderson F, Andrew ME, Billingsley G, Felder SH, Fuqua S. A community-driven model of research participation: the Jackson Heart Study Participant Recruitment and Retention Study. Ethn Dis. 2003;13(4):438–55.

Wyke S, Hunt K, Gray CM, Fenwick E, Bunn C, Donnan PT, … Boyer N. Football Fans in Training (FFIT): a randomised controlled trial of a gender-sensitised weight loss and healthy living programme for men–end of study report. Public Health Res. 2015;3(2):1–129.

Younger Onset Dementia Support Group. 2017. Retrieved 23 Nov 2017, from https://www.facebook.com/YoungOnsetDementiaSupportGroup

Ontology and Epistemology

6

John D. Hathcoat, Cara Meixner, and Mark C. Nicholas

Contents

Abstract

Health social science is an area of study wherein the methodological techniques used within the social sciences are applied to the investigation of human health. Methodological techniques, however, are not philosophically agnostic. Philosophical positions indeed matter in that they result in a range of individual

J. D. Hathcoat (✉)
Department of Graduate Psychology, Center for Assessment and Research Studies, James Madison University, Harrisonburg, VA, USA
e-mail: hathcojd@jmu.edu

C. Meixner
Department of Graduate Psychology, Center for Faculty Innovation, James Madison University, Harrisonburg, VA, USA

M. C. Nicholas
Framingham State University, Framingham, MA, USA

© Springer Nature Singapore Pte Ltd. 2019
P. Liamputtong (ed.), *Handbook of Research Methods in Health Social Sciences*,
https://doi.org/10.1007/978-981-10-5251-4_56

and societal consequences. Consequently, it is important for students and researchers interested in studying the social aspects of health to understand the role of philosophical positions within research. Philosophical positions partly consist of ontological and epistemological assumptions. Ontological issues pertain to what exists, whereas epistemology focuses on the nature, limitations, and justification of human knowledge. This chapter introduces the reader to how ontological and epistemological positions are embedded within the biomedical, biopsychosocial, and critical alternative models of human functioning. Situating each of these models in relevant vignettes, we suggest that philosophical positions serve a dual role within inquiry in that they inform, and are in some circumstances informed by, the methodological and interpretative decisions enacted by researchers.

Keywords

Ontology · Epistemology · Methods · Biomedical · Biopsychosocial · Health Social Science

1   Introduction

Research is broadly framed from vantage points that partly consist of assumptions about the underlying nature of reality, how we understand that reality, and the kind of knowledge that may be obtained from observations and investigations of social phenomena. As such, social inquiry fails to be philosophically agnostic. Positions taken toward the nature of reality, or an object of investigation, are *ontological*, whereas stances toward the meaning and process of obtaining knowledge reflect what is *epistemological*. Together, ontological and epistemological positions, even when tacit, richly inform the arc of research in the health social sciences.

As an example, let us consider anorexia as a mental health phenomenon. A researcher who views anorexia as reducible to a physiological entity residing within a person draws from a different set of ontological assumptions than a researcher who views anorexia as the manifestation of socially constructed diagnostic criteria. Each view would not only result in different methodological decisions when investigating anorexia, but these perspectives also diverge in their diagnostic, treatment, and public policy implications (Lovett and Hood 2011). Students, investigators, and practitioners of the health social sciences therefore need to understand how philosophical positions (i.e., ontology and epistemology) shape, and to some extent are shaped by, the methodological and interpretative decisions enacted by researchers.

There are numerous obstacles to overcome if one desires to learn about the role of ontology and epistemology within social inquiry. Authors from various academic disciplines have contributed to a wide body of articles on the topic. Within this literature, the same term is often used in different ways or it may be completely ill-defined. For example, the term "paradigm" has been used in the literature to indicate an overarching worldview, epistemological stance, shared set of beliefs among specialists, and exemplar for how research is done in a particular area of study (Morgan 2007). There are also a seemingly endless number of philosophical stances one may adopt; it likewise appears that new perspectives emerge frequently within the literature (e.g., Onwuegbuzie et al. 2009; Creswell and PlanoClark 2011; Denzin and Lincoln 2011; Houghton et al. 2012). In other words, there is much for a new reader to synthesize across numerous fields of study, which can seem like a daunting task.

Given these obstacles, one may question the value of taking the time to understand this line of literature. Without reflecting on such issues, one is like a ship at sea without a working navigational system. Without a navigational system – a compass, the stars, or a global positioning system – there is no awareness of direction, which could result in aimless wandering at sea. Within the context of research, ontological and epistemological considerations are our navigational systems, allowing us to become aware of our location, which in turn provides direction. We hope that, by the end of this chapter, the reader will come to appreciate the importance of navigating these considerations, as they invariably shape the direction of research studies.

We begin by discussing ontology and epistemology. This is followed by examining how these issues are connected to three models of human functioning that are frequently found within the literature. These models include the *biomedical model*, *biopsychosocial model*, and what we have labeled as *critical alternatives*. Then, we illustrate how philosophical stances serve a dual role within inquiry – to frame methodological decisions and in some circumstances to be informed by methodological and interpretative decisions. The concept of disease is used, for the purpose of illustration, for two reasons. First, the concept is highly controversial in medical discourse, which reflects differences in philosophical positions. Second, and perhaps more importantly, it seems to us that if these controversies exist in areas of research that many assume to be more certain, then how much more problematic might areas of research be that are assumed to be less certain, such as mental health and well-being? The controversies surrounding the concept of disease are applicable to mental health and well-being. With this said, examples drawn from mental health are interspersed throughout the chapter.

2 Ontology

We will begin by considering the topic of disease. What is a disease? Do diseases exist independently of our input as researchers? What conditions or states should be classified as a non-disease? These are complicated questions that have been a topic of debate among the readership of the *British Medical Journal* (*BMJ*) – one of the oldest peer-reviewed journals of medicine. Smith (2002) published an article based upon the results of a *BMJ* survey about the identification of non-diseases, which incited various debates about these concepts. From these debates, he concluded that not only is "...the notion of 'disease' a slippery one" but "health is equally impossible to define" (p. 883). Despite widespread disagreement about how to define a disease, a range of consequences may result from how healthcare professionals handle this concept (e.g., insurance reimbursement, treatment options, and so on). Though not explicitly addressed by the *BMJ*, these questions reflect ontological positions that have numerous implications within society.

Ontology is traditionally considered a branch of metaphysics aiming to describe the structure of reality (Poli 2010) or the study of being (e.g., Heidegger 1953). Historically, philosophical debates about ontology centered on the concept of realism. There are various versions of realism (e.g., Ladyman 2002); broadly speaking, a realist believes "...that entities exist independently of being perceived, or independently of our theories about them" (Phillips 1987, p. 205). Within the social sciences, a realist ontology is typically taken to mean that a phenomenon of investigation exists as an entity irrespective of our input as researchers (Hood 2013; see also ▶ Chap. 9, "Positivism and Realism"). For our purposes, we will delineate the realist from an anti-realist, defined here as the rejection of the view that a phenomenon of interest exists independently of our input as researchers.

Consider once again a disease. An ontological realist would hold that diseases are in fact "things in the world" that exist irrespective of our input as researchers. This view was illustrated by Thomas Sydenman, a mid-seventeenth-century physician, who provided an account of the history of diseases in his *Observationes Medicae* (1676). Sydenman claimed "...that diseases were distinct entities, possessing natural histories, and emphasized how disease was independent of the sufferer" (Low 1999, p. 259). Conversely, an anti-realist would reject the notion that disease exists independently in favor of a cultural construction such as this, "Each civilization defines its own disease. What is sickness in one might be a chromosomal abnormality, crime, holiness, or sin in another" (Illich 1976, p. 112). Such ontological positions are related to how one would investigate disease, the kind of knowledge believed to be obtained via such investigations, and the strategies employed to justify knowledge claims (i.e., epistemology).

3 Epistemology

Epistemology is a branch of philosophy that examines the nature, limitations, and justification of knowledge (Williams 2001). What is knowledge? What is the relationship between the knower and the known? How are knowledge claims justified? The scope of epistemology is fairly broad given that there is a range of positions to consider for each of these questions. At the risk of oversimplification, we draw upon the framework provided by Crotty (1998), which outlines three epistemic positions: (1) objectivism, (2) constructionism, and (3) subjectivism. Before discussing each view, it is first necessary to make a distinction between a phenomenon and the meaning of a phenomenon (Biesta 2010).

When asked, most people likely believe that world stuff – the material of which our universe is made of, like "matter" – would continue to exist if people ceased to exist (i.e., phenomenon). In other words, most would argue that the world does not disappear simply because no one is present to observe it. However, does this world have meaning? In order for world stuff to be meaningful, an observer must interpret such meaning (i.e., the meaning attributed to phenomenon). Consider a person who has been diagnosed with HIV/AIDS. There may be specific entities that exist irrespective of the diagnosis, but what it means to be diagnosed with HIV/AIDS could be culturally constructed. For example, the entities that compose a virus (i.e., phenomena) may exist irrespective of a diagnosis, but whether HIV/AIDS is conceived as a punishment from God, act of witchcraft or product of evolution may be culturally constructed, thus indicating that phenomena can be distinguished from meaning given to phenomena. The three general epistemic positions outlined by Crotty (1998) pertain to the meaning of phenomena.

Having established the distinction between phenomenon and the meaning of a phenomenon, an objectivist, constructionist, and subjectivist have different views toward the relationship between the knower and the known. An *objectivist* "holds that meaning, and therefore meaningful reality, exists as such a part from the operation of any consciousness" (Crotty 1998, p. 8). In other words, "diseaseness" has meaning irrespective of our interpretations, and it is an aim of science to discover this meaning. Cultural constructions of disease, under this view, may be more or less aligned with objective meaning. How we come to find meaning is rooted in positivist/postpositivist assumptions that phenomena can be understood through deductive methods, experimental approaches, and robust statistical methodologies (see also ▶ Chap. 9, "Positivism and Realism").

The *constructionist*, on the other hand, views meaning as coming "into existence in and out of our engagement with the realities of our world. There is no meaning without a mind. Meaning is not discovered, but constructed" (Crotty 1998, pp. 8–9). The constructionist does not necessarily deny the existence of a mind-independent world but instead contends that the meaning of this world is culturally, socially, historically, and politically situated. Under this view, the mind interacts with world stuff to derive the meaning of disease. Further, the constructionist upholds the possibility of multiple, coexisting meanings. To this end, methodologies associated with constructionism are often qualitative (e.g., phenomenology, grounded theory, narrative inquiry) with analyses reflecting inductive techniques (Gale and Dolbin-MacNab 2014; Hathcoat and Nicholas 2014; Nicholas and Hathcoat 2014; see also ▶ Chap. 7, "Social Constructionism").

Finally, the *subjectivist* upholds the notion that experience, and thus meaning, is independent of a fixed reality. In other words, meaning is firmly situated in culture, social interaction, and so forth. As mentioned by Crotty (1998), it is tempting under this view to state that there are no boundaries set by a mind-independent world to limit the creation of meaning. A subjectivist might, therefore, claim that the concepts of a disease and non-disease reflect but one of many possible ways to categorize, label, and impose order on the human condition. Methodologically, a subjectivist "tends toward participatory action research (PAR) approaches where the purpose of the research, analysis, and results is to involve the communities being studied and achieve emancipatory change" (Gale and Dolbin-MacNab 2014, p. 250; see also ▶ Chap. 17, "Community-Based Participatory Action Research").

Ontological and epistemological stances could be combined in different ways. We suggest, however, that not any combination of ontological and/or epistemological positions is reasonable. For example, it seems misaligned to combine ontological realism with subjectivism within a single study. In our opinion, it is probably wise to avoid seemingly contradictory ontological and/or epistemological stances within a single study. Addressing each of these complexities, however, extends well beyond the aim of our chapter, so we will refer to ontological and epistemological positions more generally as philosophical perspectives.

4 Philosophy and Models of Human Functioning

Three models are prominent within the literature when describing human functioning. These models include (1) biomedical, (2) biopsychosocial, and (3) critical alternatives. Each of these models has been informed by distinct philosophical positions and has resulted in numerous societal consequences. The biomedical model of human functioning is the oldest and most criticized yet deemed in many respects as one of the most successful for treating "physical" ailments. However, this perspective resulted in a tendency to neglect the psychological, social, and cultural aspects of human functioning that led to the biopsychosocial model. Both of these models, however, have been critiqued by scholars in the social sciences for heavily relying upon a realist ontology. For the sake of simplicity, we have labeled the third model as "critical alternatives," recognizing that such criticisms have resulted from authors with various views toward human functioning.

4.1 The Biomedical Model: Naturalism and Objectivism

The biomedical model of human functioning became prominent in the mid-nineteenth century among health professionals who sought to apply the methods of science to the diagnosis and treatment of disease. This model "...is characterized by a reductionist approach that attributes illness to a single cause located within the body and that considers disturbances of mental processes as a separate and unrelated set of problems" (Wade and Halligan 2017). Medical practice, which had been intertwined with religious ideology (Koenig 2000), started to rely upon the empirical methods of science to explain human functioning. No longer were such ailments conceived as the result of "supernatural" forces, but technological developments led to the view that diseases were instead the result of microbial organisms that attacked the body, a perspective that came to be known as the germ theory of disease (Smith 2012).

The biomedical model is consistent with the concept of naturalism, defined here as the "general view that the concept of disease reflects an objective reality about cell, organ, or system function or dysfunction" (Sisti and Caplan 2017, p. 7). This view derived from the dualistic depiction of the mind and body advocated by Descartes (Longino 1998) in which humans are conceived as an "ensouled machine" (Sisti and Caplan 2017, p. 6). The mind (nonphysical) is considered distinct from the body (physical), and diseases are located within the body. The biomedical model is, therefore, consistent with a realist ontology and an objectivist epistemology. Said differently, diseases are mind-independent entities that can be correctly classified, diagnosed, and treated. The methods of science consequently aim to predict, control, and explain human functioning.

Under the biomedical model, the application of statistical and experimental methods resulted in numerous advancements throughout the nineteenth and early twentieth centuries. Under a naturalist view, it is important to have a sense of typical or "normal" functioning in order to identify a disease, which is once again defined as a bodily dysfunction. How would one identify dysfunction without first knowing something about typical functioning? Statistics is a tool that one may employ to investigate typical patterns of a population, thus making it a viable strategy to investigate dysfunction. In fact, the employment of statistical techniques to investigate population risk factors became the basis of epidemiological research (Berkman and Kawachi 2000; see also ▶ Chap. 33, "Epidemiology").

Experimental methods were also used to develop vaccinations throughout the nineteenth and twentieth centuries, which resulted in preventing widespread outbreaks of infectious disease (Hsu 2013). Moreover, by the end of World War II, the double-blind, randomized control trial became the "gold standard" of medical research (White and Willis 2002; see also ▶ Chap. 37, "Randomized Controlled Trials"). This is evident by the fact that in 1962 the US Congress passed an act that supported the use of randomized control trials to demonstrate the quality of a drug prior to approval and by 1970 such trials were mandated by the Food and Drug Administration (Bothwell and Podolsky 2016). In other words, throughout much of the nineteenth and twentieth century, the biomedical model not only dominated the conceptualization of human functioning among healthcare professionals but also resulted in numerous benefits to society.

4.2   Biopsychosocial Model: Integrating the Psychological and Social with Biomedical

There were numerous challenges to the biomedical model that stemmed from its inability to integrate the psychological and social aspects of human health and disease (Engel 1977). For example, stress, social support, and other related variables appear to be important factors in the progression of particular forms of cancer and heart-related issues (e.g., Valtorta et al. 2016). These challenges imply that there are aspects of human functioning that fail to be captured by physiological, genetic, and/ or other forms of organic dysfunction. Said differently, human health is more than the absence of such characteristics, a view captured the World Health Organization (1948/2014, p. 1) when stating, "health is a state of complete physical, mental, and social well-being and not merely the absence of disease or infirmity." Such criticisms eventually led to the biopsychosocial model of human functioning.

The biopsychosocial model "...holds to the idea that biological, psychological, and social processes are integrally interactively involved in physical health and illness" (Suls and Rothman 2004, p. 119). This challenged the Cartesian, mind-body dualism embedded in the biomedical model. The idea that a disease is either of the mind (i.e., psychosomatic) or strictly physical is problematic given that "most psychosomatic diseases involve varied genetic and environmental determinants, and all states of health and disease are influenced to some extent by psychosocial conditions" (Berkman and Kawachi 2000, p. 4). In other words, the psychological, social, and physical intersect in a myriad of ways to contribute to human functioning.

The biopsychosocial model is perhaps best characterized as an expansion of the biomedical worldview (Wade and Halligan 2017). The biopsychosocial model does not deny the existence of physical determinants of health and disease. For example, consider the case of deafness as a biological dysfunction. Is this a disease? Under a biopsychosocial model, biological dysfunction is necessary but not sufficient to count as a disease given that a disease must include social harm (Sisti and Caplan 2017). In other words, an individual who is deaf may have a high state of well-being, thus making it questionable whether biological dysfunction alone is sufficient to be labeled as a disease (Hausman 2017). This example leads us to a range of philosophical and practical consequences.

The biopsychosocial model may be viewed as a hybrid between ontological realism and constructionism. In other words, "...we might invoke pluralistic realism – which allows for both the recognition of biologically real entities and pragmatic usages of differing definitions to meet particular needs" (Sisti and Caplan 2017, p. 17). The concept of disease consequently becomes a target that "...changes with time, depends on practice, and influences medical taxonomy" (Hoffman 2017, p. 23). In other words, there are parameters that to some extent set limits as to what could be reasonably be considered health, disease, and well-being though how each is conceived within these parameters is in many respects socially and culturally affected.

The biopsychosocial model has resulted in numerous positive consequences when compared to the biomedical model. For example, Veatch (2006, p. 589) indicated that in the 1970s, "physicians were being asked to use surgical techniques to control human behavior by destroying portions of the brain believed to be responsible for criminal and violent behavior." Such practices derive from a biomedical model and its evident ontological and epistemological foundations. The biopsychosocial model, while adhering to a scientific worldview, has led to a person-centered approach toward treatment (Wade and Halligan 2017). Therein, treatment should focus on the whole person as opposed to simply aiming to rid an individual of biological dysfunction. With this said, the biopsychosocial model of human functioning is consistent with the application of methodological techniques that aim to understand human experience (e.g., phenomenology) in conjunction with strategies that are used to assess statistical deviations from typical functioning.

4.3   Critical Alternatives: Normativism and Subjectivism

Various movements within the philosophy of science advanced criticisms that extend well beyond those raised by advocates of the biopsychosocial model. The extent to which scientific progress could be viewed as the rational accumulation of knowledge became suspect (e.g., Kuhn 1962/2012). Scientific progress, it seemed, was more disorderly, haphazard, and socially influenced than previously believed. Other challenges arose from the fact that it seemed impossible to isolate hypotheses to be tested in isolation (e.g., Quine 1953). Instead of testing hypothesis "X" in isolation, such as within an experiment, we are simultaneously testing an interrelated network of assumptions about measurement, theory, sampling, and so on. Finally, the idea that empirical observations could be separated from theoretical and value-laden presuppositions came under scrutiny (e.g., Fleck 1935/1979). Such criticisms incited various researchers within the social sciences to abandon objectivist sympathies. Scientific knowledge was instead viewed by many critics as just one of many forms of knowledge.

With respect to human health, diagnosis, and treatment, it was possible to identify historical cases in which so-called diseases or disorders served as a source of oppression. Most of these examples stemmed from behavioral or mental health conditions though in many respects a "mental health disorder" has often been viewed similarly to a disease (e.g., mental health disorders can be viewed as a biological dysfunction or the result of intersecting biological, psychological, and social conditions). Such cases include delinquency, dissidence, homosexuality, and masturbation (Hoffman 2017). For example, Englehardt (1974, p. 234) concluded the following when discussing the "disease" of masturbation:

Masturbation in the 18th and especially in the 19th century was widely…held to be a dangerous disease entity. Explanation of this phenomenon entails a basic reexamination of the concept of disease. It presupposes that one think of disease neither as an objective entity in the world nor as a concept that admits of a single universal definition: there need not, nor need there be, one concept of disease.

Masturbation was used by Englehardt to argue that values are inextricably connected with notions of dysfunction. Other commonly cited examples include a social scientist who in 1851 argued that runaway slaves suffered from a disease called drapetomania and the view that "uncivilized" people were immune to mental health problems since psychiatric issues were characterized as a maladaptive reaction to "civilization" (e.g., Whitley 2015). In other words, historically situated, moral ideologies came to be viewed by such scholars as responsible for shaping what was considered dysfunctional (e.g., Foucault 1961/1989).

These positions toward human functioning are characterized by normativism and a tendency toward epistemological subjectivism. Normativist theories view health and disease as primarily value-driven thus "…when we signify something as a disease, we are marking out something that is subjectively disvalued by society, culture, or individual preferences" (Sisti and Caplan 2017, p. 9). Masturbation was a disease because it was disvalued by the moral ideology of a particular society. An objectivist may argue that although masturbation was at one time believed to be a disease, this view was simply mistaken. In other words, the people at this period of time were incorrect. The subjectivist, on the other hand, does not view this issue so simply. For example, what criteria determine that masturbation is no longer a disease aside from a shift in cultural values? Might other so-called dysfunctions be similarly disregarded in the future because they are driven by oppressive ideologies? The subjectivist may conclude that such examples illustrate how the categories of disease and non-disease are cultural creations that in many respects both frame and limit the human condition.

What we have labeled as critical alternative views have not been widely adopted by the healthcare profession, perhaps because it is unclear how the profession would engage in diagnosis and treatment under these perspectives. Researchers adopting a critical alternative view tend to deconstruct how society shapes our concepts of dysfunction, health, and well-being. For example, many researchers have raised concerns about an increasing tendency to "medicalize" human behavior. Take attention-deficit hyperactivity (ADHD) as an example. To what extent is this disorder a function of how society has structured education? It seems that "many children with this diagnosis do not feel ill [although]…social norms for education and conduct make them sick" (Hoffman 2017, p. 20). In other words, a critical alternative view might opine that children are educated in artificially constructed environments governed by strict social norms. The subjectivist may argue that a failure to adhere to the norms that govern an artificial environment makes otherwise functional behavior appear dysfunctional (e.g., Armstrong 1999).

Nevertheless, important questions remain unanswered. What would a healthcare profession look like that adopted these philosophical positions? Might treatment reflect the philosophy of Paulo Freire (1970/2014) who, in the *Pedagogy of the Oppressed*, discussed an educational strategy in which one aims to have individuals regain their humanity through liberation? Asked differently, would the medical profession serve to "liberate" as opposed to "diagnose" individuals? Though we do not have answers to such questions, the critical alternative views sketched in this section have at least resulted in a growing awareness among the academic community of a tendency to medicalize (perhaps unwarrantedly) numerous aspects of the human condition (e.g., Veatch 2006; Murray 2014).

5 Philosophical Positions, Methods, and the Quantitative Versus Qualitative Divide

Thus far, we have argued that philosophical positions indeed matter. Different models of human functioning are informed by specific assumptions about the nature of reality, how we understand that reality, and the kind of knowledge believed to be obtained via inquiry. Each of these models also tends to result in numerous consequences with respect to how we conceptualize, diagnose, and treat human functioning within society. Within this section, we further illustrate that such positions indeed matter by addressing the relationship between philosophical stances and methodological decisions. We argue that philosophical stances serve a dual role within inquiry. Philosophical stances can inform methodological decisions; however, methodological decisions may also be more or less consistent with particular philosophical stances. In other words, philosophical stances inform, and are in some sense informed by, methodological decisions enacted by researchers. However, prior to addressing this issue, it is first necessary to clarify terminology given that there is contention within the social science literature about the nature of quantitative and qualitative inquiry.

5.1 A Brief Interlude About the Quantitative and Qualitative Divide

As previously stated, numerous movements occurred in academia throughout the 1960s and 1970s. These movements were partly informed by criticisms that were gaining traction within the philosophy of science. The picture of science as objective or value-neutral became increasingly suspect, and numerous researchers within the social sciences and education reacted to such criticisms by questioning what was viewed as the dominance of quantitative research. Perhaps the most vehement arguments for this perspective derived from Guba and Lincoln (1994) who described quantitative methods as largely deriving from particular philosophical positions that stood in contrast to alternative worldviews that informed qualitative inquiry (Lincoln and Guba 1985). Under this view, ontological positions form the foundation upon which epistemic stances emerged, which subsequently constrained methodological decisions of researchers. For example, a realist ontology was said to inform a view of knowledge that was objectivist which in turn leads to specific methods (i.e., experimental design), whereas anti-realist views may lead to methods that aim to reconstruct the meaning of an experience.

This work sets the stage for questions about the commensurability of quantitative and qualitative methods within a single study (Howe 1988; Guba and Lincoln 2005). Though largely anecdotal, we have also found that it has become increasingly common to hear individuals identify themselves as either a qualitative or quantitative researcher. However, these terms may be more misleading than helpful (see Morgan 2007; Sandelowski 2014). For example, what does it mean to be a qualitative or quantitative researcher? Is all quantitative research united by a common philosophical stance? Is there something wrong with a qualitative researcher collecting numerical data?

Although space prohibits us from addressing all of these controversies, the presence of such issues demonstrates a need to clarify terminology prior to addressing the relationship between philosophical positions and methodological decisions. Instead of referring to qualitative research or quantitative research, we will instead employ the terms quantitative and qualitative to refer to *data* (Biesta 2010). In other words, quantitative will be used to indicate numerical data, whereas qualitative will be used to indicate forms of data that are non-numerical such as texts, images, artifacts, and so on. The term *method* refers to the techniques, strategies, or procedures used to investigate a question of interest. Semi-structured interviews, focus groups, and surveys are examples of methods used to investigate questions of interest. We also remind the reader that *philosophical positions* are defined as the ontological and epistemological stances that are either explicitly or implicitly adopted within an investigation.

The distinction between data, methods, and philosophical stances allows us to draw attention to specific aspects of the relationship between philosophical stances and methodological decisions. Consider a possible relationship between philosophical stances and data. It seems counterintuitive to claim that specific epistemic stances inform the type of data that may be used in an investigation. There is nothing blatantly problematic with a subjectivist using numerical data as part of their investigation just as there is nothing problematic about an objectivist analyzing texts. Consequently, the following section focuses on the relationship between philosophical stances and methodological decisions as opposed to the kind of data that results from such decisions.

5.2   Philosophical Positions Inform Methodological Decisions

If asked, many researchers may not be able to explicitly state their ontological and epistemological positions when investigating a phenomenon. However, further probing is likely to detect their underlying philosophical assumptions. Researchers approach inquiry from a mental model. Mental models consist of "the particular constellation of assumptions, theoretical commitments, experiences, and values through which a social inquirer conducts his or her work" (Greene 2007, p. 3). Philosophical positions, though often implicit, are aspects of mental models that inform inquiry. This is similar to the position advanced by Guba and Lincoln (1994) in that it is a top-down approach, which is depicted in Fig. 1.

Figure 1 indicates that mental models are partly comprised of philosophical positions (i.e., ontology and epistemology) that in turn shape methodological decisions. Mental models are broader than philosophical positions. For example, particular commitments to social justice or theory are conceptually distinct from ontological and epistemological positions. However, under this view, philosophical positions, whether implicit or explicit, inform methodological decisions. Although we have alluded to this relationship when discussing different models of human functioning, we will discuss how this would work using a hypothetical example drawn from Lovett and Hood (2014) who discuss ADHD and the concept of malingering.

Fig. 1 Mental models consist of philosophical positions that inform methodological decisions

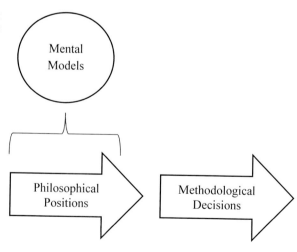

The Diagnostic and Statistical Manual of Mental Disorders, Fifth Edition, indicates that an individual must exhibit multiple symptoms, such as failing to give close attention to detail and frequently fidgeting, to receive a diagnosis of ADHD. This diagnosis is often assessed by using a variety of rating scales in which scores are given for items referring to the occurrence of such symptoms. In this scenario, malingering would occur if an individual without the disorder was motivated to present false symptoms in order to obtain the diagnosis (Suhrs et al. 2008). How would researchers with distinct philosophical stances understand the rating scale scores, the concept of ADHD, and malingering?

As indicated by Lovett and Hood (2014), ontological assumptions influence how we handle this scenario. The ontological realist would view ADHD as an entity that is independent from how we choose to operationalize the concept. When combined with an objectivist epistemology, the rating scale scores either successfully refer or fail to refer to an individual with ADHD. Malingering is compatible with this philosophical stance since it implies that some people identified as ADHD have intentionally misled the examiner. Said differently, under a realist ontology and objectivist epistemology, the scores on the rating scale allow us to say something about whether an individual possesses a real entity. Malingering is a potential concern since it suggests that we may be led to the wrong conclusion about this diagnosis.

The situation becomes more complicated when we turn to other philosophical stances. For example, the constructionist tends to adhere to ontological realism and epistemic relativism (Crotty 1998). In other words, in this case there may be a real, mind-independent entity underlying the diagnosis, but the meaning of ADHD socially created. The constructionist may, therefore, view the ADHD criteria as set of socially negotiated and/or or culturally defined concepts. When viewed in this way, there may be a common underlying experience whose meaning is constructed by individuals meeting the diagnostic criteria, though given epistemic relativism it is conceivable for other, equally legitimate constructions, to emerge. The rating scale scores could potentially be used to help us understand how individuals position themselves within such constructions. It is unclear, however, how to handle the concept of malingering from this perspective due to its reliance on the concepts of truth and falsehood. Malingering no longer consists of a discrepancy between reality and professed symptoms but may instead be viewed as a form of coherence between self and societal constructions. In other words, under this view, an individual may present the self as more or less aligned with how society has conceived of ADHD while recognizing this alignment may change given new, equally legitimate, societal constructions. Notice how in this case the collection of quantitative and qualitative data is irrelevant since the rating scale scores could presumably be interpreted differently under distinct philosophical stances.

5.3 Methodological Decisions Inform Philosophical Positions

The position that mental models, which partly consist of philosophical stances, inform methodological decisions is relatively uncontroversial. However, alternatives to this perspective, in which methodological decisions can also inform philosophical positions, have been advanced (e.g., Hathcoat and Meixner 2017). To conceptualize an alternative view, one may ask questions about the direction of the arrows in Fig. 1. Is it conceivable for the arrows to run in the other direction? Do methodological decisions frame the philosophical stances that are more or less reasonable to adopt irrespective of whether this is explicitly recognized by the researcher? It seems that the answer to this question is yes, at least under certain circumstances. Such an alternative is briefly sketched in this section since it is fairly controversial and the primary aim of this chapter is to introduce the reader to the topics of ontology and epistemology within social research.

Figure 2 illustrates a possible relationship between mental models, philosophical stances, and methodological decisions under an alternative perspective. According to Fig. 2, methodological decisions can in some sense inform philosophical positions that in turn lead to mental models. Some methodological decisions are, relatively speaking, "philosophically neutral" since they can be consistently employed from numerous philosophical perspectives. Semi-structured interviews and focus groups are examples of strategies that are relatively neutral since they can be applied by researchers who adopt objectivist, constructionist, or subjectivist epistemologies (see also ▶ Chap. 23, "Qualitative Interviewing"). Researchers adopting distinct epistemic stances would differ in how they interpret the results (e.g., an objectivist would attempt to code the data in an effort to identify the true themes), but the methods themselves do not seem to lead to any problems with a researcher who wishes to adopt one of the three epistemic stances.

Problems do seem to emerge in other situations. Consider the double-blind control trials we previously mentioned when discussing the biomedical model of human functioning. In such trials research participants are randomly assigned to either an experimental group that receives a new treatment or a control condition in which participants either receive no treatment, a placebo, or some standard treatment (see ▶ Chap. 37, "Randomized Controlled Trials"). Individuals administering the

Fig. 2 Methodological decisions inform philosophical positions that in turn inform mental models

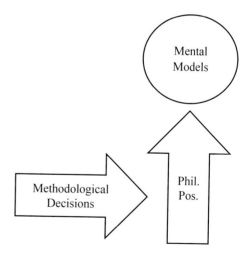

treatment are unaware of which participants are assigned to each condition. Might such methods have philosophical implications, irrespective of whether or not they are recognized by the investigators? For example, asking whether X causes Y seems to make implicit assumptions about realism. How could something nonexistent act as a cause? Other examples include the idea of parameter estimation in statistics. The concept of estimation seems difficult to defend without accepting some form of realism since there must be something to estimate for the concept to make sense (Hood 2013; see ▶ Chap. 38, "Measurement Issues in Quantitative Research").

We do not wish to imply that a realist must use experimental methods since causal questions may be investigated using alternative methodologies (e.g., Maxwell 2004). Nor do we wish to suggest that everyone who employs an experimental design must accept specific philosophical stances. Instead, there may be a range of stances that are coherent with aspects of this methodology, though some stances may be more problematic than others given apparent contradictions. In sum, under an alternative model, it is naïve to assume that methodological decisions occur in a philosophical vacuum. Failing to consider the philosophical aspects of methodological decisions may at best lead to ambiguity, lack of transparency, and/or confusion. At worst, failing to attend to the philosophical aspects of methodological decisions leads to interpretations that are altogether unwarranted, misleading, and/or incoherent.

6   Conclusions and Future Directions

When writing this chapter, we discussed with friends, family, and colleagues the controversies surrounding the concept of disease within medical discourse. Within these conversations some would ask, "Why does it matter how we conceptualize a disease?" To be candid, this question came as a surprise, which may partly be due to the fact that we were immersed within this line of literature. But, how could it *not* be important? The concept of a disease is connected to philosophical positions, resulting in a range of practical consequences for both the individual and society. In other words, the importance of ontology and epistemology was obvious to us, but it was not necessarily transparent to other people. We hope that by the end of the chapter, readers will share our sense of surprise by this question.

Ontological and epistemological considerations extend well beyond "armchair" philosophical debates. Philosophical stances indeed matter. Throughout this chapter, we have used the concept of a disease, interspersed with examples from mental health, to illustrate how ontological and epistemological considerations are embedded within models of human functioning. We have also described how philosophical stances serve a dual role within inquiry. Ontological and epistemological considerations frame social inquiry in that they provide a lens through which we investigate and interpret phenomena. Lastly, we sketched the possibility that methodological decisions can in some sense inform philosophical stances since methods themselves are often laden with ontological and/or epistemological assumptions. Nevertheless, a couple delimitations or caveats are worth mentioning.

Numerous researchers have created typologies of philosophical stances and/or theoretical perspectives such as postpositivism, pragmatism, and critical realism (e. g., Weaver and Olson 2006; Creswell 2009; Onwuegbuzie et al. 2009; Moon and Blackman 2014). We have tried to avoid creating another typology. Admittedly, we have drawn upon the work of Crotty (1998) in order to provide an overview of three epistemic positions; however, we do not want the reader to assume that these are the only positions that exist. Thus, the picture we have presented, which is also a feature of typologies, runs the risk of being overly simplistic. A second danger of typologies is that it can reinforce the position that researchers must choose between fixed, preexisting categories which fail to recognize the inherently constructed aspect of these views. Typologies also fail to consider the nuances and debates that exist within each category. With this caveat in place, the picture presented in this chapter aims to provide a useful entry point to a line of literature for the reader.

We have also not explicitly addressed other topics that are related to the philosophical stances discussed in this chapter. For example, there are debates about the aim of inquiry, which is often viewed as the idiographic and nomothetic distinction (see Robinson 2011). Is generalization possible? Are there universal laws of human functioning? What about the role of values within inquiry? There are also conceptual distinctions between health and well-being that are worth exploring (Hausman 2017). For the sake of clarity, we set such issues aside though we encourage the reader to pursue these topics as they continue to grapple with the role of ontology and epistemology within social inquiry.

In closing, we return to the metaphor of a ship being at sea without a navigational system. All too often, coursework focuses on methodology without regard for how philosophical positions provide a navigational system through which to understand social inquiry. In other words, methods are often presented as though they occur within a philosophical vacuum. This can result in aimless wandering at sea without a sense of location. Despite the obstacles to understanding this line of literature, as scholars and professors, we believe that ontological and epistemological considerations are too important to remain ignorant about. We do not expect a new reader to have figured out their navigational system by the end of the chapter. However, we do believe that at this juncture a new reader should recognize how philosophical stances can serve to navigate the vast sea of health social science research.

References

Armstrong T. ADD/ADHD alternatives in the classroom. Alexandria: Association for Supervision and Curriculum Development; 1999.

Berkman LF, Kawachi I. A historical framework for social epidemiology. In: Berkman LF, Kawachi I, editors. Social epidemiology. New York: Oxford University Press; 2000. p. 3–12.

Biesta G. Pragmatism and the philosophical foundations of mixed methods research. In: Taskakkori A, Teddlie C, editors. Handbook of mixed methods in social and behavioral research. 2nd ed. Thousand Oaks: Sage; 2010. p. 95–118.

Bothwell LE, Podolsky SH. The emergence of the randomized, controlled trial. N Engl J Med. 2016;375(6):501–4.

Creswell JW. Mapping the field of mixed methods research. J Mixed Methods Res. 2009;3(2):95–108.

Creswell JW, PlanoClark VL. Designing and conducting mixed methods approaches. Thousand Oaks: Sage; 2011.

Crotty M. The foundations of social research: meaning and perspective in the research process. Thousand Oaks: Sage; 1998.

Denzin NK, Lincoln YS, editors. The Sage handbook of qualitative research. Thousand Oaks: Sage; 2011.

Engel GL. The need for a new medical model: a challenge for biomedicine. Science. 1977;196:129–36.

Englehardt TH Jr. The disease of masturbation: values and the concept of disease. Bull Hist Med. 1974;48(2):234–48.

Fleck L. Genesis and development of a scientific fact. In: Trenn T, Merton RK, editors (F. Bradley & T. Trenn, Trans.). Chicago: The University of Chicago Press; 1935/1979.

Foucault M. Madness and civilization: a history of insanity in the age of reason (R. Howard, Trans.). New York: Routledge; 1961/1989.

Freire P. Pedagogy of the oppressed (M.B. Ramos, Trans.). New York: Bloomsbury Academic; 1970/2014.

Gale JE, Dolbin-MacNab ML. Qualitative research for family therapy. In: Mille RB, Johnson LN, editors. Advanced methods in family therapy research: a focus on validity and change. New York: Routledge; 2014. p. 247–65.

Greene JC. Mixed methods in social inquiry. San Francisco: Wiley; 2007.

Guba E, Lincoln Y. Competing paradigms in qualitative research. In: Denzin N, Lincoln Y, editors. Handbook of qualitative research. Thousand Oaks: Sage; 1994. p. 105–17.

Guba EG, Lincoln YS Paradigmatic Controversies, Contradictions, and Emerging Confluences. In: Denzin NK, Lincoln YS, editors. The Sage handbook of qualitative research. Thousand Oaks, CA: Sage Publications; 2005. pp. 191–215.

Hathcoat JD, Meixner C. Pragmatism, factor analysis, and the conditional incompatibility thesis in mixed methods research. J Mixed Methods Res. 2017;11(4):433–49.

Hathcoat JD, Nicholas M. Epistemology. In: Coghlan D, Brydon-Miller M, editors. The SAGE encyclopedia of action research ("E" entries, pp. 285–332). Washington, DC: Sage; 2014.

Hausman DM. Health and well-being. In: Solomon M, Simon JR, Kincaid H, editors. The Routledge companion to philosophy of medicine. New York: Routledge; 2017. p. 27–35.

Heidegger M. Being and time (J. Stambaugh, Trans.). New York: State University of New York Press; 1953/1996.

Hoffman B. Disease, illness, and sickness. In: Solomon M, Simon JR, Kincaid H, editors. The Routledge companion to philosophy of medicine. New York: Routledge; 2017. p. 16–26.

Hood SB. Psychological measurement and methodological realism. Erkenntnis. 2013;78:739–61.

Houghton C, Hunter A, Meskell P. Linking aims, paradigm and method in nursing research. Nurs Res. 2012;20(2):34–9.

Howe KR. Against the quantitative-qualitative incompatibility thesis or dogmas die hard. Educational researcher. 1988;17(8):10–16.

Hsu JL. A brief history of vaccines: smallpox to the present. S D Med. 2013;33(7):3–37.

Lovett BJ, Hood, SB. Comorbidity in child psychiatric diagnosis: Conceptual complications. In Perring C, Wells LA, editors. Diagnostic dilemmas in child and adolescent psychiatry: Philosophical perspectives. Oxford, UK: Oxford University Press; 2014. p. 80–97.

Illich I. Medicine is a major threat to health. An interview by Sam Keen. Psychol Today. 1976;9 (12):66–67.

Koenig HG. Religion and medicine: historical background and reasons for separation. Int J Psychiatry Med. 2000;30(4):385–98.

Kuhn T. The structure of scientific revolutions. 4th ed. Chicago: The University of Chicago Press; 1962/2012.

Ladyman J. Understanding philosophy of science. New York: Routledge; 2002.

Lincoln Y, Guba E. Naturalistic inquiry. Newbury Park: Sage; 1985.

Longino CF. The limits of scientific medicine. J Health Soc Policy. 1998;9(4):101–16.

Lovett BJ, Hood SB. Realism and operationism in psychiatric diagnosis. Philos Psychol. 2011;24(2):207–22.

Low G. Thomas Sydenham: the English Hippocrates. Aust N Z J Surg. 1999;69:258–62.

Maxwell JA. Causal explanation, qualitative research, and scientific inquiry in education. Educ Res. 2004;33(2):3–11.

Moon K, Blackman D. A guide to understanding social science research for natural scientists. Conserv Biol. 2014;28(5):1167–77.

Morgan DL. Paradigms lost and pragmatism regained: methodological implications of combining quantitative and qualitative methods. J Mixed Methods Res. 2007;1(1):48–76.

Murray M. Social history of health psychology: context and textbooks. Health Psychol Rev. 2014;8(2):215–37.

Nicholas M, Hathcoat JD. Ontology. In: Coghlan D, Brydon-Miller M, editors. The SAGE encyclopedia of action research ("O" entries, pp. 285–332). Washington, DC: Sage; 2014.

Onwuegbuzie AJ, Johnson RB, Collins KMT. Call for mixed methods analysis: a philosophical framework for combining qualitative and quantitative approaches. Int J Mult Res Approaches. 2009;3:114–39.

Phillips DC. Philosophy, science, and social inquiry: contemporary methodological controversies in social science and related applied fields of research. Oxford: Pergamon; 1987.

Poli R. Ontology: the categorical stance. In: Poli R, Seibt J, editors. Theory and application of ontology: philosophical perspectives. New York: Springer; 2010. p. 1–22.

Quine WVO. Two dogmas of empiricism. In: Quine WVO, editor. From a logical point of view. Cambridge, MA: Harvard University Press; 1953. p. 20–46.

Robinson OC. The idiographic/nomothetic dichotomy: tracing historical origins of contemporary confusions. Hist Philos Psychol. 2011;13(2):32–9.

Sandelowski M. Unmixed mixed-methods research. Res Nurs Health. 2014;37:3–8.

Sisti D, Caplan AL. The concept of disease. In: Solomon M, Simon JR, Kincaid H, editors. The Routledge companion to philosophy of medicine. New York: Routledge; 2017. p. 5–15.

Smith R. In search of "non-disease". Br Med J. 2002;324:883–5.

Smith KA. Louis Pasteur, the father of immunology? Front Immunol. 2012;3:1–10.

Suhrs J, Hammers D, Dobbins-Buckland K, Zimak E, Hughes CH. The relationship of malingering test failure to self-reported symptoms and neuropsychological findings in adults referred for ADHD evaluation. Arch Clin Neuropsychol. 2008;23:521–30.

Suls J, Rothman A. Evolution of the biopsychosocial model: prospects and challenges for health psychology. Health Psychol. 2004;23:119–25.

Sydeman T. Observationes Medicae circa Morgorum Acutorum Historiam et Curationem. London: Kettilby; 1676. As cited in Low G. Thomas Sydenham: the English Hippocrates. Aust N Z J Surg. 1999;69:258–62.

Valtorta NK, Kanaan M, Gilbody S, Ronzi S, Hanratty B. Loneliness and social isolation as risk factors for coronary heart disease and stroke: systematic review and meta-analysis of longitudinal observational studies. Heart. 2016;102(13):1009–16.

Veatch RM. How philosophy of medicine has changed medical ethics. J Med Philos. 2006;31(6):585–600.

Wade DT, Halligan PW. The biopsychosocial model of illness: a model whose time has come. Clin Rehabil. 2017;31(8):995–1004.

Weaver K, Olson JK. Understanding paradigms used for nursing research. J Adv Nurs. 2006;53(4):459–69.

White K, Willis E. Positivism resurgent: the epistemological foundations of evidence-based medicine. Health Sociol Rev. 2002;11:5–15.

Whitley R. Global mental health: concepts, conflicts, and controversies. Epidemiol Psychiatr Sci. 2015;24(4):285–91.

Williams MJ. Problems of knowledge: a critical introduction to epistemology. New York: Oxford University Press; 2001.

World Health Organization. Basic documents (48th ed.). Italy: World Health Organization; 1948/2014.

Social Constructionism

7

Viv Burr

Contents

Abstract

Social constructionism emerged in social psychology in the 1970s and 1980s, taking up many of the issues raised as part of the earlier "crisis" in social psychology and becoming a critical voice challenging the agenda of mainstream psychology. In particular, it challenged psychology's individualistic, essentialist, and intrapsychic model of the person, replacing it with a radically social account of personhood in which language is key. Viewed through the constructionist lens,

V. Burr (✉)
Department of Psychology, School of Human and Health Sciences, University of Huddersfield, Huddersfield, UK
e-mail: v.burr@hud.ac.uk

© Springer Nature Singapore Pte Ltd. 2019
P. Liamputtong (ed.), *Handbook of Research Methods in Health Social Sciences*,
https://doi.org/10.1007/978-981-10-5251-4_57

the person ceases to be a unified ensemble of stable psychological structures and traits and becomes a fluid, fractured, and changeable assemblage, distributed across and produced through social interactions and relationships. Social constructionism's critical focus has meant that its research agenda is also radically different from that of mainstream psychology and social psychology. Social constructionist research rejects the mainstream psychological experimental paradigm and turned attention to the constructive force of language and discourse, opening up new lines of research and developments in a number of methods referred to as discourse analysis.

Keywords
Social constructionism · Discourse analysis · Critical psychology · Power relations · Crisis in social psychology

1 Introduction

Social constructionism represents a radical challenge to mainstream psychology in both theory and research. It began to emerge in social psychology in the 1970s and 1980s as a recognizable body of work by those who would, today, refer to themselves as, for example, "critical psychologists" and "discursive psychologists." Social constructionism poses a challenge to psychology's individualistic, essentialist, and intrapsychic model of the person, replacing it with a radically social account personhood and is, therefore, essentially a social psychology. The taken-for-granted topics of mainstream psychology and social psychology, such as attitudes, motivation, personality, and emotion, were brought into question as structural features of the human psyche and instead seen as social constructions achieved though social interaction and language. As Craib (1997) points out, many social constructionist assumptions are fundamental to psychology's disciplinary cousin, sociology, and it is a measure of the unhelpful separation of sociology and psychology since the early twentieth century that psychologists have only recently begun to engage with social constructionist ideas.

2 Origins and Influences

Although social constructionism is a relatively new term in the social sciences, especially psychology, its ideas and practices have a longer history in disciplines such as sociology, philosophy, and linguistics. The key tenet of social constructionism is that our knowledge of the world, including our understanding of human beings, is a product of human thought, language, and interaction rather than grounded in an observable and definable external reality. Such an idea, although strange to many psychologists, is fundamental to longstanding concepts in the sociology of knowledge, such as ideology and false consciousness, which focus on how sociocultural forces construct our knowledge in particular ways.

An important contributor to social constructionist thinking has been the microsociological approach of symbolic interactionism (see Berger and Luckmann 1966). George Mead, at the University of Chicago, had developed this approach from the earlier work of Herbert Blumer and later published his ideas in *Mind, Self and Society* (1934). Mead had studied at Leipzig University under Wilhelm Wundt, who believed in the importance of sociocultural factors, such as myth, folk customs, and religion, in understanding human behavior and experience. Mead took up these ideas in his development of symbolic interactionism, which argues that people construct and negotiate identities for themselves and others through their everyday social interactions with each other. Language, as a system of socially shared symbolic meanings, is central to this constructive process. In line with this, the sociological subdiscipline of ethnomethodology, which grew up in North America in the 1950s and 1960s, aimed to research the processes by which ordinary people construct social life and make sense of it to themselves and each other, and ethnomethodology has been one of the influences on social constructionist research (see also ► Chaps. 10, "Symbolic Interactionism as a Methodological Framework," and ► 16, "Ethnomethodology").

Berger and Luckmann argue that human beings together create and then sustain all social phenomena through social practices. They argue that although our world is socially constructed through the interactions of people, it is at the same time experienced by them as if the nature of their world is pre-given and fixed; we are all born into a social world that preexists us and that, therefore, seems "natural."

In addition to these North American influences, social constructionism has also drawn on the ideas of more recent European thinkers. The historian and philosopher Michel Foucault has been highly influential, as has Ferdinand de Saussure in structural linguistics and the work of the philosopher Jacques Derrida on "deconstruction." The ideas of these writers have been extensively drawn upon in developing social constructionist thinking around the role of language in the construction of human social phenomena.

In psychology, the emergence of social constructionism is usually dated from Gergen's (1973) paper *Social Psychology as History* in which he argues that all knowledge, including psychological knowledge, is historically and culturally specific; we, therefore, must look beyond the individual and enquire into social, political, and economic realms for a proper understanding of the evolution of present-day psychology and social life. He argues that it is pointless to look for final descriptions of people or society, since the only abiding feature of psychological and social life is that it is continually changing. Social psychology, thus, becomes a form of historical undertaking, since all we can ever do is to try to understand and account for how the world and people appear to be at the present time.

Gergen's paper was written at the time of what is referred to as "the crisis in social psychology" (see, e.g., Harré and Secord 1972; Armistead 1974), which provided some of the momentum for social constructionist ideas in psychology. This crisis centered on worries about the way that social psychology's agenda was driven by the needs and motivations of powerful factions (government, the military, and commerce); social psychology as a discipline can be said to have emerged from the attempts by psychologists to provide the US and British governments during the Second World War with knowledge that could be used for propaganda and to manipulate behavior. Social psychology, therefore, had historically served, and was paid for by, those in positions of power, both in government and in industry. Its theories and research findings often seemed to bring further oppression to relatively powerless and marginalized groups (women, ethnic minorities, working class people, and those of nonnormative sexual orientation), and this operated partly through the study of human phenomena in socially decontextualized laboratory environments, since experiments ignored the real-world contexts which, it was argued, give human conduct its meaning. There was a move to attend to this social context, as well as to explore human phenomena from the perspective of psychology's "subjects" themselves rather than privilege the perspective and voice of the relatively powerful researcher. These concerns encouraged social psychologists to embrace the ideas already flourishing in neighboring disciplines, including micro-sociology, and also fed into the burgeoning call to recognize qualitative research methods as legitimate and fruitful for the discipline of psychology.

3   What Is Social Constructionism?

One of the earliest writers to describe social constructionism was Kenneth Gergen, and in his 1985 publication *The Social Constructionist Movement in Modern Psychology*, he laid out what he saw as its key features, which today remain useful orienting principles.

3.1   A Critical Stance Toward Taken-for-Granted Knowledge

Social constructionism takes a critical stance toward psychology's taken-for-granted ways of understanding the world and ourselves. Psychology has modeled itself on the natural sciences. These assume the epistemological approach of positivism, the view that knowledge comes from objective, unbiased observation of the world and that there is a true or accurate description of people, events, and things that science endeavors to reveal. Positivism assumes that the world comes to us ready-made, and our task is to discover its true nature. Within this epistemology, the experiment is the gold-standard research paradigm (see ▶ Chaps. 6, "Ontology and Epistemology," and ▶ 9, "Positivism and Realism").

But social constructionism cautions us to be suspicious of our assumptions about how the world appears to be. For example, it may seem obvious to us that there are two naturally occurring categories of human being, men and women. But social constructionism bids us to seriously question whether the categories "man" and "woman" are simply a reflection of distinct types of human being and to consider instead that they are categories constructed by people themselves in the course of social interaction. When we consider the many individuals whose sex cannot easily be decided on the basis of their anatomy, and the increasingly prevalent phenomena of transgender and gender reassignment surgery, we can begin to appreciate that being a man or a woman is much more socially negotiated that we might have imagined. The same is true of other highly socially relevant concepts, such as "race." While in former times, it may have seemed unquestionable that there exist different races of human beings, each with its own physical and psychological characteristics, research has failed to find any genetic basis for such distinctions; while human beings come in a diverse range of shapes, sizes, and colors, there is no evidence for the existence of genetically discrete races (see, e.g., Yudell et al. 2016).

We may ask why this matters and what social significance it has. The social constructionist would argue that challenging the existence of such natural categories helps us to disrupt and destabilize many of the assumptions we hold about men and women and about nonwhite peoples, assumptions that underpin inequitable practices. The familiar arguments that particular characteristics (or lack of them) are natural for women or men, or for black people, are frequently used to defend inequalities.

3.2 Historical and Cultural Specificity

The agenda of mainstream psychology is to discover universal principles of psychological functioning or "human nature." But social constructionism argues that the ways in which we commonly understand the world, the categories and concepts we use, are historically and culturally specific and depend upon where and when in the world one lives. For example, concepts of illness and disease have changed significantly over time and vary across different cultures. What we believe illness and disease to be are fluid products of our culture and history rather than fixed entities (see below). This means that we cannot assume that *our* ways of understanding are necessarily any better, in terms of being any nearer the truth, than other ways. Indeed, the concept of "truth" itself becomes problematic. Looking at knowledge this way challenges the idea of scientific progress, the idea that through science we are advancing toward a more and more accurate understanding of the physical and psychological world. Social constructionists argue that this way of thinking has led to the imposition of our own systems of knowledge upon other cultures and nations; psychology has been accused of being imperialist in its attitude toward other cultures and has colonized them, replacing their indigenous ways of thinking with Western ideas.

3.3 Knowledge, Social Processes, and Social Action Are Related

Social constructionism argues that our knowledge is not derived from the nature of the world as it "really" is; rather, people construct it between them through their daily interactions. This is why social interaction and language are of great interest to social constructionists, since it is in the course of these practices that our shared versions of knowledge are constructed. For example, what we know and understand as dyslexia is a phenomenon that has come into being through the exchanges between those who have difficulties with reading and writing, their families and friends, and others who may teach them or offer them diagnostic tests. Knowledge is, therefore, seen not as something that a person has or does not have, and is neither correct nor incorrect, but is something that people create and enact together.

Our social interactions are capable of producing a variety of possible social constructions of events. What we regard as knowledge is, therefore, one possible construction among many. But different constructions invite different kinds of action or practice from us. For example, before the growth of the Temperance movement in nineteenth-century USA and Britain, people under the influence of alcohol were seen as entirely responsible for their behavior. Therefore, punishment, such as imprisonment, was seen as an appropriate response. However, the Temperance movement represented alcohol itself, and addiction to it, as the problem. Today, addiction of various kinds, whether to drugs or to behaviors such as gambling, is constructed as a kind of illness and the alcoholic is, therefore, seen as not totally responsible for their behavior. The social action appropriate to understanding drunkenness in this way is to offer medical and psychological therapy not punishment. Constructions of the world, therefore, sustain some patterns of social action and exclude others.

4 How Does Social Constructionism Challenge Mainstream Psychology?

The three key social constructionist arguments outlined above challenge some of the equally key features of mainstream psychology and social psychology: essentialism, language as representation, and objectivity and universalism. Social constructionism also attends to power and politics, issues on which psychology has arguably been silent.

4.1 Essentialism

The model of the person at the heart of mainstream psychology represents the person as pre-existing society and social life. It has an "essence" consisting of characteristics or qualities that make it what it is, and this essence may have been produced by genes or by the influence of the environment (or a mixture of both). The social constructionist argument for cultural and historical specificity is sometimes mistakenly interpreted as just another way of taking the nurture side in the nature/nurture debate. But social constructionism is not just saying that one's cultural surroundings have an impact upon one's psychology or even that our nature is a product of environmental, including social, rather than biological factors. Both of these views are essentialist, in that they see the person as having some definable and discoverable nature, whether given by biology or by the environment, and the agenda of psychology is its discovery. Psychology has, therefore, produced a number of different theories to account for this essence, such as personality trait and type theories, which in turn give rise to instruments designed to measure personality, such as the popular Myers-Briggs Type Indicator (1943/1976), and psychodynamic theories.

The content of this individual is described by the various and competing psychological theories: personality traits and intelligence, unconscious motivations and drives, learned behaviors and habits, attitudes, and beliefs. All of these constructs are similar, however, in that they are thought of as properties of the person, whether inherent or acquired environmentally, and they are held to cause or determine our behavior. Social constructionism regards all such concepts as constructions, and like any other constructions, they have arisen in the course of social life in particular sociocultural conditions. Social constructionism is anti-essentialist and also, therefore, challenges the psychology's determinism; we cannot appeal to psychological essences to explain behavior. By contrast, social constructionism rejects the idea that the person consists of a unified and stable collection of features or characteristics. Instead, the person is viewed as fractured and distributed across social relationships; we are "produced" on a myriad of different occasions and, in different ways, dependent upon the nature of the social situation in which we find ourselves. There is, therefore, no essential "self" that describes who the person "really" is.

Psychology's essentialism often takes a reductionist form. Reductionism is the practice of describing a complex phenomenon in terms of simpler or more fundamental elements. Psychology has tended to assume that all behavior will eventually be explained in terms of biological mechanisms, and the complex social and cultural conditions that inform psychological phenomena are likewise reduced to biological causes. For example, we often encounter the idea that social inequalities and differences between women and men derive from psychological sex differences which are themselves the result of differences in hormone levels or brain structure that are said to have evolutionary origins. Developments in neuroscience have arguably led psychology even further in this reductionist direction. Neurological accounts of experience and behavior are becoming commonplace in everyday discourse, and according to Cromby (2012), we are increasingly invited to understand our experiences in terms of brain chemistry rather than social relations.

The significance of social constructionism's challenge here is that essentialism is regarded as potentially trapping people inside personalities and identities that may be restrictive and pathologizing for them, rendering psychology an oppressive practice. For example, if someone is described as psychotic and this is seen as an abiding feature of their personality, they not only face a future in which change appears unlikely but may also become subject to invasive psychiatric procedures. Essentialism also creates a tendency for psychologists to seek dispositional explanations for human behavior and to look for causes of behavior in psychological states and structures rather than in social processes. A person may be diagnosed with depression, but their depression may be constructed either as part of their personality or as a consequence of their poor living conditions; one construction leads to drug treatment and the other to improvements in housing and so on.

4.2   Language as Representation

The implicit model of language within psychology is as a vehicle for representing external or internal realities, and our representations may be accurate or inaccurate and truthful or dishonest. The accuracy and truthfulness of our linguistic accounts have assumed importance in our understanding of memory, with applications in areas such as eyewitness testimony. Memories are only one example of psychological entities that are thought to be expressed through language. Others, often captured through both quantitative and qualitative research methods, include attitudes, emotions, and subjective experience.

The position of social constructionism on the status of internal psychological conditions is not always straightforward. Some constructionists prefer to "bracket off" psychological states, much as behaviorists did in the early twentieth century; because we can have no direct access to psychological states and have only a person's verbal reports of them, behaviorists claimed that psychologists should only concern themselves with what is observable, i.e., behavior. Likewise, many social constructionists do not concern themselves with language as a representation of inner psychological states. But this is because their interest is in an important aspect of language that they feel has been ignored by psychology and that is its constructive force.

Social constructionism argues that language has practical consequences for people that should be acknowledged. For example, when a doctor declares that a person is suffering from schizophrenia, and "sections" them under the Mental Health Act, this has immediate and significant consequences for the person's freedom. It is this constructive and "performative" function of language that is of principal interest to social constructionists. It is through our discourse with each other that people are constructed as mentally ill, as masculine or feminine, and as old or as disabled, and social constructionists are interested in how and why people become constructed in the way that they do, whether through our active discourse with each other, through the small-scale linguistic turns of individual social interactions, or through the myriad linguistic and symbolic representations or "discourses" that abound in our cultural life. As I will discuss later, this focus on language leads to a radically different research agenda from that pursued by mainstream psychology.

4.3   Objectivity and Universalism

The social constructionist critique of mainstream psychology centers upon reminding psychology that its own grasp on the world is necessarily partial. It is partial both in the sense of being only one way of seeing the world among many potential ways and in the sense of reflecting vested interests.

Within the mainstream research paradigm of science, which psychology adopts, the researcher can claim truthfulness for their findings by recourse to the supposed objectivity of scientific method. The experimenter is represented as able to stand back from their own humanity and reveal the objective nature of the phenomena under study without bias and without contaminating the results with "leakage" from their own personal involvement. This vision of and value for objectivity is part of the philosophical position of realism. Realism assumes a singular and objectively describable world. It asserts that an external world exists independently of our representations of it (see also ▶ Chap. 9, "Positivism and Realism").

Psychology has also adopted from the natural sciences the assumption that there are universal laws governing the behavior of all things (including people) and that these laws can be discovered through experimental research. For example, the widely used Health Belief Model (Rosenstock 1966) describes a model of health decision-making that is assumed to explain how all people make behavioral choices, regardless of their cultural or historical location.

Within social constructionism, all knowledge is derived from looking at the world from some perspective or other and is in the service of some interests rather than others. This means that there can never be objective facts, things that are simply true for all people regardless of their time and place. Relativism, in contrast to realism, argues that the only things we have access to are our various perceptions and representations of the world, and these cannot be judged against some assumed reality for their truthfulness or accuracy. Relativists, therefore, cannot prefer one account to another on the basis of its apparent truthfulness. The dangers in insisting that there is only one truthful and, therefore, acceptable way of understanding the world and its people are arguably being played out in political arenas both internationally and more locally in issues such as religious fundamentalism, gay rights, and sexual harassment. Different constructions can only be compared in terms of their usefulness, their potential for benefitting people.

But psychology's vision of the world and people is also partial in the sense of coming from a particular cultural perspective and is, therefore, not objective. The discipline of psychology, like other academic disciplines, has been dominated by the work and interests of white, middle-class men, and this has inevitably shaped the knowledge that their work has produced. The concept of intelligence was arguably constructed by them in their own image, and it is only in relatively recent times that this has been challenged. For example, the introduction of the notion of "emotional intelligence" is an attempt to value aspects of psychological functioning that are not captured by the cognitive reasoning focus of IQ tests.

Although the tenets of social constructionism appear to lead automatically to a relativist position, some writers, often referred to as critical realists, prefer to retain some concept of a reality existing outside of our constructions of it while accepting the view that there are multiple, competing perspectives on events that are bound up with power relations (e.g., Cromby and Nightingale 1999; see also ▶ Chap. 9, "Positivism and Realism").

4.4   Power and Politics

Psychology has traditionally regarded itself as apolitical; its search for the objective facts about people seems to place it outside of the political arena. However, if we take the social constructionist view that psychology is partial and that objectivity is a fiction, it is possible to regard the discipline as instrumental in the maintenance of inequality and oppression. Through its status as a science, psychology has presented as truth representations of people that serve to legitimate inequalities. For example, through "sex differences" research, psychology has promoted the popular idea that women and men are different kinds of people and, therefore, suited to different roles in society; through its measurement and testing program and through its diagnostic classification system, it has set norms for intelligence, personality, and behavior based on white, male middle-class norms and values and has pathologized people whose behavior and experience lie outside of these prescriptions.

It is a concern with such matters that is the principle focus of critical psychology and critical social psychology (see, e.g., Fox et al. 2009; Gough et al. 2013). Critical psychologists focus on issues of exploitation and oppression and social justice. A key development has been in "postcolonial" critical psychology (see Hook 2012), which aims to problematize the discourses that help to maintain and legitimate disadvantaged Third World countries. Social constructionism and critical psychology do not, however, map onto each other directly; although much critical psychology can be said to be social constructionist in spirit, some critical psychologists would not necessarily refer to themselves as social constructionists.

Another strand to the social constructionist concern with power derives from the work of Michel Foucault. Foucault argues that the way people talk and think about, for example, sexuality and mental illness, the way these are widely represented in society, brings implications for the way we treat people. Our representations entail particular kinds of power relations. For example, we think of people who hear voices as mentally ill and refer them to psychiatrists and psychologists who then have power over many aspects of their lives. Foucault refers to such representations as "discourses," since he sees them as constituted by and operating through language and other symbolic systems. Our ways of talking about and representing the world through written texts, pictures, and images all constitute the discourses through which we experience the world.

5   Illness: A Constructionist Illustrative Example

In order to illustrate these social constructionist principles, I will use the case of "illness." The social construction of illness has now become a major perspective in medical sociology (Conrad and Barker 2010). It is generally assumed that illness is a biological matter that can be objectively determined, in line with what is termed the biomedical model of illness. However, it is by no means easy to make a judgment about whether or not a person is ill. This is because illness is not simply a biological matter – it is a social one. When we say that we or someone else is ill, we are making a judgment that only in part relates to their physical condition. Much of our judgment rests on cultural prescriptions, norms, and values surrounding our ability to perform our usual activities and upon power relations. Bury (1986) cites the work of Figlio (1982), who studied the relationship of the condition miners' nystagmus to social class and capitalism. The existence of this as a disease entity was not simply a medical matter – it was at the center of conflicts over malingering and compensation for workers. As Burr and Butt (2000) have noted, in recent times, we have seen the emergence of a number of conditions that were unknown in earlier times, for example, premenstrual syndrome and ME (myalgic encephalomyelitis), and the medical status of these is similarly problematic and infused with cultural assumptions and moral prescriptions.

In addition, it seems that the biomedical model is one that is not universal and is a fairly recent development. Medical belief systems in other cultures are often radically different from biomedicine, and in our own society, we are seeing an increasing use of alternative medicines which are often based upon belief systems quite different to biomedicine, such as homeopathy, acupuncture, and reflexology. Social constructionism argues against the view that our own predominant, biomedical view of disease is the right one and all others false. All medical belief systems operate within a culture with norms, values, and expectations that make sense of illness for people in that culture and set the criteria for what, locally, can count as illness.

The variation in ways of understanding illness that exists across cultures and across the range of alternative medicines in our own society can also be seen historically. Radley (1994) describes how, up until the end of the eighteenth century, doctors saw the patient's emotional and spiritual life as directly relevant to their state of health. With developments in the study of anatomy, it became possible to think of illnesses as things attacking the body as a system of interrelated organs, with the result that the experience of the person as a whole became irrelevant to diagnosis.

But the rise of biomedicine is not something that can be seen as simply a story of the progress of medical knowledge. It is a way of viewing the body that, it can be argued, is intimately connected to broader social developments. The study of the inner workings of the body in the anatomy laboratory took place in the context of a more general movement toward understanding the world by ordering and classifying it. Foucault (1973, 1976, 1979) has argued that such ordering and classifying, with respect to human beings, has played a key role in controlling the populace. By classifying people as normal or abnormal, mad or sane, and healthy or sick, it became possible to control society by regulating work, domestic, and political behaviors. For example, the certified mentally ill may not vote and may be forcibly confined, those who cannot obtain a sick note from their doctor may have no choice but to work and those whose parenting is seen as inadequate may be separated from their children.

6   Social Constructionist Research

It will be clear that the aim of social constructionist enquiry must leave behind psychology's questions about the nature of people and, instead, turn attention to a study of the emergence of current forms of psychological and social life and to the social practices by which they are created. Social constructionism asks how certain phenomena or forms of knowledge are constructed through language, either through the way people are positioned within wider cultural discourses or through their everyday social interaction. Research conducted from within the first of these approaches is often concerned with identifying the ideological and power effects of discourse and is referred to as Foucauldian or "critical" discourse analysis (see also ▶ Chap. 50, "Critical Discourse/Discourse Analysis"). Research focusing on the constructive work of individuals during social interaction is often referred to as discursive psychology. Confusingly, both kinds of research may be referred to as simply "discourse analysis," although much work within discursive psychology in the UK has now moved toward the adoption of conversation analysis (CA) as a research method and often uses this term instead (see also ▶ Chap. 28, "Conversation Analysis: An Introduction to Methodology, Data Collection, and Analysis"). However, Foucauldian and discursive psychological approaches to social constructionist research should not be seen as mutually exclusive; Wetherell (1998) argues that they are not in principle incompatible and that we could and should attend to both situated language use and the wider social context within which these are produced.

It would be a mistake to suggest that there are particular research methods that are intrinsically social constructionist; social constructionist research simply makes different assumptions about its aims and about the nature and status of the data collected. However, the insistence of social constructionism upon the importance of the social meaning of accounts and other texts often leads logically to the use of qualitative methods as the research tools of choice (see ▶ Chap. 63, "Mind Maps in Qualitative Research"). In practice, this has often been the analysis of interview transcripts, recordings of naturally occurring conversations, newspaper articles, and other texts of various kinds.

6.1   Foucauldian/Critical Discourse Analysis

The Foucauldian strand of social constructionism acknowledges the constructive power of language but sees this as bound up with material or social structures, social relations, and institutionalized practices. The concept of power is, therefore, at its heart. The principle form it takes today is critical discourse analysis (CDA), which is particularly associated with the work of Norman Fairclough in the UK (Fairclough 1995). Since their focus is on issues of power, these researchers are especially interested in analyzing various forms of social inequality, such as gender, ethnicity, disability, and mental health, with a view to challenging these through research and practice. The way that discourses construct our perceptions and experience can be examined by "deconstructing" texts, taking them apart and showing how they work to present us with a particular vision of the world and of ourselves and thus enabling us to challenge it (see also ▶ Chap. 50, "Critical Discourse/Discourse Analysis").

Foucauldian discourse analysis aims to identify the discourses operating in a particular area of life and to examine the implications for subjectivity, practice, and power relations that these have. The kinds of materials that may be used in such an analysis are virtually limitless; any text or artifact that carries symbolic meaning may be analyzed. So, to the extent that such things as family photographs, choices of interior décor, hairstyles, or road signs carry meanings that may be "read" by people, they may be analyzed. In the illustrative example below, Conradie (2011) used CDA to analyze a men's lifestyle magazine, *For Him Magazine (FHM)*. He asks what role such publications might have in disseminating and perpetuating gendered expectations. Conradie analyzed written articles from the magazine, examining these for the assumptions and values that appeared to be underlying these texts and for how women and men were represented within them. Some of the extracts from the magazine used by Conradie are shown below:

> Extract 56: 'Lady-pleasuring, like wiring up a stereo, requires reading the instruction book ...'.
> Extract 57: 'Learn to read the telltale physical give-aways of her body's arousal'.
> Extract 58: 'A few minutes of rub and tug [massage] and she'll be putty in your hands'.
> Extract 59: 'It might sound cheesy, but dimming the lights, playing ambient tunes and lighting a candle or two will get your lady ready for rubbing [massage]'.
> Extract 60: 'Good massage will lead directly to full sex. Guaranteed'.
> Extract 61: 'It's willingness over technique here, but it will pay off'.

Firstly, Conradie notes that the articles focused on men's sexual performance, representing sex as a skill that men should master and suggesting that successful sex depends on gaining the correct knowledge and following certain rules (e.g., Extract 56). Conradie argues that this supports an ideology in which sex is about the successful expression of masculinity. He also notes that there is no mention of any emotional or relational context of sex. Instead, it endorses a hedonistic (male) lifestyle where sexual liaisons with multiple women are accepted as the norm and women are constructed as the recipients of actions performed by the reader ("you"). Conradie concludes that *FHM* aims to encourage a hedonistic lifestyle, with relationships with women constructed as opportunities for the display of men's sexual prowess.

6.2 Discursive Psychology

The focus on social interaction and language as a form of social action introduced above has been placed center stage by discursive psychology (e.g., Edwards and Potter 1992). Discursive psychology shares the radically anti-essentialist view of the person of social constructionism, and in particular it denies that language is a representation of internal states or cognitions such as attitudes, beliefs, emotions, and memories. Its concern is to study how people use language in their everyday interactions, their discourse with each other, and how they put their linguistic skills to use in building specific accounts of events, accounts which may have powerful implications for the interactants. It is, therefore, primarily concerned with the performative functions of language.

The action orientation of discursive psychology, therefore, transforms mainstream psychology's concern with the nature of phenomena into a concern with how these are *performed* by people. Thus, memory, emotion, and other psychological phenomena become things we do rather than things we have. Discursive psychology focuses mainly on the analysis of social interactions in order to reveal the rhetorical devices that people use to construct persuasive accounts and achieve their interactional goals. The example below, taken from Dick (2013), is a brief illustrative example. Dick undertook a discursive analysis of interviews with UK policewomen about sexism in the workplace. In the following extract, "Sophie" is describing her reaction to some postcards on the wall of a CID office, showing women in scanty bikinis:

1. *S*: But I remember thinking, 'This is a professional office!
2. I've never seen naughty post cards in a professional office before'.
3. And I was foolish enough in retrospect to say, 'My Goodness me!
4. What are those doing on the wall? That's a bit off isn't it?
5. Having pictures like that in a CID office?'
6. Cos y'know, you wouldn't walk into a solicitors'
7. and see that, or most places. I mean
8. I equated that kind of calendar girl
9. stuff with garages!
10. *Me*: Absolutely! That's where I would have expected to see them!

Dick points out that Sophie has to work hard to bring off an account that persuades the hearer that sexism is objectively real. She heads off any possible suggestion that her reaction to the postcards is due to her being "prudish" or oversensitive and, therefore, prone to making accusations of sexism; she begins her account (lines 1–2) by casting the postcards as only problematic due to their location in a CID office. In lines 3–4, she frames her reaction as naïve, an example of what discursive psychologists call stake inoculation; this framing allows her to further protect herself from any claim that she might be oversensitive by constructing herself as an ordinary person who was simply perplexed by what she saw. Sophie further builds the credibility of her account in lines 5–9, introducing the idea that such a display in a public office suggests a lack of professionalism. She also uses another rhetorical device, an extreme case formulation ("or most places", line 7), to suggests that such a display is significant because it is very unusual, and in lines 8–9, she extends her initial point by comparing professional locations (like a CID office) with garages, where she would expect to see such material.

7   Conclusion and Future Directions

Social constructionist theory brings with it a reformulation of what it means to do social science research. The concepts that are the cornerstones of traditional psychology, such as objectivity and value freedom, are radically questioned. Within social constructionism, there exist approaches to research that vary in the kinds of materials they typically analyze and the conceptual tools they use to perform their analysis. A noticeable difference between them is the extent to which they are concerned with the workings of language beyond the confines of the text under analysis. Despite these differences, what they share is an understanding of language as performative and constructive, and this is what differentiates them from mainstream psychology.

Despite the rising popularity of social constructionism, there appears to have been little take-up of its ideas by the mainstream discipline. Academics continue to research and write principally within one paradigm or the other, many mainstream journals continue to favor work carried out within a traditional, positivist framework, and the agendas of the two paradigms continue to lead more or less parallel, self-contained lives.

As we are arguably living in a more and more globalized society, it may be argued that we need to see the adoption of a generally social constructionist perspective in our world view. It is no longer viable to attempt to live our lives according to our own very local "truths." In particular, we need to recognize the dangers inherent in prevailing essentialist, Western concepts of the person. The biomedical model of illness has already been successfully challenged in the field of disability, in the now widespread popularity of the "social model of disability" (Makin 1995; Hughes and Paterson 1997), which frames disability as a product of social structures and power relations rather than a quality of the disabled individual. It may be argued that similar conceptual developments are needed to help us properly understand and address other social problems. Burr and Butt (2000) argue that the proliferation of previously unknown psychological conditions that has occurred in recent decades should be seen as a socially constructed phenomenon, and the continuing rise in "mental illness" is arguably not best conceptualized through an ultimately stigmatizing, essentialist and individualist disease model.

References

Armistead N. Reconstructing social psychology. Harmondsworth: Penguin; 1974.

Berger P, Luckmann T. The social construction of reality: a treatise in the sociology of knowledge. New York: Doubleday; 1966.

Burr V, Butt T. W. Psychological distress and post-modern thought. In: Fee D, editor. Pathology and the postmodern: mental illness as discourse and experience. London: Sage; 2000. p. 186–206.

Bury MR. Social constructionism and the development of medical sociology. Sociol Health Illn. 1986;8(2):137–69.

Conrad P, Barker KK. The social construction of illness: key insights and policy implications. J Health Soc Behav. 2010;51(1):S67–79.

Conradie M. Masculine sexuality: a critical discourse analysis of *FHM*. South Afr Linguistics Appl Lang Stud. 2011;29(2):167–85.

Craib I. Social constructionism as a social psychosis. Sociology. 1997;31(1):1–15.

Cromby J. Narrative, discourse, psychotherapy- neuroscience? In: Lock A, Strong T, editors. Discursive perspectives in therapeutic practice. Oxford: Oxford University Press; 2012. p. 288–307.

Cromby J, Nightingale DJ. Reconstructing social constructionism. In: Nightingale DJ, Cromby J, editors. Social constructionist psychology: a critical analysis of theory and practice. Buckingham: Open University Press; 1999. p. 207–24.

Dick P. The politics of experience: a discursive psychology approach to understanding different accounts of workplace sexism. Hum Relat. 2013;66(5):645–69.

Edwards D, Potter J. Discursive psychology. London: Sage; 1992.

Fairclough N. Critical discourse analysis: the critical study of language. London: Longman; 1995.

Figlio K. How does illness mediate social relations? Workmen's compensation and medicolegal practices 1890–1940. In: Wright P, Teacher A, editors. The problem of medical knowledge. Edinburgh: Edinburgh University Press; 1982. p. 174–224.

Foucault M. The birth of the clinic: an archaeology of medical perception. London: Tavistock; 1973.

Foucault M. The history of sexuality: an introduction. Harmondsworth: Penguin; 1976.

Foucault M. Discipline and punish. Harmondsworth: Penguin; 1979.

Fox D, Prilleltensky I, Austin S, editors. Critical psychology: an introduction. 2nd ed. London: Sage; 2009.

Gergen KJ. Social psychology as history. J Pers Soc Psychol. 1973;26:309–20.

Gergen KJ. The social constructionist movement in modern psychology. Am Psychol. 1985;40:266–75.

Gough B, McFadden M, McDonald M. Critical social psychology: an introduction. 2nd ed. Basingstoke: Palgrave Macmillan; 2013.

Harré R, Secord PF. The explanation of social behaviour. Oxford: Basil Blackwell; 1972.

Hook D. A critical psychology of the postcolonial. London: Psychology Press; 2012.

Hughes B, Paterson K. The social model of disability and the disappearing body: towards a sociology of impairment. Disabil Soc. 1997;12(3):325–40.

Makin T. The social model of disability. Counselling. 1995;6(4):274.

Mead GH. Mind, self and society. Chicago: University of Chicago Press; 1934.

Radley A. Making sense of illness. London: Sage; 1994.

Rosenstock IM. Why people use health services. Milbank Mem Fund Q. 1966;44(3):94–127.

Wetherell M. Positioning and interpretative repertoires: conversation analysis and poststructuralism in dialogue. Discourse Soc. 1998;9(3):387–413.

Yudell M, Roberts D, DeSalle R, Tishkoff S. Taking race out of human genetics. Science. 2016;351(6273):564–5.

Critical Theory: Epistemological Content and Method

8

Anastasia Marinopoulou

Contents

Abstract

Critical theory situates science within the quest for social and political rationality. It indicates that science's normativity – which answers the question "what should science do?" – orients itself in relation to the a priori potential of society. The latter for critical theory transforms itself into concrete political vindications for science. Adorno's *Gesamtgesehen*, which differentiates from any total and, therefore, totalitarian conception of what science is, along with Horkheimer's dialectical approach to science through interdisciplinarity and Habermas' notion of communicative rationality (that emphasizes scientific dialogue) in science, finds themselves in marked contrast to the rest of modern epistemology. The chapter traces the epistemology of critical theory of the Frankfurt School through the twentieth and the twenty-first centuries via the concepts of dialectics, critique, reason, interdisciplinarity, communicative action and rationality, and their social

Some parts of the chapter are based on previous elaborations in my *Critical Theory and Epistemology: the Politics of Modern Thought and Science* (Manchester University Press, 2017). I owe particular thanks to Professor Darrow Schecter and Professor Piet Strydom for all the inspiring discussions, comments, and critique which encouraged me to reconsider many things.

A. Marinopoulou (✉)
Department of European Studies, Hellenic Open University, Patra, Greece
e-mail: anastasia.marinopoulou@gmail.com

© Springer Nature Singapore Pte Ltd. 2019
P. Liamputtong (ed.), *Handbook of Research Methods in Health Social Sciences*,
https://doi.org/10.1007/978-981-10-5251-4_58

and political function and role within modernity. The main aims of twenty-first-century epistemology of critical theory become as follows: formulate a theory of normative rationality, reclaim commitments to rational praxis, and educate the sciences to maintain dialectics as their pivotal scope and method of advance.

Keywords
Critical theory · Dialectics · Political epistemology · Normativity

1   Introduction

Critical theory situates science within the quest for social and political rationality. It indicated that science's normativity – which answers the question "what should science do?" – orients itself in relation to the a priori potential of society. The latter for critical theory transforms itself into concrete political vindications for science. Adorno's *Gesamtgesehen*, which differentiates from any total and, therefore, totalitarian conception of what science is, along with Horkheimer's dialectical approach to science through interdisciplinarity and Habermas' notion of communicative rationality (that emphasizes scientific dialogue) in science, finds themselves in marked contrast to the rest of modern epistemology (Adorno 1970; Adorno et al. 1976; Horkheimer 1988a, 1995; Habermas 1988, 2006) (For a critical account and comparison see my *Critical Theory and Epistemology: The Politics of Modern Thought and Science*).

This chapter traces the epistemology of critical theory of the Frankfurt School through the twentieth and the twenty-first centuries via the concepts of dialectics, critique, reason, interdisciplinarity, communicative action and rationality, and their social and political function and role within modernity (Habermas 1987). My aim is to answer the question "how is science evaluated?" or "who is science accountable to?". Fundamentally, I trace which questions critical theory should ask in order to measure the accountability of science and which answers it attempts to provide.

For critical theory of the twentieth century, the concept of dialectics appears significant both for society and science and marks both the first generation of Horkheimer, Adorno, and Marcuse with their critique of science, its social and political task, and the potential for interdisciplinarity and the second generation of Habermas with science's potential for public dialogue through communicative action (Marinopoulou 2008). Moreover, the chapter's arguments focus on the idea of dialectics as both a process *and* a method. Therefore, both following subchapters elaborate on the dialectical process and dialectical method, respectively. By questioning what *form of process dialectics* is, epistemology realizes that dialectical arguments are formed without losing sight of what occurs in social and political reality. For the latter reason, epistemology cannot avoid having a political character; science is socially produced and carries social and political implications. Dialectics is also *a method* because it derives from the exchange of argumentation between scientific subjects. Since dialectics has social consequences, it needs *accountability criteria* in order to be socially acceptable. Dialectics is accountable to society *because* it brings with it certain political consequences. Along the way, it also renders epistemology a scientific field that discusses the political character of scientific development.

The main aims of twenty-first-century epistemology of critical theory become as follows: formulate a theory of normative rationality, reclaim commitments to rational praxis, and educate the sciences to maintain dialectics as their pivotal scope and method of advance. Such epistemological aims would probably also advance epistemology toward realizing its political potential to influence society. As in the social and political sphere, where consensus of all participants appears important but not a condition sine qua non, the same is also valid for the sciences. It is not consensus that necessarily marks a creative scientific process; rather, it is dissensus that sciences have to promote through dialogue.

In the twenty-first century, critical theory has to prioritize dialogue that would reclaim as a process the lost honor of science, namely, to form knowledge that is accountable to society. Therefore, through dialogue the task of critical theory is to form a political epistemology. Dialogue within and about scientific arguments serves to contradict and expose mythical and dogmatic thinking. It also has the capacity to purge orthodox or absolute convictions of any arbitrary meaning. In such a manner, we come closer to understanding what constitutes the scientific sphere, philosophy, and truth and whether modern epistemology paves the way for a political epistemology in the twenty-first century. Science is neither dogma, religion, nor politics. Its basic function should be to provide a forum for open and uncoerced dialogue wherever a point of dispute arises. Science is neither a fixed understanding nor a vague assumption, but a scientific moment occurs when a question or a negation is formulated and asserted.

Critical theory's epistemological arguments are marshaled in a vehement critique of positivism (Adorno et al. 1976; Schecter 2010, 2013; Stockman 1983), which marks its claims as a reaction against rational normativity or as the new empiricist epistemology safeguarding scientific orthodoxy. By rejecting all subjective understanding according to consciousness and by prioritizing the empirical data (but what sort of empirical data? according to which criteria? the ones of time, space, historicity, or of the human senses?) as objects of knowledge, positivism questions the significance of dialectics for cognitive processes and resorts to causal explanations for its epistemological method and aims (see also ▶ Chaps. 6, "Ontology and Epistemology," and ▶ 9, "Positivism and Realism").

In an intriguing interview (http://www.youtube.com/watch?v=KyP6li6AnE0), Horkheimer states that philosophy can no longer be considered progressive because it is in the service of science, thus, distorting the task of philosophy, as well as that of science. The constant problematic, which I examine as either implicit or explicit in epistemology, is whether scientists are falling silent over what takes place in science or in the political or social sphere. It appears that philosophy has to incorporate the concerns of an epistemology orientated toward the political, the political significance of scientific research and arguments, and focus on an epistemology with a political perspective, namely, on a political epistemology. As was often the case in human history, particularly with the Newtonian scientific discoveries that pave if not open the way for the French Enlightenment, science serves as the bearer of multiple innovative changes first in the domain of research and then in society and politics.

Thus, far from any elitist approach to what science or philosophy is, philosophy has to function as the accountability bearer and evaluator of science in its relationship with society. It is neither degrading nor reductionist to consider philosophy so philosophy *among its other tasks* can and should promote political epistemology. The latter consideration entails a social and political elevation for philosophy within modernity that attributes to philosophy the normative task of social accountability regarding what science is and ought to be.

Later in the twentieth century, Horkheimer's argument found a similar elaboration in one of Marcuse's statements on science (https://www.youtube.com/watch?v= C5PU0EASi_Q). Marcuse argues that there is certainly a part of philosophy, science, and technology, which is neutral (Marcuse 1968). Nevertheless, due to its social position, science takes a firm ideological stand. Science and philosophy, by questioning what is true and valid, do not merely fill the void left by dogmatism and mythology which stretch into the realms of society and politics by creating prejudice and suppression; they set dialectics and dialogue itself at the center of their process. The negation in dialectics has to be followed by something else, too, and that can be the enlightenment of the negation, not necessarily in the form of a synthesis with the thesis but by arguments on the potential, the "other," and the alternative, both in science and society.

The ongoing innovations of modernity in science and society lay the foundation not only for an innovative science but also – and not consequently – for the promise of social progress and development. In order to understand what is modern, we probably have to affiliate it with what is scientific and to question what is dialectical. If I were to draw a hypothetical line between what is scientific and modern and separate them from the prescientific and premodern, that line would be dialectics itself.

2 The Epistemological Content of Critical Theory

The following analysis of the three major thinkers of the Frankfurt School aims to approach the basic *questions* that the first generation of the Frankfurt School expressed and attempted to answer, particularly in relation to epistemology and the social and political potential of the sciences. The first generation of critical theorists includes Max Horkheimer, Theodor W. Adorno, and Herbert Marcuse and deals extensively with the concept of dialectics in science and the potential for an interdisciplinary approach to what the scientific and the political are. The same idea of a dialectical science would be advanced by Jürgen Habermas, the leading figure of the second generation of critical theorists, with his concept of communicative action which relates to a great extent to the concepts of dialectics and the interdisciplinarity of the first generation.

For Horkheimer, Adorno, and Marcuse, there exists a recurrent epistemological concern, namely, that science contributes to the freedom of humans from social prejudices and political irrationality during modernity as well as to the formation of rationality for society. Nevertheless, science steadily transmutes rationality into irrational methodology, method, purpose, and a ruthless domination of humans over nature that culminates in the domination of man over man. Thus, the scientific crisis is, for modernity, an unavoidable dead end.

Although notably involved with articulating a certain thesis on the scientific subject and object, Horkheimer, Adorno and Marcuse deal more with eliminating the idea of a singular scientific subject and focus on dealing with three basic queries:

1. What is science?
2. What social and political meaning does science bear?
3. How is science accomplished?

Critical theory places epistemological questions at the center of its research concerns, particularly relating them to their impact on politics and society. However, when science seeks truth, critical theory does not view it as a *panacea*, nor is science said to have an exclusive claim on progress and innovation. Science is another way for society to pursue truth, rationality, and progress, but it should never serve as a deus ex machina in defense of social instrumentality, in order to establish any scientific or political authority by means of scientific works and words. Science, for critical theorists, is rarely neutral or value-free.

For critical theory, even when the objectivistic illusion prevails that everything can be assessed according to measurable facts, the knowing subject always mediates between facts and knowledge. For the epistemology of critical theory, the emphasis is placed on the conscious agent who enlightens facts by means of knowledge. Moreover, "... the empiricist 'fetishism of facts' ignored that facts were, after all, products of collectively developed modes of perception; that we only know 'mediated' facts; and that (even unconscious) theories and methods are the mediators" (Arato and Gebhardt 1998, p. 376). In addition, the knowing subject is not a meritorious individual; it is, rather, the collective subject of a scientific field, or of a whole society searching for truth by means of scientific dialogue that takes place within society, and which is also influenced by it.

From another point of view, for critical theory from the early twentieth century until the early twenty-first, science is not the docile offspring of any political ideology, nor is it the generating factor of ideological constructions or the theologian of a society eager to impose scientific authority. Authoritarianism develops in different forms. For Adorno, Horkheimer, and Marcuse, science avoids reproducing authoritarianism by declining the messianic role of social redemption (For Horkheimer the state of knowledge of a society defines what each society considers or recognizes as ideological and, moreover, converts it into a supposedly commonsense knowledge, which, because of its ideological character, is dogmatic and unwavering). It never loses sight of the concern that part of science's substantive content is to be or to become political, where the rudimentary elements of concepts, methods, and methodology are not merely the coincidental products of historical periods.

By denying the objectification of knowledge and science in a form of political ideology that establishes itself as social mythology, the critical theorists formulate an epistemological argument on the human potential for criticism and reason. They suggest that the knowing subject bearing the inherent capacity to apply dialectics in the scientific field innovates in science by extricating its social and political irresponsibility or immaturity. It is a moment of sheer Kantianism for the Frankfurt School because it maintains the a priori potential and the aim of overcoming immaturity, as well as a bold statement of surmounting the Kantian *Entwurf* for science by means of the dynamics of dialectics toward rational praxis. In Arato's words:

> Passive (non-interfering) contemplation belongs to a 'naive' stage where humans confront the world as something 'other': they have not recognized their share in its shaping, nor that the terms in which they relate to it are of their own making, nor that they are dealing with a conceptually or materially appropriated (inner or outer) nature whose terms they can change. When the idea of reason was conceived, it was to do more than regulate the relation between means and ends; it was intended to determine the ends. (Arato and Gebhardt 1998, p. 392)

For critical theory – whether of the first or second generation, and even today – a commitment to think, particularly for knowing and conscious subjects, and to formulate normative theory may potentially result in a commitment to act and generate rational praxis. This aim is bequeathed to the second generation by the first, and it remains Habermas' priority without ever ignoring commitments to normativity. The second generation, with its leading figure Jürgen Habermas, formulates the idea of a universal pragmatics based on the argument that dialogue is transformed into forms of communicative action within the scientific and renders reaching understanding and consensus among scientific participants a feasible potential.

In his epistemological concerns (if not throughout his work as a whole) Habermas opposes positivism (Habermas 1971, 1974). He attributes to the sciences themselves the responsibility for self-reflection on their transcendental capacities. Moreover, doing justice to the substance of historical processes, Habermas attempts to innovate not by rejecting the meaning of transcendental consciousness for knowledge but by combining it with the "objective life context."

One of Habermas' most notable arguments concerning the sciences is his view that the scientization of politics and social life in general is the other side of the technological control of the sciences over society. The latter takes place by means of ideologizing science, which is rendered autonomous and, thus, delegitimized by the public sphere of society. There is the question of how the colonization of the social becomes feasible by the scientific, rendering the social mute and influenced by ideologies, while the scientific is operational.

Knowledge, for Habermas, is distinct neither from life nor social process. It is not a quantifying procedure, nor is it a simplifying practice in gathering and integrating universal assumptions. Knowledge and science are not the instruments of technical control over an object of the social or over the social as an object. They bear the potential of intersubjective dialogue, which attempts to bridge the gaps between scientific fields, social forces, and research dynamics that endeavor to form social and scientific rationality. Moreover, it becomes evident both for the first and the second generation of critical theorists that normativity is not an external objective of science. Beyond what positivism suggests, science has a normative content that defines what science is, how it performs within the social sphere, and how it is accomplished. Therefore, the critical theory of all the generations attempt a threefold answer to the questions initially posed: first, that science *is* normativity and forms normative statements on the social realm; second, that there is no neutral science, the meaning of science is its social function and critique; and third, that the means to accomplish social accountability and rationality is to exercise dialectics (its negation toward the instrumental and the irrational for the first generation of critical theorists) and promote communicative action deriving from dialectics (for Habermas) so as to reach understanding.

In more detail, Habermas emphasizes that the normativity of science presupposes an ideal speech situation among participants of either individual or collective character so that potential consensus is reached and rationality is realized through the dialectics of communicative action. The rationality of science that Habermas prioritizes in his thought is the sought-after outcome of realizing theory through practice and of contradicting the irrationality of technological systems of research by means of reason, critique, and interdisciplinarity through the communicative action of the sciences.

The following schema depicts the convergence of the first and the second generation of critical theorists on their four basic epistemological concerns:

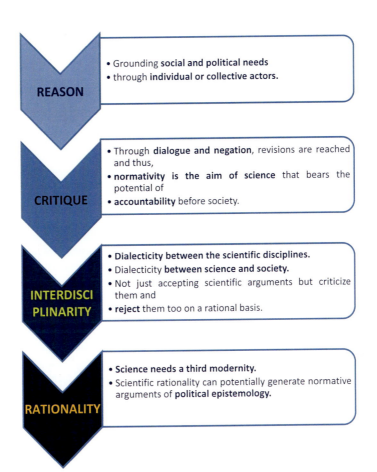

REASON
- Grounding **social and political needs**
- through **individual or collective actors.**

CRITIQUE
- Through **dialogue and negation**, revisions are reached and thus,
- **normativity is the aim of science** that bears the potential of
- **accountability** before society.

INTERDISCIPLINARITY
- **Dialecticity between the scientific disciplines.**
- Dialecticity **between science and society.**
- Not just accepting scientific arguments but criticize them and
- **reject** them too on a rational basis.

RATIONALITY
- **Science needs a third modernity.**
- Scientific rationality can potentially generate normative arguments of **political epistemology.**

The following sections will elaborate on the arguments of the schema and expand on the notion of political epistemology as the normative task of critical theory for the twenty-first century.

3   The Method of Critical Theory

The Frankfurt School's method of exploration (which refers to the first generation) is primarily based on the observation that historical and descriptive accounts might be very important methods of research. However, they still manage to miss the essence of critical theory's arguments, which entail the examination of reason and dialectics with regard to science and epistemology. The critical theory of the Frankfurt School challenges instrumentality whereby human beings also become mere instruments along the lines set out by modern science. The methodology of positivism, which the Frankfurt School attacks, focuses on sciences (in the plural) as systems of knowledge that are separate among themselves and in relation to society, thus constructing a false sense of autonomy for scientific work. Within such systems, each science or scientist aims at articulating an inner set of instructions, according to which deductive statements are produced and through which theory is dissociated from practical rules and praxis itself for the sake of producing immediate scientific results. Thus, the problematic concept of "application" arises, leaving the scientists unprepared to criticize the scientific outcome in any alternative way and unable to decide between the realm of "pure" theory and "pure" praxis as if there could ever be any theory without praxis and vice versa.

In order to deal with what constitutes science, epistemologically speaking, critical theory tackles the problem of scientific laws. The answer remains straightforward: while natural sciences facilitate the formation of scientific laws, it is rather unfeasible to expect the same degree of certainty in the humanities and social sciences. The sciences and, moreover, epistemology cannot merely be a description, nor can they be a rule, method, or methodology. They have to exert critique and influence on something else. Both are consistent with a definite method, which does not presuppose the socially autonomous formation of scientific theory and praxis but derives from society itself. If science is disconnected from what takes place within the social realm, then it is also silenced, normatively muted, and rendered socially indifferent or even redundant. For critical theory, science for the sake of science did not seem a very plausible way of scientific advancement. In fact, the reverse was the case: science becomes valid for both the scientific and the social realm when it stands critically in relation to what takes place socially and convinces people of its eternal relevance.

Science, for the first generation of the Frankfurt School, encompasses foremost the potential for critical reason, formulated because uncoerced dialogue among the sciences has taken place, thus signifying the dynamics of interdisciplinary dialectics. It is crucial for the Frankfurt School to note that dogmatism, positivism, and deterministic laws of understanding the scientific and, successively, the social create a scientific deficit whereby the sciences are unable to react to social and political crises. Science is by no means an ideological instrument. Furthermore, science is not an instrument. However, unless science deals with and directs its object of research toward the social and political questions set by society itself, and unless it reconsiders which questions it seeks to address, it *shall* become an instrument in the hands of political ideology.

Dialectics, for the Frankfurt School, is not merely a method of research; it is the subject matter of science transforming itself into interdisciplinarity and consummating dialogue among scientific disciplines. It generates dialogue with social science and posits society as the subject, namely, the agent or the acting subject, of science itself. If I could cut a long scientific story short, I would say that critical theory attempts to answer the question "Wozu noch Kritik?" ("Why more critique?") for science and gives a rather straightforward answer: "Because it is generated by dialectics." The critique that critical theory articulates and promotes in multiple arguments was (a) dialectical, meaning deriving from dialogical processes and, (b) normative, for it represented a brand-new potential for science. It is normative because it criticizes both society and science in order to reach rationality and consummate it within the social and the scientific. Thus, such a normative critique requires both epistemological and political criteria so as to answer the question "why more scientific critique?". And the answer remains timely and valid for the twenty-first century too because it appears a valid path to normativity through dialectics and to social accountability as well.

In Horkheimer's essay 'Traditional and critical theory' (Horkheimer 1972), the claim for interdisciplinarity is boldly stated and signified the scientific accountability toward truth, normativity, and rationality. Disclosing truth, for critical theory of the first generation, is a social accomplishment, achieved intersubjectively through dialectics. Science cannot afford to abandon either the social field, for the sake of some vague notion of scientific autonomy, or the dialectical process for the sake of dogmatism. In such terms, it is a clear normative turn that is to be promoted by Habermas and his furthering of the concept of dialectics into a communicative theory and action for science.

Jürgen Habermas, the philosopher and social theorist who innovated critical theory with his notion of communicative action, raises questions as to the methodology the sciences follow, the sort of logical inquiry they adopt, and the criteria they base their scientific objectives on. Habermas perceives the mode of scientific research and analysis as producing not contradiction but the interinfluence of research and analysis. However, precisely because of the mode they opt for, sciences facilitate and promote or discourage dialogue and argumentative exchange. In not being socially isolated by means of dialectics, they become major forces that shape social dialogue and rationality. Where the opposite is the case – and that is the abandonment of dialectics – they predominate by supporting and maintaining the scientization of politics.

In the same way that sciences become the authoritarian force in society by attempting to offer operational approaches to social and political problems, they also become authoritarian for their own field of research. Thus, Habermas associates science with politics by providing a concise guide to instrumental scientific methodology and the social impact of the scientific sphere as well. The scientific method, then, along with the concept of truth is established as criteria for knowledge "... only from the objective life-context in which the process of inquiry fulfils specifiable functions: the settlement of opinions, the elimination of uncertainties, and the acquisition of unproblematic beliefs ..." (Stockman 1983, p. 67).

In the preface to *Knowledge and Human Interests*, Habermas, in the same manner as Adorno's 'Subject and object', maintains that "... a radical critique of knowledge is possible only as social theory..." (Habermas 1971, p. vii). Moreover, it is essential for the clarification of the subject matter of the chapter, as well as for Habermas' work itself, to note the latter's explanation of what science is and how sciences can be considered as a unified whole composed of different fields maintaining a certain method of research.

For Habermas, science is neither the grand narrative of philosophy nor the eccentric attitude toward the "... actual business of research" (Habermas 1971, p. 4); moreover, it is neither the retreat to a philosophy of science nor the methodology of the sciences. Science is rational thought, deriving from critical consciousness, and directed toward setting dialogue into use, questioning fields of research and exploring normative concepts and social rationality. Epistemology, on the other hand, being, to an extent, the self-reflection of science, is also governed by dialectical thought and questions the social function of the scientific and the socially rational.

Being based upon the previous counterarguments of Horkheimer and Adorno to the assumed validity of commonsense knowledge, Habermas opposes the dogmatism of commonsense knowledge, which appears as the outcome of ideology, and reconstructs the "self-formative process of consciousness," as he states in *Knowledge and Human Interests*. The critique of knowledge – an aim shared by the Frankfurt School of all generations – arises when transcendental consciousness meets the socially perceived demands of science, which are then channeled into a dialectical, self-formative process.

Habermas departs from the latter notion of a dialectical self-formative process of science and introduces the notion of communicative action. He relates and designates not only social relations of uncoerced dialogue but also public communication of scientific subjects against scientific as well as social ideologies. Science is a process and aim, created by scientific dialectics and self-reflection, not the corroboration and absolute result of a socially autonomous prejudgment where science loses its social legitimacy. Science is not the empirically or methodically obligatory end; rather, it concerns itself with the dialectical means to accomplishing knowledge.

The problem and criticism of deduction reappears in critical theory with Habermas' work on epistemological interests and methodological critique. Deduction is associated with purposive-rational action, and because of its claim to methodological certainty, anticipated rules, and decisions, it entails scientific dogmatism and, thus, prejudice. In a wider critique of the symbolic processes of inference, namely, induction, deduction, and abduction, Habermas correlates the threefold schema with the instrumental approaches of pragmatism to describe the learning process by way of cumulative and quantifying criteria.

Dialectics and intersubjectivity acquire a concrete form, which combines empirical research with the "intelligible character of a community that constitutes the world from transcendental perspectives ... in a self-formative process until the point in time at which a definitive and complete knowledge of reality is attained" (Habermas 1971, p. 135). Where purposive-rational action represents scientific monism and instrumental aims toward technical control of the sciences and society, intersubjectivity represents a potential but not necessary consensus, and, most importantly, it represents the idea of a scientific dialogue that science facilitates and elaborates socially. The idea of communicative action in science becomes the first significant step toward the awareness of ignorance and then of self-formation for knowledge and social claims.

Habermas never ignores the concept of interdisciplinarity and intersubjectivity of the first generation of critical theory and aims at expanding upon its modern understanding by relating it through communicative action, either in the sciences or in society. It is one of the very rare occasions in modern epistemology that the social role of science was manifested in such a direct affiliation with its power over modern politics. Though certainly not for the first time, with Habermas' work one sees how epistemology took a political turn, avoiding the aporetic considerations of the first generation of the Frankfurt School.

Habermas maintains the political position that knowledge acquires through the elaboration of social claims and the legitimacy attributed to it within the lifeworld and the scientific public sphere. Communicative action among the sciences generated within the lifeworld and the dialogical processes in the public sphere of the sciences commits the scientific realm to redefine rational praxis through normative theory.

However, the notable accomplishment of Habermas is to *innovate the innovations of critical theory* of the first generation. He manages to extend dialectics' scientific and political dynamics into communicative action and remain loyal to how critique and interdisciplinarity can transform into a communicative rationality for science that remains accountable to society. In the following section, I analyze the scientific as well as the social and political dimensions of Habermas' *innovation of the innovative* and examine how critical theory's formulations remain a concrete normative argument of political epistemology that gives shape to the valid task of critical theory for the twenty-first century's modernity.

4   Critical Theory's Task: The Normative Turn

Apart from presenting an account of the Frankfurt School and articulating a critique on what was critical for science in the critical theory of the twentieth century, the chapter aims to present *the epistemological task of critical theory for the twenty-first century*. It remains valid and persistently timely throughout the first and the second generation and appears to continue to be the same for the epigones and any third or fourth generation of critical theorists: to criticize the social and political function of science and at the same time remain critical for science itself too.

For the first generation, to articulate a series of critical arguments on science and society is the result of dialectical procedures which are followed by participants in social and scientific dialogue. Critique is the indispensable presupposition of participation in the sphere of science and prevents any lapse into prejudice, myth, or superstition either in science or society. Critique appears more indispensable than any new scientific paradigm in order for science to progress according to critical theorists of the first generation.

The relentless critique on positivism denotes not only a rejection of the traditional scientific methods which appears insufficient to analyze modernity but the formulation of a new argument that could transcend the potential of the scientific paradigm to result in scientific revolution as well. Such a new argument on the part of critical theorists consists in the prioritization of dialectical critique as not only the scientific method but as the scientific conception that safeguards science as modern and not traditional, scientific and not a configuration of myths, as well as being accountable to society. The accountability of science toward society generates much greater and more consistent scientific innovations than the change of paradigms in scientific revolutions. It is the normative task of science to remain critical for both itself and society, and such normativity would be challenged, revisited, and reconsidered in the twentieth century many times by a number of philosophers and sociologists that perceived modernity, dialectical critique, and normativity as inimical if not opposite to the social and political function of science (For a detailed analysis of this argument see Chaps. 1, 2, 3, and 4 of my *Critical Theory and Epistemology: The Politics of Modern Thought and Science*).

Habermas moves in the same direction as the first generation of Horkheimer and Adorno and nevertheless accomplishes to innovate the innovative epistemological legacy of both previous thinkers. He theorizes critique as the task and method of a normative science and sustains Horkheimer's and Adorno's consideration that critique is constituted by the immanent negation of the traditional or, in other words, by the potential to refuse and dialectically reject what is inimical to modernity.

Such a negation would serve in critical theory throughout the twentieth century as the defender of scientific accountability to society. When, for Habermas, critique maintains and generates its validity, the latter can be accomplished without ever losing sight of science's accountability in the social sphere. The dialectical critique deriving from science leads Horkheimer to realize that scientific interdisciplinarity is essential for the profundity of such a scheme. It is only through the dialectical critique between the scientific disciplines that science can manage to innovate on their object as well as their input within the social.

Thus, Habermas realizes that not only a dialectical critique but also practice that keeps the dialectical part but forms action within society is the innovative attribute of critical theory that retains the core of the first generation's argument on interdisciplinarity but is amplified by a conjunction of the critical and its transition into practice. In this manner, communicative action in science is born, and in order to ground his innovation on a firm epistemological basis, Habermas elaborates further on the two main axes of communicative action: first, the potential force of the better argument and, second, the dynamics for an ideal speech situation which is in its turn formed by the force of the better argument.

Communicative action for Habermas is the bearer of the force of the better argument because his idea of communicative action is the development of the notion of dialectics and interdisciplinarity formulated by the critical theorists of the first generation. Dialectics generates arguments, and it also "promises" a cooperative exploration of truth along with scientific and social rationality. Furthermore, the potential for configuring and realizing the better argument is preconditioned by major aspects such as (1) the willingness and ability of participants of a dialogical critique to be accountable to the public, (2) their equivalent intentionality to include the public in a wider critique and to admit the participation of social factors within scientific dialogue, (3) to acknowledge the equal participation of all involved without undergoing outer or inner coercion of any kind and, finally, (4) the acceptance of the potentiality to reach a consensus that would include all involved parts.

Therefore, the force of the better argument is in Habermasian thought accountable to society, inclusive of negations and divergent arguments, free of any coercion or compulsion, and socially and scientifically rewarding for its ability to generate a consensus of the highest inclusive standard. Hence, Habermas realizes a threefold epistemological achievement:

(a) He innovates critical theory with the transition from interdisciplinary dialectics to the configuration of what theoretical discourse is.
(b) He grounds it on social practice.
(c) He extends it into the presupposition for an ideal speech situation that is promoted by science and applied to society and the conception of the political.

The consummation of an ideal speech situation requires the same vital aspects as the force of the better argument does: the search for truth, uncoerced expression of arguments, and the pursuit of consensus. All the previous orient toward a normative rationality that serves as the pivotal part of science and its social function in order to affect decisions and balance the asymmetries of rational critique and its contingent authoritarian applications.

Habermas' arguments on the innovation of critical theory in relation to epistemology are not a mere methodological series of propositions bound by a rationalized consistency. It is the continuation of the epistemological pursuit of a critical science that is constantly maintaining a critique by means of dialogue and which challenges social and scientific rationality by means of its normative task, namely, to maintain the political character of science to the extent that the latter criticizes and influences society. It is the pursuit for the *better* argument and the *ideal* speech situation that renders critical theory's normative task a modern task for the twenty-first century. It is the task to defend modernity through a normative science, and Habermas defends such a task by turning critical theory toward *what the normative and the rational* is in science.

Habermas' contribution to critical theory is his sheer normative turn. The quest of rationality necessitates that critical theory's epistemology present a new argument on what political epistemology is. Normative questions asked within an ideal speech situation guided by practical discourse and the force of the better argument have to be scientifically promoted and established in order to reach society. The task of a critical science is to establish normative questions in order for society to follow such a normativity. Science acts as the creator of normative challenges and attempts to answer established questions along and in cooperation with society. To an equal extent, society engenders its rationality and influences science, politics, and every other social realm but remains also in constant dialogue with the normative questions and attempted answers that science produces. The public dialogue between science and society on what the normative and the rational is remains both in Horkheimer, Adorno, and Habermas the pillar of modernity and perhaps the only hope of modernity for constant innovation in what the social, the political, and the scientific consist in.

In order to challenge the positivists but also ground modernity on a critical basis, critical theorists oscillate from the argument of a normative science to the normativity of modernity and articulate a whole series of critical arguments not on epistemological concerns but on *political epistemology* that is critical theory's contribution to the epistemology of modernity (Marinopoulou 2017).

The political epistemology of critical theory marks a lucid normative turn in modern epistemology. It is the quest for a true or false statement that renders scientific matters valid or invalid and moreover applicable or not. The example of economic theories provides epistemologists of modernity with a clear instance of critique: economic theories can be true or false not just because data and numbers are verified or not but because societies render them so, meaning applicable with evident validation and consequent validity within societies or inapplicable because societies refuse to follow numbers, indexes, or numerical indicators that can be totally verified hypothetically but fail in normative practice.

The political epistemology that critical theory promotes serves as a critique of science *and* society in a binding way that renders such a critique both dialectical and normative as well. It attributes to science the characteristic of a normative "invention" that is created when people ask normative questions. The colonization of the scientific by the social or even the political was a feature of the "old days", namely, of premodern societies. In the old days of epistemology, the prevailing anxiety among scientists was that the political will colonize the scientific sphere and take advantage of it or even worse render it its own instrument. Now, in the modern world, epistemological temptations take the form of absolute divergence between the scientific and the social. The values change: sciences appear inappropriate or even insufficient to answer the normative questions that society poses and are at a loss to produce their own, either questions or answers. Sciences produce their own "logic" which is distant if not irrelevant to social needs and remains unaccountable both to the scientific and the social spheres. In order to advance normative rationality within society, science has to consider its accountability task toward the social and in such a sense to realize its political task, namely, to be in the service of humanity's needs for dialectics, critique, and rationality within the social and the political.

For a critical theory of modernity, science needs a third modern wave of innovation. After the first one in ancient Greece and the second during the French Enlightenment, it is perhaps political epistemology that is compound of normative questions and arguments that will provide modern societies with a scientific as well as social rationality as a firm basis in order to reclaim the modern promise for a rational world.

5   Conclusions and Future Directions

For the Frankfurt School, science is a process of cognition, which contains integrity in its aim and method, as well as in its methodology. Science aspires to maintain strong bonds with social reality, but it also claims to articulate social criticism and put this into practice. On a second level of understanding, science is the process of dealing with contradictions and elaborating on them by means of dialectics.

First-generation critical theory deals with the distinction between subject and object, regarding the subject either as a collective or as an individual actor of science and society. Nevertheless, for Habermas, the object is replaced by another subject, namely, the "co-participant" in dialogue between scientific fields within the process of communicative action. Habermas attempts to find a resolution to the intricacies of the acquisition of knowledge by means of communicative action, the latter being exercised by the scientific subject that produces scientific thought through dialectics, which aims at realizing emancipatory interests. In marked contrast to the latter, when dialectical communication is muted, scientific and social elites are never far away, and knowledge for the degraded "masses" also results from the silencing of dialogue.

The main argument of the chapter concerns the undiminished significance of dialectics not only as a method but also as a process and mode of understanding for the sciences. It is an inherent characteristic of the sciences to generate and deal with contradictions, and, hence, any epistemological understanding of the scientific includes the scientific dialogue on theses and antitheses, as equally important for the evolution of any scientific argument.

For Marcuse, dialectics is the scientific stance toward what remains contentious for the sciences; or, in other words, no scientific neutrality can be accomplished as long as dialectics itself does not claim neutrality. Science is either dialectical or nonexistent. The moment in which critical theory embodies dialectics is, for Marcuse, the moment of negation of the given thesis and principles. In many references, in his work and interviews, Marcuse recognizes that the scientific constructions of modernity in the twentieth century are deprived of such embodiment of a dialectical advance and as such are subject to a scientific deficit that is concealed by means of technological progression.

Although, in many critical theorists' work, the scientific thesis represents the first advance of scientific development, namely, that there has to be an articulate and coherent position regarding scientific issues, what in essence constructs a moment of scientific reflection is the acceptance of critique and probable negation, which is encompassed either in the negation or the synthesis of the two. Critical theory then brings criticism to bear on the sciences, not from the position of the grand inquisitor, but rather from the point of view that allows for science to take an expressive position and simultaneously include the negation of a thesis and the potential synthesis, serving a self-reflective attitude toward science and its objectives.

Dialectics is a form of scientific entirety, including a certain position, its negation, and its alternative. However, the negation is not a reactionary moment within the scientific stance. It is the articulation of a critical argument that opposes a given thesis. For Adorno (http://www.youtube.com/watch?v=aIQpJNxGa90), critical theory is not a scientific field itself; rather, it constitutes the moment of reflexivity for the sciences. It can indicate that science is not a field of totality and homogeneity; on the contrary, it is a site of disagreement, negation, and opposition to the given and taken for granted.

In my understanding, science constitutes simultaneously the formation of questions and the potentiality for answers. The culture of scientific discourse that Habermas introduces and analyzes throughout his work extends from the notion of a critical theory (as originally understood by Horkheimer) to a critical science. In order to formulate a rational scientific answer, science has first to designate a concrete scientific question and a dialectical problematic.

During the late modernity of the twentieth and twenty-first centuries, critical science maintains that its aim is dialectical, in terms of producing social criticism, whereas epistemology acquires a political character. A rational reconstruction of the sciences will not become the scientific concern of modernity because philosophy occupies the position of the scientific consciousness for modern science. Rather, a rational progression of the sciences will take place when epistemology becomes political and, therefore, socially influential. The main abovementioned points raised claims that reinforce the potential accountability of science in its relationship with society. Through the formation of a normative theory, the task of political epistemology is to initiate and maintain the promise and potential of conscious subjects toward rational praxis.

References

Adorno TW. Zur Metakritik der Erkenntnistheorie, Drei Studien zu Hegel. Gesammelte Schriften, vol. 5. Frankfurt am Main: Suhrkamp; 1970.

Adorno TW, Albert H, Dahrendorf R, Habermas J, Pilot H, Popper KR. The positivist dispute in German sociology. London: Heinemann; 1976.

Arato A, Gebhardt E, editors. The essential Frankfurt school reader. New York: Continuum; 1998.

Habermas J. Knowledge and human interests. Boston: Beacon Press; 1971.

Habermas J. Theory and practice. London: Heinemann; 1974.

Habermas J. The philosophical discourse of modernity. Cambridge, MA: Polity Press; 1987.

Habermas J. On the logic of the social sciences. Cambridge, MA: Polity Press; 1988.

Habermas J. Political communication in media society: does democracy still enjoy an epistemic dimension? The impact of normative theory on empirical research. Commun Theory. 2006;16:411–26.

Horkheimer M. Critical theory, selected essays. New York: Herder and Herder; 1972.

Horkheimer M. Gesammelte Schriften, Band 3, Schriften 1931–1936. Frankfurt am Main: Fischer Taschenbuch Verlag; 1988a.

Horkheimer M. Gesammelte Schriften, Band 4, Schriften 1936–1941. Frankfurt am Main: Fischer; 1988b.

Horkheimer M. Between philosophy and social sciences, selected early writings. Cambridge, MA: The MIT Press; 1995.

Marcuse H. Reason and revolution. London: Routledge; 1968.

Marinopoulou A. The concept of the political in Max Horkheimer and Jürgen Habermas. Athens: Nissos Academic Publishing; 2008.

Marinopoulou A. Critical theory and epistemology: the politics of modern thought and science. Manchester: Manchester University Press; 2017.

Schecter D. The critique of instrumental reason from Weber to Habermas. New York: Continuum; 2010.

Schecter D. Critical theory in the twenty-first century. New York/London: Bloomsbury; 2013.

Stockman N. Antipositivist theories of the sciences. Dordrecht: D. Reidel Publishing Company; 1983.

Positivism and Realism

9

Priya Khanna

Contents

Abstract

Theory and practice of research in health social sciences involves a unique synergy of a range of quantitative, qualitative, and hybrid methodologies derived from parent disciplines of medicine, nursing, and various other branches of social sciences such as sociology and psychology. While the methodological diversity enhances the scope of research and implications of research findings, it also renders the necessity for the investigator to explicitly address the implicit theoretical stances and philosophical assumptions underpinning the evidentiary claims. Still inherent among the investigators in health social sciences is to present their evidentiary claims in binary terms of whether an intervention/

P. Khanna (✉)
Sydney Medical Program, University of Sydney, Camperdown, NSW, Australia

© Springer Nature Singapore Pte Ltd. 2019
P. Liamputtong (ed.), *Handbook of Research Methods in Health Social Sciences*,
https://doi.org/10.1007/978-981-10-5251-4_59

initiative worked or not, as opposed to why it worked and for whom. This tendency to gauge the strength of evidence in terms of objectivity and replicability seems to be emerging from the deep rooted desires for control and prediction of phenomena under investigation as opposed to meaning-making. While taking the readers on a brief journey through the emergence of history and philosophy of western science, this chapter aims to provide a deeper understanding of two major philosophical foundations of research methodologies: positivism, a theoretical stance underpinning rigor and objectivity in science and scientific method, and realism, an ontological perspective examining the truth of mind-independent reality. It is suggested that a closer inspection of emergence of scientific inquiry and its underpinnings will facilitate a better understanding of research designs and outcomes, especially for contemporary complex environments in which various initiatives in health social sciences operate.

Keywords

Positivism · Realism · Ontology · Epistemology · Research · Health · Social sciences

1   Introduction

Look round

For evidence enough. 'Tis found,

No doubt: As is your sort of mind,

So is your sort of search: you'll find

What you desire.

(Robert Browning, *Easter Day*)

No matter how the term, *"research"* is defined, regardless of discipline and of its typologies, it is ultimately a quest to deepen our understanding of the world through expanding our knowledge of it. When one looks at the more specific definitions of research, the words that often appear include: *"systematic," "investigation,"* and *"evidence."* To understand how best a systematic investigation can be undertaken to yield evidence that can be translated or adapted in other similar contexts for desired purposes, one needs to understand what really *is* meant by these terms. The first two terms are linked to the theory of action or methods of research (methodology) impacting the evidentiary claims, which in turn, are influenced by theory of how we know the object of investigation (epistemology). Closely linked to both, methodology and epistemology is the understanding of the nature of the *"reality"* or essence of the object itself (ontology). This triad of theoretical underpinnings of research can be best understood against the backdrop of historic emergence of the pursuit of knowledge and inquiry itself (see also ▶ Chap. 6, "Ontology and Epistemology").

Until Middle Ages, the world view in most of the world civilizations, especially in the Western part of the world was organic, where people lived in small, cohesive communities, and where theology guided the belief systems. The quest of knowledge, in this organic world, was focused on meaning and significance using both reason and faith. With the Scientific Revolution, the organic and spiritual worldview was gradually replaced with a "world-as-a-machine" view, and the focus of the quest shifted from meaning and significance to control and prediction (Capra 1982).

The worth of the quest of knowledge began to be gauged by the level of certainty, validity, and objectivity of scientific claims. As one travels through the journey of philosophy of science, this modernist era marks the phase of positivism – a theoretical commitment where the nature of inquiry was characterized by methodological objectivity of verifying only that is verifiable, rendering metaphysical speculations to be meaningless (Hjørland 2005). As Wight (2002) describes, positivist era laid the foundations of social science as a discipline where the principles of scientific claims governed the legitimacy of social inquiry as well.

The limitations of positivist worldview became apparent with the extraordinarily intellectual feats in modern physics such as quantum theory, as well as new wave of profound conceptions in philosophy of science which illuminated falsifiability, fallibility, and theory-laden nature of scientific knowledge. Post-positivist era also led to the emergence of alternative theoretical and methodological commitments such as interpretivism especially in social science inquiry including healthcare education which led to the popularity of qualitative and mixed-methodological research paradigms (Grant and Giddings 2002).

While we do take pride in claiming that social inquiry no longer needs to emulate its cousins in natural sciences, and the yardstick against which we assess methodological rigor is no longer limited to positivist underpinnings, the legacy of positivism still lingers on. As Crotty (1998) highlights, qualitative research, if carried out to establish evidence as facts and truths, is rooted within positivist underpinnings. For instance, an ethnographic investigation of a healthcare intervention will be usually represented as qualitative study (see ▶ Chap. 26, "Ethnographic Method"); however, if the findings are concluded in binary terms of whether "it worked or not" rather than "for whom and why," it is still underpinned under positivist framework. One can perhaps argue for the need for conclusiveness of evidence for its translation into practice, especially in biomedical sciences. The unsurpassed authority of randomized controlled trials (RCTs) as gold standard in evidence-based medicine reflects this urge for equating evidence with "Truth" (see also ▶ Chap. 37, "Randomized Controlled Trials"). This pursuit of truth or reality is an ontological issue, and as we shall see later in the chapter, understanding of reality is separate from understanding of methods of knowing the reality.

While positivism is a theoretical perspective that suggests certain epistemic commitments, the ontological position closely associated but not limited to positivism is realism. In this chapter, I shall explore major types and tenets of realism, mainly scientific realism, anti-realism, and contemporary realist perspectives such as critical realism which is compatible with constructionist epistemological frameworks.

Traversing through the history and philosophy of science, this chapter aims to provide a deeper understanding of emergence of positivist perspectives and realist ontological commitments that laid the foundations of methodological rigor in both natural as well as social sciences research. Such an understanding of historic emergence will also aid in clarifying closely related but distinct concepts such as empiricism versus rationalism versus empirical research, and empiricism versus realism.

2   Positivism

2.1   Meaning

The "positive" in positivism does not imply "good" or synonyms associated with it, but it means something that is "posited" or postulated. This is in opposition to "natural," as in natural law or natural religion, where the content emerges from the nature. On the contrary, positive religion or positive law is not arrived at by speculation, but enacted and adopted by proper authority. In context of science, and later in social sciences, what is "posited" or given from direct experiences is what is observed by scientific method as opposed to metaphysical speculations (Crotty 1998). The essence of positivism, in its various conceptualizations, is that there is a basic scientific method which is same across both natural and social sciences. Positivist school of thought believes in the "thesis of the unity of science," which means that only way by which social sciences can match the achievements of natural science in explanation, prediction, and control of phenomena being observed is by applying the methods of natural science (Lee 1991).

Positivism is not a univocal concept. Its meaning has evolved with time and can be conceptualized in three ways: positivism as a philosophical and political movement originating from French Philosopher Auguste Comte's (1798–1857) commitment to social evolution; logical positivism as well-defined philosophy of science; and methodological positivism which refers to epistemological beliefs underpinning scientific research methods and practices (Riley 2007). While positivism embodies certain epistemic commitments, it is by itself not an epistemology unless the definition of epistemology is stretched to include theoretical stances associated with theory of knowledge (Wight 2002; see ► Chap. 6, "Ontology and Epistemology"). The genesis and dominance of positivism as a philosophical and political movement, until second half of nineteenth century, is attributed to the father of modern sociology, Auguste Comte, also regarded as the founder of positivism. The origin of positivist notions, however, can be traced much earlier in the long history of early Enlightenment thinkers in the Western philosophy and in the emergence of modern science and its philosophy.

2.2   Emergence of Western Science and Positivist Notions

Capra (1982, 1992) and Russell (2013) provide a very enriching account of history and philosophy of Western science, and emergence of modern science, and scientific method. The Western science and the philosophy, into which it was originally subsumed along with theology, can be traced back to the beginning of sixth century with the Greek philosopher Thales, a native of ancient city, Miletus. The speculations of Thales, along with other two philosophers of Milesian Triad, Anaximander, and Anaximenes, can be regarded as the earliest scientific hypotheses. The late sixth century saw the influence of mathematics upon philosophy and theology with the works of Pythagoras who laid the foundations of geometry. The combination of scientific rationality, theology, and philosophy was carried on by other prominent philosophers, the most revered of which was Socrates, who was primarily concerned with critical examination of abstract concepts through dialogues. Following Socratic legacy, Plato and Aristotle continued systematic metaphysical speculations. Aristotle's influence, which was great in many different fields, was the greatest in logic and retained his authority in it throughout the Middle Ages. The first person often regarded to defend the autonomy of science as a distinct body of knowledge was St Thomas Aquinas who combined Aristotle's comprehensive system of nature with Christian theology using reason and faith. Hence, the purpose of medieval science, unlike modern science, was to understand the meaning and significance of things using both reason and faith, rather than prediction and control of phenomenon.

With the gradual decline in the dominance of the Church, and with the birth of the Renaissance, the medieval science changed radically in sixteenth and seventeenth centuries with the Scientific Revolution that began with Copernicus followed by Kepler. The end of sixteenth century saw the rise of the greatest of the founders of modern science, Galileo (1564–1642), who was the first to combine scientific experimentation using mathematical language to formulate the laws of the nature. Around the same time as Galileo was devising ingenious experiments in Italy, Francis Bacon (1561–1626) set forth empirical method of science in England primarily based on induction – combining together observed facts to form generalizations. The Baconian spirit of scientific investigation began to replace the organic view of nature with the metaphor of world-as-a-machine, and this shift in the worldview was completed by two towering figures of the seventeenth century, Descartes and Newton.

Rene Descartes (1596–1650) is usually regarded as the founder of modern philosophy, but he was also a brilliant mathematician. Descartes based his views of nature on fundamental division into two independent realms of mind (res cogitans) and matter (res extensa). The Cartesian dualism of mind as separate from matter formed the basis of the belief that world could be described objectively through analytical and rational reasoning. On one hand, the Cartesian division formed the foundations of classical physics, but on the other hand, it led to the mechanical and fragmented world view of nature as a perfect machine governed by exact mathematical laws. Limitations of the mechanistic world view, however, were not recognized until twenty-first century as we shall see later in this chapter.

The man who realized the Cartesian dream and completed the Scientific Revolution with the grand synthesis of the works of Copernicus, Kepler, Bacon, and Galileo was Isaac Newton (1642–1727). Synthesizing the Baconian empirical and inductive method with Descartes' deductive reasoning, Newton laid the foundations of classical physics in which the material particles moved in absolute space and time governed by the force of gravity. Newtonian-mechanistic model of classical physics dominated the Western scientific thought from the second half of seventeenth to the end of nineteenth century. While the Newtonian atomistic physics provided the basis of understanding the laws governing the universe, Locke developed an atomistic view of society by attempting to reduce observable societal patterns to the behavior of its individuals. He asserted that humans are born "tabula rasa" – a clean slate on which knowledge acquired through senses can be imprinted.

Within positivist notions, one can see two distinct epistemological positons – classic empiricism versus rationalism. Classical empiricism of Locke and Hume, for instance, regards observations and sensory experiences as the only method to gain knowledge; whereas rationalism, in its purest form, argues for preestablished conditions, categories, or inborn structures for obtaining knowledge. Rationalists prefer deductive rather than inductive methods for obtaining evidence. As we will see later in the chapter, the classic empiricism was combined with rationalism by a school of positivism, logical positivism (Hjørland 2005).

2.3   Social Sciences and Comtein Classical Positivism

The legacy of Descrates, Newton, and Locke was crucial to the developments in social sciences which paralleled scientific progress in natural sciences. Early Enlightenment thinkers in social sciences, such as Rousseau for instance, were concerned about the progress of humanity by conceptualizing a state-level society based on equality and agreement between the governed and those who govern. The man, however, responsible for laying out plan for mechanistic social sciences and conceived the discipline of "sociology" and coined the term itself was Auguste Comte. While his mentor, Henri de Saint-Simon may have been the originator of the positivist school of social sciences, it was Comte who elaborated the thought in series of major books. In his six-volume work, *The Course of Positive Philosophy*, Comte proposed his famous law of three stages through which the human mind, individual human beings, all knowledge, and world history is developed. The first stage is the theological stage dominated by search for the essential nature of things as explained by the existence of gods and supernatural forces. From the primitive stage of polytheism, religion evolves towards monotheism as the ultimate belief of the theological stage. This is followed by the metaphysical stage in which explanations are given in terms of abstract entities as opposed to supernatural agents. The positivist stage is the last and highest stage which involves reliance on empirical data, reason, and development of scientific laws to explain the phenomena. The law reflects Comte's idea of close association between intellectual progress and social evolution.

The second pillar of Comtein positivism is the classification of sciences into a hierarchy with mathematics and astronomy as its base, followed by physics, chemistry, biology, and culminating in sociology – the science of society. Comte's view of hierarchy of the six fundamental sciences provides justice to the diversity of the sciences without thereby losing sight of their unity (Bourdeau 2008; Bernard 2011; Ritzer and Stepnisky 2017).

Often, Comte's reductionist view of reducing all knowledge into mathematics is equated to his search for objectivism via mathematical precision. This may not be an accurate interpretation, as Crotty (1998) points out that Comte acknowledged the interdependence of human consciousness and "the social," as well as the role of theory in undertaking empirical research.

2.4 Logical Positivism

Inspired by the late nineteenth- and early twentieth-century advancements in mathematics, physics, biology, and social sciences including psychology, which were all pursued independently from philosophy, a philosophical movement called logical positivism, more suitably known as logical empiricism, originated in Austria and Germany in the 1920s (Hjørland 2005; Bernard 2011). The logical positivists, also known as the Vienna Circle, composed of mathematicians, philosophers, and physicists and primarily sought to use the rigor of mathematics for the study of philosophy, which at that time was losing its hold on natural as well as social sciences (Richard 2017).

Logical empiricists attempted to combine the classical British empiricism and continental rationalism with the method of logical analysis aiming to distinguish between analytical and synthetic propositions. They proposed that all meaningful statements are either empirical, i.e., observable and verifiable by experience (synthetic *a posteriori* propositions), or they are logical such as mathematical statements which are analytical a *priori* propositions. The statements that fall into neither of the two categories are meaningless. Science and philosophy should attempt to answer only scientifically answerable questions. In other words, no statement is meaningful unless it is capable of being verified, and this was the basis of "verification principle" – central tenet of logical empiricism, inspired by the works of Wittgenstein, an influential twentieth-century philosopher who was primarily concerned with demarcating valid and invalid use of language. The purpose of philosophy, according to logical empiricists, was to define and clarify the meaning of statements, and to distinguish statements that can be verified and from those "meaningless statements" such as metaphysical views, ethical values aesthetics and religious beliefs, which although had emotional associations but lacked cognitive meaning (Kincaid 1998; Hjørland 2005).

2.5   Post-Positivism

The influence by Logical positivism began to wane out by mid-twentieth century primarily by increasing sense of dissatisfaction with the positivist model of scientific knowledge which was perceived as unrealistic. With the revolutionary developments in physics, biology, psychology, and as well as philosophical perspectives in science, it was recognized that science is indeed a social enterprise undertaken by fallible humans (Klee 1999).

Two fundamental theories of modern physics, theory of relativity and quantum theory, transformed the image of the universe as a machine to a dynamic and holistic whole with interrelated parts. The extraordinary intellectual feat of Albert Einstein laid the foundations of collapse in the faith for dualistic thinking. The investigations of atom and subatomic particles further provided insights into matters being an abstract entity with dualist characteristic of particle as well as wave, depending upon how they are being observed, hence questioning the Cartesian division of observer and observed.

> As far as the laws of mathematics refer to reality, they are not certain; and as far as they are certain, they do not refer to reality. (Einstein (1921) in the essay 'Geometry and Experience' In: Reichenbach and Cohen (1978, p. 33))

Developments in quantum theory and the new physics, led by eminent physicists including Planck, Einstein, Bohr, de Broglie, Schrodinger, Pauli, Heisenberg and Dirac, led to significant and fundamental changes in the concepts of space, time, matter, object, and causality. The scientific community during the second half of twentieth century began to realize that the world and its entities does not exist as independent and isolated building blocks but as a unified whole with complicated web of relations between various entities, including the observer and his consciousness (Capra 1982). The twentieth-century theoretical physics is now working grand unification of four major forces of nature by combining theory of relativity and quantum physics (Dardo 2004), and this endeavor towards achieving the "final unity" parallels eastern philosophical assertions of unity in terms of soul consciousness as the very basis of existence.

Modern physics' view of seeing universe as a unified whole are echoed in social sciences by Max Weber who asserted that social realities need to be understood from the perspectives of the observed than the observer, and in totality than in isolation, thereby laying the foundations for modern interpretivism in sociology (Fox 2008).

Apart from interpretivist movements in social sciences, developments in feminism, post structuralism, critical and constructivist psychology further added to the dissatisfaction with positivist perspectives (see also ▶ Chap. 7, "Social Constructionism"). Cochran (2002) aptly articulates that positivism's main problem is not its attachment to scientific method as such but rather its commitment to what John Dewey, the leading proponent of pragmatism, called the "quest for certainty." Dewey's approach to social science focused on embracing experience is it is lived, rather than generating universal abstractions about a "real world," which is deliberately removed from everyday practice.

Parallel to the developments in natural and social sciences, "new" doctrine emerged among twentieth century philosophy of science, which although comprise of divergent views but acknowledge science as a social enterprise subject to fallibility. Prominent postmodern philosophers of science include Popper, Kuhn, Feyerabend, Lakatos, and Tolumin. This chapter briefly summaries Pooper and Kuhn's perspectives on post-positivist views of science and scientific knowledge.

Karl Popper (1902–1994) was one of the most influential philosophers of science in the post-positivist era. Although he had early associations with the Vienna Circle, his view of how scientific knowledge is developed is remarkably different from that of logical positivists. Popper substituted the idea of verifiability with the notion of falsifiability. In Popper's hypothetico-deductive scientific method, scientific theories are proposed as hypothesis which are inferred from observations but are conjectures or guesses; propositions are then deduced from these conjectures, which are then tested using observations or experimentations. Those hypotheses shown to be false must be revised or replaced. Hence, it is falsifiability of a theory and not verifiability that renders necessary condition for it to be scientific (Achinstein 2004; d'Espagnat 2006).

Crotty (1998) further explains that even if the theory has survived every failed attempt for its refutation, the theory is still "corroborated" and cannot be inferred as true as it takes only one example to falsify the theory. Popper's idea of tentativeness of scientific knowledge is in complete contrast with the objectivist epistemology of positivism.

The leading contemporary thinker, who provided a very thorough understanding of what science is, and how scientific knowledge is developed over a period of time, was Thomas Kuhn (1992–1996), an American physicist, historian, and philosopher of science. His most popular work, *The Structure of Scientific Revolution* (SSR), had an impact far and wide beyond the boundaries of philosophy of science. In SSR, Kuhn provided a model of development of scientific knowledge within a given field of inquiry in which the field evolves through a series of stages in a cyclic manner. The field evolves through a prescientific immaturity to a peaceful stable "normal science," where scientific progress is made through "puzzle-solving" activities in which most scientists are normally engaged. The package of theoretical assumptions, beliefs, rules, and concepts in which normal science works is collectively termed as "paradigm" by Kuhn. Scientific activity is carried out in line with the parameters and boundaries set by the dominant paradigm in the field of inquiry. The next stage in the scientific development comes when, in the normal science, the theoretical or experimental results do not fit the reigning paradigm (called anomalies) and the fundamental principles and assumptions of the field are called into question. This leads to scientific revolution which involves "paradigm shift" involving radical new ways of thinking and solving unsolved puzzles and eliminating anomalies.

Acknowledging the criticism of "paradigm" being insufficiently precise, Kuhn in the second edition of the book in 1970 added a postscript where he replaces the "paradigm" with the expression "disciplinary matrix" where "disciplinary" refers to the common possession of the practitioners of a particular discipline, and "matrix" is the ordered elements of various sorts, each requiring further specification (Klee 1999; Achinstein 2004). Sharrock (2002, pp. 13–14) summarizes Kuhn's antagonistic views with his logical empiricist predecessors and with Popperian logic, some which as:

There is no sharp distinction between observation and theory.

Science is not cumulative, and does not have a tight deductive structure... methodological unity of science is false as there are lots of disconnected tools used for various kinds of inquiry. Science is in time and is historical.

2.6   Legacy of Positivism and Post-Positivism in Health Social Sciences Research

In several ways, the tenets of logical positivism still have a major influence in methodological underpinnings, not only in experimental and biomedical sciences but in social science research as well. Hjørland (2005) summarizes two major influences of local positivism in contemporary research methodologies in social and healthcare sciences. Firstly, to ascertain "veracity" of research findings, a hierarchy of scientific methods has been established, under the influence of evidence-based movement, with randomized controlled trials (RCTs) as the highest form of evidence and qualitative evidence at the very bottom. This notion of assigning more weight to the evidence generated in a strict scientific method limits its sense-making and informative capacity. The dominance of RCTs as the highest form of evidence still perpetuates in health social sciences including nursing and medical education (see ▶ Chap. 37, "Randomized Controlled Trials"). The multiple or mixed methodology approach is yet to establish its credibility especially in research areas where RCTs are still dominant and multiple qualitative methods are only used as secondary sources of evidence (see ▶ Chaps. 39, "Integrated Methods in Research," and ▶ 40, "The Use of Mixed Methods in Research").

Secondly, the legacy of logical positivism still continues in much of the quantitative research which uses statistical methods to test hypotheses and to draw correlational inferences between the variables using purely mechanical and logical process. Correlation does not imply causation –- but refusal to draw causation is rooted within logical positivist's rejection of metaphysics. The emphasis on "observations" (empiricism) and "verifiability" (all observers agree on what is being verified, i.e., interrater subjectivity needs to be reduced) creates an overemphasis on reproducibility of data as opposed to its usefulness/adaptability.

Positivism, as described earlier in this chapter, is a philosophical stance that embodies objectivist commitments to its epistemology. However, understanding of epistemology is incomplete without indulging ourselves in the ontological discourse, which in the case of positivism is realism.

3 Realism

The two crucial questions that philosophers, since time immemorial, have engaged themselves in are: what is the nature of existence or reality (i.e., what "is"), and how can we know what "there is"? The latter is an epistemological question, and the former deals with ontology (see ▶ Chap. 6, "Ontology and Epistemology"). The distinction between the two is reflected in the works of two eminent physicists of twentieth century – Bohr and Heisenberg. While Heisenberg's uncertainty principle is an epistemological argument in terms of science's inability to determine location and momentum of subatomic particles simultaneously with accuracy, Bohr questions the nature of particles itself, which is an ontological issue (Crotty 1998).

Roy Bhaksar, a British philosopher, whose work on critical realism will be discussed later in this chapter, cautions us against a metaphysical dogma, which he terms as *"epistemic fallacy"*; i.e., one cannot reduce the statement about the world (ontology) into knowledge of the world (epistemology) (Bhaskar 1975). In several ways, positivism is underpinned by epistemic fallacy as ontological questions in positivism seem to be described and understood in terms of its epistemological beliefs. This creates confusion for the position of positivism itself in philosophical realm. Some scholars have regarded positivism as a form of scientific realism (Cacioppo et al. 2004); others have designated it as a "theoretical perspective" (Crotty 1998), while some other scholars have called it as a paradigm (Guba and Lincoln 1994). Epistemic fallacy also gives rise to misconceptions about the position of realism in relation to positivism.

Realism, in its simplest form, is an ontological position, which assumes that reality exists independently of our perceptions of it. Reality, in realism, refers to all that is "out there," i.e., entities, objects, forces, structures, and so on. A healthcare researcher, using a realist lens, would view the concepts of "disease" and "diseaseness" as "things in the world" independent of his or her perceptions. Realism can be contested against idealism, specifically, Kantian *transcendental idealism* where objects of human experience and their attributes do not possess any existence of themselves, outside of our mind (d'Espagnat 2006). There are several traditions and variants of notions of mind-independent reality. While some realists perceive reality to be completely independent of our perceptions such as in naïve realism, whereas others, such as critical realists, attempted to integrate realist ontology with a constructivist epistemology. This chapter covers two basic tenets of two major realist thoughts – scientific realism and critical realism.

3.1 Scientific Realism

Scientific realism is a belief about the truth in reference to scientific theories in describing the observable and unobservable entities, which primarily are regarded as "mind-independent." Scientific realism, then, is the theory that the objects of scientific enquiry exist and act, for the most part, quite independently of scientists and their activity (Bhaskar 1975).

Scientific realism can be viewed in terms of three dimensions or commitments. In ontological terms, scientific realism is committed to the mind-independent existence of the reality. The reality, which scientific theories describe, is largely independent of our thoughts or theoretical commitments. This position of metaphysical realism is contested by antirealists, primarily idealists for whom there is no reality external and independent of the mind. In epistemic terms, scientific realism is a positive attitude about the truth or approximate truth of scientific theories and theoretical claims about the observable and unobservable entities in the world. The third commitments is the sematic dimension of scientific realism, wherein the theoretical and empirical claims about the objects, events, processes and relations, and other entities, whether observable or unobservable, are to be construed as having truth values.

Human sensory capabilities render the distinction between observable (such as trees, moon, stars), and unobservable (such as distant galaxies, electrons, forces). Scientific realists typically assume that theories as described by "mature" sciences such as physics are usually true, or approximately true reference to the reality they purport to describe. If space, time, forces, atomic particles exist and have certain properties, they exist independently of our perceptions of them, and their existence is best described by scientific theories. This view is often contested by antirealists, some of which adopt an epistemically positive attitude only with respect to the observable, as we shall see later. Valid experimental arguments in favor of scientific realism were provided by Jean Perrin, French physicist who studied Brownian motion of particles suspended in liquids. Perrin's arguments can be best simplified as

Molecules exist

Molecules are unobservable entities

Therefore, unobservable entities exist. (Achinstein 2004, p. 328)

Defenders of scientific realism such as Richard Boyd and Stathis Psillos have defended a more demanding view of scientific realism. It is also claimed that scientific realism is about the aim of science wherein science provides us theories what the world is like, and acceptance of the theory involves the belief that it is true (Achinstein 2004).

A strong defense in favor of scientific realism comes from Putnam's claim that realism is the only philosophy that does not make the success of science a miracle. Putnam, initially a realist but later affiliated himself more with the transcendental idealism, in his *"no-miracle argument,"* proclaimed that the success of scientific theories in facilitating empirical predictions and explanations with remarkable accuracy can only be explained by realist explanation that the best scientific theories are true or approximately true description of the mind-independent reality. This belief in the realism is justified as a case of inference to best explanation (Boyd 1983; Bird 2006; Chakravartty 2011).

3.2 Scientific Antirealism

Any philosophical stance that opposes realism along its metaphysical, epistemic, or sematic commitments is antirealism. Antirealism comes in various forms, but the two major positions that are discussed here are instrumentalism and empirical constructivism; the other forms include historicism, social constructivism, pragmatism, and postmodern frameworks such as feminist approaches.

Instrumentalism, a form of empiricism, regards theories as instruments for making predictions. Logical empiricists were advocates of this view. As we saw in the earlier sections of this chapter, logical empiricist advocated for rationalizing the meaningfulness of the statements and felt that terms for observable entities have no meaning all by themselves, and they acquire meaning by being associated with the observables. For instrumentalists, the appropriate question is asked is not about the truth of a theory but if it is empirically adequate, i.e., if it makes accurate predictions about observable consequences (Bird 2006).

Another form of antirealism is constructive empiricism, popularized by Van Fraassen (1980), who defines it as:

> Science aims to give us theories which are empirically adequate; and acceptance of a theory involves as belief only that it is empirically adequate. (The Scientific Image 1980, p. 12)

The central tenet of *constructive empiricism* is the notion of empirical adequacy which implies that a theory is empirically adequate if its observable consequences are true. While this position is compatible with metaphysical view of realism in regards to its interpretation of a theory as true or false, its opposition for realist is epistemic as it believes in the truth of the scientific theories in terms of observables and is satisfied to claim unobservable as true or false without the need of belief in their existence (Bird 2006; Chakravartty 2011).

Constructive empiricism sometimes is also regarded as a form of instrumentalism owing to its position towards observables. One extreme side of antirealism is the view that reality is entirely dependent on our mind (transcendental idealism). A contemporary version of this form of idealism is social constructivism, which in scientific antirealist terms would imply that it is the scientists/researchers who decide what entities exist and in that process they "construct" the entities (see ▶ Chap. 7, "Social Constructionism").

In the realm of health social sciences, understandings of the theses of antirealisms may provide insights into assessing the "unobservable" competencies. In medical education, for instance, competency-based movement has created impetus on assessing not only clinical or academic competencies but nonacademic competencies (such as empathy, team-skills, interpersonal skills, and so on). While clinical skills can be directly observed and inferred (by using tools such as Direct Observation of Procedural Skills (DOPS) and Objective Structured Clinical Examination (OSCEs)), it is the assessment of nonclinical skills which is problematic. The principle of "observability," which in empirical sense demarcates theoretical and nontheoretical statements can provide some insights into how better can we understand the problems in programmatic assessment which links together various assessment of various clinical and nonclinical competencies.

3.3   Critical Realism

The most prominent form of scientific realism in the social sciences is critical realism, usually associated with the works of British philosopher Roy Bhaskar (1944–2014). In his seminal work, *A Realist Theory of Science* (1975), Bhaskar's version of critical realism (which he initially labelled as transcendental realism) seems to originate from his reflections on two issues in Western philosophy of science. Firstly, he questioned the dominant *ontological monovalence* in terms of reality being viewed as unstructured, undifferentiated, and unchanging. Secondly, he put forward the notion of *epistemic fallacy* – statements about the world (reality) cannot be reduced into statements about knowledge of the world (epistemology), because realism is a theory of *being*, and not a theory of knowledge. Bhaskar's theory, therefore, is polyvalent in terms of dimensions of knowledge, as well as reality (Bhaskar 1975).

Science, according to Bhaskar, is a product of man as a social and historic being. While science is made by man, with locus of existence being human mind, it is directed upon objects that exist independently of man. There are, therefore, two dimensions of science: social dimension wherein science is a product of human society and an objective dimension which comprises objects being studied by science. Two kinds of objects correspond to these two dimensions of science: independent or objective objects or things (physical as well as social processes) form the intransitive dimension of science, whereas objects such as theories, paradigm, models, and methods form the transitive dimension of science.

To establish the objectivity of intransitive dimension, Bhaskar viewed his version of realism as an alternative to two competing or rival philosophies – classical empiricism (knowledge a product of experience) and transcendental idealism (knowledge a product of human mind). A common element in both the rival philosophies is that the existence of reality is dependent of what can be experienced, and Bhaskar referred to this commitment as empirical realism. In this way, empirical realism rejects the intransitive dimension of science: the world of which science seeks knowledge is not independent of man; on the contrary, it can only be described in relation to man and his cognitive activity (Evangelopoulos 2013).

Bhaskar's third position, transcendental realism, regards the objects of knowledge as the structures and mechanisms that generate phenomena and the knowledge as produced in the social activity of science. These objects are neither phenomena (empiricism) nor human constructs imposed upon the phenomena (idealism), but real structures which endure and operate independently of our knowledge, our experience, and the conditions which allow us access to them. According to this view, both knowledge and the world are structured, both are differentiated and changing; the latter exists independently of the former (though not of our knowledge of this fact), and experiences and the things and causal laws to which it affords us access are normally out of phase with one another (Bhaskar 1975, p. 15). Bhaskar's view of ontology is primarily to seek the answer to the question, "what must be the world like for science (cognitive activities- sense perceptions and experimentations) to be possible?" The "world," according to Bhaskar, consists of mechanisms and not events. Mechanisms combine to generate the flux of phenomenon that constitutes the actual happenings of the world. Causal structures and generative mechanisms are intransitive, i.e., are *real* and distinct from the pattern of *actual* events they generate, which in turn are distinct from the experiences in which they are apprehended. Hence, reality, as per Bhaskar, is stratified into the realms of the real, the actual, and the empirical. Science, therefore, is a social activity directed towards understanding the generative mechanisms that underpin the observable events.

3.4 Implications of Critical Realism in Health Social Sciences Research

Critical realists retain an ontological realism in the sense of existence of a real world independent of our perceptions, and at the same time they accept epistemological constructivism and relativism, i.e., the understanding of the reality is a function of our perspectives. The relativist epistemology of critical realists legitimizes multiple accounts and interpretations of reality, and this has profound implications for health social science research methodologies including qualitative and mixed-methods research (Maxwell 2012).

Bhaskar's version of critical realism, in particular, provides not only a holistic and emergent view of ontology, its concept of generative mechanisms to explain causal relationships has led to contemporary approaches in understanding causation in the research and program evaluations especially in complex and complicated systems. One such contemporary approach in program evaluation is realist evaluation which is gaining increasing popularity in the areas of health social sciences (see also ▶ Chap. 20, "Evaluation Research in Public Health").

Realist evaluation is one of the theory-based approaches to program evaluation in complex and complicated systems such as healthcare systems. It was originally proposed by Pawson and Tilley (1997) who extended the Bhaskar's and other critical realists thoughts of causal mechanisms from ontological realm to an epistemic realm of human reasoning. Mechanisms, in realist evaluation, are combination of stakeholders' reasoning as well as resources offered by the social program or intervention to be evaluated. Mechanisms are activated when the program or initiative is placed in the context and they are responsible to generate outcomes. Mechanisms can be viewed as the underlying entities, processes, or structures which operate in particular contexts to generate outcomes of interest (Dalkin et al. 2015). Prime concern of realist evaluators is to understand the Context-Mechanism-Outcome configuration to address the question, "what works, for whom, under what circumstances, and how" (Wong et al. 2016, p. 1). Realist approach in research and evaluation is gaining increasing popularity in health social sciences, not only in program evaluation but also in undertaking systematic literature reviews (see ▶ Chaps. 45, "Meta-synthesis of Qualitative Research," and ▶ 46, "Conducting a Systematic Review: A Practical Guide").

4   Conclusion and Future Directions

What really is meant by "research"? What is meant by "evidence"? Why is the robustness of research claims still gauged in terms of "objectivity" and "replicability" of evidentiary claims? Why do our research questions are still directed towards answering "does it work", instead of "how, why and for whom."

The purpose of this chapter was not just to provide a detailed account of meaning and implications of positivism and realism as major philosophical positons, but the prime aim was to facilitate a deeper understanding of the above-stated questions which are fundamental to research in general, and health social sciences, in particular.

The chapter started with a glimpse into the history and philosophy of science, and how the organic and sense-making worldview based on meaning, faith, and reason was gradually but radically transformed into mechanistic worldview aiming for control and prediction, which further strengthened as the science progressed. This led to the birth of positivism – a theoretical position advocating for rigor and objectivity in observations, either through verification by sense-experiences or by pure logic.

While much of the quantitative research, especially in biomedical sciences, is underpinned by positivist notions in terms of emphasis on rigor, objectivity of hypotheses testing, and focus on reproducibility of the findings, qualitative research can also be underpinned by positivism if the focus is on reproducibility in terms of "what worked" instead of why and for whom. The strength of research in several areas of health social sciences, it seems, is still gauged by the generalization of evidence, where the evidence is often equated with "truth." Generalizability of research findings, however, is increasingly being recognized as a serious issue not only in qualitative but in quantitative research as well. In medicine, for instance, it has been acknowledged that a number of ostensibly robust findings of clinical trials are not replicable, and the lack of replicability of social science research including in psychology is much prevalent (Norman 2017).

While there seems to be a consensus that lack of reproducibility may be linked to methodological factors such as statistical power, degrees of freedom, and so on, there is growing body of evidence that reproducibility might be a result of *contextual* differences or "hidden moderators," as concluded by an analysis and recoding of 100 original studies in Psychology (Van Bavel et al. 2016).

The context-dependency of research is especially pertinent for complex systems such as healthcare systems, which are dynamic and emergent, with multiple and simultaneous causal strands, and where speculative contribution analysis rather than correlational attribution analysis provides better understanding to design and evaluate interventions (Moreau and Eady 2015).

These assertions seem to be highlighting the importance of meaningfulness, relevance, and sense-making pursuit of research (what works and form whom) rather than control and prediction. This also points to the relevance of critical realism in current social sciences and healthcare research. Through its acceptance of legitimate interpretations of reality, critical realism acknowledges the reality itself to be stratified and emergent rather than monovalent and static. The confluence of realist ontology with constructionist epistemology facilitates a better understanding of evidentiary claims, especially for complex and dynamic environments. This is exemplified by the growing popularity of multiple or mixed-methods research paradigm in healthcare research.

To summarize, in the postmodern era, we often seem to consider positivism and realism as "unfashionable" and "orthodox." Researchers often take pride in labelling their qualitative research as "interpretivist." The great war between qualitative and quantitative research paradigms seems to be far from over. However, a closer inspection at the very purpose of research and its emergence as a systematic quest for knowledge provides insights on how various ontological perspectives and epistemic commitments emerged with the progress in natural and social sciences. The great war, it seems, is not between qualitative and quantitative paradigm but between the replication and meaning-making, and in very ontic distinctions of *being* and *becoming*.

References

Achinstein P. Science rules: a historical introduction to scientific methods. Baltimore: John Hopkins University Press; 2004.

Bernard HR. Research methods in anthropology: qualitative and quantitative approaches. Maryland: Rowman & Littlefield; 2011.

Bhaskar R. A realist theory of science. London: Routledge; 1975.

Bird A. Philosophy of science. London: Routledge; 2006.

Bourdeau M. Auguste comte. In: Zalta NE, editors. The stanford encyclopedia of philosophy (Winter 2015 Edition). 2008. Available at: https://plato.stanford.edu/archives/win2015/entries/comte/.

Boyd RN. On the current status of the issue of scientific realism. In: Hempel CG, Putnam H, Essler WK, editors. Methodology, epistemology, and philosophy of science. Dordrech: Springer; 1983. p. 45–90.

Cacioppo JT, Semin GR, Berntson GG. Realism, instrumentalism, and scientific symbiosis: psychological theory as a search for truth and the discovery of solutions. Am Psychol. 2004;59(4):214–23.

Capra F. The turning point. London: Flamingo; 1982.

Capra F. The Tao of physics: an exploration of the parallels between modern physics and eastern mysticism. London: Flamingo; 1992.

Chakravartty A. Scientific realism. In: Zalta NE, editor. The stanford encyclopedia of philosophy (Summer 2017 Edition). 2011. Available at: https://plato.stanford.edu/archives/sum2017/entries/scientific-realism/.

Cochran M. Deweyan pragmatism and post-positivist social science in IR. Millennium. 2002;31(3):525–48.

Crotty M. The foundations of social science research. Crow Nest: Allen & Unwin; 1998.

Dalkin SM, Greenhalgh J, Jones D, Cunningham B, Lhussier M. What's in a mechanism? Development of a key concept in realist evaluation. Implement Sci. 2015;10(49):1–7.

Dardo M. Nobel laureates and twentieth-century physics. Cambridge: Cambridge University Press; 2004.

d'Espagnat B. On physics and philosophy, vol. 417. Princeton: Princeton University Press; 2006.

Evangelopoulos G. Scientific realism in the philosophy of science and international relations. Unpublished PhD thesis. London: The London School of Economics and Political Science (LSE); 2013.

Fox NJ. Post-positivism. In: Given LM, editor. The Sage encyclopaedia of qualitative research methods. London: SAGE; 2008. p. 661–664.

Grant BM, Giddings LS. Making sense of methodologies: a paradigm framework for the novice researcher. Contemp Nurse. 2002;13(1):10–28.

Guba EG, Lincoln YS. Competing paradigms in qualitative research. In: Denzin N, Lincoln Y, editors. Handbook of qualitative research. Thousand Oaks: SAGE; 1994. p. 105–17.

Hjørland B. Empiricism, rationalism and positivism in library and information science. J Doc. 2005;61(1):130–55.

Kincaid H. Positivism in the social sciences. In: Edward JC, editor. Routledge encyclopedia of philosophy (Version 1.0). London: Routledge; 1998. p. 558–561.

Klee R. The Kuhnian model of science. In: Scientific inquiry: readings in the philosophy of science. Oxford: Oxford University Press; 1999. p. 199–201.

Lee AS. Integrating positivist and interpretive approaches to organizational research. Organ Sci. 1991;2(4):342–65.

Maxwell J. What is realism, and why should qualitative researchers care. In: Maxwell J, editor. A realist approach for qualitative research. London: SAGE; 2012. p. 3–14.

Moreau KA, Eady K. Connecting medical education to patient outcomes: the promise of contribution analysis. Med Teach. 2015;37(11):1060–2.

Norman G. Generalization and the qualitative-quantitative debate. Adv Health Sci Educ. 2017;22(5):1051–5.

Pawson R, Tilley N. Realist evaluation. London: SAGE; 1997.

Reichenbach M, Cohen R. Hans Reichenbach – selected writings, 1909–1953, vol. 1. Dordrecht: D. Reidel Publishing Company; 1978.

Richard C. Logical empiricism. In: Zalta EN, editors. The stanford encyclopedia of philosophy. 2017. Available at: https://plato.stanford.edu/archives/fall2017/entries/logical-empiricism/.

Riley DJ. The paradox of positivism. Soc Sci Hist. 2007;31(1):115–26.

Ritzer G, Stepnisky J. Classical sociological theory. Thousand Oaks: Sage; 2017.

Russell B. History of Western philosophy: collectors edition. Abington: Routledge; 2013.

Sharrock WW. Kuhn: philosopher of scientific revolutions. Oxford: Polity Press; 2002.

Van Bavel J, Mende-Siedlecki P, Brady W, Reinero A. Contextual sensitivity in scientific reproducibility. Proc Natl Acad Sci. 2016;113(23):6454–9.

Van Fraassen B. The scientific image. Oxford: Oxford University Press; 1980.

Wight C. Philosophy of social science and international relations. In: Walter C, Thomas R, Simmons B, editors. Handbook of international relations. Thousand Oaks: SAGE; 2002. p. 23–51.

Wong G, Westhorp G, Greenhalgh J, Jagosh J, Greenhalgh T. RAMESES II reporting standards for realist evaluations. BMC Med. 2016;14(96):1–18.

Symbolic Interactionism as a Methodological Framework

10

Michael J. Carter and Andrea Montes Alvarado

Contents

Abstract

Symbolic interactionism is theoretical perspective in sociology that addresses the manner in which society is generated and maintained through face-to-face, repeated, meaningful interactions among individuals. In this chapter, we discuss symbolic interactionism as a methodological framework. We first provide a brief summary of interactionist thought, describing the general tenets and propositions that have defined the perspective over time. Next, we discuss methods commonly employed by symbolic interactionists, noting how the interactionist perspective informs and guides sociologists in empirical research. We discuss how symbolic interactionists employ a wide variety of methods to understand both intra- and interpersonal processes, and how methodological approaches in symbolic interactionism vary in terms of their inductive or deductive style, idiographic or nomothetic causal explanation, and quantitative or qualitative research design.

M. J. Carter (✉) · A. Montes Alvarado
Sociology Department, California State University, Northridge, Northridge, CA, USA
e-mail: michael.carter@csun.edu; andrea.montes.594@my.csun.edu

© Springer Nature Singapore Pte Ltd. 2019
P. Liamputtong (ed.), *Handbook of Research Methods in Health Social Sciences*,
https://doi.org/10.1007/978-981-10-5251-4_62

We address five main methods that are commonly used in symbolic interactionist studies: interviews, surveys, ethnographies, content analysis, and experiments. Future directions of the perspective are discussed.

Keywords

Symbolic interactionism · Research methods · Interviews · Surveys · Ethnography · Content analysis · Experiments

1   Introduction

Symbolic interactionism is a theoretical perspective in sociology that addresses the manner in which individuals create and maintain social structures (and greater society) via meaningful, symbolic communication that occurs in face-to-face encounters and in small groups. Inspired by the Scottish moralist philosophers, American pragmatist philosophers, and ideas of Charles Horton Cooley (1902, 1909) and George Herbert Mead (1934), the perspective emerged in America in the mid-twentieth century (Stryker 1980; Kuhn 1964; Blumer 1969; Carter and Fuller 2016). Symbolic interactionism is one of three main areas of inquiry and lines of research in the field of sociological social psychology (the other areas addressing *group processes* and *social structure and interaction*) (House 1977; Smith-Lovin 2001; Kelly et al. 2013; McCall 2013; Schnittker 2013). This chapter discusses the variety of methods and empirical studies that have been produced by scholars who work in the interactionist tradition.

The plan of this chapter is as follows: we first provide a brief summary of the tenets and propositions of symbolic interactionism. We then discuss basic methodological strategies that are aligned with three main variants of interactionist thought, noting how each perspective informs and guides sociologists in empirical research and how scholars use a wide variety of qualitative and quantitative methods to understand social phenomena. We then examine empirical research that has emerged over time, focusing on five main methods that are commonly used in symbolic interactionist studies: interviews, surveys, ethnographies, content analysis, and experiments. We finish with a brief discussion regarding the future of the perspective.

2   The Tenets and Propositions of Symbolic Interactionism

The term "symbolic interactionism" was coined by Herbert Blumer during his tenure at the University of Chicago, where he synthesized the work of Cooley and Mead (and others) to create a systematic framework for understanding the relationship between the individual and society. Symbolic interactionism is both a theory and method; it is particularly useful for understanding attitudes, motives, and behaviors and how individuals interpret experiences and events. It is also useful for understanding how individuals manage impressions of self and others (Goffman 1959;

Hochschild 2003 [1983]), how individuals role-play (Becker 1953; Turner 1978; Thorne 1994), how individuals cooperate and coordinate activities with others in group settings (Burke 2003; Sjoberg et al. 2003), and how individuals construct reality socially by creating shared definitions of situations (Thomas 1969 [1923]; Thomas and Thomas 1928; Berger and Luckmann 1966).

The basic tenets of symbolic interactionism are as follows: (1) individuals act based on the meanings objects have for them; (2) interaction occurs within a particular social and cultural context in which physical and social objects (persons), as well as situations, must be defined or categorized based on individual meanings; (3) meanings emerge from interactions with other individuals and with society; and (4) meanings are continuously created and recreated through interpreting processes during interaction with others (Blumer 1969).

As a perspective, symbolic interactionism developed as a reaction to the positivistic sociological theories of the day that addressed society collectively, holistically, and as a reality *sui generis* (Durkheim 1982; Parsons 2005 [1951]). At the time, the prevalent viewpoint in sociology was that external social structures and institutions impose on and constrain individuals and that any theory aimed at describing society must examine it from the "top down" by focusing on collective (macro-level) social forces rather than individual (micro-level) or social psychological processes. Blumer, who was strongly opposed to such thinking, developed symbolic interactionism as an alternative framework for understanding the social realm. Blumer's "interactionist" perspective emphasized the need for sociologists to examine society from the "bottom up" (i.e., starting at the micro-level and moving up toward the macro-level), claiming that objective "society" and external, constraining social forces are reified in collectivistic theories. Blumer's perspective was influenced by the work of Cooley (1927), who posited that there is no separation between self and society, that society exists only in the imagination, and that society is simply "an interweaving and interworking of mental selves" (pp. 200–201). Thus, in symbolic interactionism, "society" is conceived as the product of meaningful motives, gestures, and behaviors that occur in any given moment during individuals' encounters with others in specific social settings. In shifting the focus from the macro- to the micro-level of analysis, Blumer's symbolic interactionism provided sociologists with a theoretical framework that departed from over-socialized descriptions of human actors and toward an understanding of individuals as agentic, autonomous, and integral in creating their social world.

Blumer's perspective is now known to represent the "Chicago School" of symbolic interactionist thought (Musolf 2003). Other variants of symbolic interactionism emerged after Blumer, inspired by the work of Manford Kuhn and Sheldon Stryker, whose orientations have come to be known as the "Iowa" (Kuhn 1964; Couch et al. 1986) and "Indiana" (Stryker 1980; Stryker and Vryan 2003) Schools of interactionist thought, respectively (see Carter and Fuller 2016 for a summary of the distinctions among the Chicago, Iowa, and Indiana Schools of interactionist thought).

3   Methodological Divergences Across the Variants of Symbolic Interactionism

Symbolic interactionists employ a wide variety of methods to understand both intra- and interpersonal processes. Because the areas of inquiry addressed in symbolic interactionism are so diverse, methodological approaches aligned with the perspective tend to vary in terms of inductive or deductive style, idiographic or nomothetic causal explanation, and quantitative or qualitative research design (Benzies and Allen 2001; Herman-Kinney and Vershaeve 2003; LaRossa and Reitzes 2009). Even though symbolic interactionism is known for its variety of methodological strategies, it is often framed as a *pragmatic* and *qualitative* perspective (Quin et al. 1980; Weigert 1983).

Specific methodological orientations among symbolic interactionists tend to vary depending on whether one works in the *Chicago, Iowa,* or *Indiana* tradition. Those aligned with the Chicago School are known to employ a phenomenological, inductive, and interpretive approach to understanding social phenomena. Bernard Meltzer's work which utilizes methods of introspection and participant observation provides a classic example of methods commonly employed by Chicago School symbolic interactionists (Reynolds and Meltzer 1973; Musolf 2008). Methods associated with the Chicago School generally represent the mainstream framework; Blumer's version is what many (if not most) sociologists think of when one mentions the term "symbolic interactionism." More empirical studies have emerged in this tradition than in any other.

Scholars aligned with the Iowa School of interactionist thought use a logical positivist and deductive approach to the study of interaction, often addressing individual or group identity processes (Herman-Kinney and Vershaeve 2003). Empirical studies in the Iowa tradition tend to use quantitative methods to understand social behavior, often relying on surveys (such as Kuhn's "Twenty Statements Test" (Kuhn and McPartland 1954)) and laboratory experiments. By taking advantage of the laboratory as a controlled environment, experimental studies in the Iowa tradition have provided researchers with the unique ability to document the process of data collection, which (some believe) leads to more certainty that research findings are valid (McPhail 1979; Katovich 1995). The production of data that can be reexamined and validated also helps address methodological critiques regarding the unreliability of qualitative research designs.

After Kuhn's death in 1963, Carl Couch and colleagues continued work in the Iowa tradition, creating what is now known as the "New Iowa School" of symbolic interactionism (Couch 1984; Katovich et al. 2003). Methodologically (and philosophically), the New Iowa School moved away from Kuhn's logical positivist orientation and toward a more pragmatist focus, addressing lived experiences and using the laboratory to observe social transactions in dyadic relationships (McPhail 1979; Katovich 1995). Whereas the Iowa School addressed the *structure* of social interaction, the New Iowa School addressed the *elements* of interaction (Herman-Kinney and Vershaeve 2003). The New Iowa School also emphasized the importance of using third-party standpoint analysis in empirical studies, which requires a researcher to adopt the viewpoint and language of research participants when analyzing data (Diekema et al. 1996). For example, Glaser and Strauss (1964) took a third-party standpoint perspective when examining how terminal hospital patients and staff view patient conditions. They used observations to understand the awareness context of terminally ill patients (i.e., their awareness of impending death) as related to hospital staff's interactions with them, finding that staff framed their interactions with terminally ill patients with "situations of normal" interactions to shield the patient from their near death "true identity."

Sheldon Stryker (1980) and the Indiana School represent the structural symbolic interactionist perspective. Whereas Blumer's brand of symbolic interactionism views society as constantly in flux and changing from moment to moment, Stryker conceives society as stable and patterned, emphasizing the need for sociologists to consider structural conditions that exist outside the individual (Stryker 2001; Serpe and Stryker 2011). Generally, structural symbolic interactionists have more in common methodologically with the Iowa School than with the Chicago School, though those aligned with the Indiana School have used both qualitative and quantitative approaches in empirical studies (Herman-Kinney and Vershaeve 2003). Most of the research methodology aligned with structural symbolic interactionism has been developed by those who work in *identity theory* (McCall and Simmons 1978; Burke and Stets 2009; Stets and Serpe 2013) and *affect control theory* (MacKinnon 1994; Heise 2002; Robinson and Smith-Lovin 2006). Scholars working in this tradition have used participant observation (Smith-Lovin and Douglass 1992), interviews (Burke and Cast 1997), surveys (Asencio and Burke 2011; Heise and Calhan 1995), and laboratory experiments (Wiggins and Heise 1987) to understand social phenomena.

4 Common Methods Used in Empirical Studies on Symbolic Interactionism

4.1 Interviews

Interviews have long been used in social science research to learn about attitudes, beliefs, and experiences of individuals (see ▶ Chap. 23, "Qualitative Interviewing"). Interviews are conducted in a variety of ways, in face-to-face settings, by telephone, and recently online across digitally mediated domains. Interviews may be categorized in different ways, such as being structured or standardized, semi-structured or focused, and unstructured or unstandardized (Herman-Kinney and Vershaeve 2003; Babbie 2016; Serry and Liamputtong 2017). Interviews allow rich data collection and are useful in capturing the nuances of personal interpretation and biography, thus they are well-suited for studies that use a symbolic interactionist framework.

Examples of interactionist studies that employ interview methods include a study by McCabe et al. (2010), who interviewed 20 adult university staff (12 women and 8 men) for an average of 60 min to understand how people use traditional gender arrangements and stereotypes to frame their discussion of sexuality and sex, finding that people's talk about sexuality at the cultural level typically corresponds to traditional gender arrangements and stereotypes. Nugus (2008) conducted 130 semi-structured interviews on clinicians to understand the way nurses and doctors in emergency rooms carve out a unique domain for their work, by interacting and negotiating with doctors and nurses from other departments within the hospital. Nugus' study also shows how symbolic interactionism can be used as reflexive criteria for validating grounded research.

In a series of studies, Burke and his colleagues (Cast et al. 1999; Burke and Stets 1999; Burke and Harrod 2005) interviewed 207 couples to understand how masculine and feminine identity meanings change during the early years of marriage. In these studies, a variety of interview techniques were employed, including 90 min face-to-face interviews with married couples, having couples keep weekly diaries, and having couples videotape themselves while resolving an issue.

In the field of environmental sociology, Pennartz (1989) conducted interviews with households from different neighborhoods to understand their views of the environment and their opinion of inner cities. Pennartz used these interviews to identify the codes and values that operate within the consumption and production of the urban environment and to describe the function of urban elements as a potential counterbalance against the colonization processes of the lifeworld.

Symbolic interactionist studies that use interviews vary in terms of sample size, with some researchers interviewing only a few participants. For example, Day (1985) showed how symbolic interactionism is relevant for social workers in a study that analyzed interviews involving a single social worker, a client, and a team leader. In noting the different views of the people involved in the study, Day showed that it is necessary to take into account the organizational context of interviews. He noted that the interview process involves interpretation and the social construction of reality, which has implications for any study that uses interviews to collect data.

In a study that interviewed seven participants, Barton and Hardesty (2010) examined how exotic dancers define their experience of stripping and how dancers use the language of "spirituality" as a narrative resource. They found that exotic dancing is often a multifaceted experience for women and that strip bars are a more-nuanced environment than most realize. In defining dancing as a spiritual act, exotic dancers deflect the popular understanding of strip clubs. By constructing exotic dancing as a spiritual activity, stripping then becomes an unexpected and welcome source of inner power.

A final example of an interview study that uses a small sample size is provided by Curry (1993), who interviewed a single person in his study on pain and injury. By focusing on one participant named Sam who was an amateur wrestler, Curry showed how the perception of experiencing pain associated with a sport can change over time. For Sam in the early years of his career, pain and injuries were something to be ignored and endured as a test of masculinity. Over time his attitude about pain and injuries changed to be the forefront of concern and something to be avoided, as they are direct hindrances from success in one's sport. By intensely interviewing a sole participant, Curry provided a detailed account that illustrates the nature of the changing self.

4.2   Surveys

Surveys are perhaps the most common research method used in the social sciences; many studies in the interactionist tradition use some type of survey design. Surveys are appropriate for descriptive, exploratory, and explanatory research. They are especially useful for collecting data from large samples and for measuring common social psychological processes such as attitudes and beliefs. They also are useful for measuring a wide variety of demographic and psychometric variables in a short amount of time. Survey questionnaires can include closed-ended response categories, which often provide quantitative data that is conducive for statistical analysis, or open-ended questions that are appropriate for qualitative data analysis. Many surveys also have the advantage of being self-administered and thus do not have the logistical difficulties that often plague face-to-face interviews or laboratory experiments (see ▶ Chaps. 32, "Traditional Survey and Questionnaire Platforms," and ▶ 76, "Web-Based Survey Methodology").

Kuhn and McPartland's (1954) "Twenty Statements Test" (TST) (mentioned previously) is a classic example of a survey measure commonly used by symbolic interactionists. The TST is used to measure the self-concept. Known for its simplicity and ease of use, participants who receive the "test" are given a sheet of paper with the words "Who am I?" written at the top, with twenty blank lines listed below. Participants then fill in each line according to who they see themselves as (e.g., possible answers might be "a student," "a friend," etc.). Many studies have employed some variant of the TST over the past decades (Franklin and Kohout 1971; Grace and Kramer 2002).

Another example of past survey research in the interactionist tradition is provided by Burke and Tully (1977), who used semantic differential scales to measure gender role identities in school children. Students were given the survey and asked to respond to statements such as "usually boys are. . ." or "usually girls are. . .," and "as a boy (or girl) I usually am. . .." In their analysis, they found that gender role identities were unimodal and normally distributed for most participants in their sample but that approximately one-fifth of the sample had gender identities that were more aligned with the modal identity for the *opposite* sex than that of their own sex. Burke and Tully's study was groundbreaking, providing symbolic interactionists with a method for measuring identity meanings that is used to this day.

Vignettes are also commonly used by symbolic interactionists in survey designs. For example, Stets and Carter (2006, 2012) measured moral behavior by asking participants to report on how they behaved in past situations such as finding a lost wallet, having an opportunity to cheat on a test, or having an opportunity to give money to a homeless person. Reid et al. (2015) examined the importance of "setting" as a factor in shaping college students' dating and sexual behavior. They explored how students interpreted a vignette describing a casual heterosexual encounter at a party followed by a sexless dinner date, finding that rather than simply following generalized cultural scripts, heterosexual encounters of college students are guided by standardized patterns of behaviors based on the distinct settings and roles associated with each situation. A study conducted by Carter and Mireles (2016a, b) used a vignette design and administered a survey questionnaire to deaf and hard-of-hearing individuals to measure how deaf identity processes and attachment to the deaf community correlates with self-esteem and depression.

Analysis of secondary survey data is another method that is sometimes used by symbolic interactionists. For example, McPherson et al. (2006) compared the 1985 and 2004 General Social Surveys (GSS) to see if individuals' degree of closeness to other members in their social network had changed over time, showing that people had become more socially isolated over a two-decade timespan. Whether taking the form of a self-administered questionnaire or secondary data, surveys are one of the most common methods used in symbolic interactionist studies.

4.3   Ethnographies

An ethnography is a (mostly) qualitative research method that attempts to systematically investigate and then accurately represent some form of culture using an emic (i.e., "inside-out") perspective to understand social phenomena, compared to other methods (such as surveys and experiments) that generally use an etic (i.e., "outside-in") perspective. While etic approaches examine social phenomena in a detached manner, ethnographies seek to understand a culture on its own terms and from the perspective of those individuals within it (Scott and Garner 2013; see ▶ Chap. 26, "Ethnographic Method"). Symbolic interactionists are often interested in subcultures, small group norms and behavior, and role-playing. Ethnographies allow for rich descriptions of all such phenomena, so they are appealing to and commonly used by many symbolic interactionists.

In practice, "ethnography" is an umbrella term that represents a variety of research strategies (Berg and Lune 2012). Some ethnographies involve field studies or written accounts of observations (Ellen 1984; Stoddart 1986). Some involve active participant observations of a social setting (Warren and Karner 2005). Others see ethnographies as detailed descriptions of natural settings that offer no explanations (Babbie 2016). And, those who use autoethnography as a research method use their own personal experience as a basis for scholarly knowledge (Ellis and Bochner 2000; Carter 2016). Regardless of the "type" of ethnography one practices, all tend to value *depth* over *breadth* regarding data collection – data take the form of rich descriptions of events. Ethnographies are thus often considered strong on validity but weak on reliability, as some doubt the efficacy of using ethnographic data for generalizing to greater populations or generating abstract theories. Regardless, some of the most well-known and respected studies in the interactionist tradition use ethnographic methods.

Gary Alan Fine is a prominent symbolic interactionist who has used ethnographies to understand subgroup cultures and the nature of everyday social life. In *Kitchens*, Fine (1996) conducted an ethnography to understand the culture of restaurant work, showing how working conditions, time constraints, market forces, and aesthetic goals all combine to affect food that is served to customers. Fine has also used ethnographic methods to study the culture of mushrooming (Fine 1998), high school debate teams and adolescent culture (Fine 2001), and even the world of competitive chess (Fine 2015).

Robert Park, a famous sociologist from the University of Chicago said that understanding human behavior requires researchers to immerse themselves in the worlds of their subjects – to study individuals in their own terms in order to understand the symbolic meanings that those individuals themselves define as important and real (Herman-Kinney and Vershaeve 2003). Participant observation is an ethnographic field method aimed at capturing such phenomena. It is used to study small groups and homogeneous cultures. In a participant observation, ethnographers immerse themselves in a society, often living with (or spending an extended time with) the group under study, participating in daily activities and carefully observing experiences (Tedlock 2005). Data produced by participant observation are generally rich with detail and description; hence the information gathered using the method is considered to accurately reflect the views of those native to a specific group or society (Tedlock 2005).

Symbolic interactionists have long relied on participant observations to better understand the nature of social life and subcultures. As a method, it is particularly well-suited for the perspective, as it seeks to identify and understand the creation and maintenance of meanings that actors use to navigate their everyday lives. Classic participant observation studies include Liebow's (1967) *Tilly's Corner*, which documents his participation and observation of black men who spent time together on a street corner. In *Boys in White*, Becker et al. (2009 [1961]) observed young men who aspired to become physicians, documenting the difficulty of their journey toward becoming a doctor and how medical students feel about their training and the profession they will one day enter. *The Urban Villagers* portrays Herbert Gans' (1962) observations when he moved to an Italian enclave in Boston in order to immerse himself in the community and gain a better understanding of the lives of local inhabitants.

In more current research, Gottschalk (2010) used participant observation to understand how interactions in virtual spaces shape everyday life in the digital age. By participating in the virtual world *Second Life*, he documented how digitally mediated environments are both social psychological playgrounds where participants enjoy individualistic fantasies as well as virtual communities where individuals collaborate on collective projects. Rafalow and Adams (2016) observed encounters in bar settings and how patrons' use of digital communication technologies both augment the bar experience and shape social networks that may develop through interactions in such places. Harvey (2017) observed volunteers who worked to rebuild New Orleans after Hurricane Katrina, specifically the cultural performance and symbolic exchange of food among those involved in the rebuilding. She found that food was used as a way to welcome and compensate volunteers for their hard work and to celebrate progress in rebuilding the community but that over time the giving and consuming of food was renegotiated. Regardless of form, some of the most influential and recognized studies in the interactionist tradition use ethnographic methods.

4.4   Content Analysis

Content analysis is a detailed examination and interpretation of some (usually) material source that is aimed at revealing patterns, themes, and meanings that are embedded within such sources (Berg and Lune 2012; see ▶ Chap. 47, "Content Analysis: Using Critical Realism to Extend Its Utility"). Content analysis can be inductive or deductive. Inductive approaches are used in cases where no previous studies exist regarding a phenomenon (or when the understanding of a phenomenon is fragmented or unclear); deductive approaches are useful when the general aim is to test an existing theory or compare categories at different time periods (Elo and Kyngas 2007). Content analyses are common in symbolic interactionist research. They are also prevalent in health sciences, particularly in nursing studies (Elo and Kyngas 2007). Content analysis is an umbrella term that represents a wide variety of research methods, such as textual analysis, life history analysis and biographical methods, document analysis, conversation analysis, and the study of semiotics and signs (Herman-Kinney and Vershaeve 2003).

Perhaps the most famous content analysis study in the realm of symbolic interactionism is Thomas and Znaniecki's (1996) *The Polish Peasant*. In this study, the letters and diaries of Polish immigrants living in Chicago were analyzed to understand the nuances of immigrant life and the trials and tribulations of living in a different culture away from one's homeland. Other classic examples are provided by Jacobs (1967), who examined letters left by those who committed suicide, and Molotch and Boden's (1985) study of the Watergate congressional hearings, which used conversation analysis as a method to determine how power was negotiated among those involved in the political scandal. Denzin (1987) demonstrated the benefit of using semiotics (the study of signs, symbols, and the systems that create them) and symbolic interactionism together in his analysis of narrative texts of advertisements for Jack Daniel's Whiskey and Dewar's White Label Scotch to understand the political economy of signs.

In more recent research, Silva (2014) used frame analysis (Goffman 1974) to examine how anti-evolutionists neutralized the framing of their position as religious by analyzing 570 letters published in American newspapers in the months surrounding a nationally covered federal judicial decision on the legality of a school's decision to undermine evolutionary theory in a classroom. Silva showed that anti-evolutionists neutralized the framing of their position as religious through the processes of selective acknowledgement and disagreement with the problematic framing.

In her article "I am a Cheerleader, but Secretly I Deal Drugs: Authenticity through Concealment and Disclosure," Smirnova (2016) performed a content and discourse analysis of 1600 submissions to the PostSecret mail-art project and revealed how secrets are used to manage disparate social, role, and personal identities. She found that people attempt to maintain or achieve authenticity through dialectical acts of concealment and disclosure, showing that individuals keep and disclose secrets in order to maintain authenticity contextually within relationships, as well as across contexts through self-reflexive evaluations.

Other contemporary examples of content analyses includes a study by Sawicka (2016) that analyzed discussion lists for bereaved parents to understand how they collectively managed the grief that emerged due to their experience of perinatal loss/stillbirth. And, Cross (2015) analyzed magazine columns, letters, memoirs, and first person essays to discover how people form attachments to geographic locations (i.e., places), proposing that seven distinct processes interact at the individual, group, and cultural level to shape how individuals form attachments to specific places.

4.5   Experiments

While they are common in psychology, experimental methods (especially laboratory experiments) are less common in sociology. Thus, overall, symbolic interactionist studies that rely on experimental methodology are less common compared to studies that use participant observation, surveys, or interviews as methods of inquiry. However, those who work in the interactionist tradition and use experiments note the strength of the method, largely because the perspective seeks to understand social psychological processes that are often conducive for study using experimental designs.

A benefit of experiments is that they are well-suited for theory testing. Laboratory experiments offer scholars the ability to control a social environment and place subjects in a setting where stimuli can be manipulated to determine cause and effect. Some scholars use the traditional experimental design by randomly assigning subjects into experimental and control groups, pretesting each group at time 1, introducing an independent variable to the experimental group, and then posttesting both groups at time 2 to compare the effect of the independent variable. Others employ quasi-experimental designs (such as field experiments) that resemble but deviate from the traditional design in some way (see ► Chaps. 36, "Eliciting Preferences from Choices: Discrete Choice Experiments," and ► 37, "Randomized Controlled Trials").

Classic experimental studies in symbolic interactionism were conducted by those associated with the Iowa School of interactionist thought (Couch et al. 1986; Couch 1987). Notable experiments during this era (and after) include Haney et al.'s study of prison life (Haney et al. 1973), Darley and Batson's (1976) study on helping behavior, and Goldstein and Arms' (1971) study of aggression. A more recent example of symbolic interactionist studies that employ experimental designs is provided by scholars who work in identity theory (Stryker and Burke 2000; Burke and Stets 2009). Scholars who work in identity theory have used experiments to better understand how identities motivate behavior and how social structural conditions combine with identity processes to impact behavior. For example, in a study on morality, Stets and Carter (2011) used an experimental design that placed study participants in a laboratory setting where they had the opportunity to cheat to gain an advantage over others, revealing that cheating behavior is predicted depending on the meanings of one's moral identity. In related research, Carter (2013) brought participants to a laboratory and manipulated moral identity activation and different group conditions to discover how activated identities and social context influence moral behavior in situations where people are awarded more than they deserve for completing a task.

In other research, Katovich (1987) conducted a laboratory study (systemic observation) of 24 dyads that were given either a role-playing identity as a manager or potential employee and found a linking process of interpersonal situated identities and broader future-oriented concerns. Pechmann et al. (2010) used symbolic interactionist theory and conducted three experiments to examine how altering the age of models used in cigarette advertising affects whether adolescents are drawn to or deterred from smoking. They found that adolescents exhibited a boomerang effect when exposed to teen cigarette models, lowering the intent to smoke, while exposure to young adult cigarette models increased the intent to smoke.

Like all methods, experiments have both strengths and weaknesses. Some scholars who work in the symbolic interactionist tradition question the efficacy of using laboratory experiments to study micro-level social phenomena, claiming that lab settings are contrived, artificial environments that are unlike normal life experiences. Proponents of experimental methods claim that while lab settings may be unique, they aren't necessarily so distinct and removed from real experience that they offer no utility for the study of human behavior. While scholars who work in different traditions of interactionist thought may disagree on the utility of experimental methodology, there are a variety of influential studies aligned with the perspective that have used experimental designs.

5   Conclusion and Future Directions

Symbolic interactionism is approaching its centennial as a distinct theoretical perspective and method. The glut of extant interactionist studies and its continued popularity shows that it is – *de rigueur* – a leading theoretical perspective in sociology and across the social sciences. Ironically, in the past some have predicted that the perspective might wither and fade over time as a distinct sociological paradigm, as the concepts that once were unique to the perspective become more integrated into mainstream sociology (Sandstrom and Fine 2003). If the "future" is "now," it seems safe to say that such a decline has not occurred, as evidenced by the wide variety of published empirical research that continues to be aligned with symbolic interactionism. Since the framework seems to be flourishing, let us consider its future trajectory regarding research methods and epistemological strategies.

Fifteen years ago, Sandstrom and Fine (2003) predicted that in the future, symbolic interactionism would become more characterized by theoretical and methodological diversity and that the methodological differences among those in the Chicago and Iowa/Indiana traditions would begin to diminish. They also predicted that symbolic interactionists would begin to address macro-level concepts and analyze relationships among large-scale social entities. Regarding the first prediction, it seems accurate to claim that contemporary symbolic interactionists tend not to identify the school of thought that directs their study, but the common methodology likened to each school of thought is still evident in specific studies. Regarding the second prediction, a review of the recent literature reveals that studies that address macro-level or large-scale phenomena are still by far the exception rather than the rule. There are interactionist studies that address abstract or aggregate processes (Dennis and Martin 2005; Salvini 2010), but even these studies are more rooted in micro-level processes than social forces that are commonly understood to represent the macro realm. So, it seems that methodological approaches and units of analysis addressed in symbolic interactionist studies are similar to what they have been in the past. From our perspective, this is not surprising, as the perspective was originally conceived to explain micro-level phenomena, and scholars continue to find new areas of study at this level of analysis.

Regarding the future of research methods *per se*, it seems unlikely that studies conducted in the interactionist tradition will deviate much from the variety of methods described in this chapter. Interviews, surveys, ethnographies, content analyses, and experiments will continue to be methods commonly employed by symbolic interactionists (perhaps some more than others). However, these methods will evolve and will continue to be shaped by advances in technology as scholars find new ways to improve research designs and the process of data collection. Any scholar who has followed the development of symbolic interactionism over the decades has witnessed such change. For example, in only the recent past (going back 20 years or so), surveys usually took a physical form; hard copies of questionnaires were often disseminated in person or mailed to participants and then collected after completion. Response data then had to be manually entered into a computer to be analyzed (or in the distant past, analyzed by hand), often taking a team of researchers and a great deal of time. With advances in electronic technology and the creation of the Internet, surveys are now often administered via digitally mediated forums. Indeed, online surveys might now be the norm rather than the exception. Electronic survey platforms allow for quick data collection, and they eliminate the need for data entry, allowing researchers to analyze data virtually the moment they are gathered. Of course, new issues regarding reliability and validity of data collection accompany such advanced technology. Research methods in the social sciences are in a constant process of revision and refinement.

A specific example of how technology has improved survey research methods regards the experiential sampling method (ESM) (Hektner et al. 2007). ESM is aimed at measuring moment-to-moment experiences and the quality of everyday life, themes that are of central interest to symbolic interactionists. Studies using this method typically select participants and then have them respond to questions that are given to them at varying times throughout the day, during their usual routines, interactions, and encounters. For instance, an ESM study might be conducted to understand how individuals' moods and emotions change from moment to moment. Periodically, throughout the day, participants in such a study would be prompted to report their current mood state. After data are collected, a researcher can see how moods change for each participant, and a variety of analyses can be performed on the collected data. The ESM method is beneficial because it does not rely on memory recollection or past accounts of experiences, things that are commonly understood to plague survey questionnaires.

ESM methods have developed considerably with advancements in technology. Early ESM studies involved participants carrying watches or pagers that would beep or vibrate, prompting participants to stop what they were doing and answer a paper-and-pencil survey. Later studies used personal data assistant (PDA) devices (such as Palm Pilots) that allowed for responses to be recorded electronically. The advanced technology available today has greatly improved the accuracy of ESM data collection, as data can be collected in real time on aps via smartphones. These newer technologies have reduced costs and helped to eliminate logistical issues associated with ESM studies. Such strategies will likely continue to improve in the future as technology continues to improve. Methods like the ESM are especially dependent on technology. As technology continues to evolve, more and more sophisticated techniques will be used to gather data instantaneously from study participants.

Advanced technology has improved experimental methods as well. In the past, experimentalists relied on devices such as two-way mirrors for surveillance of participants in laboratories. Now experimental researchers use more sophisticated forms of surveillance, with hidden camera technology and a wide variety of instruments available that record subtle behaviors and various social processes. For example, continuing the example of studying emotions, experimentalists who study emotions have developed sophisticated technologies capable of measuring affective responses. Laboratory research documenting the relationship between emotion and brain activity has used functional magnetic resonance imaging (fMRI), electroencephalography, and infrared thermography to measure emotional reactions (Clay-Warner and Robinson 2015). As these and other technologies continue to improve, researchers who conduct experiments will have even more precise mechanisms available to study human behavior in laboratory environments. The above examples illustrate how technology continues to improve research methods used by symbolic interactionists.

The future of the symbolic interactionism seems bright, as evident by the continued interest and development of the field. As noted by the *Society for the Study of Symbolic Interaction* (SSSI) (https://sites.google.com/site/sssinteraction/), many subdisciplines within the social sciences have been influenced by symbolic interactionism, including the sociology of emotions, criminology, collective behavior and social movements, feminist studies, social psychology, communications theory, semiotics, education, nursing, mass media, organizations, and the study of social problems. In this chapter, we have discussed the variety of research methods employed by scholars who work with or align themselves with the interactionist tradition. While we have addressed many common methods that have come to define the field, our coverage is by no means exhaustive. The myriad approaches and examples of interactionist methods that have appeared in empirical studies cannot be adequately summarized in one chapter (e.g., some consider work in ethnomethodology (Garfinkel 1967) to be aligned with symbolic interactionism; we do not address this work here). As long as social scientists remain interested in micro-level social phenomena and the relationship between the self and society, symbolic interactionism will continue to hold a place among the most influential sociological paradigms. And, as the field grows in numbers and more empirical studies emerge, the research methodology associated with symbolic interactionism will continue to develop and improve.

References

Asencio EK, Burke PJ. Does incarceration change the criminal identity? A synthesis of labeling and identity theory perspectives on identity change. Sociol Perspect. 2011;54:163–82.

Babbie E. The practice of social research. 14th ed. Belmont: Wadsworth Cengage Learning Publishers; 2016.

Barton B, Hardesty CL. Spirituality and stripping: exotic dancers narrate the body Ekstasis. Symb Interact. 2010;33(2):280–96.

Becker HS. Becoming a marijuana user. Am J Sociol. 1953;59:235–42.

Becker HS, Geer B, Hughes EC, Strauss AL. Boys in white: student culture in medical school. New Brunswick: Transaction Publishers; 2009 [1961].

Benzies KM, Allen MN. Symbolic interactionism as a theoretical perspective for multiple method research. J Adv Nurs. 2001;33(4):541–7.

Berg BL, Lune H. Qualitative research methods for the social sciences. 8th ed. Upper Saddle River: Pearson; 2012.

Berger P, Luckmann T. The social construction of reality. New York: Anchor Books; 1966.

Blumer H. Symbolic interactionism: perspective and method. Berkeley: University of California Press; 1969.

Burke PJ. Interaction in small groups. In: DeLamater JD, editor. Handbook of social psychology. New York: Kluwer-Plenum; 2003. p. 363–87.

Burke PJ, Cast AD. Stability and change in the gender identities of newly married couples. Soc Psychol Q. 1997;60(4):277–90.

Burke PJ, Harrod MM. Too much of a good thing? Soc Psychol Q. 2005;68:359–74.

Burke PJ, Stets JE. Trust and commitment through self-verification. Soc Psychol Q. 1999;62(4): 347–66.

Burke PJ, Stets JE. Identity theory. New York: Oxford University Press; 2009.

Burke PJ, Tully JC. The measurement of role identity. Soc Forces. 1977;55(4):881–97.

Carter MJ. Advancing identity theory: examining the relationship between activated identities and behavior in different social contexts. Soc Psychol Q. 2013;76:203–23. https://doi.org/10.1177/0190272513493095.

Carter MJ. An autoethnographic analysis of sports identity change. Sport Soc. 2016;19(10): 1667–89. https://doi.org/10.1080/17430437.2016.1179733.

Carter MJ, Fuller C. Symbols, meaning, and action: the past, present, and future of symbolic interactionism. Curr Sociol Rev. 2016;64(6):931–61. https://doi.org/10.1177/0011392116638 396.

Carter MJ, Mireles DC. Deaf identity and depression. In: Stets JE, Serpe RT, editors. New directions in identity theory and research. New York: Oxford University Press; 2016a. p. 509–38.

Carter MJ, Mireles DC. Examining the relationship between deaf identity verification processes and self-esteem. Identity. 2016b;16(2):102–14.

Cast AD, Stets JE, Burke PJ. Does the self conform to the views of others? Soc Psychol Q. 1999;62(1):68–82.

Clay-Warner J, Robinson DT. Infrared thermography as a measure of emotion response. Emot Rev. 2015;7(2):157–62. https://doi.org/10.1177/1754073914554783.

Cooley CH. Human nature and social order. New York: Scribner; 1902.

Cooley CH. Social organization. New York: Scribner; 1909.

Cooley CH. Life and the student. New York: Alfred A. Knopf; 1927. p. 200–1.

Couch CJ. Symbolic interaction and generic sociological principles. Symb Interact. 1984;7(1): 1–13. https://doi.org/10.1525/si.1984.7.1.1.

Couch C. Researching social processes in the laboratory. Greenwich: JAI Press; 1987.

Couch CJ, Saxton SL, Katovich MA. The Iowa School. In: Couch CJ, Saxton SL, Katovich MA, editors. Studies in symbolic interaction, supplement 2. Greenwich: JAI Press; 1986.

Cross JE. Processes of place attachment: an interactional framework. Symb Interact. 2015;38(4): 493–520. https://doi.org/10.1002/symb.198.

Curry TJ. A little pain never hurt anyone: athletic career socialization and the normalization of sports injury. Symb Interact. 1993;16:273–90.

Darley JM, Batson CD. From Jerusalem to Jericho: a study of situational and dispositional variables in helping behaviors. J Pers Soc Psychol. 1976;27:100–8.

Day PR. An interview: constructing reality. Bri J Soc Work. 1985;15(5):487–99.

Dennis A, Martin PJ. Symbolic interactionism and the concept of power. Br J Sociol. 2005;56(2): 191–213. https://doi.org/10.1111/j.1468-4446.2005.00055.x.

Denzin NK. On semiotics and symbolic interaction. Symb Interact. 1987;10(1):1–19.

Diekema DA, Couch CJ, Powell JO. The third party standpoint, postmodernism, and the study of social transactions. Sociol Perspect. 1996;39(1):111–27. https://doi.org/10.2307/1389345.

Durkheim E. The rules of sociological method. New York: The Free Press; 1982.

Ellen RF. Ethnographic research. New York: Academic; 1984.

Ellis C, Bochner AP. Autoethnography, personal narrative, and reflexivity: researcher as subject. In: Denzin NK, Lincoln YS, editors. Handbook of qualitative research. Thousand Oaks: Sage; 2000. p. 733–68.

Elo S, Kyngas H. The qualitative content analysis process. J Adv Nurs. 2007;62:107–15.

Fine GA. Kitchens: the culture of restaurant work. Berkeley: University of California Press; 1996.

Fine GA. Morel Tales: the culture of mushrooming. Cambridge, MA: Harvard University Press; 1998.

Fine GA. Giften tongues: high school debate and adolescent culture. Princeton: Princeton University Press; 2001.

Fine GA. Players and pawns: how chess builds community and culture. Chicago: University of Chicago Press; 2015.

Franklin BJ, Kohout FJ. Subject-coded versus researcher-coded TST protocols: some methodological implications. Sociol Q. 1971;12:82–9.

Gans H. The urban villagers. New York: Free Press; 1962.

Garfinkel H. Studies in ethnomethodology. Englewood Cliffs: Prentice Hall; 1967.

Glaser BG, Strauss AL. Awareness contexts and social interaction. Am Sociol Rev. 1964;29:669–79.

Goffman E. The presentation of self in everyday life. New York: Doubleday; 1959.

Goffman E. Frame analysis: an essay on the organization of experience. New York: Harper and Row; 1974.

Goldstein JH, Arms RL. Effects of observing athletic contests on hostility. Sociometry. 1971;34:83–90.

Gottschalk S. The presentation of avatars in second life: self and interaction in social virtual spaces. Symb Interact. 2010;33(4):501–25. https://doi.org/10.1525/si.2010.33.4.501.

Grace SL, Kramer KL. Sense of self in the new millennium: male and female student responses to TST. Soc Behav Personal. 2002;30(3):271–80.

Haney C, Banks WC, Zimbardo PG. Interpersonal dynamics in a simulated prison. Int J Criminol Penology. 1973;1:69–97.

Harvey DC. 'Gimme a Pigfoot and a Bottle of Beer': food as cultural performance in the aftermath of hurricane Katrina. Symb Interact. 2017;40:498–522. https://doi.org/10.1002/symb.318.

Heise DR. Understanding social interaction with affect control theory. In: Berger J, Zelditch Jr M, editors. New directions in contemporary sociological theory. Lanham: Rowman & Littlefield; 2002. p. 17–40.

Heise DR, Calhan C. Emotion norms in interpersonal events. Soc Psychol Q. 1995;58(4):223–40.

Hektner JM, Schmidt JA, Csikszentmihalyi M. Experience sampling method: measuring the quality of everyday life. Thousand Oaks: Sage; 2007.

Herman-Kinney NJ, Vershaeve JM. Methods of symbolic interactionism. In: Reynolds LT, Herman-Kinney NJ, editors. Handbook of symbolic interactionism. Walnut Creek: AltaMira Press; 2003. p. 213–52.

Hochschild AR. The managed heart. Berkeley: University of California Press; 2003 [1983].

House JS. The three faces of social psychology. Sociometry. 1977;40:161–77.

Jacobs J. A phenomenological study of suicide notes. Soc Probl. 1967;15:60–72.

Katovich MA. Identity, time, and situated activity: an interactions analysis of dyadic transactions. Symb Interact. 1987;10(2):187–208.

Katovich MA. Couch the Bricoleur: using ethnographic and laboratory traditions to establish data careers. Symb Interact. 1995;18(3):283–301. https://doi.org/10.1525/si.1995.18.3.283.

Katovich MA, Miller DE, Stewart RL. The Iowa School. In: Reynolds LT, Herman-Kinney NJ, editors. Handbook of symbolic interactionism. Walnut Creek: AltaMira Press; 2003. p. 119–39.

Kelly JR, McCarty MK, Iannone NE. Interaction in small groups. In: DeLamater JD, Ward A, editors. Handbook of social psychology. 2nd ed. New York: Springer; 2013. p. 413–38.

Kuhn MH. Major trends in symbolic interaction theory in the past twenty-five years. Sociol Q. 1964;5:61–84.

Kuhn MH, McPartland TS. An empirical investigation of self-attitudes. Am Sociol Rev. 1954;19:68–76.

LaRossa R, Reitzes DC. Symbolic interactionism and family studies. In: Boss P, Doherty WJ, LaRossa R, Schumm WR, Steinmetz SK, editors. Sourcebook of family theories and methods. Boston: Springer; 2009. p. 135–66.

Liebow E. Tally's corner. Boston: Little Brown and Co.; 1967.

MacKinnon NJ. Symbolic interactionism as affect control. Albany: State University of New York Press; 1994.

McCabe J, Tanner AE, Heiman JR. The impact of gender expectations on meanings of sex and sexuality: results from a cognitive interview study. Sex Roles. 2010;62(3–4):252–63.

McCall GJ. Interactionist perspectives in social psychology. In: DeLamater JD, Ward A, editors. Handbook of social psychology. 2nd ed. New York: Springer; 2013. p. 3–29.

McCall GJ, Simmons JL. Identities and interactions: an examination of human associations in everyday life. Revised ed. New York: The Free Press; 1978.

McPhail C. Experimental research is convergent with symbolic interaction. Symb Interact. 1979;2(1):89–94. https://doi.org/10.1525/si.1979.2.1.89.

McPherson M, Smith-Lovin L, Brashears ME. Social isolation in America: changes in core discussion networks over two decades. Am Sociol Rev. 2006;71(3):353–75. https://doi.org/10.1177/000312240607100301.

Mead GH. Mind, self, and society from the standpoint of a social behaviorist. Chicago: University of Chicago Press; 1934.

Molotch HL, Boden D. Talking social structure: discourse, domination and the Watergate hearings. Am Sociol Rev. 1985;50(3):273–88.

Musolf GR. The Chicago school. In: Reynolds LT, Herman-Kinney NJ, editors. Handbook of symbolic interactionism. Walnut Creek: AltaMira Press; 2003. p. 91–117.

Musolf GR. The symbolic interactionism of Bernard N. Meltzer. Mich Sociol Rev. 2008;22:112–41.

Nugus P. The interactionist self and grounded research: reflexivity in a study of emergency department clinicians. Qual Sociol Rev. 2008;4(1):189–204.

Parsons T. The social system. Abingdon: Routledge; 2005 [1951].

Pechmann C, Pirouz D, Pezzuti T. Symbolic interactionism and adolescent reactions to cigarette advertisements. In: Campbell MC, Inman J, Pieters R, editors. Advances in consumer research, vol. 37. Doluth: Association for Consumer Research; 2010. p. 138–42.

Pennartz PJJ. Semiotic theory and environmental evaluation: a proposal for a new approach and a new method. Symb Interact. 1989;12(2):231–49.

Quin CO, Robinson IE, Balkwell JW. A synthesis of two social psychologies. Symb Interact. 1980;3(1):59–88. https://doi.org/10.1525/si.1980.3.1.59.

Rafalow MH, Adams BL. Navigating the tavern: digitally mediated connections and relationship persistence in bar settings. Symb Interact. 2016;40(1):25–42. https://doi.org/10.1002/symb.268.

Reid JA, Webber GR, Elliott S. 'It's like being in church and being on a field trip:' the date versus party situation in college students' accounts of hooking up. Symb Interact. 2015;38(2):175–94. https://doi.org/10.1002/SYMB.153.

Reynolds LT, Meltzer BN. The origins of divergent methodological stances in symbolic interactionism. Sociol Q. 1973;14(2):189–99. https://doi.org/10.1111/j.1533-8525.1973.tb00853.x.

Robinson DT, Smith-Lovin L. Affect control theory. In: Burke PJ, editor. Contemporary social psychological theories. Stanford: Stanford University Press; 2006. p. 137–64.

Salvini A. Symbolic interactionism and social network analysis: an uncertain encounter. Symb Interact. 2010;33(3):364–88. https://doi.org/10.1525/si.2010.33.3.364.

Sandstrom KL, Fine GA. Triumphs, emerging voices, and the future. In: Reynolds LT, Herman-Kinney NJ, editors. Handbook of symbolic interactionism. Lanham: AltaMira Press; 2003. p. 1041–57.

Sawicka M. Searching for a narrative of loss: interactional ordering of ambiguous grief. Symb Interact. 2016;40(2):229–46. https://doi.org/10.1002/symb.270.

Schnittker J. Social structure and personality. In: DeLamater JD, Ward A, editors. Handbook of social psychology. 2nd ed. New York: Springer; 2013.

Scott G, Garner R. Doing qualitative research: designs, methods, and techniques. Upper Saddle River: Pearson; 2013.

Serpe RT, Stryker S. The symbolic interactionist perspective and identity theory. In: Schwartz SJ, Luyckx K, Vignoles VL, editors. Handbook of identity theory and research. New York: Springer; 2011. p. 225–48.

Serry T, Liamputtong P. The in-depth interviewing method in health. In: Liamputtong P, editor. Research methods in health: foundations for evidence-based practice. 3rd ed. Melbourne: Oxford University Press; 2017. p. 67–83.

Silva EO. Neutralizing problematic frames in the culture wars: anti-evolutionists grapple with religion. Symb Interact. 2014;37(2):226–45. https://doi.org/10.1002/symb.97.

Sjoberg G, Gill EA, Tan JE. Social organization. In: Reynolds LT, Herman-Kinney NJ, editors. Handbook of symbolic interactionism. Walnut Creek: AltaMira Press; 2003. p. 411–32.

Smirnova M. 'I am a cheerleader, but secretly I deal drugs': authenticity through concealment and disclosure. Symb Interact. 2016;39(1):26–44. https://doi.org/10.1002/symb.208.

Smith-Lovin L. Social psychology. In: Blau JR, editor. The Blackwell companion to sociology. Malden: Blackwell; 2001. p. 407–20.

Smith-Lovin L, Douglass W. An affect control analysis of two religious subcultures. In: Gecas V, Franks D, editors. Social perspective in emotions. Greenwich: JAI Press; 1992. p. 217–48.

Stets JE, Carter MJ. The moral identity: a principle level identity. In: McClelland K, Fararo TJ, editors. Purpose, meaning, and action: control systems theories in sociology. New York: Palgrave MacMillan; 2006. p. 293–316.

Stets JE, Carter MJ. The moral self: applying identity theory. Soc Psychol Q. 2011;74:192–215.

Stets JE, Carter MJ. A theory of the self for the sociology of morality. Am Sociol Rev. 2012;77:120–40.

Stets JE, Serpe RT. Identity theory. In: DeLamater JD, Ward A, editors. Handbook of social psychology. New York: Springer; 2013. p. 31–60.

Stoddart K. The presentation of everyday life: some textual strategies for adequate ethnography. Urban Life. 1986;15(1):103–21.

Stryker S. Symbolic interactionism: a social structural version. Menlo Park: Benjamin Cummings; 1980.

Stryker S. Traditional symbolic interactionism, role theory, and structural symbolic interactionism: the road to identity theory. In: Turner JH, editor. Handbook of sociological theory. New York: Kluwer/Plenum Publishers; 2001. p. 211–31.

Stryker S, Burke PJ. The past, present, and future of an identity theory. Soc Psychol Q. 2000;63:284–97.

Stryker S, Vryan KD. The symbolic interactionist frame. In: DeLamater JD, editor. Handbook of social psychology. New York: Kluwer-Plenum; 2003. p. 3–28.

Tedlock B. The observation of participation and the emergence of public ethnography. In: Denzin NK, Lincoln YS, editors. The Sage handbook of qualitative research. 3rd ed. Thousand Oaks: Sage; 2005. p. 467–81.

Thomas WI. The unadjusted girl. Montclair: Patterson Smith Publishing Corporation; 1969 [1923].

Thomas WI, Thomas DS. The child in America: behavior problems and programs. New York: Knopf; 1928.

Thomas WI, Znaniecki F. The polish peasant in Europe and America. Chicago: University of Illinois Press; 1996.

Thorne B. Gender play: girls and boys in school. New Brunswick: Rutgers University Press; 1994.

Turner RH. The role and the person. Am J Sociol. 1978;84(1):1–23.

Warren CAB, Karner TX. Discovering qualitative methods: field research, interviews, and analysis. Los Angeles: Roxbury; 2005.

Weigert AJ. Identity: its emergence within sociological psychology. Symb Interact. 1983;6:183–206.

Wiggins B, Heise DR. Expectations, intentions, and behavior: some tests of affect control theory. Math Sociol. 1987;13:153–69.

Hermeneutics: A Boon for Cross-Disciplinary Research

11

Suzanne D'Souza

Contents

Abstract

Hermeneutics has long been used with huge gains in various fields of research as the underpinning paradigm. In particular, Heidegger's interpretive framework of being and becoming has influenced many a research undertaking owing to its resilience and flexibility in bringing to life the lived experience. However, despite its versatility, hermeneutics is largely overlooked in current research contexts

S. D'Souza (✉)
School of Nursing and Midwifery, Western Sydney University, Sydney, NSW, Australia
e-mail: S.D'Souza@westernsydney.edu.au

© Springer Nature Singapore Pte Ltd. 2019
P. Liamputtong (ed.), *Handbook of Research Methods in Health Social Sciences*,
https://doi.org/10.1007/978-981-10-5251-4_63

189

because of its density of ideas and verbosity. Hence, the aim of this chapter is to unpack the main tenets of Heideggerian phenomenology as embodied in *Being and Time* and make a pioneering effort to underline the implications of such a framework for cross-disciplinary research. Heidegger's facticity of being is first explicated followed by an overview of the thereness of being which comprises being in the world of care spatially and temporally. In this existential state, verstehen or understanding is an event that occurs when the three fore-closures of the hermeneutic circle – fore-having, fore-sight, and fore-conception, work collaboratively to foreground meaning. Interestingly, in this ontological circle, it is language that serves as the medium of understanding. The benefits this ontic approach offers researchers, notably, those interested in cross-disciplinary research, are manifold. The main elements of hermeneutics provide a unique, flexible framework to examine a range of complexities and open up rich avenues for a multidimensional enquiry. Hermeneutics is indeed a boon for cross-disciplinary research. Harnessing the rich potential of hermeneutics would credit researchers with a sound intellectual base, advance cross-disciplinary research, and optimize the quality of research outcomes.

Keywords

Heidegger · Hermeneutics · Ontology · Being · Becoming · Cross-disciplinary research

1   Introduction

Hermeneutics, a branch of phenomenology, has long been used with huge gains in various fields of research as the underpinning paradigm. In particular, Heidegger's interpretive framework of being and becoming, commonly referred to as interpretive phenomenology, has been influential in shaping research undertakings, owing to its resilience and flexibility in bringing to life the lived experience. Derived from a Greek word, *hermeneuein*, the term hermeneutics means the study of methods of interpretation, primarily associated with the interpretation of the Bible and other texts (Cammell 2015). Thus, Heideggerian hermeneutics is closely connected to interpretation and signals a transition from epistemology to ontology: from the concept of understanding as a mode of knowledge to a study of being.

1.1   Bracket or Not to Bracket?

Heidegger's predecessors, Husserl and Dilthey, posited that understanding the essence of human existence is a hermeneutic act that entails a study of the life world of human beings. Husserl and Dilthey privileged the subjective meaning that humans accord to the lived experience, over preconceived ideas linked to the experience, and suggested that setting aside or *bracketing* these ideas would help unveil the meaning of the lived experience (McConnell-Henry et al. 2009;

Ortiz 2009; see also ▶ Chap. 112, "Understanding Sexuality and Disability: Using Interpretive Hermeneutic Phenomenological Approaches"). However, Heidegger's strand of hermeneutics departed from these phenomenologists, in that he set much store on prior knowledge. Heidegger subscribed to the view that meaning lies latent in every situation. It is only by applying prior knowledge or the subject's world view to the situation that implicit meaning is manifested to concrete meaning (Steiner 1978; Dowling 2004). His thesis is that language is the basis for both experiencing and understanding the world (Powell 2013). The capacity to understand events in the *life world* through language, by ascribing meaning to social situations, has rendered Heideggerian hermeneutics a powerful, intellectual trajectory.

1.2   Hermeneutic Influences in Research

Given its adaptability to diverse research situations and its pliancy in bringing to life the lived experience, Heidegger's interpretive framework has influenced a sizeable amount of research in various fields, specifically health and social sciences. In the context of health, Benner (1984) has used it successfully to study the lived experience of nurses when delivering care in Australian hospital settings. In the United Kingdom, Koch (1994) employed a hermeneutic approach to investigate the relationships between elderly patients and nurses during hospitalization. There are also several hermeneutic studies in critical care (Walters 1995; Little 2000; Johnson et al. 2006), mental health (Aho 2008; Kayali and Furhan 2013), and pediatrics (Totka 1996; Sorlie et al. 2003; Olausson et al. 2006). In the field of social science, Ream and Ream (2005) drew on hermeneutics to analyze the meaning of student dwelling in learning environments, while Greatrex-White (2007) explored the experiences of nursing students studying overseas. Hermeneutic enquiry has also been conducted to analyze teacher-student relationships (Giles et al. 2012), the nature of teaching and learning (Horrocks 2006; Vu and Dall'Alba 2008; Peters 2009; Riley 2011; Jones 2011), educational programs and their emphasis on the becoming of students (Ironside 2003; Dall'Alba 2009), and curriculum development (Donnelly 2002; Slattery et al. 2007; Gibbons 2011). Nevertheless, despite its resounding success as a philosophic base, interpretive phenomenology is largely overlooked by new researchers and doctoral students.

1.3   The Current Neglect of Hermeneutics

There are three main causative factors for the current neglect of hermeneutics. Firstly, Heidegger's ideas, as expounded in *Being and Time*, his main body of work, appear to be "esoteric and forbidding" (Cammell 2015, p. 236) and considerably complex to the average reader (Schmitt 2008; Peters 2009). Moreover, according to Smith (2009), after navigating the difficult terrain of Heidegger's interpretive phenomenology, readers are left disgruntled and unsated since Heidegger fails to offer answers to the questions he raises. The evidence provided indicates that emerging researchers and doctoral students new to the research process could be discouraged by the density of Heidegger's ideas, the verbosity with which he expresses them, and their own frustration at being abandoned at the brink of being. This situation is further exacerbated by the use of certain German words that defy English translation. Another related issue is the opacity of Heidegger's ontology (Schmitt 2008; Peters 2009). To most readers, Heidegger's obsession with ontology, the study of existence, comes across as absurd and illogical, as it dwells on an abstract question – *What does it mean to be?* We take it for granted that we know what it means to physically live in the world; as such, Heidegger's emphasis on the state of *being* in the world, a vague, unsubstantial notion confounds readers. Consequently, lack of knowledge on how to relate seemingly abstract hermeneutic concepts to concrete aspects of their research deters researchers from utilizing hermeneutics as the overarching paradigm to conceptualize their study.

Additionally, the fear of misinterpreting Heidegger's philosophy also sparks methodological concerns among researchers. The bulk of hermeneutics converges upon the study of lived experience which serves as a reservoir to help people interpret and make sense of life (Huang et al. 2012; Guenther 2012; Spratling 2013). Phenomenologists view interviews as one of the primary data collection methods as respondents are able to provide a subjective account of the experience encountered from which meaning can be distilled (DePoy 2016; see also ▶ Chap. 23, "Qualitative Interviewing"). However, Paley (2014) argues that focusing on the subjective meaning of the lived experience is a misinterpretation of hermeneutics. While Paley acknowledges that human beings have a number of experiences in the world, he states that the interpretive stance on lived experience is misconstrued to include a separate layer of experience, the subjective experience, "a ghostly, subjective thing, allegedly going on at the same time" (p. 4) as the lived experience. In effect, there is no such thing as subjective experience to recall or understand because human beings find meaning in life by interacting with things around them and "get on with stuff" (p. 5). Indeed, Paley opines that interviewing is not an interpretive strategy and claims there is nothing to gain from interviews for respondents merely fabricate answers to questions. His conclusion is that phenomenological researchers should carefully consider their research design, avoid qualitative interviews, and seek other methods of enquiry.

Viewed in the light of the evidence presented by Paley (2014), it is evident that misinterpretations of Heidegger's philosophy are entirely possible. Added to this, the fear of selecting methods of enquiry incongruent with hermeneutics, which would in turn affect the trustworthiness of their study, steers doctoral students and new researchers away from hermeneutics. Being a doctoral student and a cross-disciplinary researcher, I have found myself in the same predicament: questioning the validity of interpretivism to my study, grappling with the fear of misinterpreting hermeneutics, and misapplying the philosophy. Eventually, after an in-depth study of hermeneutics, I concluded that the benefits of hermeneutics lie in the reading or – should I say – interpretation! This chapter emerged as a result of my extensive reading of hermeneutic literature in an attempt to motivate and encourage fellow researchers to pursue the less tread but most valuable path of hermeneutics. Therefore, the aim of this chapter is to unpack the main tenets of Heideggerian phenomenology, as embodied in *Being and Time*, Heidegger's monumental work, and underline the implications of such a framework for cross-disciplinary research. How to use the evidence presented here to conceptualize your cross-disciplinary study, I leave to your interpretation!

2   The Main Tenets of Hermeneutics

2.1   The Question of Being

The fulcrum of interpretive phenomenology is the essence of being. What is being? Bonevac (2014) points out that the meaning of being is "self-evident, primitive, obvious or impossible to articulate. After all, everything's doing it, all the time!" (p. 166–167). However, Heidegger elevates the meaning of being to a transcendental level: *A being is*. This response sets out being as a vague entity, for the verb "is*"* in the process of being is unknown. It is not possible to explain the "is*"* in explicit terms; yet, rich, concrete meaning is hidden within the verb (Heidegger 1978). The paradoxical characteristic of being, its tendency to be abstract and tangible at the same time, urges Heidegger to pronounce it as the poorest and the richest of concepts. It is the poorest concept because *to be*, the act of existence, is an empty, meaningless notion. On the other hand, the fact remains that beings exist and have existed since the dawn of time. If beings exist, they are not nullity – they are something, with their own rich characteristics of being. Consequently, the act of being is rich in meaning, making it a valuable concept. Cammell (2015) elaborates that beings are already a part of the world, with their own distinctive traits. They dwell in language and move along the tangent of time with a prior understanding of how to do things: birthing and raising children, forging relationships, overcoming failure, and eventually facing death. Thus, the meaningless act of being becomes rich in significance when interpreted against the backdrop of the many phases of life and the meaningful relations that beings enact with one another (Campbell 2012). In this way, Heidegger sought to move hermeneutics from the level of epistemology (a passive, cognitive understanding of a phenomenon) to ontology (an active understanding of the phenomenon by interpreting the fullness of being in the world). It is this ontic question of being that Heidegger is engrossed with, and to which I turn, to explain the *thereness* of being in the world.

2.2   The Thereness of Being

Interpretive phenomenology revolves around *Dasein* or being. *Dasein* is defined as "an entity whose Being has the determinate character of existence" (Heidegger 1978, p. 34). *Dasein*, a German word for human being, constitutes two parts: *da* referring to here or there and *sein* meaning to be or being. The two word meanings underline the ontic nature of hermeneutics – to live is to be, since existence requires human beings to be out there in the fullness of the world (Dowling 2004; Greaves 2010). According to Heidegger (1962), the world is a sort of a thing that *Dasein* dwell in. It comprises a holistic web of interrelated beings, both animate and inanimate. Bonevac (2014) notes that Heideggerian hermeneutics underlines the social nature of *being in the world*. Thus, Bonevac (2014, p. 170) maintains that "we are not, of course, alone in the world. We encounter other people, other selves, other Daseins," and when cast into the *thereness* of the world, *Dasein* or human beings forge relationships with other entities as they engage in the *everydayness* of living. They do this by drawing on the varied action possibilities present in the world to actively participate in and sustain life. In fact, *Dasein* share a symbiotic relationship with other beings, for, in the course of completing practical tasks, *Dasein* support and nurture one another (Heidegger 1962, 1978; Steiner 1978; Koch 1994; Dall'Alba 2009). Heidegger's allusion to the reciprocal nature of existence reinforces the idea that we are always there in the world in some way or the other and we use the *equipment* or entities we encounter to function in the world. All equipment are *ready-at-hand*, *in order to* complete a specific task. For instance, a pen for writing and a knife for cutting are *ready-at-hand* equipment with distinctive uses. Bonevac (2014, p. 169) contends that "their being consists of their handiness," being *ready-at-hand*. In the same way, *Dasein* have a particular purpose in life that is fulfilled by immersing themselves in the *thereness* of the world by engaged involvement with other members of the community (Greaves 2010). In the course of each engaged encounter, beings perceive their own identity in relation to other beings which makes transparent the meaning of life. Hence, shared practices and common meanings are essential to the unveiling of *Dasein*.

2.3   Being and Becoming

Heidegger pursues the idea of an individuals' capacity to become, by maximizing the potential to be. Heidegger (1962, 1978) emphasizes that the potential to be lies dormant in every being. Bonevac (2014, p. 172) writes: "As *Dasein*, I face a field of possibilities. I am located within such a field, and my location determines the possibilities that are alive for me." Heidegger (1962, 1978) reports that individuals could either be wakeful to their being and achieve optimal outcomes in life or be slumberous and let life pass by. Wakefulness to the call of being denotes active engagement in shared practices, *being with* and *being amidst* others, and apprehending the meaning of lived experiences. It is possible to apprehend the totality of life when we realize that a gap exists between the actuality (living a routine, pedestrian life) and possibility (the ability to maximize our potential) of being that prevents us from *becoming* who we are. Understanding the gap between our current state of being (merely existing) and our infinite potential to be (the capacity to live life to the fullest and achieve positive outcomes) paves the way for our *becoming* as human beings (Heidegger 1962, 1978).

Bonevac (2014) elaborates on Heidegger's theory of being and becoming by explaining that being in the world, that is, *being with* and *being amidst* others, opens up a field of possibilities for *Dasein* to use. Journeying toward the future by fully interacting with other beings, making wise choices, and converting probabilities into possibilities is being fully awake to *Dasein's* being. Dasein can raise themselves from being to becoming, from probability to possibility, by operating fully within their realm of possibilities, understanding themselves and the world they live in. In short, being wakeful to the purpose of life bestows on us the meaning of existence.

2.4   Being in the World of Care

Wakefulness to *Dasein's* being, a hermeneutic trait, draws attention to Heidegger's notion of care (*sorge*) and its connection to existentialism. Heidegger (1978) holds that care is a benevolent quality that enhances a being's selflessness. It can only be administered when beings negotiate meaning by interacting with other entities present within the world. Heidegger elaborates on two approaches to care: care as day-to-day worries (interventionist care) and care as concern or wakefulness to *Dasein's* being (emancipatory care). While the former approach is a mundane, everyday aspect of existentialism, it is the latter that elevates beings to acts of concern for others. It prompts beings to rise above themselves and learn to cope with situations by assuming a sense of responsibility and displaying sensitivity toward the needs of others (Joensuu 2012; Larivee 2014). Additionally, Heidegger maps caring on a continuum with interventionist and emancipatory care on either end. Interventionist care focuses on alleviating the anxieties of entities by arbitrating on their behalf or *leaping in* for them, as in the case of a nurse caring for a sick child or a parent looking after a child. In contrast, emancipatory care is concerned with *leaping ahead* or empowering someone with the skill to take care of themselves. It channels beings toward maximizing their future potential, and *Dasein* should aim for emancipatory care that bestows independence (Tomkins and Eatough 2013). Thus, *sorge* is critical for conducting meaningful interactions between beings in the world.

2.5   Space and Time

Being in the world also constitutes being in the world spatially and temporally. Time and space are intertwined elements in which *Dasein* dwell. Heidegger (1978) uses the terms dwelling and spatiality interchangeably. Dwelling in a house has little relation to occupying space, instead it denotes a sense of belonging; so too, human beings belong to this world and their belonging is manifested through their temporality. It is important to note that in this intellectual orientation, space and time are not perceived as measurable items; rather, emphasis is given to being in a particular place at a specific time. With regard to spatiality, Heidegger (1978) reiterates that the process of existentialism has granted human beings a predestined place in the world. Resultantly, it is not possible to situate them physically in any specific *worldspace.*

Dasein is in this world; it presences itself through active interaction and interpretation of other beings. In contrast, all other entities in this world have a physical space. It is thus possible to position them spatially by determining their proximal distance to other beings or *ready-to-hand* objects such as tools (Heidegger 1978). In a sense, through involvement with other beings (both animate and inanimate), *Dasein* give them space or *make room* for them.

Moreover, Heidegger's conception of time diverges from the generally held view of the linear flow of current time. He favors a more revolutionary outlook of time with its "endless, spatialised succession of present moments" (Orr 2014, p. 116) between which beings exist in a state of infinite possibilities. He portrays *Dasein* (human beings) as temporal beings that experience ecstatic time, while all other beings are grounded within a linear progression of time (Orr 2014). Kakkori (2013) notes that "Things in the world have duration, and this duration is given by *Dasein*" (p. 575–576). As a case in point, Kakkori offers the example of a 13-year-old vase. The vase by itself is unable to have time; only *Dasein* have the ability to afford a time duration to it, proving *Dasein's* capacity to constitute and initiate time.

Kakkori (2013) adds that the Heideggerian *ecstases* of time, the present, the past, and the future, do not follow one another but rigorously overlap each other. For instance, references to the past are in reality "the present of the past" (p. 580) because we are situated in the present but reaching out to the past. Similarly, connecting the past to the present makes the present "the future of the past" (p. 581) since viewing the present from the past makes the present the future of the past. Furthermore, Heidegger explains that although the ordinary world view of time enables us to state the time or calculate the time interval between events, it is unsuccessful in capturing the nature of time, "the now moment" (p. 573), because time is constantly advancing toward our future, our becoming. It is only by stretching ourselves *futurally* along the tangent of time and *temporalizing* our being in relation to other beings can the now moment be ontically secured (Johnson 2000; McConnell-Henry et al. 2009; Kakkori 2013). It is evident that existing spatially and temporally is a fundamental dimension of being in the world.

2.6   Verstehen

Hermeneutics is the interpretive understanding of the state of being. Heidegger (1978) posits that *Dasein* exist in a mode of endless possibilities, journeying through life and understanding themselves through the "worldhood of the world" (p. 91), which they inhabit. Beings understand and identify themselves by mediating meaning with other pre-existing entities like nature and other beings. In doing so, interpretation serves to enhance understanding (Wilson 2014). Bonevac (2014, p. 177) suggests that "*Dasein* encounters the world as already structured and already having significance in relation to *Dasein* itself." Heidegger draws on the functions of opening doors or hammering to illuminate the process of interpretive understanding. When implementing practical tasks, beings come across *ready-to-hand* tools, which include doors and hammers. In such situations, they refrain from opening and shutting the door or weighing the hammer to assess their physical characteristics. Indeed, their prior knowledge enables them to interpret something as something. They project their understanding laterally and associate doors with buildings to work out their purpose (Warnke 2011). Likewise, the purpose of the hammer is determined by relating it with other tools such as nails and boards. The working out of a tool's functionality by looking beyond it and linking it to other entities is an act of interpretative understanding or *verstehen* (Heidegger 1978; Steiner 1978; Risser 1997). Heidegger's philosophy as a whole is disposed toward *verstehen*, the ability of human beings to understand the meaning of life.

2.7   The Hermeneutic Circle

In the hermeneutic enterprise, *verstehen* occurs in a circular movement. As noted earlier, on account of beings existing amidst a matrix of interrelationships, they already possess some knowledge of the world and its entities. The potential to understand meaning lurks in every situation they encounter. By means of collectively understanding other entities and themselves, *verstehen* takes place, implying that in order to understand a part of the experience, one must understand the experience in all its entirety. This apprises us of the circular nature of understanding (Risser 1997). It also thrusts into prominence the fore-structures of understanding in this hermeneutic circle, which play a significant role in meaning attribution: fore-having, fore-sight, and fore-conception. The first fore-structure, fore-having, points to presuppositions about the entity or speculating on what the entity could be. Heidegger draws on his famous example of a hammer to explain this fore-structure. For instance, people who have never seen a hammer would speculate on its use. In doing so, their prior knowledge confers partial meaning onto the hammer by helping them realize that a hammer is a type of tool meant for a specific task. The next fore-structure, fore-sight, increases understanding by helping beings consider more information about the entity to be understood, such as working out that a hammer could be used with nails. Fore-conception completes the hermeneutic circle. It is the process of deriving more complex information about the entity. In the case of the hammer, it is construing that different types of hammers are used for diverse tasks (Heidegger 1978; Risser 1997; Johnson 2000). Overall, the hermeneutic circle involves a back-and-forth movement while making connections between the matrix of interrelations and the entity that is to be understood, in order to arrive at interpretive understanding. To sum up, interpretive understanding is manifested by "refining and piecing together ... local fore-conceptions and fore-understandings into an overall understanding" (Bonevac 2014, p. 179).

2.8   Language: The House of Being

In the hermeneutic circle, language is credited as the channel that communicates understanding. It permits beings to fulfil pragmatic activities like developing and organizing something, exchanging and debating ideas, and questioning and speculating in order to function effectively in the world (Heidegger 1962, 1978). When engaging in these functional activities, beings become oriented to the world around them through language. Language thus unites the world and enables beings to understand themselves, their world of involvement, and the purpose of life. In short, language energizes our life and activates an understanding of our potential to become. The ability of language to shelter and support beings as they fulfill their everyday tasks has urged Heidegger to refer to it as the house of being (Cammell 2015). In turn, beings dwell in language and draw on it to conclude practical transactions with other entities. Heidegger perceives language not as a coded system, but as an all pervasive being, that bestows on users a shared horizon for communication and understanding to occur (Clark 2011). The foundation of language is discourse, composed of spoken and written text. Heidegger (1978) defines a text as a "a totality of words" (p. 204) that represents the *thereness*, the diverse nature of the world. The key elements of a text are the aim of the text, its content, and the mode of communication. Research on hermeneutic textual discourse illustrates that uncovering the meaning enclosed within a text entails moving past superficial meaning, to an analysis of the relationship between the parts and the whole of the text (Neuman 2006; Ortiz 2009). In brief, language enacts meaning mediation.

The benefits this ontic approach offers researchers, principally, those interested in cross-disciplinary research, are manifold. Before I outline its' merits, some of the distinct challenges of disciplinary crossing must be acknowledged. With cross-disciplinary research, I anticipate that the selection of a suitable study design could prove problematic, since it would need to satisfy the requirements of different disciplines. It also calls for interdisciplinary knowledge on the part of the researcher and a broader set of research skills. It is alleged that there are numerous biased attitudes associated with interdisciplinary research (Aagaard-Hansen 2007), with "lack of acceptance of paradigm breaking or shifting" (Karniouchina et al. 2006, p. 274) and an excessively demanding ethics process (Aagaard-Hansen 2007) quoted as prominent obstacles. Besides these obstacles, finding a "publication outlet" for interdisciplinary research findings could be an insurmountable task (Karniouchina et al. 2006, p. 274).

3 Hermeneutics: The Springboard to Cross-Disciplinary Research

The limitations of disciplinary crossing outlined above are duly noted; yet, in higher education, the void and perceived need for interdisciplinary collaboration are acutely felt. In spite of this, literature on cross-disciplinary undertakings is disturbingly thin. What is more, it is worthwhile pointing out that while Heidegger's strand of hermeneutics is applicable to all forms of research, it holds greater promise for cross-disciplinary research. Nonetheless, remarkably, there are no traces of any hermeneutic cross-disciplinary enterprises reported in literature. I make the first attempt to call attention to this resourceful ontic approach and delineate its profitable alignment with cross-disciplinary research.

With language as the common denominator, interpretive phenomenology could cross over different fields of study and offer researchers a suitable research design that would procure a maximum amount of trustworthy data. As a research paradigm, it would afford a multidimensional view of the lived experience through the lens of ontology (see also ▶ Chap. 6, "Ontology and Epistemology"). It could bring together hitherto diverse fields such as the sciences and humanities by providing a flexible, interpretive framework that would broaden the scope of enquiry and generate new insights. What follows is an account of the hybrid benefits hermeneutics bestows on cross-disciplinary researchers.

3.1   Addressing Complex Phenomena

Employing hermeneutics allows cross-disciplinary researchers to get immersed in complex phenomena. It assists in unearthing different layers of meaning as the phenomena are examined from different perspectives. For instance, as a cross-disciplinary researcher, I undertook a research project on *Genre Mixing in Undergraduate Nursing Texts*. Using hermeneutics as the philosophy and analytical approach, the study dwelt on the research question: *What does it mean to be a student nurse?* This entailed a focus on the identity of student nurses, their goals, and their existence in the world of *sorge (care)*, that is, the manner in which student nurses are prepared to care for their patients; the relationships they form with other student nurses, tutors, literacy staff, and clinical facilitators; and the influence of these relationships on students' academic progress. Crossing the boundary from nursing to education also allowed me to conduct a language study into the becoming of the beings who deliver care (student nurses). It enabled me to research their assignment writing practices at university and the triggers and barriers to students' writing competency. Using hermeneutics as the analytic tool strengthened the cross-disciplinary endeavor for it provided the means of seeing the phenomena in its entirety and brought to light the way student beings exist spatially and temporally in the world of nursing by *making room* for the other entities they encounter such as work, family, friends, and assessment demands. Moreover, I drew on the hermeneutic circle which is best suited to addressing complex phenomena. The hermeneutic circle provides interdisciplinary researchers a platform to lay bare the research question and examine the fore-structures of understanding, by relating the parts of the phenomenon to the whole of the research experience. Hermeneutic questions that might guide researchers in this process are: *Would the aspects of the phenomena being studied contribute to answering the research question? Are there any aspects of the lived experience that might prove elusive? Have I planned for variability in the study?*

In the study on *Genre Mixing in Undergraduate Nursing Texts*, the hermeneutic circle permitted a close scrutiny of the interview data, to piece together the parts and whole of the text to generate a comprehensive picture of the phenomena in all its dimensions. It traced the *being and becoming* of student nurses as they access the range of possible support resources at their disposal, learn to overcome their writing deficiencies, write precise nursing documents, and communicate information clearly. It authorized me to study not only human behavior and the web of interrelations between people but also the language that defines human interactions, thus coinciding with the disciplines of health and social sciences. From the evidence outlined, it is possible to infer that hermeneutic cross-disciplinary interests empower researchers by stretching their range of vision and facilitating innovative solutions to the problematic under scrutiny.

3.2   Adding Depth and Breadth

In contrast to mono-disciplinary studies, a multidisciplinary endeavor founded on a hermeneutic base adds depth and breadth to the research undertaking. While hermeneutic studies in a single discipline have proved fruitful, they lack the unique richness of interdisciplinary research. The latter serves as a bridge between disciplines, fosters new ways of thinking, and extends the foci of the study over a broader area. Cross-disciplinary researchers consider the relevance of the study to their own discipline and verify data gathered by analyzing it from a cross-disciplinary perspective. This is done by actively exploring the lived experience through several modalities and in all its different aspects. Doing so negates bias and provides a fuller, richer description of the lived experience. Furthermore, positioning cross-disciplinary research within hermeneutics also shifts it from a passive, positivist form of cognitive understanding to an active, dynamic form of interpretive understanding (see also ▶ Chaps. 9, "Positivism and Realism," and ▶ "Social Constructionism"). In brief, cross-disciplinary hermeneutic studies broaden practical and scientific outcomes, make a valuable contribution to new knowledge, and attract wider exposure to multidisciplinary issues.

3.3   Providing a Selection of Interpretive Methods

Heideggerian hermeneutics helps cross-disciplinary researchers exploit the most appropriate modes of enquiry. Interpretivists use interviews, texts, and participant observation as data-gathering methods. Drawing on these methods and ensuring that they coincide conceptually with the philosophy of the study grant cross-disciplinary researchers the leverage to obtain richer and more in-depth information about the whole experience. Of the interpretive methods listed, textual analysis is frequently used because it permits researchers to listen to the speech of language in its written form. This happens when researchers listen to the story that is narrated in the text by participants of the world (see also ▶ Chaps. 47, "Content Analysis: Using Critical Realism to Extend Its Utility," ▶ 48, "Thematic Analysis," ▶ 49, "Narrative Analysis," and ▶ 28, "Conversation Analysis: An Introduction to Methodology, Data Collection, and Analysis"). In an interdisciplinary enterprise, the speech of the text could be more pronounced because things in the world speak to researchers from different perspectives and from diverse roles.

For example, an educational researcher could conduct a textual analysis on health science students' writing to understand the writing practices in health science. In this context, the disciplinary boundaries of education and health science are bridged in order to fully explore the phenomenon through diverse modalities. The data in the text could speak to the researcher in hybrid ways and unveil the writing requirements of health science students, their attitude to their discipline, and their roles and aspirations as students. All this information is revealed through the speech of the text. On the other hand, a mono-disciplinary study would only provide a flattened view of the world.

It is worth noting that multimodal, cross-disciplinary analysis calls for the researcher to be more attuned to the nuances of meaning in the text, attitudes expressed, and world views that emerge. A distinct requirement for quality textual analysis is listening to the silence of the text. Silence here refers to opinions, beliefs, attitudes, or cultural practices that are not conveyed but lie implicit in the text. It is also possible to combine heterogeneous methods such as interviews and textual analysis or interviews and participant observation in order to enrich research outcomes. Therefore, employing hermeneutic methods and conducting studies of such magnitude endow a phenomenological seeing to cross-disciplinary researchers.

4 Conclusion and Future Directions

The rapidly changing knowledge society demands novel ways of thinking and being that contribute to original research. Added to this, with technological advancement and increased complexity of issues, the need for research that opens up rich avenues of multidimensional enquiry is recognized. Most importantly, the paucity of such research in health and social sciences is increasingly noted. The multiplicity of health issues experienced by patients and their families and the academic challenges encountered by students and academics in higher education contexts warrant an exclusive approach to data gathering and analysis. Hermeneutic studies that cross disciplinary boundaries and delve into the fullness of the lived experience emerge as an asset that could power such research ventures.

A cross-disciplinary hermeneutic endeavor in tertiary education is a new concept. It is a legitimate mode of enquiry that would invigorate research practices and provide the means for researchers to capitalize on the potentiality of different disciplines and corroborate findings from an interdisciplinary perspective, through the multifocal lens of hermeneutics. However, it must be acknowledged that hermeneutic disciplinary crossing is a time-consuming and laborious process. It requires researchers to expend time understanding the core concepts of hermeneutics to avert the risk of misconstruing them and conceptually and methodologically misaligning them with their research project. Admittedly, these aspects ought to be considered when planning a hermeneutic cross-disciplinary investigation; but the benefits far outweigh any possible disadvantages in terms of time and effort.

It is envisaged that this chapter would generate sufficient interest in hermeneutics and set the trend for the convergence of hermeneutics and interdisciplinary research.

Interest in dynamic research paradigms such as hermeneutics should be fostered in undergraduate studies and nurtured right through to postgraduate research degrees. For instance, most undergraduate degrees in health and social sciences have an introductory research-based unit to orient students to the purpose and basic principles of research; help them make sense of conceptual frameworks, data collection, and analytic methods; and practically apply their research knowledge to drafting a brief research proposal. Strongly embedding hermeneutics into undergraduate coursework and continuing to scaffold it into postgraduate studies would make it more approachable as a research paradigm. Flowing on from undergraduate degrees, postgraduate research degrees too need to offer hermeneutics as an intrinsic part of the course curricula, with aspects of hermeneutics modelled by experienced researchers through workshops and hands-on interpretive analysis of data. These strategies would demystify the world of hermeneutics and make it more visible for new researchers and doctoral students grappling with methodological concerns. Allocating grants for cross-disciplinary research in higher education would also encourage more cross-disciplinary collaborations and motivate researchers to turn to multifaceted research paradigms and methods of enquiry such as hermeneutics to enhance trustworthiness and enrich research outcomes. I hope that this chapter will give fellow researchers the impetus to cross disciplinary boundaries and move from *being* to *becoming*, from the actuality to the possibility of their *being!*

References

Aagaard-Hansen J. The challenges of cross disciplinary research. Soc Epistemol: J Knowl Cult Policy. 2007;21(4):425–38.

Aho K. Medicalising mental health: a phenomenological alternative. J Med Humanit. 2008;29(4):243–59.

Benner P. From Novice to expert: excellence and power in clinical nursing practice. Menlo Park: Addison-Wesley Publication; 1984.

Bonevac D. Heidegger's map. Acad Quest. 2014;27:165–84.

Cammell P. Rationality and existence: hermeneutic and deconstructive approaches emerging from Heidegger's philosophy. Humanist Psychol. 2015;43:235–49.

Campbell S. Early Heidegger's philosophy of life: facticity, being and language. 2012. Retrieved from http://www.site.ebrary.com. Accessed 1 Sept 2017.

Clark T. Martin Heidegger. 2nd ed. New York: Routledge; 2011.

Dall'Alba G. Learning professional ways of being: ambiguities of becoming. Educ Philos Theory. 2009;41(1):34–45.

DePoy E. Introduction to research: understanding and applying multiple strategies. 5th ed. St. Louis: Elsevier; 2016.

Donnelly J. Instrumentality, hermeneutics and the place of science in the school curriculum. Sci Educ. 2002;11(2):135–53.

Dowling M. Hermeneutics: an exploration. Nurs Res. 2004;11(4):30–9.

Gibbons A. The incoherence of curriculum: questions concerning early childhood teacher educators. Australas J Early Childhood. 2011;36(1):9–15.

Giles D, Smythe E, Spence D. Exploring relationships in education: a phenomenological enquiry. Aust J Adult Learn. 2012;52(2):214–36.

Greatrex-White S. A way of seeing study abroad: narratives from nurse education. Learn Health Soc Care. 2007;6(3):134–44.

Greaves T. Dasein: a living question. In: Starting with Heidegger. 2010. Retrieved from http://www. site.ebrary.com. Accessed 25 Sept 2017.

Guenther J. The lived experience of ovarian cancer: a phenomenological approach. J Am Acad Nurse Pract. 2012;24(10):595–603.

Heidegger M. Being and time (translated by Macquarie J, Robinson E). London: SCM Press; 1962.

Heidegger M. Martin Heidegger: being and time (translated by Macquarie J, Robinson E). Oxford: Basil Blackwell; 1978.

Horrocks S. Scholarship, teaching and calculative thinking: a critique of the audit culture in UK nurse education. Nurse Educ Today. 2006;26(1):4–10.

Huang YP, Chen SL, Tsai SW. Father's experiences of involvement in the daily care of their child with developmental disability in a Chinese context. J Clin Nurs. 2012;21:3287–96.

Ironside P. New pedagogies for teaching thinking: the lived experiences of students and teachers enacting narrative pedagogy. J Nurs Educ. 2003;42(11):509–16.

Joensuu K. Care for the other's selfhood: a view on child care and education through Heidegger's analytic of Dasein. Early Child Dev Care. 2012;182(3):44–434.

Johnson ME. Heidegger and meaning: implications for phenomenological research. Nurs Philos. 2000;1:134–46.

Johnson P, St. John W, Moyle W. Long term mechanical ventilation in a critical care unit: existing in an everyday world. J Adv Nurs. 2006;53(5):551–8.

Jones A. Philosophical and socio-cognitive foundations for teaching in higher education through collaborative approaches to student learning. Educ Philos Theory. 2011;43(9):997–1011.

Kakkori L. Education and the concept of time. Educ Philos Theory. 2013;45(5):571–83.

Karniouchina E, Victorino L, Verma R. Product and service innovation: ideas for future cross-disciplinary research. J Prod Innov Manag. 2006;23:56–280.

Kayali T, Furhan I. Depression as unhomelike being-in-the-world? Phenomenology's challenge to our understanding of illness. Med Healthc Philos. 2013;16(1):31–9.

Koch T. Establishing rigour in qualitative research: the decision trial. J Adv Nurs. 1994;19:996–1002.

Larivee A. Being and time and the ancient philosophical tradition of care for the self: a tense or harmonious relationship? Philos Pap. 2014;43(1):123–44.

Little C. Technological competence as a structure of learning in critical care. J Clin Nurs. 2000;9 (3):391–9.

McConnell-Henry T, Chapman Y, Francis K. Unpacking Hedeggerian phenomenology. Online J Nurs Res. 2009;9(1):1–12.

Neuman WL. Social research methods. 6th ed. Boston: Pearson Education; 2006.

Olausson B, Utbult Y, Hansson S, Krantz M, Brydolf M, Lindstrom B, Holmgren D. Transplanted children's experiences of daily living: children's narratives about their lives following transplantation. Pediatr Transplant. 2006;10(5):575–85.

Orr J. Being and timelessness: Edith Stein's critique of Heideggerian temporality. Mod Theol. 2014;30(1):115–30.

Ortiz NR. Hermeneutics and nursing research: history, processes, and exemplar. Online J Nurs Res. 2009;9(1):1–7.

Paley J. Heidegger, lived experience and method. J Adv Nurs. 2014;70(7):1520–31.

Peters MA. Editorial: Heidegger, phenomenology, education. Educ Philos Theory. 2009;41(1):1–6.

Powell J. Studies in continental thought: Heidegger and language. 2013. Retrieved from http://www.eblib.com. Accessed 15 Aug 2017.

Ream T, Ream C. From low lying roofs to towering spires: toward a Heideggerian understanding of learning environments. Educ Philos Theory. 2005;37(4):586–7.

Riley D. Heidegger teaching: an analysis and interpretation of pedagogy. Educ Philos Theory. 2011;43(8):797–815.

Risser R. Hermeneutics and the voice of the other: re-reading Gadamer's philosophical hermeneutics. Albany: State University of New York Press; 1997.

Schmitt R. Can Heidegger be understood? Inquiry: an interdisciplinary. J Philos. 2008;10:1–4.

Slattery, Krasny K, O'Malley. Hermeneutics, aesthetics, and the quest for answerability: a dialogic possibility for reconceptualising the interpretive process in curriculum studies. J Curric Stud. 2007;39(5):537–58.

Smith C. Martin Heidegger and the dialogue with being. Cent States Speech J. 2009;36(4):256–69.

Sorlie V, Jansson L, Norberg A. The meaning of being in ethically difficult situations in paediatric care as narrated by female registered nurses. Scand J Caring Sci. 2003;17(3):285–92.

Spratling R. The experiences of medically fragile adolescents who require respiratory assistance. J Adv Nurs. 2013;68(12):2740–9.

Steiner G. Heidegger. Glasgow: William Collins Sons & Co. Ltd.; 1978.

Tomkins L, Eatough V. Meanings and manifestations of care: a celebration of hermeneutic multiplicity in Heidegger. Humanist Psychol. 2013;41:4–24.

Totka J. Exploring the boundaries of pediatric practice: nurse stories related to relationships. Pediatr Nurs. 1996;22(3):191–3.

Vu T, Dall'Alba G. Exploring an authentic approach to assessment for enhancing student learning. Paper presented at the AARE Conference, Brisbane. 2008.

Walters A. Technology and the lifeworld of critical care nursing. J Adv Nurs. 1995;22(2):338–46.

Warnke G. The hermeneutic circle versus dialogue. Rev Metaphys. 2011;65(1):91–112.

Wilson A. Application of Heideggerian phenomenology to mentorship of nursing students. J Adv Nurs. 2014;70(12):2910–9.

Feminism and Healthcare: Toward a Feminist Pragmatist Model of Healthcare Provision

12

Claudia Gillberg and Geoffrey Jones

Contents

Abstract

This chapter covers a range of topics pertaining to the ontological, epistemological, and ethical intricacies, complications, and possibilities of providing quality healthcare to women patients regardless of disability, race, ethnicity, and class by using empirical examples of certain diseases. Methodological concepts through reflections on subjectivity and objectivity are presented as contested issues, and radical objectivity, a concept comprising subjectivity, objectivity, and

C. Gillberg (✉)
Swedish National Centre for Lifelong Learning (ENCELL), Jonkoping University, Jönköping, Sweden
e-mail: claudia.gillberg@ju.se

G. Jones
Centre for Welfare Reform, Sheffield, UK

© Springer Nature Singapore Pte Ltd. 2019
P. Liamputtong (ed.), *Handbook of Research Methods in Health Social Sciences*,
https://doi.org/10.1007/978-981-10-5251-4_64

intersubjectivity, is proposed as a knowledge paradigm that allows healthcare personnel and patients to make knowledge claims that are mutually recognized as valid. Three models of healthcare, paternalistic, person-centered, and feminist pragmatist, are presented, outlining the specific problems inherent in each model of healthcare provision. The paternalistic model allows for no agency on the patients' part, elevating healthcare personnel, specifically doctors, to authoritative knowers. The person-centered model of healthcare grants some shared responsibility between healthcare personnel and patients, and some concessions are made toward patients as knowers. In the feminist pragmatist model, healthcare personnel and patients commit to equal relationships. Gender equality and gender equity are identified as insufficient tools for organizational change, and theories of professions are drawn on to deliberate about change at the systemic level.

Keywords

Feminism · Methodology · Ontology · Epistemology · Paternalism · Pragmatism

1   Feminism and Healthcare: Starting Points

Feminism and healthcare is a topic so complex that no one could ever hope to cover it satisfactorily in a single chapter. Choices regarding perspective, suggestions, selection of examples of problems, and solution-oriented proposals must be made. These are the author's choices, an author who is herself disabled by chronic disease, and is an academic. My views are colored due to my experiences of healthcare and the body I live in and with which I encounter the outside world. My ontological and epistemological points of departure are (a) women are human beings and entitled to fully participate in society, (b) "women" is not a generic term but consists of variations of being a woman and the perception by others of "woman's" multiple meanings, and (c) a belief in collaborative knowledge building.

This revelation will be understood by some as biased, subjective, and therefore invalid. Based on their understanding of valid knowledge and scientific rigor, there are ontologies and epistemologies that stand in stark contrast to the research and knowledge paradigm within which I operate (Herr and Anderson 2005). In my paradigm, feminist pragmatism, it is encouraged to reveal points of departure for critical appraisal by others and then use new/other perspectives as an enriching and fundamentally necessary contribution toward a more comprehensive understanding of healthcare. It is argued that in doing so, other forms of knowledge will be elevated with the power to reject other forms, but in this chapter I envisage the reader as unfamiliar with this line of reasoning and as malleable toward accounting for what others perceive as unscientific, anecdotal, or invalid (Minnich 2005).

It is this uniqueness of each "knower's" perspective and experience (here of healthcare) that is welcomed by feminist pragmatists exactly because multifarious perspectives, when listened to and used toward problem-solving, have the potential to and often do contribute to areas in society that are better equipped to offer solutions, benefiting more people than the few belonging to an elite whose views traditionally infuse policies, practices, and entire societal systems (Minnich 2005; Gillberg 2012). To understand this chapter is to acknowledge the existence of other paradigms than the alleged objectivity and neutrality offered in so many research disciplines, that, on closer inspection, rarely offer objectivity in terms of justice being done to all in equal measure, as research questions remain only partially or underinformed and results will not always enable a broad application of newly produced knowledge (Andrist 1997; Anjum 2016; Gillberg and Vo 2014; Minnich 2005; Reid and Gillberg 2014). Feminist pragmatists will ask questions such as for whom is knowledge produced and where "best practice" is devised. For this chapter, I have drawn on several aspects of healthcare, geographical diversity, and texts that span decades to convey some continuity for this complex and difficult topic.

2 Politically, Historically, and Culturally Embedded Healthcare

Healthcare is historically embedded in political systems and cultures with ongoing power struggles and shifting game plans coupled with ideas about privatized or public healthcare, healthcare rationing, and other rationales that have little or nothing to do with human beings' needs (Kennedy 2012; Lian 2017; Nott 2002; Sherwin 1992). Such factors perpetuate inequalities prevalent in healthcare while they also create new problems. Some inequalities can be summed up as inadequate or no access to primary healthcare, long waiting times to receive an appointment for specialist care depending on where in a country or in which country patients live, and gendered notions of patients on the part of healthcare personnel (Cody 2003; Cook et al. 2017; Evans and Mafubelu 2009; Risberg et al. 2009).

Occasionally, as occurred with AIDS and women, battles are won, though not without personal tragedies and loss of life (Epstein 1996; Shotwell 2016). Health policies, official guidelines and diagnostic criteria, curricula for medical students, access to primary and specialist healthcare, health insurance policies, funding of research and inadequate to no treatments for certain conditions can, and often do, have dire consequences for people's lives (WHO 2015; McDonnell et al. 2009).

While ontological and epistemological approaches in inquiry are culturally, ideologically, and politically informed, other factors contribute to healthcare remaining lacking for many women and men, but at the macrolevel inequalities in healthcare affect women more than men rendering chapters such as this a necessity. Differing schools of thought in medicine and organizational theories can constitute colliding ideas about how to organize healthcare in relation to notions of who, and what, the patient is. Patients can be many things to healthcare providers, which relates to academic disciplines' predominant narratives about patients but also with what constitutes a disease. It could be argued that these other factors, to a large degree, belong to the field of theories and studies of the professions and issues pertaining to power and the struggle for recognition of expertise (Freidson 1993). Such a struggle occurs both within a profession and between different professions within the same organization or system. Healthcare represents the fulcrum of such a struggle. It is also a struggle in the field of tension between professions and organizations (Andrist 1997; Butler et al. 2004; Fenton 2016a, Fenton b, Freidson 1993; Hafferty and Light 1995; Hsu 2010; Humphries et al. 2017; Klein 2004; Light 2000; Pescosolido 2013; Risberg et al. 2009; Steen 2016; Tasca et al. 2012).

3   Paternalism Within Healthcare

There are schools of thought that dismiss patients as knowers altogether, ridiculing entire patient populations, especially women patients (Fenton 2016a; Kennedy 2012; Lian and Robson 2017; cf. McEvedy and Beard 1970; Mickle 2017; Mitchell and Schlesinger 2005; Robson and Lian 2017; Staples 2017). The paternalistic school of thought deserves special mention, because it is in this paradigm that women have historically fared the worst (Fricker and Hornsby 2000; Sherwin 1992; Webster 1996, 2003; Wendell 1996). Cody (2003, p. 288) provides a definition of paternalism as:

> Paternalistic practices, wherein providers confer a treatment or service upon a person or persons without their consent, ostensibly by reason of their limited autonomy or diminished capacity, are widespread in health care and in societies around the world. . . Numerous issues surround paternalistic practices.

Cause for alarm arises due to a branching off in recent decades to extreme paternalistic schools of thought, which evidence suggests is thriving, raising the question as to why such an ontology, because it is first and foremost an ontology in my opinion, has a place in healthcare systems.

Ideally the aim should be for a relationship between the patient and medical personnel as defined by Govender and Penn-Kekana (2007, p. 4),

> A good interpersonal relationship between a patient and provider - as characterised by mutual respect, openness and a balance in their respective roles in decision-making – is an important marker of quality of care.

I would propose and argue that there are three models of healthcare provision regarding "what is" and "whose/what knowledge counts," only one of which meets the above specification, and these are:

1. **The Paternalistic Model**
   View of patient: The patient has no agency as a lay person and is therefore ignorant.
      View of knowledge and learning: Only the medical doctor knows. Patients may be able to learn from their doctors but must recognize the latter are superior. Learning is one-dimensional and one-directional. Knowledge is objective and static.
      View of medicine: What is, is, whatever the medical profession determines as true and valid. There are no margins for uncertainty on the part of the doctor.

View of impact of treatment and outcome validity: It is the patient's responsibility to comply with prescribed treatments. If the outcome is undesirable, punitive measures for the patient will ensue; since the patient has no agency, only the medical doctor's knowledge has validity.

Outcome: Recovery is determined by the medical personnel irrespective of the patient's opinion, related to continued ill-health. Patients are discouraged from seeking further help.

2. **The Neutral Patient Model**

View of patient: The patient is neutral and ascribed value as a situated, subjective knower. As such, the patient possibly holds information that will be of use to medical and other healthcare personnel.

View of knowledge and learning: Knowledge is not one-dimensional since the patient is granted agency and recognized as knowing to some extent.

View of medicine: Medicine is a fluid field of knowledge, and learning is ongoing. What is best for the patient is a semi-open question, but ultimately healthcare personnel know better, and their advice ought to be followed.

View of impact of treatment and outcome validity: Patients as subjective knowers are entitled to report back and be believed if a treatment does not yield the desired outcome. Healthcare personnel will reconsider and discuss new ways forward with the patient. Outcome validity is determined by both medical doctors and patients.

Outcome: Partially shared responsibility for improved health, if symptoms persist and patient continues seeking medical help, uncertainty ensues. The patient is likely to be treated in a paternalistic manner due to system structure.

3. **The Feminist Pragmatist Model**

View of patient: The patient is a nonconforming, knowing human being in her own right, embodying a plethora of experience and perspective that may prove useful in solving the problem at hand; in other words, patients have full agency and are regarded as equals.

View of knowledge and learning: Knowledge is best constructed from a multitude of perspectives and must have a clearly defined applicable purpose to open-mindedly address problems with a view to satisfactorily resolving them in collaboration with those affected by the problem. Learning is a process, ideally grounded in meticulously researched fact and lived experience.

View of treatment and outcome validity: A genuine interest in the patients' account of improvement or recovery. A treatment is only valid if the patient says the situation has been resolved satisfactorily.

Outcome: Based on system structure, it is unclear how far this model can accommodate patients with long-term healthcare needs. Theoretically, recovery occurs only if the patient says so and resumes pre-illness activity levels, which is acknowledged by the medical personnel.

These models are not discrete; there can be movement between models, barring the extreme paternalistic form, which arguably functions as a rigid ideology. While models are never complete or an accurate reflection of real-life situations, those detailed above might prove helpful in outlining the possibilities and obstacles healthcare personnel and patients can face. It is not easy to question one's paradigmatic base let alone to arrive at the decision that an absolutist stance in healthcare does not and cannot translate into solidarity with patients when educated in model 1, especially if one's belief in medical knowledge's absolute expertise stands firm. However, it might be possible for those located in model 1 to acknowledge that patients are complex and knowing participants in society unlike the empty vessels that a strict positivist paradigm traditionally suggests. Even the most extreme paternalistic medical practitioners today should be able to recognize that patients are a heterogenic group of people with varying backgrounds, knowledge, and scientific and medical expertise. The fact that members of the healthcare professions and scientific communities fall ill, and that they are capable of critically appraising and discussing guidelines and policies, and offers of help by assessing research designs, selection criteria, methods, outcomes, research validity, and have insider knowledge of peer review processes, and that some of these patients will engage in informal learning settings to share and disseminate their knowledge rarely register with the paternalistic groups of medical practitioners. Once a patient, regardless of educational background and professional base, the reconstruction into an infantilized unknowledgeable subject without agency categorically occurs. A dismissal of capacity and ability on the part of patients holds no value, nor do measures that do not work qualify as serious offers of medical support.

Particularly at risk for paternalistic or extreme paternalistic encounters are patients with diseases whose symptoms are either designated as vague and diffuse or whose etiology is unclear (Lian 2017; Sherwin 1992; Wendell 1996). The illogical attribution of what counts as a serious illness, e.g., multiple sclerosis (MS), versus what is considered by some as an insignificant non-illness, myalgic encephalomyelitis (ME), despite extensive symptom overlap is a decision that leaves patients, many of them women and children, living precarious lives. In marketized, neoliberal academia, research that promises allegedly cost-effective results is produced that policymakers adopt and politicians propagate.

4   Healthcare for Whom, How, and Why?

Since this is a handbook about methods in healthcare, the risk of conflating method with methodology requires highlighting. For those unfamiliar with the distinction between method and methodology, methodological issues touch upon questions of ontology, epistemology, and ethics. In its shortest form, key ontological, epistemological, and ethical questions are:

1. What is? (ontological)
2. Knowledge for whom and why? (overlapping ontological and epistemological question)
3. What counts for knowledge/whose knowledge? (feminist epistemological interlinked questions)

4. What are the consequences of 1, 2, and 3? (reflection)
5. Does the privileged/more powerful party take responsibility? (feminist ethics)

These questions are fundamental as they produce answers that transcend the current rationale for healthcare provision, indicating the type of society healthcare personnel assume they are part of and contribute to, which is particularly important.

Methodology is more than deciding on methods for inquiry. Feminism, and specifically pragmatist feminism, encompasses a problem-focused questioning of healthcare provision as there is no such thing as an unbiased researcher, clinician, or healthcare practitioner, neutral in their approach to inquiry and clinical practice. Medical research is colored by the investigator's preconceived notions and their interests (Edwards 2017; Geraghty 2016; Goldin 2016; Racaniello 2016; Staples 2017). Even rigorous and scientifically robust studies can yield results that are problematic, especially in the field of tension between the natural and social sciences, where doing science is not mathematically clear-cut and the parameters of what we need and ought to know are blurred. Many research questions that ought to have been asked and to have received funding have gone unasked, and been unfunded, leaving knowledge gaps detrimental to patient groups neglected due to their diseases lacking medical prestige (Bosely et al. 2015; Brown 2017; McEvedy and Beard 1970; Payton 2016; Risberg et al. 2009; Staples 2017; Wu et al. 2016). Consequently, there are many issues pertaining to women's health and well-being that healthcare providers are ignorant of, be that willful ignorance owing to aforementioned rationales, belonging to schools of thought that dismiss entire patient groups, or actual ignorance in terms of not knowing. This is not a new or obscure phenomenon, as the research literature indicates (Andrist 1997; Kennedy 2012; Sherwin 1992).

A tenet of feminist inquiry is the acknowledgment that the most open-minded researchers are ontologically challenged (regarding the "what is?" and therefore, also, regarding the "what is not?") as they are embedded in their cultures, academic disciplines that frame valid and to them relevant research in certain ways, and the political systems responsible for the organization of healthcare. Health policy, too, is infused with cultural notions about society's order, e.g., women's reproductive role, motherhood and women in the public sphere, and women as primary caregivers. Certain diseases are recognized in some countries but not in others or are culturally ascribed to a specific sex. To recognize that healthcare in any country is a political and cultural undertaking, influenced by notions of gender, social class, race, ethnicity, and chronic illness, and what constitutes a disability, is a vital first step toward an understanding of the place of feminism and as a logical extension, feminist-driven inquiry, in healthcare.

5   Why Paternalism Is Not Necessarily Remedied with Concepts and Practices of Gender Equality

Feminism in healthcare is partially about gender bias, but gender per se is not synonymous with feminism. To outline terms and definitions, a brief overview of terminology may be helpful (Risberg et al. 2009).

While the concepts of sameness/difference are inherent to theory, it is less clear how to define, interpret, and apply these concepts:

A. On an ontological level, issues of sameness/difference pertain to essentialism; in other words, are women and men the same or are they different, and if they are the same/different, how so? Practically speaking, this means whether or not women have the same rights as men and have or should have access to all sections of society. A relevant question in this chapter would be if we think that women should become surgeons, and if we think that, why we think this is a good idea. Is it, for instance, because they are human beings just as men are, with individual and divergent skills and interests, or because they represent other values than men and therefore have something to add?

B. On the epistemological level, discussions are about whether sameness/difference provides researchers with solid tools to ask meaningful research questions. Problems of difference/sameness are rooted in essentialism. The norm, as many claim, is still male, both in theory and clinical practice, so "sameness" can easily be mistaken "same as men"; in other words, women and men are the same provided that women behave like men and adjust to male norms. This would explain why there are women within paternalistic healthcare systems who suppress and mistreat women patients just like their paternalistic male colleagues. It is the system that is paternalistic per se. Recruiting more women into such a system does not make the system better for women patients; it remains the same, with the women adopting male norms in order to succeed. This also explains why gender awareness or even challenging gender bias does not necessarily entail fairness or inclusion, far from it.

6   Empirical Examples of Ontological and Epistemological Absences

6.1   Sick and Disabled Women's and Men's Experiences of Receiving Healthcare

In a recent study (Lian and Robson 2017), patients in Norway described the power struggle with healthcare personnel and medical doctors in terms of war metaphors. As previous studies indicate, physical symptoms inexplicable by current biomedical testing are a particularly contentious area characterized by conflict between patients and medical personnel (Bosely et al. 2015; Brown 2017; Edwards 2017; Geraghty 2016; Goldin 2016; McEvedy and Beard 1970; Racaniello 2016; Staples 2017):

We found that patients experience being met with disbelief, inappropriate psychological explanations, marginalisation of experiences, disrespectful treatment, lack of physical examination and damaging health advice. The main source of their discontent is not the lack of biomedical knowledge, but doctors who fail to communicate acknowledgment of patients' experiences, knowledge and autonomy. War metaphors are emblematic of how participants describe their medical encounters. The overarching storyline depicts experiences of being caught in a power struggle with doctors and health systems, fused by a lack of common conceptual ground. (Lian and Robson 2017)

How can such inequities and inequalities be understood?

6.2 Hysterical Women

> We should look upon the female state as being as it were a deformity, though one which occurs in the ordinary course of nature. (Aristotle, in Fricker and Hornsby 2000, p. 13).

The concept of hysteria dates to ancient Egypt though the term was used originally by Hippocrates who blamed the "disease" on uterus movement due to a lack of sexual activity in women (Tasca et al. 2012). Freud promoted the psychological model of hysteria, including men as potential sufferers, though the diagnosis is more commonly attributed to women. A more recent example is provided by ME, classified by the World Health Organization (WHO) as a neurological condition since 1969 following an outbreak at the Royal Free Hospital in 1955 (Tasca et al. 2012). In January 1970, two male psychiatrists, Colin McEvedy and William Beard, published a paper in the *British Medical Journal* that claimed, without either researcher examining any patients, that the Royal Free outbreak was a case of mass hysteria due to more women being affected: "Epidemic hysteria is a much more likely explanation...The data which support this hypothesis are the high attack rate in females compared with males." That such conjecture can be classed as data, even in 1970, is a question that directly ties into methodology and the choices doctors make to this day, thereby perpetuating discriminatory practices in healthcare (McEvedy and Beard 1970; Staples 2017; Tasca et al. 2012). It is unfortunate that such blatantly sexist, untestable opinions can be so influential on the medical profession. The paradigmatic struggle continues, with the paternalistic school of thought (demonstrated via psychogenic models with little scientific basis), practiced by both male and female medical personnel, which originated with McEvedy and Beard's work, still influencing the treatment of ME patients and sufferers of other conditions (Steen 2016).

Ascribing symptoms of organic illness in women as psychologically based "hysteria" has a long history; Freud diagnosed a 14-year-old girl as suffering with an "unmistakable hysteria," claiming she was cured following his treatment. In fact, she continued to complain of abdominal pain and died within months of sarcoma of the abdominal glands (Webster 2003).

Arguably, use of the term hysteria has led doctors to believe they have found a diagnosis for symptoms, which in fact remain unknown, meaning they may miss organic medical conditions (Webster 1996). The psychiatrist Eliot Slater conducted a study of 85 patients diagnosed with hysteria in the 1950s. Upon follow-up a few years later, nearly half these patients had died or were suffering from a significant disability. One patient, a woman suffering from severe headaches, was diagnosed with "conversion hysteria" and within 2 years had died from a brain tumor (Webster 1996). Misdiagnosis of both men and women by the medical profession is hardly unknown, but the readiness of medical practitioners to ascribe a woman's symptoms as psychological (read "hysterical") in origin is problematic and can result in serious consequences (Steen 2016).

Webster's quote is especially relevant given the rise of medically unexplained symptoms (MUS) where the paternalistic school proposes patients presenting with new symptoms be ignored and encouraged to undergo corrective or rehabilitative cognitive behavioral therapy (CBT) and similar "treatments" that often provide limited or no benefit to patients (Butler et al. 2004; Fenton 2016a; Kennedy 2012; Steen 2016; Tasca et al. 2012). The medical ethos, transmitted through medical training, introduces an inherent bias in how medical professionals approach the treatment of women compared to men, encouraging the dismissal of women's concerns regarding their health and often treating female patients in a condescending fashion (Fenton 2016a; Freidson 1993; Mickle 2017; Sherwin 1992).

The paternalistic paradigm is particularly evident when applied to women presenting physical symptoms to a doctor. Medical personnel are more likely to ascribe a woman's symptoms as emotional rather than organic in origin, resulting in misdiagnosis or the potential ignorance of serious medical conditions (Steen 2016) and poor treatment options and abuse – the patient led to believe they are responsible for their symptoms (Fenton 2016a; Mickle 2017).

For example, women are 50% more likely to be misdiagnosed following a cardiac arrest compared with men (Wu et al. 2016), which has a major impact on future health, reducing quality of life as un- or incorrectly treated heart attacks can result in irreversible damage to the organ - nearly 28,000 women die from cardiac arrest annually in the UK (Payton 2016). Women are also 30% more likely to be misdiagnosed following a stroke, and there are often significant delays in diagnosing serious conditions including autoimmune diseases, e.g., MS, which take an average of 5 years to be diagnosed, and endometriosis, which can take up to a decade (Mickle 2017).

A stark demonstration of gender variation in medical care is demonstrated through the treatment of pain (Fenton 2016; Hofman and Tarzian 2001). Women's pain is more likely to be dismissed, and female patients are less likely to be given adequate pain medication than men. In UK Accident and Emergency departments, women and men presenting with similar abdominal pain receive different treatments, men waiting an average of 49 minutes for treatment compared to 65 minutes for women; the latter also receive lower levels of pain medication independent of body mass factors (Fenton 2016; Hoffman and Tarzian 2001).

Clinical studies determined that medical personnel consider female pain more likely to be emotional in origin, their pain "all in the mind." This is alarming given evidence that doctors maintain such opinions even when clinical testing indicates the pain is real (Fenton 2016). Thus, pain in women is given a lower value than pain in men. Women experiencing chronic pain have an increased likelihood of being diagnosed with a mental health condition and prescribed psychotropic drugs (Calderone 1990; Fenton 2016; Hoffman and Tarzian 2001). This is harmful for both the mental and physical health of the patient, treating the condition as psychological may lead to serious organic conditions being undiagnosed and untreated.

Historically the ignoring of period pain, which can be as severe as that experienced during a heart attack, provides an example of the lower value placed on pain in women, "Despite affecting women and trans men around the world for days every month, the pain involved in menstruation is seldom questioned nor are serious attempts to alleviate it mentioned" (Fenton 2016).

The condition endometriosis affects 10% of women of reproductive age (approximately 176 million women globally) but, despite its debilitating effects, is often ignored or dismissed by the medical profession (Bosely et al. 2015). This situation has arisen due to the failure of the profession to investigate a "woman's issue," resulting in millions of women having their quality of life severely limited. Many GPs and specialists are ignorant of the condition leaving women to suffer for years without treatment. The UK has approximately 1.6 million sufferers, and the cost to the economy is estimated at £10.6 billion, yet research funding is limited, and women are often told to endure any pain or informed it is imaginary, "it's in your head, girl. You have got to deal with it" (Bosely et al. 2015; Butler et al. 2004; Kennedy 2012; Nott and Morris 2002).

7 Paradigmatic and Ontological Struggles Within the Medical Profession and Other Healthcare Personnel and Its Implications for Patients

Much has been said concerning hierarchies within the medical profession and the paradigmatic delineation between medical specialities as well as the relationships within the different healthcare professions (Epstein 1996; Freidson 1993; Hafferty and Light 1995; Klein 2004; Lian and Robson 2017; Light 2000). These relationships create areas of tension and power struggles within the profession, leaving patients awkwardly posited. There are problems within the healthcare professions regarding status, prestige, conceptual framing of diseases, as well as fundamental differences in how patients are approached by members of different professional groups. This must be considered when discussing changes or the possibilities of ontological shifts in health and healthcare, particularly for the following reason: points of departure are so differently evolved that some healthcare personnel are far from an ontological shift toward, for instance, person-centered healthcare ((PCH) (Anjum 2016)). At one end of the scale are paternalistic or extreme paternalistic groups, the latter opposing patient participation unless to tick boxes if a funding body requires it, while others operate within a person-centered paradigm to varying degrees and with some success, e.g., medical action researchers who actively seek to involve patients because of a genuine desire for the patient. Impact of inquiry remains unclear, as discussed in a recent study (Cook et al. 2017). Methodological issues regarding long-term effects, recovery, accountability, and genuine participation on the part of the patients require further research and possibly better frameworks for (participatory) evaluation.

In addition, the medical profession has arguably undergone a de-professionalization in recent decades owing to new public management and state-imposed limits on the profession (Bezes et al. 2012). Medical doctors' scope for action has to some extent been curtailed by new public management (NPM) (Bezes et al. 2012), but this is not the only reason for the uncertainties with which the medical professions grapple, and there are several discernible ontological movements in opposing directions, further complicating an understanding of possible ontological and epistemological changes in healthcare. The *ethical* consequences of a combination of a weakened medical profession, less time for patients and research due to more time spent on administration, and neoliberal ideology, people are responsible for their own health, as well as neoconservative forces, conjecture of the deserving versus undeserving, proactively aided and upheld by some medical schools of thought, are disadvantageous for patients presenting with "diffuse" symptoms within societies where the mainstream media no longer investigates ideological claims made by governments but instead propagates images of malingerers and benefits scroungers despite sound research clearly indicating that fraudulent behavior is minimal and an ideologically upheld myth (Duffy 2013; Stewart 2017).

8   A Problem-Solving Paradigm: Feminist Pragmatism

As for the possibilities of a transformative change toward a feminist pragmatist paradigm, ontological and epistemological trends occur on several levels, for one on the scientific level where scientific evidence is produced that debunks other findings as fatally flawed (Gillberg 2012; Minnich 2005; Reid and Gillberg 2014). Currently we see such a scientific struggle unfold between the upholders and defenders of trials that have been revealed as containing serious fallacies and flaws (Edwards 2017; Geraghty 2016; Goldin 2016; Kennedy 2012; Minnich 2005; Nott and Morris 2002; Racaniello 2016; Sherwin 1992).

At the grassroots level, ontological shifts are attempted by knowledgeable patient advocates moving in paradigms different to paternalistic hegemonies or entirely outside them. Concerning acquired immunodeficiency syndrome (AIDS), women advocates and women-specific symptoms and suffering led to revolutionary ontological, epistemological, and ethical shifts (Epstein 1996; Shotwell 2016).

9   Solutions

A solution-focused epistemology in combination with a person-centered ontology seems a sensible proposition. A feminist pragmatic approach would allow the healthcare professional to feel recognized and respected while providing room for patients to express themselves. The focus is placed on the problem that requires solving rather than on the patients' class, disability, or sex. It allows clinicians and patients to disregard ongoing power struggles between paradigms, at least to some extent, creating room for change. Furthermore, a solution-focused approach encourages medical practitioners to unpack their medical knowledge and apply their scientific curiosity.

Humility toward what we ought to know might not be known yet and humility toward another human being seeking help for the discomfort they are in, a fundamental belief in human dignity, are essential to the practice of medicine and provision of healthcare. In other words, while we cannot know what we don't know, the choice to listen to patients without prejudice *can* be made. A concession regarding the patient's knowledge about their specific reason for seeking help *can* be made.

There are further choices for healthcare personnel that could be described as feminist and can become pragmatist in addition, by concentrating on the impact they have and resulting feedback (what works/does not work), allowing for evaluation, critical debate, and renewed and improved efforts to enhance healthcare. But is PCH enough provided there is a genuine ambition to end inequalities due to sex, color, race, social background, and disability? PCH is a significant step in the right direction but may not have the capacity to bridge the gap between healthcare providers and women-specific problems concerning accessing and receiving adequate treatment (Nott 2002). For one, "person" is not a powerful enough word or concept to capture the complexities of being a woman seeking medical help, as the empirical examples above illustrate. The noun person suggests an imagined homogenous patient who responds well to drug treatments designed for a perceived average patient. While this has advantages, gender stereotyping is already built into "person." Healthcare personnel are only human and will continue to ascribe characteristics and symptoms to a sex.

PCH heralds a shift in ontological and epistemological repositioning, away from paternalistic paradigms, as it becomes impossible to apply methods which are in stark contrast to ontologies such as evidence-based medicine, as the latter effectively disregards individuals' suffering. PCH presents a viable alternative to a paternalistic healthcare model as the former requires genuine consideration of an individual's health, something that cannot be easily accommodated within a methodology that "ultimately reduces uniqueness and complexity to the sum of various averages, or derives individual propensities from statistical frequencies" (Anjum 2016).

10 A Feminist Pragmatist Discussion

A step further, then, the feminist pragmatist proposition, simple as it may seem initially, is subversive in that it enables medical professionals to reassert their expertise instead of acting as the state's extended arm in stripping patients of their right to adequate healthcare (Stewart 2017). Medical doctors who put patients' health problems first until they are satisfactorily resolved would commit an act of medical empowerment as well as empowering their patients, rejecting dogmatic administrative dictates of whichever economic rationale they must operate and negotiate their professional expertise under (cf. Freidson 1993; Pescosolido 2013; Risberg et al. 2009). This is a pragmatist stance that becomes feminist the moment medical practitioners recognize and acknowledge that women-specific healthcare is a prerequisite to resolve problems that indeed are women specific. Heterogeneity, "otherness," and embodied experience of health would become the norm in encountering "woman," be that a cis woman, an old woman, or a young woman wearing a niqab.

Inherent to feminist pragmatist thought lies the concept of radical objectivity (Minnich 2005). This ought to appeal to healthcare personnel in that radical objectivity includes subjectivities and intersubjective knowledge building. Medical doctors would be freed from the onerous roles of "knowing best" and expectations of "authoritative" knowledge.

In healthcare, what is, what is known, and what counts for knowledge are questions that are intrinsically interwoven with issues of equity and equality. As this chapter has demonstrated, to do patients justice is no easy undertaking in systems that are driven by other rationales than patients' best interests. Professional standards, guidelines, as well as healthcare personnel's own beliefs and prejudice can cause problems for patients, especially for mis- or undiagnosed women. While there are differences in the provision of healthcare globally, there are certain universal truths concerning discrimination against women seeking adequate healthcare, and to some extent these commonalities evolve regardless of women's backgrounds. Preconceived notions about women's health are still rooted in historical misconstruing, with western philosophy being as unhelpful in this respect as any other philosophy or cultural expression elsewhere in the world. Medical history and western philosophy have construed some fantastical explanatory models through time. The wandering womb being only one of many explanations as to why women are untrustworthy and intellectually inferior to men.

Knowledge, if regarded as collaborative endeavor and co-produced by applying a relational *and* systems critical approach, can potentially enable robust healthcare relationships to form (Gillberg and Vo 2014; Reid and Gillberg 2014; Cook et al. 2017). If inequalities, inequities, and power imbalances between medical doctors and patients are acknowledged and proactively addressed as problematic, the risk of diverting into more paternalistic concepts and practices of healthcare is minimized. As it makes sense to regard patients as knowledgeable and to assume they wish to live full lives, the need for paternalistic stances let alone the extreme paternalistic can in fact be considered obsolete.

For the sake of debate, however, let us assume patients have no agency and are empty vessels devoid of decision-making skills. An extreme paternalistic approach to patient care would be to tell them exactly what to do based on the medical professional's assessment of their symptoms. The assessment takes 5 to 10 minutes, and there is no patient voice. Let us assume the patient follows the doctor's orders stringently, attends therapy sessions religiously, ingests medicines unquestioningly, and then, reluctantly, even guiltily, discovers the therapy sessions do not yield the desired results or that the medicines trigger adverse events. Women are particularly vulnerable to self-blame and will endeavor to self-correct and possibly intimate improvements where there are none. They may also feel inclined to blame themselves for lacking therapeutic success and the ineffectiveness of medicines or any side effects. If they revisit the doctor's office, it will be reluctant unless their symptoms give cause for alarm. Concurrently, the underlying cause or causes for their symptoms go unnoticed, deteriorate, and, at worst, cause death. The more educated patients are, the less likely they are to seek medical help (Robson and Lian 2017), so by the time they seek help, months may have passed since the initial symptoms appeared. In addition, a long period of non-/mistreatment may occur due to misdiagnosis or no diagnosis due to the label of MUS. Once a psychogenic label like MUS appears on a patient's notes, they will receive adequate healthcare only with great difficulty, even when presenting with new symptoms (Lian and Robson 2017). It is a feminist pragmatist stance to reject such practices as impractical and unreasonable because they are impractical and unreasonable both for the affected individual and the wider societal context within which these situations unfold. The social fabric of belonging into a caring society is rendered asunder by extreme paternalistic views of women's and men's bodies, but women are more affected by such conduct (see above).

In summary, healthcare urgently requires an ontological and epistemological shift at the paradigmatic level, which is not to confuse different types of knowledge and their specific values with mythical claims about the scientific method as the only valid source of knowledge. Subjective knowledge is not "merely subjective" or inferior to "objective knowledge," and objective knowledge remains elusive while it is infused with fallacies as previously discussed (Minnich 2005). There is a need for radical objective knowledge that comprises all types of knowledge and experience, and failing to provide this for ideologically driven or paradigmatic reasons is unacceptable and detrimental to women's and men's lives (Gillberg and Vo 2014). As emphasized previously, for women, the stakes are invariably higher, and healthcare personnel have the capacity, ability, and power to alleviate suffering beyond examination rooms, patients' notes, and limited appointment times, by rethinking their professional concepts and practices, several of which have been discussed in this chapter.

References

Andrist L. A feminist model for women's health care. Nurs Inq. 1997;4:268–74. https://doi.org/10.1111/j.1440-1800.1997.tb00113.x.

Anjum R. Evidence based or person centered? An ontological debate. Eur J Person Cent Health. 2016;4(2)

Bezes P, Demazière D, Le Bianic T, Paradeise C, Normand R, Benamouzig D, et al. New public management and professionals in the public sector: what new patterns beyond opposition? Sociologie du Travail. 2012;54:1–52. https://doi.org/10.1016/j.soctra.2012.07.001.

Bosely S, Glenza J, Davidson H. Endometriosis: the hidden suffering of millions of women revealed. The Guardian. 2015. https://www.theguardian.com/society/2015/sep/28/endometriosis-hidden-suffering-millions-women. Accessed 15 Oct 2017.

Brown L. Endometriosis treatment 'unacceptable' and women aren't diagnosed quickly enough. BBC Newsbeat. 2017. http://www.bbc.co.uk/newsbeat/article/39364958/endometriosis-treatment-unacceptable-and-women-arent-diagnosed-quickly-enough. Accessed 27 Sept 2017.

Butler CC, Evans M, Greaves D, Simpson S. Medically unexplained symptoms: the biopsychosocial model found wanting. J Royal Soc Med. 2004;97(5):219–22.

Calderone KL. The influence of gender on the frequency of pain and sedative medication administered to postoperative patients. Sex Roles. 1990;23(11):713–25.

Cody WK. Paternalism in nursing and healthcare: central issues and their relation to theory. Nurs Sci Q. 2003;4:288–96. https://doi.org/10.1177/0894318403257170.

Cook T, Boote J, Buckley N, Vougioukalou S, Wright M. Accessing participatory research impact and legacy: developing the evidence base for participatory approaches in health research. Educational Action Research, 2017;25:473–488. https://doi.org/10.1080/09650792.2017.1326964.

Duffy S. A fair society? How the cuts target disabled people. Resource document, The Centre for Welfare Reform. 2013. http://www.centreforwelfarereform.org/uploads/attachment/354/a-fair-society.pdf Accessed 20 Sept 2017.

Edwards J. PACE team response shows a disregard for the principles of science. J Health Psychol. 2017;22:1155–8. https://doi.org/10.1177/1359105317700886.

Epstein S. Impure science: AIDS, activism, and the politics of knowledge. San Diego: University of California Press; 1996.

Evans T, Mafubelu D. Women and health. Today's evidence tomorrow's agenda. Report. World Health Organisation. 2009. http://www.who.int/gender/women_health_report/full_report_20091104_en.pdf. Accessed 15 Nov 2017.

Fenton S. How sexist stereotypes mean doctors ignore women's pain. The Independent. 2016a. http://www.independent.co.uk/life-style/health-and-families/health-news/how-sexist-stereotypes-mean-doctors-ignore-womens-pain-a7157931.html. Accessed 27 Oct 2017.

Fenton S. Period pain is officially as bad as a heart attack – so why have doctors ignored it? The answer is simple. Online article, The Independent. 2016b. http://www.independent.co.uk/voices/period-pain-is-officially-as-bad-as-a-heart-attack-so-why-have-doctors-ignored-it-the-answer-is-a6883831.html. Accessed 25 Oct 2017.

Freidson E. How dominant are the professions? In: Hafferty FW, McKinlay JB, editors. The changing medical profession: an international perspective. New York: Oxford University Press; 1993. p. 54–66.

Fricker M, Hornsby J. The Cambridge companion to feminism in philosophy. Cambridge: Cambridge University Press; 2000.

Geraghty K. 'PACE-gate': when clinical trial evidence meets open data access. J Health Psychol. 2016;22:1106–12. https://doi.org/10.1177/1359105316675213.

Gillberg C, Vo LC. Contributions from pragmatist perspectives towards an understanding of knowledge and learning in organisations. Philos Manag. 2014;13:33–51. https://doi.org/10.5840/pom201413210.

Gillberg G. A methodological interpretation of feminist pragmatism. In: Hamington M, Bardwell-Jones C, editors. Contemporary feminist pragmatism. New York: Routledge; 2012. p. 217–37.

Goldin R. PACE: the research that sparked a patient rebellion and challenged medicine. Sense about Science USA. 2016. http://senseaboutscienceusa.org/pace-research-sparked-patient-rebellion-challenged-medicine/. Accessed 10 Nov 2017.

Govender V, Penn-Kekana L. Gender biases and discrimination: a review of health care interpersonal interactions. Women and Gender Equity Knowledge Network. 2007. http://citeseerx.ist.psu.edu/viewdoc/download?doi=10.1.1.493.104&rep=rep1&type=pdf. Accessed 25 Oct 2017.

Hafferty FW, Light DW. Professional dynamics and the changing nature of medical work. J Health Soc Behav. 1995;36:132–53.

Herr K, Anderson GL. The action research dissertation. A guide for students and faculty. Thousand Oaks/London/New Delhi: Sage; 2005.

Hoffman DE, Tarzian AJ. The girl who cried pain: a bias against women in the treatment of pain. J Law Med Ethics. 2001;29:13–27. https://doi.org/10.2139/ssrn.383803.

Hsu J. The relative efficiency of public and private service delivery. World Health Report. World Health Organisation. 2010. http://www.who.int/healthsystems/topics/financing/healthreport/P-P_HSUNo39.pdf. Accessed 15 Oct 2017.

Humphries KH, Izadnegahdar M, Sedlak T, Saw J, Johnston N, Schenck-Gustafsson K. et al, Sex differences in cardiovascular disease – impact on care and outcomes. Front Neuroendocrinol. 2017. https://doi.org/10.1016/j.yfrne.2017.04.001.

Kennedy A. Authors of our own misfortune? The problems with psychogenic explanations for physical illnesses. South Willingham: The Village Digital Press; 2012.

Klein J. Open moments and surprise endings: historical agency and the workings of narrative in 'the social transformation of American medicine'. J Health Polit Policy Law. 2004;29:621–42.

Lian OS, Robson C. 'It's incredible how much I've had to fight.' Negotiating medical uncertainty in clinical encounters. Int J Qual Stud Health Well-Being. 2017;12. https://doi.org/10.1080/17482631.2017.1392219.

Light DW. The medical profession and organizational change: from professional dominance to countervailing power. In: Bird C, Conrad P, Fremont AM, editors. Handbook of medical sociology. Prentice Hall: Upper Saddle River; 2000. p. 201–16.

McDonnell O, Lohan M, Hyde A, Porter S. Social theory, health & healthcare. Basingstoke: Palgrave Macmillan; 2009.

McEvedy CP, Beard AW. Royal Free epidemic of 1955: a reconsideration. Br Med J. 1970;1 (5687):7–11.

Mickle K. Why are so many women being misdiagnosed? Glamour. 2017. https://www.glamour.com/story/why-are-so-many-women-being-misdiagnosed. Accessed 18 Oct 2017.

Minnich EK. Transforming knowledge. Philadelphia: Temple University Press; 2005.

Mitchell S, Schlesinger M. Managed care and gender disparities in problematic health care experiences. Health Serv Res. 2005;40(5):1489–513.

Nott SM. Body beautiful? Feminist perspectives on the World Health Organisation. In: Morris A, Nott S, editors. Well women. The gendered nature of health care provision. Aldershot: Ashgate Publishing Limited; 2002. p. 145–64.

Nott SM, Morris A. All in the mind: feminism and health care. In: Morris A, Nott S, editors. Well women. The gendered nature of health care provision. Aldershot: Ashgate Publishing Limited; 2002. p. 1–20.

Payton M. Doctors are failing to spot heart attacks in women – these are the symptoms. The Independent. 2016. http://www.independent.co.uk/life-style/health-and-families/health-news/women-heart-attacks-misdiagnosis-men-statistics-study-says-a7216661.html. Accessed 25 Oct 2017.

Pescosolido B. Theories and the rise and fall of the medical profession. In: Medical sociology on the move. New York: Springer; 2013. p. 173–94.

Racaniello V. No 'recovery' in PACE trial, new analysis finds. Virology blog. 2016. http://www.virology.ws/2016/09/21/no-recovery-in-pace-trial-new-analysis-finds/. Accessed 30 Sept 2017.

Reid C, Gillberg C. Feminist participatory action research. In: Brydon-Miller M, Coghlan D, editors. The SAGE encyclopaedia of action research. New York: SAGE; 2014. p. 343–6.

Risberg G, Johansson EE, Hamberg K. A theoretical model for analysing gender bias in medicine. Int J Equity Health. 2009;8(2009). https://doi.org/10.1186/1475-9276-8-28.

Robson C, Lian OS. 'Blaming, shaming, humiliation': stigmatising medical interactions among people with non-epileptic seizures. Wellcome Open Research. 2017. https://doi.org/10.12688/wellcomeopenres.12133.1.

Sherwin S. No longer patient. Feminist ethics & health care. Philadelphia: Temple University Press; 1992.

Shotwell A. Against purity. Living ethically in compromised times. Minnesota: University of Minnesota Press; 2016.

Staples S. During ME awareness week we revisit the toxic legacy of McEvedy and Beard. Online article. The ME Association. 2017. http://www.meassociation.org.uk/2017/05/during-me-awareness-week-we-revisit-the-toxic-legacy-of-mcevedy-and-beard-10-may-2017/. Accessed 03 Nov 2017.

Steen L. The wilderness of the medically unexplained. Resource document. theBMJopinion. 2016. http://blogs.bmj.com/bmj/2016/08/25/lisa-steen-the-wilderness-of-the-medically-unexplained/. Accessed 19 Nov 2017.

Stewart M. Cash not care, the planned demolition of the UK welfare state. London: New Generation Publishing; 2017.

Tasca C, Rapetti M, Carta MG, Fadda B. Women and hysteria in the history of mental health. Clin Pract Epidemiol Ment Health. 2012;8:110–9. https://doi.org/10.2174/1745017901208010110.

Webster R. Why Freud was wrong: sin, science and psychoanalysis. London: Harper Collins; 1996.

Webster R. Freud. London: Weidenfeld & Nicolson; 2003.

Wendell S. The rejected body. Feminist philosophical reflections on disability. New York: Routledge; 1996.

WHO. Road Map for Action (2014–2019). Integrating Equity, Gender, Human Rights, and Social Determinants into the Work of WHO. Report. World Health Organisation. 2015. http://www.who.int/gender-equity-rights/knowledge/web-roadmap.pdf?ua=1. Accessed 28 Sept 2017.

Wu J, Gale CP, Hall M, Dondo TB, Metcalfe E, Oliver G, et al. Impact of initial hospital diagnosis on mortality for acute myocardial infarction: a National Cohort Study. Eur Heart J Acute Cardiovasc Care. 2016. https://doi.org/10.1177/2048872616661693.

Critical Ethnography in Public Health: Politicizing Culture and Politicizing Methodology

13

Patti Shih

Contents

Abstract

Critical ethnography is a methodological approach to ethnographic research that is explicitly political in its epistemic and empirical focus on challenging power relations and political inequality. It has an epistemic concern about how the notion of "culture" is produced in the research process, through the assumed objective position of the researcher as the "knower" and the study participants as the "known," and thus taken for granted as a true representation of social reality. This is particularly problematic if this account of culture is integrated into public health programs as an object of intervention and decontextualized from the wider social and political structures that shape inequalities in health and healthcare outcomes. Empirically, critical ethnography attends to an expansive analysis of power relationships, by highlighting issues of social exclusion, marginalization, and injustice in its research focus. The chapter draws on my experience as a

P. Shih (✉)
Centre for Healthcare Resilience and Implementation Science, Australian Institute of Health and Innovation (AIHI), Macquarie University, Sydney, NSW, Australia
e-mail: patti.shih@mq.edu.au

© Springer Nature Singapore Pte Ltd. 2019
P. Liamputtong (ed.), *Handbook of Research Methods in Health Social Sciences*,
https://doi.org/10.1007/978-981-10-5251-4_60

novice ethnographic researcher in Papua New Guinea, which taught me the importance of understanding structural factors that contextualize the relationship between culture and health, as well as a critically reflexive research position. The chapter discusses how critical ethnography politicizes the interpretive process through a commitment to collaborative meaning-making between researchers and study participants, which enables the social utility of ethnographic knowledge for political action well after the completion of research.

Keywords

Critical ethnography · Ethnography · Epistemology · Culture · Public health research · Interventionism

1   Introduction

It is by politicising culture that we can give it back its meaning and its effectiveness. (Fassin 2001, p. 311)

A foundation of contemporary public health is the assertion that the health and well-being of individuals are contextualized by the social, economic, and political world in which they live. With a humanitarian commitment to health as a human rights and social justice issue, public health interventions aim to proactively prevent disease and improve health outcomes at the population level (Liamputtong 2016). To enable practitioners to devise strategies, policies, and practical solutions to health and healthcare challenges in targeted communities, research methods from a variety of disciplines are used to inform and examine the complex determinants of health. Ethnography, long associated with the research of cultural and social worlds, is a crucial methodology for examining the shared norms and practices of communities of people and provides a nuanced understanding of cultural factors that underscore public health issues (Pope 2005; Liamputtong 2013; see also ▶ Chap. 26, "Ethnographic Method").

However, there are practical and methodological dilemmas in the use of ethnography in public health research. In her juxtaposition of ethnographic research as the metaphor of "sitting" and the practice of public health as "doing," Stacy Pigg (2013) suggests that the lengthy immersion of researchers in the research setting and the careful production of "thick descriptions" of the social world in ethnographic methods can be dismissed by the imperative to efficiently implement solutions and focus on problem-solving in public health programs. Yet, without the thoroughness and nuances about the cultural world produced by ethnographic research, public health interventions can be implemented without thorough engagement with the local community or be less able to unravel the root causes of health problems embedded in social complexities. On the other hand, as Lisa Maher (2002) notes, ethnography traditionally values the objectivity of the researcher, who is urged to maintain a level of scientific detachment from the field and refrain from "contaminating" the analytical interpretation with their own political stance and biases. However, ethnographers uncover so many instances of health inequalities in their research that the notion of remaining politically neutral in the face of injustice and suffering is deeply unethical and counterintuitive.

There are also *epistemic* questions about the notion of "culture," one of the most central concepts in ethnography. Episteme is the study of the production of knowledge, and in ethnography this is a concern about *how we come to know* what culture means (Foucault 1972; Rabinow 1986). Critical anthropologists have long interrogated the issues associated with the production of cultural descriptions of a given community by researchers as outsiders, who are assumed to be objective in their interpretation of "culture" and produce an unproblematic account of the social reality of the people they study (see, e.g., Clifford and Marcus 1986; Abu-Lughod 1991). This has direct implications for how "culture" becomes integrated into public health interventions and into the lives of people in the studied communities, which is particularly problematic if the research conclusions do not align with the community's views and values. Research must be understood as a process that implicates the power relationship between the researcher and the study participants through the production of knowledge. Therefore, a more careful consideration is needed of the interpretive process by which the knowledge of culture is produced through ethnographic studies.

Critical ethnography is an explicitly political methodological approach to ethnographic research methods with its epistemic and empirical focus on challenging power relations and political inequality (Foley 2002; Cook 2005; Madison 2011). It has an epistemic concern about the power relationships that are produced by the research process and prompts a critique of the researcher's role as the privileged and powerful authoritative voice on "culture" in health and therefore the assumed "truth" of the knowledge made about culture as the ethnographic object of inquiry. Empirically, it analyzes the power relationships engendered within the wider structural contexts that shape health and well-being – that is, the distal social institutions such as political, economic, and legal systems that are less visible in everyday practices but are instrumental in shaping social dynamics and the allocation of social power and resources. Social inequalities are related in part to the epistemic production of knowledge about culture, because the way in which we define and research "culture". Therefore, by rendering it as an object of public health intervention can disguise the structural power distributions that underscore health inequalities. In view of this, methodology is the key site of a political revolt against current organization of power relationships in critical ethnography (Foley and Valenzuela 2008, p. 288).

This chapter examines the tradition and subsequent critical development of ethnographic research particularly in anthropology and argues how the integration of ethnography in public health thus far has been limited in applying a critical lens on these epistemic issues. In particular, it problematizes the way which culture is often oversimplified as an analytical category to explain ill health and, thus, integrated into public health programs as an object of intervention and decontextualized from the wider social and political structures that shape inequalities in health and healthcare outcomes. It draws on my experience as a novice ethnographic doctoral researcher in Papua New Guinea and illustrates how my learning about culture, marital customs, HIV risk, and the social impact of mining resonated with my personal experience as the "cultural Other" in my own writing and political voice.

2   Knowing Culture: Ethnographic Tradition and Critique

Ethnography was first used by Western colonial officials and traders to gather information about local populations and their environment under the administration of European empires in the mid-nineteenth century (Madison 2011). In the early twentieth century, anthropologists such as Bronislaw Malinowski and Franz Boas began to establish ethnography as a research methodology characterized by the lengthy immersion of researchers in their research setting, or the "field," observing and participating in the everyday "natural environment" and activities of a given group of non-Western "Others". The role of the ethnographer is to capture and write about the observed shared practices and norms of study participants, through developing interpersonal relationships and adopting the language and in some cases, clothing of the local community. As objective "outsiders" looking "inside," ethnographers are assumed to see the point of view and subjective experiences of cultural Others through their interpretation. The notion of "writing culture," quite literally the Greek lexicological derivation of ethnography (Liamputtong 2013, p. 148), was born.

The 1920s and 1930s saw further methodological development and refinement of ethnography, as it was extended to examine social marginalization and political inequalities, particularly influenced by the urban ethnography developed by the Chicago School, a group of academics from the Department of Social Science and Anthropology at the University of Chicago (Madison 2011, p. 11; Liamputtong 2013, p. 151) around the time of the Great Depression. The recognition that wider structural factors and notions of power should be integrated into the understanding of culture continued to influence methodological critiques of ethnography well into the 1960s and 1970s. Work in critical anthropology, anti-colonialism, feminism, and intersectional feminism (see for example, Clifford and Marcus 1986; Abu-Lughod 1991) critiqued the notion of "culture" as a colonial invention, constructed to reinforce cultural difference and, therefore, the social as well as political subjugation of cultural Others, and highlighted the subtle reproduction of power and knowledge embedded in cross-cultural research processes.

Critical methodologists in anthropology challenged the heart of conventional ethnography, in particular the notion that "culture" can be "known" through data collection and "written" through interpretation (Clifford and Marcus 1986). These methodologists   questioned whether objective research yields the "truth", and rejected the political neutrality of the researcher as a distant outsider able to develop scientific findings about other people's culture (Clifford and Marcus 1986; Abu-Lughod 1991; Foley and Valenzuela 2008). In fact, they suggest, the production of knowledge by researchers alone can potentially reinforce the unequal power relations between the researcher as the knower and the study participants as the known (Abu-Lughod 1991; Mertens 2007; Hesse-Biber 2010).

3   Culture as an Object of Intervention

While anthropology has had a long critical engagement with the production of knowledge and power through the ethnographic construction of the cultural Other, in the discipline of public health, "culture" and social determinants did not become a focus until the 1970s. Instead, the more privileged subdisciplines of disease control, public hygiene, and epidemiology dominated. The call to understand the role of culture in public health has been stronger in recent years, as much evidence reveals poor health outcomes are disproportionally skewed by gender, ethnicity, and social, economic, and geographic disadvantage, particularly affecting lower-income communities in developing countries of the global south and so-called "culturally and linguistically diverse" (CALD) communities in developed countries. Social scientists in public health have fought for a more central place in research and urged for an understanding of health that is informed by social research methods and sensitivity to the role of culture (Liamputtong in press).

However, the engagement of cultural research in public health remains geared toward informing the practical aims of intervention design and complementing the findings of quantitative data with qualitative explanations. From a critical ethnography perspective, the use of culture as an analytical category to explain the determinants of health can run the danger of objectifying culture as a pathological entity. For example, ethnographic research was used to explain the uneven patterns of infection among non-Western communities, gay men, sex workers, and other marginalized groups in early HIV research, which led to the development of epidemiological notions of "risk groups" based on the assumption that the shared norms and practices of a bounded community of people is central to their susceptibility to HIV infection (Glick Schiller 1992; Glick Schiller et al. 1994). By attributing culture to the practice of behaviors that lead to risk and ill health, this explanatory model essentializes a person's biological destiny to their cultural grouping (Glick Schiller 1992; Briggs 2005; Sovran 2013). The social epidemiological conclusions drawn about the group pathology of certain communities then becomes part of a wider discourse that links ill health to their cultural *difference* to those that are healthy, which further alienates people who are already marginalized and vulnerable to disease.

The influence of an epidemiological vocabulary of causation and effect tends to infiltrate the language and analytical frameworks used in public health social research. The description of culture as a social epidemiological "variable" makes culture a bounded object that can be somehow extracted or inserted into an epidemiological logical equation that explains people's behaviors and health outcomes. The black and white binary of "barriers and facilitators" is familiarly applied in numerous research projects in culture and health. When translated into interventions, cultural practices that are deemed detrimental to certain health concerns are to be problematized and challenged, but cultural practices that deemed contributing to good health are to be "fostered and promoted" (Taylor 2007; Sovran 2013). At best, culture is integrated into interventions as "cultural competence" or "culturally grounded" education (Wilson and Miller 2003) and, at worse, described as "cultural barriers" or "harmful practices" to be reformed through the rhetoric of "positive cultural change."

Culture becomes problematized as the cause of illness, as well as instrumentalized as an object of intervention (Fassin 2001; Shih et al. 2017). This is highly problematic from the perspective of critical ethnography, because the knowledge of culture becomes constructed within a positivist framework of cause and effect, conceptualized as a static object open to intervention and malleability. Moreover, when interventions are derived from research evidence produced by external researchers as outsiders and as assumed "experts" with superior knowledge about the essential characters of the culture of study participants, interventions become decisions made *for* recipient communities that do not necessarily reflect their understanding of culture.

Also concerning is that this etiologized analysis of culture is decontextualized from the wider economic and political influences, both current and historical, on cultural patterns and practices. Many of the factors affecting marginalized communities, such as economic vulnerability, violence, and racism, are collapsed into the "less threatening concept of culture" (Gregg and Saha 2006, p. 543), thus the ethnographic nuance of a contextualized temporal social reality is lost.

4   Politicizing Culture: Structural Determinants of Disempowerment and Health

Culture and structural factors are mutually implicated in producing inequality (Metzl and Hansen 2014, p. 128). The everyday experiences of health and well-being are constrained and fostered by social systems, which are in turn produced and reproduced in culture (Cook 2005). However, the role of structure in health is often disguised by the focus on culture as the interpretive lens for health-related behavior. Without understanding the complexities of historical and political power relationships, it is easy to "substitute a political economy of the disease for its cultural and behavioural interpretations" (Fassin 2013, p. 119). Therefore, a politicized understanding of culture is needed.

Indeed, studying culture shows how much culture is itself altered by a history of social inequality beyond the level of everyday behaviors and practices. A study of African-Americans' dietary patterns (Airhihenbuwa et al. 1996) revealed that a diet high in fat and salt, contributing to a range of significant chronic health conditions, was recognized by study participants as stemming from their ancestral history of slavery and economic disadvantage, yet is continued today by both higher- and lower-income African-Americans, as these food patterns are absorbed and socialized into a part of their shared understanding of a "black culture." However, the response of food patterns to structural and environmental influences remains evident today, according to the study, as it revealed that psychosocial vulnerability to institutional racism and the direct marketing of fast food to African-Americans impacted on diet and health. This suggests that successful behavioral change to healthy eating is a cultural issue that will also depend on challenging the structural vulnerability to these factors.

The collection of ethno-specific health data, whether in developed countries or globally, reinforces the assertion that people of certain ethnic origins are susceptible to different patterns of health and illness. Yet, the experience of socioeconomic discrimination along ethnic lines that are both current and historical, and relevant in both the global north and global south, is not properly accounted for in explaining the social disparities in accessing healthcare, education, housing, and other social resources pertinent to good health. The historically produced notion of ethnic difference reinforces pre-existing social hierarchies and legitimizes the division between the powerful and the powerless. This also applies for gender-specific health research and the way which women are portrayed as biologically distinct to men. Inhorn and Whittle (2001) explain that the historically male-dominated scientific knowledge of biomedicine and epidemiology amplifies women's difference to men in terms of health, when sexual difference is used to account for disease prevalence and lifestyle risk factors. In fact, historically, men and women and whites and non-whites are rarely studied together as the same group in clinical trials (Inhorn and Whittle 2001). Rather than suggesting a sense of sameness, people are differentiated and analyzed separately in populations along the blanket categories of sex and ethnicity.

Through the objective production of knowledge about cultural Others, "women, blacks and people of most of the non-West have been historically constituted as the other in the major political systems of difference on which the unequal world of modern capitalism has depended" (Abu-Lughod 1991, p. 54). The patterns of social marginalization that are linked to historical domination of the powerful over the subjugated suggest that the very construction of "culture" as an explanatory category for behavior is a way to normalize the continued marginalization of the less powerful from the "fruits of scientific and social progress," as expressed in Paul Farmer's concept of structural violence (2001, p. 79). As Farmer eloquently argues, "sickness is a result of structural violence: neither culture nor pure individual will is at fault; rather, historically given (and often economically driven) processes and forces conspire to constrain individual agency" (Farmer 2001, p. 79).

By politicizing the understanding of culture using the lens of structural inequality and its link to the production of knowledge about cultural difference, critical ethnography makes a link between its empirical and methodological concerns about unequal power relationships and social justice.

5 HIV, Marriage, and Mining in Papua New Guinea

The experience in my own research about culture and its interpretive process also taught me lessons about the importance of understanding structural factors that contextualize the relationship between culture and health.

My ethnographic research in Papua New Guinea examined faith-based HIV prevention programs and the influence of religion and culture on sexual and reproductive health. Initially, it had little focus on structural inequality. However, my fieldwork began in 2011 at the height of a mining boom. In one rural township field site, a Liquefied Natural Gas project was underway to construct a 700 km pipeline to deliver liquefied natural gas from the Southern Highlands extraction site to a processing plant in Port Moresby. It was hailed as the bastion of economic development for the country in the twenty-first century and was a significant part of everyday life in the area. Initially, I was unclear about the role and impact of mining in the big picture of healthcare and HIV. As a novice researcher, I went about observing everyday activities of my study participants, who were healthcare workers from a well-known local faith-based healthcare service, by going to clinics; watching practices; mapping outreach fieldwork, which sometimes meant hiking for hours over difficult terrain to reach remote villages; attending antenatal clinics; cooking; socializing; gardening; and going to evening prayers and attending church in the Southern Highlands community where I lived.

As I began to interview and talk more in-depth to study participants, there appeared to be a firm consensus about the cause of HIV (see expanded discussion in Shih et al. 2017). Faith-based healthcare workers deemed polygyny, the marital practice of a man marrying multiple wives and paying a bride price of material and money to the bride's family to compensate for the loss of her labor, as a "harmful tradition" and a "bad" kind of culture in the area. They believed polygyny, as a type of sexual concurrency, increased HIV risk among men and women, and the exchange of bride price further commoditized women and put them more at risk of gender-based violence and forced sex. Their assertion was reflected by policy documents, training materials, and donor-sponsored research reports used by healthcare workers, which all featured an assertive language that problematized cultural practices such as polygyny and bride price exchange as "bad" traditions that should be thoroughly discouraged to prevent further HIV spread.

However, there was also a concurrent counter-discourse about culture. The same healthcare workers also explained that in present day Papua New Guinea, traditional polygyny and bride price practices have changed, particularly with the introduction of the cash economy. Polygyny was once only available to tribal leaders, usually older men who have acquired resources such as land and pigs for bride price payment. However, with the economic opportunities provided by the mining project today, younger men were also more able to afford bride price payment. With the increased cash flow to more men and to younger men, the practice of bride price has increased and its average worth inflated. The occurrence of other forms of sexual concurrency, such as transactional sex, has also increased. Moreover, healthcare workers were most concerned that the traditional kinship relations and social obligations that were once cemented by traditional marriage and bride price exchange had been replaced by individualized forms of sexual transaction and monetary acquisition. The availability of cash meant that the cultural values of reciprocation and the social safety net that were once provided through inter-clan marriages and bride price exchange were replaced by individuals' ability to obtain financial and material wealth.

It was the loss of kinship obligations and social fabric attached to traditional practices of polygyny and bride price exchange that was lamented by healthcare workers, which they associated with cultural breakdown and lack of social cohesion. Ironically, HIV infection is blamed on polygyny as a "harmful tradition," when in reality, contemporary concurrent sexual relationships in Papua New Guinea, whether in married forms or not, have been significantly reconfigured from traditional forms due to economic and social change. In fact, polygamous practices had been problematized by Christian missionaries and colonial officials since their arrival in Papua New Guinea in the late nineteenth century, which was the basis for the further changes to marital practices today, under the influence of social and economic change. However, many types of sexual concurrency, whether in informal or married partnerships, are still labeled as "polygyny" by healthcare workers.

Having started to make sense of the intersection between the changes in the local political economy impacting marriage and sexual networks, I began to see things that I had not given much notice before – like the number of large trucks and broken roads on the region's highway, the abandoned site of a labor camp just outside town, the new airport in another township parked with giant cargo planes that looked out of place, and young men in the back of mini buses sauntering toward work camps. I noticed women were mostly excluded from such economic opportunities and that many of these young women have become much more vulnerable economically and sexually in the wake of this change.

This was a complicated story about culture that healthcare workers themselves already knew but did not articulate in the healthcare language in which they were trained. The donor-driven prevention programs they practiced mostly concentrated on discouraging individuals from participating in cultural practices such as polygyny and bride price exchange. The key focus remained on individual behavioral change and HIV awareness education as the ultimate solution to HIV risk, and not on challenging structural influences such as economic inequality and social change that can shape sexual practices. The structural complexities of HIV were drowned out by an intense focus on "culture" as the cause of illness, when in reality, so-called traditions have long been altered by almost half a century of missionization, colonization, and economic development.

6 Accounting for Reflexivity

During my fieldwork, it became apparent there were two different stories about culture: one about culture as object that causes illness and thus must be discouraged, and another about culture as subject about losing the meaning of society and kinship. In reconciling these two stories, I came to see a broader story about health that is intertwined in culture, economics, politics, and history. I realized it was not "their culture" in Papua New Guinea that I was documenting, but rather I was conveying an experience of loss. It reminded me of the opening passage of Renato Rosaldo's article (1988), in which he evoked the notion of "people without culture": We often question the authenticity of what we regard as appropriate representation of culture – temples, museums, costumes, rituals, dance; things that are beautiful, to be held and admired, not spaces left empty and hearts left broken. But, perhaps it is the reason why we find them empty and broken which most triggers critical ethnographers.

Much later after my fieldwork, I understood why I saw a story about culture and its connection to structure and history and chose to make sense of culture in this political way. Cultural dispossession and cultural fluidity was also an experience in my life and that of my own family. This is another story from another time and place, about my experience of migration and cultural dislocation. As a child, I moved from my native Taiwan to New Zealand in the early 1990s. The haunting feeling of disconnectedness from my cultural roots and making sense of my own cultural and racial difference to the New Zealand children around me will always stay. But this experience also includes the resilience of Taiwanese language and culture under attack during Japanese and Chinese occupation in my grandparents and great-grandparents' lives many decades earlier. The contestation of a rightful Taiwanese identity is passed down as family stories; my aunts tell of being fined by Chinese officials for speaking Taiwanese at school during Chiang Kai-shek's authoritarian rule; and my grandfather tells of hiding in a cellar to avoid Chinese persecution during the 228 Massacre of 1947. These are legacies of resistance and survival, shared by several generations of Taiwanese people, which propel me to assert my right to belong, speak, live, and work in a multicultural Australia. It will always be part of my research interest in culture and health and my political activism. There will also always be a constant agitation about the fluidity and uncertainty of my own identity. These are the same questions my study participants in Papua New Guinea asked themselves: *Who am I, what am I, what is my culture?*

My experience in Papua New Guinea encouraged a different way to think about researcher reflexivity. As the ethnographer of an account of cultural dispossession that contextualized health and illness, my personal familiarity with cultural fluidity triggered a sense of solidarity with my study participants and led to focusing the study analysis on a historical and structural account of power and inequality. Reflexivity is less about declaring how pre-existing experience and prior under-standing of the topic influenced how data was interpreted objectively (the "whole truth") but rather taking ownership of the researcher's subjectivity as part of the production of research knowledge (the "partial truth") (see also ▶ Chap. 26, "Eth-nographic Method"). Critical ethnography values the researcher's subjectivity in shaping the explicitly political angle that the data takes. Of course, this is not to suggest that other researchers with a different experience of culture than I would not have drawn similar conclusions about culture vis-à-vis structure, but that this was the particular way which I did, and found a specific and deeply personal pathway to discovery.

7   Blurring the Objective-Subjective Dichotomy and Collaborative Meaning-Making Through Dialogue

Reflexivity in critical ethnography challenges the strict divide between the objective and subjective positionalities taken by the researcher and study participants. This component of critical ethnography contests the dichotomy between the objective positions of the researcher as the "knower", making knowledge about the study participant as "the known", which subsequently reproduces the power relationship between them. As Madison (2011, p. 7) suggests, "positionality is vital because it forces us to acknowledge our own power, privilege and biases just as we are denouncing the power structures that surround our subjects."

To challenge the inequality stemming from the production of knowledge through the object/subject divide, a more collaborative approach to knowledge-making with study participants is required. The ethnographer entering the field and engaging with study participants is the coming together of lifeworlds. Study participants are not the cultural Other as a static representation of the ethnographic present but part of the production of ethnographic knowledge through dialogical interaction with the researcher.

However, this is not dismissive of the objective view of the research scenario. Fassin (2013) argues that the objectivity of an outsider is precisely the sharp analytical tool to examine the things that study participants themselves "cannot see or prefer to ignore" (p. 123), such as identifying and questioning the local politics and controversies which they are themselves involved and implicated in, which only the emotional distance and intellectual neutrality of strangers can offer (p. 124). Thus, the objective perspective provides a corroborative and contextualized nuance to the subjective accounts. van Meijl (2005) suggests that critical ethnographic positionality is about a degree of flexibility in shifting from the objective to the subjective *and back* by maintaining an agnostic disposition, one that can occupy both the spaces and perspective of the objective and subjective view, as both outsider and insider.

An objective but reflexive critical ethnographic account of a researcher should aim to *evoke*, rather than *explain* the lived experience of others (Foley 2002), as prescribed in the notion of bearing witness (Foley 2002). The ethnographer stands in solidarity with study participants and makes visible the injustice of a complex and often historically produced experience, yet does not intrude in making conclusions or offering solutions as an outsider. This is a stark departure from the common practice in aid and development programs where outsider-external expert-driven interventions are carried out without thorough engagement with the subjective view. Rather than intervening *on behalf* of subjects or *speaking for* participants, bearing witness is *speaking from* the ethnographic location in which study participants are situated.

There is much diversity and difference in how ethnographers chose to engage in political activism with study participants, ranging from active and deep collaboration in political movements to the more removed yet stoically provocative stance of bearing witness (Foley and Valenzuela 2008). Fassin (2013, p. 125) suggests that the "social utility" of ethnography is that it offers a critical thinking of a complex political problem, by translating it into a language that makes sense for taking political action. Thus, ethnographic research can continue beyond an academic exercise after the completion of research activities by becoming a resource for knowledge and action. I was heartened to find my research results cited by a Papua New Guinean activist website aimed at challenging the political and economic injustices of mining in their country and tweeted by a Melanesian women's interest group. After I returned to the study site to report and discuss my research with the study participants upon the completion of the project, the results were conveyed and shared in the local church community newsletter, which in turn encouraged more local healthcare workers to correspond on the issues with me. The research, I hope, will add to the body of knowledge for taking political action and make critique about the suffering of HIV as part of the injustice structural violence, as new generations of researchers are inspired by and inherit the knowledge that came before them, such as the political and critical commitment established by early ethnographers in HIV research such as Paul Farmer and Philippe Bourgois.

Returning to Pigg's (2013) metaphor of "sitting" and "doing," her argument that "sitting" is absolutely an action rings true. Ethnography diagnoses the core problems of health, not on the surface for the purpose of a superficial intervention, but at its structural and historical core. As a method, therefore, it has a much longer lasting and political potency. Indeed, the critical ethnographic method of "sitting" is by far not a passive kind of research.

8 Conclusion and Future Directions

We come to know "culture" through the ethnographer's interpretation; thus the researcher and their analytical lens are pivotal and exceptionally powerful (Madison 2011, p. 4). Critical ethnography does not seek to replace conventional ethnography – as the commitment to thorough immersion and situating knowledge in a social contextual-specificity remains its core. Rather, it cautions the dangers of an "uncritical" ethnography used in public health research that renders culture a pathologized object of inquiry and intervention. It is a critical recasting of ethnography's methodological commitment to equality and social justice, which requires challenging the researcher's role as the sole and objective producer of cultural knowledge in the "writing of culture," by transcending the objective/subjective divide in research positionality.

Likewise, critical ethnography does not seek to abandon culture as a key analytical concept in public health research but rather to approach culture more carefully in its epistemic construction and in its wider historical and structural context. As Metzl and Hansen (2014) suggest, a move from *cultural competence* to *structural competence* in healthcare recognizes that culture matters in recognizing the social diversity of communities and healthcare needs, but also that structure contextualizes the power relationships enmeshed in cultural representation of the vulnerable and marginalized and thus shapes many healthcare outcomes (see Liamputtong in press). The first step in structural competence for public health practitioners and researchers is training the critical eye to recognize structural aspects of health and illness and developing the critical voice to articulate the representation of the cultural in the complexity of structure (Metzl and Hansen 2014).

The future of critical ethnography is about being resilient to the challenges in the current climate of social sciences and health research, when qualitative research in general is already disadvantaged in funding and recognition (Denzin and Giardina 2008; Greenhalgh et al. 2016). There is increasing pressure from funding bodies for researchers to produce implementable and translatable results, with assessment and evaluation criteria measured on the deliverability of tangible results ("How many condoms were provided this month? What percentage improvement to Knowledge-Attitude-Behaviour on HIV this year?"). These benchmarks reflect what Pigg (2013, p. 128) describes as the milieu of the "neoliberal ethic of speed and efficiency that has become normalised, and moralised, in the ways global health activity makes things happen," in the expense of the rigor required to identify the root issues of vulnerability and disempowerment in health. The shrinking time and resources allocated for fieldwork reduces the thoroughness of ethnographic immersion and the capacity for relationship-building between researcher and participants and, thus, the quality of research. This increases the risk of research participants becoming simply sources of information, which makes the research process more exploitative and less equitable and collaborative. This concern was emphasized by a healthcare worker in a busy urban clinic in Port Moresby, as we were having a cup of tea in the staff room one afternoon (Shih 2015, p. 44):

> I think it's good that you're here all the time and just relaxed about it, I don't like it when some researchers come right on time at like 10am and expect us to be ready to talk to them and then leave after that. It's good you're just hanging out and eating with us.

From a logistical perspective, ethnographic methods provide ways for the researcher to generate trust and minimize the coercion and time pressure placed on healthcare workers in their participation in the research project. But, thinking about it in terms of culture and the wider context of promoting equity and sharing the production of knowledge, it elucidates a cultural discord between the Western notion of work ethic and punctuality and the local approach to what seems a chaotic, elusive yet surprisingly productive "PNG time." Has research become such a precious process that study participants do not receive some time and patience and allowed to do what they do in the very social setting that researchers are trying to understand?

The future of critical ethnography relies on researchers remaining resilient and committed to their critical voice, in both methodology and the politics of research, in the face of increasingly diminishing space and time for research.

References

Abu-Lughod L. Writing against culture. In: Fox RG, editor. Recapturing anthropology: working in the present. Santa Fe: School of American Research Press; 1991. p. 137–62.

Airhihenbuwa CO, Kumanyika S, Agurs TD, Lowe A, Saunders D, Morssink CB. Cultural aspects of African American eating patterns. Ethn Health. 1996;1(3):245–60.

Briggs CL. Communicability, racial discourse, and disease. Annu Rev Anthropol. 2005;34:269–91.

Clifford J, Marcus GE. Writing culture: the poetics and politics of ethnography. Berkeley: University of California Press; 1986.

Cook KE. Using critical ethnography to explore issues in health promotion. Qual Health Res. 2005;15(1):129–38.

Denzin NK, Giardina MD. Qualitative inquiry and the politics of evidence. Walnut Creek: Left Coast Press; 2008.

Farmer P. Infections and inequalities: the modern plagues. Berley: University of California Press; 2001.

Fassin D. Culturalism as ideology. In: Obermeyer CM, editor. Cultural perspectives on reproductive health. Oxford: Oxford University Press; 2001. p. 300–17.

Fassin D. A case for critical ethnography: rethinking the early years of the AIDS epidemic in South Africa. Soc Sci Med. 2013;99:119–26.

Foley DE. Critical ethnography: the reflexive turn. Int J Qual Stud Educ. 2002;15(4):469–90.

Foley DE, Valenzuela A. Critical ethnography: the politics of collaboration. In: Denzin NK, Lincoln Y, editors. The landscape of qualitative research. Thousand Oaks: Sage; 2008. p. 287–310.

Foucault M. The archaeology of knowledge and the discourse on language (trans: Sheridan A). New York: Pantheon; 1972.

Glick Schiller NC. What's wrong with this picture? The hegemonic construction of culture in AIDS research in the United States. Med Anthropol Q. 1992;6(3):237–54.

Glick Schiller NC, Crystal S, Lewellen D. Risky business: the cultural construction of AIDS risk groups. Soc Sci Med. 1994;38(10):1337–46. https://doi.org/10.1016/0277-9536(94)90272-0.

Greenhalgh T, Annandale E, Ashcroft R, Barlow J, Black N, Bleakley A, . . . Carnevale F. An open letter to The BMJ editors on qualitative research. BMJ. 2016; 352(i563):1–4.

Gregg J, Saha S. Losing culture on the way to competence: the use and misuse of culture in medical education. Acad Med. 2006;81(6):542–7.

Hesse-Biber SN. Mixed methods research: merging theory with practice. New York: Guilford Press; 2010.

Inhorn MC, Whittle KL. Feminism meets the "new" epidemiologies: toward an appraisal of antifeminist biases in epidemiological research on women's health. Soc Sci Med. 2001;53(5):553–67.

Liamputtong P. Qualitative research methods. 4th ed. Melbourne: Oxford University Press; 2013.

Liamputtong P. Public health: an introduction to local and global contexts. In: Liamputtong P, editor. Public health: local and global perspectives. Melbourne: Cambridge University Press; 2016. p. 1–21.

Liamputtong P. Culture as social determinant of health. In: Liamputtong P, editor. Social determinants of health: individual, community and healthcare. Melbourne: Oxford University Press; in press.

Madison DS. Critical ethnography: method, ethics, and performance. Thousand Oaks: Sage; 2011.

Maher L. Don't leave us this way: ethnography and injecting drug use in the age of AIDS. Int J Drug Policy. 2002;13(4):311–25.

Mertens DM. Transformative paradigm mixed methods and social justice. J Mixed Methods Res. 2007;1(3):212–25.

Metzl JM, Hansen H. Structural competency: theorizing a new medical engagement with stigma and inequality. Soc Sci Med. 2014;103:126–33.

Pigg SL. On sitting and doing: ethnography as action in global health. Soc Sci Med. 2013;99:127–34.

Pope C. Conducting ethnography in medical settings. Med Educ. 2005;39(12):1180–7.

Rabinow P. Representations are social facts: modernity and post-modernity in anthropology. Writ Cult Poetics Politics Ethnogr. 1986;234:261.

Rosaldo R. Ideology, place, and people without culture. Cult Anthropol. 1988;3(1):77–87.

Shih P. The biopolitics of change: a Foucauldian analysis of Christian healthcare and HIV prevention in Papua New Guinea. Unpublished PhD thesis, University of New South Wales; 2015.

Shih P, Worth H, Travaglia J, Kelly-Hanku A. 'Good culture, bad culture': polygyny, cultural change and structural drivers of HIV in Papua New Guinea. Cult Health Sex. 2017;19(9):1024–37. https://doi.org/10.1080/13691058.2017.1287957.

Sovran S. Understanding culture and HIV/AIDS in sub-Saharan Africa. SAHARA-J: J Soc Asp HIV/AIDS. 2013;10(1):32–41.

Taylor JJ. Assisting or compromising intervention? The concept of 'culture' in biomedical and social research on HIV/AIDS. Soc Sci Med. 2007;64(4):965–75.

van Meijl T. The critical ethnographer as trickster. Anthropol Forum. 2005;15(3):235–45.

Wilson BDM, Miller RL. Examining strategies for culturally grounded HIV prevention: a review. AIDS Educ Prev. 2003;15(2):184–202.

Empathy as Research Methodology

14

Eric Leake

Contents

Abstract

While a long-standing concern in psychology and philosophy, empathy is receiving increased attention in the social sciences for its importance in interpersonal relationships and its use in cross-cultural contexts. I begin this chapter with a brief history and overview of the concept of empathy as a means of understanding the perspectives and experiences of others. I then consider the features that distinguish empathy and the modes through which empathy functions. I address empathy's value across disciplines and extend the application of empathy to the health and social sciences by outlining how practices of empathy might work as a component of research, especially in consideration of different perspectives and social conditions. I apply practices of empathy to research site and participant selection, communication, collaboration, self-reflection, and the recognition of limitations.

E. Leake (✉)
Department of English, Texas State University, San Marcos, TX, USA
e-mail: eleake@txstate.edu

© Springer Nature Singapore Pte Ltd. 2019
P. Liamputtong (ed.), *Handbook of Research Methods in Health Social Sciences*,
https://doi.org/10.1007/978-981-10-5251-4_65

I advocate the practice of critical empathy, in which researchers acknowledge the biases and shortcomings of empathy while simultaneously looking to establish shared goals and interests. To conclude this chapter, I consider the continued necessity of empathy as a component of research despite empathy's limitations.

Keywords

Empathy · Psychology · Philosophy · Perspective-taking · Biases · Cultural awareness

1   Introduction

Empathy is rooted in what it means to be human. Primatologist Frans de Waal (2009) argues that empathy is not only part of human nature but part of our natural ancestry. He has observed what he considers empathy in chimpanzees and the ways that they relate to one another. Empathy all starts and coincides, de Waal argues, with maternal care. Referencing hopes for greater cooperation and social responsibility, he calls empathy "the grand theme of our time," although, admittedly, the promise of empathy now seems strained (p. ix). Even with the long-standing biological lineage detailed by de Waal, empathy is a relatively new concept, as the word *empathy* does not enter the English language until the early twentieth century when it is introduced from the German *einfühlung*, literally "feeling into" (Edwards 2013). Empathy speaks to the ways that we communicate and relate to one another, the cognitive and affective aspects of interpersonal understanding, the moves that we make in order to arrive at a fuller understanding, and the risks inherent in making such moves, loaded as they are with potential for biases.

Attention to empathy has increased in recent years across disciplines, as both an area of inquiry to itself and as a means of training practitioners. Psychologists and neuroscientists, for example, are studying the motivational nature of empathy (Zaki 2014) and how it functions in the brain, especially as related to so-called mirror neurons (Iacoboni 2009). Work in narrative medicine aims to teach physicians to be more empathic in their interactions as they listen to the stories and concerns of patients (Chen et al. 2017), social workers are being taught ways to more empathically respond to the needs of their clients for the purposes of social justice (Gerdes et al. 2011), and anthropologists are exploring the promises and limitations of empathic research practices as well as researching what empathy means in different cultural contexts (Hollan and Throop 2011). Scholars in the humanities are investigating the effects of empathic reading and writing practices (Keen 2007), while educators at all levels are considering how their teaching might increase prosocial activities and dampen bullying and other negative behaviors. And the list goes on. Some of the growing interest in empathy may be attributed to the discovery of mirror neurons, although the significance of that discovery and its potential for explaining interpersonal human interactions have been questioned (Gruber 2013). Some of the interest also likely reflects contemporary public concerns regarding social division, widening inequalities, forces of globalization, and the possibility of understanding one another across personal and cultural differences.

Empathy is useful conceptually and methodologically because of the very concerns that make it so prominent at this moment. There are significant risks to empathy, such as in empathic biases, but those risks are pre-existing; responsibly employing empathy helps to acknowledge and mitigate those risks. As will be detailed in this chapter, empathy is not a methodology to itself but a component of multiple methodologies. A greater empathic awareness highlights the position of researchers, the position of the research participants and area, and the place, social context, and relationships among them. In this chapter, I first survey interdisciplinary concepts of empathy and the significance of the term as it has developed through history and across fields. I distinguish empathy from the related concept of sympathy and break empathy into constituent modes. I then describe components of practicing empathy in research and interpersonal interactions, including the practice of perspective-taking, cultural awareness, dialogue, and the critical empathy. I acknowledge the risks and limitations of empathy, arguing in the mode of rhetorician Dennis Lynch (1998) that empathy is valuable in part because of those risks and how it brings them attention. I conclude with some thoughts on the possible future of empathy as a component of research.

2   What Is Empathy?

Much scholarship on empathy starts with just this question, of what empathy is, because empathy is an ambiguous term. To get right to it, the most complete definition of empathy, I think, is that provided by philosopher Amy Coplan (2011, p. 5): "Empathy is a complex imaginative process in which an observer simulates another person's situated psychological states while maintaining clear self-other differentiation." As Coplan explains, when she says that empathy is complex, she means that it includes affective and cognitive components. When she says that it is imaginative, she means that it requires the observers to make some assumptions concerning the other person. And when she emphasizes a "clear self-other differentiation," she does so to underscore the importance that a person keep in mind that "I am not you, and you are not me" even while using empathy as a way of approaching the understanding of somebody else.

Our contemporary concept of empathy originates in eighteenth-century German aesthetic philosophy with the term *einfühlung*. In her conceptual history of *einfühlung*, Laura Hyatt Edwards (2013) notes that the root, *fuhlung*, means "to grasp, comprehend, or know with certainty through touch" (p. 271). It was used to theorize our physical responses to art and objects. *Einfühlung* was translated to empathy and introduced to English by psychologist Edward Titchener in 1909. He uses the term to describe how concepts may be felt or acted in the "mind's muscles" (cited in Edwards 2013, p. 276). With Titchener's use, empathy transitions from aesthetic philosophy to psychology and becomes more concerned with interpersonal relations. Definitions of empathy since have proliferated, as empathy often is lumped together with sympathy and compassion. To help define empathy, it is useful to distinguish it from sympathy, one of its closest lexical relatives.

Sympathy often is used interchangeably with empathy, although empathy is currently more the vogue of the two, seeming, I assume, more sophisticated and less sentimental due to its basis in psychology. Both sympathy and empathy have similar Greek etymologies as "with" or "in" "feeling." The primary distinction is in perspective and positioning relative another. While sympathy is considered feeling *for* another, empathy is better understood as feeling *with* another. Philosopher Arne Vetlesen (1994, p. 148) distinguishes the terms by defining empathy as "humanity's basic emotional faculty, a specific manifestation of which is sympathy; being a particular feeling, sympathy is facilitated by the basic faculty of relating to others, which I term empathy." We have then a definition of empathy distinguished from sympathy in which empathy is primarily reactive to the situation of another and precedes sympathy. Empathy in this formulation is the basic way that we relate to one another prior to and including making a commitment based upon that relationship.

The concept of sympathy building upon empathy adds an important emphasis at the level of commitment leading to action. This is evident in the explanation provided by philosopher Douglas Chismar (1988, p. 257), who writes of the difference between empathy and sympathy:

> To empathize is to respond to another's perceived emotional state by experiencing feelings of a similar sort. Sympathy, on the other hand, not only includes empathizing, but also entails having a positive regard or a non-fleeting concern for the other person...A 'sympathizer' is one who goes along with a party or viewpoint, while an 'empathizer' may understand, but not agree with the particular cause.

This is a critical distinction. A researcher may empathize with a participant's beliefs or positions without necessarily supporting those beliefs or positions. Sympathy, however, results being persuaded to another's view or cause, so that in witnessing another's situation, one not only understands that situation but aligns oneself in order to amend the situation.

2.1   Interdisciplinary and Debated Approaches to Empathy

I want to collect here a few more interdisciplinary understandings of empathy in order to round out the concept. Literary scholar Suzanne Keen (2007), for example, defines empathy as "a spontaneous sharing of feelings, including physical sensations in the body, provoked by witnessing or hearing about another's condition" (p. xx); and as "the spontaneous, responsive sharing of an appropriate feeling" (p. 4). There are three important elements to this definition: that empathy is spontaneous, physical, and provoked by another's situation, in witnessing or hearing about it. Empathy works across differences and distances, which is an important quality of folklorist Amy Shuman's (2005, p. 4) theory and critique of empathy. She starts with her definition of empathy as "the act of understanding others across time, space, or any difference in experience." Psychologists and common experience tell us that we most readily empathize with those whose presence is immediate: the person on the street, our neighbors and colleagues at work, the friend, or family member. If somebody cannot be present, then the more visceral the sight or the sound and the more felt the emotion – the closer they come to being present – the more powerful the empathic identification. But presence is not a requirement, especially in an age of immediate audio and visual communication.

One of the more influential approaches to empathy is that of developmental psychologist Martin Hoffman (2000, p. 4), who defines empathy generally as "an affective response more appropriate to another's situation than one's own." Like others, Hoffman is defining empathy as the approximation of congruency between one's own and another's response to a situation or inner state or feeling. There is a shared perspective and common humanity here. This quality speaks to necessary relationships of selves and others that are always already a part of empathy. The other important quality of Hoffman's definition is his emphasis on the "affective" as characterizing the type of response one has in empathy. Through this emphasis attention to empathy can be read alongside the larger shift of attention to affect throughout the humanities and social sciences. Hoffman's focus on affect does not mean to relegate empathy to the purely emotional. Instead, he works to do just the opposite in explaining how empathy works as part of larger cognitive processes that combine principles with emotional charge. He writes of what empathy brings to cognitive processes in terms of pairing the more immediate and felt charges of empathy with the more abstract and considered workings of moral principles. This is at the center of his work on empathy, as empathy gives affective force to the support of moral principles.

The role of empathy in the development of morality is somewhat debated. Psychologist Paul Bloom (2016) has forcefully argued that empathy is a not a reliable moral guide because of the biases to which empathy is prone. He instead advocates rational compassion. Bloom defines empathy as "the act of coming to experience the world as you think someone else does" through feeling what you think other people feel (p. 16). He distinguishes between "emotional empathy" and "cognitive empathy," with his critique focused more upon emotional empathy; he views cognitive empathy as a morally neutral tool for understanding, one employed by fair-minded judges as well as psychopaths. Bloom allows that empathy is great for the arts and for intimate relationships and that it can push us to do good, "but on the whole it's a poor moral guide" (p. 2). His division of emotional (or affective) and cognitive empathy is contested, however, as many scholars consider empathy to involve both processes, as noted above with Coplan's definition. Those who promote the moral as well as social and epistemological values of empathy, such as psychologist Jamil Zaki (2017), tend to view empathy as both cognitive and affective, each part of the process counterbalancing and informing the other.

2.2 Modes of Empathy

Empathy is possible through multiple modes. Those are generally broken into two groups, the first being the nonverbal and automatic and the second being those that occur through more of a mediated process. Hoffman (2000), for example, outlines what he determines as the five modes of empathic arousal. three of which are preverbal and involuntary for the most part, those being mimicry, conditioning, and association. Two of the modes are what Hoffman (2000, p. 5) calls the "higher-order cognitive modes" of mediated association and role or perspective-taking. All of these modes are useful for a full accounting of empathy, and they can work together to reinforce one another in interpersonal encounters and in more purposeful exercises of empathy. Here, I will outline some of the most important modes of empathy for research processes in both interpersonal encounters and through mediated means.

Our earliest experiences with empathy occur through nonverbal automatic modes as infants. Empathy may be experienced as a form of affective contagion. Babies, for example, can start crying when they hear other babies crying. The babies did not hear another baby cry, determine that the other baby is sad, and then decide to share in that sad baby's emotional experience. Rather, they pick up the cry and begin to cry as well, sharing the sad baby's experience not through a purposeful perspective-taking but through physical affective feedback, in which their bodily response to their own crying produces a sense of being sad. It may be that, similar to the emotional model proposed by William James, they are not crying because they are sad but are sad because they are crying. Affect is frequently credited as being picked up physically in this sense, so that we adopt one another's mood in a somber room and pick up one another's excitement at a concert, thereby experiencing a charged but not always clearly articulated type of affective empathy.

Facial expressions and mimicry work in a similar mode to that of contagion. We often pick up the facial expressions of others, sometimes when looking at visual representations of other faces but especially in interpersonal encounters. The same can happen with body positions generally. So, if I am sitting with a group of people and many in the group are smiling, I may be more likely to smile as well. When talking with another person, I may be more likely to appear despondent if that other person appears despondent, as I mimic the other person's facial expressions. Sometimes people will even begin slightly to mouth the words that the other person is saying as the other speaks. The mimicry of these facial expressions also may provide a sense of affective feedback, so that I begin to take on not only the expressions but also some of the feelings of the other person. This happens automatically, without a person purposefully setting out to mimic another's expressions and feelings.

The mediated modes of empathy refer to those that generally depend upon intentional language use. When somebody reads or hears about another's experiences and empathizes, this is empathy through mediated means. We also experience this form of empathy when engaged with fictional narratives, whether in reading novels of watching films, which may also simultaneously engage in automatic empathy through their visual and affective cues. Two key modes for mediated empathy are association and perspective-taking. In mediated association, one person might read or hear about another person's experiences and feelings and remember when they felt likewise, drawing upon their own experiences and associating those feelings with the other person (Hoffman 2000). This is a direct communication of what somebody is feeling, but it depends not simply upon automatic visual cues but meaning as communicated through language, although that communication also can conjure simultaneous visual representations (Hoffman 2000). When perspective-taking, a person imagines what it would be like to be in the position of another. The other's feelings in this sense may be more implied, more the result of the perspective-taking, than a direct communication. Perspective-taking depends upon the imagination and drawing from one's own experiences while also attending to the experiences of somebody else. Perspective-taking can be a tricky balance between self and other. I will further detail perspective-taking later. Finally, Keen (2010, p. 80) proposes what she calls "situational empathy," a type of perspective-taking that arises out of narrative. In this case, a reader imagines themselves in the situation of a character and maybe remembers being in a similar situation. This mode of empathy depends more upon contextual factors than it does the personal experiences of another. The situation rather than the individual is the key factor here. Although Keen's focus is literary narratives, this mode of empathy may be experienced outside of literature when we rely primarily upon situational cues, rather than personal cues, in empathizing with somebody else. Such a situational empathy is weak in how it responds to the experiences of another, but it does highlight the power of particular settings and contexts.

These modes of empathy apply to multiple research situations. A researcher, for example, may find themselves picking up the feelings, moods, and expressions of others. This type of automatic empathy can make interpersonal and group research situations more comfortable, as participants feel as though they are on the same page with one another. It can also create a sense of solidarity that includes the researcher, who should take care not to use empathy and group cohesion to mislead participants in any way, such as by allowing them to misconstrue the role of the researcher. Maintaining a clear sense of role, along with a concern for the well-being of participants, is critical for the researcher. Automatic forms of empathy are helpful so long as they support those efforts. Automatic modes of empathy also highlight the importance of setting and the context for interactions in research. Some work may be more effectively done through direct interpersonal interactions rather than mediated communications. Awareness of the mediated modes of empathy can be valuable in the composition and presentation of research, as empathizing with research participants can help others better understand their situations, values, concerns, and motivations. Effective researchers are also effective communicators and may do well to learn how to responsibly employ some of the techniques of narrative empathy. Responsible empathy often is a component of the most ethical efforts in telling another person's story, thereby better serving the interests of the research participants, the researchers, and the readers.

3   Components of Practicing Empathy

Practicing empathy as researchers, and including empathy as a component of any particular research method, requires a balanced awareness of oneself, one's research participants, and the research situation. This multiple attentiveness can be difficult, as the points of awareness can be in tension. In this section, I detail the moves and considerations that may assist in practicing empathy as a researcher.

3.1   Perspective-Taking

At first, the practice of perspective-taking can seem relatively simple. We have practice in it all of the time, when we try to think about what somebody else might be thinking or feeling. Actually implementing perspective-taking, however, is much more challenging, particularly in balancing a focus on self and a focus on another.

Perspective-taking is often divided into two categories, self-focused perspective-taking and other-focused perspective-taking. Self-focused perspective-taking is that in which you imagine what it would be like if you were in another person's situation. It is the practice, as the saying goes, of putting yourself in somebody else's shoes. This is the most common form of perspective-taking and tends to be the easiest. We are pretty well equipped at imagining what we might think and how we might feel in another situation. Self-focused perspective-taking can be affectively powerful (Hoffman 2000). As we imagine ourselves in another's situation, we begin to access our emotions and start to feel how that situation might affect us. But while self-focused perspective-taking gains in affective charge, it loses in accuracy. The way that I might feel in a situation could be very different from the way that somebody else might feel in a situation, given our differences in life experiences, values, concerns, preferred outcomes, and so on. Self-focused perspective-taking makes the exercise all about me. This also leaves self-focused perspective-taking liable to what Hoffman (2000, p. 198, 59) calls "empathic overarousal" and "egoistic drift." In empathic overarousal, my sense of distress at the suffering of another, combined with my own my anxieties and emotional state, can become so intense that I shut off empathy altogether, such as by turning away from the other. In egoistic drift, while empathizing with another, I become so absorbed in my own experiences and emotional responses that I begin to focus more on myself and less on the experiences and emotions of the other person until my empathic response becomes all about me. These liabilities demonstrate the limitations of self-focused perspective-taking, although it remains useful for its emotional power. Self-focused perspective-taking falters the greater the differences between somebody else and myself. If as a researcher, I happen to be working with research respondents who are very much like myself in experiences, values, cultures, and so on, then self-focused perspective-taking could gain accuracy. That degree of similarity is unlikely, however, in most research contexts.

Other-focused perspective-taking tends to be much more accurate but less emotionally charged. For a researcher, this is a positive trade-off. Other-focused perspective-taking asks that you not imagine yourself in the place of another but instead imagine what it is like for that other person in that other person's place. To do so requires knowing more about the other person's experiences, interests, emotions, cultural contexts, desires, and anything else that makes that other person who they are. To successfully practice other-focused perspective-taking, a researcher needs to spend time talking with and getting to know the research participants. Paying attention to context clues and other details can be helpful, too. Other-focused perspective-taking makes it more difficult to make assumptions about another person, as attending to the particularities of that other person and their situation creates fewer opportunities for assuming as a way of filling in the gaps. A helpful variation of other-focused perspective-taking proposed by Hoffman (2000, p. 297) is "multiple empathizing," in which a person imagines how somebody close to them would fill in the position of another. The person might choose somebody close to them, somebody who they know relatively well, who is most like the person they want to empathize with. By empathizing with another by imagining a closer friend or relation in that other's place, one could gain some empathic accuracy alongside emotional force, since imagining somebody close to you is easier and more emotionally engaged. This could be a useful move in those situations when a researcher needs an empathic shortcut because of time or other limitations.

A successful practice of perspective-taking frequently will employ both forms, self-focused and other-focused perspective-taking, simultaneously. In doing so, one may draw upon one's own experiences and emotions while also attending to the realities of the other person. This type of dual perspective-taking is what philosopher Martha Nussbaum (2001, p. 328) calls a "twofold attention," borrowing a term from Richard Wollheim, in which one imagines both the self and another in the other's place, all the while keeping in mind the clear distinction between self and other, as underscored in Coplan's definition. The tension between self-focused and other-focused perspective-taking is never static. It is always moving as attention shifts more in one direction or another, particularly in more emotional situations. The cultivation and reliable employment of a twofold attention, one with an awareness of the tensions and shifts in either direction toward self and other, is a valuable skill for a researcher who wants to employ empathy as a component of the research process.

3.2   Cultural and Contextual Awareness

Empathy as a conscious process requires the basic assumption that we might begin to approximate the perspectives, feelings, and experiences of another. For example, in other-focused perspective-taking, how might one start to assume how another feels in a situation unless one assumes to have some shared access to what another might feel? These are always dangerous assumptions. Assumptions of commonalities are safest when one is attempting to empathize with another whose background and experiences are closest to one's own. But, even then, they are quite risky, because similarities of background and experience do not necessarily or even generally lead to shared affective states. The inherent risks of these assumptions contribute to the familiarity bias in empathy. It is generally the case, for example, that empathic identification with a sibling who shares much of one's background and experiences is easier and likely more accurate than empathic identification with someone of a different family background, from a different place, or of another generation. Some of the most significant tests of empathy, and some of the instances in which assumptions of commonalities are riskiest, often occur across the greatest differences in cultures and contexts. Because of this, empathy is a vital component of theories of cosmopolitanism, which similarly struggle with the balance of universal principles and an idea of a common humanity while simultaneously recognizing that differences cannot be neglected or erased. The resulting slogan for cosmopolitanism, offered by philosopher Kwame Appiah (2006, p. 151), is "universality plus difference." The idea of "universality plus difference" is a useful term because it demonstrates the necessary and always present tensions between the universal and the particular. These tensions are necessary because cosmopolitanism – as well as empathy – can attempt to attain validity only through the simultaneous acknowledgement of universality and of difference, two concepts that are often placed in opposition.

Empathy is of interest in anthropology both as a component of fieldwork, such as when developing an ethnography, and as an area of research itself, since conceptions of empathy vary across cultures and places. Douglas Hollan and C. Jason Throop (2011, p. 7) note the variable nature of empathy in the introduction to their collection on empathy in Pacific societies:

> Like any other form of complex human behavior, empathy emerges in an intersubjective field, partially determined by the evolved, highly social characteristic of the human species, but significantly constituted and structured as well by social, cultural, linguistic, and developmental variables.

These variables point to the challenges not only in understanding different cultural conceptions and practices of empathy but of any complex behavior and encounter. An awareness and practice of empathy should attempt to account for the "social, cultural, linguistic" and other variables that might come into play when working with diverse populations. Empathy requires cultural awareness. Researchers must question their own assumptions and focus on the lives and values of the people they are working with, not only if working with members of Pacific societies, whose ways of living may be very different from their own, but also when working with members of those communities closer to home, because with greater proximity, there is greater likelihood that significant intercultural and personal differences may be overlooked. Some of this research can be conducted in preparation for a project as researchers identify the cultural contexts they will be working in and set out to learn as much as possible about those places and people. Some of the work can only be done through immersion, contact, observation, and exchange (see also ▶ Chaps. 13, "Critical Ethnography in Public Health: Politicizing Culture and Politicizing Methodology," and ▶ 26, "Ethnographic Method"). Maria Lepowsky (2011) identifies empathy as a central concern in anthropology and ethnographic fieldwork, and she proposes methods that work toward a resolution, although empathy is always in process and never complete. Lepowsky (2011, p. 43) writes: "The core paradox of anthropological epistemology, method, and representation remains the human inability to experience full empathy with another person." As a means of addressing this, she offers "ethnographically informed narrative, the process of telling and retelling stories of intersubjective experiences in the field, at home, and in print" (p. 53). Telling and retelling stories as a mutual process, shared by researchers and research participants, can move both toward a greater cultural awareness and at least a partial empathy.

3.3   Communication, Listening, and Collaboration

At the core of any practice of empathy are effective listening, communication, and collaboration. Empathy truly realizes its value when it is practiced in connection with other people and when it becomes a joint effort. A vital figure in the development of empathic listening techniques is psychologist Carl Rogers, who pioneered the client-centered approach to therapy. Much of his practice was a practice of empathy. Rogers (1961, p. 331) considered the major barrier to communication to be the "tendency to react to any emotionally meaningful statement by forming an evaluation of it from our own point of view." Researchers, of course, should take care not to preemptively evaluate or judge a participant's statement. The communication strategy that Rogers advances is rooted in the practice of empathy as an embodied, nonjudgmental way of listening. "Real communication occurs, and this evaluative tendency is avoided, when we listen with understanding," he writes (p. 331). "What does that mean? It means to see the expressed idea and attitude from the other person's point of view, to sense how it feels to him, to achieve his frame of reference in regard to the thing he is talking about" (pp. 331–332). Empathy, for Rogers, is a means of understanding not only cognitively but also emotionally by adapting oneself to another's frame of reference. This is a powerful move both for the empathizer and the one who is empathized with. To listen with understanding, as Rogers calls it, can change the person who is empathizing. This type of listening can require courage. Rogers (1961, p. 333) advises: "If you really understand another person in this way if you are willing to enter his private world and see the way life appears to him, without any attempt to make evaluative judgments, you run the risk of being changed yourself."

It is important to stress here that Rogers is primarily referring to occasions when people are on different sides of an argument, or in therapeutic encounters, although much of the power of empathic listening – listening with understanding – remains. Rogers offers a key technique for empathic listening, what is known as the restatement principle. In this practice, people take turns restating to one another in their own words what they heard the other person say. They can only continue with the discussion once the other's point has been restated to the other's satisfaction. If I were to restate what you said and you were not satisfied with how I restated it, then I would need to try again and await your approval before we could move on. This technique also was pioneered in therapeutic contexts, but the basic move remains helpful for researchers as a way of practicing empathic listening along with the suspension of judgment. It also can be helpful when practiced in the other direction, as a means of helping researchers be sure that research participants understand the project and what is happening. Listening in this way can improve empathic accuracy and help to develop a rapport among researchers and respondents as they become collaborators in understanding their exchange.

A similar orientation is proposed in rhetorician Krista Ratcliffe's (1999) idea of "rhetorical listening." Ratcliffe's concept of listening is particularly useful in cross-cultural contexts. In defining rhetorical listening, Ratcliffe focuses on the significance of understanding, which for her "means more than simply listening *for* a speaker/writer's intent" and "also means more than simply listening *for* our own self-interested intent" (emphasis original, p. 205). Instead, for Ratcliffe, "*understanding* means listening to discourse not *for* intent but *with* intent – with the intent to understand not just the claims, not just the cultural logics within with the claims function, but the rhetorical negotiations of understanding as well" (p. 205). She asks that practitioners of rhetorical listening let themselves be immersed in other discourses, in other ways of understanding, to not try to focus on only one idea or concept in order to pull it out, for the other's purposes or your own, but instead "letting discourses wash over, through, and around us and then letting them lie there to inform our politics and ethics" (p. 205). This type of listening requires an awareness of our own cultural logics and discourses as well as those of others and the ways that they all might intersect. Ratcliffe proposes this concept in the hopes that it might work at points of discursive intersection "so as to help us facilitate cross-cultural dialogues about any topic" (p. 196). Although Ratcliffe is writing with concerns of public deliberation and social discourse in mind, her approach is valid for researchers. She warns us that when we listen too closely for intent, we can miss the larger cultural discourses that surround the communication. It is important instead to listen with intent, with the intent to appreciate not only a person's words but the worldview that gives meaning to those words.

The empathic communication and listening practices put forward by Rogers and Ratcliffe underscore the collaborative nature of empathy. As stated earlier, empathy can be most effective when it is built upon a relationship. The larger context for these moves is one in which researchers are partners with their research respondents. This partnership can be realized in many ways. It involves conferring with research participants about the research project; enlisting them in helping to develop the project; making communication a cooperative enterprise that values the listening, cultural context, concerns, and understanding of all participants; and inviting them to review how they are presented in the research and how their stories might be told, considering them as coauthors, so that they too have a voice in the project (see ▶ Chap. 17, "Community-Based Participatory Action Research"). All of these moves help to establish research practices that are more empathic, collaborative, and ethical and thereby stronger and more effective. These are especially important when working with vulnerable population, who otherwise may not have much say in how they are considered in the project (Liamputtong 2007, 2010). For people who do not have as much power as the researchers, it is imperative that institutions and, ultimately, researchers themselves make sure that they are establishing ethical and inclusive research practices. Empathic communication, listening, and collaboration are a good start for such practices.

3.4   Self-Reflection and Critical Empathy

Because of all of the biases to which empathy is prone – and because of the tensions between self- and other-identification and the tendency to erase the other to be replaced by the self in self-focused perspective-taking – it is important that any practice of empathy be paired with self-reflection and critical empathy. These moves help keep the liabilities of empathy in check, and they enhance empathic practices by offering mechanisms to review and revise empathy as well as recognize one's own interests and sociocultural position.

As attention to empathy has grown, popularly and within the academy, so have critiques of empathy and how it functions. As Shuman (2005, p. 18) writes, "empathy is almost always open to critique as serving the interests of the empathizer rather than the empathized." Shuman is interested in how personal stories are told and how they acquire meaning beyond the personal. Usually, the telling of these stories is in the interest of the teller rather than the person who first told the story from their experiences. This is a critique of empathy very much relevant to researchers, who risk promoting their own interests as they collect and tell the stories of others. As Shuman adds, empathy "rarely changes the circumstances of those who suffer" (p. 5). And so she puts forth a critique of empathy that "avoids an unchallenged shift in the ownership of experience and interpretation to whoever happens to be telling the story and instead insists on obligations between tellers, listeners, and the stories they borrow" (p. 5). Literary scholar Theresa Kulbaga (2008) similarly critiques empathy, in this case the easy empathy of some cross-cultural novels, for merely catering to the pleasures of the readers, because empathizing with distant others in ways that are comfortable can be fun. Western readers are allowed to enjoy the empathic reading experience and "remain in the realm of individual imagination, where affect remains divorced from either critical reflection or political action" (p. 517). She recommends instead that we ask "empathy to what end?" (p. 518). By questioning the conveniences of empathy, by realizing how empathy can be inviting or complacent, by acknowledging the commitments of empathy, and by asking what empathy does in a given situation, we can take a more critical approach to empathy, one that does not see empathy itself as an end but as a means toward a larger project. Researchers do well to ask themselves how they are involved in the empathic situation, how the interests of everybody are being served, what commitments for change and continuing relationships the situation proposes, and what the effects of empathy in that situation are.

The practice of critical empathy requires recognition that empathy is always an approximation and is always incomplete. By reflecting upon our own empathic processes, we can better realize the limitations of empathy and not make unacknowledged assumptions. Education professor Todd DeStigter (1999) advocates the practice of critical empathy as a way of forming relationships in communities across differences even while recognizing the limitations of empathy and how those relationships can be uneven. DeStigter (1999, p. 240) defines critical empathy, which he adopts from Jay Robinson, as:

> the process of establishing informed and affective connections with other human beings, of thinking and feeling with them at some emotionally, intellectually, and socially significant level, while always remembering that such connections are complicated by sociohistorical forces that hinder the equitable, just relationships that we presumably seek.

For DeStigter, the position of the researcher is part of the complication, because it is unequal to that of the other and is subject to sociohistorical forces. Still, those limitations do not mean that we should give up on empathic connections. Instead, they require that we be aware of the limitations and try to work with and within them. For researchers, then, the trick is not to act as though all participants, researcher included, are on equal footing but to acknowledge disparities and invite connections and collaborations across differences, making the differences themselves open to reference, critique, and, hopefully, some counterbalance as participants find points of commonality and connection.

4   Limitations and Risks of Empathy

The most significant limitations of empathy are biases. Simply put, we tend to empathize most readily with those most like ourselves. Hoffman (2000, p. 197) calls this the "familiarity bias." The greater the differences between myself and somebody else, the greater the challenge to empathize. As a corollary, if there are two people with whom I might empathize, I likely will empathize more easily and effectively with the person most like me. We see empathic failures due to familiarity biases all of the time in our public discourse, as people empathize with the victims most like themselves and downplay or ignore the claims of others. As above, researchers would do well to question their empathic impulses and to reflect upon the processes of empathy, including their own subjectivities and positions, that might lead them to more readily empathize with one person over another.

A second bias to which empathy is prone is the bias of immediacy, what Hoffman (2000, p. 197) calls the "here-and-now bias." We are more likely to empathize with somebody when that person is in our immediate vicinity, when we are witness to that person's emotions, rather than to empathize with somebody more distant. This bias is due in part to empathy's reliance upon automatic and nonverbal modes, such as body, facial, and affective cues, that are relayed through face-to-face interactions. Bloom's (2016) argument against empathy as a moral guide is based largely upon this bias. It can affect researchers in many ways, perhaps most significantly in the choice of research topics and sites of research. We are liable to investigate issues that affect those nearest us without proper consideration of what might be a more beneficial use of our time in researching issues that affect more distant populations more significantly. When weighing the ethics of research, as informed by empathy, it is important to consider not only our interactions with others in the course of research but also how we come to be interested in and to choose our research topics and sites.

An additional risk of empathy is the way that it can be self-serving and can disadvantage people with whom we are working. Because empathy generally is considered an unopposed positive value in Western societies, it is easy to use empathy as a justification for work that might serve our interests more than those of the people we are working with. So long as we are empathizing, we might unreflectingly give ourselves a pass and not question deeper motivations for and consequences of the work. The critical empathy practices addressed earlier are intended to help mitigate this. Empathy tends to flow down social power gradients.

By that I mean the more powerful tend to be in a better position to empathize with the less powerful; empathy is directed up less frequently. These differences in power can have adverse effects on the people who are empathized with, as they leave the empathic encounter reminded of their lack of social standing and power and less motivated to pursue their goals (Vorauer and Quesnel 2016; Vorauer et al. 2016). Again, the practice of a critical empathy and the development of long-term collaborations, in which participants are treated as co-creators with a voice in the research projects, hopefully can help counteract these negative outcomes. It also is worth noting that disparities in social power and standing precede empathic encounters. When practiced effectively, processes of empathy may help shine a light on some of these issues in ways that help them be addressed rather than ignoring disparities that nevertheless could still have negative implications for the work.

5 Conclusion and Future Directions

As should be clear, empathy is not a research method in itself but works as a component of multiple methods. It may be considered and practiced as a methodological framework. Part of the value of empathy is in its incorporation of the cognitive and the affective, its recognition that we think not only with the logical parts of our minds but also emotionally and with the whole body, extending even into our interactions with others and with our environments. As a component of research processes, empathy can lead to new insights and commitments and more meaningful ways of communicating. Many people are initially drawn to ideas of empathy because it seems so unquestionably positive, but empathy is not an unalloyed good. It can empower biases and lead to erroneous conclusions and problematic actions, all under the cover of positive empathy. Hence, we need practices of critical empathy.

Increased public interest in empathy speaks to the times in which we live. It seems not merely a coincidence that interest in empathy has grown at the same time that we face greater social divisions, widespread inequalities, and the challenges of engaging others near and far across more significant differences. Empathy suggests a means for how we might navigate those processes of communication, cooperation, and mutual understanding in our many interactions. Although far from perfect, empathy does offer hope that we can learn to respect and work across our differences. As we learn more and further develop our concepts and practices of empathy, our work and our relationships stand to benefit.

References

Appiah KA. Cosmopolitanism: ethics in a world of strangers. New York: W. W. Norton; 2006.
Bloom P. Against empathy: the case for rational compassion. New York: HarperCollins; 2016.
Chen P, Huang C, Yeh S. Impact of a narrative medicine programme on healthcare providers' empathy scores over time. BMC Med Educ. 2017;17(108):1–8.

Chismar D. Empathy and sympathy: the important difference. J Value Inq. 1988;22:257–66.

Coplan A. Understanding empathy: its features and effects. In: Coplan A, Goldie P, editors. Empathy: philosophical and psychological perspectives. New York: Oxford University Press; 2011. p. 3–18.

de Waal F. The age of empathy: nature's lessons for a kinder society. New York: Harmony Books; 2009.

DeStigter T. Public displays of affection: political community through critical empathy. Res Teach Engl. 1999;33(3):235–44.

Edwards L. A brief conceptual history of *Einfühlung*: 18th-century Germany to post-world war II U. S. psychology. Hist Psychol. 2013;16(4):269–81.

Gerdes K, Jackson K, Segal E, Mullins J. Teaching empathy: a framework rooted in social cognitive neuroscience and social justice. J Soc Work Educ. 2011;47(1):109–31.

Gruber D. The neuroscience of rhetoric: identification, mirror neurons, and making the many appear. In: Jack J, editor. Neurorhetorics. New York: Routledge; 2013. p. 35–50.

Hoffman ML. Empathy and moral development: implications for caring and justice. New York: Cambridge University Press; 2000.

Hollan D, Throop CJ. The anthropology of empathy: introduction. In: Hollan D, Throop CJ, editors. The anthropology of empathy: experiencing the lives of others in Pacific societies. New York: Berghahn Books; 2011. p. 1–21.

Iacoboni M. Mirroring people: the science of empathy and how we connect to others. New York: Picador; 2009.

Keen S. Empathy and the novel. New York: Oxford University Press; 2007.

Kulbaga T. Pleasurable pedagogies: *reading Lolita in Tehran* and the rhetoric of empathy. Coll Engl. 2008;70(5):506–21.

Lepowsky M. The boundaries of personhood, the problem of empathy, and 'the native's point of view' in the Outer Islands. In: Hollan D, Throop CJ, editors. The anthropology of empathy: experiencing the lives of others in Pacific societies. New York: Berghahn Books; 2011. p. 43–65.

Liamputtong P. Researching the vulnerable: a guide to sensitive research methods. London: Sage; 2007.

Liamputtong P. Performing qualitative cross-cultural research. Cambridge: Cambridge University Press; 2010.

Lynch D. Rhetorics of proximity: empathy in Temple Grandin and Cornel West. Rhetor Soc Q. 1998;28(1):5–23.

Nussbaum M. Upheavals of thought: the intelligence of emotions. New York: Cambridge University Press; 2001.

Ratcliffe K. Rhetorical listening: a trope for interpretive invention and a "code of cross-cultural conduct". Coll Compos Commun. 1999;51(2):195–224.

Rogers C. On becoming a person: a therapist's view of psychotherapy. Boston: Houghton Mifflin; 1961.

Shuman A. Other people's stories: entitlement claims and the critique of empathy. Urbana: University of Illinois Press; 2005.

Vetlesen AJ. Perception, empathy, and judgment: an inquiry into the preconditions of moral performance. University Park: The Pennsylvania State University Press; 1994.

Vorauer JD, Quesnel M. Don't bring me down: divergent effects of being the target of empathy versus perspective-taking on minority group members' perceptions of their group's social standing. Group Process Intergroup Relat. 2016;19(1):94–109.

Vorauer JD, Quesnel M, St. Germain SL. Reductions in goal-directed cognition as a consequence of being the target of empathy. Personal Soc Psychol Bull. 2016;42(1):130–41.

Zaki J. Empathy: a motivated account. Psychol Bull. 2014;140(6):1608–47.

Zaki J. Moving beyond stereotypes of empathy. Trends Cogn Sci. 2017;21(2):59–60.

Indigenist and Decolonizing Research Methodology

15

Elizabeth F. Rix, Shawn Wilson, Norm Sheehan, and Nicole Tujague

Contents

Abstract

European colonization of Indigenous nations has severely impacted the health of Indigenous peoples across the globe. Much of the burden of ill health suffered by Indigenous people today can be traced directly back to colonization. Indigenous peoples of all first world nations where colonization has occurred are experiencing epidemic proportions of chronic disease, higher levels of morbidity and mortality, and poorer health outcomes compared to non-Indigenous populations. Indigenist and decolonizing approaches to research with Indigenous peoples have emerged in recent years with the overall aim of recognition and inclusion of Indigenous epistemologies and ontologies within the western research paradigm. A significant barrier to achieving this is the disconnection between the dominant

E. F. Rix (✉) · S. Wilson · N. Sheehan · N. Tujague
Gnibi Wandarahn School of Indigenous Knowledge, Southern Cross University, Lismore, NSW, Australia
e-mail: liz.rix@scu.edu.au; shawn.wilson@scu.edu.au; norm.sheehan@scu.edu.au; nicole.tujague@scu.edu.au

© Springer Nature Singapore Pte Ltd. 2019
P. Liamputtong (ed.), *Handbook of Research Methods in Health Social Sciences*,
https://doi.org/10.1007/978-981-10-5251-4_69

biomedical approach to health and the holistic understandings of health based on Indigenist philosophies and traditional healing practices and knowledges. Conducting research that can successfully inform and improve health services and outcomes for Indigenous peoples requires a decolonizing approach where the voices of Indigenous Elders and communities are the primary informants. Integrating Indigenous ways of knowing, being, and doing with western biomedical approaches requires respect for and inclusion of Indigenous Knowledge as healing methods that have preserved community and individual well-being for thousands of years.

Keywords

Indigenous Knowledge · Colonization · Trauma aware practice · Healing culture · Dominant western paradigms

1 Introduction: What Is an Indigenist Research Paradigm?

Indigenist research respects and honors Indigenous ways of knowing, being, and doing through using methods that are informed by, resonate with, and are driven and supported by Indigenous peoples. Researchers working respectively with, and learning from Indigenous peoples aim to decolonize western research methodologies and methods in order to include Indigenous ways of seeking, analyzing, and disseminating new knowledge. In order to apply an Indigenist research paradigm to the health services sector, researchers from non-Indigenous backgrounds must firstly examine their own worldview(s) to enable them to understand that their view of the world is different than that of the Indigenous peoples with whom they are working. Indigenist research is characterized by approaches grounded in relationality and the inclusion of Indigenous ways of communicating such as storytelling or "yarning" as it is referred to in Australia (Bessarab and Ng'andu 2010; Rix et al. 2014; Wilson 2008). Applying Indigenist ways of conducting research is in accord with the United Nations Declaration on the Rights of Indigenous Peoples which states that:

> Indigenous peoples have the right to maintain and strengthen their distinct political, legal, economic, social and cultural institutions, while retaining their right to participate fully, if they so choose, in the political, economic, social and cultural life of the State. (United Nations 2008, Article 5, p. 5)

Indigenist research has been a growing body of new knowledge production over the past two decades, led by Indigenous scholars from Canada, Australia, and New Zealand. Indigenous scholar and Maori woman Linda Tuwahai Smith's groundbreaking book *Decolonizing Methodologies: Research and Indigenous Peoples* (Smith 1999) articulated Indigenous research methodologies that ensure Indigenous intellectual sovereignty of projects involving Indigenous people, interests, and concerns. A number of international scholars have built on the body of knowledge about Indigenist research methodologies. The work of Indigenous scholars in the

Americas, Africa, and Australasia has illuminated the core principles of an Indigenist paradigm for conducting respectful and safe ways for Indigenous peoples to conduct research with both their own people and other Indigenous communities. From an Indigenous methodological perspective, the entire research process must be redefined and reframed (Rigney 1997; Weber-Pillwax 2001; Martin 2008; Wilson 2008; Kovach 2010; Chilisa 2012; Kite and Davy 2015).

While Indigenous scholars from first, second, and third world nations do share methodological commonalities, researchers must be mindful not to categorize these as a homogenous group, and outsiders have no right to do so. One thing that we do share is our colonized histories and the contemporary impacts on the social, health, and political positioning of Indigenous peoples living under western governance. We are also connected via our evolving Indigenous methodologies (Walter and Anderson 2013). The collective work of these groundbreaking Indigenous scholars has built an international body of work that is now making inroads into western academic methodologies and protocols (see also ▶ Chaps. 87, "Kaupapa Māori Health Research," ▶ 89, "Using an Indigenist Framework for Decolonizing Health Promotion Research," ▶ 90, "Engaging Aboriginal People in Research: Taking a Decolonizing Gaze," and ▶ 97, "Indigenous Statistics").

2   Some History

> Everything on Earth has a purpose, every disease a herb to cure it, and every person a mission. This is the Indian Theory of Existence. (Morning Dove, Salish)

Indigenous peoples colonized by Europeans share a history of dispossession, trauma and loss of culture. The United States, Canada, Aotearoa/New Zealand, and Australia, for example, are all developed nations with common experiences of European colonization. For Indigenous peoples in these wealthier nations, seizure of land and extermination of whole communities or tribal groups were universally common and carry ongoing repercussions (Stephens et al. 2006). While the non-Indigenous populations of these countries generally enjoy high standards of living and health, Indigenous populations experience significantly poorer socio-demographic and health outcomes (Anderson and Whyte 2008). Indigenous peoples also share significantly lower life expectancy, with "epidemic" levels of chronic conditions such as cardiovascular disease, diabetes, and renal failure (King 2010; Centres for Disease Control and Prevention 2014; Australian Institute of Health and Welfare 2015).

Since the initial shock of invasion by Europeans, Indigenous peoples have been forced to give up language culture and understandings through generational impositions that separated individuals, families, and whole nations from their original knowledge systems and social structures. These traumatic events have disintegrated family and community relationships, structures, and traditional lifestyles. The profound impacts of colonization have been further compounded by the enduring history of successive failed government policies and practices (Durie 2004;

King et al. 2009). For example, the removal of Indigenous children from their families and their detention in government-controlled and Christian church-run residential schools were reinforced through government policy in both Canada and Australia throughout the nineteenth and twentieth centuries. Similar negative and traumatizing experiences when engaging with all government services further connect Indigenous peoples from the United States, Canada, Aotearoa/New Zealand, and Australia (Walter and Anderson 2013).

In her 1998 book, *Colonizing bodies*, the author Kelm explored how Canadian Indigenous peoples were not only materially affected by the colonizer's Canadian Indian policies; restricting hunting and fishing and the forced removal of children into unhealthy residential schools, traditional healing, and cultural practices were criminalized. The author discusses the impact of the use of humanitarianism and western medicine to pathologize Aboriginal bodies and inflict a monoculture of biomedical approaches in the name of assimilation (Kelm 1998). The removal of entire communities from their lands into missions and reserves was common practice, with the use of missionaries and Christian doctrine to further demonize and eliminate entire family and cultural frameworks (Kelm 1998).

Indigenous peoples in Australia and the United States experienced similar treatment under colonial policies aiming to assimilate Aboriginal peoples and destroy cultural and family structures using extreme inhumane measures, including removal of children from their families and communities.

The deliberate state-led destruction of Aboriginal communities, languages, and cultural practices has inflicted profoundly damaging levels of trauma. The systematic removal of Indigenous Australian children from their families and communities, creating the Stolen Generations, has deeply impacted social and emotional well-being, with these impacts being passed on via trans-generational trauma (Atkinson 2002). This trauma is an antecedent to Indigenous Australian peoples now suffering a huge gap in health and well-being when compared to non-Aboriginal Australian populations. Emerging theories generated by the study of epigenetics have now linked damaging environmental factors in utero and during early life and begin to explain the causation of the current epidemic of chronic disease and early mortality (Hoy and Nicol 2010). The disconnection of generations of Indigenous children from their land, families, and communities created a direct causal pathway to the current epidemic proportions of chronic disease across all first world colonized nations. Indigenous scholars stress the impact of "cultural detachment" which includes separation from people and country, loss of traditional diet, lifestyles, language, and stories on their people, evidenced in the spiralling incidence of diabetes and other chronic conditions in Indigenous populations (Sanderson et al. 2012).

3   Institutional Racism and Health: Ongoing Colonization

While we could delve further into the ongoing impact of historical trauma, it is important to recognize the ongoing nature of colonial experiences for Indigenous people. Experiencing racism is now known to have a direct physiological impact on health and well-being. Recent work has reviewed the scientific research on how racism adversely affects the health of non-dominant cultural groups. Multiple causal pathways have been identified by which racism can affect health, with institutional and cultural forms of racism being major contributors to health inequalities (Williams and Mohammed 2013). Concepts such as black inferiority and white superiority have been historically embedded in American culture and continue to impact American First Nations people today.

The link between poor health and racism for Aboriginal people was explored in depth by Australian Indigenous scholar Yin Paradies in his doctoral thesis entitled "Race, racism, stress and Indigenous health" (Paradies 2006). This epidemiological study showed strong and persistent associations between chronic stress resulting from experiences of racism and poor physical and mental health, including depression, and increased risk factors for heart disease and other chronic health conditions.

Racism is inherent within colonial government organizations and is beginning to spawn its own research. Reluctance to engage with western biomedical health services by Indigenous peoples is underpinned by a complex blend of historical, political, and economic drivers. Mainstream health services continue to lack acknowledgment and understanding of the historical trauma and racism that are the antecedents to Indigenous people's avoidance of and lack of confidence in mainstream biomedically driven health services (Larson et al. 2007). Originating at colonization and reinforced by ongoing experiences of overt individual and institutionalized racism, Indigenous people and communities remain highly suspicious of engaging with government services. In the health services context, there remains a culture of blaming and judgment of Indigenous people suffering chronic illness, despite the rapidly increasing incidence of similar patterns of "lifestyle-induced" chronic conditions in all non-Indigenous populations of first world nations (Rix et al. 2014).

Despite contemporary research showing inclusion of cultural and traditional healing methods as the way forward in addressing the serious health disparities suffered by Indigenous people, this is not translating into practice (Poche Indigenous Health Network 2016). Indigenous peoples' fear and avoidance of western healthcare systems coupled with the "one-size-fits" approach of the biomedical model remain major institutional barriers. In view of this history, it is vital that any research performed with the aim of improving health services acknowledges this by incorporating and utilizing Indigenous ways of creating new knowledge. Dissemination of findings and recommendations requires researchers to collaborate with community to ensure methods are negotiated with and approved by Indigenous Elders and community.

4   Survival and Resilience

Survival and resilience are fundamental qualities that unify colonized Indigenous peoples (Tousignant and Sioui 2009; Ramirez and Hammack 2014). Despite the deficit-based approach to "problematizing" Indigenous peoples and their health, a universal strength of all colonized Indigenous peoples is their remarkable survival and resilience, even in the face of the relentless attacks by dominant western governments, aiming to annihilate Indigenous cultural traditions and silence Indigenous voices. This shared history of extreme violence, trauma, and dispossession has, however, highlighted Indigenous peoples' abundant resilience and strengths. Indigenous peoples' ability to survive and heal from a succession of damaging and hegemonic government policies unites them (Fast and Collin-Vezina 2010).

Cultures of resistance emerge within these dominated populations fostering forms of non-compliance and aberrant behaviors that protect individuals and communities and ease the burdens of dominance. These weapons of the weak often divert and delay actions instigated by the dominant and conceal resistance because they are interpreted to be evidence of the lesser ability of the dominated. The essential tactic of avoidance is a part of resilience and also a key factor in externally applied research because approaches such as surveys may not truly reflect the Indigenous population (see also ▶ Chap. 97, "Indigenous Statistics"). In this way Indigenous populations routinely subvert research, avoid treatment, and ignore initiatives (Sheehan et al. 2009).

There is an urgent need to transform western government and policy dialogues about Indigenous peoples' health and well-being from a negative, deficit-based focus. We do not need more research that tells us how well colonialism is still working (Walter and Anderson 2013). Indigenous leaders and culturally competent and skilled healthcare professionals are stipulating a strength-and right-based approach to health policy and practice for Indigenous people (Tsey et al. 2007; Jackson et al. 2013; Neumayer 2013). Elders and Indigenous communities have been calling for Indigenous Knowledge and healing methods that have sustained Indigenous societies for thousands of years to be incorporated with western medical practices (Moodley 2005; Shahid et al. 2010).

5   Why Indigenist and Decolonizing Research in the Health Social Sciences

Indigenous Peoples must look to new anti-colonial epistemologies and methodologies to construct, re-discover and/or re-affirm their knowledges and cultures. Such epistemologies ... strengthen the struggle for emancipation and liberation from oppression. (Rigney 1997, p. 115)

Indigenous peoples have suffered a long history of having been "researched on" by western anthropologists and academics from other disciplines. As a result, many are cynical about the benefits of participating in western academic research, being wary of the colonial lens, and assumed superiority of western researchers (Prior 2006). Australian Indigenous peoples, for example, have witnessed some two centuries of being "over researched" with no prior consultation, permissions sought or any form of post research feedback or positive outcomes. There is a common recognition among many Indigenous peoples that western research paradigms contributed to their ongoing oppression. Research that is framed and supported by the very system that has dispossessed and oppressed Indigenous people has an inherent bias toward maintaining that system. Aboriginal communities in Australia have long held practices that subtly disable intrusive research (Sheehan et al. 2009).

Given this history and the urgent need to address the immense health disparities suffered by Indigenous peoples in colonized first world nations, there is an urgent need to conduct research differently. We cannot expect the same system of thinking that has caused such large health disparities to be able to envision a solution. Research must, therefore, be conducted in a way that fully captures and honors the voices and perspectives of Indigenous peoples but, more importantly, emanates from an Indigenous ontological and epistemological basis. This is Indigenist research. It ensures that Indigenous Knowledges, experience, and wisdom are captured, applied, and disseminated in ways that resonate with Indigenous ways of knowing, being, and doing. Sherwood (2010), Australian Indigenous nurse and scholar, has applied this approach to decolonizing Indigenous health services and research in her doctoral thesis "Do no harm: Decolonising Aboriginal health research." Sherwood argues that any research aimed at decolonization of Indigenous healthcare must be initiated and guided by Elders and underpinned by Indigenous critical theory and the balancing of two ways of knowing. This approach leaves no room for the "problematizing" of Aboriginal people in the healthcare context, which results in the silencing of Indigenous voices, subjugation of Indigenous Knowledges, and Indigenous peoples being viewed as the "Other" (Sherwood 2010).

All research is appropriation, and, therefore, it should be conducted in a way that ensures that both the researched and the researcher benefit (Chilisa 2012). Normalized positions of dominance must be recognized and deconstructed and deficit discourses replaced with respectful representation that leads to opportunities for sharing, growth, and learning. This approach restores hope and belief in a community's capacity to resolve challenges (Chilisa 2012). Elders and community members are indeed the experts in the health and well-being of their own people, with knowledge and expertise reaching back thousands of years prior to the evolution of the biomedical model of health and the colonization of their nations by Europeans.

Indigenist researchers have developed theoretical frameworks and research methods that are congruent with Indigenous belief systems that have been known for millennia. As the substantive theory underlying Indigenist research continues to be further articulated, Indigenous truths that are informed by Indigenist methodologies have emerged. These truths are based on our own ontological and epistemological foundations and, therefore, have the ability to envision solutions to seemingly intractable health problems. A common feature among many Indigenous peoples' ontology is that we are relational. That is to say, we do not engage in relationships, nor are we in relationships, but we are relationships. Our very being, and the nature of reality itself, is relational. We are relationships with people and communities, with the Land, with ideas, with everything. Indeed, nothing would be (or exist) without relations (Wilson 2008). Aboriginal scholar Mary Graham builds on this relational ontology with her statement "You are not alone in the world." Here Professor Graham describes the survival of a contemporary Indigenous kinship system based on relationality, despite the damage inflicted by the colonizer's urbanization of Aboriginal people using (attempted) cultural genocide. The central role of Indigenous people's connection and relationship to land is clear in her writing:

> Although Indigenous people everywhere are westernised to different degrees, Aboriginal people's identity is essentially always embedded in land and defined by their relationships to it and to other people. The sacred web of connections includes not only kinship relations and relations to the land, but also relations to nature and all living things. (Graham 2008, p. 187)

Fundamental to Indigenist research is an understanding that the researcher is not outside of reality looking in but has entered into a different set of relationships with the people and issues that they are researching. So, among the truths that emerge from an understanding of relationality is that researchers, as knowledge producers in relation, are in themselves accountable for maintaining healthy relationships with the communities, environment, and ideas that they are researching (Wilson 2008).

Reflexive practice is essential for any researcher in examining their motivations and intent in working with Indigenous people. Just as Indigenous scholars apply a critical lens to their work with their own people or other Indigenous communities, examining why and how they are doing research, close scrutiny is vital:

> If my work as an Indigenous scholar does not lead to action, it is useless to me or anyone. I cannot be involved in research and scholarly discourse unless I know that such work will lead to some change out there in that community, in my community. (Weber-Pillwax 2001)

Indigenist methodologies view the rigor and validity of research as determined through relational accountability (Wilson 2001). Researchers have relational accountability to participants, co-investigators, and the overall conduct of the study. For the non-Indigenous researcher, relational accountability is encapsulated by principles of respect, responsibility, and reciprocity (National Health and Medical Research Council 2003; Rix et al. 2014). The long-term destiny or agency of research findings can work against or in favor of the researched, and this reinforces the responsibility of the researcher (Glowszewski et al. 2012). The principles of reciprocity underpin Indigenous cultural and social identity. They relate to Indigenous rights and obligations of sharing within community (Schwab 1995). As reciprocity implies collaboration, choice, and respect, it must be deeply embedded within the research methodology and methods. Consideration must be given to where money is spent in community, what protocols are acknowledged, and how the researcher is giving back to those being researched. Opportunities to co-author research, leave knowledge, learnings and skills within the community, and respecting protocols all constitute reciprocity (Ellis 2016).

6   Indigenous Knowledge as a Protective Factor

Once we understand the relational nature of reality, we can see that Indigenous cultures are built upon complex systems of relationships and relational accountability. Therefore, Indigenous culture itself is a pathway to healthy relations and healing for Indigenous people (McDonald 2006; Aboriginal and Torres Strait Islander Healing Foundation 2014). When combined with western medicine, Indigenous Knowledge and cultural healing can provide ways for people to gain control over their lives.

> Mainstream initiatives that engage with Aboriginal cultural practice, philosophy, spirituality and traditional Aboriginal medicines are examples of how to enact the theoretical concept of Indigenous Knowledges into reality and practice. However, there are too few examples of where this is happening in a meaningful and enduring way. (Poche Indigenous Health Network 2016)

Healing for Indigenous people takes place by way of reconnection to country, family, and culture (Maher 1999; Watson 2001; Kirmayer et al. 2003). Cultural healing may be from reconnecting to Indigenous ways of knowing, being, and doing, art, dance, or simply sitting down with Elders and listening to the traditional stories passed down through the generations (Hunter et al. 2006; Aboriginal and Torres Strait Islander Healing Foundation 2013).

Any superficial acknowledgment of culture must be accompanied by a fundamental acceptance of Indigenous ontology and epistemology at the foundational level of research planning. It is not enough to window dress western practices with cultural artifacts; the issue of power imbalance of the two knowledge systems needs to be addressed. This entails collaboration, trust, and the ability to push back against our conditioned belief that the western biomedical way is the only right way. It cannot be simply acknowledging that we have two proven knowledge systems; there must be systemic change. To change the system, we must be able to understand both knowledges, value them equally, and implement what is important. It is about allowing the historically oppressed to find their voice and power and sit with equality at the table when decisions are being made. If we want true equality and healing, one side must be willing to give up the power of control.

Indigenous culture has never been static. Indigenous communities have always placed the well-being of their people and their lands at the center of their governance and have been able to evolve and embrace change with resilience. Culture is not just dance, ceremony, and language; it is a philosophy and worldview that is based on collective ways of knowing and being. Collective community is the relationship to all things from the wind and weather to the waters, lands, and its peoples. To pay lip service to culture by engaging artifacts results in a hollow attempt to show cultural safety. True healing will happen when culture is not filtered through the western ways of knowing. It is when Indigenous peoples can make decisions for the services that govern them.

7 Non-indigenous Researchers Using Indigenist Methods

Working with Indigenous peoples in a research capacity requires the non-Indigenous researcher to bracket their own worldviews and apply a critical reflexive lens to any project they contribute to (Rix et al. 2014). Reflexivity can be an effective instrument for mitigating power, class, and cultural differences in research (Bott 2010). When working with Indigenous peoples and communities, with the aim of improving health services delivery, it is crucial to proceed respectfully and remain ever vigilant of applying an epistemological approach which privileges Indigenous voices. Self-determination, addressing power imbalances, and community control must remain central to any research project performed by outsiders and non-Indigenous researchers with Indigenous communities (Rigney 1997).

According to Wilson (2007), working within an Indigenist paradigm is not limited to Indigenous researchers, just as working with a western paradigm is not restricted to researchers of "white" descent, or working with a "feminist" methodological approach is not restricted to being female. Non-Indigenous researchers are accountable to Indigenous community and elders, and it is essential that they develop mutual trusting relationships. Taking responsibility for the cultural safety of Indigenous research participants via deep consultation with elders and community to confirm that any proposed research is in accord with the needs and desires of that community is essential in the commencement of any research project.

Notions of white privilege and white guilt have emerged over roughly the same time frame as the Indigenist paradigm has been evolving. In her pivotal work, Peggy Macintosh, an American feminist and antiracism activist, described the "invisible knapsack" containing all the privileges that being a member of white middle-class society delivers (McIntosh 1990). White guilt is a concept that is defined as the individual or collective guilt felt by some white people for harm resulting from racist treatment of ethnic minorities by other white people both historically and currently (Steele 1990).

Both these concepts may form part of the non-Indigenous researcher's personal process of unpacking their own worldviews using critical reflexive practice. They are, however, of no value in the role of the non-Indigenous researcher seeking to advocate for and find creative ways to contribute to the honoring and inclusion of Indigenous Knowledges into mainstream health services research. Of more importance here is the non-Indigenous researcher's ability to develop open listening skills and bracket their own worldviews in order to view the world through the lens of the Indigenous peoples with whom they are working.

7.1 Critical Self-Reflexivity

It is of vital importance that before any first meeting or consultation with Indigenous community takes place, the non-Indigenous researcher must undertake a personal journey of critical reflexive practice in order to examine and comprehend their own position as a white person of privilege (McIntosh 1990). This process extends beyond mere awareness of colonizing histories and power imbalances, to examining one's own worldviews and the biases and assumptions that come with being a member and product of the dominant western culture. Without a genuine and ongoing process of critical examination of self, reflecting on the privileged lens through which the world is viewed, the non-Indigenous health services researcher is at risk of merely contributing to and continuing the colonization process, further embedding the dominant western paradigm that cannot provide either re-empowerment or self-determination (see also ▶ Chaps. 89, "Using an Indigenist Framework for Decolonizing Health Promotion Research," ▶ 90, "Engaging Aboriginal People in Research: Taking a Decolonizing Gaze" and ▶ 98, "A Culturally Competent Approach to Suicide Research with Aboriginal and Torres Strait Islander Peoples").

Practicing critical reflexive practice requires a toolkit. This may include, for example, regular journaling and development of strong relationships with Indigenous people as research colleagues and co-investigators. Examining one's influence on and positioning within any research performed with Indigenous communities is a crucial ingredient of preparation to work with Indigenous peoples using a relational and Indigenist methodology. It is important, however, to be aware that positioning and clarification of roles in any research endeavor are of far less importance than the non-Indigenous researcher's effectiveness as a research "instrument" who can contribute to improvements to health services in ways that resonate with and reflect the Indigenous peoples with whom they are working (Rix et al. 2014). To be successful in informing policy and positive change requires researchers to manage complex relationships between Indigenous individuals, communities and organizations, and the dominant biomedical world they are required to negotiate to access healthcare and treatment.

7.2   Supporting Indigenist and Decolonizing Methods

There are a number of important strategies that non-Indigenous researchers can use as advocates for and practitioners of Indigenist and decolonizing methodologies:

- Citing the co-creation of new knowledge as a relational exercise that cannot occur in isolation. An individual cannot own new knowledge.
- Not trying to know or understand too quickly because relational approaches with Indigenous peoples operate to an inside learning timetable; true change is paced to things other than research programs or publishing deadlines (Gnibi Elders Council 2016; Murphy 2017).
- Use of Indigenous ways of disseminating research findings, for example, storytelling, artwork, and use of metaphors.
- Publishing in health and medical journals with Indigenous Elders, community members, or research participants as co-authors.
- Publishing research results in the voice of the Indigenous participants.
- Illustrating research output such as new theory and data analysis using Indigenous artwork.
- Introducing co-authors and a little of their stories or background in publications, to enable the reader to build their own relationship with authors.

This list while by no means comprehensive provides examples of ways that non-Indigenous researchers can contribute to the use and acceptance of Indigenist methods. Academic processes often act to block or negate the influence of paradigms that challenge the dominance of western and biomedical worldviews. The inclusion of Indigenous research methods that resonate with individual participants, Indigenous Elders, and community members are becoming more commonly used and accepted within western academic conventions.

8   Conclusion and Future Directions

The purpose of this chapter is to familiarize health social scientists with the rapidly growing prominence of Indigenist and decolonizing methodologies, providing an emphasis on how these can make important and significant contributions to the health services research arena. In fact, we strongly argue that without the inclusion of Indigenous ways of knowing, being, and doing and the voices of Elders, community members, and individual Indigenous participants, there can be little or no improvement to provision of effective health services or improved health outcomes for Indigenous peoples.

The profoundly damaging impact of European colonization on Indigenous nations is the known primary driver of the current burden of ill health suffered by Indigenous peoples across many countries. Performing research with Indigenous peoples using epistemological and ontological approaches that incorporate Indigenist philosophies and traditional knowledge and practices is now known to be a genuinely decolonizing approach to research aimed at improving health and well-being.

Researchers aiming to work with Indigenist methodologies are required to use ongoing critical self-reflection to both unpack their own existing worldviews and acknowledge their privilege. Awareness of and insight into the relational and interconnected worldviews and philosophies of Indigenous peoples are vital for anyone wishing to work in this area. Any project must be approached with "respect, responsibility, and reciprocity" as the foundational principles of ethical research performed with Indigenous communities and individuals. It is also vital that researchers be mindful of the underpinning driver of Indigenist methodologies and methods: to increase the empowerment and self-determination of the Indigenous population under investigation.

In writing this chapter, we sincerely hope that more non-Indigenous researchers will be motivated and inspired to incorporate Indigenist methods and ways of knowing being and doing into any research project aiming to improve health services design and delivery for Indigenous populations. Until the western and biomedical dominance of research is broken, it is difficult to envisage how mainstream health services can successfully address the current and ongoing epidemic of chronic disease suffered by Indigenous peoples and reduce the disparities in morbidity and mortality between Indigenous and non-Indigenous populations in colonized nations across the globe.

References

Aboriginal and Torres Strait Islander Healing Foundation. Growing our children up strong and deadly – healing for children and young people. 2013. Retrieved from On-Line.

Aboriginal and Torres Strait Islander Healing Foundation. Healing Foundation STRATEGIC PLAN 2014–2017. 2014. Retrieved from http://healingfoundation.org.au/wordpress/wp-content/files_mf/1430190708StrategicPlan20142017.pdf.

Anderson I, Whyte JD. Populations at special health risk: Indigenous populations international encyclopedia of public health. Oxford: Academic; 2008. p. 215–24.

Atkinson J. Trauma trails, recreating song lines: the transgenerational effects of trauma in Indigenous Australia. North Melbourne: Spinifex Press; 2002.

Australian Institute of Health and Welfare. The health and welfare of Australia's Aboriginal and Torres Strait Islander peoples. Canberra: AIHW; 2015.

Bessarab D, Ng'andu B. Yarning about yarning as a legitimate method in Indigenous research. Int J Crit Indigenous Stud. 2010;3(1):37–50.

Bott E. Favourites and others: reflexivity and the shaping of subjectivities and data in qualitative research. Qual Res. 2010;10(2):159–73. https://doi.org/10.1177/1468794109356736.

Centres for Disease Control and Prevention. American Indian and Alaska Native heart disease and stroke fact sheet. 2014. Retrieved from.

Chilisa B. Indigenous research methodologies. Los Angeles: SAGE Publications; 2012.

Durie M. Understanding health and illness: research at the interface between science and Indigenous knowledge. Int J Epidemiol. 2004;33(5):1138–43. https://doi.org/10.1093/ije/dyh250.

Ellis JBE, Earley MA. Reciprocity and constructions of informed consent: researching with Indigenous populations. Int J Qual Methods. 2016;5(4):1–13.

Fast E, Collin-Vezina D. Historical trauma, race-based trauma and resilience of Indigenous peoples: a literature review. First Peoples Child Fam Rev. 2010;5(1):126–36.

Glowszewski B, Henry R, Otto T. Relations and products: dilemmas of reciprocity in fieldwork. Asia Pac J Anthropol. 2012;13(2)

Gnibi Elders Council. Gnibi Elder's principles. In: Gnibi Wandaran School of Indigenous Knowledge – Southern Cross University, editor. Lismore: unpublished; 2016.

Graham M. Some thoughts about the philosophical underpinnings of Aboriginal worldviews. Aust Humanit Rev. 2008;45

Hoy WE, Nicol JL. Birthweight and natural deaths in a remote Australian Aboriginal community. Med J Aust. 2010;192:14–9.

Hunter LM, Logan J, Goulet J, Barton S. Aboriginal healing: regaining balance and culture. J Transcult Nurs. 2006;17. https://doi.org/10.1177/1043659605278937.

Jackson D, Power T, Sherwood J, Geia L. Amazingly resilient Indigenous people! Using transformative learning to facilitate positive student engagement with sensitive material. Contemp Nurse A J Aust Nurs Prof. 2013;46(1):105–12. https://doi.org/10.5172/conu.2013.46.1.105.

Kelm ME. Colonizing bodies: Aboriginal health and healing in British Columbia, 1900–50. Vancouver: UBC Press; 1998.

King M. Chronic diseases and mortality in Canadian Aboriginal peoples: learning from the knowledge. Chronic Dis Canada. 2010;31(1)

King M, Smith A, Gracey M. Indigenous health part 2: the underlying causes of the health gap. Lancet. 2009;374(9683):76–85. https://doi.org/10.1016/s0140-6736(09)60827-8.

Kirmayer L, Simpson C, Cargo M. Indigenous populations healing traditions: culture, community and mental health promotion with Canadian Aboriginal peoples. Australas Psychiatry. 2003;11:S15. https://doi.org/10.1046/j.1038-5282.2003.02010.x.

Kite E, Davy C. Using indigenist and Indigenous methodologies to connect to deeper understandings of Aboriginal and Torres Strait Islander peoples' quality of life. Health Promot J Austr. 2015;26(3):191–4. https://doi.org/10.1071/HE15064.

Kovach M. Conversational method in Indigenous research. First Peoples Child Fam Rev. 2010;5(1):40–8.

Larson A, Gillies M, Howard P, Coffin J. It's enough to make you sick: the impact of racism on the health of Aboriginal Australians. Aust N Z J Public Health. 2007;31(4):322–9. https://doi.org/10.1111/j.1753-6405.2007.00079.x.

Maher P. A review of 'Traditional' Aboriginal health beliefs. Aust J Rural Health. 1999;7(4):229–36. https://doi.org/10.1111/j.1440-1584.1999.tb00462.x.

Martin KL. Please knock before you enter: Aboriginal regulation of outsiders and the implications for researchers. Teneriffe: Post Pressed; 2008.

McDonald H. East Kimberley concepts of health and illness: a contribution to intercultural health programs in Northern Australia. Aust Aborig Stud. 2006;2 86+. Retrieved from http://go.galegroup.com.ezproxy.scu.edu.au/ps/i.do?p=EAIM&sw=w&u=scu_au&v=2.1&it=r&id=GALE%7CA163262656&asid=5e1f757b5a8cd98a7702cbeae493f740.

McIntosh P. White privilege: unpacking the invisible knapsack. Indep Sch. 1990; (Winter): 31–36.

Moodley RW, editor. Integrating traditional healing practices into counseling and psychotherapy. Thousand Oaks: Sage; 2005.W.

Murphy L Yarning circle led by Lyndon Murphy Paper presented at the Sovereignty of Indigenous Knowledge Symposium, Ballina. 2017. https://www.scu.edu.au/engage/news/latest-news/2017/gnibi-hosts-sovereignty-of-indigenous-knowledge-symposium.php.

National Health and Medical Research Council. Ethical guidelines in Aboriginal and Torres Strait Islander research. 2003. 08 May 2010. Retrieved from http://www.nhmrc.gov.au/publications/synopses/e52syn.htm.

Neumayer H. Changing the conversation: strengthening a rights-based holistic approach to Aboriginal and Torres Strait Islander health and wellbeing. 2013. Retrieved from http://iaha.com.au/wp-content/uploads/2013/10/Changing-the-Conversation-Strengthening-a-rights-based-holistic-approach-to-Aboriginal-and-Torres-Strait-Islander-health-and-wellbeing.pdf.

Paradies Y. Race, racism, stress and Indigenous health. (PhD thesis), Melbourne: The University of Melbourne; 2006.

Poche Indigenous Health Network. Poche Indigenous Health Network Key Thinkers Forum: Traditional Aboriginal Healing and Western Medicine. Paper presented at the Traditional Aboriginal Healing and Western Medicine, University of Sydney 2016.

Prior D. Decolonising research: a shift towards reconciliation. Nurs Inq. 2006;14(2):162–8.

Ramirez LC, Hammack PL. Surviving colonization and the quest for healing: narrative and resilience among California Indian tribal leaders. Transcult Psychiatry. 2014;51(1):112–33. https://doi.org/10.1177/1363461513520096.

Rigney L-I. Internationalization of an Indigenous anticolonial cultural critique of research methodologies: a guide to Indigenist research methodology and its principles. J Native Am Stud. 1997;14(2):109–21. Retrieved from http://www.jstor.org/stable/1409555

Rix E, Barclay L, Wilson S. Can a white nurse get it? 'Reflexive practice' and the non-Indigenous clinician/researcher working with Aboriginal people. J Rural Remote Health. 2014;4:2679. Retrieved from http://www.rrh.org.au

Sanderson PR, Little M, Vasquez MM, Lomadafkie B, Brings Him Back-Janis M, Trujillo OV, Jarratt-Snider K, Teufel-Shone NI, Brown BC, Bounds R. A perspective on diabetes from Indigenous views. Fourth World J. 2012;11(2):57–78. Retrieved from http://search.informit.com.au/documentSummary;dn=021821765358228;res=IELIND

Schwab RG. The calculus of reciprocity: principles and implications of Aboriginal sharing. 1995. Retrieved from http://www.anu.edu.au/caepr/Publications/dP/1995_dP100.pdf.

Shahid S, Bleam R, Bessarab D, Thompson SC. "If you don't believe it, it won't help you": use of bush medicine in treating cancer among Aboriginal people in Western Australia. J Ethnobiol Ethnomed. 2010;6(1):18. https://doi.org/10.1186/1746-4269-6-18.

Sheehan N, Martin G, Krysinska K, Kilroy K. Sustaining connection : a framework for Aboriginal and Torres Strait Islander Community, cultural, spiritual, social and emotional wellbeing. Brisbane: Centre for Suicide Prevention Studies, University of Queensland; 2009.

Sherwood J. Do no harm: decolonising Aboriginal health research. (PhD thesis), Sydney: University of NSW; 2010.

Smith LT. Decolonizing methodologies: research and Indigenous peoples. Dunedin: University of Otago Press; 1999.

Steele S. White guilt. In: Steele S, editor. The content of our character: a new vision of race in America. New York: HarperCollins; 1990. p. 77–92.

Stephens C, Porter J, Nettleton C, Willis R. Disappearing, displaced, and undervalued: a call to action for Indigenous health worldwide. Lancet. 2006;367(9527):2019–28. https://doi.org/10.1016/S0140-6736(06)68892-2.

Tousignant M, Sioui N. Resilience and Aboriginal communities in crisis: theory and interventions. J Aborig Health. 2009;5(1)

Tsey K, Wilson A, Haswell-Elkins M, Whiteside M, McCalman J, Cadet-James Y, Wenitong M. Empowerment-based research methods: a 10-year approach to enhancing Indigenous social and emotional wellbeing. Australas Psychiatry. 2007;15(sup1):S34–8. https://doi.org/10.1080/10398560701701163.

United Nations. United Nations declaration on the rights of Indigenous peoples: United Nations. 2008.

Walter M, Anderson C. Indigenous statistics. Walnut Creek: Left Coast Press; 2013.

Watson C. Why warriors lie down and die: towards an understanding of why the Aboriginal people of Arnhem land face the greatest crisis in health and education. Aust J Rural Health. 2001;9(3):141–2. https://doi.org/10.1111/j.1440-1584.2001.tb00409.x.

Weber-Pillwax C. What is Indigenous research? Can J Nativ Educ. 2001;25(2):166–74. Retrieved from http://ezproxy.library.usyd.edu.au/login?url=http://search.proquest.com.ezproxy2.library.usyd.edu.au/?url=http://search.proquest.com.ezproxy2.library.usyd.edu.au/docview/230306931?accountid=14757

Williams DR, Mohammed SA. Racism and health I: pathways and scientific evidence. Am Behav Sci. 2013;57(8):1152–73. https://doi.org/10.1177/0002764213487340.

Wilson S. What is an Indigenous research methodology? Can J Nativ Educ. 2001;25:175–9.

Wilson S. What is an Indigenist research paradigm? Canadian Journal of Native Education. 2007;30(2):193.

Wilson S. Research as ceremony: Indigenous research methods. 1st ed. Winnipeg: Fernwood Publishing; 2008.

Ethnomethodology

16

Rona Pillay

Contents

Abstract

Ethnomethodology is a qualitative research methodology which has recently gained momentum across disciplines, more specifically social and health sciences. Ethnomethodology focuses on the study of methods that individuals use in "doing" social life to produce mutually recognizable interactions within a situated context, producing orderliness. It explores how members' actual, ordinary activities produce and manage settings of organized everyday situations. Practice

R. Pillay (✉)
School of Nursing and Midwifery, Western Sydney University, Sydney, NSW, Australia
e-mail: rona.pillay@westernsydney.edu.au

© Springer Nature Singapore Pte Ltd. 2019
P. Liamputtong (ed.), *Handbook of Research Methods in Health Social Sciences*,
https://doi.org/10.1007/978-981-10-5251-4_68

through everyday life is central to ethnomethodology, the methods of which produce and maintain accountable circumstances of their life activities, making use of common sense knowledge in mundane situations. Ethnomethodology originated from Garfinkel who criticized Parsons' action theory whereby Garfinkel illustrated how ethnomethodology departs from conventional social theory to develop a methodology for studying social life. Ethnomethodology draws on video-recorded data as a preferred method with detailed attention to talk-in-interaction and gestures as interaction. The rich, detailed data generated may be viewed several times over, thus demonstrating that the data is valuable and trustworthy. The concepts of indexicality, reflexivity, and accountability are central to ethnomethodology because together they illustrate meaning as a methodical accomplishment. The reflexive accountability that contributes to order and the members' local performance of shared methods to carry out a joint activity form the central values of ethnomethodology. The analytical resources of ethnomethodology have been used to produce procedural accounts of human conduct in zones like museums, classrooms, and sports. Hence health care can be explored and empirically investigated as local interactions to contribute to patient safety.

Keywords
Ethnomethodology · Indexicality · Reflexivity · Accountability · Practice and orderliness

1   Introduction

Activities and interactions in society occur through members engaging and taking part in practical activities, mostly in everyday life activities. It is here that member's interact and arranges everyday affairs in different settings like the sports field, homes, workplace, train stations, and museums. In the sociology of everyday life, be it personal or professional, events are looked at in the "real world" and how activities are organized in the everyday life, notably, the workplace for the purposes of this chapter. According to Garfinkel (2002), the primary role of ethnomethodology is to examine the social facts of the real lived case asking, what makes it accountably just – the whatness of that social fact. Ethnomethodology primarily considers the problem of order by combining a "phenomenological sensibility," as stated by Maynard and Clayman (1991), with the focus on everyday social practice. Holstein and Gubrium (2005) agree that ethnomethodology sees the social world being accomplished by how the members constitute interactional work, the activities of which produce and maintain the accountable circumstances of their lives. Health team interactions fall within the category of social interactions; hence ethnomethodology is gradually being introduced in health research. Ethnomethodology is relatively new, making a presence in just the last few decades in the world of social sciences and even more so in health research. While used effectively in other areas like museums, courtrooms, and sports, ethnomethodology in health care is a fairly recent phenomenon in health research. Therefore, this chapter calls for the introduction of readers to the use of ethnomethodology in health care to contribute to the advancement of team interaction and patient care and safety. The chapter provides assistance with its use in fieldwork, epistemological, ethical, and practical considerations for researchers. This chapter overviews ethnomethodology and a discussion of its actual research practices. The key aims of the chapter are to provide an outline of the emergence of ethnomethodology and offer a theoretical perspective and methods of ethnomethodological research. In addition, included in this chapter are strengths, advantages, weaknesses, and limitations, a brief mention of compatibility as used with other methods. A discussion of key issues and practicalities of using the method and an illustration of some examples of the use of and ways of doing ethnomethodology and advocating areas where this method could be used within health social sciences will conclude this chapter.

2   Background of Ethnomethodology

Ethnomethodology is a unique qualitative social science research methodology, known to differentiate itself from traditional presuppositions and purpose, more so in its treatment of methods and methodology. It is unique in that instead of prescribing specific research methods, it studies methods in use and suggests a collectivity's methodology as a fundamental topic (ten Have 2004). Ethnomethodologist's investigations tend to be more "empirical" than is usually the case for philosophical and humanistic scholars. They conduct case studies involving actions in selected social settings. In the study, careful attention is given to detail to describe or explain observable or at least reconstructible events. In addition, of particular importance, is that ethnomethodological studies focus on local situations of language use and practical interactions (Lynch 1999).

According to Lynch (1993), ethnomethodology and the sociology of scientific knowledge are usually considered to be subfields of sociology. Ethnomethodology is more often said to be the study of "micro"-social phenomena such as the range of "small" face-to-face interaction occurring in public places, sports fields, and workplaces and is used across disciplines. Lynch (1993) proposes that ethnomethodology be used not because of the society it investigates but due to its epistemic focus. While ethnomethodology "does not imply a unifying theory and method" (Lynch 1999, p. 220), it does hold a set of propositions that informs the practice of ethnomethodological studies. The next section delivers the emergence of ethnomethodology.

2.1   The Ethnomethodological Perspective

In this section, the birth of ethnomethodology will be detailed, together with progression of ethnomethodology. For Garfinkel (1967, p. 1) the studies of ethnomethodology look into "practical activities, practical circumstances, and practical sociological reasoning as topics of empirical study." Ethnomethodology is specific in explicating the ways, in which a group of members create and maintain a "sense of social structure" (ten Have 2004, p. 16), which is an intelligibility and accountable local social order (Garfinkel 1967; ten Have 2004). Specific attention is paid to common place activities of everyday life and seeks to learn about these as a phenomenon in its own right. Garfinkel (1967) envisioned ethnomethodology to provide a different sociology, one that would pay attention to the organization of "commonplace everyday activities" (Garfinkel 1967, p. vii). Ethnomethodology is an approach to the investigation of social life whereby the central focus is to describe *how* people coordinate ordinary social activities in organized recognizable ways and the "doing" of such activities (Garfinkel 1967). Further, Garfinkel (1967) claims that ethnomethodology is the study of actual subject matter that includes the body of common sense knowledge, procedures, and considerations whereby ordinary members of society make sense of and respond to the situation in which they find themselves. It focuses on members' situated, practical, and methodical achievement of their activities in any given setting. It is vital to note here that the fundamental recommendation "is that the activities are whereby members produce and manage settings of organized everyday affairs are identical with members procedures for making those settings 'account-able' (Garfinkel 1967, p. 1). Accountable in this sense refers to observable and reportable activities, as in situated practices of looking-and-telling. The practices performed are undertaken by members who are part of a particular setting whose "skills with, knowledge of and entitlement to the detailed work of that accomplishment-whose competence- they obstinately depend upon, recognize, use and take for granted" (Garfinkel 1967, p. 1). Ethnomethodologists are interested in the way in which members deal with issue of generality and, occasionally, with how in any particular situation generally shared notions and presuppositions can be used to make sense of the actual activity that is occurring.

2.2   Emergence of Ethnomethodology

Philosophical perspectives, including phenomenology, influenced the progression of ethnomethodology and its variants (Schutz 1972). Garfinkel is known as the "founding father" of ethnomethodology. Ethnomethodology stems from its "mother discipline" of sociology and to all social science; therefore, ethnomethodology can be traced back to a definite origin. The detailed historical background to ethnomethodology is beyond the scope of this chapter. In the mid-1950s, Garfinkel invented the term *ethnomethodology*; however, it only became known in the mid-1960s (Lynch 1993). Between 1940s and 1960s, Garfinkel was drawn to social theory, the basic problems of social order, social action, intersubjectivity, and knowledge (Heritage 1987).

The performance by the members and their interaction and construction of social activities are important to any theory of social action. Garfinkel primarily drew on the work of his PhD supervisors, Talcott Parsons and Alfred Schutz, to recommend a new approach to social theory (ten Have 2004). Despite drawing on Schutz's work, ethnomethodology is not an extension of his work (Holstein and Gubrium 2005).

Pollner (1974) refers to ethnomethodology in terms of "facts are treated as accomplishments," meaning that they are produced through members' practical activities, in comparison with classical sociology where the chief aim is to investigate "social facts" and their determinants where the focus is on the actions of the individuals (ten Have 2004). Parsons had a huge influence on the theory of action. Following on, American sociologists were introduced to a range of European theorists such as Durkheim, Weber, Marshall, and Pareto where the focus was on the voluntaristic theory of action. This work was based on the disciplines of sociology, social anthropology, and social and clinical psychology which contributed to the promotion in the development of interdisciplinary research (Heritage 1984).

Garfinkel examined the issues with the theory of action. He focused on the use of actor's knowledge and understanding within the voluntaristic theory. Garfinkel concentrated on a theoretical framework that directly captures the actions by which actors explore their circumstances and subsequently develop a plan of action. This framework would result in an account of social activity which is directly based on the analysis of the organization of the experience itself (Heritage 1984). According to ten Have (2004), ethnomethodology is very much interested and has a deep respect for the practical and accountability of the most common place of ordinary activities. Overall, ten Have (2004) clarifies that while Weber, Schutz, and Parsons examined idealized models of science and scientific rationality, ethnomethodology examines and studies the local accountability of *any* kind of practice. In so doing, ethnomethodology can show how professional practice is embedded in quite ordinary competencies, as well as elaborates on being part of a particular local version of a more generalized professional culture. Hence, ethnomethodology can be used as a research methodology in health care, including multidisciplinary health personal. Ethnomethodology leans toward the analysis of social life with the central focus being to describe how people put ordinary social activities together in orderly recognizable way while including core concepts of ethnomethodology. The core concepts are accountability, reflexivity, and indexicality.

2.3   Ethnomethodology: Core Concepts

Accountability, reflexivity, and indexicality are the core concepts of ethnomethodology and, in particular, have very special meaning in reference to ethnomethodology and Garfinkel. Garfinkel (1967), in the introductory chapter of *Studies in Ethnomethodology*, states three constituent features that are the "accountability of practical actions as an ongoing practical accomplishment." These are "(1) the unsatisfied programmatic distinction between and substitutability of objective (context free) for indexical expressions; (2) the 'uninteresting' essential reflexivity of accounts and (3) the analyzability of actions-in-context as a practical accomplishment" (Garfinkel 1967, p. 4), more commonly discussed as accountability, reflexivity, and indexicality (Pierce 1991; ten Have 2004; Koschmann 2012).

2.3.1   Accountability

Accountability in Garfinkel's terms does not relate to liability as in ordinary talk, but leans toward intelligibility or explicability. Accountability is looking at actors making sense clearly of their situations and planning action on site, locally and immediately or explicable on demand (Garfinkel 1967; ten Have 2002). Accounts are known as the ways in which members denote or explain the social situation and consist of both verbal and nonverbal reporting. One such example is people who are standing in a line at a bank teller. It is clearly seen that they are doing just that, which is demonstrated by their body position. The members may understand and respond to a question like "Are you in the queue or are you standing in line?" It is noted that the "understandability" and "expressibility" of an activity is a sensible action and, simultaneously, an essential part of that action (ten Have 2004). In addition, the term implies that the basic requirement for all social setting is that it is recognizable and accountable for what it actually represents; it is what it supposed to be (ten Have 2002). According to ten Have (2002), by the members knowingly and visibly performing their role whereby the scenes of their work are accountable, the situation is then organized where reality and meaning are conveyed. Therefore, in terms of accountability from the ethnomethodology perspective, the everyday activities undertaken by members are the methods that make those activities observable and reportable.

Garfinkel (1967) views members as possessing ordinary linguistic and interactional skills through which the accountable features of everyday life were produced. These implicated members in the production of social order, thus working to give their world a sense of orderliness. The focus became members' integral methods for accomplishing everyday reality.

2.3.2   Reflexivity

In his foundational text, *Studies in Ethnomethodology*, Garfinkel (1967, p. 1) pens:

> The following studies seek to treat practical activities, practical circumstances, and practical sociological reasoning as topics of empirical studies, and by paying to the most commonplace activities of daily life the attention usually accorded extraordinary events, seek to learn about them as phenomena in their own right their central recommendation is that the activities whereby members produce and manage settings of organized everyday affairs are identical with members' procedures for making those settings 'account-able.' The 'reflexive' or 'incarnate' character of accounting practices and accounts make up the crux of that recommendation.

Garfinkel (1967) uses reflexivity to focus on the above-specific property whereby the expression of an activity is a vital part of that activity, especially of ordinary actions. The members use the reflexivity of accounts to give meaning and sense to orderliness in a social setting. Hence reflexivity ties in with the self-explicating property of ordinary actions, thus exploring subjective social meaning through an individual's lens.

2.3.3 Indexicality

Indexicality is a key concept for ethnomethodology. Garfinkel (1967) claims that indexicality was derived from the concept of indexical expressions which appear in ordinary language. The meaning of indexicality and sense making is embedded and dependent upon the context of the local situation (Bar-Hillel 1954). For Garfinkel, indexicality is more than context-dependent and includes a sense of "particular," which refers to action in a definite situation that produces meaning or sense. The meaning or sense making is linked to particular contexts and cannot be understood outside of that situation or local circumstance (Garfinkel 1967; ten Have 2004). Further, a phrase that would make sense in one context in an interaction might seem ironic when used at another point. Actors within a particular context create indexical expressions, words, facial and body gestures, and other cues to maintain the presumption that a particular reality governs their affairs (Garfinkel 1967; Heritage 1984). For example, simple expressions like "you" can be thought of on all occasions and actions are in fact indexical, in the sense that it could hold different meanings in different contexts. Without context, it is not possible to generate meaning.

The above key concepts of ethnomethodology, namely, indexicality, reflexivity, and accountability together, have shown meaning as a methodical accomplishment. The key concepts do not stand alone but are mutually interdependent of actions and contexts, thereby producing, reflexive accountability of action thus resulting in the orderly and organized setting.

2.4 Member Categorization and Conversational Analysis as the Two Sacksian Notions

In the later development of ethnomethodology, membership categorization analysis (MCA) was a term that Sacks (1972) used in his early work in the organization of knowledge which relied on and displayed as local interactions. According to Sacks, such knowledge was organized in terms of categories of people, for example, as in "children" or in reference to the speaker, as in "my husband" and actions. A common description that Sacks mention is "The baby cried. The mommy picked it up" (Sacks 1992, p. 236), which shows the interaction of members. The beginnings of such insights and broad explications arose out of Sack's PhD research on calls to a suicide prevention center (Sacks 1972) where callers explained life situations and how they felt especially in terms of no one to look to for support (Sacks 1967). In addition, people used person categories, which took the form of a collection of categories, for example, baby, mother, and father belong to the family, which Sacks called membership categorization devices (MCD) (Sacks 1972). Moreover, in reference to gender, there are two categories, "female" and "male." Further MCDs have a team or relational implication, for example, "the doctor and patient or teacher and student." Therefore, membership categorization analysis is useful in the study of the social knowledge that people use, expect, and rely on in doing the accountable work of living together (ten Have 2004).

The other term Sacks's work included is sequential analysis which forms the basis for conversational analysis (CA). Practice in everyday life through language is central to ethnomethodology as highlighted by Pomerantz (1988, p. 361):

> Ethnomethodologists and conversation analysis seek to discover the interpretive practices through which interactants produce, recognize, and interpret their own and others' action.

See ► Chap. 28, "Conversation Analysis: An Introduction to Methodology, Data Collection, and Analysis" for more detail about this analysis method.

3 How Ethnomethodology Is Done

This section discusses the ways in which ethnomethodology research is undertaken. In social science, ethnomethodology is best known for Garfinkel's early "breaching experiments" and the use of recordings and transcripts in conversational analysis (1967).

3.1 Ethnomethodology and Common Sense Procedures

The focus of ethnomethodologically research is primarily on naturally occurring data in the everyday life practical activities versus researcher provocation data. As ethnomethodology pays attention to the study of procedures of common sense as it is used in actual practices, there are methodological problems. Members take on a practical sense rather than a theoretical awareness of their constitutive work, and in so doing, it is possible to take common sense and their constitutive work for granted, unless attention is drawn to an untoward issue (ten Have 2004). According to Zimmerman and Pollner (1971), the concern with ethnomethodology is how common sense practices and knowledge risk lose their status as an unexamined resource so as to become a "topic" for analysis. Initially, to create this scenario, an approach Garfinkel initiated was to "breach" expectations to cause "trouble" (Garfinkel 1967). The breach expectancies of actors in the midst of action of their intersubjective environments are to "deliberately modify scenic events to disappoint these attributions" (Garfinkel 1967, p. 57). Therefore, it is important to find practical solutions, which are unavoidable compromises.

There are four strategies primarily in Garfinkel's early work, namely, sense-making activities; researchers study their own sense making, closely observing situated activities; and the study of ordinary practices. Together, the above strategies tend to work best. The first three strategies make use of literal quotes as seen in Garfinkel's (1967) reports on experiments, while more recent studies, recordings, and transcripts tend to be used (Garfinkel et al. 1981; Lynch 1985). Ethnomethodology and the "visibility problem" are in part solved by the creation or selection of "strange environments," and one such environment is the workplace.

Ethnomethodology uses two phases in the research process because the use of sense making is inevitable. This means that the researcher will use "member's sensemaking practices" to understand the activities and secondly analyze the "methods used in the first phase as one's research topic" (ten Have 2004, p. 53).

3.2 Breaching

According to ten Have (2004), the term "breaching" as used by Garfinkel referred to the use of experimental demonstrations in which covert expectations were breached. In other words, it is where actors would intentionally "break" norms and rules to see the reactions of others (Garfinkel 1967). It is the violation of socially accepted common rules or norms that may be used to maintain order. It is important to note here that the term experiment does not in any way indicate cause-effect according to Garfinkel (1967), and only some were undertaken in a laboratory. Such experiments were performed by Garfinkel's students in the field. An example "Students were instructed to engage an acquaintance or a friend in ordinary conversation and, without indicating that what the experimenter was asking was in any way unusual, to insist that the person clarify the sense of his commonplace remarks" (Garfinkel 1967, p. 42). The breaching experiments can be seen as efforts to make the workings of common sense visible.

4 Data Collection Techniques: Gathering Information

Ethnomethodology is dependent primarily on video-recorded data. Below is a discussion of various data collection techniques.

Digital recordings and transcriptions dominate ethnomethodological data collection. It is very much attuned to the use of recordings of actual, naturally occurring data and social interaction, orienting to them as constitutive elements of the setting studied (Sacks 1972; Mehan and Wood 1975; Atkinson and Drew 1979; Maynard 1984; ten Have 2004). Recordings and transcriptions are clearly demonstrated in conversational analysis and used to record original events (ten Have 2004). Tape-recorded conversations allow for the detail of the actual human action, which is then subjected to close scrutiny and formal analysis (Sacks 1984). In this instance, it has taken different empirical directions and depended on whether the interactive meaning or the structure of talk is being emphasized (Holstein and Gubrium 2005). Studies that emphasize the structure of talk examine the "conversational" analysis (CA) through which meaning emerges (see ▶ Chap. 28, "Conversation Analysis: An Introduction to Methodology, Data Collection, and Analysis").

Interviews are the most common technique used to collect data in qualitative research, more so for social researchers (ten Have 2004; see also ▶ Chap. 23, "Qualitative Interviewing"). On the contrary, in ethnomethodological studies, interviews are not the method of choice to collect data case despite ethnomethodology being used to explore talk with people or listen to what people have to say. In interviews that are performed traditionally, interest is in people, like individuals, categories, or collectivities of persons, including their characteristics, value, orientations, motivations, experiences, and relations, hence most often refereeing to the properties of people. The goal here is to explore what people think, feel, or experience or a combination, and the response is formulated in their mind. On the contrary, ethnomethodology's interest is in people as *members* and as competent practitioners, because ethnomethodology is interested in order-producing *practices* and can be further narrowed to *procedures* of order production; therefore, ethnomethodology is interested in the *how* (Garfinkel 1967; ten Have 2002). Moreover, these practices are seen as specifically local and situated. Even though this order-producing practice may have general features, the overall effect is context sensitive, and the reality to be studied in ethnomethodology is a local accomplishment of member practices. Due to these issues, interviews are of limited use in ethnomethodology. Ethnomethodology studies naturally occurring situations in which practices are observable and reportable (Garfinkel 1967; ten Have 2004).

5   Ethnomethodology: Compatibility and Use with Other Methodologies

Overall, ethnomethodology is not compatible with other methodologies, even though aspects have been drawn from phenomenology. Ethnography and ethno-methodology may appear to be related to each other in complex ways whereby ethnomethodology will need ethnography, verbal depictions, and characterizations of events in particular places and times (ten Have 2004).

Ethnomethodology and conversational analysis (CA) are commonly used in organizational and workplace studies. Conversational analysis developed from ethnomethodology and analyzes interactional social behavior and talk-in-interaction (Silverman 1998; ten Have 1999). Ethnomethodology and conversational analysis share some principles according to Clayman and Maynard (1995). For example, talk and social actions are considered "indexical" because their understanding is context driven and commonly used in workplace studies (see also ▶ Chap. 28, "Conversation Analysis: An Introduction to Methodology, Data Collection, and Analysis"). Workplace studies are a naturalistic approach. It is where the details of interaction, talk, and workplace practices are explored in terms of how procedures, activities, and tasks are performed in real-time interaction, through talk and visual conduct (Heath and Luff 2000; Luff et al. 2000).

Although ethnomethodology and ethnography are not frequently combined to undertake research, according to ten Have (2004), ethnography does hold some vital virtues, especially when field recordings are used to collect data. The use of field recordings contributes to the viewing of embodied action and context character of human existence. This is evident when ethnographic studies use ethnomethodological methods to produce accounts of people's methods in everyday practical situations (Collins and Makowsky 1978). The study by Whalen and Vinkhuyzen (2000) highlights the use of ethnomethodology and ethnography, making it a multifaceted ethnomethodological ethnographical study, using transcribed phone interactions and screen displays to illustrate the findings.

6   Ethnomethodology: Weaknesses and Limitation

Ethnomethodology is not alone when it comes to weaknesses and limitations. It is noted that ethnomethodology has some essential methodological problems. According to ten Have (2004), some of the problems arise due to the fact that with ethnomethodology the phenomena of interest and sensemaking practices are hard to note in ordinary situations, because they are constitutive (intrinsic) of those very situations and unavoidably used in any research practice itself. ten Have (2004, p. 51) refers to the essential methodological problems as "the invisibility of common-sense."

As video data is the most preferred data collection method, it can be a challenge. This data collection method can be a hindrance in some cultures and in some practices in health. For example, if additional team members walk into the operating theater, unknown to them that recording is in session, the individual will need to be informed and consent obtained retrospective or the clip from the video data deleted if the individual disapproves or declines to provide consent. Ethnomethodology is also identified as lacking in a defined analysis process even though the practical events are observed and reported.

The limitation would be a challenge when using the ethnomethodological approach in cultures where faces of family members who have died may cause harm. Yet, another aspect is the limitation of the use of interview data. Interview data collected here would be a recollection of the account of an event; hence the practical action in the context of occurrence is not observed. Ethics in the breaching process is questionable.

7   Ethnomethodology: Strengths and Advantages

There are strengths and advantages to using ethnomethodology as a qualitative research process (Atkinson 1988). Ethnomethodology is a methodology that focuses on an area often overlooked in health care: the cultural influences on medical care delivery and how patients perceive and understand their interactions with their doctor, their disease, the examination room, and, possibly, with a third actor in the examination room, the computer. The data collected using video recordings are an advantage as the data can be viewed many times over by the researcher as well as other team members so as to meet rigor (Heritage and Atkinson 1984). Ethnomethodology prevents researchers depending on intuition and recollection for the analytical process but on what is actually observed. The researcher is exposed to a wide range of local interactional material and circumstance. Analytical conclusions will not arise as artifacts of intuitive idiosyncrasy or experimental design. The additional advantage with video-recorded data is that researchers hear the data and see interaction rather than read the research reports. The researcher has direct access to the data, and the data is available for public scrutiny. Tapes and transcripts have both heuristic and confirmatory functions (ten Have 2004; Heath et al. 2010).

8   Practicalities of Using the Methodology

According to ten Have (2004), any activity can be studied at real time in naturally occurring environment where order is created. Ethnomethodological studies focus on observation and note what is actually going on without the use of a traditional theoretical framework or rather by bracketing what is already known.

Like any other research design, ethnomethodology requires ethical approval, which can be a challenge due to the main data collection technique as mentioned earlier, video recording. According to ten Have (2004), research students were embarrassed to show their lack of knowledge and confidence in the use of a video camera and expected to encounter a number of objections from people who were being observed. Researchers must familiarize themselves with the use of a video camera, which can be complex as every detail is important. It is necessary to ensure that the spoken language is clearly captured. The video recordings capture interactional details like facial expressions, gaze direction, and embodied interaction. Once the recordings are completed, transcriptions of the video data must be undertaken. The researcher is required to up skill in the use of technology as editing, clips, and still photographs may be required from the large corpus of data, which can be immensely time consuming.

Access to study-specific activities may be a challenge because ethnomethodology studies the accomplishment of concrete lived orders (ten Have 2004). Ethnomethodological researchers ensure that rapport is established with organizations and that they are aware of the requirements necessary to obtain access to the complex local order-producing activities. The researchers have to choose the setting and the activities that they desire to focus on. The researcher has to consider the activities to be studied and has to construct ways for the "data" to be represented.

9   Ethnomethodology in Action

The use of ethnomethodology varies across a span of disciplines and more recently being utilized increasingly in workplace studies. Bezemer et al. (2011) performed a study that focused on how surgical work is accomplished in the operating theater. Bezemer et al. (2011) aimed to explore the interaction and organization of activities between surgeons and nurses in the operating theater and included how social interactions are used to help structure and define situations. In addition, the study included how the members' differences in knowledge are constructed and positioned. In this study, video data was recorded, and the authors used concepts from symbolic interactionism, ethnomethodology, and conversational analysis to study small clips from the video recordings. In yet another study, Svensson et al. (2009) performed a study in operating theaters on how demonstration and instruction are achieved and made accessible to trainees to witness and learn techniques and procedures in specific cases while maintaining the integrity of medical practice in highly complex and demanding procedures. The authors claim that social interaction or interaction with patients, colleagues, and other staff is the key to learning how to achieve activities when delivering health care that combines specialized activities in concert with others in ordinary everyday situation.

Ethnomethodology extended to aviation, whereby Nevile (2004) undertook a study between pilots in the cockpit of commercial aircraft. The study focused on the communication pilots engaged in with colleagues, team members, and other parties like traffic controllers during routine checks that is required in flying an aircraft. Audio and video recordings were used as data collection method, and the author drew on the analytical approaches of ethnomethodology and conversational analysis. The research demonstrated that the pilots' work activities in the cockpit are accomplished through talk.

Further the work of ethnomethodology may be applied in playing the game (Kew 1986) and young children's play (Butler 2008). Moreover, within education, ethnomethodological approaches have been used to examine how lessons get done interactionally (Mehan 1979).

10   Conclusion and Future Directions

Ethnomethodological approach has the potential to contribute to the manner in which organizational studies are done, thus contributing to research, focusing on the practical activities of the everyday among members. Ethnomethodology focuses on people's tacit resources of social action, their common sense, and interactional activities with other members. Within this approach, the researcher requires additional technical skills to enhance data collection with the use of video recording, a technique that provides a rich corpus of data. Institutional studies, more often referred to as workplace and organizational studies, benefits largely from the use of ethnomethodology to understand the *how* of practice, specifically among multidisciplinary team members to perform a joint activity. Unique to ethnomethodology are the terms accountability, reflexivity, and indexicality, especially where practices are specific to context. Ethnomethodology and the cross-disciplinary studies can contribute richly to research in various fields including health team, aviation child play, and more.

Ethnomethodology, despite being used widely currently, is still considered to be underused. In relation to future direction and applications, it is proposed that ethnomethodology be included in the research unit to provide a view of the rich data collected and varying analysis compared to traditional research methodology. In health care, ethnomethodology has the great potential to contribute greatly to enhance patient safety by observing the interaction of multidisciplinary members and the construction of an activity with a view to improving practice. In addition, ethnomethodology contributes to team interactions and potential improvement because researchers can observe how communication and interactions occur and then note room for improvement.

References

Atkinson P. Ethnomethodology: a critical review. Annu Rev Sociol. 1988;14:441–65.

Atkinson JM, Drew P. Order in court. New York: Macmillan; 1979.

Bar-Hillel Y. Indexical expressions. Mind; 1954;63(251):359–379.

Bezemer J, Murtagh G, Cope A, Kress G, Kneebone R. "Scissors, please": the practical accomplishment of surgical work in the operating theatre. Symb Interact. 2011;34(3):398–414.

Butler C. Talk and social interaction in the playground. Aldershot: Ashgate; 2008.

Clayman SE, Maynard D. Ethnomethodology and conversation analysis. In: Have PT, Psathas G editors. Situated order: Studies in the social organization of talk and embodied activities (pp. 1–30). Washington, D.C.: International Institute for Ethnomethodology and Conversation Analysis & University Press of America; 1995.

Collins R, Makowsky M. The discovery of society. London: Random House; 1978.

Garfinkel H. Studies in ethnomethodology. Cambridge, UK: Polity Press; 1967.

Garfinkel H. Ethnomethodology's program: working out Durkheim's aphorism. Boulder: Rowman & Littlefield; 2002.

Garfinkel H, Lynch M, Livingston E. The work of a discovering science constructed with materials from the optically discovered pulsar. Philosophy of the Social Sciences, 1981;11(2):131.

Heath C, Luff P. Technology in action. Cambridge: Cambridge University Press; 2000.

Heath C, Hindmarsh J, Luff P. Video in qualitative research analysing social interaction in everyday life. London: Sage; 2010.

Heritage J. Garfinkel and ethnomethodology. Cambridge, UK: Polity Express; 1984.

Heritage J. Ethnomethodology. In A. Giddens and J. Turner (eds) Social Theory Today. Cambrigde: Polity Press, 1987;224–72.

Heritage J, Atkinson JM. Introduction. In: Atkinson JM, Heritage J, editors. Structures of social action: studies in conversation analysis. Cambridge: Cambridge University Press; 1984.

Holstein JA, Gubrium JF. Interpretive practice and social action. In: Denzin NK, Lincoln SY, editors. The Sage handbook of qualitative research. 3rd ed. London: Sage; 2005.

Kew F. Playing the game: an ethnomethodological perspective. Int Rev Sociol Sport. 1986;2 (14):305–324.

Koschmann T. Early glimmers of the now familiar ethnomethodological themes in Garfinkel's The Perception of the Other. Human Studies, 2012;35(4);479–504.

Luff P, Hindmarsh J, Heath C, editors. Workplace studies: recovering work practice and informing systems design. Cambridge: Cambridge University Press; 2000.

Lynch M. Art and artifact in laboratory science: a study of shop work and shop talk. London: Routledge & Kegan Paul; 1985.

Lynch M. Scientific practice and ordinary action: ethnomethodology and social studies of science. Cambridge/New York: Cambridge University Press; 1993.

Lynch M. Silence in context: ethnomethodology and social theory. Hum Stud. 1999;22(2):211–33.

Maynard D, Clayman SE. The diversity of ethnomethodology. Annual Review of Sociology, 1991;17(1);385–418.

Maynard DW. Inside plea bargaining: the language of negotiating. New York: Plenum; 1984.

Mehan H. Learning lessons: social organisations in the classroom. Cambridge, MA: Harvard University Press; 1979.

Mehan H, Wood H. The morality of ethnomethodology. Theory Soc. 1975;2(4):509–30. Retrieved 30 Oct 2017. http://citeseerx.ist.psu.edu/viewdoc/download?doi=10.1.1.464.8004&rep=rep1&type=pdf

Nevile M. Beyond the black box: talk-in-interaction in the airline cockpit. Aldershot: Ashgate; 2004.

Pierce JF. The ethnomethodological movement: sociosemiotic interpretation. Berlin: Mouton de Gruyter; 1991.

Pollner M. Sociological and common sense models of the labeling process. In: Turner R, editor. Ethnomethodology: selected readings. Harmondsworth: Penguin; 1974.

Pomerantz A. Offering a candid answer: an information seeking strategy. Commun Monogr. 1988;55:360–73.

Sacks H. The search for Help: no one to turn to. In E. S. Shneidman, ed. Essays in self destruction. New York: Science House: 1967;203–23.

Sacks H. An initial investigation of the usability of conversational data for sociology. In: Sudnow D, editor. Studies in social interaction. New York: Free Press; 1972.

Sacks H. Notes on methodology. In: Atkinson JM, Heritage J, editors. Structures of social action: studies in conversation analysis. Cambridge: Cambridge University Press; 1984.

Sacks H. Lectures on conversation, vol. I and II. Malden: Blackwell; 1992.

Schutz A. The phenomenology of the social world. London: Heinemann; 1972. (Translation of Schütz, [1974/1932] (1967)).

Silverman D. Harvey Sacks: social science and conversation analysis. Oxford: Policy Press; 1998.

Svensson MS, Luff P, Heath C. Embedding instruction in practice: contingency and collaboration during surgical training. Sociol Health Illn. 2009;31(6):889–906.

ten Have P. Doing conversation analysis: a practical guide. London: Sage; 1999.

ten Have P. The notion of member is the heart of the matter: on the role of membership knowledge in ethnomethodological inquiry. Forum qualitative SozialResearch, Forum: 2002;3(3). Available at: http://www.qualitativeresearch.net/fqs/fqs-eng.htm

ten Have P. Understanding qualitative research and ethnomethodology. London: Sage; 2004.

Whalen J, Vinkhuyzen E. Expert systems in interaction: diagnosing document machine problems over the telephone. In: Luff P, Hindmarsh J, Heath C, editors. Workplace studies: recovering work practice and informing systems design. Cambridge: Cambridge University Press; 2000.

Zimmerman DH, Pollner M. The everyday world as a phenomenon. In: Douglas JD, editor. Understanding everyday life: towards a reconstruction of sociological knowledge. London: Routledge & Kegan Paul; 1971. p. 80–103. (1st ed. 1970).

Community-Based Participatory Action Research

17

Elena Wilson

Contents

Abstract

Community-based participatory research (CBPR) is regarded as an equitable research approach that is operationalized within a social justice framework. It has been referred to as a continuum of research approaches from action research to participatory action research. Researchers are increasingly drawn to CBPR for collaborative health research that values community participation to redress issues of health inequality arising from socioeconomic disadvantage. Distinguishing features of effective CBPR include: blurring the distinction between researchers and research participants, minimizing power imbalances, and researching in partnership with communities towards positive community

E. Wilson (✉)
Rural Health School, College of Science, Health and Engineering, La Trobe University, Melbourne, VIC, Australia
e-mail: e.wilson@latrobe.edu.au

© Springer Nature Singapore Pte Ltd. 2019
P. Liamputtong (ed.), *Handbook of Research Methods in Health Social Sciences*,
https://doi.org/10.1007/978-981-10-5251-4_87

outcomes that are sustainable beyond the life of the research. Inherent complexities of communities and partnership arrangements can, however, lead to methodological and ethical challenges for researchers. Recent studies have found this to be the case, pointing to the need for adequate training and preparation for researchers who are new to CBPR. The intention for this chapter is to provide an overview of the conceptual foundations of CBPR and practical guidance for operationalizing each phase of the research process, while raising awareness of important considerations for the researcher role, from seeking ethics approval and entering the community, to dissemination of results. Drawing on international studies, lessons learned from experienced CBPR researchers are summarized so that researchers new to CBPR can build on their understanding and strengthen their studies in the future.

Keywords

Community-based participatory research · Social justice · Equity · Participatory methods

1 Introduction

Participatory action research (PAR) and Community-based participatory research (CBPR) can be understood as collaborative approaches to research that aim to achieve social change (Israel et al. 2013; Minkler and Wallerstein 2008). While various terms are used to label collaborative research across different disciplines, there is no commonly agreed definition for research with a community participation component (Brydon-Miller 1997; Wallerstein and Duran 2010). Because of the lack of definitional clarity, the relationship between PAR and CBPR described by Minkler and Wallerstein (2008) is useful as a foundation for discussing CBPR in this chapter.

Community-based participatory research (CBPR) is described by Minkler and Wallerstein (2008) as an umbrella term that refers to an orientation to research and practice in which the focus is respectful engagement with communities while combining research with education and action for change (see also Hall 1981). CBPR is, hence, understood as a continuum of approaches to research rather than a research method. Minkler and Wallerstein (2008) suggest that CBPR ranges from action research to PAR, with participation gradually increasing from narrow participation towards full participation. Action research sits at the starting point of the continuum and involves people who are affected by a problem in practical problem solving. At the other end of the CBPR continuum, referred to by Minkler and Wallerstein (2008, p. 10) as the "emancipatory end," is the "gold standard" of full participation as the aim for researchers whose goal is to achieve social justice, particularly in redressing health disparities (Minkler and Wallerstein 2008, p. 11). A commitment to engagement and power-sharing with community partners in the research process so that communities benefit from the research (Israel et al. 2013) is present at all levels on the CBPR continuum.

Growing support for CBPR has seen it extensively used in qualitative health research and other disciplines such as geography, archaeology, and education (Minkler and Wallerstein 2008; Atalay 2012; Castleden et al. 2012; Hacker 2013). CBPR is recognized for its utility in research on health disparities that result from systemic disadvantage. It is, therefore, often used to explore issues such as poverty, racism, and forced migration in research with communities of people living with mental illness, refugees, and rural populations (Hacker 2013; Vaughn et al. 2016).

CBPR is an approach to research that is used to achieve a shift in emphasis from victimization to strengths that take into account personal lives and controlling structures (Minkler and Wallerstein 2008; Israel et al. 2013). Central to CBPR is the concept of community participation requiring a redistribution of power with the aim of producing benefits for the community (Blumenthal 2011). It is iterative and evolving and is distinguished by its emphasis on involvement of the community as co-researchers (Wallerstein and Duran 2010).

Although each CBPR project is unique (Israel et al. 2013), community participants are generally involved in decision-making from project conception to dissemination of results. CBPR includes the need to acknowledge community as variously conceptualized and to attain social equality through power sharing (Israel et al. 2013). Action and social change are, hence, promoted through cyclical, iterative, and sustainable processes (Israel et al. 2013). Despite widespread support for CBPR, there is a growing body of international literature that draws attention to methodological and ethical challenges associated with its operationalization (Banks et al. 2013). Some of these challenges will be considered in this chapter.

For researchers who aim to achieve a more equitable and just society, a CBPR approach is believed to be the most valuable (Mayan and Daum 2016). CBPR has been used successfully by researchers seeking to work collaboratively with Indigenous communities (Wallerstein and Duran 2010; Atalay 2012). Funding for health research is increasingly contingent on the inclusion of participatory approaches in the research design (Burke et al. 2013; National Health and Medical Research Council (NHMRC) 2016). Funding bodies, in national and international arenas, are increasingly requiring participatory approaches in health research. Funding of CBPR projects has been made available by agencies in the United States, such as National Institutes of Health (NIH) and Centers for Disease Control and Prevention (Burke et al. 2013), and in Australia, funding is available from the National Health and Medical Research Council (NHMRC). Researchers who apply for NHMRC funding to its Centres of Research Excellence scheme are required to demonstrate broad community participation in the research (NHMRC 2016). This is based on the NHMRC's recognition that the valuable insights provided by community participation in research contributes to its quality and direction (NHMRC 2016).

In this chapter, I introduce the conceptual foundations of CBPR with specific reference to: community participation; power equilibrium; and social justice and equity, followed by considerations for effective practice of CBPR. Aspects of the researcher role are briefly discussed with reference to partnership expectations and researcher training, after which attention turns to introducing methodological and ethical considerations for researchers wishing to use the CBPR approach.

2   Conceptual Foundations

The development of CBPR is founded on the action research approach advanced by Lewin (1946) and later the emancipatory ideals of Freire (1970). Connecting applied social science and social activism, their work developed towards a new paradigm of participatory research, (Minkler and Wallerstein 2008). CBPR evolved from the two different research traditions.

Utilitarian views of Kurt Lewin in the 1940s, rejected positivist approaches grounded in objective truths arguing rather for people's collective involvement in researching their own circumstances (Wallerstein and Duran 2008). Lewin promoted a collaborative problem-solving research cycle of planning, action, and evaluation (Wallerstein and Duran 2008). Knowledge produced in this way would be used for practical goal-oriented action. Lewin referred to this process as *action research* (Wallerstein and Duran 2008).

In his 1970 publication *Pedagogy of the Oppressed*, Paulo Freire who expressed his view of liberation through education, inspired a later research tradition based on a philosophy of transformative praxis (Wallerstein and Duran 2008). A climate of political unrest and social inequality in South America saw Freire respond by advocating love, trust, and relationality as central to the process of humanization and education towards revolutionary change (Wallerstein and Duran 2008). Freire's faith in humanity and people's role in making change was driven by a social justice agenda, leading Freire to the belief that when supported, rather than directed, by intellectuals, the powerless can take action towards their own liberation through consciousness raising (Freire 1970; Israel et al. 1998; Wallerstein and Duran 2008).

CBPR is influenced by critical theory often based on Habermasian utilitarian problem solving, normative social and cultural values, and emancipation (Habermas 1971; Wallerstein and Duran 2008). Subsequent influences were theories of feminism, poststructuralism, and post colonialism (Minkler and Wallerstein 2008). Thus, the focus of methods and goals of CBPR, influenced by social movements of the 1960s and 1970s in search of a fairer society, became social justice and challenging the idea of science and objectivity (Minkler and Wallerstein 2008; Israel et al. 2013).

Over the last two decades, the use of the term CBPR and application of this approach have grown rapidly in several fields, but most noticeably in the health field (Wallerstein and Duran 2008). The adoption of CBPR in different disciplines has resulted in several variations of the term CBPR, such as, community-partnered participatory research, participatory research, and rapid rural appraisal. The term CBPR has also been used with specific emphasis, such as tribal participatory research (Wallerstein and Duran 2008).

A number of sets of principles to guide the CBPR research partnership have been proposed (Israel et al. 1998, 2013; LaVeaux and Christopher 2009). Frequently cited principles for conducting CBPR are those developed by Israel et al. (1998). The principles were published with the caveat that "no one set of CBPR principles is applicable to all partnerships" (Israel et al. 2013, p. 8). The principles have evolved and will continue to evolve as research continues to be conducted and evaluated (Israel et al. 2017). The current CBPR principles are:

1. CBPR recognizes community as a unit of identity.
2. CBPR builds on strengths and resources within the community.
3. CBPR facilitates collaborative, equitable partnership in all research phases and involves an empowering and power-sharing process that attends to social inequalities.
4. CBPR promotes co-learning and capacity building among all partners.
5. CBPR integrates and achieves a balance between research and action for the mutual benefit of all partners.
6. CBPR emphasizes public health problems of local relevance and ecological perspectives that attend to the multiple determinants of health and disease.
7. CBPR involves systems development through a cyclical and iterative process.
8. CBPR disseminates findings and knowledge gained to all partners and involves all partners in the dissemination process.
9. CBPR requires a long-term process and commitment to sustainability.
10. CBPR addresses issues of race, ethnicity, racism, and social class and embraces "cultural humility" (Israel et al. 2017, p. 30–33).

Israel et al. (2017, p. 29) propose that CBPR is ultimately an integration of all of these individual principles and that the extent to which any combination of these principles are achieved "...will vary depending on the context, purpose, and participants involved." As the principles are considered to be situated on a continuum, they are presented as a goal towards which to strive rather than standard practice (Israel et al. 2017) as in reality, it would be rare to achieve "pristine CBPR" (Blumenthal 2011, p. 388).

3   Community Participation

Equitable participation of community partners at each stage of the research is foundational to CBPR (Blumenthal 2011). In the absence of a common definition of community participation, it has been described as a concept that emerged out of the desire to redress inequality through collective public action and attend to issues about power and control (Rifkin 2014). Arnstein's (1969) typology of community participation is based on different degrees of power sharing, represented as a ladder of participation, on which *manipulation* appears on the lowest rung, progressing through to *citizen control* at the highest level in which government power is transferred to citizens. If community participation is understood as being on a continuum, it can be applied in CBPR with the aim of achieving "maximum feasible community participation" (Buchanan et al. 2007, p. 153).

The degree of community participation in the early phase of CBPR has attracted discussion about whether CBPR projects can all be initiated by the community (Mosavel et al. 2005; Minkler and Hancock 2008). Topics for investigation in CBPR projects ideally originate in the community (Mosavel et al. 2005). For academic researchers accustomed to traditional research approaches, this represents a major paradigm shift (D'Alonzo 2010). In reality, however, time, expertise, and funding for the research study means that many CBPR projects are initiated by the researcher (Minkler and Wallerstein 2008; D'Alonzo 2010; Castleden et al. 2012). In such cases, it is essential that a genuine commitment be made to involve the community and determine in the early phase whether the proposed research topic is a priority for them (D'Alonzo 2010).

Community members sometimes participate in CBPR studies as co-researchers (also referred to as field assistants or community research workers). This type of participation directs involvement away from passive interactions between traditional notions of *subject* and *researcher* that can signal exploitation (Bromley et al. 2015). Co-researchers usually live or work in the partnering community and participate in the research with institutional researchers in a range of ways that might include translating between community members and academic researchers, assisting with recruitment, and administering surveys (True et al. 2011). Because co-researchers can bring local knowledge and cultural understanding to the partnership, they are able to facilitate access to community members through their social ties and common perspectives that link them to the community (True et al. 2011).

4   Power Equilibrium

CBPR challenges traditional power relationships between researchers and researched reinforcing partnership for mutual benefit (Wallerstein and Duran 2017). Genuine attempts at CBPR call for researchers to understand power differentials within partnerships, such as, race, gender, and class and practice a relinquishing of the traditional dominant model of power associated with research to achieve a redistribution of power across all who are involved in the research (Minkler and Wallerstein 2008). Power sharing among partners is foundational to CBPR particularly in relation to decision-making (Viswanathan et al. 2004; Minkler and Wallerstein 2008) and reflects the three interconnected goals of research, action, and education (Hall 1981; Wallerstein and Duran 2008).

Scholars emphasize the importance of researchers employing cultural humility to redress power imbalances (Tervalon and Murray-Garcia 1998). Cultural humility is understood as a lifelong commitment to egoless self-evaluation and self-critique in partnership development and maintenance based on mutual respect and trust (Tervalon and Murray-Garcia 1998; Foronda et al. 2016).

Reciprocal knowledge transfer can also lead to a greater balance of power recognizing value in the knowledge of all members of the partnership (Wallerstein and Duran 2008). Power gained through reciprocal knowledge transfer can facilitate the community's liberation from "the social structural factors that have historically silenced its voices" (Muhammad et al. 2017, p. 108) thereby enriching each partnership member's understanding of the world (Wallerstein and Duran 2008, 2017). This type of knowledge democracy has the effect of recovering knowledge that lies in the expertise found in the world "beyond academia" (Hall et al. 2015, p. 2).

5   Social Justice and Equity

CBPR is useful to researchers who seek to advance social justice goals of equity and knowledge democracy, develop respectful relationships, and influence policy for health equity (Banks et al. 2013). Braveman and Gruskin (2003, p. 254) define equity in health as "the absence of systematic disparities in health (or in the major social determinants of health) between social groups who have different levels of underlying social advantage/disadvantage – that is, different positions in a social hierarchy." Social justice goals can be achieved through a commitment to inclusive practices that promote equity of opportunity to participate. Researchers play an important role in facilitating participation so that those community members who are usually disengaged might have ways to be involved. Achieving social justice and health equity using CBPR involves sustainable outcomes, that is, members of the partnership work towards outcomes that can be sustained beyond the life of the research.

As an emancipatory approach, CBPR is often used in research seeking to address health disparities and vulnerability resulting from structural inequality and institutional racism (Muhammad et al. 2017). As Minkler (2004, p. 691) suggests, "we can approach crosscultural situations with a humble attitude characterized by reflection on our own biases and sources of invisible privilege, an openness to the culture and reality of others, and a willingness to listen and continually learn".

6   Practice of CBPR

6.1   Research Process

The increasing health disparities affecting minority groups globally render CBPR a defensible choice for researchers and communities (Chavez et al. 2008). CBPR can also accommodate the growing trend of communities that wish to participate in researching and solving their own problems (Hacker 2013). There is, however, no definitive process for conducting research using a CBPR approach.

6.2   Phases of CBPR

Rather than rigid adherence to a set of steps in a process, CBPR involves phases between which the research team and partnership members move. Israel et al. (2013) describe seven phases of conducting CBPR: Form a CBPR partnership; Assess community strengths and dynamics; Identify health priorities and research questions; Design and conduct research; Feedback and interpret findings; Disseminate and translate research findings; and Maintain, sustain and evaluate partnership. In these phases, the emphasis is on diverse partners and researchers seeking to achieve equal participation and ownership of the research. The process is circular and, therefore, iterative with some elements that can be ongoing, such as partnership maintenance (Israel et al. 2013). The following paragraphs highlight the main elements within each phase.

Form a CBPR Partnership: For community participation to be effective, partnership members must firstly define what constitutes *the community* for the purposes of the research. Defining the community can be by geographic location, health condition, common characteristics, or can be self-defined (Viswanathan et al. 2004; Hacker 2013). Identifying potential partners with which to develop operational infrastructure and establishing partnership expectations requires building trust and relationships. Groups most impacted by health disparities need to be involved in the design and implementation of strategies to address the community issues locally (Chavez et al. 2008; Israel et al. 2013).

Assess community strengths and dynamics: Each research partnership and setting is unique. Before following or adapting CBPR principles for distinct contexts, local issues need to be carefully considered (Israel et al. 2017). Therefore, it is necessary to study the community's demographic, social, and political landscape; meet with leaders of active community groups; and identify relationships and connections (Hacker 2013). Lack of trust between researchers and community members is widely acknowledged as a common difficulty (Israel et al. 1998). Building trust and building relationships is important and requires an ongoing commitment by researchers to demonstrate their trustworthiness and being sensitive to potential community impacts caused by research involvement (Israel et al. 1998; Austin 2015). A community advisory board or committee might need to be established or there might be a preestablished group in the community. This group will be important for participating and liaising with the community about the research and can contribute to the success of the project by providing guidance about community matters to the researcher (D'Alonzo 2010; Hacker 2013). Similarly, key local stakeholders can be helpful as boundary crossers for partnership success because, as local residents who are also working on the research, they can bridge the gap between the researcher and the community (Kilpatrick 2009; D'Alonzo 2010). As they are part of the community, they will have "insider" knowledge about acceptability of practices involved in the research and can help to legitimize the research by building perceptions in the community that the research and the researchers involved can be trusted (D'Alonzo 2010). These activities are time intensive and must be conducted in a participatory manner and with mutual respect within the partnership (Minkler and Wallerstein 2008; Hacker 2013). The process of going in to the community to meet people and learn the culture needs to be accounted for in research timelines (D'Alonzo 2010). Knowing the community and its cultural context can assist with tailoring recruitment methods and materials to fit the community. Recruitment might not be a one-time exercise but can be ongoing as community members dip in and out of the research as the cycle of everyday life allows.

Identify health priorities and research questions: Identifying health priorities involves researchers and partnership members working together to recognize health issues that they are sufficiently concerned about and that they want to explore and take action on (Minkler and Hancock 2008). The health priority needs to appeal to a broad range of community members and thereby be consistent with their collective vision for a healthy community (Minkler and Hancock 2008). CBPR researchers provide their skills to develop questions intended to lead to social change for improved community health and well-being (Minkler and Hancock 2008).

Design and conduct research: There is no single design or method suitable for all CBPR studies (Viswanathan et al. 2004; Wallerstein et al. 2017). Decisions about data collection methods to be used require respectful negotiation as the researcher must maintain a balance between community wishes and upholding research integrity (Hacker 2013). When choosing the research methods to be used, researchers need to consider several factors including community relations and capacity, financial resources, and time limitations (Vallianatos et al. 2015). A combination of qualitative and quantitative methods can be used in CBPR. Qualitative methods utilized can include interviews, focus groups, photovoice, and observation (see ▶ Chaps. 23, "Qualitative Interviewing," and ▶ 65, "Understanding Health Through a Different Lens: Photovoice Method").

Feedback and interpret findings: Academics and community partners are mutually dependent and with adequate training, the community can assist with collection and interpretation of data. Identifying initial themes, for example, can be facilitated through creative strategies and can lead to greater research relevance for the community (Hacker 2013; Israel et al. 2013). Research findings can be shared with the community in a culturally appropriate feedback format and setting. Making meaning from the research findings is also a collaborative activity with the participation of all partnership members (Israel et al. 2013).

Disseminate and translate research findings: Dissemination strategies can contribute to sustainable outcomes after project completion. CBPR is essentially translational, meaning that research findings lead to action applied in the community for health or environmental benefits and to build community capacity (Minkler and Wallerstein 2008; Hacker 2013; Israel et al. 2013). Commitment to action is part of the research process and involves community partners taking the lead in identifying future actions to be taken (Minkler and Wallerstein 2008; Hacker 2013). It is an important aspect of CBPR partnerships to address issues of data ownership, such as shared data ownership and how the data will be used (Minkler and Wallerstein 2008). Dissemination of results can take multiple forms and might include community forums for data presentation and discussion of recommendations, formal reports, academic publications, and conference presentations (Hacker 2013). In addition to considerations of who owns the data, questions about co-authorship and co-presentation also need to be addressed (Israel et al. 2008). There is potential for researcher and community agendas to conflict, particularly about timing of dissemination activities (Hacker 2013). Any such conflicts can be managed using open and transparent communication based on flexibility and willingness to respond (Israel et al. 2008).

Maintain, sustain and evaluate partnership: Using CBPR involves a commitment to address inequalities with partnering communities for the long-term. This phase is an ongoing process that through establishment and maintenance of trust in the partnership leads to sustainable relationships and action, often extending further than a single project or funding period (Israel et al. 2008).

7   Researcher Role

7.1   Partnership Expectations

In CBPR partnerships, between community members, organization representatives, and academic researchers, it is ideal for all members to participate fully and share equal control in decision-making (Minkler and Wallerstein 2008). Transparency about partnership expectations is needed for full and equal participation in CBPR (Hacker 2013). Hence, it is vital that the researcher's role and responsibilities for research design and conduct are clearly articulated prior to the researcher entering the community, in addition to the expectations from community members, organizations, and all involved in the partnership. Clarity about expectation help ensure that agreed pathways exist for addressing potential social, cultural, or political problems, thereby avoiding reliance on power dynamics to unfold towards a resolution (Hacker 2013; Wilson et al. 2017). Diversion from the planned process can be necessary, however, as CBPR and partnership processes are not linear but dynamic (Banks et al. 2013). Flexibility is key in avoiding partnership tensions, ensuring that agreed expectations about the researcher's role and responsibilities can be repeatedly reviewed and renegotiated as processes unfold (Banks et al. 2013).

7.2   Training

Conducting CBPR requires a high level of skill and responsibility to achieve ethical conduct of research with collectivities as well as individuals (Bastida et al. 2010; Banks et al. 2013; Carter et al. 2013). The complexities involved in managing a relationship with partner organizations and community highlight the importance of training for CBPR researchers prior to commencing contact with members of the community they are to research with. Yet, Hacker (2013) points out that the number of researchers with adequate CBPR training does not meet the increasing demand for CBPR researchers. Being armed with knowledge and practical training researchers new to CBPR have a greater opportunity for the best possible outcomes for the research. Understanding processes in theory alone are not always adequate as preparation for CBPR processes in practice (Wilson et al. 2017). Researchers need robust practical training in CBPR processes and ethics, while developing the personal awareness and skill needed for working with complex group dynamics (Chenhall et al. 2011; Castleden et al. 2012). The importance of training researchers to achieve authentic CBPR lies also in the need for developing cultural humility that builds awareness of differences in power and privilege and the hidden mechanisms perpetuating racism in social relations (Muhammad et al. 2017). Therefore, with appropriate training, CBPR researchers can achieve and better understand the disrupting role that equitable partnerships play in confronting issues such as systemic racism (Muhammad et al. 2017).

8   Methodological and Ethical Considerations

To achieve the goal of equitable participation of community partners at each stage of the research in CBPR requires greater attention to ethical considerations (Blumenthal 2011). To be noted, however, are the increasingly identified ethical issues that have coincided with the growing popularity of CBPR. The nature of ethical challenges experienced by CBPR researchers can vary between different projects and can often be attributed to the relational nature of the approach. The openness of close relationships inherent in CBPR inevitably conflicts with the duty to maintain confidentiality and anonymity of participants (Brugge and Cole 2003).

Challenges might arise for researchers who deal with CBPR projects on sensitive topics, the content of which parallels their own life circumstances (Carter et al. 2013). Ongoing evaluation and researcher self-reflection are considered essential given the frequently sensitive research content and emergent nature of relationships with communities (Banks et al. 2013).

Researching in communities with the aim of empowering people through power sharing processes can attend to social inequalities and is a guiding principle of the CBPR approach. The ideal is one in which partnerships are formed between community members, organizational representatives, and academic researchers, in which all members of the partnership participate fully and share equal control in decision-making (Minkler and Wallerstein 2008). Although there is merit in this endeavor, it can be difficult to redistribute power in CBPR to achieve an equal power base because researchers continue to be perceived as traditional power holders by community partners (Wallerstein 1999).

CBPR researchers are concerned with achieving ethical conduct of research with collectivities in addition to individuals. Ethics review processes are important, however, many research ethics guidelines are based on the moral individualism of the biomedical model and concerns about collectivities are usually only considered in terms of impacts on individuals (Campbell-Page and Shaw-Ridley 2013). Research proposals are generally evaluated by ethics committees prior to initiation of the research rendering the process misaligned to the emergent nature of CBPR (Guillemin and Gillam 2004; Lofman et al. 2004; Campbell-Page and Shaw-Ridley 2013; Wilson et al. 2017). Therefore, ethical challenges can arise in CBPR, from the conflict between the complex, social, and emergent nature of researching with communities and the operationalization and moral individualism of ethics review mechanisms (Banks et al. 2013). There is potential for unintentional harm where researchers attempt to apply requirements of misaligned models to complex social environments (Flicker et al. 2007).

9   Conclusion and Future Directions

CBPR is an overarching term for a range of collaborative research approaches – centered on community participation. CBPR is an equitable, partnership-based approach to research for social justice and health equity. Researchers using CBPR aim to address inequality by empowering communities to find solutions to their community problems. Community members are involved in decision-making throughout the phases of CBPR from selecting research questions to dissemination of results. CBPR is translational, with research findings leading to action in the community for community capacity building and health and well-being benefit. The relational nature of CBPR and its emancipatory goals are, however, thought to be at odds with the biomedical model that underpins current research ethics review practice.

In this chapter, the reader has been introduced to key features of CBPR, the phases in its operationalization, and potential challenges for consideration by researchers wishing to use this approach in future health research. As the field of CBPR continues to develop in response to institutional, community, and researcher understanding, it is important that the benefits and the challenges of CBPR for all involved remain prime considerations in research dialogue. Researchers seeking to achieve health equity and social justice will need to work more with communities in trying to solve health and social problems as global developments affect a growing number of people. CBPR offers a useful path forward for researchers responding to these pressing challenges.

References

Arnstein SR. A ladder of citizen participation. J Am Inst Plann. 1969;35(4):216–24. https://doi.org/10.1080/01944366908977225.

Atalay S. Community-based archaeology: research with, by, and for indigenous and local communities. Berkeley: University of California Press; 2012.

Austin W. Addressing ethical issues in PR: The primacy of relationship. In: Higginbottom G, Liamputtong P, editors. Participatory qualitative research methodologies in health. London: Sage; 2015.

Banks S, Armstrong A, Carter K, Graham H, Hayward P, Henry A, et al. Everyday ethics in community-based participatory research. Contemp Soc Sci. 2013;8(3):263–77. https://doi.org/10.1080/21582041.2013.769618.

Bastida EM, Tseng TS, McKeever C, Jack L Jr. Ethics and community-based participatory research: perspectives from the field. Health Promot Pract. 2010;11(1):16–20. https://doi.org/10.1177/1524839909352841.

Blumenthal DS. Is community-based participatory research possible? Am J Prev Med. 2011;40(3):386–9. https://doi.org/10.1016/j.amepre.2010.11.011.

Braveman P, Gruskin S. Defining equity in health. J Epidemiol Community Health. 2003;57(4):254–8. https://doi.org/10.1136/jech.57.4.254.

Bromley E, Mikesell L, Jones F, Khodyakov D. From subject to participant: ethics and the evolving role of community in health research. Am J Public Health. 2015;105(5):900–8. https://doi.org/10.2105/AJPH.2014.302403.

Brugge D, Cole A. A case study of community based participatory research ethics: the healthy public housing initiative. Sci Eng Ethics. 2003;9:485–501. https://doi.org/10.1007/s11948-003-0046-5.

Brydon-Miller M. Participatory action research: psychology and social change. J Soc Issues. 1997;53(4):657–66. https://doi.org/10.1111/j.1540-4560.1997.tb02454.x.

Buchanan DR, Miller FG, Wallerstein N. Ethical issues in community-based participatory research: balancing rigorous research with community participation in community intervention studies. Prog Community Health Partnersh. 2007;1(2):153–60. https://doi.org/10.1353/cpr.2007.0006.

Burke JG, Hess S, Hoffmann K, Guizzetti L, Loy E, Gielen A, et al. Translating community-based participatory research (CBPR) principles into practice: building a research agenda to reduce intimate partner violence. Prog Community Health Partnersh. 2013;7(2):115–22. https://doi.org/10.1353/cpr.2013.0025.

Campbell-Page RM, Shaw-Ridley M. Managing ethical dilemmas in community-based participatory research with vulnerable populations. Health Promot Pract. 2013;14(4):485–90. https://doi.org/10.1177/1524839913482924.

Carter K, Banks S, Armstrong A, Kindon S, Burkett I. Issues of disclosure and intrusion: ethical challenges for a community researcher. Ethics Soc Welf. 2013;7(1):92–100. https://doi.org/10.1080/17496535.2013.769344.

Castleden H, Mulrennan M, Godlewska A. Community-based participatory research involving indigenous peoples in Canadian geography: progress? An editorial introduction. Can Geogr/Le Géographe canadien. 2012;56(2):155–9. https://doi.org/10.1111/j.1541-0064.2012.00430.x.

Chavez V, Duran B, Baker QE, Avila MM, Wallerstein N. The dance of race and privilege in CBPR. In: Minkler M, Wallerstein N, editors. Community-based participatory research for health: from process to outcomes. 2nd ed. San Francisco: Jossey-Bass; 2008. p. 91–105.

Chenhall R, Senior K, Belton S. Negotiating human research ethics: case notes from anthropologists in the field. Anthropol Today. 2011;27(5):13–7. https://doi.org/10.1111/j.1467-8322.2011.00827.x.

D'Alonzo KT. Getting started in CBPR: lessons in building community partnerships for new researchers. Nurs Inq. 2010;17(4):282–8. https://doi.org/10.1111/j.1440-1800.2010.00510.x.

Flicker S, Travers R, Guta A, McDonald S, Meaguer A. Ethical dilemmas in community-based participatory research: recommendations for institutional review boards. J Urban Health. 2007;84(4):478–93. https://doi.org/10.1007/s11524-007-9165-7.

Foronda C, Baptiste DL, Reinholdt MM, Ousman K. Cultural humility: a concept analysis. J Transcult Nurs. 2016;27(3):210–7. https://doi.org/10.1177/1043659615592677.

Freire P. Pedagogy of the oppressed (trans. Ramos MB). New York: Continuum; 1970.

Guillemin M, Gillam L. Ethics, reflexivity, and "ethically important moments" in research. Qual Inq. 2004;10(2):261–80. https://doi.org/10.1177/1077800403262360.

Habermas J Knowledge and human interests (trans: Shapiro J). Boston: Beacon Press; 1971.

Hacker K. Community-based participatory research. Thousand Oaks: Sage; 2013.

Hall BL. Participatory research, popular knowledge and power: a personal reflection. Convergence. 1981;14(3):6–19.

Hall B, Tandon R, Tremblay C, editors. Strengthening community university research partnerships: global perspectives. Victoria: University of Victoria and PRIA; 2015.

Israel BA, Schulz AJ, Parker EA, Becker AB. Review of community-based research: assessing partnership approaches to improve public health. Annu Rev Public Health. 1998;19(1):173–202. https://doi.org/10.1146/annurev.publhealth.19.1.173.

Israel BA, Schultz AJ, Parker EA, Becker AB, Allen AJ, Guzman R. Critical issues in developing and following CBPR principles. In: Minkler M, Wallerstein N, editors. Community-based participatory research for health: from process to outcomes. San Francisco: Jossey-Bass; 2008. p. 47–66.

Israel BA, Eng E, Schulz AJ, Parker EA, editors. Methods for community-based participatory research for health. 2nd ed. San Francisco: Jossey-Bass; 2013.

Israel B, Schulz AJ, Parker EA, Becker AB, Allen AJ, Guzman JR, Lichtenstein R. Critical issues in developing and following CBPR principles. In: Wallerstein N, et al., editors. Community-based participatory research for health: advancing social and health equity. San Francisco: Wiley; 2017, Incorporated.

Kilpatrick S. Multi-level rural community engagement in health. Aust J Rural Health. 2009;17(1):39–44. https://doi.org/10.1111/j.1440-1584.2008.01035.x.

LaVeaux D, Christopher S. Contextualizing CBPR: key principles of CBPR meet the indigenous research context. Pimatisiwin: J Aborig Indigenous Commun Health. 2009;7(1):1–25.

Lewin K. Action research and minority problems. J Soc Issues. 1946;2(4):34–46. https://doi.org/10.1111/j.1540-4560.1946.tb02295.x.

Lofman P, Pelkonen M, Pietilae A. Ethical issues in participatory action research. Scandinavian Journal of Caring Sciences. 2004;18:333–340. https://doi.org/10.1111/j.14716712.2004.0027.x.

Mayan MJ, Daum CH. Worth the risk? Muddled relationships in community-based participatory research. Qual Health Res. 2016;26(1):69–76.

Minkler M. Ethical challenges for the "outside" researcher in community-based participatory research. Health Educ Behav. 2004;31(6):684–97.

Minkler M, Hancock T. Community-driven asset identification and issue selection. In: Minkler M, Wallerstein N, editors. Community-based participatory research: from process to outcomes. San Francisco: Jossey-Bass; 2008. p. 153–70.

Minkler M, Wallerstein N. Community-based participatory research: from process to outcomes. San Francisco: Jossey-Bass; 2008.

Mosavel M, Simon C, van Stade D, Buchbinder M. Community based research (CBPR) from South Africa: engaging multiple constituents to shape the research question. Soc Sci Med. 2005;61 (12):2577–87.

Muhammad M, Garzón C, Reyes A, The West Oakland environmental indicators project. Understanding contemporary racism, power, and privilege and their impacts on CBPR. In: Wallerstein N, et al., editors. Community-based participatory research for health: advancing social and health equity. San Francisco: Wiley; 2017.

National Health and Medical Research Council. *Statement on consumer and community involvement in health and medical research.* 2016. Canberra, Australia: Consumers Health Forum of Australia.

Rifkin SB. Examining the links between community participation and health outcomes: a review of the literature. Health Policy Plan. 2014;29(suppl 2):ii98–ii106. https://doi.org/10.1093/heapol/czu076.

Tervalon M, Murray-Garcia J. Cultural humility versus cultural competence: a critical distinction in defining physician training outcomes in multicultural education. J Health Care Poor Underserved. 1998;9(2):117–25. https://doi.org/10.1353/hpu.2010.0233.

True G, Alexander LB, Richman KA. Misbehaviors of front-line research personnel and the integrity of community-based research. J Empir Res Hum Res Ethics. 2011;6(2):3–12. https://doi.org/10.1525/jer.2011.6.2.3.

Vallianatos H, Hadziabdic E, Higginbottom G. Designing participatory research projects. In: Higginbottom G, Liamputtong P, editors. Participatory qualitative research methodologies in health. London: Sage; 2015.

Vaughn LM, Jacquez F, Lindquist-Grantz R, Parsons A, Melink K. Immigrants as research partners: a review of immigrants in community-based participatory research (CBPR). J Immigr Minor Health. 2016;19(6):1457–68. https://doi.org/10.1007/s10903-016-0474-3.

Viswanathan M, Ammerman A, Eng E, Garlehner G, Lohr KN, et al. Community-based participatory research: assessing the evidence. Evidence report/technology assessment no. 99. Rockville: Agency for Healthcare Research and Quality; 2004.

Wallerstein N, Duran B, Oetzal JG, Minkler M (Eds). Community-based participatory research for health: advancing social and health equity. San Francisco, CA: Jossey-Bass; 2017.

Wallerstein N. Power between evaluator and community: Research relationships within New Mexico's healthier communities. Social Science & Medicine. 1999;49:39–53. https://doi.org/10.1016/S0277-9536(99)00073-8.

Wallerstein N, Duran B. The theoretical, historical, and practice roots of CBPR. In: Minkler M, Wallerstein N, editors. Community-based participatory research: from process to outcomes. San Francisco: Jossey-Bass; 2008.

Wallerstein N, Duran B. Community-based participatory research contributions to intervention research: the intersection of science and practice to improve health equity. Am J Public Health. 2010;100(Suppl 1):S40–6. https://doi.org/10.2105/AJPH.2009.184036.

Wallerstein N, Duran B. Theoretical, historical, and practice roots of CBPR. In: Wallerstein N, et al., editors. Community-based participatory research for health: advancing social and health equity. San Francisco: Wiley; 2017.

Wilson E, Kenny A, Dickson-Swift V. Ethical challenges in community-based participatory research: a scoping review. Qual Health Res. 2017;28(2):189. https://doi.org/10.1177/1049732317690721.

Grounded Theory Methodology: Principles and Practices

18

Linda Liska Belgrave and Kapriskie Seide

Contents

Abstract

Since Barney Glaser and Anselm Strauss' (The discovery of grounded theory: strategies for qualitative research. New York: Adline De Gruyter, 1967) publication of their groundbreaking book, *The Discovery of Grounded Theory*, grounded theory methodology (GTM) has been an integral part of health social science. GTM allows for the systematic collection and analysis of qualitative data to inductively develop middle-range theories to make sense of people's actions and experiences in the social world. Since its introduction, grounded theorists working from diverse research paradigms have expanded the methodology and

L. L. Belgrave (✉) · K. Seide
Department of Sociology, University of Miami, Coral Gables, FL, USA
e-mail: l.belgrave@miami.edu; k.seide1@miami.edu

© Springer Nature Singapore Pte Ltd. 2019
P. Liamputtong (ed.), *Handbook of Research Methods in Health Social Sciences*,
https://doi.org/10.1007/978-981-10-5251-4_84

developed alternative approaches to GTM. As a result, GTM permeates multiple disciplines and offers a wide diversity of variants in its application. The availability of many options can, at times, lead to confusion and misconceptions, particularly among novice users of the methodology. Consequently, in this book chapter, we aim to acquaint readers with this qualitative methodology. More specifically, we sort through five major developments in GTM and review key elements, from data collection through writing. Finally, we review published research reflecting these methods, to illustrate their application. We also note the value of GTM for elucidating components of culture that might otherwise remain hidden.

Keywords

Grounded theory methodology · Qualitative research · Culture · Paradigms · Constructivist research · Objectivist research · Realist research · Situational mapping

1   Introduction

Grounded theory methodology (GTM) is a powerful approach to inductively generating theory using qualitative data. In fact, Glaser (1978, p. 2) tells us that "grounded theory is based on the systematic generating of theory from data." That said, GTM is sometimes used for other goals, for instance, when it is part of mixed methods research (Bryant 2002; Charmaz and Belgrave 2012; Belgrave and Seide forthcoming). Barney G. Glaser and Anselm L. Strauss introduced this elegant method in their groundbreaking book, *The Discovery of Grounded Theory* (1967). Since then, it has become possibly the most widely used methodology in qualitative research. GTM has spread across a wide range of fields, extending from sociology and nursing to informatics and engineering, while remaining especially valued in social scientific pursuits of theory regarding health and health care, possibly because it began there, with Glaser and Strauss' *Awareness of Dying* (1965), their study of death and dying in the hospital setting. Because the term "grounded theory" is used to refer to both the methodology and the product of its use – the theory – we use GTM here, for clarity.

GTM has grown dramatically since the publication of *Discovery*, with more explication and diversity of approach. This diversity includes differences over key features (e.g., Charmaz 1995, 2014; Corbin and Strauss 2015) whether it is a complete methodology or primarily a guide to analysis (Nagel et al. 2015) and, of course, paradigm. At times, the field can seem a bit overwhelming, even turbulent (Bryant and Charmaz 2007a), particularly to novice users of the methodology (Belgrave and Seide forthcoming). Here, we review the key elements of the methodology, from data collection through writing, sorting through the major developments in GTM. We give particular attention to (1) Glaser and Strauss's objective/realist approach, (2) Glaser's objectivist/realist/positivist work, (3) Strauss and Corbin's postpositivist/realist/interpretive explication of GTM, (4) Charmaz's and Bryant's individual but closely related constructivist/interpretive presentations, and

(5) Clarke's interpretive/constructionist/situationist extension of GTM. Finally, we review published research reflecting these methods, to illustrate their application. We note the value of GTM for elucidating components of culture that might otherwise remain hidden. Our goal is to clarify the options available, so those wanting to use GTM can select the approach that best suits their needs. We appreciate our diversity as a richness of possibilities to be celebrated (see Stern 2007; Clarke 2009; Wertz et al. 2011, for similar approaches).

2   A Brief Word on Paradigms

Distinctions between various approaches to GTM flow from differences in paradigms, which are often unmarked by authors. While "the paradigm wars" might be over, and researchers can take a pragmatist approach to paradigm (Cresswell 2009), it is valuable to examine one's underlying assumptions about what constitutes knowledge, how best to generate it, and the assumptions of the specific methods one chooses in order to reach denser, more meaningful analyses (Fielding 2008). Paradigms have methodological implications (Denzin and Giardina 2009), and researchers will be most satisfied when they work within a paradigm that matches their own. We address paradigm briefly as a guide, rather than advocacy (but see Culture below).

Kathy Charmaz (2000, 2009; Charmaz and Belgrave 2012) differentiates three main approaches to GTM, based on paradigm, postpositivist, objectivist, and constructivist, with some authors moving between paradigms over time. While following her distinctions, we use slightly different labels, reflecting how multiple authors have referred to these approaches and how authors refer to themselves, as well as to recognize that some readers might follow realist assumptions without buying into those of objectivism and vice versa (Belgrave and Seide forthcoming). GTM began in the objectivist vein, so that early explications, such as *Discovery* (Glaser and Strauss 1967) and *Theoretical Sensitivity* (Glaser 1978), provide researchers a methodology to help them uncover a knowable, real social world. They urge those using the method to be neutral and objective. Postpositivism entered the field with Strauss and Corbin's *Basics of Qualitative Research* (1990), in which the authors continue the earlier objectivist assumptions but add detailed techniques to use in analyzing data. This approach began as more prescriptive and procedural than the original, losing the earlier emergent nature of analysis (Charmaz and Belgrave 2012). In later editions of this book, Corbin provides a more interpretive approach (Charmaz and Belgrave 2012; Corbin and Strauss 2015), so that her work as a whole has been interpreted from postpositivist to constructivist (Belgrave and Seide forthcoming). Constructivist GTM (CGTM) is a newer member of the family, developed primarily by Kathy Charmaz and Antony Bryant (see Bryant 2002, 2017; Charmaz 2000, 2006, 2007, 2014; Bryant and Charmaz 2007a, b). Using constructivist GTM, researchers recognize the existence of multiple realities, treat data as mutually constructed between themselves and participants, and see their analyses as constructions of reality. This is an explicitly interpretive approach.

Situational analysis (Clarke 2009, 2015; Clarke and Friese 2007) extends this approach in a postmodern direction, with an emphasis on situations.

3   GTM: How We Do What We Do

Health social science researchers in the various GTM traditions differ in many ways. Some seek theoretical goals while others are more interested in applied work. Some follow realist/objectivist paradigms; others do interpretive/constructivist work. Certainly, health social science researchers bring a variety of data collection methods to their efforts. Despite this, there is a core of practices common to this diverse family. However, different authors specify different cores. Charmaz (2014, p. 15) lists the most commonly used by those claiming grounded theory as simultaneous, iterative data collection and analysis, analysis of actions and processes, use of comparative methods, and developing abstract conceptual categories through systematic data analysis, though full GTM also includes theoretical sampling, emphasis on theory over description, and more. Corbin and Strauss (2015) do not provide a list, per se, but include an inductive approach to ground the theory in data, constant comparative methodology, coding data and grouping similar codes into categories or themes, developing category properties and dimensions, and integrating all of this around a core category to structure one's theory. Bryant (2017) discusses multiple views of the core, arriving at a list combining characteristics and process of GTM. He argues that GTM is systematic, inductive, and comparative, it involves persistent interaction and constant involvement and is iterative and used to develop theory, and that GTM includes coding, sampling, and being open (to all possible theoretical explanations). In an effort to address all of this in limited space, we discuss the heart of GTM as Sects. 3.1, 3.2, and 3.3. Ideally, we would discuss all of these simultaneously, because we do them iteratively and concurrently, so discussion of any one requires reference to the others. However, that is not possible. We note that sometimes, research being what it is, practitioners fluctuate on some aspects of these, as well, particularly the simultaneity of data collection and analysis (see Ahmad et al. 2006; Beard et al. 2009; Bryant 2017). We present an overview of the methods used by practitioners of the key approaches. For more detail, we refer readers to the classic and recent editions of basic texts for each, including Glaser and Strauss (1967), Glaser (1978), Clark (2009), Charmaz (2014), Corbin and Strauss (2015) and Bryant (2017). Many of these provide examples that show one, concretely, what to do. Additionally, Wertz et al. (2011) and Creswell and Poth (2018) compare GTM to other qualitative methodologies, placing GTM in context. Belgrave and Seide (forthcoming) directly compare coding strategies between various GTM approaches. Finally, we write of methods in the third person plural, reflecting our view of GTM practitioners as members of a large, inclusive family.

3.1   Collecting/Constructing Data Simultaneously with Data Analysis

Although interview and observational data might be the types most typically seen in our work, GTM does not call for a specific form of data, and practitioners use a wide variety of strategies to gather theirs, including both individual and focus group interviews, document collection, observations, mixed methods, and more (Charmaz and Belgrave 2012; Corbin and Strauss 2015; see also ▶ Chaps. 4, "The Nature of Mixed Methods Research," ▶ 29, "Unobtrusive Methods," and ▶ 40, "The Use of Mixed Methods in Research"). Below, we refer to interviews, expecting readers to interpret this to include multiple forms of data. Whether we see ourselves as collecting or constructing data jointly with participants depends on the paradigm we follow. This matters more for how we view our work than for what, concretely, we do. Regardless, when using GTM, we start analyzing with the first data we have at hand, typically with some form of coding (see below), thus launching an iterative process, and we use our analysis to inform our ongoing data collection/construction, particularly via theoretical sampling, which in turn aids in explicating categories and integrating them into a theoretical framework.

3.2   Coding Data (at Multiple Levels), Constant Comparison, and Writing Memos

Coding is the most fundamental step in GTM analysis, inseparable from other steps (Glaser and Strauss 1967; Charmaz 2014; Corbin and Strauss 2015; Bryant 2017; Belgrave and Seide forthcoming). Glaser and Strauss (1967) make clear the importance of constant comparison by devoting an entire chapter to it in *Discovery*. In fact, they merely mention coding in this chapter, as they present an emergent, rather than procedural, methodology. At its most basic, coding is a process of attaching conceptual labels to data, labels that capture what the relevant data are about. Although codes are labels, coding is more than labeling. At every level, coding is conceptual; it is analysis. All major approaches to GTM contain multiple, sequentially more abstract, levels of coding. Following these stages, we begin by fracturing or taking apart the data and move toward reintegrating them into a theoretical framework. Clarke's (2009, 2015) situational analysis (SA), an extension of GTM, uses mapping, rather than coding, per se, as this approach shifts the focus to collective action. As with coding in more conventional GTM, mapping in SA begins with fracturing the data. We organize the following discussion using Charmaz' constructivist coding strategy (2006, 2014) while noting the schemes of others. Some authors provide tools to help move analysis along productively, which we also note (see Table 1).

  Whether we see ourselves as uncovering an existing social reality or working with our participants to co-construct an understanding of social life, we do the same thing for our first coding step. In *initial coding* (parallel to Glaser's and Corbin and Strauss' *open coding*), we stick close to the data, attaching labels to segments of data

Table 1 Five major approaches to coding

	Glaser and Strauss	Glaser	Strauss and Corbin	Charmaz and Bryant	Clarke
Research paradigm	Objectivist	Objectivist	Objectivist	Constructivist	Interpretive situationist, constructionist
	Realist	Realist	Realist[a]	Interpretivist	
		Positivist			
Approach to coding	Comparisons:	Substantive coding:	Open coding	Initial coding	Open coding[d]
	Incidents	Open coding	Axial coding	Open coding	Axial coding[d]
	Incidents to properties	Selective coding	Selective coding	Focused coding	Situational mapping
	Delimit theory	Theoretical coding		Varied coding strategies[b]	Social worlds
				Axial coding[c]	Arenas mapping
				Theoretical coding[c]	Positional mapping
Analytic tools		Coding families	Conditional matrix	Coding families[c]	
Early coding	Incidents	Line by line	Paragraph by paragraph	Line by line	Word by word
			Phrase by phrase	Incident by incident	Segment by segment
			Line by line	Word by word	
			Micro coding (specific strategic words)		

Table reprinted from Belgrave and Seide (Forthcoming)
[a]Interpreted variously, from postpositivist to constructivist
[b]Bryant provides varied examples of strategies for moving forward from open codes
[c]Discussed by both: neither advocated nor discouraged
[d]Overview presented, not instructional per se

to capture what is going on in those segments. Segments of data might be single words, phrases, lines, paragraphs, or incidents, depending on the approach we are using, the nature of our data and our research goals. We fracture the data in order to be able to see what is happening – the trees unobscured by the forest. Charmaz (2006, 2014) and Bryant (2017) strongly encourage coding with gerunds; these active verbs help us to focus on processes. The key in initial coding is to remain open to possibilities. In fact, Glaser and Strauss (1967) and Glaser (1978) recommend attaching multiple codes to segments of data. Whether we use in vivo codes (words used by participants) or other labels, these are abstractions. With the coding of our first data, we write our first memo(s). Memos are notes we write to ourselves in order to keep track of our analytic thoughts. Early memos tend to be short and simple notes about codes that strike us in some way. These might be codes that seem significant, unexpected, or simply interesting. They are things we might look for in future data collection/construction or probe for in upcoming interviews. After we code our second interview, we compare it to the first, seeing what is similar, what is new, what is cast slightly differently, and so on. In the third interview, we compare to the first two. These comparisons might suggest new or revised codes (and memos!), as we see codes that somehow "go together." We create new codes and recode, never losing the early ones, and write new memos. And so it goes, with constant comparison of new to data to earlier data, new codes, recoding, and more memos, always expanding our analytic ideas. Constant comparisons, at varying levels of abstraction, are at the heart of GTM.

When do we move to the next level of coding? There are no rules, per se. As we continue to collect/construct, code, and compare data, we find that some codes appear frequently. Others do not. Some codes appear to be potentially theoretically exciting and others less so. Our memos become longer and tell more of a story. These are signs that we might begin to think about moving to the next coding level and focusing our data collection/construction. The reason the previous sentence is phrased cautiously is that it can be tempting to leap ahead prematurely, something Glaser (1978) warns us to avoid. We want to collect/construct enough data to have a chance to see/find similarities, differences, theoretically interesting codes, and the like. At the same time, one of the strengths of GTM is that it helps us to focus our data collection/construction (Charmaz 2014) so that we do not spin our wheels collecting/constructing reams of redundant data that add nothing new to our theoretical analysis. Moving to the next level of coding requires taking a risk, but it is a risk that can be undone.

With *focused coding*, we move to higher levels of abstraction, combine codes into categories, and begin to **integrate categories into theoretical frameworks**. Now, we compare our *initial codes* to each other. We want to select codes that appear the most often, codes that carry much analytic power, and codes that pull together many *initial codes*. We do this in a way that helps us to construct *focused codes*, categories, and their characteristics. Charmaz (2014) suggests that we ask questions such as "Which of these codes best account for the data? What do your comparisons between codes indicate? Do your focused codes reveal gaps in the data? What kinds of theoretical categories do these codes indicate?" (2014, pp. 140–141 and 144). Gaps, characteristics of *focused codes* that are not "fleshed out," or things that we would expect to see, but do not, lead us to theoretical sampling (see below).

For Glaser (1978), the second step is *selective coding*. Here, we focus on one "core variable" and other "variables" that appear to be related to it. This "core variable" is similar to a *focused code*, in that it guides ongoing data collection and theoretical sampling. It is different in that Glaser's "variables" tap reality. Finally, using Glaser's scheme, we turn to *theoretical coding*, integrating our fractured codes by conceptualizing relationships among our substantive codes (a grouping that combines *open* and *selective codes*). Glaser provides coding families (1978) and conceptual schemes intended to add precision and clarity to our higher-level abstractions. These can be helpful but risky if we find ourselves forcing these onto our data or using them uncritically, as some contain assumptions that we might not want to include in our eventual theory (Charmaz 2014). Corbin and Strauss (2015) offer axial coding, a paradigm to use in teasing out properties and dimensions of categories and bring our fractured data back together by relating categories to subcategories, at a higher level of abstraction. Some will find this helpful, though if it is applied mechanically, it can be limiting (Charmaz 2006, 2014). In earlier versions of their work (e.g., 1990), Strauss and Corbin next had us move to *selective coding*, which is similar to Glaser's. It is also similar to their *axial coding*, only at a higher level of abstraction, but in the 2015 edition, this is subsumed as part of *theoretical integration*.

We discuss situational mapping separately because it is an extension of conventional approaches to GTM and, therefore, looks rather different. Clarke (2009, 2015) lays out three types of mapping, *situational*, *social worlds/arena*, and *positional*. These seem to parallel some of the coding steps outlined above, except that we draw social maps. In *situational* mapping, we lay out "the major human, non-human, discursive, and other elements in the research situation of inquiry and provoke analysis of relations among them" (Clarke 2009, p. 210). We do this in two steps. The first, *messy mapping*, is similar to *initial coding*, in that our goal is to capture complexity. We make working maps, open to change. Clarke provides possible categories (e.g., individual human elements, collective human elements, and spatial elements) to use for the second step of *ordered mapping*. These are analogous to the tools provided by Glaser (1978) and Corbin and Strauss (2015), intended to nudge our thinking but not to be forced on our data. *Social worlds/arena mapping* is intended for meso-level analysis. Here, we attend to social organizational, institutional, and discursive dimensions of the situation. Finally, with *positional mapping*, we focus on "major positions taken, and *not* taken, in the data," whether these are taken collectively or individually (Clarke 2009, p. 210). In this step, we move between levels of analysis, from individual to organizational. We believe situational mapping has tremendous potential for health social science, particularly as healthcare is provided in situations and organizations, all impacted by decisions made at high levels of organizations and political-economic systems.

Back to those memos – we have been writing and elaborating memos throughout this process. In our memos, if we include the relevant segment of data that gives rise to *an initial code* or is tied to a *focused code*, our analysis is physically and visibly linked to our data. This keeps us grounded in our data. As we finalize our inquiry, we see that not only have our memos helped us keep track of our ongoing analysis but much of our writing is done! Now we can clean the memos up and incorporate them into papers reporting our research. If we have included quotes from our data in our memos, we have examples at our fingertips.

3.3   Theoretical Sampling and Integration of Categories into a Theoretical Framework

As we work at advanced levels of coding, integrating categories into a theoretical framework, we typically find puzzles. Perhaps we cannot figure out why a category does not have a characteristic we think it should have, logically, or we see a glimpse of how two categories fit together, but a piece of the picture seems to be missing.

All conventional approaches to GTM call for theoretical sampling, with authors generally devoting significant space to explaining it (Glaser and Strauss 1967; Glaser 1978; Charmaz 2014; Corbin and Straus 2015); it is also raised by Clarke (2015) in reference to situational mapping. And yet, the practice is often misunderstood. We draw primarily on Charmaz (2014), who provides perhaps the clearest discussion of this aspect of integrating analysis and data collection/construction. Our goals in theoretical sampling are not to achieve some sort of representative sample, only sometimes to seek negative cases, and decidedly not to document consistent empirical patterns. Rather, we use theoretical sampling to collect/construct data to refine and elaborate our emergent *theoretical* categories, tease out their properties and ranges, and illuminate relationships between them. As we work our focused (and/or other levels of coding beyond the initial), we use abductive reasoning. That is, we make "inferential leap[s] to consider all possible theoretical explanations for [surprising or puzzling] data" (Charmaz 2014, p. 201). Then we go back to our data or collect/construct new data to use in trying out these possible explanation(s). The keys here are that we imagine theoretical explanations for our data and use data to examine these. We might already have the relevant data and simply need to look at it through a new lens. We might go back to some participants and ask them new questions. Or we might recruit new participants whose experiences or stories might hold the solution to what puzzles us. Whichever we need to do, we use our theoretical insights and reasoning to guide us in seeking new or relevant existing data. We use our theoretical sensitivity to generate insights, but ideas, categories, and concepts must earn their way into the theory (Glaser 1979; Charmaz 2014; Clarke et al. 2015).

4   Constructivist/Constructionist GTM and Culture

Culture is critically important in social scientific research in health, health beliefs, health practices, and the like (Quah 2007; Liamputtong 2018). Here, we do not use the term culture as a euphemism for ethnicity or language, much less race, though we know the literature is replete with such work. Ethnicity and race themselves, though often taken for granted as adequately summarized with labels referring to national origin, continent of origin, or language (of self or ancestors), are socially constructed, changeable, and quite complex. For instance, Linda (first author) left Cleveland, Ohio, as an ethnic, non-Hispanic white. Ethnically, she was variously assumed to be Jewish or unspecified Eastern European, while she identified as Slovak. She arrived in Miami as an Anglo, period, though one of her students frequently tells her "but you're not white," intended and taken as a compliment. Add to this that she grew up solidly working class but now finds herself identified by others as middle class. Pile on religion, and the messiness of culture, begins to become obvious. By culture, we refer to "characteristics of a group's or community's way of life" (Quah 2007) that are relevant to health issues, including such things as values; shared meanings of health and illness, life, and death; traditions, customs, and habits; understandings of power and status relationships; lifestyles; and, yes, sometimes religion and language (adapted from Quah 2007; see also Liamputtong 2018). Moreover, we recognize culture as intersectional with race, ethnicity, class, gender, sexuality, and more. Kapriskie used CGMT to explore the impact of a cholera epidemic on the everyday lives of Haitians in view of a divide between public health experts and the population regarding the effectiveness of implemented local policies (Seide 2016). Although she was born in Haiti, ongoing cultural changes required some personal and methodological adjustments. Through field-work observations and reflexivity, she (re)familiarized herself with historical and evolving stratification dynamics based on social class and skin color that permeate everyday life. With memo writing, she continuously confronted her positionality as a researcher. In Haiti, the legitimacy of her role as a scholar along with the validity of her data reposed on recognizing culturally embedded factors that one could take for granted. Examples include, but are not limited to, the status of her advisor and that of her parents. Some participants interpreted having a white research advisor as respectable, while her parents (who are bridges in large social networks) facilitated snowball sampling and authenticated her Haitian roots for some natives (see Ulysse (2006) on the othering of emigrants by native Haitians). Her self-presentation and her degree of familiarity with local sites were also relevant to managing the field-work. Occasionally, the way she dressed determined her entry in some social circles. Because of the spatial distribution of the disease, knowing the city's "hotspots" became a sine qua non to establishing rapport with participants during interviews. Each form of GTM has methodological and theoretical implications for the recognition of similar subtleties in health research.

5   Applications of GTM in Health Social Science

Having laid out the key elements of GTM, we now turn to a few studies to show the diverse application of GTM in research pertinent to health sciences. What we present here is by no means meant to be exhaustive; rather, we offer a few illustrations to complement our discussion of the GTM variants. We manually reviewed different articles' methodologies and chose those that fit with our thematic focus. Accordingly, we selected eight (8) articles drawn from multiple disciplines – each applying a variant of GTM to answer key research questions. This multiplicity speaks to the utility of the methodology to the expansion of theory and – in some cases – the explanation of components of culture.

5.1   Glaser's Objectivist/Realist/Positivist GTM

Larsson et al. (2007) explored the experiences of patients who are actively involved in their own care and their expectations of this process. Patient participation in healthcare decision-making is an integral part of the quality of services, the improvement of health outcomes, and the autonomy of patients. The team organized focus group interviews with participants recruited from four local hospitals in Sweden. All interviews were recorded and transcribed verbatim. To analyze the data, they used the constant comparative method. In their methodology, they organized their data into codes. Substantive codes conceptualized participants' description of patient participation in nursing care. The method of this study reflects the Glaserian research paradigm in several ways: the researchers conceptualized relationships among substantive codes, they used coding families in the theoretical coding stage, and they worked with the concept-indicator model, as the data (description of event, experiences, and actions from the interviews) were indicators of main concepts. The researchers generated a core category out of four interrelated categories that outlined the nurse-patient interaction, which involves series of structural and individual-level factors. For example, *Rights* is one of the four categories that emerged from data about patients' expectations of nurses' roles. According to this category, patients expect nurses to protect their rights. This need impinges on the quality of the patient-nurse interaction and is an essential part of the patient participation in nursing care. Overall, findings pointed to new discoveries and were aligned with the existing literature. Brown (2006) followed the same approach but differed in two ways. First, she explicitly drew from Glaser and Strauss's approach while providing details on her methodology. Second, she generated the shame resilience theory to explain women's experiences with the emotion of shame – a mental and public health issue for social workers. Data included more than 200 individual interviews. Just like Larsson and colleagues, Brown sought to conceptualize shame and outline its underlying mechanism, its impacts on women, and the ways in which it is managed. In accordance with Glaser's approach, both studies focused on conceptualizing and explaining patterns to clarify concepts. In the case of Brown (2006), the theory generated accounts for a phenomenon that concerns participants.

While neither of these two studies examined culture per se, they underlined the significance of participants' perspectives within specific contexts. In alignment with the Glaserian approach, there is a (artificial) boundary between researchers and participants. In this position, the studied phenomena are embedded in a reality sui generis and can be uncovered with the use of GTM as a methodological tool. Consequently, in the first study, the meaning of patient participation in nursing care is *revealed* from a patient point of view. In the second study, interviews were conducted to theoretically *determine* several aspects of women's experiences with shame.

5.2   Strauss and Corbin's Postpositivist/Realist/Interpretive GTM

Bateman et al. (2013) heavily drew from this approach while explicitly disclosing their epistemological stance, said to be parallel to that of Corbin. They created a theoretical model that describes and predicts how undergraduate medical students learn from virtual patients (VPs) and the mechanism of this human-machine interaction in light of a paucity of research on the matter. VPs are game-based learning tools and interactive computer simulations of real-life clinical cases used in healthcare education (Kononowicz et al. 2015). For this research, 48 students from one medical school completed two VP cases that the researchers designed. Subsequently, participants evaluated the tool and partook in focus group discussions. The focus group interviews were digitally recorded, transcribed, and coded to develop preliminary categories. The researchers sought to validate their interpretations of the data by constantly comparing them. As a team, they coded their data line by line with in vivo codes and then advanced to axial coding. They used emergent themes to guide their sampling method. In the selective coding stage, they established a central core category, labeled "learning from the VP," that is connected to other categories. For example, they found that learning from VP cases depends on the way VPs are constructed (how realistic, clear, and well formatted the fictitious cases used in the VPs are), the characteristics of the student and the educational institution, the experience of the student with technology, and the outcomes of this human-machine interaction. In this study, the researchers were attentive to the ways in which students' subjective experiences could be abstracted into theoretical statements about the causal relationship between learners and technology, using group interviews and observations. In addition to acknowledging the potential biases in their inquiries, the investigators convened a research team composed of different health experts and implicated them in the theory-building process in an iterative fashion. Lindsay et al. (2011) also followed this variant of GTM to *uncover* the experience of men living with chronic gout – a highly treatable, yet undermanaged and poorly diagnosed, illness worldwide. They identified three major themes linked to the experience of gout. Just like Bateman and his colleagues, this study iteratively engaged a group of professionals in the research process to analyze and validate the themes that emerged from the data. Based on their interpretive leaning, these researchers relied on experts and participants alike to define the phenomena. For example, the second set of researchers, while consulting their team of experts, found that men suffering with gout might experience wide-ranging degrees of pain with disastrous effects on their social life, their social roles, and their self-identity. Both studies diverge on one key aspect, the former explicitly indicating their theoretical stance. Bateman and his colleague (2013, p. 596) "acknowledge[d] the positivist origins of grounded theory, whilst recognizing and valuing constructivism and reflexivity in the context of [their] pragmatist theoretical orientation and training." For the latter, one can deduce their realist paradigm from their methodology.

In these studies, the notion of culture is not stressed. Still, the analyses hint at a consideration of certain cultural characteristics. In the first study, the researchers analyzed the effectiveness of a learning tool for students within an institutional context that has its own culture. Since they designed the VP cases, they recognized that individuals construct medical knowledge within the social institution. In the second study, the investigators examined a health issue among men in South Auckland, New Zealand. From a sociological perspective, we argue that lay perspectives and gender are culturally relevant. In the latter study, for instance, the gender used in the analysis is subject to specific social expectations, mode of socialization, and cultural scripts in Auckland. In both studies, researchers, more or less, treat the examined phenomena as things that exist in an independent reality, outside of the researcher and the participants' conceptual scheme. Meanwhile, they also accept the difference between lay and experts' perspectives in their methodology. To illustrate the centrality of culture in the application of this form of GTM, we refer to Aita and Kai's (2010) GT study of life-sustaining treatment (LST) practices among healthcare professionals in Japan. They applied Strauss and Corbin's GTM in their analysis and followed the same steps as the two studies reviewed in this section. Meanwhile, Aiti and Kai's desire to examine these practices was motivated by the cultural irrelevance of Anglo-American guidelines on withholding and withdrawing life-sustaining treatments (LST). This study disputed the implementation of these Western medical protocols within a Japanese context by outlining the experiences, attitudes, and perceptions of 35 Japanese physicians regarding withdrawal of mechanical ventilation and other LST. Differences in cultural beliefs and attitudes vis-à-vis withdrawal of artificial devices from dying patients pointed to the danger of globalizing certain Western medical practices and the heterogeneous conceptualization of ethics. GTM allowed for the systematic identification of barriers hindering the implementation of these guidelines for Japanese physicians working at emergency and critical care facilities in that country.

5.3   Charmaz's and Bryant's Constructivist/Interpretive GTM

Poteat et al. (2013) drew from the work of Charmaz (2006) to explore the experience of transgender patients with discrimination in healthcare access and utilization. The disentanglement of this phenomenon is essential to addressing health disparities for transgender people. The researchers analyzed typed field notes and 67 individual in-depth interview transcripts (with 55 transgender patients and 12 healthcare providers working in institutions that serve transgender patients). This study also included two "community advisory boards" composed of transgender individuals throughout the research process to "ground" the study in the community interest (23). This partnership-like approach resembles community-based participatory research (CBPR) but also respects CGTM's principle of viewing the world from the standpoint of their participants (see also ► Chap. 17, "Community-Based Participatory Action Research"). The research report emphasizes reflexivity and highlights the role of the researchers as co-authors in the research process – a key characteristic of the constructivist approach. The authors critically examined their construction of the research process as they explored how participants constructed their experiences. The interviewers wrote reflexive notes as well as general field notes after each interview. Further, one of them kept a reflexive journal to differentiate her views from those of participants – treating the research process itself as a social construction and accepting her bias and positionality as a researcher. In their analysis, the researchers found that power is a grossly overlooked element in this dyad. At this level, stigma underpins the unbalanced dynamic of power in the asymmetrical relationship between transgender patients and healthcare providers. Unlike the aforementioned studies, findings *emerged* through an interactive process between researchers and participants. A comparative method was used in all of the studies thus far; however, in this study, the data were not deemed objective. This study supplied an interpretive understanding of stigma in the healthcare provider-patient relationship and was able to highlight the impact of power in this interaction.

The method and strategies adopted from a constructivist approach offered a systematic way of delving into the pervasive practice of patient stigmatization embedded within an institution that has its own sociohistorical and cultural context. Unlike the other studies referenced, Poteat and colleagues paid attention to the nature of the interaction between members of opposing groups. In this inquiry, they studied stigma relationally and conducted a context-rich analysis of healthcare provider-patient relationship. In doing so, they tapped into the active involvement of individuals in constructing meaning from their hierarchical relationship. By placing individuals' agency and social roles in the front stage, the researchers were able to construct the meaning of stigma within a medical clinical context.

5.4   Clarke's Interpretive/Constructionist/Situationist GTM

Erol (2011) examined the medicalization of osteoporosis (OP) in Turkey and followed the situational analysis approach. Specifically, she examined how risk related to postmenopausal osteoporosis is discussed in lay and medical Turkish literatures. She collected ethnographic data, archival data (newspapers and magazines on menopause), 52 individual semi-structured interviews with hospital attendees (31 menopausal patients and 21 doctors), and three group interviews. Like conventional GTM, the analysis begun with fracturing the data and then led to their reintegration into cohesive theoretical framework (Clarke 2009). She also set up different maps (situational, relational, social worlds-arenas) to complement the analysis of the collected data. All interviews were transcribed verbatim and the first round were initially coded. When she finished gathering data, she revised those initial codes and excavated more data to further analytic ideas. Based on her findings, the construction of osteoporosis occurs at different levels. For example, biomedical explanations attributed the risks of developing OP to menopause and deemed it preventable with lifestyle changes – putting the burden solely on the individual. One of these changes entailed wearing modern clothing as opposed to traditional ones to increase women's exposure to the sun. Then, the media disseminated and reinforced these messages. Lastly, they swayed public perceptions and attitudes regarding OP – inducing anxiety among women in some cases. The framing of this health condition as a behavioral problem points to glaring cultural components. First, it depicts the risks of OP as an issue that is rooted in the habitus of women who are accustomed to wearing traditional clothes that cover their whole body. Second, it singles out women's agency (the selection of clothing which is a quintessential cultural behavior) as the primary determinant of OP risks. This established causal link is legitimized, disseminated, and cemented in cultural beliefs about OP. As the researcher underlined, an "embodied modernity is expected of the menopausal Turkish woman to counterbalance the embodied risk of osteoporosis" (p. 1496). This study outlines numerous structural (the media, the medical institution, the local clinics)- and individual-level factors contributing to the construction of osteoporosis as a gendered issue.

Central to Erol's study is the examination of discourses around postmenopausal OP in Turkey. The use of Clarke's approach to GTM enabled Erol to explain the social construction of a health condition in a culture. The researcher collected data from a variety of sources and analytic strategies to conduct a context-rich analysis on the medicalization of the risks of OP and the legitimation of this construct at different levels: macro (society), meso (organizational), and micro (individual). Situational mapping allowed for the elucidation of healthcare practices specific to postmenopausal women at risk of developing OP in Turkey.

5.5 Mixing GTM Approaches

Researchers can combine different approaches to examine a phenomenon from which they may attempt to "discover theory from data" (Glaser and Strauss 1967, p. 1). Schnitzer et al. (2011) used CGTM, Strauss and Corbin's paradigm model, and situational analysis to examine the decision process of ultra-Orthodox Jewish parents to utilize healthcare services for their children in Antwerp, Belgium. First, they drew from CGTM to analyze the interviews and coded line by line with in vivo codes. Then, following Strauss and Corbin's paradigm model, they defined links between categories and incorporated them into a procedural diagram that illustrates the help-seeking pathways of the parents. They also used three mapping techniques to explore how different people and groups (collectives), situations, contexts, and discourse influence parents' decision-making.

The application of GTM is not always the same and changes in accordance to research questions and researchers' theoretical leanings and goals. Unfortunately, not all GTM studies have a transparent and rigorous description of the methodology. This gap is not necessarily a testament to the quality of existing GTM studies. Structural constraints, like page limits in many journals, could affect the level of detail revealed. Nevertheless, we observed an epistemological issue with authors' confusing grounded theory with other methodologies. This could make the assessment of the form of GTM researchers are using a problematic endeavor. In the studies that we selected, investigators, more or less, use the core analytic tenets of GTM, outline their respective ontology, and use the technical language aligned with their approach. Given all of this, for studies in which the examination of culture is seriously intended, we recommend the use of constructivist/constructionist GTM (CGTM). The recognition by CGTM that meanings, practices, and values are socially constructed, rather than inherent in labels, and that these are changeable parallels the complexities of culture. The reflexivity demanded of the researcher by CGTM makes it ideally suited for deeper looks at culture and how it connects to health and illness; reflexive awareness of the joint construction of research can steer one toward actively listening to research participants, truly hearing them, and managing what one sees through her or his own cultural lens.

6   Conclusions and Future Directions

Clearly, GTM has much to offer health science researchers and is widely used in this field. There are multiple approaches to GTM, based on differing, often unstated, paradigmatic assumptions. Yet, these share a common core of actual research practices. Charmaz (2006, p. 9) writes: "[R]esearchers can use basic grounded theory guidelines such as coding, memo-writing, and sampling for theory development, and comparative methods are, in many ways, neutral... *how* we use these guidelines is not neutral; nor are the assumptions they bring to our research." It is best to consider one's own assumptions, in order to choose the approach best suited to one's research goals. While one can draw on multiple approaches, using helpful techniques is not clear that one can successfully mix paradigms in a single piece of research (Cresswell 2013, 2018). Moreover, readers will read into one's work, making assumptions about one's paradigm (Cresswell 2009, 2018). More troubling to readers, researchers often claim to be using GTM without providing further details, leaving them in a quandary as to what was actually done. With journal space limited, it is difficult to provide much in the way of methodological detail. That said, specifying one's approach can be helpful. Specifying some methodological detail would be even more helpful.

Culture is often a significant component of health meanings, beliefs, practices, services, and institutions, even when it is not an explicit component of one's research. It can be valuable to keep culture in mind as a sensitizing concept. This is not to advocate forcing consideration of culture onto one's data or findings but to recommend that researchers be aware that culture might be a factor in the issue being studied and/or the conduct of one's research (see Liamputtong 2010).

As indicated in our review of selected studies reflecting the application of GT methods, net of the subdiscipline, epistemologies have major implications for the research process. To account for cultural subtleties in future analyses, health science researchers may find GTM, particularly CGTM, especially useful. This variant of GTM is an alternative to the objectivist GTM which can either be impervious to culture or treat it peripherally. From this viewpoint, CGTM points to a new direction as it is ideally suited for tapping into the impact of culture on the social side of health phenomena. It is also noteworthy that attention to specific cultural practices is valuable for studying issues of social justice in healthcare, as illustrated by the work of Poteat et al. (2013) discussed above.

Regardless of the approach followed or how techniques from different approaches are mixed, GTM involves continual movement between data collection/construction and analysis and between concrete data, memos, and abstractions. This is iterative work; to lose sight of one's data is to put one's theory at risk. That said, GTM is a robust methodology, one that can be used to enter and come to understand participants' worlds. Use it in good health!

References

Ahmad F, Hudak PL, Bercovitz K, Hollenberg E, Levison W. Are physicians ready for patients with internet-based health information? J Med Internet Res. 2006;8(3):e22.

Aita K, Kai I. Physicians' psychosocial barriers to different modes of withdrawal of life support in critical care: a qualitative study in Japan. Soc Sci Med. 2010;70(4):616–22.

Bateman J, Allen M, Samani D, Kidd J, Davies D. Virtual patient design: exploring what works and why. A grounded theory study. Med Educ. 2013;47:595–606.

Beard RL, Fetterman DJ, Wu B, Bryant L. The two voices of Alzheimer's: attitudes toward brain health by diagnosed individuals and support persons. Gerontologist. 2009;49:S40–9.

Belgrave LL, Seide K. Coding for grounded theory. In: Bryant A, Charmaz K, editors. The SAGE handbook of current developments in grounded theory. London. Forthcoming.

Brown B. Shame resilience theory: a grounded theory study on women and shame. Fam Soc. 2006;87(1):43–52.

Bryant A. Re-grounding grounded theory. J Inf Technol Theory Appl. 2002;4:25–42.

Bryant A. Grounded theory and grounded theorizing: pragmatism in research practice. New York: Oxford University Press; 2017.

Bryant A, Charmaz K. Grounded theory in historical perspective: an epistemological account. In: Bryant A, Charmaz K, editors. The handbook of grounded theory. London: Sage; 2007a. p. 31–57.

Bryant A, Charmaz K. Introduction. In: Bryant A, Charmaz K, editors. The handbook of grounded theory. London: Sage; 2007b. p. 31–57.

Charmaz K. Body, identity and self: adapting to impairment. Sociol Q. 1995;36:657–80.

Charmaz K. Constructivist and objectivist grounded theory. In: Denzin NK, Lincoln Y, editors. Handbook of qualitative research. 2nd ed. Thousand Oaks: Sage; 2000. p. 509–35.

Charmaz K. Constructing grounded theory: a practical guide through qualitative analysis. London: Sage; 2006.

Charmaz K. Constructionism and grounded theory. In: Holstein JA, Gubrium JF, editors. Handbook of constructionist research. New York: Guilford; 2007. p. 35–53.

Charmaz K. Shifting the grounds: constructivist grounded theory methods for the twenty-first century. In: Morse J, Stern P, Corbin J, Bowers B, Charmaz K, Clarke A, editors. Developing grounded theory: the second generation. Walnut Creek: Left Coast Press; 2009. p. 127–54.

Charmaz K. Constructing grounded theory. 2nd ed. Thousand Oaks: Sage; 2014.

Charmaz K, Belgrave LL. Qualitative interviewing and grounded theory analysis. In: Gubrium JF, Holstein JA, Marvasti AB, McKinney KD, editors. The sage handbook of interview research: the complexity of the craft. 2nd ed. Los Angeles: Sage; 2012. p. 347–65.

Clarke A. From grounded theory to situational analysis: what's new? Why? How? In: Morse J, Stern P, Corbin J, Bowers B, Charmaz K, Clarke A, editors. Developing grounded theory: the second generation. Walnut Creek: Left Coast Press; 2009. p. 194–233.

Clarke A. Feminisms, grounded theory, and situational analysis revisited. In: Clarke AE, Friese C, Washburn R, editors. Situational analysis in practice: mapping research with grounded theory. Walnut Creek: Left Coast Press; 2015. p. 84–154.

Clarke A, Friese C. Grounded theory using situational analysis. In: Bryant A, Charmaz K, editors. The sage handbook of grounded theory. London: Sage; 2007. p. 363–97.

Corbin J, Strauss A. Basics of qualitative research: techniques and procedures for developing grounded theory. 4th ed. Los Angeles: Sage; 2015.

Cresswell JW. Research design: qualitative, quantitative, and mixed methods approaches. Thousand Oaks: Sage; 2009.

Cresswell JW. Qualitative inquiry & research design: choosing among five approaches. 3rd ed. Los Angeles: Sage; 2013.

Cresswell JW. Research design: qualitative, quantitative, and mixed methods approaches. 5th ed. Thousand Oaks: Sage; 2018.

Creswell JW, Poth CN. Qualitative inquiry & research design: choosing among five approaches. 4th ed. Los Angeles: Sage; 2018.

Denzin NK, Giardina MD. Qualitative inquiry and social justice: towards a politics of hope. Walnut Creek: Left Coast Press; 2009.

Erol M. Melting bones: the social construction of postmenopausal osteoporosis in Turkey. Soc Sci Med. 2011;73(10):1490–7.

Fielding N. Analytic density, postmodernism, and applied multiple methods research. In: Bergman MM, editor. Advances in mixed methods research: theories and applications. Thousand Oaks: Sage; 2008. p. 37–52.

Glaser BG. Theoretical sensitivity. Mill Valley: The Sociology Press; 1978.

Glaser BG, Strauss AL. Awareness of dying. Chicago: Aldine Pub; 1965.

Glaser BG, Strauss AL. The discovery of grounded theory: strategies for qualitative research. New York: Adline De Gruyter; 1967.

Kononowicz AA, Zary N, Edebring S, Hege I. Virtual patients – what are we talking about? A framework to classify the meanings of the term in healthcare education. BMC Med Educ. 2015;15:11.

Larsson IE, Sahlsten MJ, Sjöström B, et al. Patient participation in nursing care from a patient perspective: a grounded theory study. Scand J Caring Sci. 2007;21:313–20.

Liamputtong P. Performing qualitative cross-cultural research. Cambridge: Cambridge University Press; 2010.

Liamputtong P. Culture as social determinant of health. In: Liamputtong P, editor. Social determinants of health: individual, community and healthcare. Melbourne: Oxford University Press; 2018.

Lindsay K, Vanderpyl J, Logo P, Dalbeth N. The experience and impact of living with gout: a study of men with chronic gout using a qualitative grounded theory approach. J Clin Rheumatol. 2011;17:1–6.

Nagel DA, Burns VF, Tilley C, Augin D. When novice researchers adopt constructivist grounded theory: Navigating the less traveled paradigmatic and methodological paths in Ph.D. dissertation work. Int J Doctoral Stud. 2015;10:365–83.

Poteat T, German D, Kerrigan D. Managing uncertainty: a grounded theory of stigma in transgender health care encounters. Soc Sci Med. 2013;84:22–9.

Quah S. Health and culture. In: Blackwell encyclopedia of sociology. 2007. Retrieved from http://www.blackwellreference.com/subscriber/tocnode.html?id=g9781405124331_chunk_g978140512433114_ss1-7.

Schnitzer G, Loots G, Escudero V, Schechter I. Negotiating the pathways into care in a globalizing world: help-seeking behavior of ultra-orthodox Jewish parents. Int J Soc Psychiatry. 2011;57(2):153–65.

Seide K. In the midst of it all: a qualitative study of the everyday life of Haitians during an ongoing cholera epidemic. Unpublished Master's Thesis. University of Miami; 2016.

Stern PN. On solid ground: essential properties for growing grounded theory. In: Bryant A, Charmaz K, editors. The sage handbook of grounded theory. London: Sage; 2007. p. 114–26.

Strauss A, Corbin J. Grounded theory procedures and techniques. Newbury Park: Sage; 1990.

Ulysse GA. Papa, patriarchy, and power: snapshots of a good Haitian girl, feminism, & diasporic dreams. J Haitian Stud. 2006;12(1):24–47.

Wertz FJ, Charmaz K, McMullen L, Josselson R, Anderson R, McSpadden E. Five ways of doing qualitative analysis. New York: Guilford Press; 2011.

Case Study Research

19

Pota Forrest-Lawrence

Contents

Abstract

Case study research has been extensively used in numerous disciplines as a way to test and develop theory, add to humanistic understanding and existing experiences, and uncover the intricacies of complex social phenomena. Its usefulness as an exploratory tool makes it a popular methodology to employ among social scientists. This usefulness, however, has, at times, been overshadowed by several misunderstandings of and oversimplifications about the nature of case study research. Although these misunderstandings, such as the inability to confidently make scientific generalizations on the basis of a single case, may be presented as limitations, they should not detract social scientists from using case study research for certain critical research tasks. Used in such a way, case study research

P. Forrest-Lawrence (✉)
School of Social Sciences and Psychology, Western Sydney University, Milperra, NSW, Australia
e-mail: p.forrest-lawrence@westernsydney.edu.au

© Springer Nature Singapore Pte Ltd. 2019
P. Liamputtong (ed.), *Handbook of Research Methods in Health Social Sciences*,
https://doi.org/10.1007/978-981-10-5251-4_67

317

holds up considerably well to many other social science methodologies and can certainly contribute to the development of knowledge. This chapter examines case study research with emphasis on its "generality," notably what Stake proposed as "naturalistic generalization," a proxy process that enables the generalization of findings from a single case. It will interrogate the misunderstanding that case study research cannot effectively contribute to scientific development, by focusing on single-case designs. Single-case designs attempt to understand one particular case in-depth and allow for a richer understanding of the issue under investigation.

Keywords

Case study · Single case · Qualitative research method · Generalizations · Naturalistic generalizations

1   Introduction

Case study research has come to occupy a prominent place in the evolving social science landscape. Focus on this largely qualitative research method has positioned the case study as a popular "go-to" method for researchers. Its flexible approach has allowed for its extensive use in both qualitative and quantitative studies and within a vast range of disciplines including but not limited to political science, law, public health, medicine, business, social science, and education.

The versatility of the case study has positioned it at the forefront of preferred methodological approaches for many researchers. In this way, it can be used in exploratory, explanatory, and evaluation types of research to develop theory and generate and test hypotheses and can embrace varied epistemological orientations. Such features have made the case study quite an attractive research method to employ within the social sciences.

While this attractiveness has increased its popularity, it has also highlighted what some consider its many shortcomings (see Gerring 2006). One such shortcoming is the inability for a case study to make scientific generalizations. These generalizations center on notions of predictability, validity, and causation and are considered by many as the only way in which researchers can understand complex social phenomena. The perception of formal generalization as the only contributor to scientific development ignores the value of other types of generalizations that can effectively contribute to scientific knowledge and progress.

This chapter will commence with an overview of case study research, detailing some of its uses and applications. Following this summary, it then examines why it is so misunderstood as a methodology and then highlights some of its advantages. This then leads to a key focus of the chapter, generalization, by examining the utility of single-case designs as contributing to scientific development via a process proposed by Stake (1978) known as "naturalistic generalization." The chapter will conclude with a brief discussion about future directions for case study research.

2   What Is Case Study Research?

Reviewing a plethora of material and literature on case study research does little to synthesize the complementing yet, at times, contradictory definitions of a term that has come to mean different things to different disciplines. What we do know is that case study research clearly places cases *not* variables at center stage (Ragin 1992). With this in mind, the attempt here is to offer a broad-strokes overview of case study research.

Some of the earliest documented historical case studies can be traced to the works of ancient Greek writers Herodotus and Thucydides (Elman et al. 2016). Though by far, scientific revolutionist Galileo and his rejection of Aristotle's law of gravity via the process of gravitational force is considered one of the most famous case studies of the last millennium (Flyvberg 2011). Many such "case studies" have subsequently filled the pages of history books, presenting detailed overviews of complex "cases," Graham Allison's methodical case study analysis, one such contemporary example (Allison 1969). Using an explanatory single-case study, Allison examined the 1962 Cuban Missile Crisis in intricate detail. These contributions highlight specific and unique cases and offer insight into both the value of the case study and its many uses and applications.

The use of case studies according to discipline is highlighted by Thomas (2011). Here, Thomas notes that disciplines such as psychology, sociology, and education position the case study in a largely "interpretive frame." Alternatively, disciplines such as politics and business adopt a more neopositivistic epistemological stance by largely identifying variables as cases. This approach, claim Bartlett and Vavrus (2017), views the world via laws centered on cause and effect, where scientific methods prevail. Other disciplines such law and medicine largely highlight novel phenomena. Such different uses of the case study according to a researchers' epistemological stance highlight the diversity and flexibility of the case study.

2.1   A Case Study: Definitions

Such wide-ranging applications, however, do present difficulties in defining a case study (Exworthy & Peckham 2012). A plethora of definitions, explanations, insights, and understandings has made it difficult to provide a single definition of a concept so widely used. This is further compounded by the need to recognize both the researchers' methodological and epistemological stance when conceptualizing a definition of a case study (Bartlett and Vavrus 2017). The following details a number of definitions of a case study and presents ways to understand its meaning and contribution:

- A research method that involves an in-depth understanding and analysis of a case.
- A form of empirical social enquiry that interrogates social phenomena both in detail and in a real-world context (Yin 2014, p. 6).

- A "frame" that "determines the boundaries of information-gathering" (Stoecker 1991, p. 98).
- A framework that centers on answering a specific question relevant to a case (Seawright and Gerring 2008).
- An intensive examination of a single case whose aim is to illuminate a larger set of cases (Gerring 2007, pp. 20, 65).
- Cases can be anything from an event, organization, group, person, and so on. They can influence the development of public policies and practices in a broad range of fields (Duff 2014, p. 234).
- The phenomena studied are done so in their natural surroundings (Swanborn 2010, p. 15).
- Provides a multifaceted way of structuring our understanding of reality.
- Offer an explanation that holistically captures the dynamics of a period in time and of a certain social unit (Mills et al. 2010, p. 17).
- Its chameleon-like epistemological qualities allows it to embrace many different orientations (Yin 2014, p. 17).
- Can uncover patterns, establish meanings, reach conclusions, and build theory (Stake 1995, p. 67).
- Can examine causal mechanisms, test theoretical predictions specific to general models, and detail features about a specific case (Gerring 2007, p. 5).
- Allows for an intimate understanding of a phenomenon under investigation of which "the case is exemplar" (Duff 2014, pp. 236–237).
- Case study research is "not a methodological choice but a choice of what is to be studied" (Stake 2005, p. 443).

While these definitions offer some insight into the case study and, by doing so, highlight qualities that present it as an attractive research approach, it is still considered by some to exist and survive in a "curious methodological limbo" (Gerring 2007, p. 7). While this may limit its attractiveness, it still thrives because of its utility and its focus on qualitative research methods as methodological rivals to the once dominant quantitative and "positivist" models of causal explanations (Gerring 2007; see also ▶ Chap. 9, "Positivism and Realism").

2.2 Case Study: Types and Designs

Researchers typically employ a descriptive, exploratory, or explanatory case study to examine and understand the phenomenon under investigation. A descriptive case study describes the phenomenon in its context. Whyte (1955) used a descriptive case study to examine an Italian-influenced US neighborhood. Alternatively, an exploratory case study aids in the development of theories and hypotheses, whereas an explanatory case study explains how events transpire over time. It can also examine cause and effect relationships. As highlighted earlier, Allison (1969), for example, employed an explanatory single-case study to examine the 1962 Cuban Missile Crisis.

2.2.1 Multiple-Case Designs

A case study can either comprise a multiple-case or single-case design. A multiple-case design is typically selected to provide a better understanding about a collection of cases, as well as explore, test, and build theory. Stake (2006) notes that a multiple-case study examines how the subject of interest operates within different settings and within different contexts. Yeh and Hedgespeth (1995), for example, employed a comparative multiple-case study of 15 families of adolescents to examine the relationship between particular family factors and alcohol/drug abuse.

2.2.2 Single-Case Designs

A single-case design is appropriate should a researcher want to understand and examine one social phenomenon in detail (Merriam 1988). This design type also allows for an elaboration of quite complex characteristics of a single case (Duff 2014). Similarly, Stake (2005, p. 445) notes that the single-case design attempts to understand one particular case in-depth rather than what is mostly ". . . true of the many." Skar and Prellwitz (2008) used a single-case to describe how a child diagnosed with obesity understood his involvement in play activities. This method allowed them to provide a deeper understanding of the issue.

According to both interpretivists and realists, a single case is able to provide different accounts of causation, clarify theoretical relationships in a certain setting that may appear obscure, and construct theory (Mills et al. 2010). They can also offer a useful understanding and intimate knowledge of the particulars of a singular case such as a school, locality, type of food, and so on. This, according to Stake (1978, p. 6), showcases how the "truth lies in particulars."

While Stake (1995) claims that there is much one can learn that is considered general from a single case, single-case design studies are generally perceived as a poor basis for scientific generalization and, therefore, weaker than other types of research designs. This perception, however, neglects to emphasize the uniqueness of the case study and its potential as a valuable and informative research methodology.

3 Why Is It so Misunderstood?: Limitations of Case Study Research

While its varied uses may be one of its more appealing features, the case study is simultaneously criticized for its lack of rigor and representativeness (considered weaker with respect to external validity as it comprises a small number of cases) (Gerring 2007) and its inability to generalize, particularly from a single case (Simons 1986). Hamel (1993) highlights how these limitations are presented as fundamental flaws, particularly if considered in the context of the collection, construction, and analysis of empirical material. As a result, the case study has been "disregarded as a methodology" and relegated to something unscientific notably by proponents of causality via quantitative approaches, who consider this approach as the only way one can add to scientific knowledge (Flyvberg 2011).

Presented as the "weak sibling" of the social sciences methods, the case study has been further criticized as something "mere," "biased," "subjective" (though this may not necessarily be considered a weakness), and "non-generalizable" with "weak empirical leverage" that is highly "suspect" (Yin 1984; Gerring 2006, pp. 6–7). Such constructions suggest that generalizations cannot be made via case studies because they defy the laws and methods of physics, where *closed systems* are only used, meaning these systems are isolated from their environment (Bertalanffy 1973, cited in Patton and Appelbaum 2003). What Stake (1995) demonstrates, however, is that a case is both an open and an integrated system, where the rules of "closed" systems of the casual sciences *do not apply*. This is clearly something either ignored or overlooked by those who openly criticize the case study as unable to contribute to scientific knowledge and an understanding of the world.

This negative construction of the case study positions it as subordinate to natural sciences methods and secondary to studies that typically comprise large samples. Such a perception stems from a normative (empiricist) view of research – *what it ought to be, how it ought to be conducted*, and *what it ought to do*. This, according to Simons (1986), assumes a rather specific polarity, one considered quite inflexible.

This view of what research *ought to be*, and *ought to do* is illustrated in the work of Simons (2015). Here, Simons (2015, p. xi) highlights the growing perception, at least on a political level, that *evidence* produced via randomized controlled trials is the only type of evidence that should inform and influence policy (see also ▶ Chaps. 3, "Quantitative Research," and ▶ 37, "Randomized Controlled Trials"). Such an inflexible way of understanding the scope and breadth of *evidence* leaves little room for evidence produced via single cases to inform policy and political action. This is perhaps, in part, the result of the domination of the natural sciences model, based on a deductive form of reasoning that affirms anything "outside the system of explained science" to be erroneous (Stake 1995). Clearly, the use of single-case designs that adopt an inductive approach to research is at odds with this natural sciences deductive approach.

This general disregard for the case study is most pervasive in discussions on generalization. The supposed inability for case studies to generalize is, therefore, an important misunderstanding to interrogate. It affirms that there is only *one* way to generalize and that way is *not* from a single case but rather via large samples. Flyvberg (2011) explores this common misunderstanding by way of Galileo's rejection of Aristotle's law of gravity (physical theory). The process of rejection did not involve numerous observations but rather two experiments, the first, a *thought* experiment and the second, a *practical* experiment, a case study, leading to a new way of imagining and understanding physics.

Nevertheless, there still remains a general distrust of the case study in particular if it involves a single case. This is remarkable when we consider Karl Popper's (1959) notion of falsification. It involves conducting a rigorous scientific test in which a proposition is examined, and if only one observation does not fit this proposition, the proposition is either rejected or revised. Such a test renders the case study most suitable for identifying what Popper referred to as the "black swans." Looking at this scientific test, one can say that a case study is in a unique position to locate these "black swans" because of its intimate approach to the examination of that under investigation (Flyvberg 2006).

However, we must also recognize the paradox of the case study, as explored by Simons (1986). According to Simons (1986), this paradox is most visible when we consider how a case study can generate both an understanding of the particular though simultaneously generate an understanding of the universal. Case studies can, therefore, partake in both worlds, that of the general and that of the particular (Gerring 2007). This unique quality of the case study, viewed by some as a limitation, is in actuality one of its key advantages.

The many misunderstandings of the case study have generally centered on notions of generalization and subjectivity. While these are important to consider and must be recognized in discussions on the suitability of the case study as a methodology, we must also consider the many benefits of the case study. By doing so, we begin to recognize the artistry of the case study as a methodological contributor to scientific knowledge.

4   The Artistry of the Case Study: Advantages of Case Study Research

The growing shift from a variable-focused approach to causality toward a more case study approach is the result of both a creeping skepticism of standard regression techniques and the desire to preserve the details of individual cases, something commonly lost in large-N type of analyses (Gerring 2006). This has enabled the case study to emerge as quite an attractive alternative to other methods that have typically centered on causality, predictiveness, and traditional forms of generalization.

The case study contributes to our knowledge bases in rather unique and informative ways. It does so via its attention to the local situation and less so on how it represents other cases more generally (Stake 2006). It enables an investigation of complex social phenomena to "retain the holistic and meaningful characteristics of real-life events" without being reductionist (Yin 2009, p. 3). Such an inductive approach considers empirical details that comprise the object of the study.

One of the major strengths of the case study is its distinct ability to uncover the intricacies of complex social phenomena by directing attention to the local situation (Stake 2006). Unlike experiments or surveys that intentionally separate a phenomenon from its context by controlling for particular variables, the case study pays particular attention to contextual conditions. It is a useful method for researchers who seek to examine contemporary events where behavior cannot be manipulated and can also understand the unfolding case over time (Yin 2009).

The epistemological advantage of the case study is that it is a tried and tested approach that offers a rewarding way of contributing to experience by improving our understanding (Stake 1978). Meanwhile, the methodological value of the case study specific to public administration and policy analysis is its ability to draw on many sources and research designs that are largely qualitative (Marinetto 2012). This then generates a rich narrative allowing researchers a degree of methodological flexibility that encourages theoretical insights.

The value of the case study is also recognized in the theory-building process (Eisenhardt 1989). Eckstein (1975) affirms that the case study is critical to all stages of theory building though most useful when researcher theories are tested. It can add a depth and dimension to theoretical understanding and, by doing so, simplify our understanding of reality (Donmoyer 2000). Largely, the case study can be used to elaborate theory and test theory (Stoecker 1991).

The case study, thus, challenges the supremacy of the natural sciences as the only proper contributor to scientific knowledge. This is eloquently demonstrated by MacDonald and Walker (1975, p. 3) who present the case study as a valuable and unique contributor to human understanding. They poignantly state:

> Case study is the way of the artist, who achieves greatness when, through the portrayal of a single instance locked in time and circumstance, he communicates enduring truths about the human condition. For both the scientist and artist, content and intent emerge in form.

Here, MacDonald and Walker (1975) highlight the artistic flair of the case study, emphasize its intense focus on the particular, and, by doing so, position it is a strong alternative to purely quantitative methods for researchers to consider. It offers a different form of knowledge to the researcher, one that is particular, unique and intimate, and free from predictiveness and causation, though bounded by the researchers' subjectivity. The case study, thus, enables the researcher to generalize, not in the tradition sense according to the natural sciences but via the strength available in the description of the content (Stake 1995).

5   A Different Form of Knowledge: The Unshackling

The case study is a creative way to examine and understand the human condition. It allows the researcher to navigate through a process that enables a particular narrative to prevail that is outside the strict confines of empiricism. Here, we see how knowledge has the ability to be transferred without the need to aspire to formal generalization. This section explores what Stake (1978) first proposes as naturalistic generalizations, as a way for the case study to generalize via the reader's experience. It will, however, first interrogate the notion of generalization in its traditional form in order to contextualize the discussion.

5.1   Generalization

Generalization is the transference of what we know and what we learn in one situation to another. It is considered one of the most fundamental goals of scientists. It is almost always linked to terms such as randomness, sampling, statistical significance, reliability, causation, a priori, and so forth (see also ▶ Chap. 38,

"Measurement Issues in Quantitative Research"). These terms are embedded in traditional forms of research and are comforting to social scientists, largely because many consider no other suitable alternative way to discuss social phenomena (Donmoyer 2000). This unwavering conviction in the power of universal rationality centered on the pursuit of generalizable knowledge is still considered by some as the *orthodox* way to attain *real* knowledge (Schwandt and Gates 2018).

In an attempt to interrogate the view that "nomic" generalizations (centered on prediction) are the only way we can generate scientific progress, Lincoln and Guba (2000) highlight some of the major deficiencies of this type of traditional generalizability while simultaneously providing an alternative, "transferability." They point to five key problems: "dependence on the assumption of determinism," "dependence on inductive logic," "dependence on the assumption of freedom from time and context," "entrapment in the nomothetic-idiographic dilemma," and "entrapment in a reductionist fallacy (See Lincoln and Guba (2000, pp. 29–36) for a detailed overview of these deficiencies and of transferability)." These problems highlight some of the deficiencies with the popular nomothetic conceptualization of generalization.

Consequently, this reliance on nomothetic generalization has stymied progress by hindering the way we understand the purpose of generalization as one enveloped by the strict confines of empiricism. This restricted view of generalization, labeled by Donmoyer (2000, p. 46) as both "dysfunctional" and "out of step with contemporary epistemology," overstates (formal) generalizations as the only way to appropriately contribute to scientific progress.

Statistical generalization, which typically attempts to enumerate frequencies, should not be the goal of social scientists given it is not the only legitimate method of scientific inquiry. Instead, Yin (2014) proposes that the focus should be on analytic generalization. By focusing on analytical generalizations, one can create a proper case, rather than create something to be replicated repeatedly, as is the case with statistical generalization (Patton and Appelbaum 2003; see ▶ Chap. 38, "Measurement Issues in Quantitative Research").

We need to examine in more detail this idea of analytical generalization that steers away from generalizations that only refer to notions of prediction. By doing so, we begin to see the value in what Hamilton (1976) describes as a "science of the singular." While this type of science does not generalize in a propositional sense, its usefulness, because of its focus on the particular by way of narratives and others forms, allows us to learn in different ways and acquire knowledge in ways considered only decades ago as completely unscientific.

This degree of freedom (not in the statistical sense) available via the case study provides social scientists with a flexibility to produce different ways of acquiring knowledge. Eisenhardt (1989) asserts that this freedom is made possible via two ways: one, by the ability for researchers to select a range of different methods and, two, the distinctive "in-depth style" inherent in the case study researcher. Both these ways release the researcher from the shackles of strict procedure, liberate their thinking, and increase the likelihood of generating novel theory. Here, Eisenhardt (1989) allows us to consider the vast possibilities that the case study offers both qualitative and quantitative researchers.

This *unshackling* allows for different types of knowledge, such as tacit, to sit alongside other forms of knowledge and be acknowledged as key contributors to scientific progress. Naturalistic generalizations, as noted by Stake (1978, 1995), can unshackle the researcher and allow them to engage in a different, more visceral form of generalization. The attempt here is not to propose a novel approach but, rather, highlight the need to revisit naturalistic generalizations.

5.2   Naturalistic Generalizations

The case study cannot confidently make "scientific" generalizations, particularly if a researcher intends to use a single-case design as the case is based on specificities such as precise events, locations, or periods of time (Mills et al. 2010). In order to address this limitation, Stake and Trumbull (1982) coin the term "naturalistic generalization," a proxy process that enables the generalization of findings from a single-case (Stake 1978). This allows a level of generalizability previously not available to the case study, particularly single-case design studies.

5.2.1   Discussion

Naturalistic generalizations are an alternative to the more traditional forms of generalization that center on predictiveness, formality, and keeping propositions intact (Simons et al. 2003). They are quite different to explicated (nomic) generalizations where emphasis is placed on a different type of knowledge. As a result, a humanistic and subjective understanding is more prevalent and even more relevant. They develop within individuals as a product of their individualized and subjective experience with a case.

This understanding comes from tacit knowledge. This type of knowledge is described by Stake (1978, p. 6) as the knowledge of "how things are, why they are, how people feel about them, and how these things are likely to be later or in others places with which this persons is familiar." This process of generalization may be verbalized, allowing knowledge to transfer from tacit to the propositional, though unable to traverse to the empirical and even logical that typically characterize explicated (nomic) generalizations (Stake 1978). Via such tacit understandings, Stake was able to divert attention to their value in generalizing from a case (Simons 1986).

Stake (1978) claims that explanation is best suited to propositional knowledge, whereas *understanding* is best suited to tacit knowledge. Tacit knowledge is knowledge gained via experiences with objects and events, "experience with propositions about them and rumination" (Stake 1978, p. 5). Polanyi (1958) highlights how individuals harness and contain this tacit knowledge that then allows them to build new understandings of complex social phenomena. While these do not pass logical and empirical tests specific to formal generalizations, they have the ability to guide action.

Naturalistic generalizations are conclusions that provide insights into a case by reflecting on its specific and unique details. They are embedded in the experience of the reader, be it orally or via some other way (Stake 1995). There is a reliance on judgment and interpretation that enables knowledge to be transferred from one context to another (Simons et al. 2003). This is principally achieved by the reader, who reflects on the particulars and the description contained within a case. The reader then determines how particular details of the case are similar to other situations by drawing on these reflections in order to generalize. The reader then assesses whether these particulars resonate with their personal or vicarious experience and, if so, determines whether the situations are "similar enough to warrant generalizations" (Melrose 2009, p. 600).

Largely, this process enables individuals to form generalizations from their own personal experiences by adding a single case to a repertoire of cases, creating a *different* group from which to generalize. It is, however, important to note that if this process is to effectively take place, the researchers must provide thick description and enough material for the reader to determine the relevance and meaning of this information to their lives.

This process allows findings from one case study to be applied to another similar case study in order to establish useful and relevant understandings (Stake 1978). Such generalizations are then reinforced via repeated encounters (Stake 2005). They are most useful if one seeks to acquire knowledge of the particular. Such a process is described by Stake (2000, p. 22) as both "intuitive and empirical and not idiotic."

Stake (1978) argues that we must consider the point of view of the user of the generalization. Via the process of naturalistic generalization, which centers on the personal experiences of the user, cases, notably single cases, are able to build naturalistic generalizations (Lincoln and Guba 2000). This "natural experience," as purported by Stake (1978), is an effective way to add to human understanding and can be achieved via words and images of the natural experience (Lincoln and Guba 2000).

5.2.2 Example

Stake (1995) details practical points that researchers must consider in studies that encourage naturalistic generalizations. These points will assist the reader in providing an opportunity for vicarious experience. They are detailed below (Table 1):

A contemporary example where a case study researcher conducted a narrative and phenomenological case study that encouraged readers to make naturalistic generalizations is *The Wild Food Challenge: A Case Study of a Self-initiated Experiential Education Project [SEEP]*, by Graham McLaren (2015). This study focused on SEEP, a type of self-directed learning that involves a person creating specific goals, commencing the project (this is self-initiated), and following it through to completion. The SEEP was a case of a young adolescent male who attended a school that provided a "deep nature connection mentoring." Interviews with the student and his mentors revealed the students' motivations and inspirations and their impact on his own sense of self. The SEEP was presented by way of a narrative account that encouraged the reader to engage in naturalistic generalization by drawing on the particulars of the study to determine ways they can inform their personal experiences. This information may be useful to educators, mentors, and other professionals who design curriculum and educative methods that encourage teenagers to engage in such self-initiated projects on experiential education.

Table 1 A list of things to assist in the validation of naturalistic generalization

1. Include accounts of matters the reader are already familiar with so they can gauge the accuracy, completeness, and bias of reports of other matters
2. Provide adequate raw data prior to interpretation so that readers can consider their own alternative interpretations
3. Describe the methods of case research used in ordinary language including how the triangulation was carried out, especially the confirmation and efforts to disconfirm major assertions
4. Make available, both directly and indirectly, information about the researcher and other sources of input
5. Provide the reader with reactions to the accounts from the data sources and other prospective readers, especially those expected to make use of the study
6. De-emphasize the idea that validity is based on what every observer sees, on simple replication; emphasize whether or not the reported happenings could have or could not have been seen

Source: Stake (1995), The art of case study research, p. 87

5.2.3 Critique

Such a different way of generalizing has come under scrutiny over the years. One criticism is that the onus of making naturalistic generalizations is directly placed on the reader instead of the researcher (Hellström 2008; Melrose 2009). It is, therefore, up to the reader to come to their own conclusions using the thick description and context provided by the researcher. This can be viewed as incongruent to scientific induction, a process whereby the researcher, *not* the reader, develops general principles from specific observations.

Other criticisms have focused on the unoriginality of naturalistic generalizations. Hellström's (2008, p. 335) examination of naturalistic generalization concludes that to break this type of generalization from traditional notions of generalizability is unwarranted and even premature. This is largely because interpretivist types of generalizations, such as naturalistic, are "well accommodated" within the broader ambit of generalization. All types of generalizations share the objective of providing a detailed understanding by transferring knowledge from one situation to another (larger population). Therefore, Hellström claims that naturalistic generalization is no different, as it shares this goal with others forms of generalizations.

Another criticism noted by Donmoyer (2000) is that naturalistic generalization emerged as the method of choice for evaluation research and, therefore, its utility and application to other types of research is limited. While Donmoyer (2000) rightly dictates that Stake's conception of naturalistic generalization is underdeveloped, Donmoyer does little to build upon this form of generalization that has merit but requires some development. Donmoyer, however, suggest an alternative conceptualization of generalization specific to single cases, drawing on the conception of experiential knowledge.

While there is much that naturalistic generalizations can offer, and an interpretivist stance toward case studies is beneficial, we should not ignore "power relations or social structures," which Bartlett and Vavrus (2017, p. 33) claim are "underemphasized in Stake's presentation of case studies." The "politics of representation," notably "who gets to represent whom and how in a research project," is not considered by Stake who neglects to consider researcher reflexivity in the research process (Bartlett and Vavrus 2017).

However, as noted by Stake (2000), naturalistic generalization is itself a type of generalization, not in the traditional sense but a type that allows the reader to recognize similarities contained within an object both in and out of context and to be aware of this taking place. Such an engagement with the case will provide for a deeper, more meaningful understanding of the case, one that is tied to the experiences of the reader and one that engages with the reader on a more intimate level. Here, the reader is no longer a bystander, watching from the outside looking in and reading the interpretations of the researcher. Rather, the reader immerses him/herself in the case and develops his/her own interpretations of it. So, when we position naturalistic generalizations alongside nomothetic generalizations, we begin to see that knowledge should not be limited to traditional forms of causation. Instead, knowledge of the particular, knowledge gained via different methodological approaches, is valuable to human understanding of social phenomena.

6   Conclusion and Future Directions

The case study has emerged as an important methodological contributor. It allows for rich theoretical insights that are transferable over time and place. While typically considered a qualitative method, its flexibility and utility have enabled its use in both qualitative and quantitative studies. Such advantages of case study research have made it quite a favorable approach to social science and public health researchers.

While the advantages of the case study are numerous, it has also come under scrutiny, notably in terms generalization. While we must recognize that case studies, notably single cases, are not suitable for traditional (nomothetic) types of generalizations, knowledge can still be transferred even when this knowledge is not generalized in a formal sense (Flyvberg 2011).

Here, we must consider the need to challenge the traditional and rather orthodox thinking entrenched in the way we understand the utility of methods such as the case study. The dominance of natural sciences and the often uncontested approaches to scientific development they espouse have at times hindered the progress of methods beyond the strict confines of empiricism. By contesting such traditional ways of how research *ought to be* conducted, we can allow for and also encourage different types of generalizations as ways to understand that which we seek to examine. One such approach that was canvassed in this chapter was naturalistic generalizations.

We cannot ignore the power of naturalistic generalizations, not as a replacement for nomothetic forms of generalizations but what Stake (1980, p. 2) avers to in his response to criticism by Hamilton (1979), "primarily a creation of [ones's] own experimental knowing." This unorthodox approach to generalization allows the researcher to generalize according to the similarities and differences specific to their own experiences, a process Simons et al. (2003, p. 360), have referred to as "individual recipient judgement."

Naturalistic generalizations are particularly useful to researchers who want to garner the personal and vicarious experiences of their readers. They emphasize a practicality and functionality of research results that draw directly on the experiences of the reader. These accounts are completely subjective and, thus, offer a powerful way to understand the particulars and intricacies of a case directly from the readers' perspective.

A future direction for case study research specific to naturalistic generalizations is that consideration must be given to researcher reflexivity to adequately acknowledge what Bartlett and Vavrus (2017, p. 34) state as the "politics of representation." We cannot ignore that the researcher does play an important role in naturalistic generalizations principally in relation to how information is intended to be absorbed by the reader and how this information is initially represented in the study. This will allow for a more nuanced way to involve naturalistic generalizations in case study research that provides both clarity and openness.

References

Allison GT. Conceptual models and the Cuban missile crisis. Am Polit Sci Rev. 1969;63(3):689–718.

Bartlett L, Vavrus F. Rethinking case study research. New York: Routledge; 2017.

Donmoyer R. Generalizability and the single-case study. In: Gomm R, Hammersley M, Foster P, editors. Case study method- key issues, key texts. London: Sage; 2000. p. 45–68.

Duff PA. Case study research on language learning and use. Ann Rev Appl Linguist. 2014;34:233–55.

Eckstein H. Case study in theory and political science. In: Greenstein FJ, Polsby NW, editors. Handbook of political science. Reading: Addison-Wesley; 1975. p. 79–137.

Eisenhardt KM. Building theories from case study research. Acad Manag Rev. 1989;14(4):532–50.

Elman C, Gerring J, Mahoney J. Case study research: putting the quant into the qual. Sociol Methods Res. 2016;45(3):375–91.

Exworthy E, Peckham S. Case studies in health policy: an introduction. In: Exworthy E, Peckham S, Powell M, Hann A, editors. Shaping health policy: case study methods and analysis. Bristol: The Policy Press; 2012. p. 3–20.

Flyvberg B. Five misunderstanding about case-study research. Qual Inq. 2006;12(2):219–45.

Flyvberg B. Case study. In: Denzin NK, Lincoln YS, editors. The Sage handbook of qualitative research. 4th ed. Thousand Oaks: Sage; 2011. p. 301–16.

Gerring J. The conundrum of the case study. In: Gerring J, editor. Case study research: principles and practices. New York: Cambridge University Press; 2006. p. 1–13.

Gerring J. Case study research- principles and practices. New York: Cambridge University Press; 2007.

Hamel J. Case study methods. Newbury Park: Sage; 1993.

Hamilton D. A science of the singular. Urbana: CIRCE, University of Illinois; 1976.

Hamilton D. Some more on fieldwork, natural languages and naturalistic generalisation (mimeo), discussion paper. University of Glasgow; 1979.

Hellström T. Transferability and naturalistic generalization: new generalizability concepts for social science or old wine in new bottles. Qual Quant. 2008;42:321–37.

Lincoln YS, Guba EG. The only generalization is: there is no generalization'. In: Gomm R, Hammersley M, Foster P, editors. Case study method key issues, key texts. London: Sage; 2000. p. 27–44.

MacDonald B, Walker R. Case study and the social philosophy of educational research. Camb J Educ. 1975;5:2–11.

Marinetto M. Case studies of the health policy process: a methodological introduction. In: Exworthy E, Peckham S, Powell M, Hann A, editors. Shaping health policy: case study methods and analysis. Bristol: The Policy Press; 2012. p. 21–40.

McLaren G. The wild food challenge: a case study of a self-initiated experiential education project. Graham: Prescot College, Proquest; 2015.

Melrose S. Naturalistic generalization. In: Mills AJ, Durepos G, Wiebe E, editors. Encyclopedia of case study research. Thousand Oaks: Sage; 2009.

Merriam SB. Case study research in education- a qualitative approach. San Francisco: Jossey-Bass Inc; 1988.

Mills AJ, Durepos G, Wiebe E, editors. Encyclopedia of case study research, vol. 1. Thousand Oaks: Sage; 2010.

Patton E, Appelbaum SH. The case for case studies in management research. Manag Res News. 2003;26(5):60–71.

Polanyi M. Personal knowledge. New York: Harper and Row; 1958.

Popper K. The logic of scientific discovery. New York: Basic Books; 1959.

Ragin CC. "Casing" and the process of social inquiry. In: Ragin CC, Becker HS, editors. What is a case? Exploring the foundations of social inquiry. Cambridge: Cambridge University Press; 1992. p. 217–26.

Schwandt TA, Gates EF. Case study methodology. In: Denzin NK, Lincoln YS, editors. The Sage handbook of qualitative research. 5th ed. Thousand Oaks: Sage; 2018. p. 341–58.

Seawright J, Gerring J. Case selection techniques in case study research- a menu of qualitative and quantitative options. Polit Res Q. 2008;61(2):294–308.

Simons H. The paradox of case study. Camb J Educ. 1986;26(2):225–40.

Simons H. Preface. In: Russell J, Greenhalgh T, Kushner S, editors. Case study evaluation: past, present and future challenges. Bingley: Emerald Group Publishing Limited; 2015. p. ix–xvi.

Simons H, Kushner S, Jones K, James D. From evidence-based practice to practice-based evidence: the idea of situated generalisation. Res Pap Educ. 2003;18(4):347–64.

Skar L, Prellwitz M. Participation in play activities: a single-case study focusing on a child with obesity experiences. Scand J Caring Sci. 2008;22(2):211–9.

Stake ER. The case study method in social inquiry. Educ Res. 1978;7(2):5–8.

Stake ER. Generalizations. Paper presented at the annual meeting of the American Education Association, Boston. 1980

Stake ER. The art of case study research. London: Sage; 1995.

Stake RE. The case study method in social inquiry. In: Gomm R, Hammersley M, Foster P, editors. Case study method key issues, key texts. London: Sage; 2000. p. 19–26.

Stake RE. Qualitative case studies. In: Denzin NY, Lincoln YS, editors. The Sage handbook of qualitative research. Thousand Oaks: Sage; 2005. p. 443–66.

Stake RE. Multiple case study analysis. New York: The Guilford Press; 2006.

Stake ER, Trumbull D. Naturalistic generalizations. Rev J Philos Soc Sci. 1982;7(1):1–12.

Stoecker R. Evaluating and rethinking the case study. Sociol Rev. 1991;39(1):88–112.

Swanborn PG. Case study research- what, why and how? London: Sage; 2010.

Thomas G. A typology for the case study in social science following a review of definition, discourse, and structure. Qual Inq. 2011;17(6):511–21.

Whyte WF. Street corner society: the social structure of an Italian slum (original work published in 1943). 2nd ed. Chicago: University of Chicago Press; 1955.

Yeh LS, Hedgespeth J. A multiple case study comparison of normative private preparatory school and substance abusing/mood disordered adolescents and their families. Adolescence. 1995;118:413–28.

Yin RK. Case study research: design and methods. Beverly Hills: Sage; 1984.

Yin RK. Case study research: designs and methods. 4th ed. Thousand Oaks: Sage; 2009.

Yin RK. Case study research: designs and methods. 5th ed. Thousand Oaks: Sage; 2014.

Evaluation Research in Public Health

20

Angela J. Dawson

Contents

Abstract

Evaluation research is concerned with assessing the merit of health projects and programs and produces information for decision-making to improve public health. Evaluation results are critical to continuous quality improvement efforts, building organizational capacity to respond to health needs and ensuring the accountable and efficient use of resources. This chapter will introduce evaluation research to assess the outcomes of health programs and policy. The key characteristics and principles of evaluation will be examined, and the range of

A. J. Dawson (✉)
Australian Centre for Public and Population Health Research, University of Technology Sydney, Sydney, NSW, Australia
e-mail: angela.dawson@uts.edu.au

© Springer Nature Singapore Pte Ltd. 2019
P. Liamputtong (ed.), *Handbook of Research Methods in Health Social Sciences*,
https://doi.org/10.1007/978-981-10-5251-4_71

approaches can be taken in this applied area of research. Examples of process, outcome, and impact evaluation in health contexts will enable readers to:

1. Discuss approaches to evaluation using logic models and theories of change
2. Examine program/project evaluation designs to assess methodological rigor and appropriateness
3. Apply knowledge of global/national/state strategies and public health evidence to guide the development of evaluation indicators
4. Examine the culturally appropriate and ethically sound approaches in evaluation

Keywords

Program evaluation · Theory-based evaluation · Theory of change · Logical frameworks · Results-based management · Evaluation indicators · Gender-sensitive evaluation

1   Introduction

Research and evaluation are often portrayed as a dichotomy, which is not always helpful because evaluation always employs research and, therefore, evaluations are a type of research activity with different timelines and aims. Evaluation research in public health contexts is concerned with assessing the merit of a public health project or a program and produces information for decision-making. These decisions are normally about whether the intervention or set of organized activities that comprise a program should continue to be funded modified or scaled up.

Evaluation research differs from implementation research, clinical efficacy research, and operations research. Table 1 provides an overview of the features of different approach to research including evaluation research.

While often the focus of evaluation research is to improve, it can also be employed to prove that the intervention is in fact responsible for change. Delivering results for and reporting to stakeholders are features of evaluation research that is conducted with the intent to serve the information needs of stakeholders rather than curiosity-driven research. The purpose of evaluation research is, therefore, pragmatic (Patton 2008) and is part of programmatic work often comprising twenty percent or less of the resources.

Evaluation research involves the use of both qualitative and quantitative research methods and methodologies. The study design can be descriptive or experimental, while the focus can be on the effectiveness or efficiency of an intervention and/or understanding the mechanisms that help to support its implementation. According to Habicht et al. (1999, p. 11), evaluations are conducted to determine "plausibility, probability, or adequacy" of interventions. However, all evaluation research in the field of health is applied and part of a cycle of planning, implementing, and assessing interventions that focus on changing people lives including the realization of their rights and improving health outcomes. This may also involve the evaluation of

Table 1 Overview of the features of different types of research

Type of research	Evaluation research	Implementation research	Translational research	Clinical efficacy research	Operations research
Characteristics					
Assess a program implementation	✓				
Assess a program effect	✓				
Identify factors that facilitate implementation effectiveness		✓			
Develop strategies to achieve effective implementation		✓			
How can evidence be applied in practice to affect health outcomes			✓		
Examine how a therapy works on a health outcome				✓	
Construct data-based models for decision-making					✓

behavioral change and institutional change including the organization of components of health systems requiring operational change.

Learning is a key feature of evaluation research described by the European Union (2013, p. 17) as a process of learning through systematic enquiry what public programs and policies have achieved and understand how they perform to better design, implement and deliver future programs and policies.

2   Theory-Based Evaluation

Underpinning all evaluation research is a theory or a conceptual analytical model that provides a way of structuring analysis in an evaluation. A theory is a collection of assumptions and hypotheses that are empirically testable or that are logically connected. In the literature, theory-based evaluation can be found as early as the 1930s (Coryn et al. 2011) and was further developed by key figures such as Chen (1990) and Weiss (1995). Today, theory-based evaluation is commonplace and an integral part of local, national, and international public health practice.

In line with an evidence-based approach to quality public health, we must ensure that our programs are underpinned and guided by principles of public health programming and that evaluation is not an ad hoc enterprise. Theory helps enhance our understanding of complex situations taking into consideration specific contextual factors. Two types of theory can be identified:

- Explanatory theory that helps to identify factors that a health program might try to change
- Change theory that helps us to develop range of intervention strategies to address correct variables in appropriate combination with appropriate emphasis and in evaluation to assess whether all the right components are in place

Theory, therefore, provides a meaningful way for framing or prioritizing evaluation questions. It also provides a guide to the design and execution of the evaluation as well as the interpretation and application of the reported findings. An underpinning theory also allows programs to be generalizable to the larger population and/or transferable to other similar contexts by identifying successful elements and outcomes that can be predicted or anticipated enabling an understanding of what works and why.

A number of organizations including the expert consensus process undertaken by the Agency for Healthcare Research and Quality in the UK (Foy et al. 2011) have called for evaluation research to be integrated into the health program structure from the beginning of the planning phase to build understanding of change. This enables the team to identify which outcomes are key to the program's success and select which ones should be the focus of the evaluation.

Theory in evaluation is often driven by evaluation practice, and many of the theories used have been found to been unsubstantiated by empirical studies (Coryn et al. 2011). Despite this, theory is important to the structure, planning, design, and implementation of the program and execution of the evaluation, and more research is required to deliver exemplars of theory use in evaluation practice.

2.1   The Theory of Change (ToC)

The theory of change (ToC) approach in evaluation is underpinned by concepts of "how and why the program will work" (Weiss 1995, p. 66) and is widespread in public health evaluations (Breuer et al. 2016). ToC as a term in evaluation emerged from social change movements and the work of the Aspen Institute on Community Change. Weiss, who was a key member of this group, described the need to articulate the assumptions upon which each of the steps in a program is based in order to make the change process explicit. ToC is a causal model that explains the complexity of this change by revealing the conceptual framework that explains the causal relationships between program activities and the immediate, short-term and long-term outcomes.

Evaluation theory, therefore, seeks to determine what changes have taken place at each level goal being change at many levels:

- Changes in people's lives such as the achievement of their rights and improvements in health status
- Change in the culture and organization of institutions including their values, the services they provide, legal status, and their performance
- Changes in behavior such as attitudes and practices
- Change in the ways in which products and services are delivered involving improvements in knowledge and skills and cost- and time effectiveness

Despite there being a lack of a definition of what ToC is, there is agreement on the important considerations that comprise a ToC (Vogel 2012). These considerations include an explanation of the:

- Context of the initiative, i.e., the sociocultural, political, and environmental conditions, the current state of the problem the initiative is aiming to influence
- Long-term change or impact that the initiative is aiming for
- Process or stages of change expected that will lead to the desired long-term change
- Assumptions about how these changes might occur
- Outputs that are conducive to the desired change in the specific context
- Diagrammatic summary that outlines the change

The process of developing a ToC is usually collaborative and begins with establishing what the far-reaching outcomes or impact will be as the result of a program that are often expressed in terms of the health or social impact (see Fig. 1). This is then mapped to what can be achieved in a long term such as changes in the health outcomes of a defined population and then to the immediate effects of the program upon the beneficiaries themselves. The assumptions or preconditions required to achieve the desired change at each stage are laid bare in a ToC including the contextual factors that may influence these necessary preconditions. The ToC development may also include the design of indicators to assess the change achieved through the program implementation and the evidence required to verify this.

There is considerable literature to guide public health practitioners to develop their own theory of change. This includes guidance from the United Nations (Rogers 2014), philanthropic foundations (Reisman et al. 2004), universities (Taplin et al. 2013; University of Kansas 2017), community organizations (Australian Communities Foundation 2015), and networks (De Silva et al. 2014a).

Theories of change are usually expressed graphically and in a temporal fashion from left to right. Outcomes are noted along the hypothesized causal pathway that is required to achieve the anticipated impact. There are a number of examples in the literature of these diagrams including some in the area of mental health: a theory of change for peer counseling for maternal depression in Goa, India

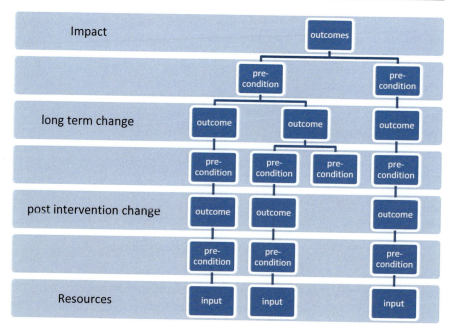

Fig. 1 Theory of change (ToC) mapping

(De Silva et al. 2014a), the program for improving mental health care cross-country summary theory of change (Breuer et al. 2015), ToC approach to develop a mental health care in a rural district in Ethiopia (Hailemariam et al. 2015), and in adolescent health (Van Belle et al. 2010; Weitzman et al. 2002). A worked example is provided (see Fig. 2) from the community case management (CCM) project in Indonesia. CCM is a community-based service delivery model designed to address childhood illnesses such as diarrhea, pneumonia, and malaria, particularly in resource-poor settings (Marsh et al. 2012; Setiawan et al. 2016). Here, readers can see the interventions as they pertain to political buy-in, resourcing, and capacity building and the effect upon treatment and care outcomes, service use, health status, and costs.

While Connell and Kubisch (1998) call for credible, achievable, and testable theories of change, Breuer et al. (2016) have developed a useful framework that can be used to report on ToCs in public health evaluations. This consists of four elements outlined below that serve to guide those wishing to develop their own ToC.

- Clear definition of the ToC
- Description of the ToC development process (methods including stakeholder involvement)
- Summary of ToC in diagrammatic form
- Mapping of the ToC to the evaluation questions, indicators used for assessing the program's success, methods of data collection analysis, and data interpretation at various time points including during and after the program implementation.

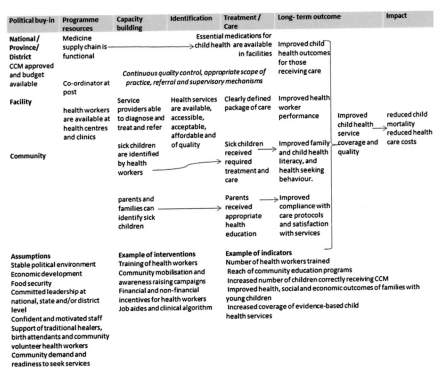

Political buy-in	Programme resources	Capacity building	Identification	Treatment / Care	Long-term outcome	Impact
National / Province/ District CCM approved and budget available	Medicine supply chain is functional		→child health	Essential medications for are available in facilities	Improved child health outcomes for those receiving care	
Facility	Co-ordinator at post	*Continuous quality control, appropriate scope of practice, referral and supervisory mechanisms*				
	health workers are available at health centres and clinics	Service providers able to diagnose and treat and refer	Health services are available, accessible, acceptable, affordable and of quality	Clearly defined package of care	Improved health worker performance	Improved child health service coverage and quality
Community		sick children are identified by health workers		Sick children received required treatment and care	Improved family and child health literacy, and health seeking behaviour.	reduced child mortality reduced health care costs
		parents and families can identify sick children		Parents received appropriate health education	→ Improved compliance with care protocols and satisfaction with services	

Assumptions	Example of interventions	Example of indicators
Stable political environment	Training of health workers	Number of health workers trained
Economic development	Community mobilisation and	Reach of community education programs
Food security	awareness raising campaigns	Increased number of children correctly receiving CCM
Committed leadership at	Financial and non-financial	Improved health, social and economic outcomes of families with
national, state and/or district	incentives for health workers	young children
level	Job aides and clinical algorithm	Increased coverage of evidence-based child
Confident and motivated staff		health services
Support of traditional healers,		
birth attendants and community		
volunteer health workers		
Community demand and		
readiness to seek services		

Fig. 2 Theory of change for community case management in Indonesia

2.2   ToC and Classic Change Theories

Theories of change, however, are not rooted in one philosophical traditional; they are pragmatic and can be strengthened by adding theories such as those from sociology or psychology. These theories can be inserted to explain change at various levels and at selected time points either before, during, and after the program implementation.

Some theories focus on understanding the individual factors that influence health behavior, such as knowledge, attitudes, beliefs, and personality traits. For example, Ramsey and colleagues (Ramsay et al. 2010) have used the theory of planned behavior to examine the implementation of a knowledge translation intervention to improve the diagnostic test requesting behavior of general practitioners. A specially designed survey was used to gauge how the intervention affected the attitudes of GPs toward requesting certain tests, their beliefs about others behavior, and perceptions of how easy or difficult it would be to undertake a new regime including the associated contextual factors that would hinder of facilitate this change (Ramsay et al. 2010). Other theories help to clarify processes between individuals and groups such as family, friends, peers, and colleagues to explain social identity, support, and roles. A post-implementation evaluation of a workplace educational program to promote exercise (Amaya and Petosa 2012) used a survey based quasi-experimental design to show the effect of learning by observing others in a social context.

Evaluations that examine changes in communities can use theory to understand how organizational factors such as rules, regulations, and policies affect health or the effect of social norms and networks. The diffusion of innovations theory has been applied in an evaluation of the dissemination of best practice guidelines in substance abuse treatment. The evaluation mapped the effect and rate of the uptake of the guidelines through social networks on health professional knowledge and awareness of the guidelines, how persuaded they were to change their practice, decisions taken toward change, and implementation of the guidelines in services (Hubbard and Hayashi 2003).

Finally, ecological theories attempt to understand the multiple levels and see change in health behaviors, care, services, and policy in terms of a complex system of interrelated factors. The California Healthy Cities evaluation framework sought to measure change at five levels: individual, civic participation, organizational, interorganizational, and community (Kegler et al. 2000). Bauer (1999) employed an ecological model of community organizing to evaluate a capacity and advocacy initiative for residents to impact on public health policy and training of public health professionals.

2.3   Realistic Evaluation Theory

Realistic evaluation is concerned with an examination of the underlying mechanisms and contextual factors that trigger change (Pawson and Tilley 1997). Many evaluation studies have developed a model of change based upon realistic theory to explain what aspects of the intervention bring about change, the extent of this, and the associated contextual circumstances. In Australia, Schierhout et al.'s (2013) evaluation of a continuous quality improvement process in Indigenous health services was able to identify what worked from whom and in what contexts. Similarly Byng et al.'s (2008) evaluation of a multifaceted intervention to improve the care of people with long-term mental illness was able to develop a context-specific, mechanism-based explanations for health-care effectiveness. Realistic evaluation is an iterative process that gradually reveals patterns of outcomes to determine how the program works rather than a focus on what worked.

3   Frameworks to Guide Evaluation Research

3.1   Logic Models and Results-Based Frameworks

Logic models and the more a detailed form known as the logical framework or the logical framework approach (LFA) are tools designed to plan and evaluate programs and describe the goals and resources of an initiative or organization. These tools give less attention to the complex political, sociocultural, economic, and organizational processes that underpin change in health and health care; rather they focus on the implementation of a program. LFAs are useful to plan evaluations and employed as a metric to understand the aims, plan methods, and indicators for measurement. Theory can be added to strengthen the explanation. Figure 3 lists the logic levels alongside examples of evaluation questions, the indicators employed to measure success, the means through which these indicators are verified, and the underpinning assumptions upon which this change is based.

Spearheaded by USAID, the logic framework (LF) was adopted by many donor agencies and applied across international health settings. In the late 1990s, the UN system adopted the results-based management (RBM) approach in its major agencies. RBM evaluation grew out of the logical framework approach and is a management strategy that focuses on defining results based on appropriate analyses, monitoring progress, identifying and managing risks, capturing lessons learned, and reporting on results achieved and resources involved. The WHO now employs a results framework to monitor the implementation of the organization's program budget, activities, and outputs against its performance according to the achievement of the sustainable development goals (WHO 2017). This approach identifies the monitoring and results-based evaluation phases as well as the responsibility of the WHO Secretariat and member states and partners for accountability and results.

While the diagram at Fig. 4 represents one chain, programs are made of multiple chains that require evaluation. RBM is composed of a series of results chains (see Fig. 5) that, like a logic model, is a simplified picture of an intervention designed in

Logic model	Logic framework evaluation questions	Performance Indicators	Means of Verification	Assumptions
Impact	**Goal** To what extent have unplanned pregnancies been reduced?	Measures of goal Achievement used for evaluation	Various sources of information; methods used	Assumptions concerning Goal-purpose linkages
Outcome	**Purpose/overall objective** What increase is there in the use of family planning (FP)?	End-of-project status –to assess purpose achieved Used for project completion and evaluation	Various sources of information; methods used	Assumptions concerning the purpose/goal linkage.
	Component objectives 1. How has knowledge of FP been increased? Is there an increase in the acceptance of FP services? 2. Has the quality of FP counselling and services improved?	Measures of the extent to which component objectives have been achieved for review and evaluation.	Various sources of information; methods used	Assumptions concerning the component objective/purpose linkage.
Output	**Output /results** 1. Increased availability of educational materials 2. Improved FP supervisory system	Measures of the quantity and quality of outputs and the timing of their delivery. Used for monitoring and review.	Various sources of information; Methods used	Assumptions concerning the output/component objective linkage
Process	**Activities** 1. Community mobilization activities 2. Mass media campaign 3. Train health workers 4. Quality improvement process	Implementation/work program targets. Used during monitoring.	Project data, other sources of information	Assumptions concerning the activity/output linkage.
Input	**Inputs/ Resources** Money staff, time, political support			

Fig. 3 Logic framework for the evaluation of a community-based family planning program

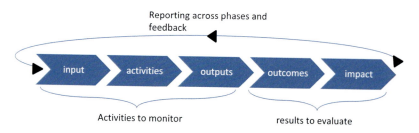

Fig. 4 Results-based management approach to evaluation research

response to a health issue or problem and articulates the logical relationships between the resources invested, the activities, and the stages of changes that result, also known as impact.

The logic model approach and RBM have been criticized for being too focused on a top-down and linear approach that minimizes the characteristics and expertise of people and the interaction of contextual factors on change. However, the strength of this approach is the articulation of the causal connections between conditions that need to change to reach the impact goal. A theory of change can express the assumptions that underpin the results framework.

3.2   Other Frameworks to Guide Evaluation

In the literature, there are many other conceptual models and frameworks that can guide evaluation. The PRECEDE-PROCEED (Predisposing, Reinforcing and Enabling Constructs in Educational Diagnosis and Evaluation-Policy, Regulatory, and Organizational Constructs in Educational and Environmental Development) model was designed for health promotion planning, and evaluation (Green and Kreuter 2015) has been employed in many public health interventions to evaluate workplace interventions (Post et al. 2015) to individual chronic disease programs (Azar et al. 2017). The Re-Aim (Reach, Effectiveness, Adoption, Implementation, Maintenance) framework (Glasgow et al. 1999) is another useful tool to structure evaluations of individual (Belkora et al. 2015) and community (Jenkinson et al. 2012) and partnership (Sweet et al. 2014) initiatives.

The Centers for Disease Control and Prevention in the United States has developed a Framework for Program Evaluation in Public Health (CDC 1999). This framework summarizes the key elements of evaluation and proposes a six-stage cycle comprised of engaging stakeholders, articulating the program and the evaluation design, gathering credible evidence, justifying the conclusions reached, and sharing lessons learned. This is coupled with standards for effective program evaluation that have been applied in public health disease control programs (Logan et al. 2003).

More recent frameworks include Proctor et al.'s (2011) eight conceptually distinct outcomes for potential evaluation: acceptability, adoption (also referred to as uptake), appropriateness, costs, feasibility, fidelity, penetration (integration of a

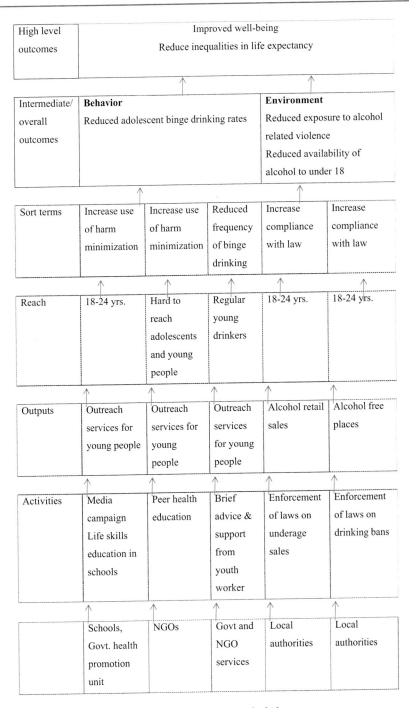

Fig. 5 Results chains for an evaluation of an adolescent alcohol program

practice within a specific setting), and sustainability (also referred to as maintenance or institutionalization). This has been largely applied in implementation research such as the population-based care program for those at risk for delirium, alcohol withdrawal, and suicide harm (Lakatos et al. 2015). Finally, another potentially useful approach to evaluation design is ten steps to making evaluation matter outlined by Sridharan and Nakaima (2011) that add considerations from the realist tradition including sustainability and learning considerations.

4   Purpose and Phases of Evaluation Research

Evaluation may be shaped by the purpose for which it is designed as well as the time frame in which it is executed. As Habicht et al. (1999) suggest, the purpose of an evaluation research can be to establish plausibility, adequacy, or probability. If the aim is plausibility, then the focus will be on designing the evaluation to reveal best how a program achieved its expected objectives and that the change that occurred during the process can potentially be attributed to the program activities. If the aim is to determine adequacy, then this will be an evaluation that seeks to establish if the program goals were achieved. However, an evaluation with the goal of determining probability will most likely employ an experimental design to demonstrate that improved health outcomes or impact is directly attributed to the program activities.

In addition to this, there are several phases or stages of a program implementation where evaluation research can be undertaken as outlined in Fig. 6. This can proceed the design and implementation of a health program or intervention so that baseline data can be collected to not only inform the design of the intervention but also provide a yardstick for measuring change. The next phase of evaluation might involve piloting or testing aspects of the intervention to ensure feasibility, appropriateness, or fit. This process may involve some modification of the intervention and provide additional baseline data. Implementation or process evaluation is known as "real-time" evaluation and involves the regular collection and reporting of information to track whether activities are being implemented and immediate results are achieved as planned (Moore et al. 2015). Theory can be useful to structure this evaluation (Ramsay et al. 2010). Post-implementation reviews take place immediately after rather than during the implementation of an intervention, while outcome evaluation and impact evaluation map short-term and longer-term change, respectively. Outcome and impact evaluation are often termed summative evaluation and

Time

Needs assessment/ baseline assessment	Feasibility formative/pilot assessment and review	Implementation evaluation or process evaluation	Post-implementation review	Outcome evaluation	Impact evaluation

Fig. 6 Phases or stages where evaluation research can be undertaken

aim to answer specific questions about performance of the activities. They are concerned with answering how and why questions linked to plausibility or causality.

There are a number of useful guides to these various types of evaluations in public health contexts provided by departments of health (ACI 2013), international non-government organizations (IFRC 2011), and the United Nations (UNDP 2009; WHO 2013).

Different types of evaluations may be undertaken across these phases that draw on both qualitative and qualitative evidence. Pre-intervention evaluations may comprise assessments of health needs that involve surveys of or interviews with community members or from existing statistical health data, desk reviews of existing reports and policy documentation, or financial audit and risk assessments of the context into which a program or policy may be implemented. Economic evaluations including cost-effectiveness assessment and cost-benefit analysis can be undertaken across all phases alongside quantitative analysis and qualitative evaluations involving observations of behavior, key informant interventions, and participatory processes.

5 Evaluation Research Designs

Selecting the study design for an evaluation depends on the purpose of the evaluation. The purpose will determine the stage or phase where evaluation activities are carried out and the type of evaluation. For example, an evaluation that aims to understand whether the budget was allocated effectively or the performance of health professionals during the implementation of a program may involve systems to monitor the finances or standards over a specific time frame. Other evaluation activities might involve an examination of changes in knowledge or behaviors such as the uptake of contraception. These activities could be part of a process evaluation and employ a quasi-experimental pre-and post-intervention design. Such activities contrast with experimental longitudinal designs where causal links are sought to identify if the program demonstrated an impact on health outcomes of the beneficiaries or the larger population. Impact and outcome evaluation may also involve mixed methods combining, for example, ethnography involving the data collection from in-depth interviews and observation with survey- and/or population-based surveillance data. Table 2 identifies some characteristics and examples of experimental, qualitative, and mixed methods evaluation designs in public health. However, it is possible that an evaluation of a program could be comprised of all or some of these designs and methodologies.

6 Evaluation Indicators

An indicator is a variable that provides accurate and reliable evidence about the achievement of a specific result. Indicators should be observable, well-defined, measurable, and agreed upon. They can be both qualitative and quantitative and are at all levels of the program logic or results chain. Indicators that make up a process evaluation usually involve the regular collection and reporting of data to

Table 2 Study designs and methodologies for evaluation research

Study design	Explanation and example
Experimental and quasi-experimental evaluation designs	
Randomized control trial	The health program's impact is the outcome of interest. Common form involves one group being randomly assigned to receive the intervention, and the other receives no intervention or usual treatment (see also ▶ Chap. 37, "Randomized Controlled Trials"). Useful when intervention is introduced in small population in highly structured manner, see in the case of the evaluation of a mindfulness program (Hou et al. 2014). Limited by high resource implications and does not necessarily reflect how interventions will work beyond the experiment
Quasi-experimental, comparison group design	May involve a study of a group before and after receiving an intervention. A comparison group could be included. See an example in the evaluation of an urban health initiatives (Weitzman et al. 2002)
Economic evaluation	Statistical measurement of the inputs and outcomes of an evaluation to examine the costs and consequences of an initiative. Sinha et al. (2017) undertook a cost-benefit analysis of a program involving women's groups facilitated by community workers to reduce neonatal mortality in rural India.
Qualitative evaluation	
Ethnography	This methodology involves the study of culture using observation, in-depth interview, and field notes. It involves the researcher spending long periods in the field studying knowledge systems of groups of people (see ▶ Chaps. 13, "Critical Ethnography in Public Health: Politicizing Culture and Politicizing Methodology," ▶ 26, "Ethnographic Method," and ▶ 27, "Institutional Ethnography"). Ethnography has been applied in the formative evaluation of infant feeding initiatives (Young and Tuthill 2017)
Mixed methods	
Participatory evaluation	An approach that engages stakeholders in design, planning, and undertaking the evaluation with the goal of improving skills and ensuring more responsive health care and services (see also ▶ Chap. 17, "Community-Based Participatory Action Research"). One example from mental health involves consultation with consumers, community people, and providers to contextualize and validate the findings from case studies (Lea et al. 2015)
Realist evaluation	Theory-driven evaluation to determine the contextual mechanisms that enable the successful achievement of program outcomes. Qualitative and quantitative approaches are employed according to what best answers the questions. Pragmatic design visible in an evaluation of continuous quality improvement in primary health care (Schierhout et al. 2013)
Developmental evaluation	An approach that is responsive to context by allowing constant adaption and enables the gathering of real-time data. Suits complex situations, for example, the evaluation of social change in communities (Patton et al. 2016)

monitor whether results are being realized as planned and to identify problem areas and possible solutions. Such indicators are often found in processes of continuous quality improvement efforts and require an operational definition.

Indicators focused on assessing the achievement of results in outcome, and impact evaluations are analytical efforts to answer specific questions about performance of program activities. There are generally concerned with answering questions concerning why the intended outcomes were or were not realized and how the results were achieved. Such indicators are designed to determine the probability of a program to health and social outcomes over time or the causal contributions of activities to results to confirm a hypothesis.

In Table 3, we can see that the evaluation questions outline in Fig. 3 have been formed into objectives that have been further qualified by indicators across the various evaluation levels. These indicators relate to the provision, utilization, coverage, and impact of health services as well as the legal and social environment. Other indicators can include:

- Improved health outcomes
- Increased use of health facilities
- Extension of quality health services
- Development of human resources for health
- Improved legal environment
- Achieve gender equality

Other indicators could include:

- Improved economic productivity.
- Improved social capital that includes the use of social networks to improve health; this includes the facilitation of cooperation and mutually supportive relations in communities to reduce social isolation, improve well-being and harness the skills and talents of individual, and increase access to employment and education opportunities.
- Improved cultural capital education (knowledge and skills) that provides advantage in achieving a higher social status in society (Table 3).

There is considerable generic guidance on developing quality indicators for evaluation in general; they should be valid, reliable, precise, timely, and comparable. Table 4 defines these attributes using indicators from a family planning evaluation as an example.

7   Considerations in the Development of Indicators

One of the issues of evaluation research is ensuring that everyone involved is applying the same assessment framework to the measurement of outcomes. An operational definition of each indicator is, therefore, required so that those

Table 3 Examples of evaluation objectives and indictors at impact, outcome, and impact levels

Impact evaluation	Impact evaluation indicator
Impact objective Reduce adolescent fertility	Adolescent birth rate (aged 10–14 years; aged 15–19 years) per 1,000 women in that age group reduced by three quarters in country x by 2030
Outcome evaluation	**Outcome evaluation indicator**
Overall outcome evaluation objective 1 Increase adolescent use of modern methods of contraception	Contraceptive prevalence rate in province X increased by x
Overall outcome evaluation objective 2 Improved social and policy environment for contraception and sexual and reproductive health and rights	Institution of laws and regulations that guarantee women aged 15–19 years access to sexual and reproductive health care, information, and education
Component 1 outcome evaluation objective Increased uptake of adolescent contraception services	% of new clients and return of clients
Component 2 outcome evaluation objective Improved quality of contraceptive counseling and services for adolescents	% of sites adhering to adolescent friendly standards
Component 3 outcome evaluation objective Increased access to contraception services	% satisfaction
Component 4 outcome evaluation objective Increase availability of contraceptive commodities	% of functional procurement and distribution in the supply chain
Component 5 outcome evaluation objective Increase in female adolescent reproductive health decision-making	Proportion of female adolescents who make their own informed decisions regarding sexual relationships, contraceptive use, and reproductive health care
Output evaluation	
Component 1 output evaluation objective Increased adolescent knowledge and acceptance of modern methods of contraception and service location	% of adolescents with knowledge of available services and commodities Positive attitudes toward contraception and increased expressed demand
Component 2 output evaluation objective Improved health workers contraceptive counseling skills	% of staff trained and assessed as competent
Component 3 output evaluation objective Appropriate clinic opening hours, timeliness of consultation, and appropriate staffing numbers	% of facilities with minimum staffing norms (List of minimum staffing defined)
Component 4 output evaluation objective Health centers stocked with low-cost essential RH commodities	% of facilities without 7-day stock outs of essential drugs (list of essential drugs defined)
Component 5 output evaluation objective Increase in contraception services at health clinics	% of services delivering evidence-based contraceptive services, care, and information to adolescents

Table 4 Attributes of quality indicators

Attribute	Example of indicator
Valid	Participants will recall/describe at least three modern methods of family planning
Reliable	The indicator above could be used and classified as reliable if in pretesting different people (interviewer and participants) demonstrated a consistent understanding of the term "modern." If not, then validity may be affected since different people may understand different methods as modern
Precise	The indicator must be able to be clearly defined. In this case, a predefined list of modern family planning methods should be able to be produced. The indicator must be precise so that the answers can be clearly assessed
Timely	Change in this indicator could be expected to be within a short time frame. However, if the evaluation sought to measure change in family size in a 2–3-year project, it will not be possible to observe such an indicator within the time frame of the project
Comparable	Knowledge of three modern family planning methods should be comparable across various populations. It should be straightforward to make a comparison between men's and women's knowledge of contraception. However, if an intervention-specific indicator was selected, for example, if we wanted to know how many modern methods that adolescents who are peer health educators can list, this is only useful for that group of people but could not be applied to other groups

involved in collecting data can assess the achievement of the indicator in a standard manner. This also requires that the evaluation design is rigorous and aligned with best practice efforts that provide comparable data on changes over time. A protocol is also required to guide data collection, as well as standard tools to collect such data. Piloting or testing indicators in the field with the proposed data gathering tools is useful to ensure that all issues can be addressed before the rollout.

Another area to consider when developing indicators is how they might best connect with existing measures and could be integrated across the health system to provide a useful picture of change. Indicator designs can, therefore, benefit from being aligned with global/national/state strategies and public health evidence. This enables comparability, and although they may need to be field-tested for the unique context of your evaluation, they will already be quite sturdy. For example, countries may already have goals and measures by which they would like to reduce the adolescent fertility rate in line with their Sustainable Development Goals (SDG) targets and measures. This indicator could be inserted at Table 2 to specific the impact evaluation objective and indicator. Other SDG target and goals may also be relevant here such as existing country indicators to achieve gender equality and empower all women and girls.

Some indicators such as the measurement of community participation will require extensive consultation to ensure that what is measured is appropriate and sound. For example, several indicators may be required to evaluate community engagement in a participatory action and learning health initiative. The list below outlines many indicators that could be included in an evaluation.

- No. and % of activities that had a record of community participation
- No. and % activities where community members were involved in identifying the problem or issue
- No. and % activities where community members were involved in determining strategies (deciding what to do) about the problem or issue
- No. and % activities where community members were involved in implementing the strategy (doing the work)
- No. and %. activities where community members were involved in evaluating the results of the work

However, the measures of these indicators will be dependent on the capacity of the community to participate including the skills and knowledge of the people, the strength of the community organizations, and stability of the political and economic context. It is necessary in an evaluation to have buy-in from all sectors, particularly the community to ensure success.

8   Culturally Appropriate, Gender-Sensitive, Ethically Sound Evaluation

Engaging stakeholders including health professionals, decision-makers, and community members before, during, and after evaluation research is essential to ensure that the evaluation questions and indicators are relevant and appropriate and that data is ethically collected. It is critical to include sex, gender, culture/ethnicity, and age categories for data collection as this helps to identify norms, values, attitudes, and behaviors that may affect health and the impact of a program. Gender norms, for example, can be a basis for discrimination and bias. Gender norms around early marriage can work to a girl's disadvantage by preventing their engagement in education and fulfilling employment and predispose them to early childbearing and associated death and disability.

While sex-disaggregated data (data that are collected, analyzed, and reported for men and women and boys and girls separately) is useful, gender-sensitive indicators can be effective in measuring gender or social differences between the sexes. These indicators can measure changes in status, roles, expectations, and norms pertaining to people based on what gender they are or identify themselves as. Gender-sensitive indicators vary in complexity, with some requiring elaborate data collection or analytic methods. Examples of such indicators could be: the proportion of people (disaggregated by sex) who can make decisions about their own health care/health care for their children, or the proportion of people (disaggregated by sex) who experienced physical violence from an intimate partner in the last 12 months. As many of these indicators require the collection of sensitive data, consent and ethical processes are mandatory as is the case of all evaluation research where the results are to be published. However, many evaluations are internal processes that do not require ethical approval. The collection of data against gender-sensitive indicators may also require the employment of field workers of the same gender, culture, and religion to ensure that participants are comfortable in responding.

Effectively engaging stakeholders as equal partners facilitates ownership over the evaluation process and outcomes to ensure that modifications to the program are made during implementation evaluation and lessons transferred in policy and practice.

Thought needs to be given to who should be involved and how this might contribute to the effect of the actual intervention. For example, engaging men in discussions about how the outcomes of a maternal health program might be evaluated or how the results can be applied may increase husband's participation in birth preparedness, a known factor to improve maternal health outcomes. Involving men may also facilitate women's access to facilities in cases where men's approval must be given and finances may be required to travel to a health clinic. Training and involving community midwives in collecting data as part of a maternal health evaluation at village level may provide the most up-to-date information on women who are pregnant in rural situations where data collection is poor.

9   Conclusion and Future Directions

The goal of evaluation research is utilization in policy and practice to improve quality of life. A balance must be, therefore, achieved between quality data and rigorous processes and ensuring that there is ownership and involvement of all stakeholders so that change and health improvement can be actioned.

> In the end, the measure of our success will not be predicated on the number of evaluations done, or stored within a database, or even solely upon the quality of the findings…Our success will depend on our ability to use evaluation findings to strengthen our efforts and sharpen our decision-making." (USAID 2011, pp. Rajiv Shah, Administrator, Preface)

Success in evaluation is not always communicated past the reports to funders due to budget and time constraints. However, while sharing lessons learned in peer-reviewed literature is important, so too is the dissemination of evaluation results in the form of practice or policy options briefs for decision-makers. Such dissemination formats help to make evaluation findings accessible and organizations accountable for the resources used. Documenting and sharing evaluation knowledge are, therefore, key to institutionalizing health improvement efforts.

Institutionalizing data-informed decision-making derived from evaluation research is likely to become a key part of future practice with technology playing a central role. Instead of establishing systems to collect data, evaluators are likely to become more involved in data mining and data linkage activities using existing sources that will enable real-time evaluation across multiple sites and countries. The internet may accommodate an increased and participatory approach to evaluation research where citizens and stakeholders can offer comments, contribute data, and undertake analyses. This will facilitate evaluations that capture and respond to the sociocultural diversity in society locally and globally. Evaluation research processes are also likely to become more transparent with activities taking place in online open access platforms that enable learning to be easily accessed and shared. With these changes may come challenges that could affect the independent nature and quality of evaluation research that standard education and the professionalization of the field can help to keep in check.

References

ACI. Understanding program evaluation an ACI framework, agency for clinical innovation. Chatswood, NSW Department of Health; 2013.

Amaya M, Petosa R. An evaluation of a worksite exercise intervention using the social cognitive theory: a pilot study. Health Educ J. 2012;71(2):133–43.

Australian Communities Foundation. Theory of change. Fitzroy: Australian Communities Foundation; 2015. Viewed 17 Jan 2018, http://www.communityfoundation.org.au/about-acf/theory-of-change/.

Azar FE, Solhi M, Nejhaddadgar N, Amani F. The effect of intervention using the PRECEDE-PROCEED model based on quality of life in diabetic patients. Electron Physician. 2017;9(8):5024–30.

Bauer G. Developing community health indicators to support comprehensive community building initiatives: A case study of a participatory action research project. (Dr.P.H.), University of California, Berkeley, California; 1999.

Belkora J, Volz S, Loth M, Teng A, Zarin-Pass M, Moore D, Esserman L. Coaching patients in the use of decision and communication aids: RE-AIM evaluation of a patient support program. BMC Health Serv Res. 2015;15:209. https://doi.org/10.1186/s12913-015-0872-6.

Breuer E, De Silva M, Shidaye R, Petersen I, Nakku J, Jordans M, Fekadu A, Lund C. Planning and evaluating mental health services in low-and middle-income countries using theory of change. Br J Psychiatry. 2015:s1–8. https://doi.org/10.1192/bjp.bp.114.153841.

Breuer E, Lee L, De Silva M, Lund C. Using theory of change to design and evaluate public health interventions: a systematic review. Implement Sci. 2016;11:63. https://doi.org/10.1186/s13012-016-0422-6.

Byng R, Norman I, Redfern S, Jones R. Exposing the key functions of a complex intervention for shared care in mental health: case study of a process evaluation. BMC Health Serv Res. 2008;8:271. https://doi.org/10.1186/472-6963-8-274.

CDC. Framework for program evaluation in public health. CDC Evaluation Working Group, Centers for Disease Control and Prevention, No. RR-11. 1999. https://www.cdc.gov/mmwr/preview/mmwrhtml/rr4811a1.htm.

Chen HT. Theory-driven evaluations. Thousand Oaks: Sage; 1990.

Connell JP, Kubisch AC. Applying a theory of change approach to the evaluation of comprehensive community initiatives: progress, prospects, and problems. New approaches to evaluating community initiatives, 1998;2(15–44):1–16.

Coryn CL, Noakes LA, Westine CD, Schröter DC. A systematic review of theory-driven evaluation practice from 1990 to 2009. Am J Eval. 2011;32(2):199–226.

De Silva M, Lee L, Ryan G. Using theory of change in the development, implementation and evaluation of complex health interventions A practical guide. London: The Centre for Global Mental Health & the Mental Health Innovation Network; 2014a.

De Silva MJ, Breuer E, Lee L, Asher L, Chowdhary N, Lund C, Patel V. Theory of change: a theory-driven approach to enhance the Medical Research Council's framework for complex interventions. Trials. 2014b;15:267. https://doi.org/10.1186/745-6215-15-267.

EU. EVALSED: the resource for the evaluation of socio-economic development. European Commission. 2013. http://ec.europa.eu/regional_policy/en/information/publications/evaluations-guidance-documents/2013/evalsed-the-resource-for-the-evaluation-of-socio-economic-devel opment-evaluation-guide.

Foy R, Ovretveit J, Shekelle PG, Pronovost PJ, Taylor SL, Dy S, Hempel S, McDonald KM, Rubenstein LV, Wachter RM. The role of theory in research to develop and evaluate the implementation of patient safety practices. Qual Saf Health Care. 2011;20(5):453–9.

Glasgow RE, Vogt TM, Boles SM. Evaluating the public health impact of health promotion interventions: the RE-AIM framework. Am J Public Health. 1999;89(9):1322–7.

Green L, Kreuter M. Health program planning: an educational and ecological approach, vol. 4. New York: McGraw-Hill Higher Education; 2015.

Habicht J-P, Victora C, Vaughan JP. Evaluation designs for adequacy, plausibility and probability of public health programme performance and impact. Int J Epidemiol. 1999;28(1):10–8.

Hailemariam M, Fekadu A, Selamu M, Alem A, Medhin G, Giorgis TW, DeSilva M, Breuer E. Developing a mental health care plan in a low resource setting: the theory of change approach. BMC Health Serv Res. 2015;15:429. https://doi.org/10.1186/s12913-015-1097-4.

Hou RJ, Wong SY-S, Yip BH-K, Hung AT, Lo HH-M, Chan PH, Lo CS, Kwok TC-Y, Tang WK, Mak WW. The effects of mindfulness-based stress reduction program on the mental health of family caregivers: a randomized controlled trial. Psychother Psychosom. 2014;83(1):45–53.

Hubbard SM, Hayashi SW. Use of diffusion of innovations theory to drive a federal agency's program evaluation. Eval Program Plann. 2003;26(1):49–56.

IFRC. Project/programme monitoring and evaluation (M&E) guide. International Federation of Red Cross and Red Crescent Societies, Geneva; 2011.

Jenkinson KA, Naughton G, Benson AC. The GLAMA (Girls! Lead! Achieve! Mentor! Activate!) physical activity and peer leadership intervention pilot project: a process evaluation using the RE-AIM framework. BMC Public Health. 2012;12(1):55.

Kegler MC, Twiss JM, Look V. Assessing community change at multiple levels: the genesis of an evaluation framework for the California Healthy Cities Project. Health Educ Behav. 2000;27(6):760–79.

Lakatos BE, Schaffer AC, Gitlin D, Mitchell M, Delisle L, Etheredge ML, Shellman A, Baytos M. A population-based care improvement initiative for patients at risk for delirium, alcohol withdrawal, and suicide harm. Jt Comm J Qual Patient Saf. 2015;41(7):291–AP3.

Lea S, Callaghan L, Eick S, Heslin M, Morgan J, Bolt M, Healey A, Barrett B, Rose D, Patel A. The management of individuals with enduring moderate to severe mental health needs: a participatory evaluation of client journeys and the interface of mental health services with the criminal justice system in Cornwall. National Institute of Health Research, Southampton, No. 3.15. 2015. https://www.ncbi.nlm.nih.gov/books/NBK285789/.

Logan S, Boutotte J, Wilce M, Etkind S. Using the CDC framework for program evaluation in public health to assess tuberculosis contact investigation programs. Int J Tuberc Lung Dis. 2003;7(12):S375–S83.

Marsh D, Aakesson A, Anah K. Community case management essentials: treating common childhood illnesses in the community. In: A guide for program managers. Washington, DC: CORE Group Save the Children BASICS MCHIP; 2012.

Moore GF, Audrey S, Barker M, Bond L, Bonell C, Hardeman W, Moore L, O'Cathain A, Tinati T, Wight D. Process evaluation of complex interventions: Medical Research Council guidance. BMJ. 2015;350:h1258. https://doi.org/10.1136/bmj.h258.

Patton MQ. Utilization-focused evaluation. 4th ed. Thousand Oaks: Sage; 2008.

Patton MQ, McKegg K, Wehipeihana N. Developmental evaluation exemplars: principles in practice. New York: Guilford Press; 2016.

Pawson R, Tilley N. Realistic evaluation. London: Sage; 1997.

Post DK, Daniel M, Misan G, Haren MT. A workplace health promotion application of the Precede-Proceed model in a regional and remote mining company in Whyalla, South Australia. Int J Workplace Health Manag. 2015;8(3):154–74.

Proctor E, Silmere H, Raghavan R, Hovmand P, Aarons G, Bunger A, Griffey R, Hensley M. Outcomes for implementation research: conceptual distinctions, measurement challenges, and research agenda. Adm Policy Ment Health Ment Health Serv Res. 2011;38(2):65–76.

Ramsay CR, Thomas RE, Croal BL, Grimshaw JM, Eccles MP. Using the theory of planned behaviour as a process evaluation tool in randomised trials of knowledge translation strategies: a case study from UK primary care. Implement Sci. 2010;5:71. https://doi.org/10.1186/748-5908-5-71.

Reisman J, Gienapp A, Langley K, Stachowiak S. Theory of change a practical tool for action, results and learning. Seattle: Organizational Research Services, Annie E. Casey Foundation; 2004. http://www.aecf.org/resources/theory-of-change/.

Rogers P. Theory of change, methodological briefs: impact evaluation, vol. 2. Florence: UNICEF Office of Research; 2014.

Schierhout G, Hains J, Si D, Kennedy C, Cox R, Kwedza R, O'Donoghue L, Fittock M, Brands J, Lonergan K. Evaluating the effectiveness of a multifaceted, multilevel continuous quality improvement program in primary health care: developing a realist theory of change. Implement Sci. 2013;8:119. https://doi.org/10.1186/748-5908-8-119.

Setiawan A, Dignam D, Waters C, Dawson A. Improving access to child health care in Indonesia through community case management. Matern Child Health J. 2016;20(11):2254–60.

Sinha RK, Haghparast-Bidgoli H, Tripathy PK, Nair N, Gope R, Rath S, Prost A. Economic evaluation of participatory learning and action with women's groups facilitated by Accredited Social Health Activists to improve birth outcomes in rural eastern India. Cost Effectiveness and Resource Allocation, 2017;15(2), https://doi.org/10.1186/s12962-12017-10064-12969.

Sridharan S, Nakaima A. Ten steps to making evaluation matter. Eval Program Plann. 2011;34(2):135–46.

Sweet SN, Ginis KAM, Estabrooks PA, Latimer-Cheung AE. Operationalizing the RE-AIM framework to evaluate the impact of multi-sector partnerships. Implement Sci. 2014;9(1):74.

Taplin D, Clark C, Collins E, Colby D. Theory of change a series of papers to support development of theories of change based on practice in the field. New York: Center for Human Environments; 2013.

UNDP. Handbook on planning, monitoring and evaluating for development results. New York: United Nations Development Programme; 2009. http://www.undp.org/eo/handbook.

University of Kansas. Developing a framework or model of change. Kansas: Center for Community Health and Development, University of Kansas; 2017. Viewed 17 Jan 2018, http://ctb.ku.edu/en/4-developing-framework-or-model-change.

USAID. Evaluation learning from experience: USAID evaluation policy. Washington, DC: U.S. Agency for International Development; 2011.

Van Belle SB, Marchal B, Dubourg D, Kegels G. How to develop a theory-driven evaluation design? Lessons learned from an adolescent sexual and reproductive health programme in West Africa. BMC Public Health. 2010;10:741. https://doi.org/10.1186/471-2458-10-741.

Vogel I. Review of the use of 'theory of change' in international development review report. UK Department of International Development, London; 2012.

Weiss CH. Nothing as practical as good theory: exploring theory-based evaluation for comprehensive community initiatives for children and families. In: Connell JP, Kubisch AC, Schorr LB, Weiss CH, editors. New approaches to evaluating community initiatives: concepts, methods and context. Washington DC: The Aspen Institute; 1995. p. 65–92.

Weitzman BC, Silver D, Dillman K-N. Integrating a comparison group design into a theory of change evaluation: the case of the urban health initiative. Am J Eval. 2002;23(4):371–85.

WHO. WHO evaluation practice handbook. Geneva: World Health Organization; 2013.

WHO. WHO's results framework. Geneva: World Health Organization; 2017. Viewed 21 Oct 2017, http://www.who.int/about/who_reform/change_at_who/results_framework/en/#.WfLd4rVx2 Uk www.who.int/about/resources_planning/WHO_GPW12_results_chain.pdf.

Young SL, Tuthill E. Ethnography as a tool for formative research and evaluation in public health nutrition: illustrations from the world of infant and young child feeding. In: Chrzan J, Brett J, editors. Research methods for anthropological studies of food and nutrition: food research. New York: Berghahn Books; 2017.

Methods for Evaluating Online Health Information Systems

21

Gary L. Kreps and Jordan Alpert

Contents

Abstract

This chapter will examine the rationale, strategy, and methods for systematically evaluating the effectiveness of online health information systems. Evaluation research will be framed as an essential activity for designing, refining, and sustaining robust health information systems. The best health information system evaluation research programs should include: (1) formative evaluation research activities, such as needs analysis and audience analysis, for designing responsive and appropriate systems; (2) process evaluation research activities to assess how well health information systems work with users, primarily through use of message testing, system usage analysis, and user feedback systems; as well as (3) summative evaluation research activities to assess the influences of the health information systems on important health outcomes, including costs and benefits. We will describe the use of multiple research methods as part of multi-methodological designs for conducting health information system evaluation.

G. L. Kreps (✉)
Department of Communication, George Mason University, Fairfax, VA, USA
e-mail: gkreps@gmu.edu

J. Alpert
Department of Advertising, University of Florida, Gainesville, FL, USA
e-mail: jordan.alpert@jou.ufl.edu

© Springer Nature Singapore Pte Ltd. 2019
P. Liamputtong (ed.), *Handbook of Research Methods in Health Social Sciences*,
https://doi.org/10.1007/978-981-10-5251-4_111

These methods will include examinations of the applications of content analysis, interviews, focus groups, usability tests, cost-benefit analysis, user feedback systems, unobtrusive measures, the critical incidents method, and field experiments for evaluating health information systems.

Keywords

Evaluation research · Health information systems · Formative evaluation · Process evaluation · Summative evaluation · Multimethodological research · Audience analysis · Needs analysis · Cost-benefit analysis

1   Introduction

Powerful new online health information technologies (HITs) have been touted as holding tremendous promise for enhancing the delivery of health care and promoting public health (Kreps 2011a, 2015). The tremendous growth and widespread adoption of online health information systems, such as health care system portals, health information websites, and online support groups, has the potential to transform the modern health care system by supplementing and extending traditional channels for health communication (Kreps 2015). The use of new health information technologies can enable broad dissemination of relevant health information that can be personalized to the unique information needs of individuals (Neuhauser and Kreps 2008). These ehealth communication channels can provide health care consumers and providers with the relevant health information they need exactly when and where they need the information (Krist et al. 2016).

Unfortunately, many of the enthusiastic predictions about the amazing contributions of digital health programs for promoting public health have not reached fruition and the great potential of health information systems has resulted in limited returns (Kreps 2014a, b). Too often, health information technologies fail to communicate effectively with users due to problems with the design of messages and the usability of the technologies (Neuhauser and Kreps 2003, 2008, 2010; Kreps 2014b). There is a long way to go for digital health information systems to reach their incredible potential. To enhance the quality of online health information systems, rigorous evaluation research needs to guide the design and refinement of these systems (Kreps 2002, 2014a, c; Alpert et al. 2016b). It is critically important to conduct regular, rigorous, ongoing, and strategic evaluation of health communication programs to guide development, refinement, and strategic planning (Rootman et al. 2001; Green and Glasgow 2006).

Failure to engage in careful and concerted evaluation research is likely to doom the success of online health communication systems (Kreps 2002, 2014a). Evaluation research answers important questions about the specific influences online health communication programs have on different audiences, identifying which audiences are paying attention to the programs and what they are learning from the programs (Kreps and Neuhauser 2013; Kreps 2014a). Evaluation data can identify when online health communication programs are having any unintended influences, including boomerang and iatrogenic (negative) effects (Rinegold 2002; Cho and Salmon 2007). Poorly designed health communication programs have had negative influences on key audiences, such as with the infamous National Youth Anti-drug Media Campaign, which instead of combating youth drug abuse, actually increased interest in using illegal drugs by at-risk youth (Hornik et al. 2008). Well-conducted evaluation research can help explain why some programs work and what parts of these programs work the best (Kreps 2014a).

2   Formative Research

Formative evaluation research is conducted prior to the introduction of health information systems to guide the design of these programs (Kreps 2002, 2014a). Formative evaluation helps health information system designers answer key questions about the goals and purposes of the programs they are developing. Revealing formative evaluation data can clarify what system designers want to accomplish with specific communication programs, which audiences they want to reach and influence, and what they want audience members to do in response to the health communication programs. Formative data can provide essential information to system designers about the audiences they want to reach, such as what health issues audience members are likely to be interested in, what audience members currently know about key health issues, and which messages are likely to make sense to and resonate with different audiences. Formative evaluation data can be used to establish measurable goals and outcomes for online health information programs. Formative data can be used to establish baselines for establishing current levels of knowledge and health activities to track over time. Formative evaluation research can guide adoption of relevant theories and intervention strategies to guide development and implementation of health information systems. Furthermore, good formative evaluation research can also help ensure that communication programs are sensitive to unique audience needs, cultural orientations, literacy levels, and expectations (Neuhauser and Paul 2011; Kreps 2014a).

There are two primary and interrelated forms of formative evaluation research that are critically important to the design of health information systems: needs analysis and audience analysis (Kreps 2014a, c). Needs analysis is conducted to help systems designers develop a full understanding of the scope of health issues, health behaviors, and current knowledge about the health issues confronting different audiences. Needs analysis data help system designers focus on the most relevant health issues with their programs and provide audiences with the most useful and up-to-date health information. It helps system designers determine the gaps between what is currently happening related to specific health issues within different communities and what needs to happen to promote health.

Needs analysis data can often be collected through use of multiple research methods, such as secondary analysis of existing data, such as reviewing relevant epidemiological studies about disease incidence and outcomes, surveys about health issues confronting different communities, public and private health utilization records, and research reports about best practices for addressing specific health issues (Kreps 2011b, 2014a; Alpert et al. 2016a). Sometimes, when there is insufficient research that has already been conducted about specific health issues and trends, new data should be collected to fully evaluate the health issues within specific communities. New needs analysis data can be collected with surveys, interviews, and observations. Both quantitative and qualitative needs data can help health information systems designers develop a full understanding about relevant health issues. Situation analysis is a form of needs analysis that examines the history and extent of specific health issues within communities, focusing on how widespread the health issue is, who the health issue affects, how the issue has been treated, and what recommendations have been made for ideal ways to address the health issue (Rootman et al. 2001). Channel analysis is a form of needs analysis that focuses on examining the current health information systems in effect within communities and how effective these channels have been in disseminating relevant health information (Rootman et al. 2001). SWOT analysis (strengths, weaknesses, opportunities, and threats) is a needs analysis framework that focuses on identifying and analyzing the internal and external factors that can have an impact on addressing community health issues (van Wijngaarden et al. 2010). Needs analysis is essential for helping systems developers understand the nature of the health issues that their health information systems are designed to address. It also indicates the kinds of information that is needed to address different health issues (Kreps 2014a).

Audience analysis is another form of needs analysis that focuses on providing information about the different key populations that health information systems designers want to reach and influence (Kreps 2014a). Audience analysis should provide system designers with information about which groups of people are at greatest risk for different health threats, what they currently know about key health threats, and what they need to know. It tells system designers what beliefs, attitudes, and values key audiences hold relevant to the health issues to be addressed, how the audiences have responded to health issues in the past, what channels of communication they use for accessing health information, and how effective these channels have been at providing them with accurate, relevant, and up-to-date health information. Audience analysis also provides data about relevant communication characteristics of different key audiences, such as the primary languages they use (are they native English speakers or do they use another language), their health literacy levels, their levels of trust in different information sources, and their receptivity to information about different health issues. Audience analysis data are essential for guiding the design of health information systems.

Audience analysis data help system designers segment the most relevant and homogenous audiences for different health information systems, so the information systems can be designed to be meaningful and influential for these populations. This means that health information systems are often best designed for specific audiences and one size does not necessarily fit every audience (Kreps 2012). Audience analysis data are typically collected with interviews, focus groups, and surveys to gather self-report information from different populations. Sometimes key documents are analyzed, such as websites, online posts, letters, and newspapers through content analysis to examine key audience beliefs and attitudes. Secondary analysis of relevant surveys, such as the Health Information National Trends Survey (HINTS), can also provide relevant audience analysis data (Hesse et al. 2005; Finney Rutten et al. 2011). In addition, observational data can provide insightful audience analysis data for guiding design of health information systems. Formative evaluation research provides rationale and direction for the design of health information systems that address important health issues and provide relevant and up-to-date health information and reflects the unique cultures, communication orientations, and health information needs of intended audiences (Kreps 2014a).

As social media has become a vital channel for online health communication campaigns, it can also be leveraged as a powerful tool during formative evaluation research. In general, web-based tools like social media have advantages over in-person methods because physical barriers can be overcome, searchable content makes it convenient to find specific information, and the medium encourages interactivity (Chu and Chan 1998). Social media reaches large and specific audiences and among public health departments, 60% employ at least one social media application, with nearly 90% using Twitter and 56% utilizing Facebook (Thackeray et al. 2012). Facebook is the most popular social network for individuals, including 62% of adults 65 and older who have joined (Greenwood et al. 2016). Other applications like Twitter, Pinterest, and Instagram are gaining popularity and tend to be used more by online adults ages 18–29 (Greenwood et al. 2016).

Eliciting feedback or input from users can be collected from reviewing websites, blogs, or social networking groups. For instance, collecting needs or audience analysis from a population segment that has experience with a particular health topic can be accomplished by posting a question on a Facebook wall to trigger a discussion (Neiger et al. 2012). This method is particularly effective to gather insights from hard-to-reach populations or when stigmatized issue is concerned. For instance, a discussion was created on a popular social networking website to understand teenagers' HIV prevention strategies (Levine et al. 2011). This technique enabled the researchers to capture the exact language used by teenagers and was a convenient and low cost means of collecting valuable insights.

3   Process Evaluation Research

To ensure that online health information systems achieve their health communication goals, it is important to test key program components. This is known as process evaluation research (Moore et al. 2014). Health information programs are carefully assessed to determine their suitability for addressing specific health issues. User responses are tracked to determine whether health information programs are working well with different audiences. Tests are often conducted to determine the adequacy of message strategies used and the communication channels deployed to disseminate health messages. Field tests are conducted to determine how well the intervention programs have been implemented in key settings. User responses to programs are tracked over time, especially after refinements are made to the programs. Programs are tested to determine how acceptable and usable they are for key audience representatives. These tests often generate user recommendations for refining program features that can be implemented to improve intervention programs. Process evaluation can identify strategies to improve the quality and delivery of online health systems.

Process evaluation data can be collected with user-response systems, such as questionnaires, interviews, or focus groups, which ask representative program participants about their experiences using the health information system, as well as their evaluations of the strengths and weaknesses of program components. These tools are sometimes referred to as user satisfaction surveys. The Critical Incident Method is a sophisticated qualitative user-response system for process evaluation that asks representative users about the best and worst elements in health information systems, leading to in-depth recommendations for emphasizing the strongest parts of health information systems and refining the weaker elements of the information systems (Alpert et al. 2016c).

Message testing experiments are also often used to assess user responses to messages, such as how much they liked the messages, how informative, how believable, and how influential the messages were. Respondents are typically asked for suggestions for revising the messages to make them clearer, more interesting, and more influential. A/B testing is a message testing strategy that compares two versions of a webpage or application against each other to determine which one performs better. Sometimes eye-tracking tests are conducted to determine which messages respondents focus on and which messages they find most arousing. There are also standardized text analysis programs that are used to assess readability levels of system content, such as the CDC's Clear Communication Index (Alpert et al. 2016a).

Usability tests are often conducted to determine how well different representative users can navigate online health information systems (Nielsen 1999; Kreps 2014a). In these usability tests, representative system users are asked to demonstrate how they use the system, showing how they can find specific information on the system. Researchers often will ask respondents to comment on how easy or difficult it is for them to find information and navigate the system during the usability tests, inviting respondents to suggest better ways to design the information system to make it easier to use. The data provided from the usability tests can be very revealing about hidden system flaws and about strategies for refining system design. For instance, an exercise simulation called BringItOn aimed to increase physical activity for health, recovery, or rehabilitation purposes (Albu et al. 2015). Usability testing included heuristic expert analysis and think-aloud verbal protocol. Heuristic analysis is when an expert, in this case, a software engineer, evaluates an application and compares it to the industry's best practices (Nielsen 1994). Think aloud verbal protocol is when an individual describes their decision-making criteria during a problem-solving task (Fonteyn et al. 1993). Based on these methods, BringItOn revised the software to better fit the needs and wants of participants.

In addition to usability tests, system usage data are tracked to identify who uses the information system, how often they use the system, and how much time they spend interacting with the system (Kreps 2002). Tracking data can often be collected unobtrusively through analysis of system use and billing records. Also, website usage metrics and surveys can be used to measure levels of reach and engagement (Nguyen et al. 2013). This type of process evaluation was utilized in The FaceSpace Project and metrics provided objective data about audience characteristics and timing of their engagement. Survey data explained users' online behavior and sexual behavior, while team meeting notes kept records of the challenges associated with conducting a sexual health promotion using social media (Nguyen et al. 2013). While usage data are interesting, it is often necessary to question users directly to find out why they use the system, how well the system works for them, and whether the information they accessed from the system influenced their health decisions, behaviors, and outcomes (Webb et al. 1972; Kreps 2014a). Process evaluation research is critically important for tracking user responses to health information systems over time and for providing evidence for refining system components to meet the needs of system users (Kreps 2002).

4 Summative Evaluation Research

Summative evaluation research is used to measure overall influences and outcomes from online health information systems. Summative research is conducted after the information system has been in use for a substantial period of time to document the positive and negative influences the information system has had on addressing health issues. Many of the evaluation research methods that were used in conducting both formative and process evaluation research on the information system are conducted again to compare system performance over time. By comparing baseline (pre-test) data on audience member's beliefs, attitudes, knowledge, behaviors, and health status with outcomes (post-test) data on these same factors, a quasi- experimental field test can be conducted to assess changes that have occurred during use of the health information system. These changes can be compared to measures of comparison groups that did not have access to the health information system to illustrate whether changes that occurred with the test group were related to system use. The summative evaluation data that are collected can provide important measures of the overall utility of online health communication programs for addressing important health issues and promoting public health (Nutbeam 1998; Kreps 2002, 2014a).

Summative data should examine overall patterns of program use, user satisfaction with programs, message exposure and retention from the programs, changes in key outcome variables (such as learning, relevant health behaviors, health services utilization, and health status) related to the intervention, as well as economic analyses of program costs and benefits (cost-benefit analysis). Summative research also identifies strategies for sustaining the best intervention program. Strong summative evaluation data can be very influential in determining the overall value of the health information systems, directions for improving systems, and securing support for program sustainability and institutionalization (Kreps 2014a).

A way of bolstering summative evaluation of online health communication campaigns is to utilize social media and track web analytics, or key performance indicators (KPI). KPIs are metrics that assess pre-established goals of a social media campaign (Sterne 2010). Metrics such as the number of clicks, shares, mentions, and followers can be used to gauge a variety of KPI's, like interaction and awareness. Other KPIs include exposure, or the number of times content on social media is viewed; reach, the number of people who have contact with the social media application; and engagement, which is participation in creating, sharing, and using content (Neiger et al. 2012). Based on a campaign's goals, KPIs should be identified and defined during the formative and process evaluation research stages. To monitor KPIs, typically a social media performance dashboard is used, which is an insight tool that monitors media performance and provides guidance for program enhancement and optimization (Murdough 2009).

Social media provides a wealth of information that could be evaluated both quantitatively and qualitatively. Summative evaluation dashboards can be used to evaluate reach, discussions, and general outcomes. Reach focuses on several factors, including the volume of mentions, where mentions are occurring (e.g., Twitter, social networks, blogs, discussion forums), and the social influence of individuals discussing the issue (Murdough 2009). Discussions identify the main topics or themes, the tone of discussions (e.g., positive or negative) and whether sentiment has changed (Murdough 2009). In an example of a qualitative approach utilized social media to inform the public about food safety, comments from Facebook walls were monitored, which provided a rich source of how individuals interacted with the information and took action or changed behavior (James et al. 2013). Quantitatively, different forms of engagement were measured in a Facebook study focusing on chronic disease (James et al. 2013). Metrics included the number of "likes" and "shares" to determine low, moderate, or high levels of engagement (Rus and Cameron 2016).

5   Conclusion and Future Directions

Evaluation research should be an indispensable part of the development and refinement of every online health information system (Rootman et al. 2001; Kreps 2014a). It is critically important to utilize user experience in designing and refining health information systems; this is known as participatory or user-centered design, and it not only provides important insights into audience experiences with information systems, it also encourages user involvement with the information systems (Neuhauser 2001; Neuhauser et al. 2007; Neuhauser and Kreps 2011, 2014). The best health information systems are designed to reflect the experiences and insights of system users (Neuhauser et al. 1998).

Evaluation researchers should carefully identify available sources of audience analysis data. What do we already know about key audiences? Are there natural sources of information about key events that can inform evaluation efforts, such as medical billing records, public records, or message transcripts? Health information system designers should build in user-response mechanisms into every online health communication intervention program to provide user feedback about program use.

Researchers should carefully identify existing sources of data about key audience attributes and behaviors to use as benchmarks for later comparisons after use of health communication programs, or establish new data collection measures to establish key baselines and track use over time. Usability tests should be conducted regularly to determine the effectiveness of communication programs for different groups of users. Researchers should work closely with key representatives from targeted audiences to conduct user-centered design and community participative evaluation research (Neuhauser et al. 2007). Data from evaluation research should be applied to refining and improving all digital health communication programs.

References

Albu M, Atack L, Srivastava I. Simulation and gaming to promote health education: results of a usability test. Health Educ J. 2015;74(2):244–54.

Alpert J, Desens L, Krist A, Kreps GL. Measuring health literacy levels of a patient portal using the CDC's Clear Communication Index. Health Promot Pract. 2016a. https://doi.org/10.1177/1524839916643703.

Alpert JM, Krist AH, Aycock BA, Kreps GL. Applying multiple methods to comprehensively evaluate a patient portal's effectiveness to convey information to patients. J Med Internet Res. 2016b;18(5):e112. https://doi.org/10.2196/jmir.5451.

Alpert JM, Krist AH, Aycock BA, Kreps GL. Designing user-centric patient portals: clinician and patients' uses and gratifications. Telemed e-Health., Ahead of print. 2016c. https://doi.org/10.1089/tmj.2016.0096.

Cho H, Salmon CT. Unintended effects of health communication campaigns. J Commun. 2007;57(2):293–317.

Chu LF, Chan BK. Evolution of web site design: implications for medical education on the internet. Comput Biol Med. 1998;28(5):459–72.

Finney Rutten L, Hesse B, Moser R, Kreps GL, editors. Building the evidence base in cancer communication. Cresskill: Hampton Press; 2011.

Fonteyn ME, Kuipers B, Grobe SJ. A description of think aloud method and protocol analysis. Qual Health Res. 1993;3(4):430–41.

Green LW, Glasgow RE. Evaluating the relevance, generalization, and applicability of research: issues in external validation and translation methodology. Eval Health Prof. 2006;29(1):126–53.

Greenwood S, Perrin A, Duggan M (2016) Social media update 2016. Retrieved 16 March 2017, from http://www.pewinternet.org/2016/11/11/social-media-update-2016/

Hesse BW, Nelson DE, Kreps GL, Croyle RT, Arora NK, Rimer BK, Viswanath K. Trust and sources of health information. The impact of the internet and its implications for health care providers: findings from the first Health Information National Trends Survey. Arch Intern Med. 2005;165(22):2618–24.

Hornik R, Jacobsohm L, Orwin R, Piesse A, Klton G. Effects of the National Youth Anti-Drug media campaign on youths. Am J Public Health. 2008;98(12):2229–36.

James KJ, Albrecht JA, Litchfield RE, Weishaar CA. A summative evaluation of a food safety social marketing campaign "4-day throw-away" using traditional and social media. J Food Sci Educ. 2013;12(3):48–55.

Kreps GL. Evaluating new health information technologies: expanding the frontiers of health care delivery and health promotion. Stud Health Technol Inform. 2002;80:205–12.

Kreps GL. The information revolution and the changing face of health communication in modern society. J Health Psychol. 2011a;16:192–3.

Kreps GL. Methodological diversity and integration in health communication inquiry. Patient Educ Couns. 2011b;82:285–91.

Kreps GL. Consumer control over and access to health information. Ann Fam Med. 2012;10(5). Available at: http://www.annfammed.org/content/10/5/428.full/reply#annalsfm_el_25148

Kreps GL. Evaluating health communication programs to enhance health care and health promotion. J Health Commun. 2014a;19(12):1449–59.

Kreps GL. Achieving the promise of digital health information systems. J Public Health Res. 2014b;3(471):128–9.

Kreps GLR. Epilogue: lessons learned about evaluating health communication programs. J Health Commun. 2014c;19(12):1510–4.

Kreps GL. Communication technology and health: the advent of ehealth applications. In: Cantoni L, Danowski JA, editors. Communication and technology, Volume 5 of the Handbooks of communication science, p. 483–493, (Schulz PJ, Cobley P, General Editors). Berlin: De Gruyter Mouton Publications; 2015.

Kreps GL, Neuhauser L. New directions in ehealth communication: opportunities and challenges. Patient Educ Couns. 2010;78:329–36.

Kreps GL, Neuhauser L. Artificial intelligence and immediacy: designing health communication to personally engage consumers and providers. Patient Educ Couns. 2013;92:205–10.

Krist AH, Nease DE, Kreps GL, Overholser L, McKenzie M. Engaging patients in primary and specialty care. In: Hesse BW, Ahern DK, Beckjord E, editors. Oncology informatics: using health information technology to improve processes and outcomes in cancer care. Amsterdam: Elsevier; 2016. p. 55–79.

Levine D, Madsen A, Wright E, Barar RE, Santelli J, Bull S. Formative research on MySpace: online methods to engage hard-to-reach populations. J Health Commun. 2011;16(4):448–54.

Moore G, Audrey S, Barker M, Bond L, Bonell C, Cooper C, Hardeman W, Moore L, O'Cathain A, Tinati T, Wight D, Baird J. Process evaluation in complex public health intervention studies: the need for guidance. J Epidemiol Community Health. 2014;68:101–2.

Murdough C. Social media measurement: It's not impossible. J Interact Advert. 2009;10(1):94–9.

Neiger BL, Thackeray R, Van Wagenen SA, Hanson CL, West JH, Barnes MD, Fagen MC. Use of social media in health promotion purposes, key performance indicators, and evaluation metrics. Health Promot Pract. 2012;13(2):159–64.

Neuhauser L. Participatory design for better interactive health communication: a statewide model in the USA. Electron J Commun. 2001;11(3 and 4):43.

Neuhauser L, Kreps G. Rethinking communication in the e-health era. J Health Psychol. 2003;8: 7–22.

Neuhauser L, Kreps GL. Online cancer communication interventions: meeting the literacy, linguistic, and cultural needs of diverse audiences. Patient Educ Couns. 2008;71(3):365–77.

Neuhauser L, Kreps G. Ehealth communication and behavior change: promise and performance. J Soc Semiot. 2010;20(1):9–27.

Neuhauser L, Kreps GL. Participatory design and artificial intelligence: strategies to improve health communication for diverse audiences. In: Green N, Rubinelli S, Scott D, editors. Artificial intelligence and health communication. Cambridge, MA: AAAI Press; 2011. p. 49–52.

Neuhauser L, Kreps GL. Integrating design science theory and methods to improve the development and evaluation of health communication programs. J Health Commun. 2014;19(12): 1460–71.

Neuhauser L, Paul K. Readability, comprehension and usability. In: Communicating risks and benefits: an evidence-based user's guide. Silver Spring/Bethesda: U.S. Department of Health and Human Services/Food and Drug Administration; 2011.

Neuhauser L, Schwab M, Obarski SK, Syme SL, Bieber M. Community participation in health promotion: evaluation of the California wellness guide. Health Promot Int. 1998;13(3):211.

Neuhauser L, Constantine WL, Constantine NA, Sokal-Gutierrez K, Obarski SK, Clayton L, Desai M, Sumner G, Syme SL. Promoting prenatal and early childhood health: evaluation of a statewide materials-based intervention for parents. Am J Public Health. 2007;97(10):813–9.

Nguyen P, Gold J, Pedrana A, Chang S, Howard S, Ilic O, et al. Sexual health promotion on social networking sites: a process evaluation of the FaceSpace project. J Adolesc Health. 2013;53(1):98–104.

Nielsen J. Usability engineering: Amsterdam: Elsevier; 1994.

Nielsen J. Designing web usability: the practice of simplicity. Indianapolis: New Riders Publishing; 1999.

Nutbeam D. Evaluating health promotion – progress, problems, and solutions. Health Promot Int. 1998;13:27–44.

Ringold DJ. Boomerang effects in response to public health interventions: some unintended consequences in the alcoholic beverage market. J Consum Policy. 2002;25:27–63.

Rootman I, Goodstadt M, McQueen D, Potvin L, Springett J, Ziglio E, editors. Evaluation in health promotion: principles and perspectives. Copenhagen: WHO; 2001.

Rus HM, Cameron LD. Health communication in social media: message features predicting user engagement on diabetes-related Facebook pages. Ann Behav Med. 2016;50(5):678–89.

Sterne J. Social media metrics: how to measure and optimize your marketing investment. Hoboken: Wiley; 2010.

Thackeray R, Neiger BL, Smith AK, Van Wagenen SB. Adoption and use of social media among public health departments. BMC Public Health. 2012;12(1):242.

van Wijngaarden JDH, Scholten GRM, van Wijk KP. Strategic analysis for health care organizations: the suitability of the SWOT-analysis. International journal of health planning and management. 2010. Available at: https://www.researchgate.net/profile/Jeroen_Wijngaarden/publication/45094861_Strategic_analysis_for_health_care_organizations_the_suitability_of_the_SWOT-analysis/links/541fc9860cf203f155c25f28.pdf

Webb EJ, Campbell DT, Schwartz RD, Sechrist L. Unobtrusive measures: nonreactive research in the social sciences. New York: Rand McNally & Company; 1972.

Translational Research: Bridging the Chasm Between New Knowledge and Useful Knowledge

22

Lynn Kemp

Contents

Abstract

The failure to translate health research findings into practice costs lives. Less than 20% of research on the efficacy of new interventions or practices finds its way into ongoing clinical practice, and it takes between 15 and 20 years for this translation to occur. Translational research involves a series and combination of methods to achieve the nonlinear process of progressing basic scientific discovery to a

L. Kemp (✉)
Translational Research and Social Innovation (TReSI) Group, School of Nursing and Midwifery, Ingham Institute for Applied Medical Research, Western Sydney University, Liverpool, NSW, Australia
e-mail: lynn.kemp@westernsydney.edu.au

P. Liamputtong (ed.), *Handbook of Research Methods in Health Social Sciences*,
https://doi.org/10.1007/978-981-10-5251-4_72

healthcare intervention, to the assessment of efficacy of that intervention for health outcomes in trial groups, to the determination of effectiveness of the intervention in the broader population, and finally to the sustainable adoption of the effective practice at population scale. More simply put, translational research is the movement of basic science into human research and human research into healthcare practices: the former sometimes referred to as translational research and the latter as implementation research. This chapter will provide some clarity to the complex labeling and conceptualizing of translational and implementation research and their methodological frameworks including the characteristics and key procedures of research methods that facilitate quality and timely translation of interventions and programs, including hybrid and reflexive research designs, diffusion and dissemination research, and decision-making and policy research.

Keywords
Translational research · Hybrid trial designs · Pragmatic trials · Decision-making tools · Fidelity research · External validity · Collaboration

1   Introduction

To him who devotes his life to science, nothing can give more happiness than increasing the number of discoveries, but his cup of joy is full when the results of his studies immediately find practical applications. (Louis Pasteur cited in Brownson et al. 2012, p. 3)

Increasingly, research funders, the public, and policy-makers are demanding that the significant investments made to conduct health innovation research result in measurable patient and/or population improvement and show a return on research investment (Brownson et al. 2012; Institute of Medicine 2013). Internationally, assessments about quality in research are adding engagement and impact measures to traditional measures such as publication in academic journals. For example, the Australian government, through its National Innovation and Science Agenda, has piloted assessment of engagement and impact of university research to its Excellence in Research for Australia (ERA) assessment (Australian Government 2017). Some writers have suggested that researchers engage in what they term "Designing for Dissemination and Implementation" (D4D&I), which "refers to a set of processes that are considered and activities that are undertaken throughout the planning, development, and evaluation of an intervention to increase its dissemination and implementation potential" (Brownson et al. 2012, p. 34). These initiatives are designed to address concerns that too little research results in benefit to patient care, and on average it takes 17 years for this small proportion of research to be ultimately translated and implemented in healthcare settings (Green et al. 2009; Morris et al. 2011).

Further, once translated, there is considerable evidence from many fields that the quality of implementation is poor and that programs are subject to reduced fidelity and poorer outcomes (Bopp et al. 2013; Moore et al. 2013). Programs and interventions loose effectiveness due to "drift," defined by Aarons et al. (2012, p. 2 of 9) as "a misapplication or mistaken application of the model, often involving either technical error, abandonment of core and requisite components, or introduction of counterproductive elements." They are also subject to dilution. For example, evaluation of the US implementation of 35 various early childhood home-visiting programs by Daro et al. (2014) showed that retention at 12 months ranged from 3.9% to 73.0%, the proportion of families receiving the full dosage ranged from 5.3% to 26.4%, and the proportion receiving 80% of dosage ranged from 41.2% to 51.6%. A study of five program models implemented nationally in the USA found that only 53% of families were retained until the child was aged 12 months and less than 20% of enrolled families received the recommended number of visits (Latimore et al. 2017). Earlier evaluations found that no matter what home-visiting model was implemented, families typically receive about half of the models' intended dose (Gomby 1999).

There is concern that to date translational research has both lacked investment and been poorly conducted. As the field has emerged, there has been a plethora of models and tools to support translational research; however, this has contributed to confusion rather than adequate investment and quality science. In addition, there has been a lack of attention to translation research in the health social sciences, both in terms of the translation of social interventions (as opposed to clinical interventions) and despite the social sciences being critical to the success of translational research.

This chapter will clarify the language of translational research, provide a social intervention translational research parallel for the oft-presented and more widely understood clinical translation methods, and highlight the role of health social science in translational research.

2   What Is Translational Research

> Translational research means different things to different people but it seems important to almost everyone. (Woolf 2008, p. 211)

Translational research developed alongside the evidence-based medicine movement, as a framework to more systematically detail the research processes of moving discoveries and clinical innovations from "bench-to-beside." The website of the US National Institutes of Health (2007) has offered the following definition:

> Translational research includes two areas of translation. One is the process of applying discoveries generated during research in the laboratory, and in preclinical studies, to the development of trials and studies in humans. The second area of translation concerns research aimed at enhancing the adoption of best practices in the community. Cost-effectiveness of prevention and treatment strategies is also an important part of translational science.

The framework for clinical translation was thus initially conceptualized as having two types (Institute of Medicine 2013):

- Type 1 translation of basic science (e.g., cell and chemical studies) to humans: classically the creation of new drugs or technologies and their initial human studies
- Type 2 practice-based research: conducting efficacy (trials in very controlled settings) and effectiveness (trials in real-world settings) studies

Type 2 translational research was "divided" into T2 and T3, to be inclusive of the additional processes required for broader practice-based research to support dissemination and adoption of innovative clinical treatments, detailing:

- T2 translation to patients: covering systematic review and guideline development
- T3 translation to practice: with dissemination and implementation research

Facing criticism that these "types" were still not descriptive of the full process from basic research to patient trials, and particularly failure of the framework to include broader population adoption, a more recent framework describes five (5) types of clinical translational research (T0–T4), within two domains (Institute of Medicine 2013):

- Translational from basic science to human studies:
  T0: Basic science research
  T1: Translation to humans
  T2: Translation to patients
- Translation of new data into the clinic and health decision-making:
  T3: Translation to practice
  T4: Translation to community

Figure 1 shows these frameworks and how they equate, together with details of the forms of research in each type. Not depicted in the figure is the reflexive dynamic that should exist across all the types of translational research. Rather than a unidirectional movement from T0 to T4, the learning from subsequent translation should inform earlier types. It is this lack of reflexivity that can contribute to the conduct of trials that test treatments or interventions that are either not relevant to or unable to be implemented in the real world. This can be the result of the failure of clinical research to consider the importance of delivery processes and contexts.

Clinical research has a prime focus on internal validity, that is, being able to "prove" that the intervention *alone* is responsible for the outcomes produced: context and processes are "noise" that needs to be controlled (Blamey and Mackenzie 2007). However, in a reflexive translational research framework, they are considered an interactive part of both the intervention and the research processes that are necessary to conduct "translatable" research. Box 1 provides an example of nonreflexive and reflexive clinical research, highlighting the enhanced translatability of the reflexive research.

Figure 1 also depicts the research funding investment in the different types of research. Despite increasing recognition of the importance of translational research,

Fig. 1 Phases of translational research

	Type 1		Type 2	
BENCH	**Translation to humans**	**BEDSIDE**	**Practice-based research**	**PRACTICE**

CLINICAL INTERVENTION RESEARCH

| Basic research | Case Series Phase 1 and 2 clinical trials | Human clinical research Controlled observational studies Phase 3 clinical trials | Phase 3 and 4 clinical trials Observational studies Survey research | Clinical practice Delivery of recommended care to the right patient at the right time |

	T2	**T3**
	Translation to Patients	**Translation to Practice**
	Guideline development Meta-analyses Systematic reviews	Dissemination research Implementation research

Translation from basic science to human studies			Translation of new data into the clinic and health decision-making	
T0	**T1**	**T2**	**T3**	**T4**
Basic science research	**Translation to humans**	**Translation to patients**	**Translation to practice**	**Translation to community**
Pre-clinical and animal studies	Proof of concept Phase 1 clinical trials	Phase 2 clinical trials Phase 3 clinical trials	Phase 4 clinical trials and clinical outcomes research	Population-level outcomes research
Defining mechanism, targets, and lead molecules	New methods of diagnosis, treatment, and prevention	Controlled studies leading to effective care	Delivery of recommended and timely care to the right patient	True benefit to society

Current research investment — Research investment needed

SOCIAL INTERVENTION RESEARCH

Translation from basic theory and evidence to human studies			Translation of new practices/programs into the clinic/community, health decision-making and implementation research	
T0	**T1**	**T2**	**T3**	**T4**
Theoretical research	**Translation to humans**	**Translation to clients**	**Translation to policy and practice**	**Translation to community**
Evidence and theoretical reviews	Proof of concept and pilot studies	Hybrid research models and pragmatic trials	Adoption studies Comparative effectiveness studies	Action research Population-level implementation and outcomes research
Defining mechanism, outcomes, and theory of change	New methods of diagnosis, treatment, and prevention	Controlled studies leading to effective care	Delivery of recommended and timely care to the right clients	Sustainable, quality adoption and outcomes

Box 1 Nonreflexive and reflexive clinical research

Nonreflexive	Reflexive (Designing for Dissemination and Implementation (D4D&I))
Researchers at the New Drug Institute have developed a new drug treatment (T0, T1) and are conducting their phase 2 clinical efficacy trial (T2). With a strong focus on internal validity, in order to prove the efficacy of their new drug, they are running a randomized controlled trial to test the *drug* with their target population in the outpatient rooms at the local specialist hospital	Researchers at the New Drug Institute have developed a new drug treatment (T0, T1) and are conducting their phase 2/3 clinical efficacy and effectiveness trial (T2). With a focus on both internal and external validity, the researchers have asked health providers and patients what they need and what would work for them (T4). In conducting the trial, they have tested the *drug and the delivery mechanisms* and they importantly included a range of patients who are representative of those who would use the drug (T3)
A few years later, the trial results show that the *drug* is efficacious and results are published. Some further years later, the drug has been in use across the country; however, questions are being raised about its effectiveness as health providers and patients are not seeing the results that the trial suggested should be expected. Research is conducted that shows that providers and patients are reluctant to try the new drug and that there are problems with the drug ordering and delivery processes, so patients are not using the drug with the needed consistency	A few years later, the trial results show that the *drug and its delivery mechanisms* are effective and it is widely adopted and in effective use in the community

it remains of considerable concern that the majority of research investment is in T0 through T2 research, despite, as noted by Woolf (2008, p. 212):

> [P]atients might benefit even more – and more patients might benefit – if the health care system performed better in delivering existing treatments than in producing new ones.

3   Translation in Health Social Science Research

To date, translational research has been largely considered the domain of clinical research. The development and implementation of social interventions in health, for example, health promotion and behavioral interventions, has not been subject to the same scientific rigor as clinical research, with assumptions that social interventions are done with "good intentions" and so science is redundant. The aphorism "the road to hell is paved with good intentions" is pertinent here, as social interventions, which are often implemented on a community or population scale, can have significant and generational benefits or harms. For example, the social intervention of placement of children of unwed mothers into care, based on evidence [assumptions] and good intentions that life chances for children are better in families with married parents, has resulted in generational trauma for which many governments have subsequently apologized.

The lack of rigorous science in social health intervention research has been mirrored by a lack of funding for such research. There is increasing recognition of the need for science in social health that is equivalent to that used in clinical research. Frameworks for rigor and methods for specific phases in the development, trialing, and implementation of social health interventions are now beginning to be developed, and greater investment is needed. Figure 1 shows the social health intervention parallels for the types of translations research T0 through T4, within the two domains of:

- Translational from basic theory and evidence to human studies:
  T0: Theoretical research
  T1: Translation to humans
  T2: Translation to clients
- Translation of new practices/programs into the clinic/community, health decision-making, and implementation research:
  T3: Translation to practice
  T4: Translation to community

In addition to this redefinition of the types of translation research to encompass social, rather than clinical, interventions, health social science has a considerable contribution to make in T3 and T4 translational research. The use of social science is fundamental to translation to practice and community of both clinical and social interventions.

Best practice research studies, strategies, and frameworks for translational research in social interventions, and the use of social science methods to enhance T3 and T4 translation, regardless of the intervention type, will be discussed in the remainder of this chapter. The example of the development, testing, and widespread implementation of the Maternal Early Childhood Sustained Home-visiting (MECSH®) and Volunteer Family Connect (VFC) programs will illustrate the components and process of each type of health social science translation research (for more information about MECSH, see http://www.earlychildhoodconnect.edu.au/home-visiting-programs/mecsh-public/about-mecsh, and for more information about VFC, see http://www.volunteerfamilyconnect.org.au/).

4   Translation from Basic Theory and Evidence to Human Studies

This section describes methods for T0, T1, and T2 translational research for social and behavioral intervention, which parallels the basic science and trial research for clinical interventions. Here, the focus is on providing guidance on research methods that are more appropriately used by those developing and trialing social and behavioral interventions, particularly where the context is a key component of the intervention, and both internal validity (the ability to confidently attribute outcomes to the studied intervention) and external validity (the ability to generalize the outcomes of the studied intervention to other situations and people) are important.

4.1   T0: Theoretical Research

The health social science equivalent of basic clinical research is evidence and theoretical reviews to explicate the desired outcomes and the mechanisms that will achieve them through an explicit (and testable) theory of change. Theory of change states the hypothesized links between an intervention program context, its activities and anticipated outcomes. An intervention's theory of change is grounded in a thorough understanding of the literature, both theoretical and empirical, and may also include exploratory studies where contextual understanding is needed, or there are gaps in the literature. Connell and Kubisch (cited in Blamey and Mackenzie 2007, pp. 445–446) state that:

> A theory of change approach would seek the agreement from all stakeholders that, for example, activities A1, A2, and A3, if properly implemented (and with the ongoing presence of contextual factors X1, X2 and X3) should lead to outcomes O1, O2 and O3; and if these activities, contextual supports, and outcomes all occur more or less as expected, the outcomes will be attributable to the interventions.

The MECSH program drew together the best available evidence on the importance of the early years, children's health and development, the types of support parents need, parent-infant interaction from 98 prior studies of early childhood home visiting interventions, and holistic, ecological, and partnership theory-based approaches to supporting families to establish the foundations of a positive life trajectory for their children. A contextual exploratory study was conducted with users (families) and providers of 45 early childhood programs in the study region that includes some element of home visiting. These data were analyzed and integrated to develop a hypothesized theory of change for subsequent testing.

MECSH®
Maternal Early Childhood
Sustained Home-visiting

3 CO-DEPENDENT ASPECTS OF INTERVENTION				
1. CONTEXT *Create the conditions*	2. TRUST RELATIONSHIP	3. RESPONSE		
Integrated into normal activities	(institutional ↔ familiar)	Psychosocial	Instrumental	Education
Integrated in environment	Reliable	- Affirmation	- Information made accessible	- Adaptive parenting/ attachment skills
Predictable	Non-authoritarian	- Normalising	- Resources	- Parentcraft skills
Opportunistically identifying needs	Back-up, safety net	- Empowerment	- Linking	- Child development
Flexible	Agreed boundaries/expectations	- Reflecting behaviour	- Practical help	- Life skills
Accessible	Persistence, continuity	- Different perspective	(transport, child minding)	- Health promotion
	Support	- Problem solving	- Advocacy	
	Shared experiences	- Goal setting	- Accompanying	
	Confidante			

CAPACITY TO DELIVER/RESPOND TO INTERVENTION (mediating layer)		
CLIENT (mother/family)	VISITOR	INSTITUTIONAL
Resilience Stage of change	Training Skills	Staffing Programmatic Networks
Skills Personal & family strengths	Experience	Funding number, length, Reputation
Support	Support/supervision	Resources duration of visits) Goals/values

5 CORRELATED OUTCOMES				
1. SOCIAL RESOURCES *Generalised and institutional trust*	2. SOCIAL WELL-BEING	3. DEMONSTRATED KNOWLEDGE	4. EMOTIONAL WELL-BEING	5. ADAPTABILITY
Increased appropriate use of services and programs	Improved networks	Parentcraft	Reduced anxiety/stress	Improved problem solving
Increased community participation	Improved friendship relationships	Adaptive parenting	Reduced negative feelings	Resourcefulness
	Improved family relationships	Appropriate developmental expectations	Increased confidence	Control
		Life skills	Increased self esteem	
		Health behaviours		

4.2   T1: Translation to Humans

Translating the developed theory of change into an implementable and testable inter-vention or program involves examination and analysis of the context, participants, and actors (or potential actors) as well as logistic and infrastructure characteristics that would be needed to support any program activity that would ensue. The importance of context is well documented in research, with a number of processes proposed to analyze contexts such as stakeholder interviews, community mapping, and pilot intervention studies. In his advocacy for increased practice-based implementation research in the field of health promotion, Green (2001) makes a particular call for a rise in attention to "setting-level social contextual factors." Green notes that if these issues were addressed in the design of programs as well as in attempts to measure and report efficacy, it would greatly advance the current quality of research and our knowledge base.

> The MECSH program developers were a team of academic and clinical researchers, practitioners, and service managers, who undertook a co-design process to develop the program and document program manuals and curricula. A piloting period tested family engagement processes and practitioner competencies to implement the program as designed (Kemp et al. 2005, 2006; Kardamanidis et al. 2009).

4.3   T2: Translation to Clients

High-quality translational research should aim to ensure that intervention and programs that are provided for clients and communities are effective. Health social scientists should be encouraged to conduct comparison trials of developed interventions (the term "comparison" is used rather than "controlled" to indicate that those receiving the intervention will be compared to those not receiving the intervention, within an envi-ronment that is not "controlled"). The clinical model of randomized controlled trials with high levels of internal validity, as discussed above, however, is not a comfortable fit for many social scientists, as they attempt to control for rather than incorporate the real world (see also ▶ Chap. 37, "Randomized Controlled Trials"). This is the core difference between clinical efficacy and effectiveness studies, which are described by Flay (1986, p. 455) as "efficacy trials are concerned with testing whether a treatment or procedure does more good than harm when delivered under optimum conditions, effectiveness trials are concerned with testing whether a treatment does more good than harm when delivered via a real-world program." The next sections describe some types of study designs that can incorporate internal and external validity, and quality causal comparison and context, and so are more suitable for studies of social health interventions.

4.3.1   Hybrid Research Models

Hybrid research models seek to test both efficacy and effectiveness by investigating the efficacy of an intervention in a real-world setting and/or within the context of an established service delivery system (Atkins et al. 2006). The intervention and the context are understood to be related in a reflexive process where the knowledge of how an intervention works in the real world can help identify the efficacious components of the intervention and these components can be formulated to be effective within the real-world context (Atkins et al. 2006). Hybrid research capitalizes on the "external validity strengths of service research (e.g., 'embeddedness' of the study in usual care settings, with usual care clinicians) and the rigorous measurement methods developed in interventions research" (Garland et al. 2006, p. 37). This is the purposeful reflexive combination of "evidence-based practice" and "practice-based evidence" (Thase 2006; Green 2008).

Within the real-world background, most study forms that support causal conclusions can be conducted, including randomized trials, interrupted time series, wait-list/step-wedge, controlled before and after studies, and quasi-experimental designs. Designs other than those requiring individual randomization, for example, community randomization, or randomizing small groups to commence the intervention at different times (dynamic wait-list/step-wedge) can be especially important in underserved populations and low-resource settings and where it may be practically or culturally unacceptable or not feasible to individually randomize (Glasgow et al. 2005).

> The MECSH study conducted a hybrid individually randomized comparison trial, with intervention delivery embedded in the established
> service system. Both the intervention and the control group were encouraged to use the service system, which reflexively engaged in development to accommodate the new practice intervention. The MECSH program randomized trial simultaneously maintained a strong focus on internal validity, seeking demonstration of a clear causal pathway from the intervention to outcome through individual randomization of study participants into intervention and usual care groups while also focusing on external validity by trialing the intervention for a wide range of participants drawn from the population and delivered within the existing service system (Atkins et al. 2006; Flay et al. 2005; Hohmann and Shear 2002; Kemp and Harris, 2012). The MECSH intervention included both practice and service system change and required that the research be conducted in a participatory way, where the researchers worked with the service managers and practitioners to develop and understand the intervention and the service context within in which it is based and to solve problems as they arose.

4.3.2 Pragmatic Trials

Pragmatic trials are a rigorous method for assessing effectiveness, that is, the degree of beneficial effect of interventions in real-world conditions, answering the question "Does this intervention work under usual conditions?"(Thorpe et al. 2009). Godwin et al. (2003, p. 2) contend that "pragmatic trials inform practitioners and health care planners on the most clinically effective and cost effective treatments." The Pragmatic-Explanatory Continuum Indicator Summary (PRECIS) tool was developed by Thorpe et al. (2009) to assist researchers in identifying and quantifying elements of the research design and the extent to which these reflect highly controlled or real-world conditions, distinguishing between explanatory (ideal conditions) and pragmatic (real-world conditions) trials.

The PRECIS tool consists of ten domains identified that critically distinguishing pragmatic trials from explanatory trials: (1) participant eligibility criteria, (2) flexibility of experimental intervention, (3) experimental intervention-practitioner expertise, (4) flexibility of the comparison intervention, (5) comparison intervention-practitioner expertise, (6) follow-up intensity, (7) primary trial outcome, (8) participant compliance, (9) practitioner adherence to study protocol, and (10) analysis of the primary outcome. The tool results in a "hub and spoke" or "spider map" plot where the explanatory end is central and the pragmatic end is the edge. Each spoke is a domain that is marked using a 0–4 rating scale (0 = highly explanatory, 4 = highly pragmatic). The PRECIS tool is promoted by the CONSORT Work Group on Pragmatic Trials and should become widely adopted for planning and reporting studies.

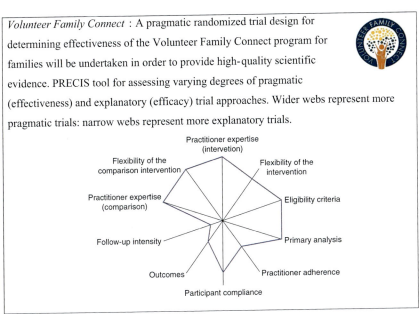

Volunteer Family Connect : A pragmatic randomized trial design for determining effectiveness of the Volunteer Family Connect program for families will be undertaken in order to provide high-quality scientific evidence. PRECIS tool for assessing varying degrees of pragmatic (effectiveness) and explanatory (efficacy) trial approaches. Wider webs represent more pragmatic trials: narrow webs represent more explanatory trials.

5   Translation of New Practices/Programs into the Clinic/ Community, Health Decision-Making, and Implementation Research

Consideration of and researching the processes for translating effective programs into standard practice and population-scale implementation has recently emerged as the critical final phase of the translational research journey that starts with basic theoretical or scientific research. There are debates about whether this phase of translation is more appropriately labeled "implementation science," but for the purposes of this chapter, the nomenclature of T3 and T4 translational research will be used.

T3 and particularly T4 translational research is an area that is still developing and requires significantly greater research investment in both clinical and social intervention research. Social science research methods are particularly important in T3 and T4 translation, regardless of whether the intervention being translated is a clinical or social/behavioral intervention. In particular, reflexive consideration of the questions of fit, reach, and feasibility of adoption and sustainability should inform the designs and processes of T0, T1, and T2 translational research.

5.1 T3: Translation to Policy and Practice

T3 translation to policy and practice describes research conducted to support policy-makers, practitioners, and health and social care decision-makers to understand, choose, and adopt effective interventions and/or programs.

Comparative effectiveness research (CER) are processes for conducting and/or synthesizing research that allow comparison of the benefits (and harms) of different interventions or programs, as they are applied to real-world settings. CER studies directly compare two or more alternative interventions, rather than comparing an intervention to a control or comparison condition, as in T2 translational research. CER can also be used to explore different ways of implementing the same intervention, for example, different service designs for delivery. Using CER processes and tools, efficacious and effective interventions, once identified and assessed, are compared and prioritized. There are a number of tools that can be used.

5.1.1 Hexagon Tool

This tool for evaluating and comparing evidence-based programs addresses six key considerations in program implementation (see Fig. 2): need, fit, evidence, readiness, resources, and capacity (Blase et al. 2013). Each key consideration is scored using a 5-point rating scale, with high-level scoring a 5, medium-level scoring a 3, and a low-level scoring a 1. Midpoints can be used and are scored as 2 or 4. Points for each evidence-based program can then be totaled and used as a means of comparison in this optional analysis.

Need refers to whether the issue being addressed is identified as significant, including social perceptions of need and data indicating need.

Fit indicates how the program aligns with agency, community, and state priorities. It also refers to how the program fits within current organizational structures and community values.

Evidence is composed of what outcomes are evidence-based, the strength of this evidence, and how effective the program is. It also includes considerations relating to population similarities.

Readiness for replication refers to whether the evidence-based program is able to be implemented as is. This involves considerations relating to operational format and expertise availability.

Resource availability requires technology and data capabilities, staffing and training, and administrative support.

Capacity to implement refers to workforce qualifications and the sustainability of the program.

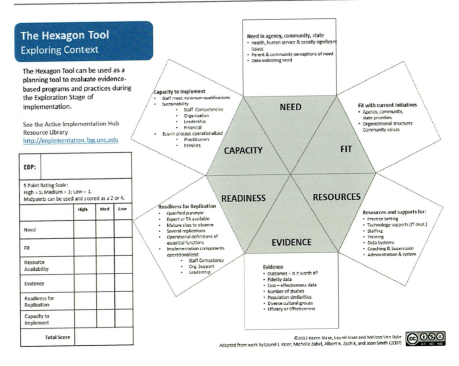

Fig. 2 The hexagon tool

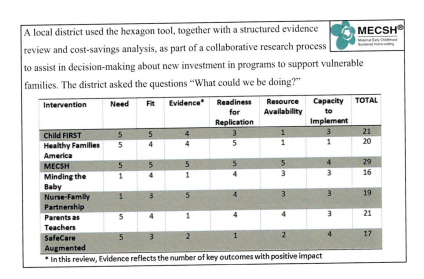

A local district used the hexagon tool, together with a structured evidence review and cost-savings analysis, as part of a collaborative research process to assist in decision-making about new investment in programs to support vulnerable families. The district asked the questions "What could we be doing?"

Intervention	Need	Fit	Evidence*	Readiness for Replication	Resource Availability	Capacity to Implement	TOTAL
Child FIRST	5	5	4	3	1	3	21
Healthy Families America	5	4	4	5	1	1	20
MECSH	5	5	5	5	5	4	29
Minding the Baby	1	4	1	4	3	3	16
Nurse-Family Partnership	1	3	5	4	3	3	19
Parents as Teachers	5	4	1	4	4	3	21
SafeCare Augmented	5	3	2	1	2	4	17

* In this review, Evidence reflects the number of key outcomes with positive impact

5.1.2 RE-AIM Model

The Reach, Effectiveness, Adoption, Implementation, and Maintenance (RE-AIM) model (RE-AIM 2017) provides a set of guidelines to support selection and evaluation of programs and policies designed to have a public/population health impact. The guidelines require questioning about:

- Reach: will/does the program reach a large and representative proportion of the target population, particularly underserved and those most in need?
- Effectiveness: will/does the program produce robust positive effects in the target population, with minimal negative effects?
- Adoption: is it feasible to implement the program in the context and in real-world and less ideal contexts?
- Implementation: can the program be consistently implemented with reasonable costs?
- Maintenance: is the program likely to support long-term and sustainable improvement?

5.1.3 Practical, Robust Implementation, and Sustainability Model (PRISM)

The Practical, Robust Implementation, and Sustainability Model (PRISM) developed by Feldstein and Glasgow (2008) guides decision-makers through systematic consideration of four elements:

- Program (intervention): from both an organizational perspective (including readiness, strength of evidence base, usability, and adaptability) and patient perspective (including patient centeredness and choice, service and access, and burden)
- External environment: including payor satisfaction, regulatory environment, and community resources
- Implementation and sustainability infrastructure: performance (fidelity) data, training and support, and sustainability
- Recipients: from both an organizational perspective (including organizational goals, management, and leadership) and patient perspective (including demographics, competing demands, and knowledge and beliefs).

5.1.4 Decision Analysis

Decision analysis with microsimulation modeling (DA) is a method that is often combined with an economic analysis to support health decision-makers to choose interventions or programs. Simulation techniques, commonly using computer modeling, are used to explore various implementation scenarios over time modeling, for example, the impacts on outcomes of differing programs, human resources and financial investment, budget constraints, and contextual and system impacts. This allows consideration of a number of "what-if" scenarios, supporting holistic and complex decision-making (Institute of Medicine 2013).

5.2 T4: Translation to Community

The final, often neglected, stage of translation is the sustainable and quality implementation of interventions or programs in the real-world, community environment.

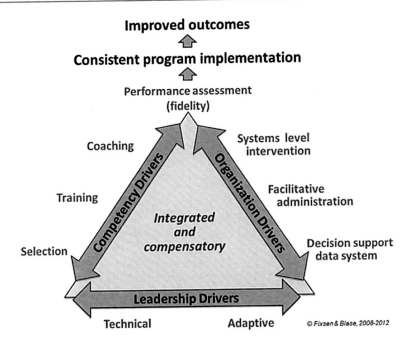

Fig. 3 The "Fixsen triangle"

In this context, the term "community" is used to denote all those in the population for whom the intervention was designed to benefit and can refer to communities of patients, clients, or populations.

Fixsen and Blase, in their seminal works on implementation research (Fixsen et al. 2005; Bertram et al. 2015), summarize the elements needed for effective translation to community in the so-called Fixsen triangle (see Fig. 3). Two primary forms of translational research are suggested by this model: business research methods and fidelity research, both detailed below.

5.2.1 Business Research Methods

T4 translation to community research requires the extensive use of business research methods, including understanding marketing and distribution systems, organizational and leadership research, forecasting and economic analyses, and research on training and workforce and system competency to deliver interventions or programs with sustainable quality to achieve population and community outcomes. As discussed by the Institute of Medicine (2013, p. 215), this area of research is often "unassigned, underemphasized, and underfunded. If they are undertaken at all, it is usually only as one of many responsibilities of someone who may lack the training or resources to do it well."

Figure 4 provides an overview of some of business research questions that require answering in translational research in order to promote quality and sustainable implementation of new or revised programs or policies. Such business research would intimately engage the deliverers and users of the program being translated and should

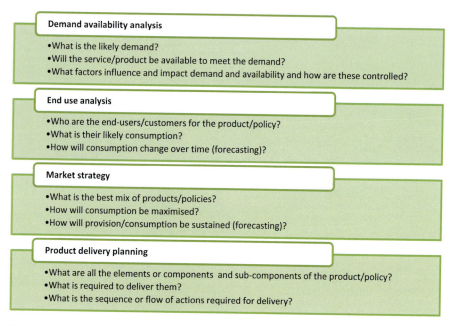

Fig. 4 Business research questions

result in practical applied knowledge arising from collaboration with practitioners and decision- and policy-makers. Research approaches and methods may include quantitative and survey methods (e.g., analysis of user data and need assessment, surveys of practitioners, training assessments), qualitative methods (e.g., policy analysis and provider and user interviews and focus groups), case studies, and, increasingly, mixed methods.

5.2.2 Fidelity Research

Implementation fidelity, or treatment integrity, "refers to the degree to which an intervention or programme is delivered as intended" (Carroll et al. 2007, p. 1 of 9). Achieving a balance of fidelity and adaptation is one of the key debates in implementing evidence-based programs beyond the research paradigm. Fidelity similarly defined by Mowbray et al. (2003, p. 315) as "the degree to which delivery of an intervention adheres to the protocol or program model originally developed" and adaptation, that is, modification of a program to suit a particular context (Castro et al. 2004), are generally viewed as conflicting or competing drivers (Aarons et al. 2012; US Department of Health and Human Services 2002). Regardless of the decision made about ways of managing this conflict through limiting or managing adaptation in order to maintain treatment integrity (see Kemp 2016 for a discussion of ways of managing the conflict), a structured and systematic program is needed to monitor fidelity and adaptation.

Daro and colleagues (2014) usefully describe two forms of fidelity to be monitored:

- Dynamic fidelity: adherence to the way or processes of intervention/program delivery
- Structural fidelity: adherence to the intervention/program content

The ability to identify these elements and establish appropriate performance indicators for monitoring is reflexively and inherently dependent upon the clear articulation of theory of change (T0), outcomes effected by the intervention/program (T1 and T2), and effective T3 research identifying and assuring organizational, leadership, and competency drivers of quality implementation.

The MECSH program in population-wide implementation requires that participating organizations and communities establish a sustainable data system for three-monthly monitoring fidelity against established performance indicators. They work with the MECSH International Support Service to interpret and improve performance.

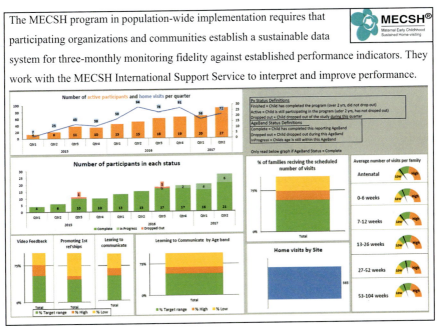

6   Critical Elements of Translational Research

Across all types of translational research are a set of critical elements needed to ensure that the translation process is timely and, indeed, achieved, that is, to ensure that a greater proportion of research is translated in a shorter time frame than the 17-year average currently experienced (Morris et al. 2011). These are:

• Attention to external validity
• Understanding both whether a program/intervention is effective and why
• Collaboration and community engagement

6.1   Attention to External Validity

As noted throughout this chapter, translational of research is aided by maintaining attention to external validity throughout all research phases. In particular, researchers conducting trials in T1 and T2 translation phase should systematically consider and document:

- The community need or issue that the research is addressing and whether this is of public health concern
- The context of and participants in the study and how representative these are of the expected end-user community
- The processes and delivery components of the intervention and how replicable these are likely to be in community-based delivery
- The potential costs and benefits (and also potential negatives) of the intervention in community delivery (Green and Glasgow 2006)

6.2   Understanding Both Whether Program Effective and Why

Randomized controlled trials, both in conduct and reporting, are focused on demonstrating that the studied intervention produced the desired outcomes. Rarely are questions of "why" or how the intervention worked explored or documented. It is uncommon for specifics of intervention/program content to be published (often for commercial confidence reasons), and, even when described, disclosed detail usually reports expected content rather than content actually delivered (Gomby 2007), although more recently greater detail is being provided, aided somewhat by the phenomenon of online publication and the capacity to include greater details and appendices to journal articles. For example, the Building Blocks early childhood home-visiting trial publication in the Lancet included a large appendix detailing the intervention provided and reported families generally received the expected content focus as a proportion of visiting time; however, details of specific content were not provided (Robling et al. 2016).

Still absent from the literature, however, is publication of the theory of change that should be developed in the T0 phase; in the absence of publication, it is not possible to know whether this is because there was no developed theory of change or a failure to publish. Where published the documentation states the expected mechanisms; rarely published is the end users' experience or understanding of how or why the tested intervention or program worked. In the absence of such documentation, it is not possible for both those conducting the research and importantly potential community providers of the intervention to fully understand what is supposed to be happening and what actually does happen over the course of an intervention or program, and little is known about the program processes necessary for success.

Addressing this deficit in conduct and/or reporting that is inhibiting of translation requires greater use of multi-method trial designs, specifically including qualitative research exploring both the intervention/program deliverer and recipients' understandings of the intervention and how and why it is having the impact documented. Such knowledge would considerably aid those engaged in T3 and T4 translation through documentation of:

- What the benefit of the intervention is, from both the provider and recipient perspective
- Why and how the benefit was realized
- And for whom

Such knowledge would considerably aid the "act of converting program objectives into actions, policy changes, regulation, and organization" (Green and Kreuter 2005, p. G-5).

The MECSH program has a documented program and logic model, developed using the theoretical model shown in the T0 MECSH example provided above. This model was constantly tested throughout the MECSH research, using multiple methods including qualitative research with providers and clients, and detailed quantitative data on program delivery, with such testing published throughout (see Kemp et al. 2005, 2006; Kardamanidis et al. 2009). The final testing of the model explored families' understanding of why and how the program had longer-term benefit, using qualitative research. The families described impacts (see below) that were consistent with the program and logic model (Zapart et al. 2016).

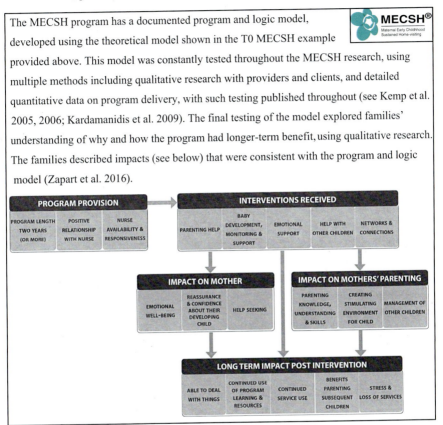

6.3   Collaboration and Community Engagement

Collaboration, with the "industry" who will be the end provider and the community who will be the end users of the intervention/program, in the design and conduct of all types of research within the translation T0–T2 pathway has the potential to significantly accelerate the translation of research findings into practice. As noted by Brownson et al. (2012, p. 192) "the importance of community and other stakeholder participation for improving the quality and relevance of research has long been acknowledged. With the growing interest in closing the 'chasm' between research and practice and more effectively eliminating health disparities, the potential benefits of participatory approaches for dissemination and implementation of research findings are increasingly being be considered." Industry collaboration and community engagement in all stages of research are key metrics being included in the new assessments of quality in research (Australian Government 2017).

7   Conclusion and Future Directions

There are three places for health social science in translational research. Firstly, applying the structured and rigorous T0–T4 translational research framework well developed for clinical interventions to social and behavioral interventions can result in improved interventions that are effective and implemented with quality. Secondly, the use of health social science research methods throughout all phases of translation can improve the "translatability" and speed of translation of both clinical and social/behavioral interventions. The concept of Designing for Dissemination and Implementation (D4D&I) captures the idea that consideration of the end users' requirements (both provider and community) should be integral to all phases of intervention/program development and trials. Finally, T3 and T4 translation phases are predominantly conducted through the use of social science methods, supporting healthcare decision-making and high-quality community-scale implementation.

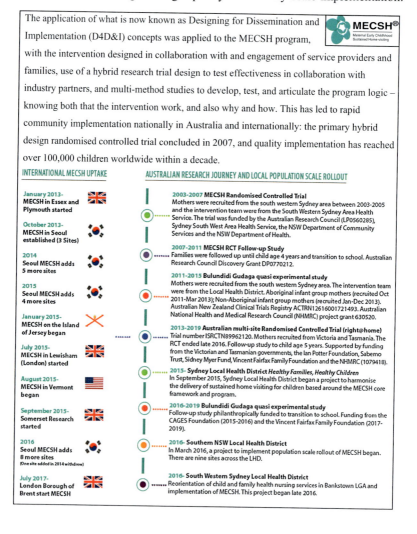

Health social science is critical to increasing the likelihood of and rapidity with which interventions/programs are successfully disseminated and implemented, as discussed throughout this chapter, and evidenced in the MECSH and VFC program examples throughout. Future use of the parallel translation science framework for social/behavioral interventions, and increased use of social science methods in all types of translational research, will ensure that the focus of research effort is increasingly on community impact.

References

Aarons GA, Green AE, Palinkas LA, Self-Brown S, Whitaker DJ, Lutzker JR, . . . Chaffin MJ. Dynamic adaptation process to implement an evidence-based child maltreatment intervention. Implement Sci. 2012;7(1):32.

Atkins MS, Frazier SL, Cappella E. Hybrid research models: natural opportunities for examining mental health in context. Clin Psychol Sci Pract. 2006;13(1):105–8. https://doi.org/10.1111/j.1468-2850.2006.00012.x.

Australian Government. National Innovation and Science Agenda: assessing the engagement and impact of university research. 2017. Retrieved from https://www.innovation.gov.au/page/measuring-impact-and-engagement-university-research

Bertram RM, Blase KA, Fixsen DL. Improving programs and outcomes: implementation frameworks and organization change. Res Soc Work Pract. 2015;25(4):477–87.

Blamey A, Mackenzie M. Theories of change and realistic evaluation: peas in a pod or apples and oranges. Evaluation. 2007;13(4):439–55.

Blase K, Kiser L, Van Dyke M. The hexagon tool: exploring context. 2013. Retrieved from http://implementation.fpg.unc.edu/resources/hexagon-tool-exploring-context

Bopp M, Saunders RP, Lattimore D. The tug-of-war: fidelity versus adaptation throughout the health promotion program life cycle. J Prim Prev. 2013;34(3):193–207.

Brownson RC, Colditz GA, Proctor EK. Dissemination and implementation research in health. Oxford: Oxford University Press; 2012.

Carroll C, Patterson M, Wood S, Booth A, Rick J, Balain S. A conceptual framework for implementation fidelity. Implement Sci. 2007;2(1):40.

Castro FG, Barrera M, Martinez CR. The cultural adaptation of prevention interventions: resolving tensions between fidelity and fit. Prev Sci. 2004;5(1):41–5. https://doi.org/10.1023/B:PREV.0000013980.12412.cd.

Daro D, Boller K, Hart B. Implementation in early childhood home visiting: successes meeting staffing standards, challenges hitting dosage and duration targets. 2014. Retrieved from Chapin Hall.

Feldstein AC, Glasgow RE. A practical, robust implementation and sustainability model (PRISM) for integrating research findings into practice. Jt Comm J Qual Patient Saf. 2008;34(4):228–43.

Fixsen DL, Naoom SF, Blase KA, Friedman RM, Wallace F. Implementation research: a synthesis of the literature. 2005. Retrieved from Tampa.

Flay BR. Efficacy and effectiveness trials (and other phases of research) in the development of health promotion programs. Prev Med. 1986;15(5):451–74.

Flay BR, Biglan A, Boruch RF, Castro FG, Gottfredson D, Kellam S, . . . Ji P. Standards of evidence: criteria for efficacy, effectiveness and dissemination. Prev Sci. 2005;6(3):151–73.

Garland AF, Hurlburt MS, Hawley KM. Examining psychotherapy processes in a services research context. Clin Psychol Sci Pract. 2006;13(1):30–46. https://doi.org/10.1111/j.1468-2850.2006.00004.x.

Glasgow RE, Magid D, Beck A, Ritzwoller D, Estabrooks P. Practical clinical trials for translating research to practice: design and measurement recommendations. Med Care. 2005;43(6):551.

Godwin M, Ruhland L, Casson I, MacDonald S, Delva D, Birtwhistle R, . . . Seguin R. Pragmatic controlled clinical trials in primary care: the struggle between external and internal validity. BMC Med Res Methodol. 2003;3:28.

Gomby DS. Understanding evaluations of home visitation programs. Future Child. 1999;9(1): 27–43.

Gomby DS. The promise and limitations of home visiting: implementing effective programs. Child Abuse Negl. 2007;31(8):793–9.

Green LW. From research to "best practices" in other settings and populations. Am J Health Behavior. 2001;25(3):165–178.

Green LW. Making research relevant: if it is an evidence-based practice, where's the practice-based evidence? Fam Pract. 2008;25:i20–4.

Green LW, Glasgow RE. Evaluating the relevance, generalization, and applicability of research. Issues in external validation and translation methodology. Eval Health Prof. 2006;29(1):126–53.

Green LW, Kreuter MW. Health program planning: an educational and ecological approach. 4th ed. New York: McGraw-Hill; 2005.

Green LW, Ottoson JM, Garcia C, Hiatt RA. Diffusion theory and knowledge dissemination, utilization, and integration in public health. Annu Rev Public Health. 2009;30:151–74.

Hohmann AA, Shear MK. Community-based intervention research: coping with the "noise" of real life in study design. Am J Psychiatry. 2002;159(2):201–7. https://doi.org/10.1176/appi.ajp.159.2.201.

Institute of Medicine. The CTSA Program at NIH: opportunities for advancing clinical and translational research. 2013. Retrieved from Washington, DC.

Kardamanidis K, Kemp L, Schmied V. Uncovering psychosocial needs: perspectives of Australian child and family health nurses in a sustained home visiting trial. Contemp Nurse. 2009;33(1): 50–8.

Kemp L. Adaptation and fidelity: a recipe analogy for achieving both in population scale implementation. Prev Sci. 2016;17(4):429–38. https://doi.org/10.1007/s11121-016-0642-7.

Kemp L, Harris E. The challenges of establishing and researching a sustained nurse home visiting programme within the universal child and family health service system. J Res Nurs. 2012;17(2): 127–38. https://doi.org/10.1177/1744987111432228.

Kemp L, Anderson T, Travaglia J, Harris E. Sustained nurse home visiting in early childhood: exploring Australian nursing competencies. Public Health Nurs. 2005;22(3):254–9.

Kemp L, Eisbacher L, McIntyre L, O'Sullivan K, Taylor J, Clark T, Harris E. Working in partnership in the antenatal period: what do child and family health nurses do? Contemp Nurse. 2006;23(2):312–20.

Latimore AD, Burrell L, Crowne S, Ojo K, Cluxton-Keller F, Gustin S, . . . Duggan A. Exploring multilevel factors for family engagement in home visiting across two national models. Prev Sci. 2017; 1–13. https://doi.org/10.1007/s11121-017-0767-3.

Moore JE, Bumbarger BK, Cooper BR. Examining adaptations of evidence-based programs in natural contexts. J Prim Prev. 2013;34(3):147–61.

Morris ZS, Wooding S, Grant J. The answer is 17 years, what is the question: understanding time lags in translational research. J R Soc Med. 2011;104(12):510–20.

Mowbray CT, Holter MC, Teague GB, Bybee D. Fidelity criteria: development, measurement, and validation. Am J Eval. 2003;24(3):315–40. https://doi.org/10.1016/S1098-2140(03)00057-2.

National Institutes of Health. Definitions under Subsection 1 (Research Objectives), Section I (Funding Opportunity Description), Part II (Full Text of Announcement), of RFA-RM-07-007: Institutional Clinical and Translational Science Award (U54). 2007. Retrieved from http://grants.nih.gov/grants/guide/rfa-files/RFA-RM-07-007.html

RE-AIM. 2017. Retrieved from http://re-aim.org/about/applying-the-re-aim-framework/

Robling M, Bekkers MJ, Bell K, Butler CC, Cannings-John R, Channon S, . . . Torgerson D. Effectiveness of a nurse-led intensive home-visitation programme for first-time teenage mothers (building blocks): a pragmatic randomised controlled trial. Lancet. 2016;387(10014):146–55. https://doi.org/10.1016/S0140-6736(15)00392-X.

Thase ME. A tale of two paradigms. Clin Psychol Sci Pract. 2006;13(1):94–8. https://doi.org/10.1111/j.1468-2850.2006.00010.x.

Thorpe KE, Zwarenstein M, Oxman AD, Treweek S, Furberg CD, Altman DG, . . . Chalkidou K. A pragmatic-explanatory continuum indicator summary (PRECIS): a tool to help trial designers. J Clin Epidemiol. 2009;62:464–75.

US Department of Health and Human Services. Finding the balance: program fidelity and adaptation in substance abuse prevention: a state-of-the-art review. Rockville: Substance Abuse and Mental Health Services Administration, Center for Substance Abuse Prevention; 2002.

Woolf SH. The meaning of translational research and why it matters. JAMA. 2008;299(2):211–3.

Zapart S, Knight J, Kemp L. 'It was easier because I had help': mothers' reflections on the long-term impact of sustained nurse home visiting. Matern Child Health J. 2016;20(1):196–204. https://doi.org/10.1007/s10995-015-1819-6.

Qualitative Interviewing

23

Sally Nathan, Christy Newman, and Kari Lancaster

Contents

Abstract

Qualitative interviewing is a foundational method in qualitative research and is widely used in health research and the social sciences. Both qualitative semi-structured and in-depth unstructured interviews use verbal communication, mostly in face-to-face interactions, to collect data about the attitudes, beliefs, and experiences of participants. Interviews are an accessible, often affordable, and effective method to understand the socially situated world of research participants. The approach is typically informed by an interpretive framework where the data collected is not viewed as evidence of the truth or reality of a

S. Nathan (✉)
School of Public Health and Community Medicine, Faculty of Medicine, UNSW, Sydney, NSW, Australia
e-mail: s.nathan@unsw.edu.au

C. Newman · K. Lancaster
Centre for Social Research in Health, Faculty of Arts and Social Sciences, UNSW, Sydney, NSW, Australia
e-mail: c.newman@unsw.edu.au; k.lancaster@unsw.edu.au

© Springer Nature Singapore Pte Ltd. 2019
P. Liamputtong (ed.), *Handbook of Research Methods in Health Social Sciences*,
https://doi.org/10.1007/978-981-10-5251-4_77

situation or experience but rather a context-bound subjective insight from the participants. The researcher needs to be open to new insights and to privilege the participant's experience in data collection. The data from qualitative interviews is not generalizable, but its exploratory nature permits the collection of rich data which can answer questions about which little is already known. This chapter introduces the reader to qualitative interviewing, the range of traditions within which interviewing is utilized as a method, and highlights the advantages and some of the challenges and misconceptions in its application. The chapter also provides practical guidance on planning and conducting interview studies. Three case examples are presented to highlight the benefits and risks in the use of interviewing with different participants, providing situated insights as well as advice about how to go about learning to interview if you are a novice.

Keywords

In-depth interviews · Semi-structured interviews · Qualitative interviewing · Interview study design · Interview methodology · Interview method

1   Introduction

Interviewing is often described as the "foundational" method of qualitative research, with good reason (Liamputtong 2013). Using verbal communication and face-to-face conversation as a means to share our experiences and views of the world around us is fundamental to human social life (Serry and Liamputtong 2017). With a strong grounding in the social convention of conversation, then, interviewing offers an accessible and effective mechanism for accessing the socially situated life worlds of individuals and groups and recording a rich source of knowledge on the distinctive views and perspectives of our informants (Braun and Clarke 2013).

However, there is an art and a science to qualitative interviewing which risks being underestimated by this representation of the interview as an "intuitive" social process. Therefore, this chapter aims to introduce the novice interviewer to the range of benefits, challenges, and strategies that we have learned which are associated with qualitative interviewing in our work as social researchers in various areas of health and medicine. We have deliberately imagined a reader who is new to qualitative interviewing, but has a fairly robust understanding of the complexities of methodological decisions, and is fuelled by a curiosity about what is possible to be achieved through interviewing methods.

The chapter is divided into sections which discuss what qualitative interviewing can offer as a research method, how to effectively plan for an interview, and how to conduct the interview well, to achieve the best possible experience and outcome for both interviewer and participant. These are accompanied by three case studies, which draw on our own experiences in using qualitative interviewing in our work, which provide more situated insights into the challenges and the rewards of using this method.

2   Why Use Qualitative Interviewing?

Qualitative interviewing is employed today as a primary method across the social and health sciences and in many other related fields (Minichiello et al. 2008). But why is this the case? While interviewing can also be used to collect quantitative data, with the researcher using a structured tool which limits answers, what exactly are the benefits and limitations of using interviewing for qualitative analysis (Bryman 2016)? The simplest answer is that qualitative interviewing permits us to broaden or extend our understanding of what is known about a specific set of issues or experiences (Silverman 2017). Qualitative interviewing is, therefore, typically informed by an interpretive research framework which does not view the accounts provided by research participants as evidence of truth or reality nor as generalizable beyond the specific context in which they are originally provided (Crotty 1998). Instead, qualitative interviewing aims to curate historically and culturally specific insights into how the subjective experience of a unique social world is viewed and how those perspectives come to be knowable in particular ways (Davies 2007).

In most contexts, a key principle which underpins the use of interviewing is an interest in capturing the "richness" and "breadth" of perspectives on the topic of interest, which supports the use of techniques which prioritize open-endedness and open-mindedness (Braun and Clarke 2013). By this, we mean that interviewing is best employed as an exploratory research method for answering questions which hold few, if any, prior assumptions about the views and experiences that participants will share. Researchers must remain genuinely open to being surprised by what they hear and to be willing to actively pursue further insights from participants regarding the subject under investigation. Indeed, attempting to understand the viewpoint of the participant is critical to this method, both in terms of understanding the purpose and in conducting interviews well.

There are distinctive traditions within interviewing practice and key differences between these (Punch 2005; Bryman 2016). For example, in the *oral history interviewing* tradition, the aim is to document personal accounts to build a picture of the lived experience of important or under-recognized moments in history. These interviews are usually not recorded anonymously, as the accuracy of the account is essential to the history-recording aims of the method. In the *in-depth interviewing* tradition, the researcher is very "hands-off," sometimes commencing the interview with only one question and building the interview around the stories the participant shares, rather than any predetermined topics. Finally, in the *semi-structured interviewing* tradition, arguably the most common approach, there is a balance between the interests of the researcher and participant, with the questions moving across a range of topic domains, but still permitting the interviewer to remain responsive and flexible in asking further questions about emerging topics and stories (see Serry and Liamputtong 2017).

Across these distinctive approaches, there are shared advantages and disadvantages of qualitative interviewing methods (Bryman 2016). Focusing first on the advantages, we know that qualitative interviewing is valuable for exploring topics on which there is little known, or when the issues are particularly complex, given they can be used to pursue a deep understanding of an issue. We also know qualitative interviewing works well as a flexible and responsive method which can be adapted to suit each context and conditions and the needs and preferences of individual participants. Because the emphasis is placed on understanding how participants give meaning to their experiences, this flexibility also increases the likelihood of achieving the quality and depth of perspective that is sought through this method.

There are also practical advantages of using qualitative interviewing (see next section), which holds appeal for researchers operating in constrained resource environments. Interviewing is a relatively cheap method to employ, with costs mainly focused on employing skilled interviewers, reimbursing participants for time and costs and translating audio recordings into analyzable transcripts. Minimal specialist equipment is required, and even novice researchers can conduct a quality interview, once they have a good understanding of the key principles and practices (see "Case Study 1: Learning How to Interview – Christy Newman"). Finally, there are many advantages to using interviews to capture the perspectives of people who have experienced disempowerment or marginalization or who, due to physical illness or disability, may have difficulty participating in other types of research. Interviews can be conducted in people's homes or other "safe" spaces, and if confidential, participants should be reassured that their courage in sharing their stories will be justified by the care and attention paid by ethical researchers to the research process and to really recognizing and learning from the participant's account.

Case Study 1: Learning How to Interview – Christy Newman

As a qualitative social researcher, interviewing represents both the everyday stuff, and the greatest reward, of my work. I have conducted over 200 interviews myself, exploring a wide range of experiences relating to health, sexuality, and relationships, with an incredibly diverse range of people. I have also had the opportunity to "teach" interviewing skills at both the undergraduate and doctoral student levels, including preparing junior and doctoral researchers to record over 300 of their own qualitative interviews. I am also the Editor in Chief of a journal which publishes health sociology from around the world, and we receive manuscripts reporting data collected in the form of qualitative interviews more than any other method. These opportunities have led me to reach a number of conclusions about the exceptional breadth of affordances and complications associated with putting this method into practice and the key traps that novices need to keep in mind when learning how to interview.

The first challenge is to learn how to prepare interview guides and practices which will support the novice interviewer to both answer the research question and remain flexible and responsive to what comes up during the exchange with

(continued)

Case Study 1: Learning How to Interview – Christy Newman (continued)

a research participant. The idea of an "open question" is familiar to most, but it holds a particular meaning in the context of a research interview. Learning how to craft an interview question so that it avoids permitting a yes/no answer, but also directs the participant toward the issues and ideas you are interested in exploring, can take quite some work. Learning from experienced researchers is helpful, but finding your own, unique way of posing questions can probably only be developed through trial and error. You need to learn how to feel alright with sounding a little bit stupid when you have to ask repeated follow-up questions, or sit with an awkward silence, without jumping in or moving on as we would in a social relationship.

The second challenge builds on these complexities by requiring skilled interviewers to learn how to juggle multiple priorities in the moment. Not only must you proficiently manage the technical dimensions of an interview – recording equipment, sound quality, pace, and timing – you must also attend to maintaining a meaningful and sustained rapport with the participant through eye contact, body language, verbal reassurance, and responsiveness to changing needs and preferences. Adding in the challenge of staying focused on the topic, while also pursuing unexpected leads, means novice interviewers can feel they are being asked to learn something as complex as circular breathing! Students who have had the opportunity to practice interviewing in my courses almost always confess that they had expected this skill to be more "intuitive" and felt quite discomforted by how difficult they found it in reality. Nonetheless, with practice and attention to becoming more proficient with each of these dimensions, it is possible to relax into the experience and really start to enjoy the intensive interaction that a research interview makes possible.

The third trial in learning how to interview relates to managing your post-interview responsibilities. Students can often find it difficult to know how to record their observations of the interview experience, in the form of field notes or reflexive comments, and can benefit from reading a range of examples of the kind of sensory, interpersonal, ethical, and empirical observations which they should ideally be seeking to capture. Transcribing and deidentifying are always shockingly time consuming for students, who were unaware how many choices are available to them in translating the spoken word into a usable form of written text. Nonetheless, gaining these experiences is essential to becoming a good interviewer, because it is only when you work with the final version of a qualitative interview transcript that you understand what you are aiming to achieve from the beginning.

My approach to supporting others in learning how to interview is to try to convey the incredible complexity and privilege that can come from engaging in this method. There is an incredible sense of intimacy, generosity, and connection that you can achieve in many one-to-one interviews, which is not

(continued)

> **Case Study 1: Learning How to Interview – Christy Newman** (continued)
> something I've experienced as much with other methods. As you become
> more senior in the academic world, there is an expectation that one will
> increasingly delegate research tasks such as interviewing to others. But per-
> sonally, this aspect of my work is one of the most rewarding, and I hope very
> much to be able to be interviewing people for the rest of my working life.

However, there are also a range of limitations of interviewing, and some inherent
challenges and complexities which mean this method needs to be carefully appraised
before employing. The most important issue to resolve is whether you want to know
what people actually do, rather than what they *say* they do. Interviews capture an
account of an experience, attitudes, or a view or feeling but cannot document
practices and behaviors in everyday life, as do observational methods, for example.
Another potential limitation relates to the accuracy of people's accounts. Even if
participants believe they are providing you with a "true" recollection of an event or
experience, all memory is shaped by cognition, culture, and context. There is a very
clear – albeit often unconscious – motivation to present a more socially acceptable
account in the context of research than may be ideal for answering some questions,
particularly when exploring sensitive or taboo subjects Suggets refer to "Case
Study 2: Interviewing the Unenthusiastic – Sally Nathan". This has added complexities when interviewing "up," given there may be additional reasons for someone in a
position of power to manage their position carefully, even in a confidential interview
(see ▶ Chap. 126, "Researching Among Elites"). As a related point, given qualitative interviewing aims to achieve depth and insight from participants, the expectations of the interviewer may need to be tempered when working with people for
whom this way of thinking or speaking is not familiar or easy (see "Case Study 3:
Interviewing "Elites" – Kari Lancaster").

> **Case Study 2: Interviewing the Unenthusiastic – Sally Nathan**
> In-depth interviews are commonly used in research with "vulnerable groups"
> and about sensitive topics, such as people's drug taking behaviors, and the
> experience of traumatic events, such as domestic violence. Vulnerable groups
> can be those whose behavior, experience, or power in society may make them
> more at risk of harm from research (Liamputtong 2007). They can sometimes
> also be less than enthusiastic about your research study as their life priorities,
> such as finding a home, or a job, or just getting through the day are more
> pressing.
>   As with all research, these participants must be given assurances that their
> identity will be protected. Researching sensitive topics and with vulnerable
> groups using in-depth interviews, however, requires extra vigilance. The slow

(continued)

Case Study 2: Interviewing the Unenthusiastic – Sally Nathan (continued)
building of trust and rapport which often underlies the consent process for vulnerable participants may mean more time in the field to become known to the participants (Foster et al. 2010).

A major group of participants I have been doing research with for more than a decade are young people aged 13–18 years in residential drug and alcohol treatment (Foster et al. 2010; Nathan et al. 2011). This body of research has been focused on examining young people's experiences of treatment and related services, and their life pathways, before and after treatment. Unsurprisingly, they are sometimes less than enthusiastic about many parts of the process. First is having another person ask them questions. They have already been asked questions by counselors, maybe also the police, and family and community services. They are also wary of signing forms and some have poor literacy. Then recording them can be the final straw. One young person turned the recording device on and off every few minutes as if it was admissible in court. Another said more at the end of the interview as we were walking back to the service, but the recording device was off, and the passing traffic would have made it impossible to capture the audio. Sometimes a young person will say yes to participating, but they are not really engaged. They do not want to share very much about themselves and answer in one-word responses despite open-ended questions and probing for more depth.

Together with my team, we have worked hard to get the most from the interviews we do with young people using a variety of strategies. We have trialed many locations. We have trained a peer interviewer who now comes to many of the interviews with another member of the team. She shares about her experiences briefly at the beginning and as appropriate, during the interview. It becomes more of a three-way conversation, and the data we collect is often a co-creation of meaning between us as a small group but focused on the young person's experiences. It is not necessarily "the truth," but the approach we have taken has yielded rich an interesting data, and the process seems to be more comfortable for the young person. There is a participation continuum in in-depth interviews from being neutral to active as an interviewer, and my stance in this and related studies has often lent toward the latter.

In our interviews, we have also strived to come from a strength-based perspective, not just problematize their lives. However, we have found the participants were not used to saying anything "good" about themselves. The sharing by the peer interviewer often opened up this discussion about what was good in their life. We have also trialed ways to move beyond talk using more participatory methods, such as drawing, and this is an important approach to consider in researching with young people.

In reviewing the data we have collected, we are asking what we are finding: Are they performing an "identity" (Rhodes et al. 2010; Riessman 1993) for us

(continued)

Case Study 2: Interviewing the Unenthusiastic – Sally Nathan (continued)

as a research team, whether consciously or unconsciously? Is social desirability a factor at play? For example, some young people may not want to admit to certain practices, like injecting or using "ice" or crystal methamphetamine due to stigma. Alternatively, they may "drug rave" and overstate their drug and alcohol using practices.

Despite its challenges, working with this group of young people is revealing and rewarding. Sometimes, we are the first people who have asked them about their lives without judgment or to provide treatment. We do, however, need to question what is the data we are collecting and what can it tell us to make a material difference to the lives of these young people.

Case Study 3: Interviewing "Elites" – Kari Lancaster

When we think about using interviewing methods in the health social sciences, we often think about the complexities of research involving vulnerable participants with lived experience of disease, trauma, or disability or studying sensitive topics like sexuality, terminal illness, or mental health. There has been a focus in the method literature on how to manage issues of confidentiality, anonymity, and power dynamics to ensure that vulnerable participants are appropriately protected, respected, and not placed at risk of harm. However, notions of "vulnerability" and "power" are not always straightforward nor are "patients" the only kinds of participants we might encounter.

In my research, I have studied health policy processes in the highly politicized domains of illicit drug use and blood-borne virus transmission. This research has involved interviewing government policy-makers, clinicians, professors, experts, as well as advocates holding senior roles in non-government organizations. These participants are by no means people we would ordinarily think of as "vulnerable." In many cases, they occupied important public positions and possessed decades of professional experience and knowledge. They were educated, articulate, and had a good understanding of research. But, in my experience, assumptions about what constitutes "vulnerability" and "power" can get us into trouble. We need to be careful in approaching every interview encounter.

There is a body of literature which has grappled with the issues associated with interviewing "up" or interviewing "elite" participants (e.g., Lancaster 2017; Morris 2009; Neal and McLaughlin 2009; Ostrander 1995). Some scholars suggest that "elite" participants may be difficult to recruit because their professional positions create barriers to researcher access. Others argue that "elite" participants might seek to exert too much control over the interview

(continued)

Case Study 3: Interviewing "Elites" – Kari Lancaster (continued)

process or that it may be difficult for the interviewer to develop rapport, especially if the interviewer is younger, female, relatively junior in status, or inexperienced in corporate culture. This literature emphasizes that "elite" participants are "used to being in charge, and they are used to having others defer to them" (Ostrander 1995, p. 143). What is striking is the way that "power" is seen as fixed and given, which leaves little room to better understand the vulnerabilities of participants usually seen as "elite" and consider how these might play out in interviews.

It is important to challenge the idea that certain participants are "in possession" of power, while others are necessarily always already "vulnerable" (Lancaster 2017). "Elite" participants are not a homogenous group, and thinking of them as such obscures crucial differences and the fact that every interview encounter depends on context and a process of co-production between the interviewer and the participant. Seeing power dynamics as fixed (as assumptions about the category of "elite" participants tends to do) limits discussion about how else we might think about and be responsive to encounters with various kinds of participants (Neal and McLaughlin 2009) and the ways such dynamics will shift and change.

Despite their "elite" status, in my experience interviewing policy-makers, clinicians, and other professional experts, there are a range of vulnerabilities and sensitivities which must be considered. Issues of anonymity and confidentiality are complex to overcome when small sampling frames are used, especially when the individuals involved occupy public positions. Moreover, participants might know each other through ongoing work and collaborations and discuss the study between them. Being attuned to these particular challenges is important so as not to expose participants to risk of harm such as retaliation from others involved in political processes, potential job loss, or, indeed, compromising the ongoing viability of policies and programs. While these may not be the kinds of concerns usually associated with protecting "vulnerable" participants, these sensitivities needed to be managed, and participants reassured to maintain trust throughout the study. However, these are not easy choices to make. Choosing not to report sensitive material or "keeping secrets" (Baez 2002) can help produce and perpetuate problematic power dynamics and, therefore, must be seen as a political choice which shapes the interview data and research process.

The major practical drawback of interviewing is that this method can be time consuming, particularly when researchers reach the stages of transcription and data analysis. In addition, while the method is teachable, as are all research methods, some of the more "intuitive" aspects of good interviewing practice require both aptitude and experience to execute them well (see "Case Study 1: Learning How to Interview – Christy Newman"). A genuine rapport with the research participant is crucial but can be quite difficult to achieve in some contexts (see "Case Study 2: Interviewing the Unenthusiastic – Sally Nathan", and "Case Study 3: Interviewing "Elites" – Kari Lancaster"). Researchers need to be willing and able to reflect on their own personal and social backgrounds to understand how they are shaping the interaction and how they may come to figure in the interview data. Nonetheless, with preparation, reflection, and practice, anyone with a genuine interest in understanding the experiences of others can learn how to employ, and to love, qualitative interviewing methods.

3 Planning for Interviews

As we intend to make this chapter more practical to students and novice researchers, we will use the first person in our discussions in this and following sections. Once you have decided that qualitative interviews will enable the study to collect the kind of data you need to answer your research questions, the next step is to plan and design the project to ensure it is achievable within the available timeframe and resources. However, when planning a qualitative study of any type, the researcher must be open to the reality that their research plan may need to be adapted in the field. Being flexible in your approach is not only acceptable but often necessary, to ensure rich data is collected that answers the research questions (Braun et al. 2017; Patton 2015).

The reasons that what you actually do in the field may differ from what was planned are many and varied. The researcher may be able to predict some of these potential changes in the planning stages, but many of them cannot be predicted. For example, you may have overestimated the number of participants that can be recruited, or you may need to recruit them differently than planned.

In planning, the researcher needs to address question such as: Who should I interview? What questions do I want to ask them? How will I recruit them? Where will I conduct the interviews? How will I record the conversation? What ethical issues need to be considered? Each of these questions requires attention to the nature of the method and reasons for choosing interviews (see Sect. 2). Importantly, your research questions should be a central consideration in the planning of an interview study. In short, good planning will increase the likelihood firstly that the study will obtain ethics approval and secondly that you will collect rich and informative data to fully answer the research questions. Interview studies with limited planning often result in "thin" data which does not provide novel insights to the field of inquiry.

3.1 Sampling or Selecting Participants to Interview

When selecting participants for interview, the researcher may choose many characteristics that need to be included in the sample, depending on the objectives of the research. For example, a researcher may decide to study participants of an age, gender, location (city or country), profession, ethnicity, health status, and so on. The researcher needs to decide what characteristics might matter in the experiences or for the phenomenon that is the focus of the study. Deciding who will be interviewed and why is best answered by thinking back again to the study research questions. Sample size is often much less important than the "make-up" of the sample (Ritchie 2001).

Sometimes it is useful to begin with maximum variety sampling (Liamputtong 2013; Patton 2015), gathering data from a range of different people, such as, in a study of homelessness, or to focus in on one particular group, such as young people who are homeless. The researcher may also hone in on a group following initial interviews. Give yourself time to review and listen to early interviews to inform the sample composition and shape the direction of later interviews (see section on "Doing Interviews").

Different approaches to selecting participants are often called sampling frames (Liamputtong 2013; Patton 2015). These are usually divided according to probability and non-probability sampling. "Logic of probability" sampling is common to quantitative research to permit generalization to population, whereas "logic of non-probability" or purposive sampling is to select information-rich cases for in-depth understanding not generalization (Liamputtong 2013; Patton 2015). Snowball sampling, where participants suggest someone else to interview, is also common in interview studies, particularly with more hidden populations, such as sex workers (Liamputtong 2013). The number of participants is usually not large in a qualitative interview study, but the depth of the data and the time with each participant is likely to be long. The most important principle is to ensure that the data collected is sufficient to answer the research question and provide something of value to the field, such as a new way of thinking about a problem or issue. This can often only be determined in the field. It is also important to ensure, as much as possible, that the research fairly tries to explore the range of views of your chosen participant group or is transparent about the views it does and does not explore (Glaser and Strauss 1967).

There is no magical number for an interview study; it is topic and context dependent. There are good qualitative studies with only a few participants and some with 30 or more. A common term in the method literature is "saturation" (Mays and Pope 2000; O'Reilly and Parker 2013). The term originated in grounded theory research but has now become increasingly employed to justify sample size across diverse paradigms, unfortunately as though it can stand as a measurable marker of adequacy (O'Reilly and Parker 2013). This concept of "saturation" is often used uncritically as a shortcut way to explain sample size, with many authors stating "data were collected until no new themes emerged." This use of the term is highly problematic. Transparency about how sampling was undertaken (and why) is a far better marker of sample size adequacy for a particular study than the use of the blanket term "saturation" (Liamputtong 2013; O'Reilly and Parker 2013).

3.2   Recruitment, Access, and Ethics

Working out how you will gain access to participants is a key question to ask at the planning stage and includes a range of ethical and consent considerations. Many ethics committees require "arms-length" recruitment (National Health and Medical Research Council 2007), that is, someone other than the researcher being involved in inviting the participants to be part of the study so that they do not feel "coerced" to participate. This can be easier said than done. Sometimes it is less ethical to have a clinician, for example, recruit patients for a study, as the patient may feel more coerced by their treating clinician than an unknown "researcher." Recruitment requires careful planning to be both feasible and ethical (see ▶ Chap. 5, "Recruitment of Research Participants"). Confidentiality and informed consent also need to be addressed, as well as procedures to prevent potential harmful consequences to participants (see ▶ Chap. 106, "Ethics and Research with Indigenous Peoples").

Think carefully about the timing (day and time) of interviews to ensure they will be most suitable for participants and the expected duration of the interview. Face-to-face interviews are the most common and are more likely to help establish rapport (Bryman 2016), but phone and video interviews are also options to consider if participants are geographically spread (Irvine 2011). Also consider whether the interview needs to occur after an event, such as following discharge from hospital. Where to conduct the interview is the next consideration. It needs to be somewhere that is comfortable for the participant but also where you can capture the conversation without too much background noise. Interviewing someone in their home is an option in some studies, but issues relating to the safety of the interviewer need to be carefully managed. You also may decide you are not the most suitable person to conduct the interviews, and a peer interviewer can be trained to assist (see "Case Study 2: Interviewing the Unenthusiastic – Sally Nathan").

3.3   Planning and Piloting Your Approach

In-depth interviews are usually very unstructured, and sometimes can consist of only one question, followed by probes to explore issues of relevance to answer your research questions (Gillham 2000; Hesse-Biber and Leavy 2011) (for further discussion of probes, see section below). Semi-structured interviews usually involve the development of an interview guide. This involves developing and planning questions in advance, and, on the whole, similar wording will be used when asking questions of different participants (Bryman 2016). Although the questions or topics for discussion are planned in advance using the interview guide, the interviewer should still be flexible and responsive, giving the participant plenty of freedom and scope to respond, asking additional questions that pick up on ideas introduced by the participant and should not rigidly following the order of the questions prepared (Bryman 2016). Thinking about not only the topic or content of the interview but also different styles of questions is important when constructing an interview guide. For example, introductory or opening questions should allow the participant to respond at great length, get them talking, and help the researcher to understand better what participants *themselves* see as important in relation to the topic being studied (Serry and Liamputtong 2017, p. 44). Other types of questions, like structuring questions, can help redirect the focus of the interview and assist the participant to move on to a new line of questioning, helping to close off the previous discussion (Serry and Liamputtong 2017, p. 45) (for more discussion of questioning, see section below).

As discussed earlier, qualitative interviews are focused on the participant's frame of meaning and their worldview, so the interviewer is often led by the issues of relevance to the participant addressed in their response to the opening question. Nonetheless, the interviewer needs to keep in mind the aims of the research and steer the participant back to the topic of the study if things go offtrack. Probing is, therefore, a critical tool to employ in in-depth interviewing (see next section for discussion of probing questions). Piloting for in-depth interviewing is, therefore, a bit more difficult than in other types of interviews, such as semi-structured, as the way an interview unfolds will be participant dependent. However, test out the opening question and possible probes with colleagues or someone from the participant group if feasible. A key learning from many years of interview research by the chapter authors is the importance of getting to concrete examples or experiences by asking the participant to tell a story or give a more specific account, to move the participant beyond generalities, such as "I really felt let down." Key questions generally relevant to most studies are:

- Can you tell me more about that incident?
- Can you tell me about a specific time that you felt like that?
- Do you have a story about <topic> that stands out as important?

Even if you have only a few questions to pilot, the use of audio or video recording equipment requires practice. Coming back from an interview that was rich and rewarding to find nothing on the recording device is a researcher's worst nightmare and more common than you may think. Choose a recoding method appropriate to the participants and context. Methods could include written field notes or electronic recording with audio or video devices. If the interviewer will be doing the interviews on their own, then practicing the skill of using the equipment and note-taking while remaining focused on the participant and encouraging them to tell their story as well as keeping to time is essential.

4   Doing Interviews

Having selected participants for recruitment, arranged access, piloted the approach, and planned your interview questions, you are ready to enter the field and start your interviews. There is no single right way to conduct a qualitative interview (Kvale 2007; Minichiello et al. 2008). As we noted earlier, the key to a good interview is often the ability to be adaptive and responsive to socio-political contexts and the needs of individual participants. Depending upon the epistemological, ontological, and methodological commitments of the study, the interviewer may see her role as gathering and documenting a rich account of the participant's knowledge, experiences or thoughts on a topic, or, alternatively, see herself as an active participant in the co-construction of such accounts or subject positions. We concur with Minichiello et al. (2008) that "the interview is neither an objective nor subjective method but that its essence is *inter-subjective interaction*" (Minichiello et al. 2008, p. 78). It is from this standpoint that we provide some general guidance for thinking about how to successfully engage in the interview process and for developing techniques and skills.

4.1 Interview Practice: Listening, Questioning, and Probing

Conducting successful interviews involves practical elements such as good planning and preparation, as well as interpersonal elements such as rapport. Good interviewing practice, therefore, requires the development of a range of skills and attributes. Building on Kvale's (1996) criteria, Bryman (2016) outlines key elements for successful interviewing. The interviewer must be *knowledgeable* (familiar with the interview's topic and prepared), use *structuring* throughout (to give purpose and direction to the interview), be *clear* (ask simple, understandable questions), be *gentle* (give participants time to think and allow for silences), be *sensitive* (listen attentively and empathetically), be *open* (remaining flexible and responsive to the participant's interests and emphasis), employ *steering* (to get to what they want to discover), stay *critical* (being prepared to challenge participant's inconsistencies), *remember* what has been said, *interpret* (clarify and extend what a participant says), maintain *balance* between saying too much or too little (directing the interview while allowing for flexibility and not exerting too much control), and, finally, be *ethically sensitive* to ethical aspects of interviewing (including confidentiality and anonymity). Underpinning this comprehensive list are a number of key qualities to be fostered: listening, flexibility, and nonjudgment (Bryman 2016).

While interviewing is much more complex than an everyday conversation, the interviewer must also work to help the participant feel relaxed and comfortable and ensure a conversational social interaction. Listening, building rapport, and styles of questioning will all contribute to this dynamic.

An interviewer's listening skills are crucially important. Minichiello et al. (2008) describe different modes of listening including *listening analytically* and *listening as support and recognition*. Maintaining concentration and focus throughout the interview is essential. While it might sound simplistic, it is important to enter an interview feeling refreshed and not tired or socially drained. This might mean planning space and time before and after an interview and not conducting multiple interviews back-to-back in one sitting, especially if the research involves talking about sensitive or difficult subjects. Being alert and listening attentively to the participant's language, emotions, attitudes, inferences, arguments, perceptions, ideas, obfuscation, contradictions, and needs throughout the interview, while simultaneously maintaining balance and control over the direction and structure of the interview, and internally beginning the process of analytic interpretation, is a juggling task which requires energy, focus, and presence.

As an intersubjective interaction, a good interview requires the building of trust between the interviewer and the participant (Johnson 2001). *Rapport* is a term often used when discussing this aspect of interviewing technique, but it is rarely defined with any specificity. Minichiello et al. (2008, p. 82) helpfully describe this process as generating "a *productive interpersonal climate.*" Building rapport requires more than the innate social skills of the interviewer. Like all aspects of interviewing, it is important to develop the skills required to be actively involved and present in the interaction. Rapport is about interpersonal interaction but also relates to ethical sensitivity. Building rapport involves ensuring that the participant is comfortable and at ease in your company. This involves both verbal and body language. Eye contact is important, as is being observant of both your own body language and the participant's. If a participant's body language changes, this might be an indication that they are anxious about responding to questions or have become aware of something in the interview environment which is causing them discomfort. Maintaining open body language, smiling and offering verbal reassurance through "listening noises" can put the participant at ease and communicate that you are interested in what they have to say, encouraging in-depth responses to questions rather than closing them off.

Rapport is also an important consideration when thinking about the order in which you will ask questions of the participant. Starting with easier, open-ended questions will help to get the participant talking and comfortable, before you approach more sensitive or difficult topics.

How you approach the process of asking questions will in some ways depend on the study design, the chosen style of interview, and preparation and piloting of the interview guide (see earlier section). Despite the possible variability in interviewing styles, in qualitative interviewing, there are a few general principles to consider in relation to questioning technique. Whether you are using a semi-structured interview guide or a more open-ended in-depth interviewing approach, you will need to be responsive and flexible, pursuing different lines of enquiry and drawing out detail from the participant. Kvale (1996) and Bryman (2016) categorize different types of questions: introducing questions, follow-up questions, specifying questions, direct questions, indirect questions, structuring questions, silence, interpreting questions, and probing questions (to which we shall return below) (see also Serry and Liamputtong 2017, pp. 44–45, for a helpful list of question types). When asking questions of your participant, it is important to give them time to reflect and respond and answer in their own way. Do not jump in to silences, but rather practice being comfortable in them and giving the participant space to think and speak. When formulating questions, it is important that they are worded in such a way that they elicit open-ended and not closed (or yes/no) responses. The way that questions are asked should generally be neutral and nonjudgmental. The question should not anticipate any response or position. Never assume you know what the participant's view or experience might be, and try to avoid offering examples which might shape the participant's response or reveal your own assumptions about a topic. It is also important that questions are clear and simple, rather than rolling multiple questions into one (Rubin and Rubin 2012).

The interview guide will guide the shape, substance and structure of the main interview questions or topics, but these questions will also be supplemented with follow-up questions and probes to ensure a successful interview. Probes "help regulate the length of answers and degree of detail, clarify unclear sentences or phrases, fill in missing steps, and keeping the conversation on topic" (Rubin and Rubin 2012, p. 164). Being skilled in these questioning techniques can help the interviewer to remain flexible and be confident when moving away from the structure of the interview guide and generate a richer understanding of what the participant is saying. Rubin and Rubin (2012) describe different types of probing questions. *Continuation* probes encourage the participant to keep talking about the topic without necessarily knowing what will be said next ("And then what happened?"). *Elaboration* probes are slightly different in that they ask the participant for more detail about something which has already been said ("Can you tell more about that?"). *Attention* probes signal that you are interested and will also encourage elaboration ("That's interesting"). *Clarification* probes resolve ambiguity or ask the participant to explain something you found confusing ("Would you mind saying that again? I am sorry I didn't quite follow.") *Steering* probes can be used to bring the conversation back to the topic of concern, if the interview has diverged down a different path. Probes can also be used in anticipation of interpretation and analysis, for example, seeking *evidence* to determine what may be more important or using *sequence* probes to clarify the order of events. Of course, not all probes are verbal questions. A smile or a nod, a hand gesture, or leaning forward in anticipation of a detailed answer will demonstrate to the participant that you are interested in hearing more, whereas a confused or quizzical look may encourage a participant to elaborate or clarify their meaning.

In addition to being attentive to what is being said, and listening analytically, the interviewer must also maintain focus on the pragmatic aspects of the interview process throughout. How much time is left? Have all the topic areas or questions been covered? Is the recording device on and working throughout the interview? Is the space still comfortable – temperature, air flow, noise, and privacy? How can the interview be wrapped up well at the end in such a way that the interview feels complete and the participant is happy to conclude? Both verbal and nonverbal cues can also be used to close the interview well and signal to the participant that the process is coming to an end. It is important that the interviewer allows plenty of time to end the interview and that it is not closed off abruptly. Giving the participant an opportunity to add further thoughts and reflection and wind down the process can be done by saying "I think we've come to the end of my questions," "You've given me a lot to think about," or "We're almost out of time" or asking questions like "Is there anything you like to add?" Positioning the participant as the expert in the research can also help end the interview positively and generate final reflections, for example: "Is there anything else that you think might be helpful for us to think about?" "Have I missed anything?" Never turn the recording device off prematurely. It is often at this point of winding down the interview that participants will have more to say. At the end of the interview, thank the participant for their time, and when leaving the interview setting, be warm and respectful of the fact that this has been an intimate social interaction. Ending the interview well is a particularly important consideration for studies which involve follow-up interviews. The number of contacts you have with participants will depend on your study design but also has implications for the interview process. A one-off interview may enable people to reveal sensitive experiences more readily as they may feel more "anonymous" in the process. On the other hand, multiple interviews may help build trust and rapport which may assist in gaining a deeper understanding of someone's experience but may lead to boundary blurring for the researcher and difficulties in ending the research and leaving the field (Dickson-Swift et al. 2008).

4.2   After the Interview

Keeping a research diary (or recording what are called "field notes") is an important practice that can help the interviewer process the interaction which has just taken place and assist later with data analysis. After each interview, spend some time writing notes to record your insights. This can also be done via an audio recording soon after the interview and later transcribed as field notes. It is best to record reflections as soon as possible after you complete the interview, so your recollection of the interview is fresh. These notes should cover analytic, practical, and personal elements. Reflect on your own role as an interviewer: How did the interview go? What went well and what could be improved in your approach? Could you have probed differently at points? What happened that could have been handled differently? Did the participant raise issues or perspectives you had not yet thought about in this study? If so, is there anything that you need to follow up in your reading or raise with participants in subsequent interviews? On a practical level, make some notes about the participant and the setting where the interview took place. What was the participant wearing? Was the setting noisy and busy or quiet and comfortable? Were you interrupted? Did you have any technological problems that need to be resolved before your next interview? Are there other things to follow up? This process can form the basis of an audit trail and promote reflexivity, both essential to rigor in qualitative research. An audit trail is a clear account of the research process and includes documentation of steps taken, for example, how access was gained, how mistakes and surprises were addressed, and how data were collected. Reflexivity is about documenting and reflecting on how your beliefs and values, and who you are as a person, such as your sex, age, and other characteristics, including any shared experience with your participants, may have impacted on their interaction with you in the interview. Reflexivity also comes into play when you are making meaning or interpreting your data in analysis.

As you finish the interview and leave the field with your digital recording in hand, there are a range of important analytic and ethical choices to be made regarding data preparation (McLellan et al. 2003). Analysis, coding, and considerations related to transcription and data management need to be considered (see ▶ Chaps. 47,

"Content Analysis: Using Critical Realism to Extend Its Utility," and ► 48, "Thematic Analysis").

In applying for ethics and planning your study, it is important to think about how findings will be fed back participants. Checking the data or account collected with the participant is sometimes advocated in qualitative research texts and "quality criteria" and is often called *member checking* (Tong et al. 2007). Planning to send transcripts to participants can be useful in checking that someone's experience has not been misrepresented but can also be problematic in that it assumes that data collected are "truth," whereas it is a time-situated telling of experience. It also assumes that the participant has the time and is interested in seeing the transcript. If member checking is employed, the participant seeing this "telling" in black and white at another time may result in significant changes to the transcript or challenges regarding "control" over data (see "Case Study 2: Interviewing the Unenthusiastic – Sally Nathan"). Importantly, no study uses all the data in every transcript, and this is another reason to consider an alternative approach. Another option can be to feedback a summary of key issues raised in the interview or across a number of interviews to see if there is anything the participant would like to add to the issues raised. This could even become a second interview for further data collection. This approach is more about building a shared understanding of the findings and relates to the idea of co-construction of meaning. Presenting back to a group of participants and having a discussion is another way to feedback findings and again collect further data to inform your research.

5   Conclusion and Future Directions

In this chapter, we have reviewed what we see as the key issues to be resolved before using qualitative interviewing as a research method, to ensure they are anticipated and addressed during and after the interview is conducted. We hope this is both illuminating of the complexities involved in this method and encouraging for novice interviewers who have much to gain from immersing themselves in learning how to become a good interviewer. To conclude, we want to emphasize that we see qualitative interviewing as both complex and teachable, both intuitive and rule-bound, and it is these contradictions which add to the richness and surprise inherent in this method. Braun and Clarke (2013, p. 80) have so beautifully described in their description of the experience of the interviewer:

> A qualitative interviewer is not a robot, precisely programmed to conduct every interview according to a set of inviolate rules. Rather, a qualitative interviewer is a human being, with a distinctive personal style, who uses their social skills, and flexibly draws on (and in some cases, disregards) guidance on good interview practice to conduct an interview that is appropriate to the needs and demands of the research question and methodological approach, the context of the interview and the individual participant.

While there is already a substantial methodological literature on interviewing, this quote also points to one of the future directions we see emerging in this field: the role of digital technologies in qualitative interviewing practice. As discussed, interviews can be recorded online, as well as in person, but the increasing opportunities to record interviews virtually are likely to reveal a whole new range of issues associated with the quality, ethics, and security of digital interviewing (Braun et al. 2017). In addition, there is an increasing recognition of the value of participatory and arts-based methods in research (Leavy 2015), and many issues to be examined regarding how interviewing might support and be challenged by these explicitly co-constituted approaches to research practice.

References

Baez B. Confidentiality in qualitative research: reflections on secrets, power and agency. Qual Res. 2002;2(1):35–58. https://doi.org/10.1177/1468794102002001638.

Braun V, Clarke V. Successful qualitative research: a practical guide for beginners. London: Sage Publications; 2013.

Braun V, Clarke V, Gray D. Collecting qualitative data: a practical guide to textual, media and virtual techniques. Cambridge: Cambridge University Press; 2017.

Bryman A. Social research methods. 5th ed. Oxford: Oxford University Press; 2016.

Crotty M. The foundations of social research: meaning and perspective in the research process. Australia: Allen & Unwin; 1998.

Davies MB. Doing a successful research project: using qualitative or quantitative methods. New York: Palgrave MacMillan; 2007.

Dickson-Swift V, James EL, Liamputtong P. Undertaking sensitive research in the health and social sciences. Cambridge: Cambridge University Press; 2008.

Foster M, Nathan S, Ferry M. The experience of drug-dependent adolescents in a therapeutic community. Drug Alcohol Rev. 2010;29(5):531–9.

Gillham B. The research interview. London: Continuum; 2000.

Glaser B, Strauss A. The discovery of grounded theory: strategies for qualitative research. Chicago: Aldine Publishing Company; 1967.

Hesse-Biber SN, Leavy P. In-depth interview. In: The practice of qualitative research. 2nd ed. Thousand Oaks: Sage Publications; 2011. p. 119–47

Irvine A. Duration, dominance and depth in telephone and face-to-face interviews: a comparative exploration. Int J Qual Methods. 2011;10(3):202–20.

Johnson JM. In-depth interviewing. In: Gubrium JF, Holstein JA, editors. Handbook of interview research: context and method. Thousand Oaks: Sage Publications; 2001.

Kvale S. Interviews: an introduction to qualitative research interviewing. Thousand Oaks: Sage; 1996.

Kvale S. Doing interviews. London: Sage Publications; 2007.

Lancaster K. Confidentiality, anonymity and power relations in elite interviewing: conducting qualitative policy research in a politicised domain. Int J Soc Res Methodol. 2017;20(1):93–103. https://doi.org/10.1080/13645579.2015.1123555.

Leavy P. Method meets art: arts-based research practice. New York: Guilford Publications; 2015.

Liamputtong P. Researching the vulnerable: a guide to sensitive research methods. Thousand Oaks: Sage Publications; 2007.

Liamputtong P. Qualitative research methods. 4th ed. South Melbourne: Oxford University Press; 2013.

Mays N, Pope C. Quality in qualitative health research. In: Pope C, Mays N, editors. Qualitative research in health care. London: BMJ Books; 2000. p. 89–102.

McLellan E, MacQueen KM, Neidig JL. Beyond the qualitative interview: data preparation and transcription. Field Methods. 2003;15(1):63–84. https://doi.org/10.1177/1525822x02239573.

Minichiello V, Aroni R, Hays T. In-depth interviewing: principles, techniques, analysis. 3rd ed. Sydney: Pearson Education Australia; 2008.

Morris ZS. The truth about interviewing elites. Politics. 2009;29(3):209–17. https://doi.org/10.1111/j.1467-9256.2009.01357.x.

Nathan S, Foster M, Ferry M. Peer and sexual relationships in the experience of drug-dependent adolescents in a therapeutic community. Drug Alcohol Rev. 2011;30(4):419–27.

National Health and Medical Research Council. National statement on ethical conduct in human research. Canberra: Australian Government; 2007.

Neal S, McLaughlin E. Researching up? Interviews, emotionality and policy-making elites. J Soc Policy. 2009;38(04):689–707. https://doi.org/10.1017/S0047279409990018.

O'Reilly M, Parker N. 'Unsatisfactory saturation': a critical exploration of the notion of saturated sample sizes in qualitative research. Qual Res. 2013;13(2):190–7. https://doi.org/10.1177/1468794112446106.

Ostrander S. "Surely you're not in this just to be helpful": access, rapport and interviews in three studies of elites. In: Hertz R, Imber J, editors. Studying elites using qualitative methods. Thousand Oaks: Sage Publications; 1995. p. 133–50.

Patton M. Qualitative research & evaluation methods: integrating theory and practice. Thousand Oaks: Sage Publications; 2015.

Punch KF. Introduction to social research: quantitative and qualitative approaches. London: Sage; 2005.

Rhodes T, Bernays S, Houmoller K. Parents who use drugs: accounting for damage and its limitation. Soc Sci Med. 2010;71(8):1489–97. https://doi.org/10.1016/j.socscimed.2010.07.028.

Riessman CK. Narrative analysis. London: Sage; 1993.

Ritchie J. Not everything can be reduced to numbers. In: Berglund C, editor. Health research. Melbourne: Oxford University Press; 2001. p. 149–73.

Rubin H, Rubin I. Qualitative interviewing: the art of hearing data. 2nd ed. Thousand Oaks: Sage Publications; 2012.

Serry T, Liamputtong P. The in-depth interviewing method in health. In: Liamputtong P, editor. Research methods in health: foundations for evidence-based practice. 3rd ed. South Melbourne: Oxford University Press; 2017. p. 67–83.

Silverman D. Doing qualitative research. 5th ed. London: Sage; 2017.

Tong A, Sainsbury P, Craig J. Consolidated criteria for reporting qualitative research (coreq): a 32-item checklist for interviews and focus groups. Int J Qual Health Care. 2007;19(6):349–57. https://doi.org/10.1093/intqhc/mzm042.

Narrative Research

24

Kayi Ntinda

Contents

Abstract

Narrative research aims to unravel consequential stories of people's lives as told by them in their own words and worlds. In the context of the health, social sciences, and education, narrative research is both a data gathering and interpretive or analytical framework. It meets these twin goals admirably by having people make sense of their lived health and well-being in their social context as they understand it, including their self-belief-oriented stories. Narrative research falls within the realm of social constructivism or the philosophy that people's lived stories capture the complexities and nuanced understanding of their

K. Ntinda (✉)
Discipline of Educational counselling and Mixed-methods Inquiry Approaches,
Faculty of Education, Office C.3.5, University of Swaziland, Kwaluseni Campus, Manzini,
Swaziland
e-mail: kmntinda77@gmail.com; kntinda@uniswa.sz

© Springer Nature Singapore Pte Ltd. 2019
P. Liamputtong (ed.), *Handbook of Research Methods in Health Social Sciences*,
https://doi.org/10.1007/978-981-10-5251-4_79

significant experiences. This chapter presents a brief overview of the narrative research approaches as forms of inquiry based on storytelling and premised on the truth value of the stories to best represent the teller's life world. The chapter also discusses data collection, analysis, and presentation utilizing narrative analysis. In doing so, this chapter provides illustrative examples applying narrative-oriented approaches to research in the health and social sciences. The chapter concludes by outlining the importance of narrative research to person-centric investigations in which the teller-informant view matters to the resulting body of knowledge.

Keywords

Collaboration · Lived experience · Intersubjectivity · Life world · Narrative inquiry · Storytelling · Meaning-making

1   Introduction

Narrative research or inquiry is one of the more recent qualitative methodologies that focuses on life stories as the essence of people-oriented sciences. As a research inquiry, narrative approaches endeavor to attend to the ways in which a story is constructed, for whom and why, as well as the cultural discourses that it draws upon (Bochner 2007; Trahar 2009). Narrative research is based on the premise that people understand and give meaning to their lives through the stories they tell (Andrews et al. 2013; McMullen and Braithwaite 2013). In doing so, people utilize narratives to compose and order their life experiences. Through the use of story forms, people account for and give meaning or significance to their lives (Bleakley 2000).

Among the early proponents of narrative research include Connelly and Clandinin (1990) who proposed to put the person back to the center of research inquiry ensuring that people's voices are not lost in translation. The two main elements comprising this approach are participants' account of a particular experience and the exploration of meaning embedded in the participants' stories. The focus on particular experiences is from the presumption that lives are bounded by events which vary in significance to the people involved. Exploration of personal meaning refers to the fact that meanings are evolving and persons may recognize some meanings and not others.

Narrative research is increasingly used in studies of health, education, and social sciences practice for its unique value to representing social phenomena in its full richness and complexity as well as providing a particularly generative source of knowledge about meaning individuals ascribe in their daily social contexts (Clandinin and Connelly 2000; Riessman 2008; Clandinin et al. 2009; Clandinin 2013). Richness and complexity are from the multiple layers of meanings people impute to their life worlds. Generativity of knowledge is from how constructing a meaning leads to new and deeper meanings than at the start. There is a growing interest and utilization of narrative research among the social science-related disciplines and also for its benefit in sensemaking, communication, learning/change, identity and identification (Errante 2000; Rhodes and Brown 2005; Atkinson and Delamont 2006; Spector-Mersel 2010) as discussed below.

The chapter considers narrative research as both a data collection and interpretive or analytical framework to understand people's sensemaking and life choices. First, the chapter begins with definitions of narrative research and when narrative research can be used. Next, the chapter presents types of approaches to data collection, analysis, and interpretation in the narrative research tradition. In this regard, this chapter presents and discusses illustrative examples of use of narrative research in the health and social sciences. Finally, the chapter concludes with suggestions for importance of narrative research in health sciences and future directions.

2 Narrative Research: Definition and Scope

As previously noted, narrative research (also referred to as narrative analysis) is a family of approaches which focus on the stories that people use to understand and describe aspects of their lives from the stories they tell (Riessman and Quinney 2005; Kim and Latta 2009). The term "narrative" carries multiple meanings and is used in a variety of ways by different human or social science disciplines. It is mostly used synonymously with "storytelling," although a distinction is made by some researchers between narrative as an account by an individual of their own experience and storytelling as it is related by others. Narrative research allows for comprehending, describing, and acting within the frame of the storyteller experiences; the story is how we make sense of the world (Clandinin and Connelly 2000). Several types of narrative inquiry have been proposed: relational, lived, therapy, and autobiographical.

2.1 Relational Narrative

This refers to shared intersubjectivity between the researcher and participant for understanding of the phenomena under study based on their construction of meanings authentically (Murphy and Aquino-Russell 2008). Relational narrative may be useful in the discipline of nursing to assist individuals who experience health inequalities to clarify their values, and, in becoming more fully their authentic selves, community members who typically feel powerless in the public space may act with confidence in influencing the distribution of healthcare resources. For example, Gadow (1999) indicates that in engagement in a relational narrative or intersubjectivity, nurses may be present to patients as they clarify their values and, therefore, transform experiences of disease and suffering into experiences of personal development. Relational narrative has been utilized for understanding response to treatment by patients attending nursing care. For example, Tsianakas et al. (2012) explored the relative value of surveys and detailed patient narratives in identifying priorities for improving breast cancer services as part of quality improvement process. In this case, patients' narratives revealed "relational" aspects of patient experience. Those identified by the survey typically related to more "functional" aspects and were not always sufficiently detailed to identify specific improvement actions. Patients' experiences have become central to assessing the performance of healthcare systems globally and are increasingly being used to inform quality improvement processes of which relational narrative is ideally suited.

2.2   Lived Narrative

This type of narrative seeks to engage participants through telling stories about their lived lives with no presumptions about the importance of specific experiences (Connelly and Clandinin 2006). The typical characteristic common in lived narrative inquiry is negotiating into the relationships, research purposes, transitions, and how researcher and participants are going to be useful in those relationships. For instance, Jeon et al. (2010) reported the findings of a systematic narrative review of qualitative studies concerning people's experience of living with chronic heart failure, aiming to develop a wide-ranging understanding of what is known about the patient experiences. The review identified the most prominent impacts of chronic heart failure on a person's everyday life including social isolation, living in fear, and losing a sense of control. Thus, lived narratives have been identified as a common strategy through which, for example, patients with chronic heart failure can manage their illness through sharing experiences and burdens with others.

2.3   Narrative Therapy

This is used in counseling and psychology premised on the fact that mental health healing is a personal matter and with unique meanings constructed around people's everyday lives. Narrative therapy is based on the construct that there is no single "truth" (Nwoye 2006). For instance, Heidari et al. (2016) and Kim and Park (2017) explored the effectiveness of narrative therapy as a person-centered therapy on happiness and death anxiety of elderly people and on people with dementia. Findings revealed that person-centered narrative therapy has a positive effect on increasing happiness and reducing death anxiety. Findings also indicated that person-centered narrative therapy can reduce agitation, neuropsychiatric symptoms, and depression and improve the quality of life. As a therapy, retelling narratives enable people to create new meanings to get rid of disabling stories harmful to their happiness.

2.4   Autobiographical Narrative

This refers to a special form of narrative inquiry which seeks to understand people's stories as autobiographies within a cultural context (Bruner 2004). For instance, Appel and Papaikonomou (2013) explored how three culturally diverse South African women constructed death and bereavement. The three diverse cultures were Tswana, Islamic Muslim, and Afrikaans. The themes of "mourning procedures and practices," "bereavement behavior," "sociopolitical context," and "private and public display of grief" were identified as valuable areas for clinical practice and future research. Likewise, Wood et al. (2006) explored the narratives of older children in their teens, who have experienced parental AIDS-related illness and death in six cities in Zimbabwe. Findings indicate that, even though many orphaned teenagers desire direct communication with adults about parental illness and death, adults themselves, whether the sick parent, other relatives in the household, or a caregiver following parental loss, are often ill-equipped to identify and manage children's distress positively. Research data generated by autobiographical interviews are usually regarded and analyzed as monological narratives drawn from autobiographical memory of the participants as shown above. If one specific style of narrative research catches the researcher's interest, then the research ought to focus on the discipline-based literature to guide their research efforts.

3   Narrative Research as a Methodological Framework

As a methodological framework, narrative research is defined by the following three aspects: temporality, sociality, and place (Connelly and Clandinin 2006; Clandinin and Huber in press). Temporality, sociality, and place stipulate the three elements of inquiry and serve as a conceptual framework. Elements are dimensions which need to be concurrently explored in conducting narrative inquiry. Temporality refers to events under study which are in temporal transition (Connelly and Clandinin 2006). Focusing attention temporally leads researchers toward the past, present, and future of people, places, and events under study. It is important to attend to temporality in narrative inquiry as quality of experience through time is viewed as narrative.

Sociality entails paying attention to both personal and social conditions by narrative researchers. Personal conditions involve "feelings, hopes, desires, aesthetic reaction and moral disposition of the inquirer and participants" (Connelly and Clandinin 2006, p. 480). Social conditions refer to conditions under which people's experiences and events are taking place. These social conditions are typically understood in a way through culture, social, and language narratives (Craig and Huber 2007). The relationship between the researcher and participant's lives is another aspect that is important to observe under narrative research as the researcher cannot detach him/herself from the inquiry relationship.

Place is about the specific concrete, physical, and topological boundaries of place where the inquiry and events take place (Connelly and Clandinin 2006, p. 480). Of essence to observe under this element or commonplace is that all events take place some place (Connelly and Clandinin 2006). Connelly and Clandinin (1990) contend that experience is narratively constructed and narratively lived in the research process. An important characteristic of narrative inquiry or research rests in a view of the research process as relational. During the research process, the researcher and participant work collaboratively in constructing meaning of the phenomena as experienced by the participant and researcher. Meaning is constructed through negotiation and collaboration between the participant and the researcher (Clandinin and Connelly 2000). According to Clandinin and Huber (in press), attending to experience through inquiry in all three elements is, somewhat, what differentiates narrative inquiry from other methodologies. Attending to the commonplaces or elements allows for narrative inquirers to study the complexity of relational composition of people's lived experiences both inside and outside of an inquiry.

The importance of attending to all three elements is very critical in narrative research. Clandinin and Connelly's (2000) framework for narrative research highlights three specific conditions of experience: interaction, continuity, and situation. Experience is perceived as involving people in relationship with others and their environment. They indicate that for an individual to be able to comprehend experience, the individual ought to consider the personal, social, and temporal elements of experience and also the context in which experience takes place (Connelly and Clandinin 2006).

3.1   Data Collection

There are various methods of data collection that can be utilized in narrative research as the researcher and participants enter into a collaborative partnership. Data can be collected in many forms such as autobiographical writing, documents such as class plans and bulletins, journals, field notes, interview transcripts, observations, storytelling, letter writing, pictures, metaphors, and personal philosophies. However, most narrative studies commonly use interviews as a key research tool (see also ▶ Chap. 23, "Qualitative Interviewing"). Data are transcribed, and then transcripts of these interviews are made available to the participant for further discussions, and these form part of the narrative record. The participant then interprets his or her own biography as a series of causal, meaningful events. Journals kept by participants also form a source of data in narrative research. Data gathered are analyzed through narrative analysis with the aim to provide evidence from experience described (Kim 2006; Clandinin and Rosiek 2007).

3.2   Narrative Analysis

Narrative analysis refers to a number of procedures for interpreting of the narratives generated in research: formal structural and functional (Clandinin and Connelly 2000; Frank 2002; Ellis and Bochner 2005). Formal structural means of analysis entails exploring how a story is structured, how it is developed, and where the story starts and ends. Functional analysis focuses on what the narrative is "doing" or what is being conveyed in the story (e.g., moral tale or a success story) (Freeman 2007; Clandinin and Huber in press). In the analysis of the narrative, the researcher tracks sequences, chronology, stories, or processes in the data, acknowledging that most narratives have backward and forward nature that ought to be unraveled in the analysis (Creswell 2012; Zulu and Munro 2017). Narrative meaning is transferred at different levels (for instance, textual level that is suitable for hermeneutic analysis, informational content level that is ideal for content analysis, or interpersonal level that could be subjected to conversational analysis).

Furthermore, narrative analysis has its own methodology which entails analysis of data in search for individual narrative accounts by the researcher (mainly have to do with commonalities in and across texts), narrative threads (core emerging themes), and temporal/spatial themes (present and future contexts) (see also ▶ Chap. 49, "Narrative Analysis").

3.3 Narrative Interpretation

Narrative interpretation is concerned with meaning-making and construction. It seeks to understand social action in which people attach subjective meaning (Crotty 1998; Currie 2010). In narrative interpretation as meaning-making, knowledge and meaning are acts of understanding people's lives on their own terms (Herman 2009; Caracciolo 2012). For instance, narrative interpretation entails exploring why narratives carry the meanings they do in the social, cultural, economic, and political context in which they are produced. That is when the application of social theory is necessary in that it takes into consideration the wider context of the reality presented. In order to achieve the intersubjective understanding of narratives, digital technologies can be used as a contemporary channel of communication that can support but cannot exclusively reproduce the social context in which the research occurs. The use of different tools to record participants' narratives can enhance the credibility of narrative research interpretation (see ▶ Chap. 2, "Qualitative Inquiry"). It can enable for suitability of dialogue with the research participants during extended periods of time and, thus, allow for cross-checking for consistency of participant's accounts during different platforms in which they interact with the researcher.

Narrative interpretation seeks to go behind and inside data to identify hidden meanings and not just accepting data at face value. The ordering is critical to the narrative interpretation, and whether it is genuinely a time ordering or not is less critical. Scientific examples of this ordering issue abound (Denzin and Lincoln 2000; Cochran 2007). It is important to study all data in order to obtain a complete picture.

Data analysis in narrative research includes four stages: (1) preparing the data, (2) identifying basic units of data, (3) organizing data, and (4) interpretation of data as suggested by Newby (2014). Preparing the data entails grouping it in a form that can be manipulated. Identifying basic units of data involves categorization procedure as categories of significance to the research issue are constructed and named. Organizing data is a chronological procedure in which associations between units are built, evaluated, and maybe rejected for the whole procedure to start again. This grouping of data can be done at different levels. Basic data units are aggregated into first-level groupings and first level to second level and so on as long as the data and interpretation permit it. Interpretation of data is about aggregating data into meaningful groupings as interpretation or from implicit understanding. For instance, Ntinda (2012) sought to unravel conceptions of ability aspects among educational consumers using a narrative approach. Findings indicated people's narratives in areas of assessment domains: learning readiness, aptitude, personal development, community norms, socialization, and guidance and counseling. The study also provided an opportunity for participants and researchers to co-construct knowledge disconcerting the power dynamics between outside experts and local community insiders about appropriate support for Botswana school learners by providing critical aspects of ability they perceived learners needed support in.

4   Applications to Health and Social Sciences Research

In recent years, there has been an increasing interest in narrative ways of knowing, resulting in a rise in using narrative research methods and techniques (Clandinin and Rosiek 2007; Riessman 2008; Spector-Mersel 2010). Even though narrative research has long tradition in disciplines such as anthropology, counseling, history, and psychology (Connelly and Clandinin 2006), the turn to narrative in health is relatively new. Narrative research has been used in health to help researchers understand patients, medical personnel-patient relationships, or other issues such as personal identity and culture. Health and healthcare issues can be expressed through the narrative process (Wang and Geale 2015).

The concept of narrative inquiry was first utilized by Connelly and Clandinin (1990) as an approach to describe personal stories of teachers. Clandinin and Connelly's (2000) approach to narrative research is rooted in Dewey's philosophy of experience. Thus narrative inquiry is a means of "understanding and inquiring in to experiences through collaboration between the researcher and participants, over time, in a place or series of places, and in social interaction with milieus" (Clandinin and Connelly 2000, p. 20). Every individual has his or her own story, and some research studies are designed to collect and analyze the stories of participants (e.g., when we study the experiences of parents of children with schizophrenia). Similarly, narrative research can be used to enhance teaching and learning for students and educators in health and social science-related disciplines. This has been through placing experiences of students and educators at the center of curriculum development. Stories are at the core of narrative analysis, whether the stories be of illness (Frank 2000), stories of participants in programs (Gibson 2012), or stories of students (Wang 2017). How to make sense of stories and, more specifically, the texts that tell the stories is at the core of narrative research.

The study on the effects of narrative career facilitation on the personal growth of a disadvantaged student is an example of using narrative research application (Maree et al. 2010). It emerged from researcher's desires to share career story of a gifted 20-year-old, male student from a poor economic background called Lebo, an undergraduate student from South Africa enrolled in education. Narrative career facilitation starts when the facilitator creates a safe atmosphere within which the client, which is being respected throughout, is invited to tell his/her story (Eloff 2002). Stories do not only simply describe a person's life but also compose one's view of oneself as a human being. The stories that are told eventually become a person's frame of himself/herself (Eloff 2002). The concept of his motivation to study was challenging for researchers and allowed for more critical investigation.

> As you can see, I have always been dictated and motivated learner. I always wanted to study further and get a good job so I can take care of myself and my family. I knew from an early age that I was clever and that I could do it. [Lebo] (Maree et al. 2010, p. 408)

This response was very important in helping the researchers make meaning of the motivation for Lebo to study education. Following the intervention, the student evidenced an improved future perspective and a more positive academic self-image. The process of narrative career facilitation had a positive effect on the overall personal growth of the student.

Ngazimbi et al. (2008), in their article, "Counseling caregivers of families affected by HIV/AIDS: The use of narrative therapy," assert that narrative therapy has a potential to enhance counseling interventions with African caregivers who are caring for family members diagnosed with HIV/AIDS. A caregiver may tell a counselor that caring for a member diagnosed with HIV/AIDS is challenging because the community believe that the infected person is immoral (which is related to the stigma associated with HIV/AIDS). Even though the client might not necessarily agree with this assumption, he or she does not challenge it. In this case, the caregiver is not mentally healthy due to the inability to challenge what society says about caring for people living with HIV/AIDS. Thus, narratives have also be utilized to understand vulnerability among carers of people living with HIV and AIDS and have indicated improvement in the quality of life of the carers with regard to burden of care, isolation, and stigma.

5   Advantages and Disadvantages of Narrative Research

Advantages of narrative research include the following: it is easy in getting people to tell their story, it gains in-depth data, participants are willing to reveal self and account reflection, the revelation of truth, and the provision of a voice for participants (Creswell 2012; Newby 2014). In using narrative research, it is fairly easy to get people to tell stories, since most people are usually pleased to share a story about themselves and one wants to report their story. Gaining in-depth data (thick description) is possible since this often occurs with ease in narrated events. In using the narrative approach to present findings, researchers can access rich strata of information that give a more in-depth understanding of the specifics of the participants' viewpoints. The knowledge gained from narrative research can provide the reader with a detailed understanding of the subject matter and further insight on how to apply the stories to their own context (Savin-Baden and Niekerk 2007). Moreover, individuals have a habit of not hiding truths when telling stories, or if they attempt to, it mostly becomes obvious in thorough data interpretation. In carrying out narrative studies, researchers form a close bond with participants where participants may feel that their stories are heard and important (Creswell 2012). Furthermore, the approach allows for bridging the gap between research and practice.

The disadvantages of narrative therapy include the difficulty of establishing the role one assumes in the inquiry. For example, Ellis and Bochner (2000) assert that if one is a storyteller rather than a story analyst, then their goal becomes therapeutic rather than analytic. This role is usually difficult to negotiate in narrative research. Stories can be challenging to understand in terms of the relationship between the storytelling in the interview and story-making in the presentation of data. It is often difficult to decide the relationship between the narrative account, the interpretation, and the retold story. The negotiations of data interpretation and presentation of data can be problematic.

Also setting boundaries to stories can be difficult in five aspects: (1) who authors the account (e.g., the researcher or the participant), (2) the scope of the narrative (e.g., an entire life or an episode of life), (3) who provides the story, (4) the kind of conceptual framework that has influenced the study (e.g., critical or constructivist), (5) and whether or not all these elements are included in one narrative.

6   Conclusion and Future Directions

Even though narrative research shares features of other qualitative research approaches such as the social focus in ethnography and the focus on experience in phenomenology, it is the simultaneous exploration of all the three elements: temporality, sociality, and place that shape and make narrative research/inquiry. Experience can be narratively created and narratively lived. The distinctive feature of narrative research is dependent on the view of the research process as relational. Narrative inquiry is collaborative and informant centered. Meaning is constructed through negotiation and collaboration between the participant and the researcher. The narrative research design adopted for health science studies provides opportunities to probe deeply into complexities surrounding health-related research. Narrative research is not simple storytelling; it is a method of inquiry that uses storytelling to uncover nuances around people's lived experiences. The process of storytelling, a key element in narrative research, provides the opportunity for dialogue and reflection, each intertwined and cyclical. The narrative approach can be adopted for health-related studies as it provides a unique opportunity to explore researcher/participant relationship over time in health sciences and to place health in the context of participants' lives. Employing a specific narrative research approach requires close attention to the "fit" of the research question/context with the particular method under consideration: relational, lived, therapy, and autobiographical. While sharing some narrative commonalities, each of the various methods enables the emergence of unique analytic and interpretive perspectives about stories relevant to health research and practice.

With advances in word-recording softwares, narrative research will become a mainstay approach from easy access to construct personal and experiential stories with reliable systematic scoring making for greater confidence in research findings and their transportability in that data will be more objectively analyzed so as not to only rely on subjective interpretation of researchers. Narrative research will also continue to present as a useful tool for empowerment which is important for social construction of reality from the perspective of participants. In the discipline of health, for instance, patients' experiences will continue to be essential to assessing the performance of healthcare systems worldwide and are increasingly being utilized to inform quality improvement processes of which relational narrative is particularly appropriate. There is a growing need for research on person-centered narrative therapy which is a holistic and integrative approach designed to maintain well-being and quality of life for people with mental and health issues, as it includes the elements of care, the individual, the carers, and the family for which narrative research is ideally suited.

References

Andrews M, Squire C, Tamboukou M, editors. Doing narrative research. London: Sage; 2013.

Appel D, Papaikonomou M. Narratives on death and bereavement from three South African cultures: an exploratory study. J Psychol Afr. 2013;23(3):453–8.

Atkinson P, Delamont S. Rescuing narrative from qualitative research. Narrat Inq. 2006; 16(1):164–72.

Bleakley A. Writing with invisible ink: narrative, confessionalism and reflective practice. Reflective Pract. 2000;1(1):11–24.

Bochner AP. Notes toward an ethics of memory in autoethnographic inquiry. Ethical futures in qualitative research: decolonizing the politics of knowledge. 2007;197–208.

Bruner J. Life as narrative. Soc Res. 2004;71(3):691–710.

Caracciolo M. Narrative, meaning, interpretation: an enactivist approach. Phenomenol Cogn Sci. 2012;11:367–84.

Clandinin DJ. Engaging in narrative inquiry. Walnut Creek: Left Coast Press; 2013.

Clandinin DJ, Connelly FM. Narrative inquiry: experience and story in qualitative research. San Francisco: Jossey-Bass; 2000.

Clandinin DJ, Huber J. Narrative inquiry. In: McGaw B, Baker E, Peterson PP, editors. International encyclopaedia of education. 3rd ed. New York: Elsevier; in press.

Clandinin DJ, Rosiek G. Mapping a landscape of narrative inquiry: borderland, spaces and tensions. In: Clandinin DJ, editor. Handbook of narrative inquiry: mapping a methodology. Thousand Oaks: Sage; 2007. p. 35–76.

Clandinin DJ, Murphy MS, Huber J, Orr AM. Negotiating narrative inquiries: living in a tension-filled midst. J Educ Res. 2009;103(2):81–90.

Cochran L. The promise of narrative career counselling. In: Maree K, editor. Shaping the story: a guide to facilitating narrative counselling. Pretoria: Van Schaik; 2007. p. 7–19.

Connelly FM, Clandinin DJ. Stories of experience and narrative inquiry. Educ Res. 1990; 19(5):2–14.

Connelly FM, Clandinin DJ. Narrative inquiry. In: Green JL, Camilli G, Elmore P, editors. Handbook of complementary methods in education research. 3rd ed. Mahwah: Lawrence Erlbaum; 2006. p. 477–87.

Craig C, Huber J. Relational reverberation: shaping and reshaping narrative inquires in the midst of storied lives and contexts. In: Clandinin DJ, editor. Handbook of narrative inquiry: mapping a methodology. Thousand Oaks: Sage; 2007. p. 251–79.

Creswell JW. Educational research: planning, conducting, and evaluating quantitative. 4th ed. Upper Saddle River: Prentice Hall; 2012.

Crotty M. The foundations of social research: meaning and perspectives in research process. London: Sage; 1998.

Currie G. Narratives and narrators: a philosophy of stories. Oxford: Oxford University Press; 2010.

Denzin NK, Lincoln Y. The landscape of qualitative research: theories and issues. Thousand Oaks: Sage; 2000.

Ellis C, Bochner AP. Autoethnography, personal narrative, reflexivity: researcher as subject. In: Denzin NK, Lincoln YS, editors. Handbook of qualitative research. 2nd ed. Thousand Oaks: Sage; 2000. p. 733–68.

Ellis C, Bochner AP. Autoethnography, personal narrative, reflexivity: researcher as subject. In: Denzin NK, Lincoln YS, editors. Handbook of qualitative research. 2nd ed. London: Sage; 2005. p. 733–68.

Eloff I. Narrative therapy as career counselling. In: Maree K, Ebersohn L, editors. Lifekills and career counselling. Sandton: Heinemann; 2002. p. 129–38.

Errante A. But sometimes you're not part of the story: oral histories and ways of remembering and telling. Educ Res. 2000;29(2):16–27.

Frank AW. Illness and autobiographical work: dialogue as narrative destabilization. Qual Sociol. 2000;23(1):135–56.

Frank AW. Why study people's stories? The dialogical ethics of narrative analysis. Int J Qual Methods. 2002;1(1):109–17.

Freeman M. Autobiographical understanding and narrative inquiry. In: Clandinin DJ, editor. Handbook of narrative inquiry: mapping a methodology. Thousand Oaks: Sage; 2007. p. 120–45.

Gadow S. Relational narrative: the postmodern turn in nursing ethics. Sch Inq Nurs Pract. 1999; 13(1):57–70.

Gibson M. Narrative practice and social work education: using a narrative approach in social work practice education to develop struggling social work students. Practice. 2012;24(1):53–65.

Heidari F, Amiri A, Amiri Z. The effect of person-centered narrative therapy on happiness and death anxiety of elderly people. Asian Soc Sci. 2016;12(10):117–26. https://doi.org/10.5539/ass.v12n10p117.

Herman D. Basic elements of narrative. Chichester: Wiley-Balckwell; 2009.

Jeon YH, Kraus SG, Jowsey T, Glasgow NJ. The experience of living with chronic heart failure: a narrative review of qualitative studies. BMC Health Serv Res. 2010;10(77):2–9. https://doi.org/10.1186/1472-6963-10-77.

Kim JH. For whom the school bell toll: conflicting voices inside an alternative high school. Int J Educ Arts. 2006;7(6):1–19.

Kim JH, Latta MM. Narrative inquiry: seeking relations as modes of interactions. J Educ Res. 2009;103(2):69–71.

Kim SK, Park M. Effectiveness of person-centered care on people with dementia: a systematic review and meta-analysis. Clin Interv Aging. 2017;12:381–97. https://doi.org/10.2147/CIA.S117637.

Maree JG, Ebersöhn L, Biagione-Cerone A. The effect of narrative career facilitation on the personal growth of a disadvantaged student – a case study. J Psychol Afr. 2010;20(3):403–11.

McMullen C, Braithwaite I. Narrative inquiry and the study of collaborative branding activity. Electron J Bus Res Methods. 2013;11(2):92–104.

Murphy N, Aquino-Russell C. Nurses practice beyond simple advocacy to engage in relational narratives: expanding opportunities for persons to influence the public space. Open Nurs J. 2008;2(40):40–7.

Newby P. Research methods for education. 2nd ed. New York: Routledge; 2014.

Ngazimbi EE, Hagedorn WB, Shillingford MA. Counselling caregivers of families affected by HIV/AIDS: the use of narrative therapy. J Psychol Afr. 2008;18(2):317–23.

Ntinda K. Constructing a framework for use of psychometric tests in schools: a consumer-oriented approach. Unpublished doctoral dissertation. Botswana: University of Botswana; 2012.

Nwoye A. A narrative approach to child and family therapy in Africa. Contemp Fam Ther. 2006; 28(1):1–23.

Rhodes C, Brown AD. Narrative, organizations and research. Int J Manag Rev. 2005;7(3):167–88.

Riessman CK. Narrative methods for the human sciences. Los Angeles: Sage; 2008.

Riessman CK, Quinney L. Narrative in social work: a critical review. Qual Soc Work. 2005; 4(4):391–412.

Savin-Baden M, Niekerk LV. Narrative inquiry: theory and practice. J Geogr High Educ. 2007; 31(3):459–472.

Spector-Mersel G. Narrative research: time for the paradigm. Narrat Inq. 2010;20(1):204–24.

Trahar S. Beyond the story itself: narrative inquiry and autoethnography in intercultural research in higher education [41 paragraphs]. Forum Qual Soc Res. 2009;10(1):Art. 30. http://nbn-resolving.de/urn:nbn:de:0114-fqs0901308.

Tsianakas V, Maben J, Wiseman T, Robert G, Richardson A, Madden P, Griffin M, Davies EA. Using patients' experiences to identify priorities for quality improvement in breast cancer care: patient narratives, surveys or both? BMC Health Serv Res. 2012;12(271):2–11. https://doi.org/10.1186/1472-6963-12-271.

Wang CC. Conversation with presence: a narrative inquiry into the learning experience of Chinese students studying nursing at Australian universities. Chin Nurs Res. 2017;4:43–50.

Wang CC, Geale SK. The power of story: narrative inquiry as a methodology in nursing research. Int J Nurs Sci. 2015;2(2):195–8.

Wood K, Chase E, Aggleton P. 'Telling the truth is the best thing': teenage orphans' experiences of parental AIDS-related illness and bereavement in Zimbabwe. Soc Sci Med. 2006; 63(7):1923–33.

Zulu NT, Munro N. "I am making it without you, dad": resilient academic identities of black female university students with absent fathers: an exploratory multiple case study. J Psychol Afr. 2017;27(2):172–9.

The Life History Interview

25

Erin Jessee

Contents

Abstract

In this chapter, I explore the "best practices" and core values with which researchers should align when conducting life history interviews to elicit information about an individual's past and present lived experiences. Drawing primarily on literature from the multidisciplinary field of oral history, I outline the process of determining in which circumstances life history interviews might be beneficial for addressing a research question and how life history interviews are typically designed, conducted, and analyzed. I also examine the challenges that can arise when conducting life history interviews, particularly when investigating sensitive subject matter or working in conflict-affected settings, for example. In the process, I reflect on over a decade of fieldwork in post-genocide Rwanda and Bosnia, wherein discussions of the past are often highly politicized and researcher fatigue – particularly related to the recent atrocities – is common. This provides a starting point for discussing how the best practices for life history interviewing may need to be adapted to ensure that they remain culturally and politically appropriate in different settings. Taken together, the chapter provides readers

E. Jessee (✉)
Modern History, University of Glasgow, Glasgow, UK
e-mail: erin.jessee@glasgow.ac.uk

© Springer Nature Singapore Pte Ltd. 2019
P. Liamputtong (ed.), *Handbook of Research Methods in Health Social Sciences*,
https://doi.org/10.1007/978-981-10-5251-4_80

with a foundation for deciding where life history interviews might enhance their research, and how to adapt current best practices on life history interviewing to suit their research needs and maintain a high ethnic standard in their fieldwork when documenting intimate details about participants' lives.

Keywords

Life history · Interview · Ethics · Methodology · Intersubjectivity · Memory

1 Introduction

The *life history interview* – a common means of documenting an individual's account of their life – is one of several types of interviews used by scholars and practitioners from a range of disciplinary backgrounds (Yow 2014). As a methodology, it is valued for its ability to amplify individual actors' voices and privilege their insights on historical events in tandem with those of the researcher and other relevant sources. For this reason, practitioners stress that the life history interview is distinct from the autobiographical narrative and other related methods aimed at revealing people's lived experiences (see ▶ Chaps. 23, "Qualitative Interviewing," ▶ 24, "Narrative Research," and ▶ 30, "Autoethnography"). The interviewer's presence, and their efforts to guide the interviewee's reflections, however minimal, as well as the interviewer's efforts to bring the resulting narrative into conversation with a broader body of academic literature, mean that practitioners understand the life history interview to be a co-creation between the interviewee and the interviewer. Oral historians refer to this collaborative process as *sharing authority* – "a complex, demanding process of social and self discovery" (Frisch 1990, p. 112) based on "a reimagination of the past that is being shared in a joint moment between the narrator and interviewer" (Abrams 2016, p. 27). As such, practitioners do not typically regard the life history interview as an objective source of information about the past. Instead, they engage with its subjective and intersubjective nature to explore the meaning that past has for an individual at a particular moment in time and what this can tell us the production of historical knowledge surrounding a given topic. Thus, they approach the narratives they elicit with a critical eye (as should always be the case when working with oral or written sources), anticipating that it will encapsulate a blend of fact and fiction that is unique to the narrator and the time when the interview was conducted (see, e.g., Passerini 1979, 2007; Portelli 2016; Lummis 1988; Abrams 2016).

For these reasons, the life history interview is arguably best suited to those research questions where a comprehensive understanding of an interviewee's subjective worldview is desirable. Life history narratives are particularly beneficial for revealing "the tangle of relations" and symbiotic interactions that exist between an individual's memories and those memories that circulate in the broader cultural circuit in which individuals are embedded – referred to most commonly as *collective memory* (Basu 2011, p. 33; see also, Summerfield 2004). It can facilitate a more nuanced understanding of how public and private narratives of a historical event can evolve over time in response to a range of personal, political, cultural, and social factors, resulting in selective remembering and forgetting that greatly shapes what people know and understand about the past, alongside their lived experiences of past events. And the reverse is also true. As people try to make sense of their past experiences by sharing their stories publicly, their narratives can also gradually influence what the broader public holds understands to be "true" about historical events and actors. By understanding this symbiotic relationship and its impact on personal and collective understandings of the past, practitioners can – in the spirit of renowned anthropologist Michel-Rolph Trouillot (1995) – offer insights on the production of history under different governments and within different communities, for example, as well as at different points in time.

Despite these potential strengths, however, the life history interview is not without complications and potential pitfalls. Of particular importance, life history interviews can be quite time-consuming and often involve long-term engagements with participants and their communities that over time can pose challenges for researchers and their analysis, as well as for participants. This is particularly true when working in crisis- or conflict-affected communities, among other highly politicized research settings, as well as with people whose memories may have been impacted by traumatic or distressing experiences or whose narratives the researcher finds repellent (see, e.g., Blee 1993; Thomson 2010; Adler et al. 2011; Jessee 2011; Field 2012; Cave 2016). Similarly, the commonly cited "best practices" for life history interviewing often need to be adapted to address the specific needs of individual researchers and participants, as well as the specific cultural and political contexts in which the interviews will occur (Jessee 2017a). For these reasons, researchers and practitioners should carefully consider whether life history interviews are personally, culturally, and politically appropriate for their research projects, and in instances where they decide to proceed, revisit their ethical and methodological frameworks throughout their fieldwork, and adapt their methodology as necessary to ensure it remains safe and ethical for the people and communities with whom they are working (see also ▶ Chaps. 106, "Ethics and Research with Indigenous Peoples," and ▶ 108, "Ethical Issues in Cultural Research on Human Development").

In the following sections, I will discuss how to determine whether life history interviews might be appropriate for a given research project. I will then outline the core values and practices that practitioners should use as a starting point for their research. Having covered the basic principles of life history interviewing, I will then reflect on a decade of experience conducting life history interviews in Rwanda and Bosnia to illuminate the potential pitfalls of this particular method and highlight the need for careful, ongoing adaptation of our best practices as our research progresses. In doing so, I will draw primarily upon literature from the multidisciplinary field of oral history, within which the life history is but one type of interview that practitioners may use. Taken together, the chapter will provide readers with a foundation for maintaining high ethnic standards in their fieldwork when documenting intimate details about interviewees' lives and their insights about the past and present.

2 Approaching the Life History Interview: Preliminary Considerations

In considering whether the life history interview might be appropriate methodology to employ for a research project, a good starting point is to consider whether the project might benefit from the wealth of data and insight that can emerge from engaging with people's firsthand experiences and inherited memories. Early practitioners recognized that the life history interview can be essential for developing an understanding of historical events and actors that consists not just of factual statements but also the narrator's "memory, ideology and subconscious desires" (Passerini 1979, p. 84). They likewise recognized that by using interviews and other oral sources to complement the historical record, they could serve a social purpose. For example, in 1988, oral historian Paul Thompson (2016, pp. 34–35) noted that:

> [u]ntil the present century, the focus on history was essentially political: a documentation of the struggle for power, in which the lives of ordinary people, or the workings of the economy or religion, were given little attention except in times of crisis... This was partly because historians, who themselves belonged to the administering and governing classes, thought that this was what mattered most... But even if they had wished to write a different kind of history, it would have been far from easy, for the raw material from which history was written, the documents, had been kept or destroyed by people with the same priorities. The more personal, local, and unofficial a document, the less likely it was to survive. The very power structure worked as a great recording machine shaping the past in its own image.

For this reason, the life history interview – and the practice of oral history more generally – was conceptualized by many early practitioners as a powerful means of challenging the elite power structures that had previously controlled the production of history in many settings around the world. Thompson focuses on documenting the life histories of working class people to enhance understanding of working class families and communities. However, early efforts to address this power imbalance are not limited to working class communities. Oral historian Daniel Kerr (2016) has highlighted the pioneering work of popular educators like Myles Horton, Septima Clark, Ella Baker, and Paolo Freire, who starting in the 1930s were actively conducting research among oppressed communities in the United States. Their common goal was to evoke the life experiences of the working classes, people of color, and members of lesbian, gay, bisexual, transgender, and intersex (LGBTI) communities, among other people they perceived as vulnerable to political violence. They then disseminated the resulting narratives to the public to promote enhanced understanding of the structural oppression and discrimination that undermined the social vitality of these marginalized communities. Taken together, the "radical" social roots of the life history interview – at least in terms of their application across the Global North – become clear, as does the rationale underlying practitioners' focus on the life history interview as a means of "democratizing history," revealing the multiple truths of people's lived experiences and complicating the grand historical narratives that might otherwise exclude them from the historical record, giving rise to a popular "history from below" (see, e.g., Samuel 1976).

However, in deciding whether the life history interview might be an appropriate method for a given research project, it is important to assess the project's legal and ethical merits and whether it is personally, culturally, and politically appropriate for participants. As a starting point, researchers should familiarize themselves with their discipline's best practices surrounding "informed consent," according to which the researcher provides potential recruits with enough information up-front to ensure that they clearly understand "the facts, implications, and future consequences" of their participation in the research project, enabling them to make an informed decision as to whether or not they want to participate (OHS 2012). In instances where the desired participants are below the age of consent or have impaired judgement – for example, due to severe learning disabilities, dementia, or intoxication – it may be impossible for them to give informed consent, requiring special protocols to be introduced, such as the acquisition of informed consent from a legal guardian.

Additionally, due to the method's goal of evoking detailed life stories – often documented over multiple interview sessions stretched over days, weeks, or even months – participants often become highly visible within the research project, posing potential challenges to the researcher's ability to preserve participants' confidentiality, where requested (OHS 2012). Researchers should consider whether in visiting with participants – particularly when the researcher is a visible outsider in the communities where they conduct fieldwork, or where family, friends, neighbors, and local officials are likely to take an active interest in the comings and goings of newcomers – they will be able to maintain people's confidentiality, if participants request it. Similarly, efforts to anonymize participants' contributions can be easily stymied. The personal details that often emerge during life history interviews can make it possible for participants' friends, families, and other intimates to identify which pseudonym the researcher has used in reference to their contributions to their writing, presentations, and exhibits, even in the absence of photographs and video and audio recordings (Jessee 2011; see also ▶ Chaps. 106, "Ethics and Research with Indigenous Peoples," ▶ 107, "Conducting Ethical Research with People from Asylum Seeker and Refugee Backgrounds," and ▶ 108, "Ethical Issues in Cultural Research on Human Development").

It is also important to consider up-front the labor we are requesting of our participants in conducting lengthy interviews, particularly if we are not offering to reimburse them for their time. While some people might be eager for an opportunity to speak at length about their lives and not expect or need compensation, others might find that the researcher's requests place them in a difficult position personally or financially, by taking time away from their daily subsistence activities and employment, for example (Liamputtong 2007, 2010). While researchers may have valid ethical concerns about paying participants for their life histories – key among them the fear that the promise of money might coerce people into consenting to participate in research projects against their best interests (see, e.g., Russell et al. 2000; Head 2009) – organizations like the American Anthropological Association warn researchers against exploiting participants and call upon them to "compensate contributors justly for any assistance they provide" (AAA 2012). Furthermore, where researchers deem financial compensation to be inappropriate or infeasible, there are a host of other options for "giving back" to the participants and communities who feature in our research, from engaging in volunteer work to ensuring we work with participants to identify "research-as-intervention" opportunities that can facilitate meaningful positive change in our participants' daily lives (Swartz 2011, p. 50).

The researcher's schedule and budget is another key concern related to the life history interview. Due to the lengthier nature of the life history interview and the large amounts of data it can generate, researchers should carefully consider up-front how many life history interviews they can reasonably complete given the time frame for their research project and the funds available to them. This involves consideration of not only the interviews themselves but costs associated with travel to and from the interviews and potential financial and in-kind compensation for participants and the time involved in transcribing or summarizing, editing, annotating, and analyzing the interviews in preparation for their eventual dissemination in the various formats desired by the researcher and their collaborators. There may also be costs and additional time requirements associated with the different software researchers decide to use in editing and annotating their interviews, or making them available via online repositories and archives, all of which should be assessed prior to the start of interviews and budgeted for accordingly.

3 Conducting Life History Interviews: Best Practices

Once the researcher has addressed these preliminary considerations and decided to proceed with life history interviews, there are a number of reliable resources they can use to familiarize themselves with the "best practices" for conducting and analyzing interviews. Particularly notable, the American Oral History Association (OHA 2009) and Oral History Society (OHS 2012) offer valuable guidelines for conducting interviews that range from the aforementioned legal and ethical concerns to more practical concerns related to drafting the necessary consent and recording agreement forms and creating an appropriate interview guide for your research project. A number of oral history centers around the world also offer introductory workshops and templates that newcomers to the field can use to prepare themselves. For example, in addition to running regular training workshops, the Centre for Oral History and Digital Storytelling at Concordia University in Montréal, Canada, has an impressive online "toolbox" that includes free webinars on the basics of oral history interviewing, transcription, and walking interviews as well as sample consent forms, among other valuable resources (COHDS 2018). Similarly, the Columbia Center for Oral History Research in New York offers workshops and short courses but has also drafted a series of research guides that are freely available on its website, including one that specifically addressed the challenges posed by using oral history methods to document and interpret conflict and related human rights abuses (CCOHR 2018).

However, one of the most accessible and thorough sources of information on best practices surrounding the interview is oral historian Valerie Yow's *Recording Oral History: A Guide for the Humanities and Social Sciences* (2014). In addition to detailing all aspects of what she terms the "in-depth interview," Yow includes a series of appendices that practitioners can use to create their own sample interview guide, which outlines the questions they intend to ask, legal release forms specific to the United States, archive release forms for the long-term preservation of the interviews, participant information sheets for informing potential participants about the researcher's project and their rights within it, indexing forms for annotating the resulting recordings, and citation guidelines. In terms of the practical documentation that researchers should have in hand in preparation for starting the interview phase of their research, *Recording Oral History* is an essential source of information based on Yow's decades of experience working in academic and public settings.

To this end, most guidelines recommend that first-time interviewers gain some kind of training in oral history prior to recruiting potential participants and take the time to ensure that all necessary paperwork and related procedures for establishing informed consent, arranging for long-term preservation of the interviews, and other key elements of their project are in place. In instances where researchers are working in foreign contexts, in conflict-affected settings, or on potentially sensitive subject matter, they may want to pursue additional forms of training, such as Basic First Aid, Hazardous Environment Training, and trauma counseling. They should also make sure they clearly understand their legal and ethical responsibilities as outlined by their host institutions, which likely requires some form of institutional ethics approval, but may also be involving completing a risk assessment form for insurance purposes in the event the researcher plans to work abroad. Researchers should also take care to familiarize themselves with any code of ethics, laws, or research approval processes specific to the communities and nations in which they will be working (Fujii 2012). For example, in order to conduct human subject research in Rwanda at present, foreign researchers are required to arrange various forms of approval including at minimum ethics approval from the Rwanda National Ethics Committee, a research permit from the Ministry of Education, and a research visa from the Rwanda Directorate General of Immigration and Emigration. They may also require institutional approval from other government ministries or community-based organizations in order to secure the necessary supporting paperwork from in-country partners in their research, as all research in Rwanda must be collaborative in order to ensure the knowledge being gained benefits Rwandans, as well as foreign researchers (Jessee 2012).

Most guidelines similarly recommend that researchers prepare an interview guide that details the specific questions and themes they intend to address. The interview guide is typically understood as a backup, rather than a strict guide to which the interviewer and interviewee must adhere. To this end, the life history interview may be quite different from other types of interviews. Generally speaking, the life history interview should be directed by the interviewee, with the interviewee speaking in as little or as much detail as they feel is necessary to narrate those events and experiences they feel are most relevant. However, interviewers may still find it helpful to prepare an interview guide to get them thinking in advance about the kinds of topics they want to discuss. And in the event that the interviewee is reluctant to take the lead in the interview, the guide can be a helpful point of reference for kick-starting the conversation until the interviewee feels more comfortable.

In drafting a life history interview guide, many practitioners advocate a chronological approach. For example, Yow's sample interview guide begins with prompting the interviewee for some basic background information related to their birthplace, parents' names and occupations, and family's cultural life before shifting to specific memories related to the different life stages – the interviewee's childhood, adolescence, and key life events, such as marriage and raising children – until the interviewee reaches the present (Yow 2014). This makes for a logical narrative format in many Euro-American settings, though it is important to be mindful that in some communities, "good storytelling" may look quite different (Cruikshank 1990; Krog, Mpolweni, and Ratele 2009). Furthermore, adopting a chronological approach may prompt interviewees to revisit their memories in a way that is distressing or culturally inappropriate. Oral historian Amy Tooth Murphy (2014) has noted that in her life history interviews with British lesbians, her efforts to encourage a chronological format in the interviews often forced the women she interviewed to revisit painful and distressing memories that they might otherwise have chosen to avoid or reflect upon at a different point in the interview.

I encountered a similar phenomenon in my fieldwork in post-genocide Rwanda and Bosnia-Herzegovina. While interviewing people whose lives had been negatively affected by genocidal violence, I noticed that participants often seemed to find it jarring and painful to start with their childhood, reflecting on memories of deceased loved ones and a way of life that no longer existed, even in cases where they demonstrated remarkable resilience in their post-genocide lives. Thus, I quickly adapted my life history interview guide to start with an open-ended request: "Tell me about your life." Participants may still have chosen to talk about distressing memories or experiences – indeed, people's personal experiences of genocide were typically focal points of the interviews I conducted – but at least participants were able to broach these memories on their own terms and in their own time (Jessee 2017b, pp. 83–84). Furthermore, encouraging participants to take the lead in the interview granted me a sense of which experiences they found possible to narrate. Simultaneously, it afforded me a sense of which topics participants might find incommunicable, unbearable, or irretrievable, and where I might need to proceed with special caution or sensitivity, or avoid altogether (Greenspan 2014). In settings where people are negotiating trauma, defined as a range of psychological or psychosomatic symptoms that temporarily interfere with the "normal functioning" of their mind or nervous system (Rothschild 2011, p. 18), or post-traumatic stress disorder, described as long-term "clinically significant distress or impairment of an individual's social interactions, capacity to work or other important areas of functioning" resulting from "exposure to actual or threatened death, serious injury or sexual violation" (American Psychiatric Association 2013), this approach can provide participants with a crucial degree of control over the interview. This, in turn, can make for a more positive and less distressing experience than might otherwise be the case if the researcher insists upon the more common chronological format or otherwise attempts to adhere to a strict interview guide or questionnaire.

To this end, the life history interview can be a more intimate form of interview than participants typically expect in settings where shorter, thematic interviews aimed at understanding a clearly defined phenomenon are the norm. Where time permits, non-recorded pre-visits with potential participants – perhaps combined with open discussion of the project and its intended outcomes as guided by the participant information sheet – can be valuable for helping to establish positive rapport between the interviewer and the interviewee. It provides the interviewee with an opportunity to ask questions and address any concerns they may have about the project or the researcher. The OHA's "principles and best practices" guidelines (2012) note that the interviewee should have a clear understanding of their rights surrounding the interview, how the interview will be used within the broader project including copyright agreements, access restrictions, and expected forms of dissemination, and whether or not they would like their confidentiality to be maintained in the documentation, preservation, and dissemination of any data that results from their participation. Given the pre-visit typically occurs off-the-record, it can be a good time to map any topics that participants are keen to discuss and conversely any subjects that they consider off-limits for the interview. The pre-visit is also an ideal time to address any potential harms that could affect participants as a result of the specific subject matter being discussed. For example, if it is possible that the interviewee might – in the course of reflecting on their life – discuss past or present crimes or other potentially sensitive subjects that places the researcher in a legally or ethically problematic position wherein they may feel obligated to inform the police, the researcher should mention this potential risk and take steps in conversation with the interviewee to ensure this risk is minimized, as much as is possible. In my work with convicted perpetrators of the 1994 genocide in Rwanda, I always informed the people I interviewed up-front that it may not always be within my power – due to the likelihood of surveillance within the prisons – for me to guarantee them confidentiality, and so to be especially careful of discussing crimes for which they had not already been convicted. This was a primary reason why so many of the convicted perpetrators I interviewed spoke about atrocities they "observed" within their communities, often without indicating their role in these atrocities or who else might have been involved (Jessee 2017b, pp. 149–188).

The necessity of such caution is demonstrated by the recent controversy surrounding Boston College's "Belfast Project," which used oral history methods to document "The Troubles" in Northern Ireland. The project's participants included former leader of the Irish Republican Army (IRA), Brendan Hughes. As part of his interviews, Hughes allegedly discussed his role in organizing the Bloody Friday attack, during which the IRA detonated approximately 19 car bombs across Belfast in an hour on 21 July 1972, killing 9 and injuring an estimated 130 civilians (BBC News 2017). He also allegedly disclosed information about the criminal actions of Sinn Féin's leader, Gerry Adams, who had controlled a paramilitary squad that was allegedly responsible for kidnapping, murdering, and disappearing perceived enemies of the IRA. Knowledge of these disclosures prompted the Police Service of Northern Ireland to submit a legal bid in 2011 to gain access to all of the Belfast Project interviews. The project's lead researcher, Anthony McIntyre, received a subpoena demanding he give the police the relevant interviews, and upon refusing found himself embroiled in a legal battle (McDonald 2016). This, in turn, prompted panic throughout the oral history community related to whether practitioners can, in such extreme circumstances, genuinely protect participants from legal investigation and prosecution where police suspected that participants have discussed their own or others' criminal actions. The OHS issued a formal statement in 2014 that condemned the Police Service's efforts to violate participants' confidentiality within the Belfast Project but simultaneously noted that the case should serve "as a warning not only to oral historians, but to all those engaged in collecting historical data about criminal activity or allegations of criminal offences" (OHS 2014). This warning is particularly well-considered by those engaging in life history interviews, again due to the detailed information and lengthy narratives we often elicit from participants, which can make it easier for people to be identified even in the absence of a name, date of birth, or other information.

As the life history interview takes shape, practitioners encourage participants to take whatever time is necessary to fully develop their life story. Ideally, the researcher should not need to ask many questions, but merely follow the interviewee's lead. Where the interviewee requests that the researcher ask questions to help guide them through the interview, these questions should be open-ended to encourage reflection, rather than "yes" or "no" responses from the interviewee (Yow 2014, p. 79; Ritchie 2015, p. 92). Likewise, the interviewer can use follow-up questions and prompts to direct the interviewee and demonstrate interest. However, in the life history interview, these kinds of interruptions should, ideally, be minimized to empower the interviewee and avoid skewing the narrative with the researcher's opinions (Ritchie 2015).

The life history interview may need to be conducted over several sessions in the event that the interviewee has a great deal to say, or just as likely, may be brief or perfunctory in the event that participants feel they do not have much of importance to say on-the-record, for example. Practitioners generally regarded it as polite to set an approximate time limit in advance, to which they then try to adhere to avoid exhausting themselves and the interviewee (OHA 2009). Likewise, the resulting narrative may be a carefully rehearsed version of events, a spontaneous reaction in the moment to the interviewer and their interests, or a blend of both. Oral historian Lynn Abrams (2016, pp. 64–86) has written extensively about the important interplay between the subjectivity – an individual's sense of self – of the interviewer and interviewee, and how this may influence any interview's course and content, among other interactions between researchers and participants. She encourages practitioners to be especially mindful and honest about the way that gender, heritage, age, class, profession, and other intersecting facets of the researcher's and interviewee's identities may influence the interview (see also, Bouka 2015). Sociologist Kathleen Blee's work (1993, 2002) among women members of the Ku Klux Klan is particularly helpful for those researchers who work among so-called "unloved" participants – "individuals who are immersed in events or subject matter marked by conflict and controversy, and from whom other often seek to distances themselves due to the morally reprehensible nature of their actions" (Jessee 2017b, p. 152; see also, Fielding 1990). While ideally researchers hope to establish positive rapport with participants and conduct insightful interviews marked by "engaged and sympathetic interaction," Blee (1993, p. 327) recognizes this may not always be possible or desirable.

To this end, practitioners like oral historian Alexander Freund (2013) advocate paying particular attention to not just the words spoken in an interview but the silences that become apparent. It can be difficult for interviewers to know where to draw line between probing silences to determine why they might exist – a case of genuine forgetfulness on the part of the interviewee, for example, that is not indicative of a conscious plan to withhold what might otherwise be important information for the researcher – versus respecting the participant's privacy, where discussion might cause them emotional distress or embarrassment, for example. Freund (2013) has called upon researchers to develop an "ethics of silence" that recognizes a participant's silence as an act of agency that should be prioritized above the researcher's desire to document "the whole story" or to amplify the experiences of people who might otherwise have been excluded or underrepresented in the historical record. Indeed, Freund reminds us that silence – particularly when manifested in secrets, taboos, and misinformation – can be a powerful weapon for subaltern communities. For this reason, he recommends taking the time to document silences in the interview in the transcripts and summaries we produce to make them visible in the historical record. He likewise encourages practitioners to ask questions about silences – ideally during an unrecorded session after the interview, where possible – to get "the story behind the story." This does not mean forcing participants to answer questions on topics they have clearly expressed a desire to avoid, but rather asking them why they wish to avoid the topic, rather than the researcher making assumptions as to their motives and potentially colonizing the silence with their interpretation (see also, White 2000).

Once the interviewee concludes their life history, the researcher may want to ask some follow-up questions in order to address any lingering research interests that went undiscussed or to otherwise achieve some balance between the project's broader objectives and the interviewee's narrative (OHA 2009). Oral historian Donald Ritchie (2015, p. 108) advocates having a prepared "wrap-up question" that encourages the interviewee to reflect back on their life as a whole. It can also be helpful to debrief with the interviewee, where time permits, to invite them to reflect on the interview experience itself. Before finally ending the session, the researcher should briefly revisit the consent and recording agreement forms to make sure that the interviewee is comfortable with the conditions that surrounded their interview and that any changes to the interviewee's wishes are clearly documented (OHS 2012). The final versions of these forms should be archived alongside any data that results from the life history interview.

4   After the Interview: Analyzing and Disseminating Life History Interviews

This leads to considerations of how best to proceed after the life history interview. Where participants have requested confidentiality, researchers should immediately ensure that all personally identifying information has been edited out of any interview recordings and related materials that will end up in an archive. This includes any transcripts or time-coded summaries they might produce. It may also be desirable, where time, funding, and participant interest permits, to vet these documents with individual interviewees to ensure they are happy with the final content of the interview and satisfied that their confidentiality is being maintained. This is particularly good practice in settings where the political climate surrounding the topics discussed in the life history interview can change rapidly and unexpectedly, as is often the case in conflict-affected communities or when dealing with politically controversial subject matter.

As researchers begin to analyze the life history interviews they have conducted, in addition to the aforementioned importance of silences, they should be aware of the potential for what anthropologist Antonius Robben (1996) has termed "ethnographic seduction." Robben (1996, p. 74) defines ethnographic seduction as "a complex dynamic of conscious moves and unconscious defenses" through which an interviewee attempts to "influence the understanding and research results of their interviewers." He argues that ethnographic seduction is a common element of any human subject research that can significantly impact both the content and course of an interview or ethnographic encounter, as well as how researchers analyze the resulting data. This risk may be heightened surrounding the life history interview, the intimate and prolonged nature of which may foster deeper emotional connections between the interviewer and interviewee. For this reason, Robben cautions researchers to be especially mindful of "frequent thoughts about the interviewee, possibly accompanied by depression, dreams and fantasies, as well as slips of the tongue and the compulsion to talk to others about the interviewee" as signs that ethnographic seduction could be interfering with the researcher's ability to analyze and understand participants' narratives in a balanced and critical manner (Robben 1996, p. 99; see also Jessee 2017a, p. 338).

I constantly navigated this phenomenon in my work among convicted perpetrators of the 1994 genocide in Rwanda, many of whom insisted they were victims, even as they discussed the ways they contributed to the torture, murder, and mutilation of their ethnic Tutsi compatriots. At times, their claims to victim status seemed genuine as they described, for example, being coerced by armed militia who threatened to kill them if they did not agree to join in the massacre of Tutsi civilians in their communities. But in other instances, their claims seemed disingenuous – cloaked in a morally reprehensible understanding of the past and the sense that the atrocities they perpetrated where somehow justified by their nation's history or perceived historical injustices visited upon the Hutu majority by the Tutsi-dominated monarchy that had ruled the nation prior to independence in 1962 (Jessee 2018). As such, I had to be keenly aware of how my efforts to build positive relationships with convicted perpetrators might influence my understanding of the genocide and Rwandan history and by extension, my research outcomes – at times, it seemed more so than other interviewees and informants. However, Blee (2017, p. 16) has noted that any relationships that emerge in the course of fieldwork – from gatekeepers who provide permits and contacts to the specific individuals who are ultimately interviewed – will shape a researcher's theorizing "by affecting not only what researchers can access but what they notice or find puzzling and what they regard as significant in a research setting."

To this end, the process of analyzing life history interviews can be a lengthy endeavor and one that involves multiple listenings to the recordings and revisiting associated data in order to glean a thorough understanding of not only what the interviewee has said but the deeper meaning behind their words. As indicated in the introduction, the ideal outcome is for the researcher to develop a nuanced understanding of the meaning the past has for individuals in the present, as informed by their lived experiences and the broader cultural circuit in which they are embedded, as well as the myths they rely upon to give meaning to their lives. This is typically accomplished through a combination of reconstructive analysis, whereby the life history is used to reconstruct an approximation of a participant's lived experiences, and narrative analysis, which "identifies and then explains the ways in which people create and use stories to interpret the world" to create a "storied past" (Abrams 2016, p. 106). As part of this, practitioners may choose to analyze a single life history in conversation with a broader body of literature around particular events or kind of experience, for example. However, a more common approach is to bring multiple life histories into conversation by using excerpts that offer "thick description" on the topics being explored with the goal of identifying commonalities and outliers on a wide range of phenomenon from physical and mental health challenges in the aftermath of trauma to reconstructing working class lives in different settings around the world (Geertz 1973).

In terms of dissemination, these days the sky is the limit. Typical academic outcomes include theses and dissertations, as well as academic articles and books aimed at a specialist audience. But the life history interview is increasingly acknowledged as an integral way of demonstrating how policies, ideas, and events take on different meanings over time in response to a wide range of personal, social, political, and historical factors. In the process, life history interviews can amplify the voices of people whose perspectives are typically absent or obscured in the historical record in many contexts. For these reasons practitioners are increasingly experimenting with ways of making these interviews publicly accessible to ensure they have impact beyond academia. Across the physical and social sciences, and arts and humanities, researchers are using digital media to make the life histories they document and analyze accessible to the public in different ways – from museum exhibits and educational documentaries to video games and graphic novels. The rapidly expanding field of digital storytelling is particularly provocative and has captured the interests of publics around the world, as evidenced by the range of online workshops and related educational materials aimed at teaching people to engage in digital storytelling activities in their communities, as well as the vast number of digital storytelling projects and organizations that have taken shape around the world in the last 25 years (Lambert 2013; see also ▶ 74, "Digital Storytelling Method").

Digital storytelling is not without its critics, however. Freund (2015), among others, has warned against digital storytelling of the kind championed by organizations like the US-based multinational and transcultural phenomenon, StoryCorps. Freund (2015, p. 96) argues that StoryCorps "frequently aligns itself with (or appropriates) oral history, reinforces neoliberal values of competitive individualism and thus depoliticizes public discourse" in its efforts to bring people together around what appear to be positive, healing, and empowering narratives of personal transformation and social change. The Smithsonian's National Air and Space Museum's senior curator, Roger Launius (2013, p. 31), shares Freund's criticism, noting that the homogenizing story of America as "one nation, one people" promotes nostalgia for a mythical past in which the American people were "all one." Freund (2015, pp. 108–109) argues that the resulting "consensus history" is often invoked with particularly dangerous consequences by the political and social right, who use it to "silence citizen critique" of the state by situating hardship and failure as a matter of individual responsibility. For this reason, he advocates that practitioners resist the "vortex of storytelling" by ensuring that they adequately historicize the life history interviews and related materials that they analyze, and in dissemination take care "not to be mesmerized by the emotional power of the storytelling phenomenon or by the economic success of the storytelling industry" (Freund 2015, p. 130, 131). Such warnings are particularly salient in conflict-affected communities and other highly politicized settings where the international community, state-level actors, and other parties may have a stake in encouraging people to adopt and adhere to an idealized "single story" that might not accurately represent ordinary people's lived experiences or might serve to silence those people who experiences contradict the desired single story (Adichie 2009).

5 Conclusion and Future Directions

Taken together, this chapter has offered readers a starting point for thinking through the benefits and challenges that surround the life history interview as a methodology. It has demonstrated the democratizing potential of the life history interview as a means of engaging with people's intimate life experiences and the meaning they attribute to their past as a result of the cultural circuit in which they are embedded and bringing this highly personal interpretation into conversation with the typically elite-dominated historical record. But it has also explored the potential limitations of the life history interview, particularly in contexts where government surveillance makes it difficult to ensure participants' confidentiality and safety, as well as the well-being of others involved in the research project. Finally, it has pointed readers to key readings on the current "best practices" that life history interviewers typically adhere to in their research and offered a preliminary overview of key debates regarding the analysis and dissemination of the resulting life histories. The overall goal is to promote a high ethical standard in the elicitation and dissemination of life history interviews, regardless of the audience for which they are intended.

Despite the range of literature and related guidance on the use of life history interviews, the method is changing rapidly and engaging an ever-widening range of individuals and communities around the world. While practitioners have been excellent to date at articulating the ethical and legal responsibilities to which they should adhere, and analyzing the intersubjective nature of the life history interview and the way that individual and collective memories can shape life histories in different contexts, it is crucial that practitioners continue to reflect on the method's "best practices" and speak about the ways – both overt and subtle – that they adapt these practices in their research. This is particularly true of practitioners who come from diverse backgrounds, whose experiences of conducting life history interviews as people of color, members of religious minority communities, and gender or sexual minorities, for example, may be substantially different from those of the cisgender, heterosexual, Judeo-Christian, white majority in the Global North that typically dominates the literature (Berger Gluck and Patai 1991; Bouka 2015; Fobear 2016). It is similarly important for researchers who are working across national, regional, and linguistic boundaries or in crisis- or conflict-affected communities to discuss the particular challenges and ethical and legal responsibilities that they negotiate in using life history interviews to evoke people's experiences and opinions. And finally, as we increasingly experiment with digital media as a means of disseminating the life histories and related information that we evoke, honest reckonings with the benefits and limitations of the various technologies that are available could only help improve our understanding of which options might be personally, culturally, and politically appropriate in different contexts.

References

Abrams L. Oral history theory. 2nd ed. New York: Routledge; 2016.

Adichie C. The danger of a single story. TEDGlobal. 2009. https://www.ted.com/talks/chimamanda_adichie_the_danger_of_a_single_story. Accessed 16 Feb 2018.

Adler N, Leydesdorff S, Chamberlain M, Neyzi L, editors. Memories of mass repression: narrating life stories in the aftermath of atrocity. New Bruinswick: Transaction Publishers; 2011.

American Anthropology Association. Principles of professional responsibility. AAA Ethics Blog. 2012. http://ethics.americananthro.org/category/statement/. Accessed 10 Feb 2018.

American Psychiatric Association. Posttraumatic stress disorder. 2013. http://www.dsm5.org/Documents/PTSD%20Fact%20Sheet.pdf. Accessed 14 Feb 2018.

Basu L. Memory dispositifs and national identities: the case of Ned Kelly. Mem Stud. 2011;4 (1):$32#33–41.

BBC News. What are the Boston tapes? 2017. http://www.bbc.co.uk/news/uk-northern-ireland-27238797. Accessed 14 Feb 2018.

Berger Gluck S, Patai D, editors. Women's words: the feminist practice of oral history. New York: Routledge; 1991.

Blee K. Evidence, empathy, and ethics: lessons from oral histories of the clan. J Am Hist. 1993;80 (2):596–606.

Blee K. Inside organized racism: women in the hate movement. Berkeley: University of California Press; 2002.

Blee K. How field relationships shape theorizing. Sociol Methods Res. 2017;1–34. https://doi.org/10.1177/0049124117701482.

Bouka Y. Researching violence in Africa as a Black woman: notes from Rwanda. Research in Difficult Settings Working Paper Series. 2015. http://conflictfieldresearch.colgate.edu/wp-content/uploads/2015/05/Bouka_WorkingPaper-May2015.pdf.

Cave M. What remains: reflections on crisis oral history. In: Perks R, Thomson A, editors. The oral history reader. 3rd ed. New York: Routledge; 2016. p. 92–103. 2015.

Centre for Oral History and Digital Storytelling. Toolbox. 2018. http://storytelling.concordia.ca/toolbox. Accessed 14 Feb 2018.

Columbia Center for Oral Historical Research. CCOHR Services. 2018. http://www.ccohr.incite.columbia.edu/services-resources/. Accessed 14 Feb 2018.

Cruikshank J. Life lived like a story: Life stories of three Yukon Native elders. Vancouver: University of British Columbia Press, 1990.

Field S. Oral history, community, and displacement: imagining memories in post-apartheid South Africa. New York: Palgrave; 2012.

Fielding N. Mediating the message: affinity and hostility in research on sensitive topics. Am Behav Sci. 1990;33(5):608–20.

Fobear K. Do you understand? Unsettling interpretative authority in feminist oral history. J Fem Scholarsh. 2016;10:61–77.

Freund A. Toward an ethics of silence? Negotiating off-the-record events and identity in oral history. In: Sheftel A, Zembrzycki S, editors. Oral history off the record: toward an ethnography of practice. New York: Palgrave Macmillan; 2013. p. 223–38.

Freund A. Under storytelling's spell? Oral history in a neoliberal age. Oral Hist Rev. 2015;42 (1):96–132.

Frisch M. A shared authority: essays on the craft and meaning of oral and public history. New York: SUNY Press; 1990.

Fujii LA. Research Ethics 101: Dilemmas and Responsibilities. PS: Political Science & Politics. 2012;45(04):717–723.

Geertz C. The interpretation of cultures. New York: Basic Books; 1973.

Greenspan H. The unsaid, the incommunicable, the unbearable, and the irretrievable. Oral Hist Rev. 2014;41(2):229–43.

Head E. The ethics and implications of paying participants in qualitative research. Int J Soc Res Methodol. 2009;12(4):335–44.

Jessee E. The limits of oral history: ethics and methodology amid highly politicized research settings. Oral Hist Rev. 2011;38(2):287–307.

Jessee E. Conducting fieldwork in Rwanda. Can J Dev Stud. 2012;33(2):266–74.

Jessee E. Managing danger in oral historical fieldwork. Oral Hist Rev. 2017a;44(2):322–47.

Jessee E. Negotiating genocide in Rwanda: the politics of history. New York: Palgrave Macmillan; 2017b.

Jessee E. Beyond perpetrators: complex political actors surrounding the 1994 genocide in Rwanda. In: Smeulers A, Weerdesteijn M, Hola B, editors. Perpetrators of International Crimes. Oxford: Oxford University Press, forthcoming 2018.

Kerr D. Allan Nevins is not my grandfather: the roots of radical oral history practice in the United States. Oral Hist Rev. 2016;43(2):367–91.

Krog A, Mpolweni N, Ratele K. There was this goat: investigating the truth commission testimony of Notrose Nobomvu Konile. Scottsville: University of KwaZulu-Natal Press, 2009.

Lambert J. Digital storytelling: capturing lives, creating community. 4th ed. New York: Routledge; 2013.

Launius R. Public history wars, the 'one nation/one people' consensus, and the continuing search for a usable past. OAH Mag Hist. 2013;27(1):31–6.

Liamputtong P. Researching the vulnerable: a guide to sensitive research methods. London: Sage; 2007.

Liamputtong P. Performing qualitative cross-cultural research. Cambridge: Cambridge University Press; 2010.

Lummis T. Listening to history: the authenticity of oral evidence. Totowa: Barnes & Noble Books; 1988.

McDonald H. Boston college ordered by US court to hand over IRA tapes. 2016. https://www.theguardian.com/uk-news/2016/apr/25/boston-college-ordered-by-us-court-to-hand-over-ira-tapes. Accessed 14 Feb 2018.

Oral History Association. Principles and best practices. 2009. http://www.oralhistory.org/about/principles-and-practices/. Accessed 10 Feb 2018.

Oral History Society. Is your oral history legal and ethical? 2012. http://www.ohs.org.uk/advice/ethical-and-legal/2/. Accessed 10 Feb 2018.

Oral History Society. Oral history society statement on the Boston College Belfast Project. 2014. http://www.ohs.org.uk/documents/OHS_Statement_Boston_College_Belfast_Project_May2014.pdf. Accessed 14 Feb 2018.

Passerini L. Work, ideology, and consensus under Italian fascism. Hist Work J. 1979;8:82–108.

Passerini L. Memory and utopia: the primacy of intersubjectivity. London: Routledge; 2007.

Portelli A. What makes oral history different? In: Perks R, Thomson A, editors. The oral history reader. 3rd ed. New York: Routledge; 2016. p. 48–58.

Ritchie D. Doing oral history. 3rd ed. Oxford: Oxford University Press; 2015.

Robben A. Ethnographic seduction, transference and countertransference in dialogues about terror in Argentina. Ethos. 1996;24(1):71–106.

Rothschild B. Trauma essentials: the go-to guide. New York: W.W. Norton & Company; 2011.

Russell M, Moralejo D, Burgess E. Paying research subjects: participants' perspectives. J Med Ethics. 2000;26(2):126–30.

Samuel R. Local history and oral history. Hist Work J. 1976;1:191–208.

Summerfield P. Culture and composure: creating narratives of the gendered self in oral history interviews. Cult Soc Hist. 2004;1(1):65–93.

Swartz S. 'Going deep' and 'giving back': strategies for exceeding ethical expectations when researching amongst vulnerable youth. Qual Res. 2011;11(1):47–68.

Thompson P. The voice of the past: oral history. In: Perks R, Thomson A, editors. The oral history reader. 3rd ed. New York: Routledge; 2016. p. 33–9.

Thomson S. Getting Close to Rwandans since the Genocide: Studying Everyday Life in Highly Politicized Research Settings. Afr Stud Rev. 2010;53(03):19–34

Tooth Murphy A. The continuous thread of revelation: chrononormativity and the challenge of queer oral history. Scottish Oral History Centre seminar series. 2014.

Trouillot M. Silencing the past: power and the production of history. Boston: Beacon Press; 1995.

White L. Speaking with vampires: rumor and history in colonial Africa. Berkeley: University of California Press; 2000.

Yow V. Recording oral history: a guide for the humanities and social sciences. 3rd ed. New York: Rowman & Littlefield; 2014.

Ethnographic Method

26

Bonnie Pang

Contents

Abstract

This chapter explores the use of a range of ethnographic methods within qualitative methodology. Alongside introducing the building blocks of ethnographic methods, with a focus on reflexivity, participant-observation, fieldwork, and visual and sensory methods, it will draw upon my experiences in studying young Chinese Australians' lived experiences in health and physical activity. My experiences in the field provide insights into the potentials and perils of using some of these ethnographic methods and issues navigating research ethics. I emphasize the advantages of an approach that allows for innovative and unique interactions between the researcher and research participants within the

B. Pang (✉)
School of Science and Health and Institute for Culture and Society, University of Western Sydney, Penrith, NSW, Australia
e-mail: b.pang@westernsydney.edu.au

© Springer Nature Singapore Pte Ltd. 2019
P. Liamputtong (ed.), *Handbook of Research Methods in Health Social Sciences*,
https://doi.org/10.1007/978-981-10-5251-4_81

participants' real-life environments. The use of innovative ethnographic methods encourages thinking and practice beyond traditional modes of enquiry and beyond understanding the participants' lived experiences through texts and numbers alone.

Keywords

Ethnographic methods · Participant-observations · Reflexivity · Visual methods · Sensory methods · Lived experiences

1 Introduction

Ethnographic methods fall within the broader category of qualitative methodologies and are commonly used within health-related research (Berg 2004; Liamputtong 2013; Willis and Anderson 2017). It entails a variety of research techniques that aim to understand human actions, thoughts, and behaviors. Ethnographic methods can be used to gain further understanding of the research issues lurking behind the scenes of surveys and quantitative methods. They are well suited for understanding local points of view. Local voices – in particular, those of minority backgrounds, who may be linguistically diverse and disadvantaged, and whose voices have been undermined in research – are relatively difficult to access through traditional survey methods (e.g., Pang et al. 2015; see also ▶ Chap. 13, "Critical Ethnography in Public Health: Politicizing Culture and Politicizing Methodology"). This chapter introduces the main concerns in using various ethnographic methods (with a focus on reflexivity, fieldwork, observations, visual methods, and sensory methods). I also examine the potentials and perils of doing reflexive, embodied, and emplaced research in the field, offering a combination of hands-on examples and my own research experiences on how to carry out these methods effectively. Drawing upon examples from my ethnographic research with young Chinese Australians in health and physical activity, I examine some of the distinctive characteristics of fieldwork and interactions with participants in the environment. In doing so, I argue that the success of conducting ethnographic research methods require a relational, flexible, and reflexive informed approach. I also put forward the need to conduct sensory ethnographic research that gives more consideration to the multiple senses and embodied dimensions in an in-depth understanding of people's lived experiences. The chapter ends by highlighting future directions for ethnographic research methods in qualitative health and physical activity research.

Individual methods that are commonly used within an ethnographic research study include participant observation, interviews, and surveys. More contemporary forms of ethnographic methods include visual, digital, sensory, and spatial approaches. These forms of ethnographic methods can be very valuable in gaining a deeper understanding of a particular social group or a specific problem. The use of a range of traditional ethnographic methods is well covered in existing research methodology books (e.g., Liamputtong 2013, 2017). Recent ethnographic research frequently involves the use of digital, visual, and audio technologies in the practice of such methods (see Pink 2007, 2015). Other less conventional methods may entail, for example, arts-based methods and collaborations between the researcher and the researched by producing a video, writing a song, or inviting the participants to reflexively engage in an everyday or designed activity (Liamputtong and Rumbold 2008; see chapters in the ▶ Chap. 61, "Innovative Research Methods in Health Social Sciences: An Introduction" section of the handbook).

In using various ethnographic methods, there is a need to follow a set of practices that explores social life as the result of an interaction of structure and agency in people's everyday practice and experiences. Several factors may compromise the trustworthiness of the research data collected from participants, including participants' fear of giving "incorrect" answers, power imbalances, and a lack of rapport and trust between the researcher and researched (O'Reilly 2012). Social researchers often enter one of the three different forms of power relations in their ethnographic work (Tuck and McKenzie 2015). Reciprocal relationships are those in which the researcher and participants are in similar social positions and have relatively equal benefits and costs in participating in the research. Asymmetrical relationships are those in which there are significant differences in the social positions of the researcher and participants; the researcher may have relatively better access to cultural and economic resources and/or be in a position of greater power than the participants. Power exists in all social interactions, and social research cannot be excluded. In responding to potentially exploitative relationships, Dowling (2015) proposes two strategies that researchers can draw on to minimize power imbalances. First, researchers can involve participants in the design and conduct of the research, for example, by asking participants to set the research questions that need to be examined. Researchers can also invite participants to verify or "talk back" to the researchers' interpretations throughout the research process.

Since ethnographic research takes place in the real world with human beings, there are a number of ethical concerns to be aware of before, during, and after the study. For example, researchers must convey clearly to the participants the rationale of the research, develop ongoing rapport throughout the research process, and have an ethical exit strategy (Morrison et al. 2012, p. 200). McCorkel and Myers (2003) comment that "the assumptions, motivations, narratives, and the relations which are part of the researcher's backstage" are often invisible in the practice of qualitative research. Information about the situatedness of the knower, the context of discovery of the knowledge, and the relation of the knower to the subjects of the study are important in the production of legitimate knowledge (McCorkel and Myers 2003). Authenticity and fairness are regarded as a unique and core characteristic of interpretivist research (Schwandt 2000). The main concern of ethnographic research as a form of interpretivist research is to show the diverse "realities" of the participants' voices. Researchers need to "be there" and represent fully their participants' lives in relation to the research questions. This is what Jackson (1995, p. 163) suggests as "the authenticity of ethnographic knowledge depends on the ethnographer recounting in detail the events and encounters that are the grounds on which the very possibility of this knowledge rests."

2   Reflexivity

The use of reflexivity in ethnographic research serves to ensure researchers have systematically and rigorously conducted their methodology and their self as an instrument of data collection and representation. This is because ethnographic researchers are part of the social world in which they study, which in turn raises concerns over issues of subjectivity when attempting to demonstrate the trustworthiness of their research findings. Finlay and Gough (2003) describe reflexivity as a process which researchers conduct thoughtful, self-aware analysis of the intersubjective dynamics between researcher and the researched. Researchers are required to be aware of ongoing dynamics and power relations in the research process and to reflexively draw on appropriate ethnographic methods to apply in the research with the participants. There are some possible questions to reflect on throughout the research process and to modify the process where appropriate (Dowling 2015):

- Would you be doing anything differently based on the participants' response to your methods?
- Could you justify your actions to others?
- Are you presenting what you heard and saw or what you expected to hear and see?
- Are you reproducing stereotypical representations?

Given that these concerns are often undermined and ignored in qualitative research, it is important to highlight some of them when conducting various ethnographic methods backstage. Below is an example of how I conducted reflexivity in my research (Pang 2016):

> I have conducted an ethnographic research study with 12 young Chinese Australians aged 11–15 and their Health and Physical Education teachers in two Brisbane schools. The research was based on fieldwork and observations, in-depth interviews, and drawing-elicitation methods, and it provided rich data on the students' lived experiences in their physical activity lives. Research argues that the essentialist notion of being a complete insider or outsider as a researcher is problematic (Fletcher 2014). Some of the central questions about the insider/outsider debate in researching with participants, especially with those who are different to us include, e.g., who can be a "knower" (Sparkes 2002), and what are the relationships between "self and other" and the "self-as-other" (Fletcher 2014). In response to this debate of insider and outsider in the research process, I concluded that my phenotype, interests, age, class, gender, and country of origin all shaped the questions that I posed to the participants, my interactions with the participants, and the focus of my observations in schools. It seems that a similar racial identity to the young Chinese students' may add a dimension of diversity and depth to the data collected (e.g., in discussing my English accent and linguistic capacity as well as issues of racism and microaggressions) and rapport built in this research process (e.g., in discussing my different lifestyles in Australia and Hong Kong and my career choices as a Chinese person) that might not have surfaced if this research had been conducted by White researchers born in Australia. The rich and/or different data are a result of how the Chinese students perceived me as a "Chinese young academic," "family member," and/or "Chinese with good English," and how I interacted with the students during the fieldwork.

This reflexivity process is important to overcome what is usually referred to as the researchers' bias in the positivistic paradigm (Davies 2008). When researchers conduct ethnographic methods, reflexivity allows them to acknowledge human subjectivity in the research process. The participants are indeed a constructivist outcome of the intersubjectivity between the researcher and the researched (Finlay and Gough 2003).

3 Traditional Forms of Ethnographic Methods

Traditional approaches to ethnographic research endeavor to collect information from the field through interviews, fieldwork, and observations. These traditional forms of ethnographic methods aim to unravel the contextual meanings of the everyday practices in their participants' natural settings. The focus of such methods is to obtain a comprehensive understanding of a specific problem or question. Fieldwork and observations are discussed in detail in the following sections.

3.1 Fieldwork

Fieldwork is an essential attribute of ethnographic methods. O'reilly (2012) defines fieldwork as a form of inquiry that requires a researcher to be immersed personally in the ongoing activities of participants in the research. Fieldwork requires the researcher, being the main instrument of data collection, to have a clear intent apart from merely a presence in the field (O'reilly 2012).

In recording field notes, the researcher records not only his or her notes on the community or context but also his or her feelings and emotions regarding the field experiences. Goffman (1989, p. 125) notes that field research involves "subjecting yourself, your own body and your own personality, and your own social situation, to the set of contingencies that play upon a set of individuals, so that you can physically and ecologically penetrate their circle of response to their social situation, or their work situation, or their ethnic situation." These personal experiences should be analyzed in the ongoing fieldwork process with respect to the context and researchers themselves, as well as the insider and outsider perspectives, or the space between being an insider and outsider (Dwyer and Buckle 2009).

In the process of conducting fieldwork, we therefore need to be cautious of the significance of understanding ongoing meanings and how researchers interpret them. Ontologically, human beings are the primary focus of the study, and they construct multiple realities that are complex, multifaceted, differently expressed in specific contexts, and continually undergoing changes and transformations. Epistemologically, in order to gain an understanding of such realities, the aim of conducting ethnographic methods is not to begin with predetermined hypotheses to be proved or disproved as objective fact, but with open-ended exploratory questions to learn as much as possible about those realities. This process enables the researcher to describe these etic realities (the perspective of an outsider looking in) and the connections between them and the emic view (the insider perspective) of participants (Guba and Lincoln 1994; see also ▶ Chap. 6, "Ontology and Epistemology").

3.2   Participant Observation

Participants might respond to interviews with what they should do (the norm), and thus, observation provides a complementary means to reveal the enacting culture through their "patterns of behaviour" (Bernard 2017). Accurate observations are fostered by systematic and scrupulous attention to detail. Observation is immersing oneself into the actual world of culture in study. This suggests that the ethnographic method is connected to naturalism, in which the social world is studied as far as possible in its natural state (Atkinson and Hammersley 2007). Observations also provide first-hand encounters with the research interests (Merriam and Tisdell 2015) and allow researchers to discover complex interactions in natural social settings (Marshall and Rossman 2014). Bernard (2017) suggests that the focus of observations should include a clear outline in the research design of the boundaries and participants for, and the arrangements of, observations. The ethnographic fieldwork continuum ranges from direct nonparticipant observation to participant observation. An outline of a range of research roles in observations is provided below (Bernard 2017):

- Complete observer (e.g., a researcher with no physical involvement in a sport by remotely observing the sport)
- Observer-as-participant (e.g., a researcher being a newcomer to bodybuilding by being part of the group)
- Participant-as-observer (e.g., a researcher who is familiar with bodybuilding but seeks to understand this sport in a new light)
- Complete participation (e.g., a researcher living in a new city to understand a particular group of people's health experiences)

There are different stages of participant observation, including the choice of site, accessing the site, navigating field relations, recording data on site, and analysis of observation data (Atkinson and Hammersley 2007). In choosing a site for participant observation, one might choose to focus on a familiar setting and attempt to shed new light on the experience central to the research focus. In the case of unfamiliar sites, researchers should have a good understanding of the chosen community before entering the site for research purposes (DeWalt and DeWalt 2011). Doing fieldwork in a familiar place can be as challenging as it is in an unfamiliar one. The point is, therefore, to strive for balance in choosing a research site that enables the researcher to be both an insider and an outsider (DeWalt and DeWalt 2011).

After identifying an appropriate site, a crucial issue in gaining access to the site is to identify the gatekeepers in order to recruit targeted participants in the site. The difficulty of this process depends on whether the site is a private or public place and how sensitive is for researchers to conduct research on the topic at their site (DeWalt and DeWalt 2011). Generally, gaining access is more challenging when the place is more private, the researcher is not already an insider on the site, and the topic under examination is sensitive. In navigating the field, researchers take their bodies with them on the site. That is, we are embodied subjects, and what we wear and how we conduct ourselves can be a key marker of who we are and how we are perceived by the participants on the site (Atkinson et al. 2001). Here is an example of how I conducted my observations and field notes during my research:

> On the spot field notes, including sketched drawings and jotted notes, were recorded by hand during the day and transferred to the computer afterwards. More recently, I have used iPads/ iPhones to jot notes in the field and import them to the computer folders. Field notes included three parts: The first part was an accurate and detailed description of participants' behaviors and appearance (e.g., what the participants were doing) and the physical state of the environment (e.g., what was the place like?). In contrast, the second part included my personal observations and emotional feeling about the event. Sometimes, photos of the place were included to remind me how the place looks and how I feel about the place. And the third part included snippets of thoughts that were linked to the theoretical concepts.

Analyzing the results of observation data will be different depending on the research purpose. For example, observations that involve counting require a structured research design. This may include the presentation of descriptive statistics and how they relate to the context of the research question. When observation is related to understanding the context, analysis will focus on the text and descriptions about the place and people. This will include finding the meaning of the data and drawing on software such as NVivo to organize and interpret the data (Bryman 2016).

4   Innovative Forms of Ethnographic Methods

It is not uncommon to find research projects that only draw on interviews as their main source of ethnographic knowledge. However, these more conventional qualitative interviewing methods have been argued by researchers to inadequately represent the lived experiences of participants in their specific contexts (e.g., Rapley 2004). Indeed, as Pink (2015) notes, the relationship between what *is* said in interviews and knowledge that is not articulated in this way means that there is a need for other methods to reveal further information from participants. Visual methods and sensory methods are discussed in detail in the following sections.

4.1   Visual Methods

The "visual-turn" (Rose 2016) has brought an increasing growth of image-based research in the health, education, and social science disciplines. We are living in a world where "ocularcentrism" (an increasing saturation of images) has a powerful impact on how we experience our lives (Rose 2016). Visualization is a way of interpreting, and thus, interpretation can influence our ways of seeing (Bustle 2003). In using visual methods, researchers are able to represent knowledge in different ways. For example, visual methods can be used to analyze cultural representations on social media so that alternative narratives can be understood. Images have the potential to stimulate reflections that words alone cannot (Schwartz 1989; see also ▶ Chap. 61, "Innovative Research Methods in Health Social Sciences: An Introduction"). Various visual methods relevant to this discussion include visual ethnography (Pink 2007), visual narrative (Carrington et al. 2007), visual sociology, image-based research, and photo elicitation (Harper 2002; Clark-Ibáñez 2007), among others (see also ► Chap. 65, "Understanding Health Through a Different Lens: Photovoice Method").

Pink (2007, p. 40) highlights that there are three different approaches in using visual methods: "(a) examining pre-existing visual representations, (b) making visual representations, and (c) collaboration with social actors in the production of visual representations." Visual approaches allow us to ask questions about health and physical activity research such as: *What is being seen/not seen by the participants and how is it socially and culturally shaped? How are we made to see/not see? How are we allowed to see/not see?* Visual methods allow researchers to probe participants' lived experiences and act as a medium through which emotions and sensitive topics can be expressed more clearly than using linguistic communication alone (Rich and O'Connell 2012).

Harrison (2002) distinguishes between the visual as resource and the visual as topic. The visual as resource means that the use of visual forms is used to explore a particular research question (e.g., the use of drawings to examine how people construct meanings around physical activity). This may also include the use of visual methods to collect data from participants (e.g., the researcher co-producing sketch-mapping with the participant to understand how the participant's environment has an impact on his or her physical activity participation, as in Azzarito 2010). In other studies, researcher-produced photographs can be used alongside participant interviews (e.g., Hill and Azzarito 2012). These methods, as resources, allow the researcher to enter the participants' world visually and move beyond the traditional hierarchical relationships between the researcher and researched in positivist research methods (Harrison 2002). The use of visual methods as topic suggests the use of visual forms as the subject of analysis (e.g., analyzing the media representations of young people in body size in online media, as in Millington and Wilson 2012). Below, I expand on photo-elicitation (visual methods as resource), and its use alongside interviews and demonstrate its potential and perils in ethnographic research.

Photo-elicitation is a form of visual method that allows participants' experiences and meanings related to the research topic to be shared simultaneously and discussed during traditional interviews. Photo elicitation can be used during interviews, which allows participants to discuss dimensions of their social world that may be ignored or taken for granted (Clark-Ibáñez 2007). Photo-elicitation interviewing was originally used by to study migration and the participants' everyday lives in relation to technological and economic changes. Photo-elicitation involves using photographs to evoke participants' memories and discussion during interviews and to explore their lived experiences (Harper 2002). Lived experiences are sometimes difficult to articulate, but participants can use photographs to anchor their thoughts on past, present, and future in relation to the research topic. An image produces multiple readings and messages, which are, according to Schwartz (1989), intrinsically ambiguous. However, this ambiguity should not be considered a limitation or disadvantage. On the contrary, when used to aid participants' communication of complex messages, it may contribute to the generation of richer data.

Photo-elicitation has also been shown to reduce the power imbalance between the researcher and researched (Harper 2002). When images are produced by the researcher, some level of distance between the participant and the researcher is generated. However, when participants bring or take their own photos, they are more able to direct the discussion, which is useful when exploring sensitive issues. Harper (2012) asserts that the world that is seen and represented visually is different from the world that is represented through words, and as a result, the former connects to different realities than more conventional research methods. In visual methods, drawings and photographs are not just passive images but symbolic representations of a structural order of things in society. For researchers, the use of visual methods uncovers the meanings embedded in these images. Below is an example of using Culture grams (Chang 2008), drawing, and sketch mapping with youth participants:

> The various drawing methods were useful communication tools to elicit the participants' voices, which reflected their diverse sociocultural surroundings. Culture gram (Chang 2008) was used to help capture the young people's meaning of their cultural identity during the second and last interview. In understanding who they are, responses on their religion, ethnicity, interests, hobbies, strengths and weaknesses, and physical activity experiences were elicited. During the interviews, students were asked to draw or write down their responses to the questions on a piece of drawing paper alongside answering the interview questions. Students were also asked to circle the three most important characteristics that best represented themselves. This method enabled the students and me to visually see the responses and, thus, allowed the students to describe the possible interconnections among the different topics. In addition, preserving this vivid and visual detail of their commentaries allowed us to recapture the students' responses with reference to their own cultural gram after a year's time in the last interview. As for eliciting the research data, it helped to visually note that these young people's cultural practices were diverse and multicultural. Their drawings showed a colourful array of their experiences, capacities, and engagements. The data helped to challenge as well as respond to the binary understanding of "East" or "West" cultures. The use of a culture gram was new to research for understanding young Chinese people's engagement in physical activity, specifically in demonstrating their subjectivities in relation to their surrounding contexts.
>
> In particular, the visual representation of drawings assisted in eliciting students' interview responses, as it allowed the young people and me to refer to the interconnectedness of the topics under discussion. The use of mapping their neighborhoods in Australia and overseas also provided an anchor for the young people, the readers, and me to visualise possible social and environmental differences and thereby understand and discuss why and how these young people live across two different cities or countries. By looking at two maps (i.e., Brisbane and their overseas contexts), these young people were better able to compare and contrast the facilities, weather, and spaces that were conducive to their physically active lifestyles. In addition, the drawings allowed us to visualize the perceived functionality of the facilities and the distance between the sites. For example, a number of the participants' drawings showed accessible recreational facility sites, such as basketball courts, parks, and shopping malls, while others suggested there were risks in using these sites, such as busy roads, steep slopes in the parks, and the intense sunshine.

4.2   Sensory Methods

We learn through our senses, and this sensoriality influences how we understand and represent others' lives. Pink (2009) discusses the use of the sensory approach to understand the complexity of people's meanings in the contemporary social world. For example, researchers have been concerned with their own sensory embodied experiences in relation to researching their participants (Downey 2005). Sensory perceptions are central to researchers' encounters in fieldwork, including the sociality and materiality of the research. Some important groundwork on sensory methods includes the sensual and affective dimensions in space (Thrift 2004), perceptions and the environment (Ingold 2000), and cultural differences and senses (Howes 2005). In conducting sensory research, the modern Western five-sense sensorium can offer useful analytical categories through which to understand embodied knowledge and practice (Pink 2015). Researchers have focused on using various combinations of sight, sound, smell, taste, and texture to understand and represent participants' lived experiences. For example, a cold breeze will have an impact on how people feel about the social and physical environment and shape how they act in relation to others (as in Sunderland et al. 2012). Research has also drawn on other methods – for example, go along or walk-along interviews (Carpiano 2009; see ▶ Chaps. 72, "Walking Interviews," and ▶ 73, "Participant-Guided Mobile Methods"), participatory video documentaries (Pink 2007), and photovoice narratives (Baker and Wang 2006; see ▶ Chap. 65, "Understanding Health Through a Different Lens: Photovoice Method") – to access local knowledge and sensory data from the research and participants. These methods are often participant-led and allow participants to share their experiences more naturally with the researcher.

Pink (2007) notes that ethnographic practice entails our multisensorial embodied interactions with others and the social, material, discursive, and sensory environment. Sensory researchers believe that human experience is mediated through the body and therefore our senses. This embodied experience is how we make sense of ourselves, others, and things in a place (Howes 2005; Pink 2009; see ▶ Chap. 61, "Innovative Research Methods in Health Social Sciences: An Introduction"). Therefore, human experience is not only embodied but also emplaced. As Merleau-Ponty (1964) describes, place is central to gathering people's experiences, histories, and things. In particular, sensory geographers considers how the specificity of place can only be understood through a zone of entanglement (Ingold 2008) constituted through lived bodies and things (Casey 2001). Research questions that can be used in sensory ethnographic methods in health and physical activity research include: *What does it feel to live and do physical activity here?* and *How can the rich sensory ethnographic accounts of lived experiences be used to influence health planning in this community?*

5   Opportunities and Perils of Using Ethnographic Methods

Based on the literature and research examples in the previous sections, there are some conclusions that we can make in relation to the opportunities and perils of using ethnographic research methods. One of the main advantages associated with ethnographic research is that it allows researchers to identify issues that arise during their ongoing interactions with participants (Atkinson et al. 2001). When conducting other types of studies that are not based on fieldwork and observations or more generally interactions, there is a lesser chance to introduce new topics within the broader research question under examination. This is because researchers are encouraged to exercise their reflexivity during fieldwork and to probe further questions or change the direction of their research topics in response to the participants' responses and behaviors. Another main advantage of ethnographic research methods is that they aim to represent the detailed and authentic experiences of the insider. As a result of this intersubjective nature, an ethnographic study can be very useful in revealing in-depth and multisensory data about the participants.

One of the main criticisms of ethnographic research is the amount of time and effort that are required to produce trustworthy results (Atkinson and Hammersley 2007). The use of interviews, observations, and fieldwork within an ethnographic study requires a certain amount of time, and the results tend to take longer to generate. This is because understanding of a cultural practice takes time to unfold, and it takes time for the researcher and participants to make meaningful connections (Liamputtong 2010, 2013).

As stated above, the conduct of ethnographic research methods requires the researcher to be a participant in the observation within the environment. There are two main potential risks with using ethnographic methods. First, ethnographic researchers need to be skilled to minimize potential pitfalls, including the "thickness" of the data collected and potential bias, as well as a lack of reflexivity in data collection or analysis (Atkinson et al. 2001). It is also vital that the participants be willing and open with the researcher in discussing their lived experiences. As such, the quality of researchers themselves and their role in study design play an important part in eliciting responses from participants. As discussed, one of the most important criteria in ethnographic research methods is the researcher. This means an ethnographic researcher is critical to a study's success. Based on my research with young Chinese Australians in physical activity, I offer a few tips to researchers who may be considering using an ethnographic research methods approach to their study:

- Do not choose ethnographic methods or qualitative research only because you cannot do statistics.
- Do not expect there to be standard procedures.
- Do not regard your initial research plan as sacred.
- Do not leave analysis till the end.
- Do not take things or people for granted.

6   Conclusion and Future Directions

Ethnographic research methods entail an eclectic approach. They involve making connections between people's complex lived experiences and their ever-changing structural environments. In this chapter, I have suggested several key ethnographic research methods and examples of how to carry them out in research. I have also noted that human beings are both embodied and emplaced subjects, and that experiences, history, and things interact and accumulate in a place. In using ethnographic methods, researchers are interested in delving into the entanglement of these histories, cultures, and lived experiences. Words alone are sometimes insufficient to shed light on the complexity of people's lived experiences in contemporary society (Banks 2008). Visual and sensory methods provide alternative means of conducting ethnographic research with participants. Working with images, unlike working with texts alone, encourages research to move towards sensory methods (Pink 2007). One of the tasks of ethnographic researchers is to develop an awareness and reflexivity of how they become involved not only in understanding participants' embodied feelings, experiences, and practices but also in reflecting on their co-involvement in the research, the place where the research takes place, and the interactions between the researcher and the researched. The success of ethnographic research requires a relational, flexible, and reflexive informed approach that acknowledges our multiple senses and embodied dimensions in relation to understanding people's lived experiences.

Rapidly advancing technology and increased globalization require innovative ethnographic methods to capture the socially networked and ever-changing living environments. Digitalization provides researchers with unprecedented opportunities for understanding and exploring people's lives. The increased use of social media as means of sharing information is creating opportunities for health and physical activity researchers to examine their participants and the data they generated online. This emergence of online ethnographic research methods such as netnography (Kozinets 2015) has allowed researchers to collect data on online communities and social media spaces such as blogs and twitters (see ▶ Chap. 75, "Netnography: Researching Online Populations"). Such approach moves beyond the limitations of quantitative survey methods that relies upon a participant's memory, and therefore increases the trustworthiness of the research findings (Costello et al. 2017). The emergence of new forms of digital cultures and methods will continue to shape the nature and practice of ethnographic research. For example, the use of "ethnomining," a way of joining ethnography and data base mining, attempts to collect big data in an online environment which is relatively cost-effective than traditional ethnographic research approaches (Varis 2016). Nonetheless, as Kozinets (2015, p. 97) reminds us, "the key element is not to forget the participative, reflective, interactive and active part of our research when using the communicative function of social media and the internet," which aligns with the purpose and practice of ethnographic research methods.

References

Atkinson P, Hammersley M. Ethnography: principles in practice. 3rd ed. London/New York: Routledge; 2007.

Atkinson P, Coffey A, Delamont S, editors. Handbook of ethnography. London: SAGE; 2001. p. 1–7.

Azzarito L. Ways of seeing the body in kinesiology: a case for visual methodologies. Quest. 2010;62(2):155–70.

Baker TA, Wang CC. Photovoice: use of a participatory action research method to explore the chronic pain experience in older adults. Qual Health Res. 2006;16(10):1405–13.

Banks M. Using visual data in qualitative research. London: SAGE; 2008.

Berg BL. Methods for the social sciences. Boston: Pearson Education; 2004.

Bernard HR. Research methods in anthropology: qualitative and quantitative approaches. Maryland: Rowman & Littlefield; 2017.

Bryman A. Social research methods. 5th ed. New York: Oxford University Press; 2016.

Bustle LS, editor. Image, inquiry, and transformative practice: engaging learners in creative and critical inquiry through visual representation, vol. 203. New York: Peter Lang Pub Incorporated; 2003.

Carpiano RM. Come take a walk with me: the "go-along" interview as a novel method for studying the implications of place for health and well-being. Health Place. 2009;15(1):263–72.

Carrington S, Allen K, Osmolowski D. Visual narrative: a technique to enhance secondary students' contribution to the development of inclusive, socially just school environments – lessons from a box of crayons. J Res Spec Educ Needs. 2007;7(1):8–15.

Casey ES. Between geography and philosophy: what does it mean to be in the place-world? Ann Assoc Am Geogr. 2001;91(4):683–93.

Chang H. Autoethnography as method. London/New York: Routledge; 2008.

Clark-Ibáñez M. Inner-city children in sharper focus: sociology of childhood and photo elicitation interviews. In: Stanczak GC, editor. Visual research methods: image, society, and representation. London: SAGE; 2007. p. 167–96.

Costello L, McDermott ML, Wallace R. Netnography: range of practices, misperceptions, and missed opportunities. Int J Qual Methods. 2017;16(1):1–12.

Davies CA. Reflexive ethnography: a guide to researching selves and others. New York: Routledge; 2008.

DeWalt KM, DeWalt BR. Participant observation: a guide for fieldworkers. Walnut Creek: Rowman Altamira; 2011.

Dowling. Methods of critical place inquiry. In: Tuck E, McKenzie M, editors. Place in research: theory, methodology, and methods, vol. 9. New York/London: Routledge; 2015.

Downey G. Learning capoeira: lessons in cunning from an Afro-Brazilian art. Oxford: Oxford University Press; 2005.

Dwyer SC, Buckle JL. The space between: on being an insider-outsider in qualitative research. Int J Qual Methods. 2009;8(1):54–63.

Finlay L, Gough B. Reflexivity: a practical guide for researchers in health and social sciences. Oxford: Blackwell Science; 2003.

Fletcher T. 'Does he look like a Paki?' An exploration of 'whiteness', positionality and reflexivity in inter-racial sports research. Q Res Sport, Exerc Health. 2014;6(2):244–60.

Goffman E. On fieldwork. J Contemp Ethnogr. 1989;18(2):123–32.

Guba EG, Lincoln YS. Competing paradigms in qualitative research. In: Denzin NK, Lincoln YS, editors. Handbook of qualitative research. Thousand Oaks: SAGE; 1994. p. 105–17.

Harper D. Talking about pictures: a case for photo elicitation. Vis Stud. 2002;17(1):13–26.

Harper D. Visual sociology. New York: Routledge; 2012.

Harrison B. Seeing health and illness worlds–using visual methodologies in a sociology of health and illness: a methodological review. Sociol Health Illn. 2002;24(6):856–72.

Hill J, Azzarito L. Representing valued bodies in PE: a visual inquiry with British Asian girls. Phys Educ Sport Pedagog. 2012;17(3):263–76.

Howes D. Empire of the senses. Oxford: Berg Publishers; 2005.

Ingold T. The perception of the environment: essays on livelihood, dwelling and skill. London: Routledge; 2000.

Ingold T. Bindings against boundaries: entanglements of life in an open world. Environ Plan A. 2008;40(8):1796–810.

Jackson M. At home in the world. Durham: Duke University Press; 1995. p. 163.

Kozinets R. Netnography: redefined. London: SAGE; 2015.

Liamputtong P. Performing qualitative cross-cultural research. Cambridge: Cambridge University Press; 2010.

Liamputtong P. Qualitative research methods. 4th ed. South Melbourne: Oxford University Press; 2013.

Liamputtong P. Research methods in health: foundations for evidence-based practice. 3rd ed. Melbourne: Oxford University Press; 2017.

Liamputtong P, Rumbold J. Knowing differently: arts-based and collaborative research methods. New York: Nova Publishers; 2008.

Marshall C, Rossman GB. Designing qualitative research. 5th ed. London: SAGE; 2014.

McCorkel JA, Myers K. What difference does difference make? Position and privilege in the field. Qual Sociol. 2003;26(2):199–231.

Merleau-Ponty M. The primacy of perception and other essays on phenomenological psychology, the philosophy of art, history, and politics. Illinois: Northwestern University Press; 1964.

Merriam SB, Tisdell EJ. Qualitative research: a guide to design and implementation. San Francisco: Wiley; 2015.

Millington B, Wilson B. Media analysis in physical cultural studies: from production to reception. In: Young K, Atkinson M, editors. Qualitative research on sport and physical culture. Bingley: Emerald Group Publishing Limited; 2012. p. 129–50.

Morrison ZJ, Gregory D, Thibodeau S. "Thanks for using me": an exploration of exit strategy in qualitative research. Int J Qual Methods. 2012;11(4):416–27.

O'reilly K. Ethnographic methods. London/New York: Routledge; 2012.

Pang B. Conducting research with young Chinese-Australian students in health and physical education and physical activity: epistemology, positionality and methodologies. Sport Educ Soc. 2016;1–12.

Pang B, Macdonald D, Hay P. 'Do I have a choice?' The influences of family values and investments on Chinese migrant young people's lifestyles and physical activity participation in Australia. Sport Educ Soc. 2015;20(8):1048–64.

Pink S. Doing visual ethnography. London: SAGE; 2007.

Pink S, editor. Visual interventions: applied visual anthropology, vol. 4. Oxford: Berghahn Books; 2009.

Pink S. Doing sensory ethnography. London: SAGE; 2015.

Rapley T. Analysing conversation. In: Seale C, editor. Researching society and culture. London: SAGE; 2004. p. 383–96.

Rich E, O'Connell K. Visual methods in physical culture: body culture exhibition. In: Young K, editor. Qualitative research on sport and physical culture. Bingley: Emerald Group Publishing Limited; 2012. p. 101–27.

Rose G. Visual methodologies: an introduction to researching with visual materials. London: SAGE; 2016.

Schwandt TA. Three epistemological stances for qualitative inquiry: interpretivism, hermeneutics, and social constructionism. In: Handbook of qualitative research, vol. 2. Thousand Oaks: SAGE; 2000. p. 189–213.

Schwartz D. Visual ethnography: using photography in qualitative research. Qual Sociol. 1989;12(2):119–54.

Sparkes A. Telling tales in sport and physical activity: a qualitative journey. Champaign: Human Kinetics; 2002.

Sunderland N, Bristed H, Gudes O, Boddy J, Da Silva M. What does it feel like to live here? Exploring sensory ethnography as a collaborative methodology for investigating social determinants of health in place. Health Place. 2012;18(5):1056–67.

Thrift N. Intensities of feeling: towards a spatial politics of affect. Geogr Ann Ser B. 2004;86(1):57–78.

Tuck E, McKenzie M. Place in research: theory, methodology, and methods, vol. 9. New York/London: Routledge; 2015.

Varis P. Chapter 3 Digital ethnography. In: Georgakopoulou A, Spilioti T, editors. The Routledge handbook of language and digital communication. London/New York: Routledge; 2016. p. 55–68.

Willis C, Anderson K. Ethnography as health research. In: Liamputtong P, editor. Research methods in health: foundations for evidence-based practices. 3rd ed. Melbourne: Oxford University Press; 2017. p. 121–37.

Institutional Ethnography

27

Michelle LaFrance

Contents

Abstract

Institutional ethnography (IE), a form of critical ethnography introduced to the social sciences in the late 1990s by Canadian sociologist Dorothy J. Smith, poises researchers to uncover how "work" (a concept defined generously) is co-constituted within institutional environments. The IE approach reframes institutional sites as dynamic shape shifters that use texts to mediate, organize, and lend value to the social practices of diverse and knowing individuals. Workplaces and practices can be said to reproduce the broader spheres of influence, prestige, and value that structure society at large. As such, IE seeks out the (often implicit and/or erased) connections between work processes and institutional discourses, revealing how work is coordinated across time and space. More plainly, the

M. LaFrance (✉)
George Mason University, Fairfax, VA, USA
e-mail: mlafran2@gmu.edu

© Springer Nature Singapore Pte Ltd. 2019
P. Liamputtong (ed.), *Handbook of Research Methods in Health Social Sciences*,
https://doi.org/10.1007/978-981-10-5251-4_82

methodology uncovers *how things happen* – how institutional discourse compels and shapes practice(s) and/or how norms of practice speak to, for, and over individuals. IE research offers opportunities for more situated and finely grained understandings of the sites where we work, the people we work most closely with, the generative power of institutional texts and discourse, and the ways that our participation in work then gives material face to the institutions that govern the social world.

Keywords

Institutional ethnography · Ethnography · Feminist methodology · Cultural materialism · Standpoint · Material relations · Institutional discourse · Textual analysis

1   Introduction

Institutional ethnography (IE) is a form of ethnographic inquiry introduced to the social sciences in the 1990s by Canadian sociologist Dorothy J. Smith. Smith's career work critiques traditional models of social science research; these traditions relied upon positivist paradigms and universalist models of empirical observation and disinterested analysis, processes that many ethnographers claim overdetermine and reify research participants and the social realm (1974). Generalized understandings of institutions erase the actualities of lived experience for real people; Smith (2005) argues, flattening important disjunctions of doing, knowing, and being. The model of ethnography Smith has developed instead draws upon principles of feminist cultural materialism to focus the researcher's eye upon the unique personal experiences and practices of individuals. Feminist research methods draw on the notion that the personal is political and strive to bring forward stories of the marginalized, hidden, and overlooked as they affirm differences among people (DeVault and Gross 2012). Cultural materialists hold that the material actualities of a culture (social structures, infrastructure, economics, technology, the political, ideology) exert enormous power on groups and individuals; these structural forces (the macro) play an undeniable role in the organization of everyday life (the micro).

Institutional ethnographers locate a unique standpoint and engage in the research practice of "looking up from where [they] are" (Smith 2006, p. 5), a means of uncovering the highly situated and personal experiences and practices of individuals. This move is similar to what Laura Nader (1969) calls "study[ing] up," a process of inverting the power differentials that often inform ethnographic research. The goal of IE inquiry is to reveal how actual lives take shape as a process of negotiating social relations. Seeking to uncover the relationships between highly personal experiences and practices of active individuals and their everyday (material) contexts, according to Smith, reveals any number of stories and experiences that are often otherwise erased, elided, or ignored.

Ethnography is a ubiquitous methodology in many fields. Defined by Linda Brodkey (1987, p. 25) as "[t]he study of lived experience," by Laura Nader (2011,

p. 211) as a mix of descriptive and theoretical research that seeks "an emic perspective" (or "the insider's point of view," and humorously by Clifford Geertz (1998) as "deep hanging out," ethnography offers an adaptable and reflexive means by which to explore the complex and highly networked topoi of everyday life. Originating from the field of anthropology, where it often focuses on participant observation and the intensive study of a community or social environment, ethnography offers a sense of richness and specificity that other forms of research, particularly those that seek patterns in human behavior or that view social organization from a disinterested distance, may not (see also ▶ Chaps. 13, "Critical Ethnography in Public Health: Politicizing Culture and Politicizing Methodology," and ▶ 26, "Ethnographic Method"). Institutional ethnographers often position themselves as social activists whose research uncovers the experiences of actual people who carry out their work in neoliberal, so hierarchical and highly prescribed, environments.

While traditional ethnographers are often interested in what is happening – how what people are doing in a site offers insights into a particular social order, for instance – the IE project sets out to uncover *how things come to happen*, noting that "[p]eople participate in social relations, often unknowingly, as they act competently and knowledgeably to concert and coordinate their own actions with professional standards" (Campbell and Gregor 2002, p. 31). The methodology investigates how individuals within a location co-create the dynamics and processes that give the site its unique character. To understand what is actually happening within institutional sites, institutional ethnographers ask how experiences and practices are co-constituted in the moments that knowing and active individuals negotiate social, professional, and institutional systems as they carry out their "everyday/everynight" lives (Smith 2005).

Much recent institutional ethnography has been invested in exploring the ways people carry out their "work," a concept defined generously (Griffith and Smith 2014). In IE, "work" denotes a series of coordinated practices within a local setting that an individual routinely puts time and energy into. It is through work that institutions coordinate the experiences and practices of individuals, particularly in "corporations, government bureaucracies, academic and professional discourses, [and] mass media," highly structured social complexes that have an inordinate power over the ways people go about their everyday lives (Smith 2005, p. 10). The work at the center of an IE study might be paid labor, but it might also be forms of invisible work, such as the running of a household, unrecognized and/or unacknowledged activities associated with paid labor, or the activities of clients, patients, or other members of a community. Workplaces and practices are sites that reproduce the broader spheres of influence, prestige, and value that structure society at large. As such, IE seeks out the (often implicit and/or erased) connections between work processes and institutional discourses, revealing how work is coordinated across time and space.

The IE framework holds that work experiences and practices take shape in relationship to local materialities; work is always socially coordinated, rule-governed, and textually mediated. An example would be the ongoing historical construction of household chores as the work of women. Another example might be the problem-solving processes patients must go through in order to receive paid treatment under a particular insurance plan. Using IE to study the work that people carry out makes visible the values, practices, beliefs, investments, and belongings that circulate below more visible or dominant institutional discourses and ideals of the everyday. To trace these mutually constitutive relationships, IE focuses the researcher on the practices people engage in, the decisions they make, and how their negotiations of values, policy, procedure, labor hierarchies, and work systems take on a particular shape. On the one hand, these everyday work experiences and practices are a matter of choice, personal forms of identification, and taste; on the other hand, because institutions are material locations and social relations have material implications, these everyday doings are often highly circumscribed by active social and professional norms. The doing, knowing, and being of people bring these tensions into visibility, making the institution itself legible for study (Smith 2001). Institutional ethnography, then, reveals the material actualities – what people do to negotiate the everyday – at the heart of our institutional existences.

2   Key Terms for IE Inquiry and Analysis

Several distinctive analytic moves are central to IE, lending focus to the institutional ethnographer's development of a dynamic and evolving protocol, data collection activities, and the analysis and interpretation of data: *problematic, institutions, ruling relations, standpoint, social coordination, institutional discourse*, and *Work*. Two concepts are central to the analysis of social sites that institutional ethnographers enact: "standpoint" and "ruling relations." With both terms, Smith asks us to think about how the spectrum of social forces organize actual people, as well as how institutions regulate and bind individuals to ideals of practice. These moments of ethnographic inquiry collapse distinctions between broader discursive forces (such as professional and institutional discourse), beginning with the understanding that our everyday lives are discursively constituted. Individuals are unique, and knowing, but also act from places of shared identity, local belonging, professional alignment, and personal investment – their "standpoint." "Ruling relations" are "that extraordinary yet ordinary complex of relations...that connect us across space and time and organize our everyday lives" (Smith 2005, p. 8). An IE project seeks to empirically trace the connections between these two points of understanding, noting that there is always a relationship between the "micro" and the "macro" elements of the sites under study (DeVault 2008, p. 4). The other terms offer insights into IE as a research practice and additional fine-grained understandings of how ethnographers may come to uncover elements of institutions, especially how their discourses coordinate people across time and space.

2.1   Problematic

An IE project begins with a problematic, a process of "research and discovery," according to Smith (2005, p. 227). A problematic suggests a direction and/or a loose set of boundaries for an investigation, as it takes into account how experience and practice are situated within and contoured by discourse. A problematic is not necessarily a "problem" in an organization, such as the problem of overtime for medical staff. Instead, a problematic exposes an overlap of competing ideals or calls attention to where institutional discourse and the particularities of lived experience refuse and resist one another. An example problematic might be a question, such as: What is the relationship between the policies of insurance companies and the language use of health-care providers when speaking with patients? A focus on a problematic entails the recognition that not all people experience a site in the same way and that practices, particularly work practices, will take shape in any number of ways as the efforts of knowing people are coordinated under the influence of institutional discourse, professional expertise, and personal predilection. The uneasy moments at the center of any workplace story often suggest that people within an institution are differently organized in relation to their daily work. Thinking of these differences in relation to the "problematic" of a workplace reaffirms that some practices within institutions will always be scripted for individuals, but that individuals will also actively negotiate and renegotiate these points of institutional contact in highly personal ways.

2.2   Institutions

"Organized around a distinctive function, such as education, health care, and so on" (Smith 2005, p. 225), institutions are complex rhetorical, social, and material entities, which house any number of diverse rule-governed, hierarchical, and textually mediated workplaces. The challenge for the institutional ethnographer is to recognize the dynamic and generative nature of the institution as a social entity. IE supports this move by conceptualizing individuals as unique and knowing, while emphasizing that institutions function as "shape shifters" (LaFrance and Nicolas 2012, p. 131), social constellations that take shape relationally around the distinctive needs and roles of the individuals who engage them. Institutions encompass multiple, dynamic experiences and a proliferation of practices. The institution is a site of dialogic and multivocal belongings, where actual people make visible their unique understandings of the institution and their roles within it.

Most people tend to have a "general macro-level idea" in mind when thinking about or discussing institutions and large formalized organizations (LaFrance and Nicolas 2012, p. 131). That is, we share a collective understanding of these sites based on common preconceptions and experiences in and around them (Smith 2005, p. 160). Think, for example, of a community hospital. The patient of a community hospital will have a different viewpoint of the hospital than will a medical professional employed by the hospital. One medical professional will have a different set of interactions with the hospital from other medical professionals who work in different areas of that hospital. As they interact with different people, different times, and locations, they will literally experience the hospital differently. Staff who do not carry out medical procedures will have a different set of working conditions and practices from professionals with other job designations. The family and friends who visit a patient will have a different view from those who are employed or are attended by the hospital. The private hospital that is only a few minutes away will have a different (if also somewhat coincidental and similar) group of people seeking its services, sets of conditions for those who are employed by its offices, and discursive understanding of its mission or goals. It is easy to take these differences for granted; on their surfaces, these locations of work may seem quite similar. But the differences of experience, practice, and work are crucial to recognize and flesh out with specificity if we are to understand how institutions recreate the social order in line with regimes of power, prestige, and authority. Indeed, the institution of the community hospital can be thought to shape shift – taking on different qualities, providing different services, and responding differently to needs – based upon the unique standpoints of those who access this site. And, the relationship of the patient to the hospital will shape shift again once the patient has left the hospital or no longer needs its services.

2.3   Ruling Relations

As people act with purpose and knowledge, they act in coordination with professional standards and the expectations of organizations, colleagues, and employers. It is in this term that we see IE's most explicit nod to Marxist cultural materialism; as Marx (1852, 1913, p. 9) is known to have said: "Men make their own history, but they do not make it as they please; they do not make it under self-selected circumstances, but under circumstances existing already, given and transmitted from the past." This means that we must be careful to acknowledge that social relations do not simply happen to people; they are not accidents but rather the product of historical moments and the particularities of location. Working conditions and daily routines bear traces of ideology, history, and social influence. Akin to powerful social or workplace norms, ruling relations draw on complexes of power, authority, and labor – expertise, marginality, influence, and decision-making. This social landscape coordinates how actual people carry out their particular daily practices as they negotiate the everyday. As people act with purpose and knowledge, they act in concert with the expectations of organizations, communities, and employers. Participating in these forms of social organization naturalizes the multitudes of practices that imbue a site, entrenching certain practices into the cultural fabric, making them "just how it's done," "common sense," and/or easily taken for granted.

Ruling relations make themselves visible when we see over time and space how the work of one person (or a small group of people) bears similarities to the work of others in other locations. When we see researchers and practitioners share vocabulary, philosophies, routines, and regimes, or when those we interview independently tell the same stories, reflect on the same moments, discuss the same issues, or offer a shared sense of purpose, ruling relations are coming into visibility.

2.4   Standpoint

Smith's early career work aligned with other feminist thinkers who posed "standpoint theory" as a challenge to the universal or "pure" knowledge and positivism of the sciences and social sciences. Smith's (1974, p. 22) arguments in "Women's Perspective as a Radical Critique of Sociology" critiqued masculinist models of sociology, which excluded women's experiences and perspectives from "methods, conceptual schemes, and theories." These models privileged abstraction, objectivity, and a disembodied subject, according to Smith, not only dominated, and so structured, mainstream sociological thought, but relegated women and the concerns of women "outside and subservient to this structure" (p. 26). The resulting paradigm erased and marginalized differences of experience, being, and knowing, restricting what could be known, studied, and understood. Other standpoint theorists, such as Haraway (1988), argue that knowledge is "situated" (1988), or as Harding (2004, p. 3) explains that "the social order looks different from the perspective of [the] lives and or struggles [of those marginalized by the social order]." Recognizing all knowledge as the product of a particular epistemological framework allows us to understand a richer range of experiences (Haraway 1988). Beginning IE projects from the unique standpoint of someone who carries out their work within an institutional setting is one way to come to know the particularities of experience and practice that give the institution its face. An embrace of standpoint allows us to uncover and tell the stories of people whose lived experiences may otherwise be elided, erased, or ignored.

2.5 Social Coordination

Social mechanisms sanction practices, granting some legitimacy; in this way, the social order comes to coordinate *doing, knowing, and being*. Drawing upon Marx and Engels' critique of the German ideologists, Smith (2005, p. 65) grounds IE in the argument that the social and people's activities mutually inform one another:

> The social might be conceived as an on-going historical process in which people's doings are caught up and responsive to what others are doing; what they are doing is responsive to and given by what has been going on; every next act, as it is concerted with those of others, picks up and projects forward into the future.

Institutions in the IE framework, as such, are social entities created in the moments that individuals take up particular practices toward specific ends – institutions serve as "networks" or "complexes" that serve a distinctive function (education, government, health, entertainment), perpetually constructed and reconstructed (socially coordinated) through the active participation of individuals. This framework focuses ethnographers on tracing actualities of experience, especially "distinctive relational sequences" – *or how work gets done* – as these reveal the ways local cooperative efforts respond to and reinscribe broader economies of value (Smith 2005, p. 54). IE's focus on the social nature of institutions acknowledges that the work of individuals is coordinated in alignment with (or resistance of) a spectrum of different values, notions of expertise, ideals of belongings, and other factors. These complex interrelations between the individual and the social realm are always dynamic, evolving, and mutually constitutive within a site (Campbell and Gregor 2002).

2.6   Institutional Discourse: Texts, Textual Mediation, Boss Texts, and Institutional Circuits

Institutional discourse and texts are not just sources of information but shapers of experience and practice. Smith (2001, p. 100) writes that formalized discourses and texts have an "architectural significance" within organizations; institutions coordinate individuals across time and space through the vehicle of written and visual texts. To say this differently, the variety of different communications central to work and participation in the social order rhetorically influence what people do, sanction experience, grant agency and authority, and privilege certain practices over others. "Institutional discourse" operationalizes, organizes, and controls workplaces and how people carry out their work, creating generalizations, and so a sense of continuity, across individuals, practices, and sites (Smith 2005, p. 225).

IE's emphasis on texts "emerges from empirical observation as well as from theory; it comes from the insight that technologies of social control are increasingly and pervasively textual and discursive" (DeVault 2008, p. 6). The power of texts particularly arises out of their replicability – they persist over time and space – and exhibits a seemingly fixed nature. As Smith (2001, p. 160) writes:

> Texts and documents make possible the appearance of the same set of words, numbers or images in multiple local sites, however differently they may be read and taken up. They provide for the standardized recognizability of people's doings as organizational or institutional as well as for their co-ordination across multiple local settings and times.

"Boss texts" carry a certain type of authority, determining experience and practice "in such a way that an institutional course of action can follow" (Griffith and Smith 2014, p. 12). As texts carry ideas, language, and elements of persuasion between individuals (even those with little personal interaction), they subsequently transfer ideals of practice and affiliation across sites. Likewise, through texts and textual practices, individuals are enabled to recognize, organize, and respond to processes of social coordination. "Boss texts" particularly act as forms of "institutional circuits," which create ideals of accountability, professionalism, and disciplinarity, as they regulate – and often standardize – practice, mediating idiosyncrasies and variability in local settings. An example of a boss text would be the patient review form. The blank template is produced by medical professionals and their office staff; this form is used to evaluate patient needs, the services provided, and an array of other procedures from paperwork to medically oriented interactions.

2.7   Work and Work Processes

Much recent IE research has focused on the "front line" of public sector employee work, with the concept of "work" defined very generously. Because the term "work" is so comprehensive, it is somewhat difficult to define; but Griffith and Smith (2014, p. 11) note that:

In those institutional settings where services are provided to clients, we should remember that, using the "generous" conception of work, those who are served are also working; they put in time and energy and are active in actual local settings as they engage with or are caught up in an institutional process.

More generally, Smith (2005, p. 229) uses the term to refer to "anything that people do that takes time, effort, and intent." As an analytic lens that focuses institutional ethnographers on what people do, "work" is one site where the actualities of experience and practice can be readily traced in the interplay between individual and broader systems of value.

Devault (2008, p. 6) explains that work *processes* are "[o]rganizational strategies. . . [that] highlight and support some kinds of work while leaving other tasks unacknowledged, to be done without recognition, support, or any kind of collective responsibility." These processes ground the work of multiple individuals in conceptions of practice, providing the opportunity for the writing researcher to trace individual and institutional values in action. As "distinctive relational sequences" – *or how work gets done* – these processes reveal the ways local cooperative efforts respond to and reinscribe broader economies of value (Smith 2005, p. 54). We can understand "work" (in an office, in a clinic, one-on-one with a patient or client), then, as both a social collaboration and a product of uniquely personal understandings, preferences, identifications, and affiliations with and within particular institutions and disciplinary and professional identities. Devault (2006, p. 295) argues that institutional work life (experience and practice) is more tightly organized, regularized, and so coordinated, than in "households or family groupings, for instance," where texts and discourses more closely align through institutional logics and other social mechanisms.

Uncovering how work processes take shape – who determines what will happen and how – reveals the influences, hierarchies, and organizing factors at work upon individuals as they go about their daily activities. The process of asking for disability accommodations is an excellent example of the power of work processes: individuals must fill out quite particular forms (typically approved in advance by a number of offices and/or individuals), with very particular types of information to be included. The types of information included on the forms will often index closely to the individual's personal history. But, whether a person can be considered disabled or "qualifies" to claim a particular disability is a situation defined by governmental and health agencies; the criteria for consideration has been determined by medical, legal, governmental, and professional bodies in separation from the on-the-ground experiences of people who are living a unique array of embodied actualities. In many cases, the forms that document these experiences must be submitted for review to an office or committee of professionals (depending upon the local protocol), and the individual's abilities and experiences are evaluated against a number of preestablished (and sometimes hidden) factors and criteria. Each of these steps is typically highly prescribed, so that a clinic, office, or an employer is in alignment with state employment disability and discrimination laws, local culture, and other expectations associated with the site. Of course, having a procedure for determining disability does not mean that these processes are always entirely transparent, fair, or clear. The relationships individuals may have to these processes are crucial to note, as these will reveal the components of experience: the ways legal status, social status, and other factors, such as the personal philosophies, motives, and mutable practices, may become slippery as they move from a conceptual realm to be applied to a specific case.

Each of these interdependent terms related to the activities and practices people carry out as they go about their work focus the institutional ethnographer on powerful indicators of hierarchy, authority, and belonging within systems of work and labor, as they coordinate the practices of unique individuals across time and space. These key terms and analytic moves within the larger framework of IE allow us to explore the problematics of work from the standpoint of those who do the work, seeing experience and practice that may not be visible form other vantages.

3   Data Collection

Data collection for the IE project comprises observations, interviews, collection of documents, and other artifacts for analysis (see ▶ Chaps. 23, "Qualitative Interviewing," ▶ 26, "Ethnographic Method," and ▶ 29, "Unobtrusive Methods"). The institutional ethnographer seeks to bring to light an assemblage of experiences and practices through the course of a study; sites of study are recognized as locations of dialogic and multivocal belongings. The practices studied, the interactions, access of services, the creation of documents, and the carrying out of work, are read as moments of negotiation, where actual people make visible their unique understandings of the institution and their roles within it. The narrative produced presents a dynamic and multilayered understanding of lived experience and practice within the institution, often focusing on how people make choices, access resources, and participate in established routines.

Wright and Rocco (2016) argue that the IE project generally unfolds in two stages. Stage one begins with "fully understanding and developing the research [p]roblematic" (p. 28), by getting a sense of the contradictions and disjunctions of lived experience that are central to a particular set of social interactions. This process of observing the general experiences and practices of participants establishes the scope and specific focus of the study, as it offers the institutional ethnographer a sense of how things are happening, the ways the hierarchies of work coordinate how people do what they do, and other important aspects of the location. Once the institutional ethnographer has established an understanding of the standpoints central to the site, then the process of "looking up," previously described, comes into play – what Wright and Rocco describe as "stage two."

"Looking up from where you are" enables the institutional ethnographer to observe the particularities of a location, a recognition that how people are positioned within a site will often dramatically impact not only what people do but how they do it. Smith (2006, p. 5) argues that personal experiences (and so, work practices) express a social order; the institutional ethnographer then seeks an understanding of how the particularities of the social within a location then coordinate how people do what they do and influence their experiences and practices. DeVault and McCoy (2006, p. 20) describe the process of IE inquiry in similar steps: "(a) identify[ing] an experience, (b) identify[ing] some of the institutional processes that are shaping that experience, and (c) investigat[ing] those processes in order to describe analytically how they operate as the grounds of experience." As institutional ethnographers study a site, they begin to get a sense of the "language, thinking, concepts, beliefs and ideologies" that constitute a site (Luken and Vaughan 2005, p. 1604); these elements of the social realm are all clues to how the social takes shape in a specific setting. That is, how is the fluid constellation of values, identities, belongings, hierarchies, and claims to authority coming together to give a site, and so experience, practice, and work, a unique character.

IE as a framework for inquiry gains its rigor and systematicity from this process of "looking up" and teasing out the relationships between the individual and the social realm. As Campbell and Gregor (2002, p. 29) explain: "Analytically, there are two sites of interest [to the institutional ethnographer] – the local setting where life is lived and experienced by actual people and the extra-local that is outside the boundaries of one's everyday experience." When data collection activities have resulted in a reliable body of data, the institutional ethnographer will begin to analyze how individuals speak of and engage in their daily practices, thinking about how participant responses and practices reveal the ongoing coordination of activity. Variations, disjunctions, disagreements, or absences may reveal themselves in the rationales enabled by this process of "looking up," as these complex moments tell a story about the ways in which personal experience and work practices have been reflexively contoured by the material and discursive conditions of the site (Campbell 2003, p. 4). The institutional ethnographer explicates the confluence(s) of individual experience, work practice, and dominant discourses at each location, demonstrating "an empirical bridge between local and particular processes," as it brings to light the ways in which individuals actively negotiate the "social relations that order everyday existence" (Luken and Vaughan 2005, p. 1604). Brotman (2000, p. 109) has described this process more succinctly as establishing a "gaze on the macro structure from the micro level."

4   Conclusion and Future Directions

Institutional ethnography extends and reframes the work of the ethnographer, posing a line of inquiry that uncovers the relationships between the social realm and how people do what they do. Some have argued that ethnography's explicit attention to lived experience and the social nature of practice offer a rich and varied understanding of the everyday that cannot be captured from disinterested forms of empirical study. Ethnographic study has been a go-to for researchers who hope to offer a human face to social issues and to offer a personal understanding of how public policy, social norms, and cultural experiences unfold for real people in real time and space.

Even so, a number of critical challenges to ethnography have been given voice since the mid-1990s. Hammersly (2006, p. 2), for instance, has called into question the "[legitimate] claim [that ethnography factually] represent[s] an independent social reality." Others, such as Lubet (2018), question the veracity of the evidence collected by ethnographers. Arguing that ethnography suffers from a lack of fact checking, cross-examination, and the presentation of counterevidence, Lubet suggests that ethnographers must be more conscientious about whether their research narratives present reliable and *factual* representations. A number of others have claimed that ethnography relies too much upon the assumptions and observations of a single researcher (Kawulich 2005), cannot be replicated (LeCompte and Goetz 1982), and has not yet established a clear and systemic taxonomy of research practices (Wall 2015). These particular arguments retrace long-standing arguments in the social sciences, humanities, and the tradition of qualitative research about the nature of interpretation, analysis, and representation in research activities, how knowledge is constructed, and how study of the social achieves reliability.

In response to these critiques, ethnographers have adapted and evolved their stances, seeking models such as IE which have a central set of scalable and adaptable, but also somewhat regularized, analytic moves and a shared objective. These conversations gesture to the importance of situating chosen methods and methodologies firmly in an ontological and epistemological reflexivity, offering models of research that are positioned within areas of our research interest but that also methodologically extend and deepen our understandings of research practice as a local and grounded endeavor. IE does not seek to pose the researcher as an independent and so impartial observer, but rather to reveal that how people talk about work, experience, and practice is itself a negotiation of the norms of power and persuasion in a site of study.

Institutional ethnography is keenly attuned to helping researchers uncover aspects of experience and practice that other methodologies might not. Through its focus on the individuals carrying out the work of their institutions, the IE framework enables us to answer current calls in many fields to uncover how what we do in our everyday lives and as workers is coordinated by ideological and political discourses. IE enables us to systematically study the hierarchical systems of labor, professional systems of value, and notions of expertise and prestige that structure our local actualities. Through its lenses we are able to uncover stories that are often elided and that speak to the important actualities of everyday life. Cumulatively, these studies contribute to a broad picture of the discursive and organizational regimes of our lives. Ruling relations are so pervasive and so often "naturalized," most of us are not always aware of how these social forces have structured our lives, our work, and our relationships. IE allows us to explore the dynamic facets of the everyday and to discover how what we do as workers, as people with full lives, is coordinated in alignment with the lives, the practices, and the work of others.

While IE has deep roots in the field of Sociology, it has now spread to a number of other fields: nursing, psychiatry, public health, occupational therapy, and many others. Those who use IE are continuing to evolve the ways that the framework for discovery supports a range of research endeavors related to uncovering the hidden experiences of work and institutional knowing, being, and doing. Because so much about how people carry out their social lives is undergoing radical change in the twenty-first century, the result of transformations in innovation, technology, the shrinking of the public sphere, and the nature of organizational life, those interested in how actual people are negotiating these emerging contexts have found IE an invaluable tool. Many who study the global economy and neoliberal logics that drive commerce and other social forms of organization have commented upon the rise of accountability, measurement, efficiency, and adaptability as pressures structuring the global workforce. IE is being used to study how workers are coordinated by these emergent landscapes of labor. Future work with IE in many fields will continue to explore how people negotiate the everyday experience within these dynamic and global contexts.

References

Brodkey L. Academic writing as social practice. Philadelphia: Temple University Press; 1987.

Brotman SL. An institutional ethnography of elder care: understanding access from the standpoint of ethnic and "racial" minority women. PhD diss, University of Toronto; 2000. http://www.collectionscanada.gc.ca/obj/s4/f2/dsk2/ftp03/NQ50005.pdf.

Campbell M. Dorothy Smith and knowing the world we live in. J Sociol Soc Welf. 2003;30(1):3–22.

Campbell M, Gregor F. Mapping social relations: a primer in doing institutional ethnography. Walnut Creek: AltaMira Press; 2002.

Devault ML. What is institutional ethnography? Soc Probl. 2006;53(3):294–8.

Devault ML. People at work: life, power, and social inclusion in the new economy. New York: New York University Press; 2008.

DeVault ML, Gross G. Feminist qualitative interviewing: experience, talk, and knowledge. In: Hesse-Biber SN, editor. Handbook of feminist research: theory and praxis. Thousand Oaks: Sage; 2012. p. 206–36.

DeVault ML, McCoy L. Institutional ethnography: using interviews to investigate ruling relations. In: Smith DE, editor. Institutional ethnography as practice. London: Rowman & Littlefield; 2006. p. 15–43.

Geertz C. Deep hanging out. The New York review of books. October 22, 1998. http://www.nybooks.com/articles/1998/10/22/deep-hanging-out/.

Griffith AI, Smith DE. Under new public management: institutional ethnographies of changing front-line work. Toronto: University of Toronto Press; 2014.

Hammersly M. Ethnography: problems and prospects. Ethnogr Educ. 2006;1(1):3–14.

Haraway D. Situated knowledges: the science question in feminism and the privilege of partial perspective. Fem Stud. 1988;14(3):575–99.

Harding S. Introduction: standpoint theory as a site of political, philosophic, and scientific debate. In: Harding S, editor. Feminist standpoint theory reader: intellectual and political controversies. New York: Routledge; 2004. p. 1–15.

Kawulich BB. Participant observation as a data collection method. FORUM Qual Soc Res. 2005;6(2).

LaFrance M, Nicolas M. Institutional ethnography as materialist framework for writing program research and the faculty-staff work standpoints project. Coll Compos Commun. 2012;64(1):130–50.

LeCompte MD, Goetz JP. Problems of reliability and validity in ethnographic research. Rev Educ Res. 1982;52(1):31–60.

Lubet S. Interrogating ethnography: why evidence matters. New York: Oxford University Press; 2018.

Luken PC, Vaughan S. '… Be a genuine homemaker in your own home': gender and familial relations in state housing practices, 1917–1922. Soc Forces. 2005;83(4):1603–25.

Marx, K. The eighteenth Brumaire of Louis Bonaparte. New York: International publishers; 1852; 1913.

Nader L. Up the anthropologist: perspectives gained from 'studying up'. In: Hymes D, editor. Reinventing anthropology. New York: Random House; 1969. p. 284–31.

Nader L. Ethnography as theory. HAU J Ethnogr Theory. 2011;1(1):211–9.

Smith DE. Women's perspective as a radical critique of sociology. Sociol Inq. 1974;44(1):7–13.

Smith DE. Texts and the ontology of organizations and institutions. Stud Cult Organ Soc. 2001;7(2):159–98.

Smith DE. Institutional ethnography: a sociology for people. Walnut Creek: AltaMira Press; 2005.

Smith DE. Institutional ethnography as practice. New York: Rowman & Littlefield; 2006.

Wall S. Focused ethnography: a methodological adaptation for social research in emerging contexts. FORUM Qual Social Res. 2015;16(1).

Wright UT, Rocco TS. Institutional ethnography: a holistic approach to understanding systems. Int J Adult Vocat Educ Technol. 2016;7(3):27–41.

Conversation Analysis: An Introduction to Methodology, Data Collection, and Analysis

28

Sarah J. White

Contents

Abstract

Conversation analysis is a qualitative research methodology with roots in sociology, and, in particular, ethnomethodology. Over the past 50 years, it has developed not only within sociology but across the fields of linguistics, anthropology, and psychology. In health care research, conversation analysis has been successfully applied

S. J. White (✉)
Macquarie University, Sydney, NSW, Australia
e-mail: sarah.white@mq.edu.au

© Springer Nature Singapore Pte Ltd. 2019
P. Liamputtong (ed.), *Handbook of Research Methods in Health Social Sciences*,
https://doi.org/10.1007/978-981-10-5251-4_107

in researching interactions in primary care, surgery, pediatrics, and psychotherapy, to name a few examples. Conversation analysis allows the researcher to analyze the structures of interaction at a micro level, focusing on how the participants make sense of each other in conversation through shared interactional norms. In this chapter, I begin by surveying the history and development of conversation analysis. I consider methods of data collection and explore aspects of analysis in everyday conversation and in institutional interaction. I review key conversation analytic research in health care and consider its application and use for health care researchers.

Keywords
Conversation analysis · Qualitative · Interaction

1   Introduction: Why Use Conversation Analysis

Conversation analysis is the detailed microanalysis of talk-in-interaction, examined in order to provide insight into the structures of action that are usually (or normatively) oriented to by conversational participants. If the goal of the research is to understand how people are doing things using talk, like seeing a doctor, conducting a multi-disciplinary team meeting, or working in an operating theatre, conversation analysis allows researchers access to do so through its detailed and methodical approach.

The structure of conversation ensures that the production of social actions is both achievable and intelligible. Much of everyday sociality, both personal and institutional, is manifest through conversation (Heritage 2004). As Silverman (2001, p. 161) emphasizes, "conversation is the primary medium through which social interaction takes place." Children are socialized through conversation (Heritage 1984), and, following Schegloff (1987), Clayman and Gill (2004, p. 589) note that it is "the primordial site of human sociality and a fundamental locus of social organization in its own right." Not only is social interaction of the mundane variety managed through conversation, but so are the institutions that make up society – government, law, education, and health.

Communication is an integral part of the delivery of health care (Drew et al. 2001). From clinical handover (e.g., Jorm et al. 2009; Roger et al. 2016) and working in teams in the operating room (e.g., Mondada 2016; Yule et al. 2006) to eliciting patients' concerns (e.g., Robinson 2006) and delivering diagnoses (e.g. Maynard 1992), much of health care is performed through talk-in-interaction. It cannot be emphasized enough how central communication is to the delivery of safe and effective patient care.

There have been two major methodological categories used in researching clinician-patient interactions: process analysis and the microanalysis of discourse (Heritage and Maynard 2006a, p. 2). Conversation analysis (henceforth CA), which belongs to the latter group of these methodologies, has been applied to primary care interactions (e.g., Heritage and Maynard 2006b), surgical interactions (e.g., Hudak et al. 2009; White et al. 2014), psychotherapy (e.g., Peräkylä 2008), nursing care (e.g., Jones 2003), physiotherapy (e.g., Parry 2004), mental health (e.g., Pino 2016), and much more. Arguably, the most effective way to understand what is occurring in a conversation is to record it and analyze it, and CA allows you to do just that.

CA does not focus on the *why* of social action, but on the *what* and the *how* (Clayman and Gill 2004), which differentiates it from other microanalytic methodologies. It is directed at finding patterns in conversational structure and understanding and explaining their logic (ten Have 2007), describing existing structural patterns to which participants orient their production and understanding of talk-in-interaction. In short, CA is "centrally occupied with describing the procedures and expectations through which participants produce and understand ordinary conversational conduct" (Heritage 1984, p. 245) and, as applied to institutional talk, "how ordinary talk is adapted or modified to accomplish specialized tasks and achieve the visibility of these social contexts, and how participants orient to institutional identities and entitlements" (Gill and Roberts 2013, pp. 575–576).

In this chapter, I begin by contextualizing this methodology through an exploration the origins, the fundamental assumptions, and methodological foci of CA. This is followed by two sections on how to do CA, separated into data collection and data analysis. To finish, I examine the application of CA, what we have learnt so far from the method, how it can be used, and what the future directions are.

2 Methodology

2.1 Beginnings and Development of CA

Conversation analysis was developed during the 1960s and 1970s, primarily by Harvey Sacks with Emanuel Schegloff and Gail Jefferson (Heritage 1984). Some of Sacks' theorizing about how to study everyday social life can be traced to the work of his theoretical predecessors, Goffman and Garfinkel and, in particular, a field of sociology called ethnomethodology (Heritage 1984; Maynard 2012). Yet, his development of CA was new and remarkable. While not having a particular interest in language per se, Sacks found organization in the apparent chaos of conversation, demonstrating an underlying orderliness that had previously been thought impossible to describe (Heritage 1984). The focus of Sacks (and of CA in general) was form over content – the machinery of talk that allows participants to produce social actions (Silverman 1998). Sacks's unique thinking formed a new theoretical framework, aiding in the development of a new way of understanding conversation.

In the seminal paper, *A simplest systematics for the organization of turn-taking for conversation* (1974), Sacks, Schegloff and Jefferson describe and analyze how people take turns in conversation, evidencing that talk is locally managed and structurally organized through norms that govern conversational practice. This paper, which has been cited over 15,000 times, also established the systematic and scientific method of CA. Unfortunately, Sacks had an early death, leaving it to others to continue the development of this field, in particular Schegloff and Jefferson, the latter developing the widely used transcription systems used in CA (Jefferson 2004). The transcription notations used in this chapter are based on Jefferson's system (see transcription notation at the end of the chapter).

Conversation analysis has since developed considerably within a number of fields, including linguistics, anthropology and psychology (Gardner 2004) and has had substantive and methodological influence over these and other disciplines (Heritage 1984; Maynard 2012), as the advantages of examining naturally occurring data have become apparent. While CA is congruent with other observational research methodologies (Clayman and Gill 2004), it goes beyond such observation, as it provides analysts with the opportunity to repeatedly observe interactional phenomena. The advantage of this is the ability to discover fine details of interactions that would remain hidden without this methodology and it has been shown that such fine detail can have huge effects (e.g. Heritage et al. 2007). Aspects of CA can be found across different methods, such as the close analysis of recorded data and its "quantification" in building collections of interactional phenomena (Clayman and Gill 2004).

2.2   Fundamental Assumptions and Methodological Foci

In the recent collection, *The Handbook of Conversation Analysis*, Stivers and Sidnell (2013, p. 2) describe five key stances that coalesce to create CA: "(i) its theoretical assumptions, (ii) goals of analysis, (iii) data, (iv) preparation of data for analysis, and (v) analytic methods." Throughout this chapter, I explore each of these, showing how they work together to create the distinctive methodology of CA.

In analyzing conversation, the analyst is assuming that conversation is orderly and that orderliness is able to be studied (Sidnell 2013). When people talk, they are orienting to a set of norms so that they can construct the conversation together. As participants in a conversation, they can take turns, they can respond in a way that makes sense, and they can repair the conversation if it stops working, among many other things. The analyst approaches conversation with a view to the participants' understandings and orientations to the talk, rather than attempting to describe motivation, to which we have no access (Mondada 2013, p. 42). The focus is how interaction and intersubjectivity are achieved through the structures of talk.

There are three central methodological foci in CA (Silverman 2001): the structural organization of talk; its sequential organization; and the empirical grounding of its analysis. Each of these, briefly described here, encapsulate the principal findings of CA and the ways in which such findings are evidenced.

2.2.1   Structural Organization

The structural organization of talk is accomplished through the turn-taking rules (Sacks et al. 1974). These "rules" provide a normative guide that participants follow so as to maintain order within conversation; turns are managed by the participants in the conversation turn-by-turn with reference to these rules (Heritage 1984). While these are not necessarily consciously followed, they become apparent to speakers when the "rules" are broken. In the case of turn-taking, this would be, for example, when a speaker is interrupted by another party. When there is a momentary "break down" in the turn-taking organization, such as an interruption, this is when participants generally become aware that the structure exists as it is not being followed. It then may (or may not) be addressed within the conversation. Analyses of such "deviant cases" are often used to demonstrate the existence and architecture of the structure of the conversation in question (ten Have 2007).

The structural organization of talk is further managed through a system of repair. Repairs are not concerned with the correction of "errors" in conversation, but instead aid the flow (or progressivity) of conversation when there are difficulties of speaking, hearing or understanding (Schegloff 2007). If there were no system of repair, there would be no way for conversations to progress if there were any such problems, thus completely halting conversation or at least making it mutually unintelligible.

2.2.2   Sequential Organization

The conversational structures and practices which are described by CA are those that make social interaction and mutual understanding possible (Heritage 2004). Conversation analytic research has demonstrated that context is created and maintained at a local level by the participants and that the creation of meaning is reliant on the sequential environment of the talk (Heritage 2004). Sequential organization forms the basis of understanding in conversation through this turn-by-turn building of context. Participants in conversation operate under the assumption that what is said relates to what has been said just prior (unless something is said to show participants that what is being said is not to be understood with reference to the prior talk (Sacks et al. 1974)), thus creating a contextual environment for mutual understanding and intersubjectivity (Heritage 1984). Heritage (1984) regards turns as both context-shaped and context-renewing; that is, turns are delivered with reference to the previous turn and they create a context for any subsequent turn. Thus, when analyzing conversation, it is essential to have access to the surrounding talk to the utterance(s) under examination, otherwise the analysis will be limited (Silverman 1998).

2.2.3   Empirical Ground of Analysis

Any claim in conversation analytic research must be supported by actual examples found in natural conversation. Therefore, CA is rigorous in the collection of data and its analyses (Clayman and Gill 2004). This is why conversation data is audio or video recorded to ensure the empirical soundness of the analysis. Heritage (1984, p. 237) notes that "it can be difficult to treat invented or recollected sequences as fully persuasive evidence for analytic claims." Invented examples of talk based on the intuition of analysts are not a reliable source of data (Clayman and Gill 2004). On the other hand, recorded naturally occurring data exists independently of the analyst's intervention and gives access to conversational practices akin to those experienced by the participants themselves (Clayman and Gill 2004).

2.3   Institutional Interaction

In studying institutional interactions, CA provides access to the normative structures and constraints to which participants orient their talk as they work toward common institutional goals. By using CA, it is possible to understand how participants modify mundane conversational practices to achieve institutional outcomes. Conversation analytic research has found that in institutional interactions the practices of ordinary talk are modified and specialized for "task-oriented institutional contexts" (Clayman and Gill 2004, p. 592).

According to Drew and Heritage (1992, p. 22), there are three key concepts that differentiate institutional interactions from mundane conversation:

1. Institutional interaction involves an orientation by at least one of the participants to some core goal, task, or identity (or set of them) conventionally associated with the institution in question. In short, institutional talk is normally informed by *goal orientations* of a relatively restricted conventional form.
2. Institutional interaction may involve *special and particular constraints* on what one or both of the participants will treat as allowable contributions to the business at hand.
3. Institutional talk may be associated with *inferential frameworks* and procedures that are particular to specific institutional contexts.

There have been various methodologies used in the analysis of institutional interactions. These range from coding (e.g., Roter 1977) to CA (e.g., Drew and Heritage 1992; Heritage and Maynard 2006b), with many other methods in between. Quantitative methodologies have proved useful in this area, particularly when combined with qualitative research (e.g., Heritage et al. 2007); however, qualitative research gives more in-depth insight into the structures of the interactions. By using CA to research institutional interactions, we can see how everyday conversational practices are employed and modified for institutional purposes (Heritage 2004). It shows how members "invoke a particular context for their talk" (Silverman 1998, p. 171). Because CA is a very detailed method of analysis, it provides a comprehensive examination of the interaction and as such is a useful methodology in the study of institutional interactions. Through CA, analysts can, for example, explore how the roles of the professional and the layperson are co-constructed through the interaction, and how and why participants structure their conversational turns in the way they do.

In CA, as noted above, analysts use recorded data and, therefore, the details of institutional interaction are not lost as they are in observations, interviews, and experimental research (Heritage 1984). It involves the analysis of actual observable occurrences rather than invented or reported ones (ten Have 2007). Recorded data also has the advantage that the data are "neither idealized nor constrained by a specific research design or by reference to some particular theory or hypothesis" (Heritage 1984, p. 238). By using CA to study institutional interaction, the identities are made relevant and observable in the aims and activities of the participants involved (Drew and Heritage 1992).

Having explored the background and assumptions of CA, including its data-driven approach, I now move to describe the two primary aspects of conducting conversation analytic research – data collection and data analysis.

3   Data Collection

3.1   Data Collection in Health Contexts

Prior to recording interactional data, work must be done to adequately address the ethical, practical, and relational aspects of data collection (Parry et al. 2016). Collecting interactional data in a health care setting requires building trust with those from whom you are collecting it and that can be assisted through careful consideration of how you will ensure the privacy and security of the data (see also ▶ Chap. 106, "Ethics and Research with Indigenous Peoples"). In planning your research, you can consider the following questions:

- Is this the best method for what you are wanting to understand?
- What "controls" will you include in your study design (e.g., participant profession, level of training, interaction type, etc.)? (Robinson and Heritage 2014)
- How will privacy be maintained? Will the data be anonymized or de-identified and how will this affect the way in which it is presented?
- How will the data be stored securely?
- How will the data be managed? (Jepson et al. 2017)
- Who will have access to the data? What are the governance protocols?
- What are you wanting to capture?

It is advised that you engage in proto-analytic ethnography (Mondada 2013, p. 42) prior to recording to gain a sense of what should be recorded. This involves observing the environment and interactions that you intend to record in to familiarize yourself, which will assist in defining the context of what you will be recording and the scope of the project. CA is well suited to collaborative research, thus involving participants who are familiar with these aspects (such as clinicians and/or patients) in the study design (and even the analytic process) can assist you in your planning.

3.2   Recording

The naturalistic stance of CA influences data collection, with a preference for capturing as much of the interaction as unobtrusively as possible (Mondada 2013) and video recording provides other contextual information beyond that which can be elicited from audio recording alone (ten Have 2007). The approach of CA prohibits the analyst from inventing and manipulating data and ensures a strong, empirical, and accountable basis for any conclusions drawn from it (Schegloff 1988). The increasing concerns around the ethical collection of data has meant that in most places anyone being recorded must be told in advance, though the various laws regarding this differ by jurisdiction. This means that the data might be affected by the consciousness of the participants that their conversation is being recorded. However, as Mondada (2013, p. 34) argues "(c)ontrary to what is often suggested..., the camera, although permanently present, is not omni-relevant for participants, and moments in which they do orient to it can be identified and studied." More specifically, the natural, unconscious structures in conversation seem to be generally unaffected; the effects of using video and audio recording equipment are minimal on the conversational structures that are the focus of CA as the effects are generally limited to content rather than form (Clayman and Gill 2004).

There are a number of key considerations in the collection of video data as discussed by Mondada (2013, pp. 39-41):

- Perspective choices
  - Field size
  - Camera placement
  - Focus
  - Static vs. moving shots
- Technical choices
  - Cardioid vs. omnidirectional microphones
  - Lenses
  - Level of portability (with reference to setting)

These considerations form part of your study design and should be evaluated throughout the data collection process. Mondada (2013) and ten Have (2007) both provide practical details of the data collection process.

3.3   Transcription

Once the data is collected, the analyst then transcribes it. It is important to remember that transcriptions are not the data itself, but are used to make the original data accessible for in-depth research (ten Have 2007). The way a conversation is transcribed can affect the interpretation of the data as each transcriber hears and transcribes different elements of a conversation (Silverman 2001), so it is important to have the audio and/or video accessible during the analytic process.

Transcripts make the recordings more accessible to repeated analysis by visualizing the talk (and sometimes non-verbal action) into written text (Clayman and Gill 2004). Ideally, analysts should do at least some transcribing of data themselves (Clayman and Gill 2004), whether it be one or two whole transcripts from a collection or the fine-tuning of basic transcripts made by others (I have used both processes in my own research). The number of hours of conversation, the number of interactions recorded, and the focus of the analysis all affect how much will be transcribed.

The use of detailed nature of transcription in CA is motivated by the idea that no detail, no matter how minute, can be dismissed as being "disorderly, accidental or irrelevant" (Heritage 1984, p. 241). The transcripts preserve sequential detail of talk (Hepburn and Bolden 2013), including overlaps, pauses, and continuers, which, given the analyst is concerned with describing how people co-construct conversation, is important to maintain in the representation of the recorded data.

Transcriptions can also include nonverbal aspects of the conversation, included through transcriber commentary presented in double parentheses or through a set of symbols (Hepburn and Bolden 2013). To see what a CA transcript looks like, consider Excerpt 1. In this interaction, the patient is waiting in the preoperative area for surgery. The surgeon has come in to discuss the procedure. This excerpt is from the end of this discussion.

Excerpt 1 MQ-CARM12–14 (White 2015)

1	S:	$we'll see i'll see if there's there's still$
2		any evidence. alright, (.) well look we'll fly
3		into it.
4	P:	okay =um (.) tanya said can (.) one of you guys
5		give her a ca:ll [when it's done.]
6	S:	[yeah:: i can.] and i
7		should have her numbah? [but >but=
8	P:	[yeah
9	S:	=you have it on the top of y[ah
10	P:	[>it's on the
11		system.<
		((omitted 30 seconds regarding post-operative call))
12	S:	alright, (0.2) so (1.0) fly into it. (.)
13		[(very good)]
14	P:	[(do a good job)] heh [heh heh]
15	S:	[okay:] thank you.

In Excerpt 1, different details within the transcript can be seen, such as the square brackets which indicate where overlapping talk between speakers occurs (e.g., lines 5 and 6), pauses measures to the tenths of seconds (e.g., line 12), and turn-final pitch movement shown through different punctuation marks (e.g., line 3). Creating such a transcript takes time and through the repeated exposure to the data that is required in the very detailed transcription and by being forced to listen in much more detail than usual, one begins to notice different phenomena within the data (ten Have 2007). This makes transcription stand somewhere between being part of the data collection and the data analysis in its role as an analytical tool in CA. In the following section, I further explore the analytic process.

4   Data Analysis

There are two primary ways to approach interactional data: unmotivated looking and analytic keys.

4.1   Unmotivated Looking

The first of these involves the analyst noticing phenomena through repeated exposure to data either through transcription or through multiple plays. These inquiries are unmotivated insofar that they do not look for a specific feature but "discover" what features are present within that particular interaction and are open to discovering new features as well. Analysts "approach data without a specific agenda in mind at the outset, and thus remain open to previously unexplored practices of interaction" (Clayman and Gill 2004, p. 596). While the methodology of CA encourages analysts to start their inquiries with "unmotivated looking" (ten Have 1999), the nature of CA nowadays means that analysts have already been exposed to the theories and findings of previous research and are, thus, influenced by these theoretical notions when beginning their research. Entirely unmotivated looking is an unattainable ideal (Clayman and Gill 2004) and, as ten Have (1999) notes, it would be impractical to ignore the conceptual apparatus already built by CA research over the past several decades. However, we can view unmotivated looking as an open-minded and inductive approach to analysis.

4.2   Systematic Analysis

Another way of analyzing data within CA is systematic analysis which involves transcribing a sequence and then analyzing it systematically using previous principal findings in CA to "unpack" the sequence. This assists the analyst in finding patterns in the data. Many of the more recent studies have used the principal findings of CA to help begin their research and identify patterns in conversation. Ten Have (1999, pp. 107–108) provides an analytic package as an example of the systematic analysis of a natural recording using a detailed transcript. To summarize ten Have's method:

(a) Analyze a selected piece of data systemically, working turn-by-turn, explicating the use of the following "organizations" of conversational structure:
   - Turn-taking
   - Sequence organization
   - Repair
   - Turn construction/design
(b) Formulate general observations about the specific piece of data, taking note of features of particular interest.

4.3   Validity

No matter which method of analysis (or combination of methods) is used, all analyses in CA are data-driven and come from naturally occurring interactions. Ten Have (1999, p. 103) summarizes "three distinct elements" from Schegloff (1996), which relate to the assumptions and foci described earlier in the chapter, that are ideal in the empirical account for conversation analytic explications of actions:

1. "A formulation of what action or actions are being accomplished"
2. "A grounding of this formulation in the 'reality' of the participants"
3. An explication of how a particular practice, i.e., an utterance or conduct, can yield a particular recognizable action.

Both of the analytic approaches described above allow for the validity of claims through these empirical accounts. Peräkylä (2011, p. 415) summarizes the different ways in which validation is achieved in CA as: "[T]ransparency of analytic claims; validation through 'next turn'; deviant case analysis; questions about the institutional character of interaction; the generalizability of conversation analytic findings; [and] the use of statistical techniques." While not all of these aspects are present or used in every analysis, the inclusion of these within the analysis is central to the validity of claims made (c.f. Peräkylä 2011 for a detailed explanation of validity in CA research; Sidnell 2013).

4.4   Interactional Phenomena and Building Collections

Sidnell (2013) describes the analytic process as moving from noticing a potential phenomenon occurring more than once to then deliberately searching for and collecting examples of that phenomenon. In doing so, the analyst focuses on the position and composition of the phenomenon, casting a wide net initially to ensure nothing is missed. This approach to collection building works for both unmotivated looking and more systematic analysis using keys.

In creating a collection of a particular phenomenon, "one should include not only those that appear to be clear instances of the phenomenon in question, but also less clear boundary cases in which the phenomenon is present in a partial or imperfect form, as well as negative or 'deviant' cases where the phenomenon simply did not occur as expected" (Clayman and Gill 2004, p. 601). Deviant cases can often prove the "rule," or systematic practice, that is under observation as participants will often orient to normative practices in deviant cases. Such cases strengthen the analytic explication of a phenomenon by broadening its scope and clarifying its boundaries (Clayman and Gill 2004). When the normative orientations of participants are not adhered to, participants will often account for such deviations; this is because this framework of normative orientations also means deviations are normatively accountable (Heritage 1984).

To exemplify the analytic process, let us revisit Excerpt 1. As noted above, this is an excerpt from a conversation between a surgeon and patient. In this excerpt, the surgeon is beginning to close the consultation, which is a preoperative discussion.

From an overall structure perspective (Robinson 2013; White et al. 2013), we can see this is nearing the end (or possible end) of the consultation. This is not only because we have access to the full recording, but also because the participants are orienting to the end of the conversation. There is evidence for this beginning in lines 2–3, where the surgeon announces that the next activity will start soon by saying "we'll fly into it." We can also see that the patient orients to this being the closing as he pauses the closing by bringing up something new that he wanted to mention prior to the end of the conversation, that his wife wanted the team to call her after the surgery (lines 4–5). This starts a longer discussion at the end of which the surgeon partially recycles his announcement of moving to the next activity with "fly into it" in line 12. The patient accepts this and the conversation ends (for more analysis of this excerpt, see White 2015).

As an analyst, you could consider the whole interaction with the aim of noticing a phenomenon and building a collection. Alternatively, you could choose from the outset to focus on part of interaction, such as closings, to understand how they are constructed across a data set. You could even use what is known about a phenomenon to develop a broader trial, such as Robinson and Heritage (2014) describe (more on intervention studies below).

4.5   Furthering the Analysis

There are two routes to consider in furthering and presenting the analysis: single case analysis and analyzing practices of action (Robinson 2007). These are not alternatives, but rather the latter builds on the former (Schegloff 1993). The analysis still involves the close, detailed analysis of talk and requires an inductive and rigorous approach to ensure quality of the analysis.

A single case analysis involves the description of a single case or series of single cases when a phenomenon has been identified. Although it is referred to as "single case," often examples from other cases from a collection will be referenced in that analysis (Raymond 2017). Single case analyses are useful for smaller collections, such as for newly described phenomena or rare phenomena (see also ▶ Chap. 19, "Case Study Research"). Analyzing practices of action involves larger collections of the phenomenon that often also include borderline and deviant cases as well as a significant core collection.

CA, in some respects, bridges a divide between qualitative and quantitative research. From its early days, there has been an emphasis of building collections of phenomena, allowing for the identification of context-independent practices. As Sidnell (2012, p. 90) notes, "we want to identify a phenomenon that happens often enough to allow for a collection to be made." There is continued debate in the field as to how many instances of a phenomena in a collection are required to identify a conversational practice (Albert 2017). Schegloff (1996) suggests 60 instances within a core collection, while Robinson (2007), following psychological research, notes that a core collection of a practice (excluding borderline or deviant cases), needs at least 87 cases for statistical analysis.

One reason this debate continues is that numbers are in some ways irrelevant in CA. This relates the concept of "order at all points," where since the participants are able to make sense of each other, then it follows that they are able to do so due to adherence to the same norms (Sidnell 2013). Thus, choosing where to look or what to look at can also be considered unimportant, as "if. . . the way people organize their talk-in-interaction is 'orderly'. . . then it does not matter very much which particular specimens one collects to study that order" (ten Have 1999, p. 50). That is, any episode of talk-in-interaction will elucidate something about the orderly, normative structures of conversation. However, when using CA to consider a particular institutional setting, for example, an analyst deliberately restricts their data sample to those from that setting so as to concentrate on how the structures of conversation are used in that setting (Silverman 1998). Deciding whether you approach it as a single case analysis or analyzing practices of action would usually occur during the analytic process, depending on what you find.

5   Application

In the final section of this chapter, I describe some of the ways in which CA has already been used to improve communication in health care and consider future directions.

5.1   Key Findings in Health Care

Around 40 of the last 50 or so years of conversation analytic research has involved a specific focus on health care interactions. The research, as noted in the introduction to this chapter, has covered a wide breadth of clinical interactions and continues to do so. In their chapter on CA in medicine in *The Handbook of Conversation Analysis* (2012), Gill and Roberts identify three streams of research in CA in health care (pp. 578–9):

- Physician-patient interactions
- Interactions between patients and other health care professionals and paraprofessionals
- Interactions among health care professionals

The aims of such research, according to Gill and Roberts, is "to understand and document *what* social actions and activities are accomplished by participants in medical encounters and *how* participants use interactional resources and sense-making practices to accomplish their goals, with the aim of identifying recurrent practices of interaction" (p. 577; italics in original). There have been hundreds of studies applying CA to clinical interactions and this research has resulted in findings that allow us to better understand how patients participate in decision-making, how clinicians ask questions, how people work in teams, to name just a few. Heritage (2011), for example, identifies three areas in which findings from CA research could be used to improve clinician-patient interactions and from this recommends training to assist clinicians in understanding how conversational norms can impact the clinical encounter. What CA studies into communication in health care show is that through these interactions the participants are demonstrating their roles and identities (e.g. patient, doctor, nurse), establishing relationships, and accomplishing the activity at hand (Gill and Roberts 2012, p. 581).

5.2   Applied CA

The term applied CA has been used to describe both CA that involves focusing on specific interaction types or environments (as opposed to analyzing the talk without reference or specific consideration of the context) and to intervention-type studies. Antaki (2011) describes six types of applied CA studies (pp. 3–9):

- Foundational applied CA
- Social-problem applied CA
- Communicational applied CA
- Diagnostic applied CA
- Institutional applied CA
- Interventionist applied CA

Some research streams cover several of these, such as work that began as diagnostic CA which has moved into interventionist applied CA. In recent years, there has been significant progress in using CA as a diagnostic tool, particularly in dementia and seizure clinics. Reuber, a neurologist and conversation analyst, demonstrated that differences in conversational profiles of patients (i.e., the way in with patients talk about their problem) could be used to differentiate between patients presenting with epileptic seizures and those presenting with non-epileptic seizures (Reuber et al. 2009). Since then, diagnostic linguistic features in patient talk have been researched (Ekberg and Reuber 2016; Jones et al. 2016), taught (Jenkins et al. 2015; Jenkins and Reuber 2014), and developed as a diagnostic tool (Mirheidari et al. 2017) in relation to both seizure presentations and dementia.

In interventionist CA, there have been developments relevant to health. Stokoe (2011) has developed a more formal approach to using CA for training in Conversation Analytic Roleplay Method (CARM). Video-taped consultations have been used in medical school training since the early 1980s (Gill and Roberts 2012); however, CARM provides a more structured approach. This involves identifying "trainables" through CA that are then presented in a workshop format along with training on how conversation works. The workshops are created for specific professional groups or workplaces, ensuring that the trainables identified are relevant to the work of the participants.

Robinson and Heritage (2014) describe an approach to study design that factors in the possibility of quantification from the beginning. In this approach, CA is used in both the preintervention phase to identify the phenomenon to be used in the intervention and in the intervention phase as a way of evaluation and in dissemination. This is particularly useful for interventional research, such as in Heritage et al. (2007). If you are using an interventional approach, designing it as such from the beginning is important.

6   Conclusion and Future Directions

There are many avenues available for future CA in health care including researching different aspects of health care interactions across a range of professions as well as interventional and diagnostic applications, such as those described above. Gill and Roberts (2012, p. 589) argue that "recognition is growing that CA is a crucial resource for medical educators, practitioners and others whose aim is to improve the quality of medical care and relationships among participants in medical encounters."

The considerable contribution of CA can already be evidenced in its use within clinical communication training, including the leading text on communicating with patients (Silverman et al. 2013). As health care communication training relies heavily on simulation, there is significant scope not only for more CA-based training such as CARM for group training and remediation, but also for critiquing and improving simulation through a better understanding of its authenticity (White and Casey 2016). The use of CA in assessing communication might also be further explored and tested (Kelly 2009).

Looking at the changing nature of health care might lead some to explore less researched areas such as the integration of technology into the interaction, other types of team interaction, or how training and supervision is conducted. This might also include using historical data from larger databases (Jepson et al. 2017) to analyze the way in which consultations have changed over time or to track patients through the health system (e.g., Barton et al. 2016), comparing and contrasting their interactions.

Conversation analysis allows analysts to meticulously analyze conversation in order to understand the architecture of talk. It is increasingly being used to analyze doctor-patient interactions (Heritage and Maynard 2006b). The strength of the methodology lies not only in its use of naturally occurring data, but also in the replayability of the data, allowing it to be viewed and reinterpreted by other analysts. While other methodologies may also use recorded data, the microanalysis of CA develops an intricate understanding of the processes of talk-in-interaction. Although CA is congruent with other observational research methodologies (Clayman and Gill 2004), it goes beyond such observation as it provides analysts with the opportunity to repeatedly observe interactional phenomena. Through CA, we can see how we get things done through talk, that is, what is going *on*, which can help us understand and improve when things go *wrong*.

Appendix: Transcription Notation

The transcription notations that are used in this research are taken from ten Have (1999, pp. 213–214) and Gardner (2001, pp. xi–xxi). These are based on the Jeffersonian transcription system.

Sequencing	
[	A single left bracket indicates overlap onset.
]	A single right bracket indicates the point at which an overlap terminates in relation to another utterance.
=	Equal signs, one at the end of one line and one at the beginning of the next, indicate no gap between the two turns. This is called latching.
>	A carat bracket is used within a speaker to indicate no gap between a speaker's turn constructional units.
Intervals	
(0.0)	Numbers in parentheses indicate elapsed time in silence by tenth of seconds. This works within a turn, a turn constructional unit or between speakers. For example, (2.1) is a pause of 2 s and one tenth of a second.
(.)	A dot in parentheses indicates a tiny gap of less than 0.2 s within or between utterances.
Prosodic features of utterances	
word	Underscoring a word or part thereof indicates some form of stress.
::	Colons indicate prolongation of the immediately prior sound. Multiple colons indicate a more prolonged sound.
-	A dash indicates a cutoff.
w-w-word	Stuttering is indicated by a repetition of the stuttered sound connected by hyphens.
*	An asterisk around an utterance or part thereof indicates creaky voice.
$	A dollar symbol around an utterance or part thereof indicates smiley voice.
.	A period indicates a stopping fall in intonation.
,	A comma indicates a slightly rising, continuing intonation.
?	A question mark indicates a rising intonation.
¿	A "Spanish question" mark indicates stronger rise than a comma but weaker than a question mark.
_	An underline symbol after the word indicates a level pitch contour.
x:x	An underlined colon within a syllable indicates that the intonation within the syllable falls then rises.
xx:	An underlined second letter within a syllable followed by a nonunderlined colon indicates that the intonation within the syllable rises then falls.
	The absence of an utterance-final marker indicates some sort of "indeterminate" contour.
↑	An upward arrow indicates a marked shift into higher pitch in the utterance-part immediately following the arrow.
↓	A downward arrow indicates a marked shift into lower pitch in the utterance-part immediately following the arrow.
WORD	Upper case indicates especially loud sounds relative to the surrounding talk.
·word	Staccato talk is indicated by a bullet prior to the utterance-part.

(*continued*)

Sequencing	
°word° °°word°°	Utterances or utterance-parts bracketed by degree signs are relatively quieter than the surrounding talk. Very quiet talk is indicated by two degrees signs on each side.
<word>	Left/right carats bracketing an utterance or part thereof indicate slowing down as compared to the surrounding talk.
>word<	Right/left carats bracketing an utterance or part thereof indicate speeding up as compared to the surrounding talk.
.hhh	A dot-prefixed row of "h's indicates an in breath.
hhh	Without the dot, the "h"s indicate an out breath.
w(h)ord	A parenthesized h, or a row of hs within a word, indicates breathiness, such as can be heard in laughter and crying.
Transcriber's doubts and comments	
( )	The length of empty parentheses indicates the length of talk that the transcriber was unable to hear. Empty parentheses in the speaker designation column indicate inability to identify a speaker.
(word)	Especially dubious hearings or speaker identifications are indicated by parentheses around the utterance, utterance-part, or speaker designation.
(( ))	Transcriber descriptions are indicated by double parentheses.

References

Albert S. How to explain #EMCA collections to quants [Twitter moment]. 2017. Retrieved from https://twitter.com/i/moments/900684578066759680

Antaki C. Six kinds of applied conversation analysis. In: Antaki C, editor. Applied conversation analysis: intervention and change in institutional talk. London: Palgrave Macmillan; 2011.

Barton J, Dew K, Dowell A, Sheridan N, Kenealy T, Macdonald L, Stubbe M. Patient resistance as a resource: candidate obstacles in diabetes consultations. Sociol Health Illn. 2016;38(7):1151–66.

Clayman SE, Gill VT. Conversation analysis. In: Byman A, Hardy M, editors. Handbook of data analysis. Beverly Hills: Sage; 2004. p. 589–606.

Drew P, Heritage J, editors. Talk at work. Cambridge: Cambridge University Press; 1992.

Drew P, Chatwin J, Collins S. Conversation analysis: a method for research into interactions between patients and health-care professionals. Health Expect. 2001;4:58–70.

Ekberg K, Reuber M. Can conversation analytic findings help with differential diagnosis in routine seizure clinic interactions? *Commun Med.* 2016;12(1):13–24. https://doi.org/10.1558/cam.v12i1.26851.

Gardner R. When listeners talk: response tokens and listener stance. Amsterdam/Philadelphia: John Benjamins Publishing Company; 2001.

Gardner R. Conversation analysis. In: Davies A, Elder C, editors. The handbook of applied linguistics. Oxford: Blackwell; 2004. p. 262–84.

Gill VT, Roberts F. Conversation analysis in medicine. In: Stivers T, Sidnell J, editors. The handbook of conversation analysis. Chichester: Wiley; 2013. p. 575–92.

Hepburn A, Bolden GB. The conversation analytic approach to transcription. In: Stivers T, Sidnell J, editors. The handbook of conversation analysis. Chichester: Wiley; 2013. p. 57–76.

Heritage J. Garfinkel and ethnomethodology. Cambridge: Polity Press; 1984.

Heritage J. Conversation analysis and institutional talk: analyzing data. In: Silverman D, editor. Qualitative research: theory, method, and practice. 2nd ed. London: Sage; 2004. p. 222–45.

Heritage J. The interaction order and clinical practice: some observations on dysfunctions and action steps. Patient Educ Couns. 2011;84(3):338–43. https://doi.org/10.1016/j.pec.2011.05.022.

Heritage J, Clayman S. Talk in action: interactions, identities, and institutions. West Sussex: Wiley-Blackwell; 2010.

Heritage J, Maynard DW. Introduction. In: Heritage J, Maynard DW, editors. Communication in medical care: interaction between primary care physicians and patients. Cambridge: Cambridge University Press; 2006a. p. 1–21.

Heritage J, Maynard DW, editors. Communication in medical care: interaction between primary care physicians and patients. Cambridge: Cambridge University Press; 2006b.

Heritage J, Robinson JD, Elliot MN, Beckett M, Wilkes M. Reducing patients' unmet concerns in primary care: the difference one word can make. J Gen Intern Med. 2007;22(10):1429–33.

Hudak P, Clark S, Raymond G. In the shadow of surgery: treatment recommendations and the institutionality of orthopaedic surgery. Paper presented at the 2nd international meeting on conversation analysis and clinical encounters, Plymouth. 2009.

Jefferson G. Glossary of transcript symbols with an introduction. In: Lerner GH, editor. Conversation analysis: studies from the first generation. Amsterdam: Benjamins; 2004. p. 13–31.

Jenkins L, Reuber M. A conversation analytic intervention to help neurologists identify diagnostically relevant linguistic features in seizure patients'; talk. Res Lang Soc Interact. 2014;47 (3):266–79. https://doi.org/10.1080/08351813.2014.925664.

Jenkins L, Cosgrove J, Ekberg K, Kheder A, Sokhi D, Reuber M. A brief conversation analytic communication intervention can change history-taking in the seizure clinic. Epilepsy Behav. 2015;52(Part A):62–7. https://doi.org/10.1016/j.yebeh.2015.08.022.

Jepson M, Salisbury C, Ridd MJ, Metcalfe C, Garside L, Barnes RK. The 'One in a Million' study: creating a database of UK primary care consultations. Br J Gen Pract. 2017;67(658):e345–51. https://doi.org/10.3399/bjgp17X690521.

Jones A. Nurses talking to patients: exploring conversation analysis as a means of researching nurse–patient communication. Int J Nurs Stud. 2003;40(6):609–18. https://doi.org/10.1016/S0020-7489(03)00037-3.

Jones D, Drew P, Elsey C, Blackburn D, Wakefield S, Harkness K, Reuber M. Conversational assessment in memory clinic encounters: interactional profiling for differentiating dementia from functional memory disorders. Aging Ment Health. 2016;20(5):1–11. https://doi.org/10.1080/13607863.2015.1021753.

Jorm C, White SJ, Kaneen T. Clinical handover: critical communications. Med J Aust. 2009;190 (11):S108–9.

Kelly A. Articulating tacit knowledge through analyses of recordings: implications for competency assessment in the vocational education and training sector. In: Wyatt-Smith C, Cumming JJ, editors. Educational assessment in the 21st century: connecting theory and practice. Dordrecht: Springer Netherlands; 2009. p. 245–62.

Maynard DW. On clinicians co-implicating recipients' perspective in the delivery of diagnostic news. In: Drew P, Atkinson J, editors. Talk at work. Cambridge: Cambridge University Press; 1992. p. 331–58.

Maynard DW. Everyone and no one to turn to: intellectual roots and contexts for conversation analysis. In: The handbook of conversation analysis: Wiley; 2012. p. 9–31.

Mirheidari B, Blackburn D, Harkness K, Walker T, Venneri A, Reuber M, Christensen H. Toward the automation of diagnostic conversation analysis in patients with memory complaints. J Alzheimers Dis. 2017;58(2):373–87. https://doi.org/10.3233/JAD-160507.

Mondada L. The conversation analytic approach to data collection. In: Stivers T, Sidnell J, editors. The handbook of conversation analysis. Chichester: Wiley; 2013. p. 32–56.

Mondada L. Operating together: the collective achievement of surgical action. In: White SJ, Cartmill JA, editors. Communication in surgical practice. Sheffield: Equinox; 2016. p. 206–33.

Parry RH. The interactional management of patients' physical incompetence: a conversation analytic study of physiotherapy interactions. Sociol Health Illn. 2004;26(7):976–1007. https://doi.org/10.1111/j.0141-9889.2004.00425.x.

Parry R, Pino M, Faull C, Feathers L. Acceptability and design of video-based research on healthcare communication: evidence and recommendations. Patient Educ Couns. 2016;99 (8):1271–84. https://doi.org/10.1016/j.pec.2016.03.013.

Peräkylä A. Conversation analysis and psychoanalysis: interpretation, affect and intersubjectivity. In: Peräkylä A, Antaki C, Vehviläinen S, Leudar I, editors. Conversation analysis and psychotherapy. Cambridge: Cambridge University Press; 2008. p. 170–202.

Peräkylä A. Validity in research on naturally occurring social interaction. In: Silverman D, editor. Qualitative research. 3rd ed. London: Sage; 2011. p. 365–82.

Pino M. Knowledge displays: soliciting clients to fill knowledge gaps and to reconcile knowledge discrepancies in therapeutic interaction. Patient Educ Couns. 2016;99(6):897–904. https://doi.org/10.1016/j.pec.2015.10.006.

Raymond CWC. Remember also that single case analyses bring findings from other *collections* to bear on the analysis of the single case. [Tweet]. Retrieved from https://twitter.com/ChaseWRaymond/status/899656562242863104. 2017.

Reuber M, Monzoni C, Sharrack B, Plug L. Using interactional and linguistic analysis to distinguish between epileptic and psychogenic nonepileptic seizures: a prospective, blinded multirater study. Epilepsy Behav. 2009;16(1):139–44. https://doi.org/10.1016/j.yebeh.2009.07.018.

Robinson JD. Soliciting patients' presenting concerns. In: Heritage J, Maynard DW, editors. Communication in medical care: interaction between primary care physicians and patients. Cambridge: Cambridge University Press; 2006. p. 22–47.

Robinson JD. The role of numbers and statistics within conversation analysis. Commun Methods Meas. 2007;1(1):65–75. https://doi.org/10.1080/19312450709336663.

Robinson JD. Overall structural organization. In: Stivers T, Sidnell J, editors. The handbook of conversation analysis. Chichester: Wiley; 2013. p. 257–80.

Robinson JD, Heritage J. Intervening with conversation analysis: the case of medicine. Res Lang Soc Interact. 2014;47(3):201–18. https://doi.org/10.1080/08351813.2014.925658.

Roger P, Dahm MR, Cartmill JA, Yates L. Inter-professional clinical handover in surgical practice. In: White SJ, Cartmill JA, editors. Communication in surgical practice. Sheffield: Equinox; 2016. p. 333–54.

Roter D. Patient participation in the patient-provider interaction: the effects of patient question asking on the quality of interaction, satisfaction and compliance. Health Educ Behav. 1977;5 (4):281–315.

Sacks H, Schegloff EA, Jefferson G. A simplest systematics for the organization of turn-taking for conversation. Language. 1974;50(4):696–735.

Schegloff EA. Between micro and macro: contexts and other connections. In: Alexander JC, Giesen B, Munch R, Smelser NJ, editors. The micro-macro link. Los Angeles: University of California Press; 1987. p. 207–34.

Schegloff EA. Goffman and the analysis of conversation. In: Drew P, Wootton A, editors. Erving Goffman: exploring the interaction order. Oxford: Polity Press; 1988. p. 9–135.

Schegloff EA. Reflections on quantification in the study of conversation. Res Lang Soc Interact. 1993;26(1):99–128.

Schegloff EA. Confirming allusions: toward an empirical account of action. Am J Sociol. 1996;102 (1):161–216.

Schegloff EA. Sequence organization in interaction, vol. 1. Cambridge: Cambridge University Press; 2007.

Sidnell J. Basic conversation analytic methods. In: Stivers T, Sidnell J, editors. The handbook of conversation analysis. Chichester: Wiley; 2013. p. 77–99.

Silverman D. Harvey sacks: social science and conversation analysis. Cambridge/New York: Polity/Oxford University Press; 1998.

Silverman D. Interpreting qualitative data. 2nd ed. London: Sage; 2001.

Silverman J, Kurtz SM, Draper J. Skills for communicating with patients. 3rd ed. Oxford: Radcliffe Publishing; 2013.

Stivers T, Sidnell J. Introduction. In: The handbook of conversation analysis. Chichester: Wiley; 2013. p. 1–8.

Stokoe E. Simulated interaction and communication skills training: the 'Conversation-analytic role-play method'. In: Antaki C, editor. Applied conversation analysis. Basingstoke: Palgrave Macmillan; 2011. p. 119–39.

ten Have P. Doing conversation analysis. London: Sage; 1999.

ten Have P. Doing conversation analysis. 2nd ed. London: Sage; 2007.

White SJ. Closing clinical consultations. In: Busch A, Spranz-Fogasy T, editors. Sprache in der Medizin [Language in Medicine]. Berlin: De Gruyter; 2015. p. 170–87.

White SJ, Casey M. Understanding differences between actual and simulated surgical consultations: a scoping study. Aust J Linguist. 2016; 1–16. https://doi.org/10.1080/07268602.2015.1121534.

White SJ, Stubbe MH, Dew KP, Macdonald LM, Dowell AC, Gardner R. Understanding communication between surgeon and patient in outpatient consultations. ANZ J Surg. 2013;83(5): 307–11. https://doi.org/10.1111/ans.12126.

White SJ, Stubbe MH, Macdonald LM, Dowell AC, Dew KP, Gardner R. Framing the consultation: the role of the referral in surgeon-patient consultations. Health Communication. 2014;29(1): 74–80. https://doi.org/10.1080/10410236.2012.718252

Yule S, Flin R, Paterson-Brown S, Maran N. Non-technical skills for surgeons in the operating room: a review of the literature. Surgery. 2006;139(2):140.

Unobtrusive Methods

29

Raymond M. Lee

Contents

Abstract

Unobtrusive methods use ways of collecting data that do *not* involve the direct elicitation of information from research participants. They are useful in situations where it might be dangerous or difficult to question respondents directly or where, for one reason or another, using self-report methods will not yield reliable information. Three main sources of unobtrusive data can be distinguished: traces, documentary records, and direct nonparticipative observation. Each of these is discussed with examples of their use, as is the increasing use of unobtrusive data acquired online. The ethical challenges associated with the use of unobtrusive methods are identified, and the issues involved in their generation are discussed.

R. M. Lee (✉)
Royal Holloway University of London, Egham, UK
e-mail: r.m.lee@rhul.ac.uk

© Springer Nature Singapore Pte Ltd. 2019
P. Liamputtong (ed.), *Handbook of Research Methods in Health Social Sciences*,
https://doi.org/10.1007/978-981-10-5251-4_85

Keywords

Unobtrusive measures · Traces · Running records · Episodic records ·
Observation · Online methods · Ethics · Triangulation · Ethics

1   Introduction

There is a recurrent joke in Laurent Binet's (2017) satirical novel *The 7th Function of Language* in which a character on first meeting another person immediately produces a detailed account of that person's biography and way of life based, Sherlock Holmes-like, on observing small details such the presence of a signet ring or the rumpled state of their shirt. The joke would not have been lost on a group of social scientists at Northwestern University in the 1960s who met regularly and made it a game to come up with ever more outlandish research methods. The group – Eugene Webb, Donald T. Campbell, Richard Schwartz, and Lee Sechrest – went on to write a best-selling book, *Unobtrusive Measures* (1966), that set out the results of their deliberations. In the playful spirit that animated their original discussions, they explicitly encouraged researchers to seek out novel, creative, and innovative ways of collecting data rather than reaching automatically for the standard methods then in use, predominantly self-report methods based on questionnaires and interviews, both quantitative and qualitative. The difficulty with such methods, Webb and colleagues argued, was that they were *reactive*. People changed their behavior because they knew they were being studied or tailored their responses to questions in order to create favorable impressions of themselves in the eyes of an interviewer. Unobtrusive methods, by contrast, removed precisely those elements from the situation.

Webb et al. (1966) identify three sources of nonreactive data: traces, documentary records, and observation. Some alternative approaches have been proposed. Emmison et al. (2012) distinguish between two-dimensional visual sources, images, signs, and representations and the like, and three-dimensional sources, such as settings, objects, and traces, and lived and living forms of visual data, i.e., the built environment, human bodies, and interactional forms. Lee (2000) has proposed recasting Webb et al.'s typology to make clearer how particular measures are generated. To this end, traces become "found data," observational methods yield "captured data," and documents "retrieved data" (p. 13). Fritsche and Linneweber (2006) make the point that reactivity might be better seen as a continuous rather than a dichotomous variable. They set out a typology in which examples of different methods are distinguished in terms of levels of reactivity, on the one hand, and levels of participant awareness, on the other. Arbitrary though it is, Webb et al.'s distinction is retained here for expository purposes.

2   Traces

As we move through the world, we leave behind traces of our passage. Those traces, "wear, tear and rubbish" (Emmison et al. 2012, p. 132), provide evidence of the actions and processes that produced them. Webb et al. distinguish between two kinds of trace measure: "erosion measures" produced when a surface is worn away and "accretion measures" produced where material is added to the environment usually by abandonment or adornment. That floor tiles around a popular exhibit at Chicago's Museum of Science and Industry needed to be replaced more frequently than those around others that were less popular (Webb et al. 1966) has become an almost quintessential, and much cited, example of an erosion measure. Looking at wear on library books can give an indication of how often they have been read (Webb et al. 1966; Abbott 2014), while Emmison et al. note that shiny patches on metal statues provide evidence of the extent to which people have made contact with the statue in some way, by sitting on it, for example (2012).

Litter has creative possibilities. Picasso's famous sculpture, *Tête de Taureau*, is made from the seat and handlebars of a discarded bicycle. Litter can also be data. New York City banned smoking in parks and on beaches in 2011. Using a before-and-after design, Johns et al. (2013) evaluated the effect of the law by using, among other methods, an audit of litter left on parks and beaches. The ban seems to have produced a significant reduction in smoking litter for beaches and playgrounds but not parks. Interestingly, Johns et al. attribute this pattern to an environmental property of parks; it is harder to clean litter from grassy areas than from areas that are paved.

Webb et al. (1981, p. 21) describe graffiti as the "example par excellence'" of accretion data. The presence of graffiti has been used to map territoriality, as in Ley and Cybriwsky's classic (1974) study of gang distribution in Philadelphia. The content of graffiti can also be a useful measure of attitudes. When a hurricane approaches, people board up windows and doors with plywood, which is then often graffitied. Alderman and Ward (2008) point to how hurricane graffiti allowed residents to express defiance, desperation, concern, or contempt for politicians in the face of an impending disaster. The longevity of certain kinds of graffiti might be significant, as might its absence. Wilson (2014) used the time taken to remove racist graffiti appearing in a small Australian city as a measure of local authorities' willingness to meet their legal obligations to remove offensive material.

Latrinalia, bathroom graffiti, has been extensively studied. Much interest has focused on how graffiti found in male and female restrooms on university campuses differs in content. Generally, differences seem to be in line with traditional stereotypes, with men being more aggressive and sexually focused and women being more supportive and willing to offer advice. There are signs, though, of a broadening both of context and conceptualization. Trahan (2016) notes the extension of latrinalia studies to contexts such as bars and restaurants and a greater willingness to explore how issues of sexuality intersect with gender (Rodriguez 2016).

Garbage provides an important, if underused, resource for tracking social activities. Founded around 1973 at the University of Arizona by William Rathje, the Garbage Project, as it became known, aimed to apply techniques from archaeology to the study of modern-day waste (Rathje and Cullen 2001). Very much in the spirit of Webb et al.'s work, Rathje and Murphy (2001) point to recurrent findings from Garbage Project studies of a disconnect between self-reports of consumption behavior based on interviews and estimates of derived from studies of discarded packaging. Householders misreported alcohol consumption, amounts and types of food consumed and discarded, and the frequency with which they disposed of hazardous waste. Comparisons suggest that self-reports were affected by a variety of factors, including difficulties with recall and an apparent desire to portray the respondent in a positive light.

Webb et al. (1966, pp. 43–46) propose that one might improve the reliability of trace measures through "controlled erosion" or "controlled accretion," that is, manipulating the properties of materials such that erosion or deposit could be more accurately calibrated or situations such that traces are left behind. A carpet might be brushed periodically, for example, to make it easier to see the amount of traffic passing over it. The rate at which leaflets for or against a particular candidate or controversial issue left on car windscreens were discarded when drivers returned to their cars provides a measure of levels of support for the issue or candidate and potentially overcomes biases that can occur when individuals are asked in an interview to express an opinion about a controversial matter (Cialdini 2011).

Although, as Zeisel (1984) points out, thought needs to be given to how trace observations are recorded, obtaining trace data poses few operational challenges. Physical traces can be recorded quickly and at low cost, and relatively large volumes of data can be accumulated in a fairly short time. Data collection requires little cooperation from those within the setting and poses little inconvenience to them. Few ethical difficulties arise in collecting trace data. Personally identifiable material is sometimes encountered in garbage, but in the context of a trace-based study, little reason exists for it to be recorded, retained, or used (Rathje and Cullen 2001). Traces are easy to count and relatively durable. This makes it easy to track them over time or by location, opening up possibilities for longitudinal analysis or between-site comparison.

There are some rather obvious disadvantages to the use of trace data (Lee 2000). Some activities leave no trace or obliterate those already there. As a consequence, the estimates produced are usually conservative. Surfaces can be repaired, disrupting the continuity of data already there. Trace measures are not completely free of response sets or patterns of selectivity in the data. The extent of wear on a surface will depend on its physical properties. Hard surfaces wear less readily than soft ones. A surface, once degraded, wears faster than it did before. Since wear or deposit is rarely a speedy process, accumulating evidence from trace data can take some considerable time. One rarely has population data that would make it possible to calculate a rate for some particular measure.

Traces are "inferentially weak" as Bouchard (1976) puts it. What one can infer from them is limited. Zeisel (1984), writing from the point of view of someone interested in environmental design, for example, argues against the use of the term "accretion measure" precisely on the grounds that Webb et al.'s definition ignores the intention behind the addition of a particular object in an environment. Emmison et al. argue that in the absence of material from more conventional sources or a "strong sociological imagination," research using unobtrusive methods can produce "banal and largely descriptive inferences" of little interest beyond their immediate context (2012, p. 151).

3   Running Records

For Webb et al. (1966), "running records" are documentary sources produced on a reasonably regular basis in a form broadly comparable from one time period to the next, thus allowing the possibility of longitudinal analysis. By contrast, "episodic records" are archival materials, discrete in themselves and not explicitly time-ordered. Running records are highly diverse; no entirely satisfactory or comprehensive typology or listing exists. They include, for example, official statistics of various kinds: actuarial records; administrative data; government documents emanating from executive, legislative, and judicial sources; and so on. They also encompass a vast range of materials from the mass media, not just news stories but editorials, cartoons, obituaries, advertisements, personal advertisements, wedding announcements, and advice columns.

Uses of running records are as diverse as the records themselves. Tracking the popularity of baby names from birth records allowed Lieberson (2000) to analyze processes associated with cultural innovation. Wiid et al. (2011) explored political sex scandals through an analysis of political cartoons, noting that the eventual outcome tended to depend on who came to be portrayed as the "loser" in the popular imagination. Running records can be used to explore issues difficult to tap using other methods. Using information from local newspaper articles, Rowe et al. (2011) examined incidents where people with dementia went missing. The material allowed them to explore the antecedents of going missing, the characteristics of the people concerned, and the circumstances in which people were subsequently found dead or alive.

"Lonely hearts" advertisements have long been a source for understanding presentations of self and partner selection preferences (see, e.g., Jagger 1998). A fairly extensive literature now exists that examines such issues in relation to age, gender, race, and sexuality. The advice literature has been heavily mined in a number of fields, including the sociology of emotions, despite reservations by historians about how far guidance translates into practice (Lees-Maffei 2003). Many kinds of running record have moved online, making them more accessible. Yampolskaya (2017) has noted that possibilities created by the shift from paper to electronic records have encouraged policy-makers, researchers, and funders to take greater interest in the research uses of administrative data. Typically large, diverse, and comprehensive with low rates of attrition, administrative datasets provide a cost-effective source of nonreactive data on populations in contact with official bodies.

Although they have many advantages, one needs to be vigilant about the constraints, restrictions, and hidden fallibilities that surround running records. Such records are socially situated products; what is recorded is not independent of the processes involved in producing the record. With administrative data, the extent to which material collected is central to the administrative task itself is likely to have a direct effect on the quality of the information collected (Yampolskaya 2017). Trends shown in the data might result from changes in record-keeping practices. Careful judgments, in other words, might need to be made about how measures are expressed, combined, or aggregated.

4   Episodic Records

Episodic records have a more discontinuous form than running records and are "usually not part of the public record" (Webb et al. 1966, p. 88). They encompass, for example, maps, product packaging, architectural drawings, postcards, letters, diaries, suicide notes, regulatory handbooks, records of legislative debates, sales records, institutional records, inscriptions on cemetery markers, and more besides. The range and diversity of such records makes summary difficult (though see Scott 1990). For present purposes, only a few kinds of different types of record will be discussed in detail.

Once a rather devalued genre, "documents of life" (Plummer 1983, 2001) include sources such as life history interviews, letters, diaries, memoirs, and the like that document processes by which individual lives come to be creatively recounted and interpreted. Although perhaps not best thought of as "unobtrusive" in the normal sense of the term, they often have considerable power to illuminate aspects of lived experience not always well captured by more conventional methods. Jones (2000), for example, describes a somewhat unusual instance where an individual who had kept a diary over many years subsequently made it available, unsolicited, to a researcher. A particularly valuable feature of the diary was that it contained contemporaneous accounts of how the diary-keeper had experienced and responded to various medical consultations, including a cancer diagnosis (see also ▶ Chap. 83, "Solicited Diary Methods"). Documents of life are most readily found in archives and reflect the processes by which archival materials come to be selectively preserved, deposited, and catalogued. Accordingly, they tend to be more readily available for elites or for times past, a circumstance that might limit their applicability for some researchers.

Latterly, researchers have been encouraged to make wider use of data obtained under freedom of information legislation and to broaden the scope of requests beyond the law enforcement and security agencies that have traditionally been the targets of access to information requests (Greenberg 2016). Greenberg argues that research based on freedom of information requests has become easier and more fruitful over time. The processes involved have become routinized; sources are now commonly created in releasable, usually digital, formats; and researchers and officials have a greater awareness and understanding of the possibilities and limitations of relevant legal and operational frameworks. As well as a source of data in its own right, Greenberg (2016) suggests that disclosed material can be used to cross-validate data from other sources and as the basis for case study research on organizations, large bureaucracies. However, it is important to recognize the limitations of such data. Officials can thwart disclosure by self-censoring or by keeping information off the record.

Self-report methods have rather serious limitations in research focusing on wrongdoing, particularly where those involved are relatively powerful. In such cases, researchers can use documents that have come into the public domain through the activities of whistleblowers, the deliberations of investigative commissions or tribunals, or court cases. Public health researchers and others have studied previously secret internal tobacco company documents made public as a result of legal settlements (see, e.g., Fooks et al. 2011). Emails released following the collapse of the Enron Corporation have been extensively studied by researchers interested in email as a communicative medium (Janetzko 2017), while Lee (2010) was able to use tobacco industry documents dealing with the marketing of cigarettes to trace in part the history of focus group methodology.

5   Observation

Observation might be preferred to interviewing in busy or noisy environments that make speech difficult, where participants are engrossed in their activities or where participants are unwilling or unable to report on their own behavior. What you look like, what you wear, how you move and behave, your speech, your posture, your gestures, how you arrange yourself in space, and how you interact with others can all be seen, recorded, inspected, and analyzed. Aspects of physical appearance – hair, clothing, jewelry, tattoos, and the like – often convey information about culture or status. Meaning is frequently conveyed nonverbally, through gestures or often quite subtle expressions of deference and demeanor. Related to this, people use time, space, and duration to convey social meaning through such things as seating arrangements or the length of time someone is kept waiting for a meeting. Direct observation of such things, "simple observation" as Webb et al. (1966, p. 112) describe it, can often be accomplished easily and in a nonreactive manner. Of course, simplicity can be deceptive. Even straightforward observation is vulnerable to distorting factors. What is to be observed needs to be visible and accessible. It can be affected by the extent to which those studied are unaware or unconcerned about being observed, as well as by observer fatigue, inattention, distraction, or narrowness of focus. What is seen at any given point might not be representative, while the presence or absence of those in the setting might fluctuate in some nonobvious but systematic way.

Sometimes, where behavior occurs infrequently or is relatively indiscernible, the period of observation might need to be prolonged before relevant data emerges. In these circumstances, "contrived observation" is sometimes used. There are by now a number of well-known instances of contrived observation based on the manipulation of activities such as driving behavior, help-seeking, the return of lost objects, and the provision of goods and services. Compared to other unobtrusive methods, contrived observations can be vulnerable to experimenter effects, although they also permit a degree of experimental manipulation of the situation (Fritsche and Linneweber 2006).

Reiss (1971) has argued for an approach he calls "systematic social observation" that uses explicit procedures for observation and recording designed to ensure that findings are replicable. This approach has affinities with earlier observational studies of children by pioneering female researchers such as Charlotte Bühler, Dorothy Swaine Thomas, and Ruth Arrington, small group research, and the development of observational methods by human ethologists (McCall 1984). However, much of methodological development of systematic social observation took place in the

1970s in the context of research on police-citizen encounters, and such methods, which make use of trained observers, have remained important in criminology (see, e.g., Mastrofski et al. 2010). The overt presence of researchers raises obvious concerns about reactivity, dealt with in this tradition by the observers themselves monitoring in a close and careful way their possible influence on the interaction observed.

5.1 Observational Sampling

Since behavior flows continually through time and space, sampling is necessary in observational research. Martin and Bateson (2007) identify four broad strategies. The observer can record whatever seems relevant (ad libitum sampling), though this can encourage a focus on the visible or unusual at the expense of more routine or subtle behavior. In focal sampling, one observes, for a specified period of time, a particular sample "unit" such as an individual or pair of individuals and records all instances of relevant behavior occurring during that time. This can produce a bias toward visible behavior since the focal unit might not always be in sight of the observer. Scan sampling involves sweeping the behavior of a subject or group at regular intervals and recording what is happening at that particular point. It can be difficult to record more than a few categories of behavior when scan sampling, and conspicuous individuals or behaviors can be noticed more readily and, therefore, to be overrepresented. With behavior sampling, a group or setting is observed in its entirety. Each time a particular behavior occurs, its occurrence and the identity of the sample element involved are recorded.

Martin and Bateson identify two methods for recording behavioral data: continuous (or "all occurrences") recording and time sampling. Continuous recording produces a record of how often and for how long behaviors of interest occur, with start and stop times being recorded. While precise, the observer is often able to attend to only a few categories of behavior limiting the method to some degree. With time sampling, observations are recorded periodically at random intervals, a procedure that in theory is more reliable than continuous sampling since it allows more categories to be measured and more of the subjects present in the setting to be studied. When sampling, behavior measurement accuracy needs to be set against the ease with which measures can be obtained. The former implies short sample intervals, the latter long ones. Choosing an interval will often be a matter of trial and error and/or judgment or need to be determined on the basis of a pilot study.

Two types of time sampling can be distinguished: instantaneous sampling and one-zero sampling. In the former, the observation period is divided up into short sample intervals. At the precise moment each sample point is reached, the behavior of interest is recorded as having occurred or not. Instantaneous sampling does not capture well rare, brief, or inconspicuous events. In one-zero sampling, the observer records at each sample point whether or not the behavior of interest occurred during the preceding sample interval. One-zero sampling can produce biased results. Behavior is recorded no matter how often it appears or for how long it occurs, and events clustered at particular times tend to be undercounted compared to those that are spaced out evenly across the whole observation period. The method can, though, be useful in studies of intermittent behavior difficult to capture with either continuous recording or instantaneous sampling methods.

5.2 Hardware Data Capture

How to capture and preserve for analysis of what is observed has prompted a recurrent interest in recording and data-logging devices. Manual methods for recording observational data have the virtue of low cost and simplicity. Their relative inflexibility encouraged researchers in many fields to turn toward hardware-based methods that promised accuracy, ease of use, and adaptability. More recently, the advent of handheld tablets and smartphones has opened up a range of newer applications including geo-spatial mapping and data linkage to audio, video, and physiological data (see, e.g., Wessel 2015).

6 Ethics and Unobtrusive Methods

Ethical concerns arise with unobtrusive methods in relation to informed consent, the principle that people should participate in research of their own volition having been fully appraised of what their participation will entail, and with their express agreement having been obtained and documented (see ▶ Chap. 106, "Ethics and Research with Indigenous Peoples"). Since nonreactive methods, by definition, avoid direct engagement with research participants, it would seem impossible for that principle to be upheld.

Trace data, documentary research, and some kinds of observational study, such as those taking place in public settings, raise few substantial ethical issues. People in public settings act in the knowledge that their behavior is observable and open to scrutiny. Although visible, they are not directly identifiable. In many situations, sporting events come to mind, gaining permission to observe from those present would in any case be almost impossible. Issues arise most clearly in studies using contrived observation, where the researcher changes the setting to invoke a response or to make some pattern of behavior more visible or where concealed hardware is used to record people's activity. The former involves deception, while photographs, video, audio, and, latterly, geo-locational data potentially invade people's privacy and can allow individuals to be identified.

While the responsibility for ethical conduct lies firmly with the researcher, many countries require researchers to have prior approval for their research from an ethics committee or review board. Research using unobtrusive methods might need to be very carefully framed in submissions for ethical approval, balancing in particular wider societal benefits from the research against possible harm to participants and suggesting, where possible, remedial strategies such as obtaining consent after the event (King et al. 2013). A wider and controversial issue that arises in this context concerns how far, as Page (2000) argues, the principle of informed consent has become entrenched within regulatory frameworks to the extent that it is now difficult to study socially relevant topics such as stigmatization or discrimination using methods that rely on offering people disguised opportunities to act in stigmatizing or discriminatory ways without apparent sanction. How far certain kinds of study have been "chilled" in the manner Page suggests is difficult to assess. A decline in the number of observational studies reported in journals in a number of fields has been attributed in part to the impact of ethical regulation (see, e.g., Giuliani and Scopelliti 2009), although the role of other factors, changing theoretical trends, for example, cannot be ruled out. Focusing specifically on research using visual methods, Wiles et al. (2012) argue that, while concerns about ethical regulation are widespread among social researchers in the UK, it is difficult to find evidence that the system has deterred researchers from using particular methods. They do conclude, however, that "subtle but significant self-censorship" has been one outcome of ethical governance. A number of researchers who participated in their study reported, for example, being increasingly cautious in relation to issues such as the anonymization of visual materials, the ownership of data, and the dissemination practices as a result of ethical committee concerns.

7   Unobtrusive Data Online

Online methods are transforming social research, including unobtrusive methods. It is difficult to arrive at a satisfactory classification of online data sources, but some different types can be distinguished. Increasingly available in online repositories and archive are large volumes of what might be called "retrievables" – episodic and running records such as official documents, historical records, secondary data, statistical material, and so on. These can be distinguished from online materials directly harvested from sources such as webpages, blogs, online forums, and the like. Beyond this are the digital traces that computer systems produce to document their own operations and that are captured in log files (Janetzko 2017). Finally, one can mention "paradata" (Couper 1998), the auxiliary data generated as people complete surveys online, the time taken to answer a question, for example, which provide, among other things, a source of unobtrusive information about data quality and survey operation.

Online user-generated material is not necessarily free of self-presentational elements, but blogs, websites, forums, and online discussion groups have been widely used as sources of unobtrusive data (see ▶ Chap. 77, "Blogs in Social Research"). Seale et al. (2010) have made interesting use of data from online forums for cancer sufferers. They were able to compare forum data with detailed qualitative interviews from a sample of cancer sufferers. They note that while the interviews were rich in biographical and contextual data, they were shaped by respondents' self-presentational concerns and largely retrospective. The large volumes of material extracted from online forums provided, by contrast, direct and contemporaneous information about people's experiences of illness and treatment, and there was more openness about sensitive matters. Clearly, this is not to suggest the superiority of one data source over the other but to point to their complementarity.

Social media data has become a focus of interest in recent years, often under the rubric of "big data," and provides many examples of topics that can be studied unobtrusively. Some topics – fads, fashions, rumors, and protests are examples – are interesting, precisely because their evanescence, fluidity, and impermanence make them difficult to study by conventional means but accessible to online study. Zubiaga et al. (2016), for example, used Twitter data relating to a variety of news events in order to examine how people spread, supported, or denied rumors associated with an event.

The size and complexity of "big" datasets mean they need to be analyzed using computational techniques such as data mining, and there are technical issues associated with their acquisition. Bright (2017) provides an introduction to methods for accessing social media data, including the use of APIs (application programming interfaces), programmatic methods for extracting data generated by a particular provider. The use of APIs can be problematic. They are provided by the social media platform itself, and their use is governed by the provider's commercial interests rather than the needs of researchers. This means what is available through the API can be limited or might change in unanticipated ways that cut off certain avenues of inquiry or make longitudinal analysis difficult. In addition, as Bright points out, the technical aspects of the procedures involved have implications for the skilling of social researchers and the future content of research training yet to be resolved.

Although the accessibility of archival and similar materials has been enhanced by their move online, difficulties remain. While online material can generally be searched easily, keyword searching can make finding nonobvious materials more difficult, while for some sources, the variable quality of optical character recognition used in the digitization process can create problems (Abbott 2014). Online research raises ethical issues. Research conducted online often extends beyond national boundaries raising concerns about ethical governance. In some instances, on online forums, for example, the boundary between public and private is not always clear. Particularly in relation to social media, large volumes of data, techniques for combining datasets, and individuals with fairly unique combinations of attributes mean that anonymity can be compromised. One consequence can be a degree of suspicion that encourages users to be careful about self-disclosure or to provide wrong or misleading information (Janetzko 2017). On the other hand, as Buchanan (2016) points out, the possibility of data breach can make problematic the obtaining of informed consent since users are typically not well informed about policies and practices governing reuse of their data, which, in any case, are subject to change. Issues of informed consent also arise in self-report studies where paradata are collected (Couper and Singer 2013).

8   Triangulation

Webb et al. (1966) were critical not only of reactive methods. The problem with traditional methods was "that they are used alone" (p. 1). Drawing on Donald Campbell's work on "convergent" and "divergent" validity (Campbell and Fiske 1959), and the use of "triangulation" as a methodological strategy, Webb et al. argued that greater faith could be placed in a study's results when they had been obtained using different methods. As they put it, "if a proposition can survive the onslaught of a series of imperfect measures, with all their irrelevant error, confidence should be placed in it" (1966, p. 3). Denzin (1970) drew on and extended Webb et al.'s notion of triangulation to refer more generally to the combining of quantitative and qualitative methods in a single study. Subsequently, the issue became controversial. Disputes raged about how far differing epistemological assumptions rendered different methods incompatible rather than complementary (Bryman 2006). Over time, a somewhat broader view gained traction. This saw triangulation in terms of the implicative juxtaposition of different methods understood reflexively in relation to their respective epistemological foundations (Fielding 2010). In all of this, the potential contribution of unobtrusive methods to mixed methods studies was somewhat lost. Also lost, arguably, was Webb et al.'s insistence that researchers would do well to foster methodological imagination, creativity, and ingenuity. Recently, however, many health social science researchers have attempted to employ more creative and innovative methods when working with marginalized and vulnerable people (see chapters under the ▶ Chap. 61, "Innovative Research Methods in Health Social Sciences: An Introduction" section).

9   The Generative Problem

Serendipity obviously has an important role to play in the generation of unobtrusive measures. The classic example of wear on floor tiles as a measure of the popularity of a museum exhibit is a case in point. Sometimes, variables or research settings are sufficiently well-tailored to a particular problem that their use commends itself. For example, Lieberson's (2000) use of data on children's names to explore processes associated with cultural innovation allows innovations in the realm of culture to be studied independent of organizational or institutional influence, while the organizational processes that surround the registration of baby names are likely to have little effect on the naming process itself, providing data relatively free of artifact. Borrowing a method or technique from somewhere else can also be useful (Abbott 2004). Garbage studies can be seen, for example, as involving the adaptation of archaeological techniques to the study of contemporary consumer culture. What is relatively lacking in the unobtrusive methods literature is systematic guidance to help researchers identify specific methods, approaches, or data sources relevant to a particular research problem.

Although Webb et al. (1966) emphasized creativity, it is not clear that this can be summoned at will. Webb et al.'s (1981) later attempt to develop a "generative taxonomy" of unobtrusive measures proved unwieldy. Subsequently, Lee (2000, Chapter 1) suggested an alternative approach, never fully developed, in which Webb et al.'s typology of data *sources* – traces, running records, episodic records, and observation – is recast as the search for heuristics associated with *modes of data acquisition*, "finding," "capturing," and "retrieving" (on heuristics in the social sciences, see Abbott 2004).

One heuristic designed to capture aspects of routine, mundane, or fleeting aspects of social life or personal conduct is to look for ways of making a setting or situation perceptually, normatively, or culturally problematic. Researchers have used a wide range of strategies to draw out hidden or less obvious features of social situations. Some involve shifting the capabilities of normal human perception. Changing the time base for observation or the depth of focus through which some setting or activity is viewed is one example. Contriving, provoking, and disrupting tactics have commonly been used: the use of props such as lost letters or "breaching experiments" (Garfinkel 1967) in which some norm is deliberately violated in order to see it more clearly is a further example.

A heuristic useful to the identification of physical traces might be to consider how the physical properties of objects are inadvertently implicated in their social use. Webb et al. (1966) tend to see this implicitly in quasi-economic terms. Traces are evidence of production, consumption, demand, and supply. The extent to which production is implicated in consumption triggers their interest in garbage, litter, and so on. Conversely, one can consider what consumption is implicated in production. In practice, Webb et al. (1966) frame this as a question about how demand is naturally calibrated, leading them to consider measures such as abrasion on surfaces. A further and somewhat different heuristic considers the performative opportunities that objects offer directing one to consider graffiti, inscriptions, and the like.

Although it forms only a very brief part of their discussion and they had reservations about using such materials, Glaser and Strauss (1967, pp. 167–8) enjoined researchers to be alive to the potential for using "caches" of documents (see also Ralph et al. 2014). Treating such caches metaphorically as repositories of voices waiting to be heard suggested to Glaser and Strauss parallels between library and archival work and the processes of locating and assessing the usefulness of informants and respondents in field research. With this in mind, asking "How far are there documents to be found that might be treated as if they were informants or respondents sought out on the basis of their developing theoretical relevance to the topic in hand" might form a useful heuristic for identifying and evaluating running and episodic records and other documentary sources. Beyond this, the processes involved in identifying, locating, and retrieving documents often have quite specific heuristics associated with them to do, for example, with useful places to look for material, when to lightly scan materials as opposed to intensively mining them, when to start or stop searching, and so on (a detailed account can be found in Abbott 2014).

10   Conclusion and Future Directions

Webb et al.'s (1966, p. 34) commitment to what they themselves described as "oddball measures" could be seen as a gentle satire on then dominant approaches which unthinkingly applied standard methodological solutions to research problems and treated interviews and questionnaires as methods of choice without regard to their weaknesses. Alvesson and Sandberg (2013) have argued that today increasing specialization often encourages a style of "boxed-in" research that is narrow, fragmented, rigid, and defensive. They propose instead a broad range of strategies for encouraging "box-breaking" research. In this context, Webb et al.'s insistence on thinking against the grain of the conventional has a continued resonance.

It would be an exaggeration to say that unobtrusive methods have needed periodically to be rediscovered in the social sciences, but psychologists at least have on occasion needed to be reminded of their usefulness (see, e.g., Reis and Gosling 2010). In fields such as sociology, measurement-based approaches to unobtrusive data were decisively challenged by the cultural turn emerging in the 1970s that treated traces, documents, and the visual as topics in their own right subject to interpretive and critical interrogation and understanding. Nevertheless, the Australian researcher Michael Emmison (2010, p. 243) has pointed to "obvious methodological affinities" between newer approaches to the visual and to material culture and "an older and these days some – what neglected and unfashionable – branch of social research, the use of unobtrusive or non-reactive measures." Like Webb et al., these newer approaches took "a rather sideways glance at traditional sources of data in social research" (Lee 2000, p. 7) by opening up to serious study aspects of social life often regarded as peripheral, insignificant, or taken-for-granted. They also drew on work in the humanities where elicited data was little used but where sources historically deprecated or marginalized within mainstream social science were important. One might also argue, without overstating the case, that a playful and irreverent stance toward potential sources of data resonates to a degree with certain kinds of postmodern sensibility.

It is possible to exaggerate the case against self-report methods. If you want to know what somebody thinks, it is probably good to ask them. An argument can be made, however, that interviews too often become the conventional rather than the appropriate methodological choice (Alvesson 2003). In difficult or dangerous situations, interviewing is not always appropriate (Lee 1995). Not everything can be verbalized. The recall of past behavior and discussions of sensitive matters is problematic in the interview, which in any case tends to be time and place-bound. To the extent that they do encourage playful and creative approaches, unobtrusive methods force researchers to think beyond methods that are familiar or routine. Simplicity and accessibility are also advantages of unobtrusive measures; they rarely require great technical or technological sophistication and are widely adaptable to many kinds of research situation.

As the twenty-first century progresses, the methodological repertoire is likely to look quite different than it does today. New sources of data will come on-stream, new skills will be necessary, and new technological affordances will likely make it easier to combine data from diverse sources (Fielding 2014). Such trends are likely to enhance the importance of unobtrusive data, even if some of the forms that data will take are not yet apparent to us. Whatever happens, however, the oblique, the playful, and the creative are still likely to be important. Imagination is no bad thing.

References

Abbott A. Methods of discovery: heuristics for the social sciences. New York: WW Norton & Company; 2004.

Abbott A. Digital paper: a manual for research and writing with library and internet materials. Chicago: University of Chicago Press; 2014.

Alderman DH, Ward H. Writing on the plywood: toward an analysis of hurricane graffiti. Coast Manag. 2008;36(1):1–18.

Alvesson M. Beyond neopositivists, romantics, and localists: a reflexive approach to interviews in organizational research. Acad Manag Rev. 2003;28(1):13–33.

Alvesson M, Sandberg J. Has management studies lost its way? Ideas for more imaginative and innovative research. J Manag Stud. 2013;50(1):128–52.

Binet L. The 7th function of language. London: Harvill Secker; 2017.

Bouchard TJ Jr. Unobtrusive measures: an inventory of uses. Sociol Methods Res. 1976;4(3): 267–300.

Bright J. Big social science: doing big data in the social sciences. In: Fielding NG, Lee RM, Blank G, editors. The SAGE handbook of online research methods. London: SAGE; 2017. p. 125–39.

Bryman A. Paradigm peace and the implications for quality. Int J Soc Res Methodol. 2006; 9(2):111–26.

Buchanan EA. Ethics in digital research. In: Friese H, Rebane G, Nolden M, Schreiter M, editors. Handbuch Soziale Praktiken und Digitale Alltagswelten. Wiesbaden: Springer Fachmedien Wiesbaden; 2016. p. 1–9.

Campbell DT, Fiske DW. Convergent and discriminant validation by the multitrait-multimethod matrix. Psychol Bull. 1959;56:81–105.

Cialdini RB. Littering as an unobtrusive measure of political attitudes: messy but clean. In: Arkin RM, editor. Most underappreciated: 50 prominent social psychologists describe their most unloved work. New York: Oxford University Press; 2011.

Couper M. Measuring survey quality in a CASIC environment. Proceedings of the Survey Research Methods Section of the American Statistical Association. 1998; 41–49.

Couper MP, Singer E. Informed consent for web paradata use. Surv Res Methods. 2013;7(1):57–67.

Denzin NK. The research act in sociology: a theoretical introduction to sociological methods. London: Butterworth Ltd.; 1970.

Emmison M. Conceptualizing visual data. In: Silverman D, editor. Qualitative research. London: Sage; 2010. p. 233–49.

Emmison M, Smith P, Mayall M. Researching the visual. London: Sage; 2012.

Fielding NG. Mixed methods research in the real world. Int J Soc Res Methodol. 2010; 13(2):127–38.

Fielding NG. Qualitative research and our digital futures. Qual Inq. 2014;20(9):1064–73.

Fooks GJ, Gilmore AB, Smith KE, Collin J, Holden C, Lee K. Corporate social responsibility and access to policy élites: an analysis of tobacco industry documents. PLoS Med. 2011;8: e1001076. Retrieved from https://doi.org/10.1371/journal.pmed.1001076

Fritsche I, Linneweber V. Nonreactive methods in psychological research. In: Eid M, Diener E, editors. Handbook of multimethod measurement in psychology. Washington, DC: American Psychological Association; 2006.

Garfinkel H. Studies in ethnomethodology. Englewood Cliffs: Prentice-Hall; 1967.

Giuliani MV, Scopelliti M. Empirical research in environmental psychology: past, present, and future. J Environ Psychol. 2009;29(3):375–86.

Glaser BG, Strauss AL. The discovery of grounded theory: strategies for qualitative research. Chicago: Aldine; 1967.

Greenberg P. Strengthening sociological research through public records requests. Soc Curr. 2016; 3(2):110–7.

Jagger E. Marketing the self, buying an other: dating in a post modern, consumer society. Sociology. 1998;32(4):795–814.

Janetzko D. Nonreactive data collection online. In: Fielding NG, Lee RM, Blank G, editors. The SAGE handbook of online research methods. London: SAGE; 2017. p. 76–91.

Johns M, Coady MH, Chan CA, Farley SM, Kansagra SM. Evaluating New York City's smoke-free parks and beaches law: a critical multiplist approach to assessing behavioral impact. Am J Community Psychol. 2013;51(1–2):254–63.

Jones RK. The unsolicited diary as a qualitative research tool for advanced research capacity in the field of health and illness. Qual Health Res. 2000;10(4):555–67.

King EB, Hebl MR, Morgan WB, Ahmad AS. Field experiments on sensitive organizational topics. Organ Res Methods. 2013;16(4):501–21.

Lee RM. Dangerous fieldwork. Thousand Oaks: Sage Publications; 1995.

Lee RM. Unobtrusive methods in social research. Buckingham: Open University Press; 2000.

Lee RM. The secret life of focus groups: Robert Merton and the diffusion of a research method. Am Sociol. 2010;41(2):115–41.

Lees-Maffei G. Introduction: studying advice: historiography, methodology, commentary, bibliography. J Des Hist. 2003;16(1):1–14.

Ley D, Cybriwsky R. Urban graffiti as territorial markers. Ann Assoc Am Geogr. 1974; 64(4):491–505.

Lieberson S. A matter of taste: how names, fashions, and culture change. New Haven: Yale University Press; 2000.

Martin P, Bateson P. Measuring behaviour: an introductory guide. 3rd ed. Cambridge: Cambridge University Press; 2007.

Mastrofski SD, Parks RB, McCluskey JD. Systematic social observation in criminology handbook of quantitative criminology. In: Piquero AR, Weisburd D, editors. Handbook of quantitative criminology. New York: Springer; 2010. p. 225–47.

McCall GJ. Systematic field observation. Annu Rev Sociol. 1984;10(1):263–82.

Page S. Community research: the lost art of unobtrusive methods. J Appl Soc Psychol. 2000; 30(10):2126–36.

Plummer K. Documents of life: an introduction to the problems and literature of a humanistic method. London: Allen & Unwin; 1983.

Plummer K. Documents of life 2: an invitation to a critical humanism. London: Sage Publications Ltd.; 2001.

Ralph, N., Birks, M., & Chapman, Y. (2014). Contextual positioning: using documents as extant data in grounded theory research. Sage Open, 4(3). Retrieved from https://doi.org/10.1177/2158244014552425

Rathje WL, Murphy C. Rubbish!: the archaeology of garbage. Tuscon: University of Arizona Press; 2001.

Reis HT, Gosling SD. Social psychological methods outside the laboratory. In: Fiske ST, Gilbert DT, Lindzey G, editors. Handbook of social psychology. Hoboken: Wiley; 2010. p. 82–114.

Reiss AJ. Systematic observation of natural social phenomena. Sociol Methodol. 1971;3:3–33.

Rodriguez A. On the origins of anonymous texts that appear on walls. In: Lovata TR, Olton E, editors. Understanding graffiti: multidisciplinary studies from prehistory to the present. London: Routledge; 2016. p. 21–31.

Rowe MA, Vandeveer SS, Greenblum CA, List CN, Fernandez RM, Mixson NE, Ahn HC. Persons with dementia missing in the community: is it wandering or something unique? BMC Geriatr. 2011;11(1):28.

Scott J. A matter of record: documentary sources in social research. Cambridge: Polity Press; 1990.

Seale C, Charteris-Black J, MacFarlane A, McPherson A. Interviews and internet forums: a comparison of two sources of qualitative data. Qual Health Res. 2010;20(5):595–606.

Trahan A. Research and theory on latrinalia. In: Ross JI, editor. Routledge handbook of graffiti and street art. London: Routledge; 2016. p. 92–102.

Webb EJ, Campbell DT, Schwartz RD, Sechrest L. Unobtrusive measures: nonreactive research in the social sciences. Chicago: Rand McNally; 1966.

Webb EJ, Campbell DT, Schwartz RD, Sechrest L, Grove JB. Nonreactive measures in the social sciences. 2nd ed. Boston: Houghton Mifflin; 1981.

Wessel D. (2015). The potential of computer-assisted direct observation apps. Int J Interac Mob Tech, 9(1). Retrieved from http://online-journals.org/index.php/i-jim/article/view/4205

Wiid R, Pitt LF, Engstrom A. Not so sexy: public opinion of political sex scandals as reflected in political cartoons. J Public Aff. 2011;11(3):137–47.

Wiles R, Coffey A, Robison J, Prosser J. Ethical regulation and visual methods: making visual research impossible or developing good practice? Sociol Res Online. 2012;17(1):8. Retrieved from http://www.socresonline.org.uk/17/1/8.html

Wilson JZ. Ambient hate: racist graffiti and social apathy in a rural community. Howard J Crime Justice. 2014;53(4):377–94.

Yampolskaya S. Research at work: administrative data and behavioral sciences research. Fam Soc: J Contemp Soc Serv. 2017;98(2):121–5.

Zeisel J. Inquiry by design: tools for environment-behaviour research. Cambridge: Cambridge University Press; 1984.

Zubiaga, A., Liakata, M., Procter, R., Hoi, G. W. S., & Tolmie, P. (2016). Analysing how people orient to and spread rumours in social media by looking at conversational threads. PLoS One, 11(3). Retrieved from https://doi.org/10.1371/journal.pone.0150989.

Autoethnography

30

Anne Bunde-Birouste, Fiona Byrne, and Lynn Kemp

Contents

Abstract

Autoethnography is a branch of ethnography that enables a practitioner to also be a researcher and vice versa. While ethnography is concerned with the descriptive documentation of the sociocultural relationships within a given research environment, the researcher remains an observer of the situation under study. Autoethnography enables the researcher to maximize her (his) personal involvement with the action. The researcher's lived experience is an integral part of the

A. Bunde-Birouste (✉)
School of Public Health and Community Medicine, UNSW, Sydney, NSW, Australia
e-mail: Ab.birouste@unsw.edu.au

F. Byrne · L. Kemp
Translational Research and Social Innovation (TReSI) Group, School of Nursing and Midwifery, Ingham Institute for Applied Medical Research, Western Sydney University, Liverpool, NSW, Australia
e-mail: f.byrne@westernsydney.edu.au; Lynn.Kemp@westernsydney.edu.au

© Springer Nature Singapore Pte Ltd. 2019
P. Liamputtong (ed.), *Handbook of Research Methods in Health Social Sciences*,
https://doi.org/10.1007/978-981-10-5251-4_86

learning; her engagement with the context, stakeholders, and processes, along with her reflections on that engagement, is paramount to the autoethnographic methodology. Autoethnography is considered to have two clear branches: emotive and analytic. Emotive autoethnography seeks to bring the readers to an empathetic understanding of the writer's experience. Analytic autoethnography allows for the researcher's engagement in the situation to be included in the analysis, adding to the theoretical understanding of the social processes under study by making more interpretive use of available data. Analytic autoethnography is, therefore, particularly useful for the design phases of community-based action research in areas such as community development, health promotion, and social work. This chapter will provide an overview of methods involved in autoethnography, with focus on analytic autoethnography as an "action-oriented" method for social science researchers. Advantages and limitations will be discussed and illustrated with lived experience from the authors' study of complex community interventions.

Keywords

Autoethnography · Health promotion · Translational research · Analytic reflexivity · Crystallization, practice-based research, program design

1   Introduction

I don't want to just be watching, I want to be doing...
(Bunde-Birouste 2013, p. 13)

This quote exemplifies the frustration that many practitioner-oriented public health/health promotion/community development-oriented professionals may feel when they are faced with the need to engage in researching their interventions. It may also represent the feelings that practitioner-turned-researchers may feel as they try to grapple with finding a research approach which suits them. This is actually a direct quote from Anne Bunde-Birouste's own doctoral thesis, reflecting her circumstances of being pushed in order to keep her job at the university where she was teaching and managing an exciting research program on rebuilding health systems in post-conflict settings (Bunde-Birouste 2013).

In seeking to positively integrate these demands, Anne, supported by her mentor Lynn Kemp, came to find out about autoethnography, which is the subject of this chapter. Autoethnography is a research method that enables a practitioner to be the researcher and vice versa – the researcher to engage in the doing. The discovery of autoethnography was exciting for both Anne and Lynn; they had grappled with what they found to be extremely complex world of phenomenology, hermeneutics, narrative inquiry, and were extremely frustrated that to research what we wanted to, according to those methods, we needed to distance ourselves. Even ethnography was only about observing – Anne wanted to be doing!

Autoethnography is especially useful for those professionals who are not used to, or are uncomfortable with, research because they cannot see themselves having to distance themselves from the action – and wondering why their engagement is not "allowed." Rather than [merely] accompanying the processes involved in the research, the autoenthographer engages actively in both research and program/intervention. This is not to imply that autoethnography does not involve methods, protocols, and rigor. As we will show across the following pages, autoethnography is not just about the researcher, and there are specific methodological parameters to be respected and applied.

The world of research has greatly evolved since the days when only quantitative research was judged to have merit within the positivist tradition. As noted by Wall (2006, p. 147), "the postmodern era has made it possible for critical theories to emerge and take hold in academic inquiry and to open up the possible range of research strategies by the rise of a variety of qualitative methods." It is not the purpose of this chapter to debate the place of autoethnography within what Denzin and Lincoln (1994) refer to as the fifth moment of research (Denzin and Lincoln 1994; Duncan 2004). This chapter will provide an overview of methods involved in autoethnography, with a focus on analytic autoethnography as an "action-oriented" method for social science researchers. In this chapter, we will set the scene as to what autoethnography is, clarify what it is not, and propose where it is most useful. Advantages and limitations will be discussed and illustrated with lived experience from the authors' study of complex community interventions.

2   Placing Autoethnography

The theory and methods of ethnography has been described in **critical ethnography** and **ethnographic method**, but to briefly reiterate, ethnography is the systematic exploration of people and cultures where the researcher situates him/herself within the community under study. The goal is "to learn from the people (the insiders) what counts as cultural knowledge (insider meanings)"(Green et al. 2012, p. 309). Ethnographic research enables a detailed, often termed "thick," in-depth description of the culture under study (Liamputtong 2010). Ethnography incorporates "graphy" which refers to the duality of both the research processes and the written representation of that process: that is, it comprises processes beyond "ology" referring to a branch of knowledge. In ethnographic research, however, even though the researchers situate him/herself within the community under study, the researcher is still an observer; she/he may be "inside" the community but not really an insider (for a useful discussion of insider and outsider positions in research, see Corbin Dwyer and Buckle 2009).

2.1   Automethodologies

"Auto"- methods are those that go beyond situating the researcher as an "insider" (Corbin Dwyer and Buckle 2009), that is, "when researchers conduct research with populations of which they are also members" (Corbin Dwyer and Buckle 2009, p. 58; see also Kanuha 2000; Asselin 2003), to the inclusion of, and indeed focus on

Fig. 1 Methodological
components

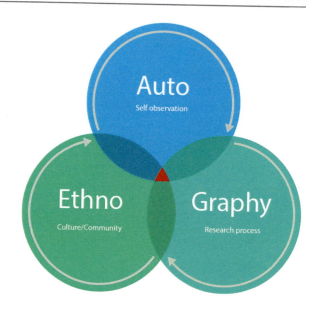

the researcher's first-person experiences of both the object (culture/practice/process) of interest and the process of knowledge discovery and creation (Pensoneau-Conway and Toyosaki 2011). Automethodologies are formed by the intersection of three components: "auto," the self; "graphy," the research process; and the epistemological frame, for example, "ethno," knowledge of culture, community, and social worlds (see Fig. 1).

There are a number of emerging automethodologies, reflecting the combination of insider researcher positioning and the epistemological position of the research/ researcher. Figure 2 provides three examples of automethodologies to highlight key components and, critically, to help distinguish autoethnography, the subject of this chapter, from other automethodological forms: autobiography (self [auto], writing [graphy] about their life[bio]), autophenomenography (self [auto], writing [graphy] about experiences or processes [phenomenon]), and autoethnography ([self], writing [graphy] about a cultural or social world [ethno]). Pensoneau-Conway and Toyosaki (2011) provide useful examples of other forms of automethods.

Our focus here is on research defined by Creswell (2008, p. 3) as "research is a process of steps used to collect and analyze information to increase our understanding of a topic or issue," with steps being drawn from posing questions, making arguments, gathering empirical evidence, and answering the question through empirical claims derived from the evidence. We hence exclude autobiography as a genre of popular literature, as an account of life, usually presenting anecdotal evidence, which does not seek to answer a question or make an argument. Autobiography can, however, provide a data source for research in what can be termed autobiographical research (Coffey 2004). This form of autobiographical research differs from our purpose here, which is first-person insider research.

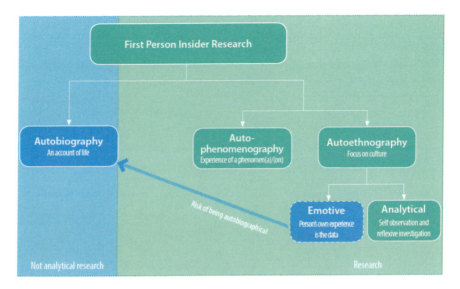

Fig. 2 Automethodologies

Autophenomenography "is about recording what is actually said or happens in a given situation without direct manipulation" (Hasselgren and Beach 1997, p. 197), through self-observation of naturally occurring routine interactions where the author is both the researcher and participant (Allen-Collinson 2011). The researcher engages in both the first-order experience of the phenomenon and the second-order reflection on the nature of that experience (Hasselgren and Beach 1997). The researcher then describes the ways that the phenomenon is experienced (Linder and Marshall 2003; Cibangu and Hepworth 2016).

2.2   Autoethnography

Autoethnography situates itself within the overarching field of ethnography, and in early development, the term referred to the study of a group to which the reflexive observer belongs (Maréchal 2010). The term evolved, however, with scholar Vryan (2006, p. 47) describing autoethnography as "a way to conduct traditional ethnography with a significantly enhanced role for the researcher: the researcher is visible, a 'strong member.'" Autoethnography, as a method of data collection and analysis, extends beyond (just) the narrative as it seeks to abstract and explain. In other words, autoethnographers should expect to be involved in the construction of meaning and values in the social worlds they investigate. The research is reflexive involving self-observation within a social or cultural world to which they necessarily belong (Pensoneau-Conway and Toyosaki 2011).

Autoethnography has developed a number of "branches" which can essentially be divided into those that seek to be evocative or emotive and analytic autoethnography.

Evocative or emotive branches of autoethnography present ideographic case studies, life stories, or autobiographical performances of first-person narrative aiming to share in an expressive or emotional way the subjective experience of the cultural or social (Maréchal 2010). These methods will primarily or solely draw evidence from the self and reproduce images of the self.

In contrast, the use of analytic autoethnography as a research approach should include systematic methodology in data collection and a diversity in types of data (Ellis et al. 2010). The autoethnographer should include data in forms of interviews, focus groups, and document analysis and could include media analysis if pertinent (see also ▶ Chaps. 75, "Netnography: Researching Online Populations," and ▶ 29, "Unobtrusive Methods"). This moves the analytic autoethnographic approach beyond frequent criticism of evocative or emotive autoethnography (Atkinson 2006) that it is just researchers being "being self-indulgent, narcissistic, introspective, and individualized" (Wall 2006, p. 155). This focus on and analysis clearly positions analytic autoethnography away from autobiography and into research. Analytic autoethnography goes beyond the "me" of the researchers and makes more use of the data. It brings in interpretive analysis. Rather than simply explaining "what is going on," the analytic autoethnographer seeks to refine and add to the theoretical understanding of the social processes under study (Anderson 2006; Vryan 2006).

Analytic autoethnography is consistent with Ellis' conception of autoethnography where the researcher's gaze moves back and forth between an outward focus and looking inward as change agent (Ellis 2004; Ellis and Bochner 2006). It is this view of analytic autoethnography as being disruptive and effecting change that makes analytic autoethnography particularly useful for design, planning, and implementation of community-based public health and health promotion initiatives (Bunde-Birouste 2013).

2.2.1 Analytic Autoethnography

Anderson (2006) designates five features key to analytic autoethnography: complete member researcher, analytic reflexivity, narrative visibility of researcher's self, dialogue with informants beyond self, and commitment to theoretical analysis. The first attribute of being a complete member researcher means that the researcher has a dual role. She/he is a member of the group under study and a researcher. Acosta et al. (2015) use the term "practitioner-researcher" which represents many professionals in this day and age. The analytic autoethnographer purposefully engages in the action as well as systematically studying it, through observation and documentation, interviews, focus groups ,and other data gathering. It is important as well to emphasize that the autoethnographer also engages in analysis and ensures that there is rigorous analysis. Anderson (2006) proposes that there are two types of complete member researchers: they are either "opportunistic" or "convert."

Opportunistic complete member researchers are either born into the group they are studying, are there by chance circumstances (e.g., through illness), or have acquired intimate familiarity with the group through occupation or some kind of lifestyle participation. Convert complete member researchers begin their belonging through a purely research-oriented situation but become "converted" due to the complete immersion and membership of the group during the research (Anderson 2006).

Although the autoethnographic methodology involves being a complete member of the group, the researcher has the additional role of being the social science researcher – thus belonging to another group as well. Anderson (2006) cautions as to the tensions that this can pose for the researcher, to be able to manage the documentation, observation, and analysis while at the same time being a member of the group. This multiple focus can evoke tension within the researcher – as the researcher must manage the multitasking. This phenomenon is further represented in the second characteristic of analytic autoethnography – analytic reflexivity. Atkinson et al. (2003, p. 62) contend that "autoethnographic data is situated within the personal experience and sense making of the researcher; s/he is part of the research process, an integral part of the story as it unfolds and that s/he shares." As an active participant, rather than a passive observer, the autoethnographer then should expect to be involved in decision-making and potentially engage in divisive issues (Anderson 2006). This again may take time for the researcher to come to terms with feeling initially that they can actively engage in decisions.

The third defining characteristic of analytic autoethnography, according to Anderson (2006), is what he calls the "narrative visibility of researcher's self": the researcher must not only be visible, active, and reflexively engaged during the research but must be highly visible within the text. As authors, analytic autoethnographers frame their accounts with personal reflexive views, situating their data within their personal experiences and sensemaking (Bunde-Birouste 2013). Here, again the auto-ethnographer is different from the others engaged in the processes, be they design development, planning, or implementation. The autoethnographer's engagement goes far beyond the experience of the moment because she/he is more fully engaged in the experience through the analysis, which involves also including their own lived experience in the data and analysis of it as integral to the whole. Further than engagement is the "textualizing" of this particular place of the researcher. The challenge for the autoethnographer in relation to this third characteristic is to include the self-reflexivity, and recount it, but without becoming the only focus of the study, which scholars in the field refer to as "self-absorption."

To avoid this latter pitfall, the fourth feature Anderson shares in analytic auto-ethnography involves "dialogue with informants beyond self" (Bunde-Birouste 2013). This crucial element involves a conscious commitment to engage with others along the journey. This feature significantly differs from traditional ethnography. Conscientious engagement with the experience as captured through the different data is needed to reach beyond the sole experience of the researcher to "make sense of the complex social worlds of which we are only a part" (Atkinson et al. 2003).

Just as the first four elements are logically linked and build upon each other, the fifth feature of autoethnographic study involves "commitment to theoretical analy-sis" (Anderson 2006). This is one of analytic autoethnography's distinguishing characteristics. The autoethnographer needs to refine, to add theoretical understand-ing to the use of empirical data, to gain insight beyond that of simple narrative description.

3   Doing Autoethnography

3.1   Design Questions

We would suggest that the first step to using an autoethnographic approach would not be dissimilar to many other research projects, which is taking time to figure out what you want to do and why and, particularly in the case of automethods, by whom and, subsequently, how best to do it. Critical to keep in mind is the integration of purposeful and deep reflexivity across all phases of action and research. This reflexivity is perhaps the most distinguishing characteristic of autoethnography and sets it apart from other research methods (Duncan 2004; Wall 2006; Acosta et al. 2015).

The following graphic presents some critical domains of design questions for autoethnography, which are described further below (see Fig. 3). Section 4 provides a case study that details how these design decisions determined and impacted on an autoethnographic study by Anne Bunde-Birouste (ABB) of the development of a complex community health promotion program.

3.1.1   The Context (Setting)

The importance of context, always significant in any research endeavor, has specific meaning with regard to autoethnography. In an autoethographic study, context is not only referring to the physical setting and cultural context in which the study is implemented, but importantly it concerns how the researcher came to be there, the how, the why (Anderson 2006; Vryan 2006). As discussed earlier in this chapter, what the setting means to the researcher, how it influences them, the history that the researcher(s) brings to it: their past experiences, their disciplinary back grounds, world views, and personal experiences, all contribute to the richness of the learning. Logically the same context is relevant for members of a collective autoethnographic study (see case study example Sect. 4.1.1).

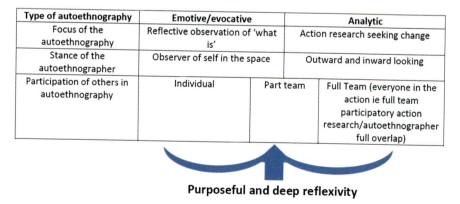

Type of autoethnography	Emotive/evocative	Analytic	
Focus of the autoethnography	Reflective observation of 'what is'	Action research seeking change	
Stance of the autoethnographer	Observer of self in the space	Outward and inward looking	
Participation of others in autoethnography	Individual	Part team	Full Team (everyone in the action ie full team participatory action research/autoethnographer full overlap)

Purposeful and deep reflexivity

Fig. 3 Design domains

3.1.2 The Why

The focus, or purpose, of the study needs to be clear to the autoethnographer: Is the study being done to observe what "is" (usually associated with evocative/emotive autoethnography) or to catalyze or accompany action, hence to affect change? Moving along the design continuum, the engagement of the autoethnographer can go from being a "self-observer," focusing on what is happening to him/her but not driving any change, to engaging to drive some of the action, being progressively more involved in what is happening to others as well as self, to fully driving the action (see case study example Sect. 4.1.2). Autoethnography is increasingly being considered a practical and pragmatic approach to take in combination with various action research approaches, including the diverse forms of participatory action research (Acosta et al. 2015).

For instance, is this research the foundation for program design? Or, is the objective of the research to accompany and assess program implementation and impact, hence designed to accompany an intervention in an action research framework? Is the research retrospective; is the study being done to observe and document an initiative, and if so for what purposes? The aims and objectives of the research will influence how the autoethnography is set up and whether it would be an individual research project, driven and implemented by a single autoethnographer or whether it is a variant of collective autoethnography (see Sect. 3.1.4 below). The answers to these questions will influence where the research or researchers are placing themselves and what data collection and analysis methods are to be used.

3.1.3 The What

The autoethnographer's stance determines, and is determined by, the object of the research, the what: Is the self the sole object or the self and others? In the former, as per emotive/evocative autoethnography, the self is the object. In the latter, the autoethnographer is both outward and inward looking, noting that an autoethnographer cannot be only outward looking as self-observation underpinned by reflexivity is a distinguishing characteristic.

When conducting action-focused autoethnography, taking an outward and inward looking stance, and regardless of whether autoethnographic research is used in a design phase, planning or implementation of a program or intervention, intentional planning of the phases of research elements, and linking to planning of the program/action components are essential. Ideally, the same care in planning the different phases of action will be accompanied by similar care in planning the steps to the research, which will be carried out in parallel. It is important to set out the autoethnographic stance before beginning the action, in all phases from prepping the research, to analyses, writing and sharing it. The various phases of action need to be considered, planned, and underpinned by preparatory research. Similarly, the various forms of data and data gathering will need to be considered, arranged, and scheduled; in other words, they need to be based on previous research, sequential, structured, and purposeful (see case study example Sect. 4.1.3).

3.1.4   The Who

One of the attractive characteristics of autoethnography is that it can be done either solo or with a group of stakeholders (Acosta et al. 2015). The choice of this will depend on where the researcher/practitioner is placing him or herself and as such is one of the very first and crucial decisions the researcher should make, as the ensuing research design will depend on it. In such cases, there are multiple scenarios. It is helpful to consider the participation design domain of solo to full group auto-ethnographic participatory action research along the lines of a continuum (see Fig. 3 above). Some autoethnographic research will be solo; in other words, the autoethnographer, alone, is designing, driving, and analyzing the research him- or herself. In this case, although there may be other actors involved in the initiative, or action that is being researched, the autoethnographer manages the various research stages, data collection, and analysis alone.

In some instances, the autoethographer drives the research action and analysis yet also integrates and supports a team-based approach, in which the reflexivity is done solely by the autoethnographer, but she/he integrates learning from involvement of stakeholders along the way. This is not to imply that the full community or action team has stance of the autoethnographer; they can be involved, yet the autoethnographer is still the solo researcher. Progressing further along the continuum from solo to group is the scenario where a single researcher drives the research yet engages "collectively and cooperatively within a team of researchers" for the stages of analysis (Chang et al. 2013; Acosta et al. 2015). As such, the group, or team, begins to be involved in the autoethnographic process. Chang et al. (2013) call this "type" of autoethnography as "collaborative autoethnography," while Acosta et al. (2015) call it collaborative and analytic autoethnography (CAAE) in action research inquiry. This team approach can remain somewhat "driven" by a single auto-ethnographer in that one person takes ultimate responsibility for moving things forward and final processing of analysis (see case study example Sect. 4.1.3). Finally, at the far end of the spectrum is the full group autoethographic process, wherein all members are fully autoethnographically engaged in all stages of the research.

In the following sections, we will focus on analytic autoethnography that is focused on change and both outward and inward looking, which may be conducted by an individual autoethnographer or engage a team.

3.2   Data Gathering

As noted earlier, use of analytic autoethnography as a research approach should include a systematic methodology in data collection and a diversity in types of data (Ellis et al. 2010). The data sources will be in different forms, mixing query, reflections, observations, and straight noting of activity. There will be the personal involvement of the researcher-practitioner and "classic" elements of ethnography such as participant observation and specific and significant diarizing of activities (see ▶ Chap. 26, "Ethnographic Method"). In addition to these researcher-/practitioner-specific elements, there are the additional data gathering mechanisms to include: interviews, focus groups, and document/media analysis. We suggest that communication elements such as meeting minutes and e-mails can also be brought into the data pool. These bring in information additional to autoethnographer's notes and contribute to corroborating findings or ensuring trustworthiness of the data.

From this collective, compiled analysis, the full story emerges, and with it the contribution to learning unfolds (Bunde-Birouste 2013). In examining results from different methods of data gathering, the metaphor of triangulation is the common term currently used to corroborate findings and ensure trustworthiness of the data gathering methods (Denzin and Lincoln 1994). For autoethnographic studies, we prefer a technique that more fully recognizes the variety of facets involved in any given phenomenon – crystallization (Richardson 1994; Wall 2006). Acknowledging that there are "far more than three sides through which to view the world," the crystallization approach is particularly appropriate for autoethnographic studies where methods used are multiple and thus often complex (Richardson 1994, p. 522). Crystallization, as explained by Richardson (1994), deconstructs the traditional idea of validity without losing structure, providing a deepened and complex understanding of the phenomena. This approach allows for the diverse, sometimes minimally structured, data sources, recognizing that they will all contribute to a better understanding of the whole. Each source of data does not have to be comprehensive in and of itself – but must be comprehensive across data sources (Kemp et al. 2008). The rigor and structure will come in the analysis, as opposed to the sources (Bazeley and Kemp 2012; Bunde-Birouste 2013).

3.3   Data Analysis

As noted in Sect. 2.2.1 analytic autoethnographic research involves multiple layers of analysis to interrogate the data to find the "hidden story" within new learning (Bunde-Birouste 2013). As with any qualitative research, there are multiple and progressive steps to data analysis and discovering the depth of the story that the research is telling. The progressive steps of analysis enable interrogation of the complexity, separating the background noise to delve deeply into the "story," discovering underlying themes, and exploring the findings to understand their different facets (Bunde-Birouste 2013).

A critical part of autoethnography is the hard thinking about how to "represent" the data. The choice of how, where and why the story is told, will frame the representation of the data and the telling of the story. To enable engagement with the story the data is telling, autoethnographic analysis is often presented in some form of analogy, commonly employing literary or visual metaphors. As the steps in data analysis progress in an iterative fashion, often the choice of representation format comes in early in the process. This is perhaps best understood by reading the excerpts from our case study (see case study example Sect. 4.2).

4   Introduction to Case Study Vignettes

Our case study is from Anne Bunde-Birouste's PhD research, "Kicking goals for social change." This is a story of how a vision turned into a viable social innovation program using football to support refugee youth and families feel welcome in Australia (see www.footballunited.org.au). The study examined the challenges and processes involved in the design and development phases of a program that uses football as a vehicle to contribute to building social inclusion in complex sociocultural settings in urban areas of Western countries, such as Australia, that largely consist of refugees from fragile and conflict-impacted areas. This is also a story of how a practitioner came to reconcile herself with research by using an auto-ethnographic approach. Finally, this is a story that proposes that health promotion consider the change agent through the lens of social innovation as a better fit for the complex world of community-based health promotion. The autoethnographic methodology allows for appreciation of the researcher as a social innovator. Significantly, the tale contributes new learning into responsive program design in what is increasingly referred to as complex, community-based health promotion interventions and presents the value of autoethnography as an effective method for research in health promotion, one currently underappreciated and underused.

4.1   Design Questions

4.1.1   The Context

Anne was working in the area of health and peace building, with particular emphasis on post-conflict health sector redevelopment. Through this work, she became highly sensitized to the dramatic plight of refugees, asylum seekers, and humanitarian immigrants. In a previous life, she had witnessed the power of football (aka soccer) to bring people together:

> I was among more than 350,000 people gathered to celebrate the French National Team's World up victory in 1998. My then little 8 year old son and I were among those hundreds of thousands, and were completely and utterly awestruck not only by the number of fans, but by the indescribable shared euphoria of all of them, and even more so by their diversity. Old and young, rich and poor, black, brown and white, all races and ethnic groups brought together by the amazing victory on the football (soccer) field. Ecstatic people crowding on the Champs-Élysées to celebrate the multi-racial French victory team, at a time when the country was significantly frayed by racial tensions.
>
> During my professional life as an international practitioner in community-based health promotion and community development I also witnessed the passion and popularity that the "World Game" had across the entire globe. There is not a community where one does not see children, youth and adults kicking balls around, playing on makeshift fields, often barefoot.

4.1.2   The Why

These two phenomena, Anne's professional expertise and the popular passion for football (including her own), combined into a vision that she wanted to explore:

Could she find a way to use football to "bring people together" and help newly arrived refugee and humanitarian immigrants settle into Australia? A great idea – a vision. But how to develop this vision into a viable program? Of course, community-based, action research seemed the way to go according to her theoretical background. Yet, she struggled with how to remain outside the action when she felt that the:

> vision was so powerful, I could literally see the programs happening. After considerable exploration of methods in the area of action inquiry and reflective practice, and pondering about how best to approach the research, my supervisor suggested that the approach best fit for her was autoethnography, which allowed me to be part of the research process, an integral part of the story as it unfolded and that I would share.

4.1.3 The What and the Who

In setting out to explore if Anne's vision that the passion that the world game of football could bring people together and help newly arrived feel welcome and belong, Anne traversed several pathways in her autoethnographic journey. Committed to a participatory, responsive approach to her research, Anne followed the classic path of engaging with various stakeholder groups across all phases of her research, initially attempting to work consistently with a core steering committee. As challenges emerged and pilot efforts came and went, it became clear that her role was that of the driver, the change agent, and sole autoethnographer: the main character in her cast of dozens.

> As the initiator and driver, I was constantly juggling with the desire to maintain principles of consultation and participatory decision-making versus the need to just get on with the development. I was hesitant to be too directive, feeling that I might undermine the concept of participatory community building. This consultative process was frequently not understood by the community members or even the Steering Committee (SC), who often wondered why I continued querying and discussing, rather than taking the lead in decision-making. I realized that they were not understanding what I was trying to do, which prompted the understanding that at times a pragmatic approach was needed. A driver was needed; I had that role and also needed to be a leader in taking a decision and moving the group forward.

Through these experiences, Anne realized that the fitting design for her study was analytic autoethnography, focused on action research seeking change, led by herself as the individual sole ethnographer but taking and outward and inward looking stance (Fig. 4) (Bunde-Birouste 2013).

4.2   Data Analysis and Representation

The choice of metaphor for this study was that of literary analysis. Literary analysis goes beyond a surface reading of a text, story, or, in this analogy, a theater play. As in the analysis of a play, the setting or context within which the story takes place

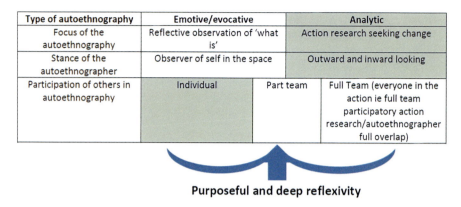

Type of autoethnography	Emotive/evocative	Analytic	
Focus of the autoethnography	Reflective observation of 'what is'	Action research seeking change	
Stance of the autoethnographer	Observer of self in the space	Outward and inward looking	
Participation of others in autoethnography	Individual	Part team	Full Team (everyone in the action ie full team participatory action research/autoethnographer full overlap)

Purposeful and deep reflexivity

Fig. 4 Case study example of design domain choices. (Source: Bunde-Birouste 2013)

includes not only time and location but it also includes the multiple levels of social reality present. A play is an evolving story with multiple stages and many acts. A play mirrors life, which is ever changing – as was the experience of Football United's development. Stories evolve, as does life, and as do health promotion interventions and necessarily the processes to design them. A play may have many actors and, like analytical autoethnography, is not a "one-man show." The analysis involved needs to go beyond that of the self. Within this construct comes the challenge to reproduce the views of the other actors.

A theater play has a number of different parts that combine to tell the story: the plot, subplots, characters, theme(s), tone, and stage. As within any tale, it can be viewed superficially or examined more deeply. A theater review requires objective analysis beyond the apparent and involves more than a simple plot summary – it must be grounded in the production itself. Anne's ethnographic study was indeed grounded in the action from initial conception and enabled her to see beyond the superficial – to explore the depth of what is really going on as I moved from a good idea to build a viable program concept.

The initial experience of a play is often summarized by critics in the *plot synopsis*. This is the general impression when you first experience the play: you view the stage, "feel" the overall story, gain an overall impression of the subject matter and genre (in this case community-based health promotion), and see the things you expect to see. Correlating this stage to autoethnographic study, it relates to the initial immersion in reviewing data following the fieldwork.

The initial steps to discover underlying themes and explore the characters to understand their different facets bring in character notes and further plot analysis to unearth the various story lines. To enable her to move beyond the plot synopsis, Anne needed to take a number of progressive steps, as summarized in the following table (Table 1) and detailed further below.

The plots synopsis development provided a good sense of things that a "standard" production might involve. A more profound appraisal of the play was needed to begin to look deeper into the characters, to understand their various facets, and what

Table 1 Theater review and steps for analysis

Theater review	Steps for analysis
Plot synopsis	Coding according to genre: exploration of data with community-based health promotion elements in mind; data entered and coded in free nodes in NVivo
Character notes	Clarke's framework for situational analysis (2003) used to delve deeper into the data to explore beyond the surface; framework elements recorded in tree nodes in NVivo
Plot analysis	Multiple methods to interrogate the data to find the "hidden story" within new learning; interrogating the complexity; separating the background noise

they contribute to a deeper meaning to the experience. This next level of analysis required a review of the initial data analysis with a more in-depth manner of coding; the choice of Clarke's (2003) framework for situational analysis is used for this step.

Clarke's framework analysis was appropriate as it enables a researcher to draw on multiple data sources including, among other things, participant observation, key informant interviews, focus groups, and any other data sources the researcher deems relevant (Clarke 2003). The method allows the researcher to draw together action, structure, context, history, and agency for analysis of complex situations. Clarke's framework was particularly suited to Anne's preliminary level of analysis given that it draws on ecological frameworks to map the key human and nonhuman elements involved in the processes under study. This mapping of connections enables a further and deeper analysis and according to Clarke, enhances the analysis as it allows the social side to be added to the more individualistic analysis, which is the center of ethnographic, narrative, and other forms of interpretive phenomenology, thus providing for a full situation analysis, which is sensitive to the complexities, multiplicities, and contradictions, and taking them seriously. It is a particularly useful method of analysis for complex, community-based autoethnographic research. Again, multiple levels of analysis within this phase enabled me to come up with more significant understanding of the characters and deeper meaning to the forces at play. The resulting character notes were extensive and shed an interesting light on the overall story, yet they only told part of the tale.

To understand further required a plot analysis, which would pull all the various elements of the story together by delving into context, character analysis, and story lines, with the intent to unearth the deeper, hidden story or stories. These subliminal stories may either highlight new learning or perhaps shed new enlightenment on old issues.

Trying to find the "hidden stories" in her production proved to be more of a taxing process than Anne would ever have imagined. She ended up referring to this very long stage of interrogating the data as "drudging through the analysis." She reviewed the framework analysis, reviewed relationships, developed further tables, matrixed learning, and checked for cross-referencing, pulling out issues, querying them, and then doing it all over again and again and again. To be honest, it was exhausting! However, it all did end with an intriguing plot analysis, which produced significant learning as to situation, processes, managing complexities, and future potential.

The resulting learning is presented in the "Reviewers" critique, which brings together the results from the various stages of analysis and, by applying a final "filter," provides us with what might be commonly referred to as discussion and conclusion in a more traditional presentation of research results. Anne found this a fascinating step, one much more interesting than the fatiguing multiple layers within the previous analysis stages.

One thing that plagues those working in community-based programs and health promotion in general is the complexity; managing this complexity in research, and particularly design phase, research can be overwhelming. The challenge is to work through the complexity; filter out the background, offstage, and audience sounds; and focus on recurring bits to understand how the various elements and forces come together. Complexity scholars Funnell and Rogers (2011) advise to focus on the characteristics that are most relevant in developing a program theory to decipher the complexity.

If we can succeed with an effective filtering mechanism, one might consider that the complex becomes merely complicated; the chaotic feeling of being overwhelmed (feeling unable to act) becomes one of analysis, understanding, and then managing. We need to unpack the complexity and in doing so will often find that we move in a sense back from complex to complicated. Anne's own attenuation filtering process (see Fig. 5) was the application of the editing framework to the draft sections of the character notes, where she queried each section to explore: What are the forces, what did they do, how did they move the story forward, and what do we conclude from it?

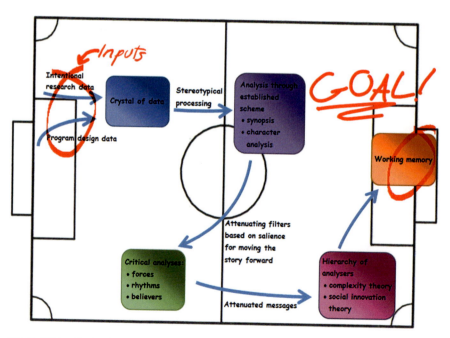

Fig. 5 Taking the inertia out of complexity (ABB attenuation filtering model)

The resulting "story lines" provided the final results in the reviewer's critique, which in her case were two: the value of analytic autoethnography in health promotion research and particularly the importance of it to underpin the design phases of programs or interventions and that community-based health promotion involves working with a set of complex social innovation processes, which require the practitioner, if she/he is to be effective, to be a social innovator. Because the mechanisms of transforming the vision to action often seem elusive, the social innovator can be particularly aided through an autoethnographic approach. Through analytic autoethnography, the social change processes are underpinned by a robust research technique which enables analysis and processing – allowing us to apply the learning for future social change endeavors.

5   Conclusion and Future Directions

Autoethnography, particularly analytic autoethnography, is an emerging method that supports the practitioner-researcher to be subjectively and fully engaged within the action being studied. We propose that it is especially useful for studying community-based health promotion as a set of complex social innovation processes, where the genesis of the social change is a desire to change the "what is." The autoethnographic method can have significant implications for health promotion research, in particular, the practice-based research that is increasingly sought from theorists. As we look for more practice-based research to inform policy and theory, an autoethnographic approach is particularly effective. It can empower the practitioner to become the researcher. Autoethnography acknowledges the researcher's experience and expects that it will inform the research. Yet, it is not just about the researcher, and there are specific methodological parameters (Anderson's five features; refer to Sect. 2.2.1) in analytic autoethnography that ensure the researcher engages in a high level of dialogue with informants and in commitment to theoretical analysis. As such, this is an excellent method to provide the foundation for future practice-based research.

Acknowledgments The authors gratefully acknowledge support from Adjunct Professor, Dr Patricia Bazeley, Research Support P/L, Translational Research and Social Innovation group Western Sydney University for her guidance, editorial contributions and critical feedback.

References

Acosta S, Goltz H, Goodson P. Autoethnography in action research for health education practitioners. Action Res. 2015;13(4):411–31.
Allen-Collinson J. Intention and epoché in tension: autophenomenography, bracketing and a novel approach to researching sporting embodiment. Qual Res Sport Exerc Health. 2011;3(1):48–62.
Anderson L. Analytic autoethnography. J Contemp Ethnogr. 2006;35(4):373–95.
Asselin ME. Insider research: issues to consider when doing qualitative research in your own setting. J Nurses Staff Dev. 2003;19(2):99–103.

Atkinson P. Rescuing autoethnography. J Contemp Ethnogr. 2006;35(4):400–4.

Atkinson PA, Coffey A, Delamont SS. Key themes in qualitative research: continuities and change. Walnut Creek: AltaMira Press; 2003.

Bazeley P, Kemp L. Mosaics, triangles, and DNA: metaphors for integrated analysis in mixed methods research. J Mixed Methods Res. 2012;6(1):55–72.

Bunde-Birouste A. Kicking goals for social change: an autoethnographic study exploring the feasibility of developing a program that harnesses the passion for the World Game to help refugee youth settle into their new country. Unpublished PhD thesis, University of NSW, Sydney; 2013.

Chang H, Ngunjiri FW, Hernandez KC. Collaborative autoethnography. Walnut Creek: Left Coast Press; 2013.

Cibangu SK, Hepworth M. The uses of phenomenology and phenomenography: a critical review. Libr Inf Sci Res. 2016;38(2):148–60.

Clarke AE. Situational analyses: grounded theory mapping after the postmodern turn. Symb Interact. 2003;26(4):553–76.

Coffey A. Autobiography. In: Lewis-Beck MS, Bryman A, Liao TF, editors. The SAGE encyclopedia of social science research methods. Thousand Oaks: Sage; 2004.

Corbin Dwyer S, Buckle JL. The space between: on being an insider-outsider in qualitative research. Int J Qual Methods. 2009;8(1):54–63.

Creswell JW. Educational research: planning, conducting, and evaluating quantitative and qualitative research. 3rd ed. Upper Saddle River: Pearson; 2008.

Denzin N, Lincoln Y. Handbook of qualitative research. Thousand Oaks: Sage; 1994.

Duncan M. Autoethnography: critical appreciation of an emerging art. Int J Qual Methods. 2004;3(4):28–39.

Ellis C. The ethnographic I: a methodological novel about autoethnography. Walnut Creek: AltaMira Press; 2004.

Ellis CS, Bochner AP. Analyzing analytic autoethnography: an autopsy. J Contemp Ethnogr. 2006;35:429–49.

Ellis C, Adams T, Bochner A. Autoethnography: an overview. Forum: Qual Soc Res. 2010;12(1):10.

Funnell S, Rogers P. Purposeful Program Theory: Effective use of theories of change and logic models. San Fransisco: John Wiley & Sons; 2011.

Green JL, Skukauskaite A, Baker WD. Ethnography as epistemology. In: Arthur J, Waring M, Coe R, Hedges L, editors. Research methods and methodologies in education. 1st ed. Los Angeles: Sage; 2012.

Hasselgren B, Beach D. Phenomenography – a "good-for-nothing brother" of phenomenology? Outline of an analysis. Higher Educ Res Dev. 1997;16(2):191–202.

Kanuha VK. "Being" native versus "going native": conducting social work research as an insider. Soc Work. 2000;45(5):439–47.

Kemp L, Chavez R, Harris-Roxas B, Burton N. What's in the box? Issues in evaluating interventions to develop strong and open communities. Community Dev J. 2008;43(4):459–69.

Liamputtong P. Performing qualitative cross-cultural research. Cambridge: Cambridge University Press; 2010.

Linder C, Marshall D. Reflection and phenomenography: towards theoretical and educational development possibilities. Learn Instr. 2003;13(3):271–84.

Maréchal G. Autoethnography. In: Mills AJ, Durepos G, Wiebe E, editors. Encyclopedia of case study research. Thousand Oaks: Sage; 2010.

Pensoneau-Conway SL, Toyosaki S. Automethodology: tracing a home for praxis-oriented ethnography. Int J Qual Methods. 2011;10(4):378–99.

Richardson L. Writing: a method of inquiry. In: Denzin N, Lincoln Y, editors. Handbook of qualitative research. Thousand Oaks: Sage; 1994. p. 519–29.

Vryan K. Expanding analytic autoethnography and enhancing its potential. J Contemp Ethnogr. 2006;35(4):405–9.

Wall S. An autoethnography on learning about autoethnography. Int J Qual Methods. 2006;5(2):9.

Memory Work

31

Lia Bryant and Katerina Bryant

Contents

Abstract

Memory work is a methodology and method first introduced by Frigga Haug and others in Germany and appeared in academic publications in the 1980s. As an approach to data collection, memory work involves writing a memory in the third person in relation to a question or theme. The methodology and method was used with groups of women to examine power relations through writing and analyzing specific situations. Memory work is an approach that enables emotions to come to the fore, particularly emotions that are not easily voiced. Through processes of writing in the third person and time for analysis and rewriting, the approach provides distance and space for the emotional and sensory to emerge. Memory work facilitates the discovery of the tangible and intangible aspects of sensations that may not emerge from other qualitative methods like semi-structured interviews. In this chapter, the research processes for using memory work are outlined

L. Bryant (✉)
School of Psychology, Social Work and Social Policy, Centre for Social Change, University of South Australia, Magill, Australia
e-mail: Lia.Bryant@unisa.edu.au

K. Bryant
Department of English and Creative Writing, Flinders University, Bedford Park, SA, Australia
e-mail: katie.bryant@live.com.au

© Springer Nature Singapore Pte Ltd. 2019
P. Liamputtong (ed.), *Handbook of Research Methods in Health Social Sciences*,
https://doi.org/10.1007/978-981-10-5251-4_88

and contextualized in relation to working with individual participants and groups. Processes for analyzing memories are explained and examined as are the ethical dimensions of writing, sharing, analyzing, and publishing memories that have arisen from groups or participants.

Keywords

Memory work · Collective analysis · Ethics · Reflexivity · Social transformation · Living historically

1   Introduction

I can only note that the past is beautiful because one never realizes an emotion at the time. It expands later, and thus we don't have complete emotions about the present, only about the past. (Virginia Woolf)

Memory and perceptions of place, people, and events may come into being through a constellation of, or singular, visual images or scenes, language, and/or sensory perceptions and emotions. Understanding how memories work has been the foci of many disciplines over time including history, psychology, and philosophy. As postmodern theories of memory came to the fore, increasingly conceptualizations of memory were understood as fragmented, nonlinear, and not about remembering "how it really happened" (Haug 2008, p. 538). Theorizing memory as temporally created and recreated which is "true" for the individual – that is, telling a story about an event, experience, place, or emotion – dislodges circular debates about memory and its relation to truth.

Memory work is a methodology and method first introduced by Frigga Haug and others in Germany and appeared in academic publications in the 1980s. As an approach to data collection, memory work involves writing a memory in the third person in relation to a question or theme. It can be used in conjunction with other approaches including photography, diary entries, interviews, or focus groups. Initially, memory work was constructed as a tool and theoretical device to examine female sexualization (Haug 1987). Haug and others critiqued psychological and sociological knowledges which prioritized abstract theorizing to understand women's experiences that resulted in "flattening the multiplicity and diversity of experiences" (Stephenson 2005, p. 35). Haug was seeking to develop an approach that would challenge the "opposition between objective, transcendent theory and subjective, bounded experience" thereby "returning to the experience side of the divide" (Stephenson 2005, p. 35). In this way, Haug and colleagues were using a grounded methodology using experience and subjective experience to develop theory. By centralizing the perspective of the subject, Haug and others disturbed the notion of the objective researcher and the participant as the subject of research. In memory work, the researcher becomes both the object and subject of research.

Haug and others (1987) developed memory work as a collective strategy that "was written with and for the feminist movement" (Haug undated, p. 2). The methodology and method was used with groups of women "to provide understandings of relations of power through writing and analyzing specific situations in their recollections" (Bryant and Livholts 2014, p. 285). Collectivity in the approach was important as it provided the basis for a discussion of individual experiences in relation to broader social issues. For Haug et al., the purpose of memory work was to move beyond documenting inequality. The memory work project was to provide a process for social transformation. They "wanted to do academic work which had an explicit political value...[and] enabled identification of an intervention..." (Stephenson 2005, p. 35).

Although first developed as a collective method (Haug et al. 1987; Crawford et al. 1992), over time memory work has also been used as autoethnography (Bryant and Livholts 2013) and with individual participants (Onyx and Small 2001; Shea et al. 2016). For example, Widerberg (1999) used memory work as autoethnography focusing on sexualization of the body. Her analysis of her memories showed how experiences of sexual harassment during her school years shaped her choices and participation in higher education. Shea et al. (2016) used memory work with individual participants to examine younger women's identities while negotiating motherhood and transitioning to adulthood. Participants were asked to write a memory on the positive experiences of becoming a younger mother and another on a challenging experience. Shea and colleagues (2016) found that younger mothers while internalizing discourses about younger mother's as "dysfunctional" also resisted these, drawing on feelings of pride and resilience. Memory work enabled participants to reflect on their experience of mothering alongside dominant discourses enabling "emotions to come to the fore, with fear and uncertainty being overcome by pride, self-belief and determination" (Shea et al. 2016, p. 851).

Henriksson et al. (2000) and Bryant and Jaworski's (2015) work are examples of studies using collective memory work to explore the "everyday" and "personal" emotions and experiences of doctoral students in the academy. Through writing memories, doctoral students identified that thesis writing and the production of knowledge were imbued with emotions of risk, fear, and shame despite hegemonic understandings that privilege rationality and using technical skills to producing theses (Bryant and Jaworksi 2015). These studies brought into focus the embodied aspects of study that conflict with the structures, cultures, and process in neoliberal universities when studying for a doctoral qualification.

Memory work, while used collectively with individuals or by a researcher as autoethnography, cannot be simply categorized as a collective and/or individual method (Bryant and Livholts 2007). As a collective method, individual members are writing and also analyzing memories individually and constructing knowledge about themselves and the social world. In this sense, while a collective method, memory work is also an individual method. In research that involves individuals, memories are often analyzed by participant-researchers and academic researchers and as such, also creates collectivity.

In this chapter, the research processes for using memory work are outlined and contextualized in relation to working with individual participants and groups. While the steps are sequential, analysis is ongoing with theorizing and analysis often occurring in tandem. Processes for analyzing memories are explained and examined as are the ethical dimensions of writing, sharing, analyzing, and publishing memories that have arisen from groups or participants.

2   When to Use Memory Work as a Research Approach

Memory work has been used across multiple disciplines including nursing, sociology, psychology, social work, human geography, and gender studies. In health and associated disciplines, it has been used to research a wide range of topics including pain (Gillies et al. 2004), lived experiences of HIV (Stephenson 2005), young people's experience of contraception (Harden and Willig 1998), and gender and caring (Bryant and Livholts 2013; Livholts and Bryant 2013).

Memory work is an approach that enables emotions to come to the fore, particularly emotions that are not easily voiced. Through processes of writing in the third person and time for analysis and rewriting, the approach provides distance and "space for the emotional and sensory to emerge in ways that may not occur or may not be allowed for in for example, a face to face semi-structured interview" (Bryant 2015, p. 11). Simply put, writing memories allows focused attention to detail. Other qualitative methods are less likely to involve descriptions of color, taste, and sound – the surrounding thoughts and images that make memories. Memory work facilitates the discovery of the tangible and intangible aspects of sensations (Mason and Davies 2009). The approach assists in examining the intangible, shifting, and hard to articulate concepts like identity or awareness of one's body. Gillies et al. (2004, pp. 111–112) argued that when using memory work to write about sweating and pain, they generated rich descriptions of their bodies as well as psychological and emotional experiences. Writing memories brought greater attention to experiential aspects of pain. In terms of the tangible, writing memories, according to Jansson et al. (2008, p. 236), allows "focusing on the 'banal' and tangible everyday practices, [and] comes across as a very rewarding method to grasp deeply naturalized structures that are difficult to discern in other forms of empirical materials, just because they appear so unimportant and unproblematic." Moreover, researchers using memory work have argued that writing memories brings forth new unexpected knowledge which is able to be viewed as "social and political. . .[therefore] plac[es] the blame and responsibility beyond [the] individual. . ." (Jansson et al. 2008, p. 238).

As memory work is a method rich in description, it can bring the past more firmly into the present. This brings forth previously unarticulated thoughts or emotional pain. Consequently, memory work as an approach requires skilled facilitation of the group or individual participants, preorganized pathways to access counseling or other services, and enough time for group sessions to provide support to members.

3   Doing Memory Work

Haug's intention was that the stages and process of memory work remain flexible and adaptable, and, therefore, memory work has been adapted in various ways (Stephenson and Kippax 2008). Crawford et al. (1992) suggest the following four steps are useful in creating a theme, writing a memory, undertaking analysis, and interpreting in relation to theory.

1. Identify a theme or question.
2. Write a memory in the third person in as much descriptive detail as possible avoiding biography and interpretive comments.
3. Analyze memories individually and within the group. New questions are proposed, and new memories are written or original memories rewritten.
4. Memories and analysis of memories are examined in relation to relevant social theories.

Step 1: Setting a Topic

The first step in doing memory work involves identification of a theme and/or question. Identifying the right theme is important, and adequate time should be allocated for the group to identify a theme or question of significance. As a researcher, you may set a theme or question. The question or theme needs to be written in everyday language. Haug (undated, p. 3) argues that "it is important not to pose the question in scientific or analytical terms since memories will not emerge when the appeal to them takes the form of language that is not in the vernacular. For example, 'A time when I was afraid' is common language to which everyone can relate." Further, the theme or question chosen will need to be broad enough to trigger memories while avoiding well-rehearsed memories.

Academic researchers develop their topic from experience, bodies of literature, or theory. Jansson and others (2008, p. 232) provide a clear example of how topics have been derived by researchers stating:

> For instance, Widerberg (1999) started memory work on knowledge and gender by asking why she fells so uneasy in university contexts. . .In a project carried out by Bronwyn Davies et al. (2001). "Becoming Schoolgirls", the group started by discussing different theoretical concepts as a way to inspire and awake memories, and later decided to write memory stories of school experiences that actualize certain concepts.

Bryant and Livholts (2013) derived their topic for memory writing on gender, care, and the telephone by identifying key themes from the literature about telephone use and care work. Their memories were written on the theme "Where are you?" which was a common question reportedly asked by (mobile) telephone users in studies of telephone use and care (Garcia-Montes et al. 2006).

There has been discussion on how the topic chosen for writing memories influences the production of knowledge (Jansonn et al. 2008). Davies et al. (2001) have argued that if theoretical concepts drive the writing of memories, there is the danger of reproducing knowledge and everyday understandings of the topic being researched. Haug and others, recognizing that memories could simply repeat what is known, suggest that the problem under investigation should be displaced. For Haug et al. (1987), displacement of a problem means, firstly, transferring the conceptual problem to a specific situation. That is, if the concept is gendered work, then the situation might be "motherhood" or "becoming a teacher/academic etc." Secondly, displacement of the problem involves dislocating the topic from the theoretical theme being studied, for example, Haug et al., when studying women's sexuality, turned their focus instead to women's bodies and hair growth. Jannson et al. (2005, pp. 232–233) explain:

> In Female Sexualization (1987) Haug et al. describe how the group wanted to write about women's sexuality. One of their first memory stories tells [us] about sexual assault. This story is written in a language the participants find familiar. It is a story about subordination and about women and girls lacking a sexuality of their own. The story is 'located at the centre of the discourse in which what we understand as sexuality is produced" (Haug et al. 1987:74). The group does not find anything they consider is "new" in...[these] kind of stories; instead they see how modes of already existing explanations and understandings are reproduced by themselves in their own stories. This in turn leads them to focus their memory stories on different parts of the female body. Now the stories do not primarily deal with sexuality, but for instance, hair growth in "wrong" places or with leg posture.

Dislocating and transferring the problem in relation to creating a topic enables different results to emerge, and as such the topic or question becomes a critical part of the methodology requiring elaboration and justification.

Step 2: Writing a Memory

During the second step, each member is required to write a detailed memory in the third person. Haug (1987) argues that writing in the third person created necessary distance to explore emotions and sensory aspects of memory. In their experience of writing memories, Jansson and others (2008, p. 235) found that "writing about oneself as 'she' enables us to approach this 'she'-person with greater empathy and understanding: it is a form of textual distance that makes it possible to stay near 'her' and take her experiences fully seriously, in a way that is more difficulty when using 'I'." Further, writing in the third person "is partly a way to facilitate remembering," and some scholars suggest it makes the corporeal more readily available in memory and facilitates writing that captures the sensory and emotional aspects of experience (Jansson et al. 2008, p. 236). As such, memories should be as detailed as possible, stating emotions, color, smell, sounds in relation to an event, action, or episode. Frost et al. (2012, p. 234) suggest "they should be concrete and detailed, avoiding interpretation, biography and explanation."

Memories should also be written in one attempt without stopping for corrections or self-censorship (Jansson et al. 2008). Writing without censorship is not always easy, requiring the writer to allow herself to be vulnerable and sharing thoughts and experiences that she may not have shared with others or has yet to share fully with herself. Having said that, writing memories is best done without censorship, and it may be appropriate at times for the memory work group to place caveats on what is shared. For example, when teaching memory work to university students or colleagues, the purpose is not to develop theory, and as such, asking group participants to withhold details that may be upsetting to themselves or others is appropriate.

Examples of Memories
Topic: Loss of Faith (Frost et al. 2012)

Memory 1
Her car was hit. They sent her the bill, and when that didn't work, they tried talking to her in growling tones. She felt small, reminded of girlhood. She tried to tell herself she wasn't small; she was being made to feel that way. She called the insurance agency, making her voice sound as steady as she could. They invoiced her, and a year later, the claim is pending.

Memory 2
She is slow to rise. Caught between sitting and standing. Moments after she has stood, it registers in her mind that this is no longer automatic. As she moves bent but standing, her knees feel stiff. Rigid. Not a part of her. As she walks down the corridor, she catches a glimpse of herself in the reflection of the glass pane which separate one office from another. She is folded forward with right shoulder jutting out, her hip bones uneven. Her eyes catch her looking. She is surprised by what she sees. Always that quick half smile. She attempts to straighten and the same thought visits her on these occasions – not at work.

Step 3: Analysis and Rewriting
The third step involves members reading each other's memories and analyzing them in terms of contradictions and patterns. This process draws the individual's attention to "moments" that are often fragmented and nonlinear to enable examination of aspects of our personal lives in the context of societal structures and power relations (Haug 2008). Take the example memories we provided above. Each author analyzed their own memory and their colleague's memory by seeking themes, patterns, and differences or non-patterns. For example, the author of Memory 1 reflected that her memory indicated that gender was hierarchical and when dealing with a large organization, gender and age intersect to produce differential power relations. Her colleague's analysis was similar suggesting that power was being practiced through authority and control, creating feelings about being small and not in charge. The author of Memory 2 in analyzing her memory reflected that her memory brought forth the "ideal" construct of worker that did not fit with an aging body in the workplace. Her colleague noted the liminal space between sitting and standing which reflected the lack of full acknowledgement of the body at work and how social constructions of disability linked to shame are reflected in the memory.

Apart from a broad identification of themes, analysis also involves a close focus on language used in the writing of the memory. Haug (undated, p. 14) suggests that the "first step is to break from the realm of the conveyed meaning, and distance ourselves for the work of deconstruction." For Haug (undated, p. 15), distancing

Table 1 Analysis of language memory 1

Verbs	Linguistic peculiarities	Named emotions	Conveyed emotions	Other people
Hit	Growling	Felt small	Girlhood	They
Talking			Made to feel = conveys pressure	Insurance company
Tried			Steady voice	

involves asking explicit questions about the language used in the text, and she suggests that all verbs be written down, all "linguistic peculiarities" be recorded (e.g., lightening in my head), emotions that have been named or conveyed be considered, and whether other people have been written into the memory should be recorded. If others are written into the memory, is there narration about them, their feelings or have they been "faded out" (undated, p. 23)? In relation to Memory 1, the following Table 1 indicates an analysis of language.

Placing words in a table shows how language is used and "how. . . [the author] constructs herself through language" (Haug undated, p. 16). Haug suggests that using a table like the one above will bring new observations to analysis of memories and in particular show how the writer leaves an impression for the reader. In Memory 1, for example, the words in the table show the author is being taken back to the past, to other memories of powerlessness.

In relation to both memories, analytical notes were shared at each stage of analysis and further analysis took place asking each other questions about specific words or themes. Shared meanings were developed and agreed upon. For Memory 1, the central theme was institutional gendered power and in Memory 2 the interrelationships between emotions and the social construction of the "disembodied" worker. Analysis is, however, not a linear process but as Crawford et al. (1992, p. 49) suggest happens with several readings and is recursive, that is:

> The collective reflection and examination may suggest revising the interpretation of the common patterns and the analysis proceeds by moving from individual memories to the cross-sectional analysis and back again in a recursive fashion.

However, there are some assumptions in memory work about the nature of collective analysis. There is the assumption that group must reach a consensus during analysis, leaving unanswered how a diversity of views can be incorporated into analysis (Onyx and Small 2001). Stephenson (2005) has suggested that an emphasis on collectivity diminishes the examination of different and challenging ideas and, as a consequence, may also diminish some voices and stories and limit theorizing.

Analyzing memories collectively may hold a series of other challenges as a-nalysis may cause tension, conflict, and/or distress to an individual and/or the group. In their account of reflecting on their experiences of doing memory work and collective analysis, Frost and others (2012, p. 235) explain:

In our debates and questions, we have had to manage our different epistemological orientations in a careful and respectful manner and discuss and reflect on intra-personal conflicts and interpersonal dynamics.

Further, Frost and others (2012, p. 241) provide concrete experiences of tensions when analyzing collectively and provide an example where a member of the group was not present when analysis took place feeling that her experiences were "cast aside" in how her language in the memory was analyzed.

Rewriting Memories

While Haug (1987) suggests rewriting memories, memory work groups often do not fulfill this requirement (Stephenson and Kippax 2008). In our case, the original memories were modified rather than rewritten. Minor modification of words occurred to enable reader clarity.

Haug et al. (1987, p. 245) show that the purpose of rewriting memories is to achieve a "a shift in one's perception of one's self and social reality" where "the individual's active engagement in the process carries the sense of agency that comes with being an active participant in the analysis of one's own experience." Frost et al. provide an example whereby a participant in rewriting her memories gained agency as Haug and others hoped. Frost et al. (2012, p. 244) explain: "Her loss of faith [became] a moment of realization that she had outgrown something, which could be felt as empowering and not just as disappointing."

Step 4: Applying Social Theories

Memory 2, when analyzed collectively, revealed an analysis of the body at work in a corporate structure and can be theorized in relation to knowledge on embodiment and work or social constructions of aging and/or disability. Taking embodiment and work, for example, Acker (1990) has argued that organizations reproduce embodied norms through structures and processes shaping organizational cultures which privilege some bodies over others. When analyses of both memories are read side by side, there is a common analytical thread around age, the body, and institutions that reflect through vision and voice dominant discourses about younger and older bodies, their power, and worth. While analysis in relation to social theory raises hegemonic discourse, the point as Stephenson (2005, p. 38) argues:

> ...is not to identify the imprint of hegemonic discourse in people's experience, but to gain a better understanding of the available and emerging processes of appropriation and their effects, not only on lived experience, but on the development of a particular discourse.

Therefore, our analysis about embodiment and work for us as researcher-participants assisted us as individuals to understand the effect of discourses that may not have been apparent to us in our daily lives. In turn, this awareness provides an opportunity on a daily basis to resist or challenge self-perceptions and organizational cultures and structures. It is in this way that memory work can be transforming for individuals, as well as structurally. Haug and others have called this process of theorizing experiences as "living historically," that is, "what they had previously thought to be natural sequences of their lives started to appear as historically constituted avenues for interpreting and managing the material and social realities in which they were immersed" (Stephenson 2005, p. 38).

4   Memory Work and Social Transformation

Writing memories involves introspection, while analysis is likely to involve reflective and reflexive action. It is through reflexivity that transformation occurs, and as such, the concept requires interrogating. The concept of reflexivity has been understood in multiple ways (Finlay 2002; Pillow 2003). In relation to the possibility of being transformed by one's own memory work, reflexivity involves self-scrutiny in relation to the written text, an introspective assessment of one's experience, and assessment of social phenomena gained from the writer's insight. Collective transformation in relation to writing and analyzing within a memory work group is more likely to occur from intersubjective reflexivity (Bryant 2015). Intersubjective reflexivity is a process where the self is co-constructed through dialogue and the multiple perspectives of other subjects (e.g., Bahktin 1981). While collective analysis and transformation may derive from subjects drawing meaning in relation to other subjects, equally difficulties may arise in reaching consensus impacting on the possibility for collective transformation.

Stephenson (2005, p. 44) drawing on Haraway (1988) brings attention to memory groups as groups consisting of "networks of actors, and choruses of multitudes..." Different perspectives, practices, and experiences raise the complexity of how to achieve collective understanding and/or transformation. Worldviews are shaped through our location which is situated historically, politically, socially, geographically, and culturally informed by social signifiers like age, class, sexuality, and ethnicity. Our situatedness provides the possibilities for seeing and limitations of what we see. Multidimensional ways of seeing are inherent within subjects as well as across populations of subjects. Frost et al. (2012, p. 234) discuss the interconnection between multidimensionality and collectivity in an academic memory work group stating that "despite this positive engagement and genuine sense of embarking upon a journey, we were also aware of how our different intellectual backgrounds and individual frames of reference were likely to pose problems."

5   Ethical Considerations

Cadman et al. (2001, p. 76) believe that feminist memory work, when it allows group members to interpret the data as a whole, can bring "about some positive change in the participants and in the world." Yet memory work that addresses trauma can also evoke distress within participants. As Fraser and Michell (2015, p. 325) argue, "care must also be taken to recruit participants and generate discussions that do not require participants to re-live their traumatic experiences in front of a non-therapeutic group.

If conducted with sensitivity, respect and adequate expertise, memory-work groups can foster levels of trustworthiness, even solidarity, among participants and researchers." To this point, topics that are overtly traumatic for participants should be avoided (Fraser and Michell 2015).

Fraser and Michell (2015, p. 326) argue that it is key that when conducting memory work, participants are "people first and foremost, rather than sources of data." Researchers have a duty of care for participants' well-being and as such, should employ skilled facilitators and refer participants to support services after the session (Fraser and Michell 2015). In view of this, informed consent is a key aspect of conducting memory work ethically (Fraser and Michell 2015). Participants should receive written documentation about the project as well as being informed of their rights (especially their right to withdraw from participating at any time) and responsibilities to other participants (Fraser and Michell 2015; see also ► Chap. 106, "Ethics and Research with Indigenous Peoples").

Paid participation also creates complexities as it may induce a participant to become involved, tainting their free consent (Fraser and Michell 2015). For example, in a memory work study conducted by Beddoe and Jarldorn, the researchers revised $100 participant gift cards to $60 as the ethics committee felt that $100 acted as "too much of an inducement to participate" (Fraser and Michell 2015, p. 328).

As Newton (2017, p. 98) writes, the key to consent is that it is ongoing: "Participants will invariably have a different understanding about the project at the beginning of the research from their understanding of it at the end. . ." As members of the group cannot anticipate what they will write in their memories, post-consent is an important ethical consideration. Post-consent forms enable group members to have time to reflect and consider what they have written and reconsider their willingness for their memories to be published.

Newton (2017, p. 98) argues that best practice for participant and researcher well-being is when researchers "reconnect with participants a little while after the research. . .to ask them of their later feelings about the research experience." Using memory work as a method, therefore, facilitators of the group (usually researchers) will need to ensure that the time for group meetings is adequate to discuss processes and emotions as well as write and analyze memories. Given the emotional nature of writing and reading memories, it is also important that researchers facilitating the group provide space after meetings to check in with individuals and allow time for those who wish to stay back to talk about their personal experiences.

Upon publication of memory work studies, further ethical considerations such as privacy and anonymity arise. Glenda Koutroulis (2014, p. 81) discusses that despite changing names in her study to procure anonymity for her participants, one participant was concerned that her actions and descriptions would reveal her identity and create "profound implications" for her family. Publishing memories requires decisions about what to include and exclude when it comes to stories about others. In research submitted for publication, if the material is sensitive, consent will need to be obtained from those who appear in the memories or with the memory writer's consent sensitive information or revealing details require deleting from the text (Bryant and Livholts 2013).

6 Conclusion and Future Directions

Memory work has been used in a variety of ways given the "openness" of the process since its inception by Haug and others (1987). It is an approach that can be used as autoethnography, with individual participants or in groups. The key benefits of memory work are that it taps into recalled experiences bringing to the fore experience as sensory, emotional, and embodied. Writing memories provides detail to a fragmented moment in time that might not come into being with more commonly used qualitative methods like face-to-face open or semi-structured interviews. Analyses of memories involve thematic analysis, as well as a linguistic analysis (see ▶ Chaps. 28, "Conversation Analysis: An Introduction to Methodology, Data Collection, and Analysis," and ▶ 48, "Thematic Analysis"). Combined, these forms of analyses highlight central themes which underlie what has been said and an understanding of how the author of the memory linguistically positions herself. Central to analyzing memories is relating the themes which emerged to social theories in order to uncover taken for granted understandings in subject's lives. For Haug and others, taken-for-granted knowledge is part of "living historically" and analyses of living historically open avenues for transformation to the self, group members, and, through the agency of subjects potentially, social structures.

Memory work still remains somewhat on the margins of academic research, and as such, there is a scarcity of literature on the processes and outcomes of using memory work with different populations and subpopulations of people. Is memory work an appropriate tool when co-researching with, for example, older people or people from different class backgrounds (Onyx and Small 2001)? How useful is memory work in capturing the memories of people from diverse ethnic backgrounds whose memories might unfold in multiple linguistic structures? Would the recording of oral memories using a similar process be more useful for people who prefer to express themselves in ways other than writing? If oral memories are replaced with written ones, will the processes and outcomes change?

The "openness" of memory work provides opportunities for researchers to explore how the method might be used according to different settings and with diverse populations. Memory work remains a particularly useful method for focusing on a fragmented moment in time and "making us see things in a new way" (Widerberg 1999, p. 158). As Widerberg (1999, p. 160) suggests, "if we live our lives in episodes . . .The overall plot of the life-history that makes up. . .these episodes is something we cannot know until afterwards. Remembering is therefore not only a recounting of the past, but also a reinterpretation." Memories bring the past to the present and through this process old memories become new ones. Memory work enables us to expand our understanding of our lives and the social worlds we inhabit.

References

Acker J. Hierarchies, jobs, bodies: a theory of gendered organizations. Gend Soc. 1990;4(3): 139–58.

Bakhtin MM. The dialogic imagination: four essays, vol. 1. Austin: University of Texas Press; 1981.

Bryant L. Taking up the call for critical and creative methods in social work research. In: Bryant L, editor. Critical and creative research methodologies in social work. Farnham: Ashgate; 2015. p. 1–26.

Bryant L, Jaworski K. Women supervising and writing doctoral theses. Lanham: Lexington Books; 2015.

Bryant L, Livholts M. Exploring the gendering of space by using memory work as a reflexive research method. Int J Qual Methods. 2007;6(3):29–44.

Bryant L, Livholts M. Location and unlocation: examining gender and telephony through autoethnographic textual and visual methods. Int J Qual Methods. 2013;12:403–19.

Bryant L, Livholts M. Memory work and reflexive gendered bodies: examining rural landscapes in the making. In: Pini B, Brandth B, Little J, editors. Rural femininities. Plymouth: Lexington Books; 2014. p. 181–94.

Cadman K, Friend L, Gannon S. Memory-workers doing memory-work on memory work: exploring unresolved power. In: Small J, Onyx J, editors. Memory-work: a critique. Sydney: School of Management, University of Technology; 2001.

Crawford J, Kippax S, Onyx J, Gault U, Benton P. Emotion and gender: constructing meaning from memory. London: Sage; 1992.

Davies B, Dormer S, Laws C, Rocco S, Lenz Taguchi H, McCann H. Becoming schoolgirls: the ambivalent project of subjectification. Gend Educ. 2001;13(2):167–82.

Finlay L. Negotiating the swamp: the opportunity and challenge of reflexivity in research practice. Qual Res. 2002;22:209–30.

Fraser H, Michell D. Feminist memory work in action: method and practicalities. Qual Soc Work. 2015;14(3):321–37.

Frost A, Eatough V, Shaw R, Weille KL, Tzemou E, Baraitser L. Pleasure, pain and procrastination: reflections on the experience of doing memory-work research. Qual Res Psychol. 2012;9(33):231–48.

Garcia-Montes JM, Kaballero-Munos D, Perez-Alvarez M. Changes in the self resulting from the use of mobile phones. Media Cult Soc. 2006;28(1):67–82.

Gillies V, Harden A, Johnson K, Reavy P, Strange V, Willig C. Women's collective constructions of embodied practices through memory work: Cartesian dualism in memories of sweating and pain. Br J Psychol. 2004;43(1):99–112.

Haraway D. Situated knowledges: the science question in feminism and the privilege of partial perspective. Fem Stud. 1988;14(3):575–99.

Harden A, Willig C. An exploration of the discursive constructions used in young adults' memories and accounts of contraception. J Health Psychol. 1998;3(3):29–445.

Haug F, et al. Female sexualization, translated from the German by Erica Carter, London: Verso; 1987.

Haug F. Philosophy and overview of memory work. In: Hyle AE, Ewing MS, Montogmery D, Kaufman JS, editors. Dissecting the mundane: international perspectives on memory work. Lanham: University Press of America; 2008. p. 3–16.

Haug F. Memory-work as a method of social science research: a detailed rendering of memory-work method, a research guide. (undated). Retrieved from http://www.friggahaug.inkrit.de/documents/memorywork-researchguidei7.pdf.

Henriksson M, Jansson M, Thomsson U, Wendt Hojer M, Ace C. In the name of science. A work of memory. Kvinnovetenskaplrg Tidskr. 2000;21(1):5–25.

Jansson M, Wendt M, Åse C. Memory work reconsidered. Nord J Fem Gend Res. 2008;16(4):228–40.

Koutroulis G. Memory-work: a critique. Ann Rev Health Soc Sci. 2014;3(1):76–96.

Livholts M, Bryant L. Gender and the telephone: voice and emotions shaping and gendering space. Human Technology, 2013;9(2):157–170.

Mason J, Davies K. Coming to our sense? A critical approach to sensory methodology. Qual Res. 2009;9(5):587–603.

Newton VL. 'It's good to be able to talk': an exploration of the complexities of participant and researcher relationships when conducting sensitive research. Women's Stud Int Forum. 2017;61:93–9.

Onyx J, Small J. Memory-work: the method. Qual Inq. 2001;7(6):773–86.

Pillow W. Confession, catharsis or cure? Rethinking the uses of reflexivity as methodological power in qualitative research. Int J Qual Stud Educ. 2003;16(2):175–96.

Shea R, Bryant L, Wendt S. Nappy bags instead of handbags: experiences of young motherhood. J Sociol. 2016;52(4):840–55.

Stephenson N. Living history, undoing linearity: memory-work as a research method in the social sciences. Int J Soc Res Methodol. 2005;8(1):33–5.

Stephenson N, Kippax S. Memory work. In: Willig C, Stainton-Rogers W, editors. The Sage handbook of qualitative research in psychology. Thousand Oaks: Sage; 2008. p. 127–46.

Widerberg K. Alternative methods – alternative understandings: exploring the social and multiple 'I', through memory-work. Sociologisk Tidsskr. 1999;2:147–61.

Traditional Survey and Questionnaire Platforms

32

Magen Mhaka Mutepfa and Roy Tapera

Contents

Abstract

Platforms for administering surveys have evolved in the past 20 years, and increasingly electronic platforms are utilized by many research programs. Historically, paper-and-pencil interviewing was the norm, and in recent years computer-assisted interviewing has been adopted by many. This chapter considers the questionnaire as a mode of data collection, the use of survey methods that employed paper and pencil, their merits, and limitations. It also compares traditional and online surveys, reasons for their wider adoption,

M. M. Mutepfa (✉)
Department of Psychology, University of Botswana, Gaborone, Botswana
e-mail: magen.mhaka@yahoo.com; mutepfam@ub.ac.bw

R. Tapera
Department of Environmental Health, University of Botswana, Gaborone, Botswana
e-mail: Taperar@ub.ac.bw

© Springer Nature Singapore Pte Ltd. 2019
P. Liamputtong (ed.), *Handbook of Research Methods in Health Social Sciences*,
https://doi.org/10.1007/978-981-10-5251-4_89

and potential into the future. The discussion also considers ethics liabilities and strengths of data management in which either data collection platforms are used. Researchers should choose the most appropriate method depending on the topic, goals of the study, geographic region, timeframe, and budget. The issues highlighted above may need exploring through methodological research.

Keywords

Traditional surveys · Questionnaires · Online surveys · Representativeness · Social desirability · Interviewers

1   Introduction

Surveys were defined as several quantitative and qualitative research strategies or procedures used to systematically collect data from a sample through some form of invitations or appeals, such as face-to-face interviews, telephone interviews, or mail questionnaires (self-administered questionnaires) (Ponto 2015). Historically, surveys were administered by mail, and telephone, and also face-to-face (F2F). There are several ways to conduct research and collect information, but one way that is reliable is a survey, if well planned. More recently tablets, personal computers, and smartphones are being used for data collection. Internet-based surveys have become more popular than traditional surveys because of lower costs and faster modes of processing (Szolnoki and Hoffmann 2013; Kramer et al. 2014; Callegaro et al. 2017) (see also ► Chap. 76, "Web-Based Survey Methodology").

Nonetheless, traditional surveys are more appropriate in rural and remote areas because 60% of the people living in these areas still do not have Internet access (UN News Centre 2015; Egan 2016), suggesting online surveys exclude the bulk of the population. Traditional surveys reduce the demographic discrepancies as people residing in the country can also be recruited as participants in research, increasing the validity of most studies. The UN News Centre (2015) also reported that broadband Internet had not reached billions of people living in low- to middle-income countries (LMIC), including 90% of people living in poverty-stricken countries. Information computer technology (ICT) still has to be availed to more rural and remote areas, particularly LMIC. The implication is that researchers should be able to select the most appropriate approach depending on the geographic area.

This chapter describes the utility of the traditional survey research method using paper-and-pencil tools. The questionnaire as a mode of data collection is discussed. It then considers the merits of both survey approaches (traditional and online) regarding response rate, social desirability, and sensitive questions. Finally, ethics of data management with either approach are considered. Both traditional surveys and ICT-based surveys are used in cross-sectional and longitudinal studies. Longitudinal studies include trend, cohort, and panel. Traditional surveys will be described below.

2 Descriptions of Traditional Surveys

Traditional surveys are used to provide more data on respondents, from basic demographic information (e.g., age, education) to social data (e.g., causes, activities). Survey design involves the planning of the whole survey project and the outlining of steps to take when conducting research. These are the steps that start from the formulation of the survey goals to the interpretation of the survey results. Depending on the existing state of knowledge about a problem being studied, different types of questions may be asked which require the use of different study designs (Varkevisser et al. 2003) (see examples given in Table 1).

Careful survey design can help researchers obtain the responses they want. For instance, a survey design should be limited to the extent necessary for respondents to understand questions or to stimulate the response so as to reduce measurement errors. Questionnaire design is a multistage process that requires attention to several details.

2.1 Traditional Survey Tools

The questionnaire is the primary data collection instrument in social, health, epidemiological, and other areas of research. Thus, researchers should take cognizance of question wording and sequencing, the appearance of the questionnaire, the mode of

Table 1 Survey types and examples of research questions

State of knowledge of the problem	Type of research questions	Type of survey
Knowing that a problem exists but knowing little about its characteristics or possible causes	What is the nature/magnitude of the problem?	Cross-sectional surveys
	Who is affected? How do the affected people behave?	
	What do they know, believe, and think about the problem and its causes	
Suspecting that certain factors contribute to the problem	Are certain factors indeed associated with the problem? (e.g., Is lack of preschool education related to low school performance? Is low-fiber diet related to carcinoma of the large intestines?)	*Analytical (comparative) studies*
		Cross-sectional comparative surveys
		Case-control studies
		Cohort studies
Having established that certain factors are associated with the problem: desiring to establish the extent to which a particular factor causes or contributes to the problem	What is the cause of the problem?	Cohort studies
	Will the removal of a particular factor prevent or reduce the problem? (e.g., smoking cessation, provision of safe water)	

Adopted from Varkevisser et al. (2003)

administration, and enhancing response rates to get the most out of data collection. Collection of data in traditional survey research has been laid on different methods which are the questionnaire, interview, panel survey, observation, and telephone interview. The characteristics of the target population, resources available, and sensitivity of the topic of interest determine the method of data collection chosen.

Market research has relied on explicitly given answers to survey questions for insight into people's behavior (Schoen and Crilly 2012). This is because business needs quick answers in order to make business decisions. Explicit methods comprise questionnaires, semantic differential, focus groups, in-depth interviews, thinking aloud, and rating scales. While explicit methods divulge information on motivations, values, and rational explanations, implicit methods divulge hidden and difficult information that is hard to verbalize (Nosek et al. 2011). Using both explicit and implicit methods enables comprehensiveness of research data.

The traditional survey tools will be discussed as follows: the format, validity and reliability, effect of mode of administration on data quality, and the potential biases. Traditional survey tools have their merits and limitations when compared to online survey tools. The survey tools, specifically the questionnaire, will be discussed in the section that follows.

2.1.1   Questionnaire Format and Design

The layout and design of the questionnaire have an impact on data collection (response and completion rates); therefore, they should be well planned (Bowling and Windsor 2008; McColl et al. 2001). The social exchange theory and theories of perception and cognition should be used during the physical design of a questionnaire to enable researchers to get the most from data collection (McColl et al. 2001). Social exchange theory explains the weighing of potential benefits and losses (cost-benefit analysis) obtained from a questionnaire. A good questionnaire design is one in which the benefits outweigh the costs. With regard to theories of perception and cognition, researchers take cognizance of how people may view, process, and interpret the questionnaire so as to evoke response. For instance, a highlighted section in a questionnaire stands out (attention) and allows participants to focus their interest.

Designing of traditional questionnaires should thus be customized to local circumstances and to improving data quality. An easy-to-use questionnaire reduces measurement error and the potential for nonresponse error of the research participant. Questionnaire design issues inclusive of providing a PDF version and careful use of design elements may also affect data reliability (Callegaro et al. 2017; van Gelder et al. 2010). Questionnaires should thus be explicit about the required data and data format (see Table 2).

To develop a good structured traditional questionnaire, the following steps must be followed: state the hypothesis, outline the analysis plan, and list the variables to be measured. The data analysis plan must be structured in terms of specific objectives and show the statistical tests to be used and types of variables. Nonetheless,

Table 2 Be clear about the data format

Unspecified format	Specified format
Name: Roy Tapera?	Surname: Tapera
Name: Tapera Roy?	First name: Roy
Date: 7/11/17	Date: 11/7/2017 (**dd/mm/yy**)
11/7/17	
November 7, 2017	
Age: 65	Age: 65 years 4 months
(What? Years or months)?	

online questionnaire data formats and methods often comprise attitude response scales, for instance, the Likert scale or semantic differential rating, as well as open-ended or multiple-choice questions (Devine and Lloyd 2012). These online questionnaire programs are more user-friendly than traditional questionnaires. Forced-choice formats, item nonresponse, and "don't know" answers were less prevalent in online questionnaires compared to paper-and-pencil questionnaires (Marleen et al. 2010). The order of questions is also said to improve response rate. Further, partial responses in online questionnaires may be used to identify survey questions that were difficult to answer. When identified the questions may be amended to increase response rates and decrease item nonresponse.

Online questionnaires also have the advantage of fewer errors in data entry and coding because data are electronically entered and may easily be transformed into formats that are easy to analyze (Marleen et al. 2010; Callegaro et al. 2017). Online questionnaires may have skip patterns to hide irrelevant follow-up questions which is not possible with traditional paper-and-pencil questionnaires. Furthermore, visual and audio aids and pop-up windows may also be added to simplify responding by providing additional information. However, the additional features increased download time and contributed to nonresponse, thereby reducing the sample size (Marleen et al. 2010). Despite the advantages, the online questionnaires have their limitations.

2.1.2 Reliability and Validity of Traditional Survey Data

Administration of surveys should consider the aims of the study, the population under study, and the resources available to enhance the validity and reliability of the study (Wyatt 2000; McColl et al. 2001). The primary objective of a survey should be to collect reliable, valid, and unbiased data from a representative sample in a timely manner and without resource constraints. The process of responding to questions and modes of data collection has different effects on the validity and quality of data collected during research. The validity and reliability of a questionnaire are affected when the response rate is low.

Online surveys mostly use convenient samples and, according to Egan (2016), the "opt-in" bias, in which those who participate in online survey polls are already predisposed to taking them over a random sample, is lower in traditional surveys. In addition, the likelihood that participants' demographics may differ among respondents and non-respondents is higher when the response rate to a study is low.

Traditional ratings are consistently more favorable than online survey ratings (Taylor et al. 2009; van den Berg et al. 2011; Liljeberg and Krambeer 2012), because of the low response rate. For instance, websites like "TripAdvisor" used by a minority of hotel guests are generalized, with results showing similar and conflicting findings and extreme variability. Low response rate affects the precision (reliability) of the survey, resulting in study bias, and weak external validity (generalizability) of the survey results (McColl et al. 2001; Bowling 2005).

Online questionnaires, in addition to poor questionnaire design, have larger amounts of measurement error than the traditional methods of data collection (Wyatt 2000; Brigham et al. 2009; Callegaro et al. 2017), as a result of participants not scrolling down to find all questions or answer choices and slow reading speed. However, the quality of data of online questionnaires on sensitive and private questions, such as anthropometry (Touvier et al. 2010), perceived health status (Graham et al. 2006), oral contraceptive history (Rankin et al. 2008), and substance use (Brigham et al. 2009), was high. This is mainly due to the anonymity of online questionnaires (see ▶ Chap. 76, "Web-Based Survey Methodology").

Data on diet, health-related quality of life (QOL), and weight were found to be reliable in both online and traditional questionnaires (Marleen et al. 2010). Several psychological and psychiatric clinical and research scales, for example, the Edinburgh Postnatal Depression Scale (Spek et al. 2008), have been validated for Internet administration, and the results were reported to be slightly different from paper-and-pencil administration results, confirming the validity and reliability of the scales.

2.1.3   Effects of Mode of Administration on Data Quality

In selecting the mode of questionnaire administration, the availability of an appropriate sampling frame, anticipated response rate, potential from sources of bias other than nonresponse, time availability, financial budget, and other resources (e.g., equipment) should be considered. The different modes of administration comprise method of contacting participants, the media of delivering the questionnaire, and the administration of the questions (McColl et al. 2001; Bowling 2005). The traditional questionnaire uses the traditional paper-and-pencil interview (PAPI) and self-administration modes. The two modes impact the data quality differently (see Table 3); thus, the most appropriate mode should be selected.

Data quality is defined in terms of survey response rates, questionnaire item response rates, the accuracy of responses, absence of bias, and completeness of the information obtained from respondents. The mode of questionnaire administration has effects on (i) non-measurement errors (survey design, sampling frame and sampling, nonresponse, and item nonresponse) and (ii) measurement errors (survey instrument and data collection processes) (McColl et al. 2001; Bowling 2005; Bowling and Windsor 2008). Both the traditional and online questionnaires have their strengths and weaknesses.

2.1.4   Potential Biases of the Three Traditional and Online Surveys

Table 3 gives a summary of potential biases by mode of questionnaire administration of traditional and online questionnaires. However, the table should be interpreted

Table 3 Summary of potential biases by mode of questionnaire administration

Potential for	Face-to-face interviews	Telephone interviews	Self-administered, postal	Self-administered, programmed, electronic
More complete population coverage for sampling	**High**	Low	**High**	Low
Cognitive burden	**Low**	Great	Great	Great
Survey response	**High**	Low	Medium-low	Low
Item response/completion questionnaire	**High**	Low	Low	Low
Question order effects	Low	Low	High	**Low**
Response choice order effects	**Moderate**	High	High	High
Recall bias	Low	Low	High	High
Social desirability bias	High	High	**Low**	**Low**
Yes-saying bias	High	High	**Low**	**Low**
Interviewer	High	High	–	–
Length of verbal response/amount on information	**High**	Low	–	–
Willingness to disclose sensitive information	Low	Low	**High**	**High**
Respondents preference for mode of administration	**High**	Low	Low	Moderate

Adapted from Bowling (2005, p. 284)

with caution because findings are not always consistent, are not always based on experimental designs, and often have different topics (Bowling 2005). The table reveals that the traditional F2F interview questionnaire administration is the best mode of data collection (see bold text in table) as it comprises fewer biases when compared to the other modes.

The table also gives a summary of what is under discussion in this chapter, for instance, non-measurement errors and measurement errors.

2.2   Non-measurement Errors

Non-measurement errors comprise survey design, sampling, and response rate. Sample selection bias is lower in traditional questionnaires since sample selection is not limited to those with internet only. Nonetheless, health-related literature on the differences between responders and non-responders is inconsistent or inconclusive in most literature (McColl et al. 2001). Traditional surveys will be discussed in terms of representativeness.

2.2.1   Analysis of Traditional Surveys in Terms of Representativeness

The low statistics of internet coverage (e.g., 2% found in some sub-Saharan African countries) (UN Centre 2015) and other LMIC expose the lack of representativeness that may be portrayed by using online surveys in these low-income countries. However, the samples should be determined mainly by the type of study and validity of the method used. Szolnoki and Hoffmann (2013) compared different sampling methods in wine consumer research. The authors posited that depending on the topic, goals of the research, and the budget, all kinds of survey methods are being used to collect consumer data for research in the wine industry. A F2F and a telephone survey (comprising 2000 F2F and 1000 telephone respondents) were compared to online surveys [(quota sampling, 2000 participants) and snowball sampling (3000)] using identical questions. The F2F, telephone, and quota sampling methods were representative of the sociodemographic structure of the whole population with regard to the selected demographics of the quota sample. The snowball online survey had the least representative sample, suggesting snowballing technique was an inappropriate method for this particular study.

In the wine consumer study, traditional surveys were found to be more favorable than online surveys (Szolnoki and Hoffmann 2013). In addition, behavioral characteristics of consumers were delivered in the following order starting with the best: F2F, telephone interviews, and online quota surveys. In another previous online survey comprising 1,586 guests, 3% reported having posted a review about their August hotel stay on TripAdvisor (world's largest travel site) or a similar website (Brandt 2012). Three percent is a very small number for it is far off 10%, the most representative sample in research studies. Furthermore, only 12% of the respondents in the same study had posted a review regarding a hotel stay in the previous 12 months. The findings on TripAdvisor suggest the traditional sample did not use social media to share their experiences; therefore, the sample was not representative of the general population.

In other studies, one on air pollution in a national park by Taylor et al. (2009) and a German study by Liljeberg and Krambeer (2012) that also compared the two modes, it was found that the online sample response rate was much lower than the traditional telephone method even after controlling demographic variables. Sample sizes for online surveys may also be too small to support statistical analysis and inference (Brandt 2012). Online surveys are usually preferred by younger people, the educated, and people of high socioeconomic status (Blasius and Brandt 2010; Walker 2012). However, social media surveys enable participants to evaluate other people's responses by market researchers, which may lead to social desirability and acquiescence bias. Influence by other social media participants may be controlled by using other methods of research.

Previous findings (Fricker et al. 2005; Hoogendorn and Daalmans 2009; Taylor et al. 2009; Brandt 2012) suggest that online surveys cannot replace traditional surveys but could be used as a supplement. However, according to Szolnoki and Hoffmann (2013), proponents for traditional surveys, F2F surveys, still deliver the most representative results, and telephone surveys may be a good alternative although they advised using a larger sample. The findings were consistent with those of the wine consumer research.

Thus, researchers should be aware that sampling error and nonresponse error distort survey results by compromising representativeness. Wright, this volume, posited that if reminder emails and easy-to-use web questionnaire formatting are used in online surveys, a diverse sample of participants may be obtained to improve representativeness (see ▶ Chap. 76, "Web-Based Survey Methodology"). Response rates to different modes of questionnaire administration vary by topic and in particular, complexity issues. For instance, traditional surveys may have a lower response rate than self-administered, postal, and telephone questionnaires for sensitive questions, which give rise to measurement errors.

2.3   Measurement Errors

The data collection process involves an interaction among the questionnaire, the respondent, and, in the case of F2F and telephone interviews, the interviewer (McColl et al. 2001; Bowling 2005; van den Berg et al. 2011). The interaction is affected by the pacing of the interview and the control over the order of questions by the interviewer. It is also affected by social desirability, acquiescence bias, interviewer bias, and response-choice order. Although traditional surveys have advantages of representativeness, they have higher social desirability response bias than Internet surveys. Measurement errors can jeopardize results; thus, they should be taken into account during research proceedings. The section below discusses social desirability acquiescence bias of traditional surveys.

2.3.1   Analysis of Traditional Surveys in Terms of Social Desirability and Acquiescence Bias

In traditional surveys, participants are more likely to consider social norms, morality, and ethnic values when responding to questionnaires which may result in social desirability bias. Social desirability is more prevalent in traditional survey interviews (Szolnoki and Hoffmann 2013; Zhang et al. 2017) but lower in self-administered questionnaires (Bowling and Windsor 2008; Morales-Vives et al. 2014). However, some previous researchers have reported no differences between interviewer versus self-completion modes and type of response (e.g., McColl et al. 2001).

Social desirability is a strength for online surveys as participants do not feel pressured to overreport desirable behaviors or underreport undesirable behaviors in online questionnaires (Bowling and Windsor 2008; Szolnoki and Hoffmann 2013; Callegaro et al. 2017; Zhang et al. 2017). Taylor et al. (2009) found evidence of social desirability in a study where participants agreed to pay higher rates to curb air pollution than the online sample. Further, perceived health status is likely to be exaggerated, and substance use, religion, politics, and sexual behavior are likely to be underestimated in traditional F2F or telephone interviews than in self-administered questionnaires and online surveys (Bowling and Windsor 2008; Liljeberg and

Krambeer 2012; Szolnoki and Hoffmann 2013; Callegaro et al. 2017; Zhang et al. 2017).

Unlike in traditional surveys, online responders feel less concerned about how they appear to others because feelings of anonymity and privacy are higher. Thus, researchers should consider using online surveys when the research requires sensitive information or use of traditional self-administered questionnaires where online surveys are not feasible. In addition, the use of well-trained interviewers can reduce the bias. Further, assurances of confidentiality and anonymity may reduce social desirability in traditional interview questionnaires. Furthermore, checking responses against known "facts," indirect questioning, correlation of responses with social desirability measures, and randomized response techniques, which are feasible with large populations, may also reduce social desirability (Bowling 2005; Morales-Vives et al. 2014). Questions that have explicitly enunciated and identifiable options may also resolve the bias issue (see Table 2).

In addition to social desirability, acquiescence bias may also determine whether participants are going to answer truthfully particularly in traditional surveys. Traditional interview questionnaires have a higher agreement bias or "yes saying" (acquiescence bias) than self-administered questionnaires and online questionnaires. Age was also found to have effects on social desirability and acquiescence bias, both of which increase with age (Morales-vives et al. 2014). It is culturally easier to agree with others than disagree. However, Bowling (2005) posited that the acquiescence bias may be reduced by switching the order of responses periodically in a measurement scale (e.g., from "strongly agree–strongly disagree" to "strongly disagree–strongly agree"). In addition to acquiescence bias, interviewer bias also impacts F2F interview surveys.

2.3.2 Interviewer Bias

The three major sources of interviewer bias may be the interviewer (prejudices or asking leading questions), the respondent (may lie or evade questions), and the interview situation itself (physical and social setting). Interviewers may have a negative impact with regard to questionnaire administration. The interviewers vary in their ability to appear or sound neutral, listen, probe adequately, and use techniques to aid recall and record responses (Bowling 2005; Callegaro et al. 2017). Evasiveness ("don't know" replies or no reply) was also found to be more common in paper-and-pencil self-administered questionnaires. Despite the interviewer differences, the negative impact can be minimized through careful training and monitoring of interviewers and analysis of responses by interviewers (to check for interviewer bias). Despite the criticism leveled against interviewers, they have several advantages in questionnaire administration.

A well-trained interviewer can help increase response and item response rates, maintain motivation with longer questionnaires, probe for responses, clarify ambiguous questions, help participants with enlarged show cards of response-choice options, use memory jogging techniques for aiding recall of events and behavior, and control the order of the questions (Bowling 2005; Callegaro et al. 2017). Further, Bowling posited that interviewers may also be trained to follow complex question routing and skipping instructions. In addition, the personal F2F interview (auditory channel) is regarded as the least burdensome (Bowling 2005) since the participant is only required to have basic verbal and listening skills and ability to speak the same language as in the questionnaire, whereas self-administered questionnaires demand literacy and ability to follow routing instructions. This suggests the efficacy of interviewers in data collection.

The interviewers also ensure participants are motivated, answer, and record all responses correctly, which increases response rate despite the criticism they face. It is also easier to convince participants of the legitimacy of the study in person.

2.3.3   Response-Choice Order: Primacy and Recency Effects

Participants responding to self-administered questionnaires were found to select the first response option presented in a questionnaire (primacy effects). In contrast, the final response option is offered first in oral questions in F2F or telephone interviews because participants still remember and, where agreeable, are likely to select that option (recency effects) (Bowling 2005; Bowling and Windsor 2008). These tendencies lead to response-choice order effects (see Table 3). Recency effects were found to be more common among older people in Bowling's (2005) studies. Researchers should take cognizance of these effects when collecting data. The section below compares traditional and online surveys that have not been discussed in the chapter.

3   Comparison of Traditional Surveys and ICT-Based Surveys

Previous studies concurred that both traditional and online surveys are generally comparable in terms of validity and reliability and that psychological and communication measures are similar (Touvier et al. 2010; van den Berg et al. 2011). However, Zhang et al. (2017) disagree with this assertion. The assertion could be logical in high-income countries although debatable in LMIC. Internet connectivity is available mostly in Asia and the Middle East (Republic of Korea, 98%; Saudi Arabia, 94%) (UN News Centre 2015), which make up the top ten countries with Internet. Further, lowest levels of Internet access were reported to be mostly in sub-Saharan Africa, with connection available to less than 2 % of the population in Guinea, Somalia, Burundi, and Eritrea. Smartphone penetration in LMIC was also reported to be very low, for instance, in India, coverage was 22.4%, Bangladesh 5.2%, and Uganda 4% (UN News Centre 2015). The model used to assess availability of smartphones took into consideration the economic progression, demography, online population, and inequality (Poushter 2017). Further, 40 nations were found to have smartphones in high-income countries (e.g., South Korea: 88%) (Pew Research Centre 2015), making it logical to conduct online surveys in these high-income countries and traditional surveys in the poor countries.

Most recent studies on elderly caregivers in low-income countries (Mhaka-Mutepfa et al. 2014; Aransiola et al. 2017; McKoy Davis et al. 2017) in West African countries, Jamaica, and Zimbabwe had problems accessing rural and remote populations. The authors revealed that the elderly with low income, less education, and living in non-metropolitan areas (rural) had no access to phones and Internet; therefore, it was difficult to make appointments for interviews. This kind of scenario is reported in LMIC where online platforms are not always available. Furthermore, online platforms (e.g., Facebook, Pinterest, and Snapchat) may be unavailable in certain countries (e.g., China) for legal or political reasons.

Nonetheless, Touvier et al. (2010) and van den Berg et al. (2011) found that online participants were comparable to respondents taking part in traditional surveys in terms of age, gender, income, education, and health status. The samples included caregivers that resided in urban areas with some possessing different demographics (Mhaka-Mutepfa et al. 2014). Although traditional surveys were found to have higher coverage in most studies, the administrative costs were much higher (2.5 times more) compared to online surveys (Hohwü et al. 2013; Szolnoki and Hoffmann 2013; Kramer et al. 2014). The cost for paper-based questionnaires used by Hohwü et al. (2013) was twice that of the online questionnaire surveys, confirming that online surveys are less costly. Therefore, there is a need for researchers to take cognizance of these differences to enable selection of the most appropriate survey method.

Online surveys were also not as representative or as projectable as traditional surveys (Hoogendorn and Daalmans 2009; Taylor et al. 2009). Hence, online surveys should be treated as a supplement to, rather than a substitute for, traditional surveys (Brandt 2012), even though they are cheap and readily available. In Germany, for example, 4.7% of all Internet users were found to be registered in some kind of web panel, and the response rate of these panels was about 20%, suggesting that only 1% of the Internet users in Germany could be reached by web panel surveys (Liljeberg and Krambeer 2012). Considering that Germany is a high-income country with such a low coverage, what can we make of low-income countries?

The discussion around effectiveness of online and traditional surveys has become synonymous with the nature/nurture controversy, whereby saying one is better than the other may be incorrect. Each of these methods has great merits, making it difficult to decide whether a good study is determined by the type of survey used. As of now, the authors believe that both traditional and web-based surveys are useful in research and complement each other. In addition, researchers should examine their target population carefully before selecting the most suitable method. Marleen et al. (2010) and Joan Lewis, a global consumer and market knowledge officer at Procter and Gamble, were some of the proponents of complementarity of both methods. Online surveys should be considered as alternatives to traditional surveys and as a complementary mode of data collection in research (Fricker et al. 2005; Hoogendorn and Daalmans 2009; Marleen et al. 2010; Brandt 2012).

Odds ratios were used to estimate differences in response rates between four modes of data collection: paper version questionnaire only, paper and web questionnaire, web questionnaire only, and web questionnaire and an incentive in the five Nordic countries (Denmark, Finland, Iceland, Norway, and Sweden) (Hohwü et al. 2013). The paper mode had a higher response rate (67%), and the three other modes had lower response rates. Lower response rates to online questionnaires were also found in other studies (Kongsved et al. 2007; Zuidgeest et al. 2011), although online surveys had higher response rate for younger respondents with a mean age of 30 years (van den Berg et al. 2011). This finding supports the idea that although online questionnaires have lower response rates than traditional questionnaires, the response rates have specific demographics, for instance, the highly educated and undergraduate students (Greenlaw and Brown-Welty 2009; Poushter 2017). However, the younger respondents in van den Berg et al.'s study reported a preference for the paper-based survey questionnaire (83%) to the online one.

Despite traditional surveys having their merits, there were controversies between the F2F and telephone surveys. Previous researchers (Szolnoki and Hoffmann 2013) found differences between telephone and F2F methods (see Table 3) with regard to random digital dialing (RDD), good geographical coverage, personal interaction, and cost. The merits and limitations of both approaches to research will be discussed below.

4   The Merits and Limitations of Traditional and Online Surveys

Traditional survey methods have several key strengths which should be noted to enable selection of the most appropriate research method. The choice of using an online or traditional survey method depends on the topic, objectives, timeframe, and the budget at the researcher's disposal (McColl et al. 2001; Szolnoki and Hoffmann 2013). The types of questions and information needs also play a pivotal role in selection of the method or data sources to use for research depending on the most appropriate (Brandt 2012: see Table 1). Traditional surveys are reported to be explicitly structured, flexible, and adaptable. Design researchers use interviews and self-report questionnaires to measure consumer response to products regardless of the limitations of these "explicit" self-report methods (Schoen and Crilly 2012). Implicit methods have thus been developed to try and overcome self-report biases and obtain a more automatic measure of attitudes.

Another strength for traditional surveys is that participants are observable and environmental changes may be detected while administering them. Personal interaction and control within the traditional survey environment are also an advantage. These advantages may hamper the use of online surveys. Marleen et al. (2010) argued that although online surveys may be an attractive alternative, epidemiological research was still scarce because of major concerns with selective nonresponse and reliability of the data obtained. Further, reluctance to use online surveys because of safety and confidentiality issues may also be of concern.

However, despite the strengths mentioned for traditional surveys, they also have their own limitations. Multimedia elements, for instance, videos, that can provide more information to participants cannot be used in traditional surveys. Traditional surveys are also more time consuming because they involve setting up interviews and lengthy collection of data and sifting through the results manually. In addition, people who participate in traditional surveys may feel obliged to suit the interviewers' time and cannot pace the interview, whereas Internet surveys can be saved and continued at a later date. However, premature termination cannot be prevented in online surveys.

A strength for online surveys is that busy people who decline telephone or F2F interviews may be willing to take surveys popping up on their computer screens (Kellner 2004). In Mhaka-Mutepfa et al.'s (2014) study, it was difficult to get busy people to sit down for interviews for a paper-and-pencil survey. The investigators eventually gave up after turning up for appointments several times to no avail. Nevertheless, relying on modes that require initiative from participants (like the pop-ups) may result in selective samples which raise concerns about nonresponse bias (Cooper 2011). Selection bias results from self-selection although traditional modes of data collection (e.g., the questionnaire) have shown little bias resulting from nonparticipation. Avidity bias (those with greater interest in the survey) may also be a factor especially in online surveys.

Another limitation for traditional surveys is unwillingness by participants to respond to sensitive questions, which will be discussed in the next section.

5   Ethics in the Use of Sensitive Information and Questions

Sensitive questions are situational and depend on the design features of the survey (Tourangeau and Yan 2007; Callegaro et al. 2017). Participants' reluctance to disclose sensitive information is increased in F2F surveys, particularly on illicit substance use and sexual and criminal behavior. Interviewers administering traditional surveys may provide more information to participants, but the participants are less likely to report the sensitive information. Previous researchers have reported that participants prefer online questionnaires to traditional questionnaires if research comprises sensitive questions (Tourangeau and Yan 2007; Mhaka-Mutepfa et al. 2014; Callegaro et al. 2017). For instance, in a traditional survey in Zimbabwe (Mhaka-Mutepfa et al. 2014), participants did not respond to the interview question on "How satisfied are you with your sex life?" in the WHOQOL-BREF questionnaire. This became a limitation because most of the grandparent carers in the study reported not having sex and those having sex were embarrassed to discuss sexual behaviors. Deleting the item from the WHOQOL-BREF scale would have reduced the validity of the measure.

A study on substance use and other stigmatized behaviors using traditional methods and Internet surveys was done (Newman et al. 2002). They found that people are more likely to respond to socially unacceptable behavior online than in traditional surveys (Tourangeau and Yan 2007; Liljeberg and Krambeer 2012; Szolnoki and Hoffmann 2013). The same authors reiterated that online surveys are more user-friendly where there is fear of embarrassment with exposure of weakness, failure, or deviancy to a stranger. In Marleen et al.'s (2010) study, the Internet response rate was higher than the F2F interview, because of the sensitive nature of their study. However, Wright (2017) and Marleen et al. (2010) posited that some participants were still uncomfortable responding to sensitive online questions because of the belief that investigators may use people's IP address or other information to identify respondents, thereby compromising safety and confidential issues.

Despite the limitations caused by sensitive questions, traditional surveys enable a broad understanding of people's attitudes and feelings particularly when done alongside implicit surveys. It is possible to determine whether participants are overstating or understating their attitudes and feelings. True feelings may also be hidden in response to sensitive questions if participants are not convinced of confidentiality issues (e.g., abortion, homosexuality, HIV status) and are skeptical about how their data is going to be used. The techniques used for reducing item response in social desirability bias may be used to reduce item response when responding to sensitive questions. Reassurances on confidentiality and anonymity could alleviate lack of confidence in responding to sensitive questions.

6   Strategies to Reduce Measurement Errors: Recommendations for Practice

The following recommendations are suggested to reduce measurement errors during research:

- A pilot test or focus group in early stages of questionnaire development to trial the tool and aspects of the data collection protocol should be done. This helps researchers to better understand how people think about an issue or comprehend a question.
- A pilot study might give advance warning about where the main research project could fail, where research protocols may not be followed, or whether proposed methods or instruments are inappropriate or too complicated. Information on how the testing environment affects performance may also be gathered. Focus groups in the early stages of questionnaire development may also help.
- Data should be double-checked and verified.
- Statistical procedures may be used to adjust for measurement error.
- F2F surveys were found to be the most advantageous mode of data collection; therefore, careful training and monitoring of interviewers and analysis of responses by interviewers should be done to minimize the interviewers' negative impact on participants and prevent accidental introduction of errors.
- Triangulation of data across multiple measures may lead to more accurate findings.
- Enforcing the sustainable development goal (SDG) that stresses access to Information Computer Technology (ICT), particularly broadband Internet, may speed up development and human progress, resolve the digital divide, and develop knowledge societies (UN News Centre 2015), thereby increasing access to ICT. This will increase the use of online surveys in rural and remote areas not in the so distant future.

- Different modes of questionnaire administration were found to affect the quality of the data collected suggesting that all questionnaire users need to be cognizant of the potential effects of the mode of administration on their data. Although some previous researchers (e.g., Callegaro et al. 2017) reported that participants prefer online questionnaires to traditional questionnaires if research comprises sensitive questions, researchers should consider their sample, type of research, and demographics when choosing the mode of data collection. The biasing effects of mode of questionnaire administration have important implications for research methodology, the validity of the results of research, and the soundness of public policy developed from evidence using questionnaire-based research (Bowling 2005).

7   Conclusion and Future Directions

Despite the fact that advances in ICT have enabled researchers to move away from traditional surveys, the historical surveys still play an important role because of unavailability of Internet services in most LMIC. The chapter discusses the merits and limitations of using traditional surveys in research. The advantages of representativeness in populations that are not easily accessible were also discussed. In addition, the problems with social desirability and sensitive questions were also discussed. However, the discussion strengthens the view that researchers should move away from comparing social media and traditional data sources but should be able to select the most appropriate method, which is determined by the type, topic, goal, budget, timeframe, and geographic site of the research.

Each of the modes of data collection (traditional and online) methods has its comparative strengths and limitations. It is common that alternative methods and data sources may result in different or at times conflicting tales of accounts of data, suggesting the complementarity of the two methods. Paper-based surveys certainly still have their place in survey research and will always have. Nevertheless, continued advances in technology may increasingly diminish that demand, but researchers should choose the best survey method depending on appropriateness. Whatever method is used for data collection must fulfill the researcher's requirements.

References

Aransiola JO, Akinyemi AI, Akinlo A, Togonu-Bickesteeth F. Grandparenting in selected West African countries: implications for health and hygiene behaviours in the household. GrandFamilies: Contemp J Res Pract Pol. 2017;4(1):195. http://scholarworks.wmich.edu/grandfamilies/vol4/iss1/11.

Blasius J, Brandt M. Representativeness in online surveys through stratified samples. Bull Méthodologie Sociol. 2010;107:5–21.

Bowling A. Mode of questionnaire administration can have serious effects on data quality. J Publ Health (Oxf). 2005;27(3):281–91. https://doi.org/10.1093/pubmed/fdi031.

Bowling A, Windsor J. The effects of question order and response-choice on self-rated health status in the English Longitudinal Study of Ageing (ELSA). J Epidemiol Community Health. 2008;62(1):81–5. https://doi.org/10.1136/jech.2006.058214.

Brandt R. Website vs. traditional survey ratings. Do they tell the same story? Mark Res. 2012. http://www.thevoicecrafter.com/files/107673625.pdf.

Brigham J, Lessov-Schlaggar NC, Javitz SH, Krasnow ER, McElroy M, Swan EG. Test-retest reliability of web-based retrospective self-report of tobacco exposure and risk. J Med Internet Res. 2009;11(3):e35. https://doi.org/10.2196/jmir.1248.

Callegaro M, Manfreda KL, Vehovar V. Mario callegaro discusses web survey methodology. Slovenia, University of Ljubljana: SAGE Research Methods; 2017.

Cooper MP. The future of modes of data collection. Publ Opin Q. 2011;75(5):889–908.

Devine P, Lloyd K. Internet use and psychological well-being among 10-year-old and 11-year-old children. Child Care Pract. 2012;18(1):5–22. https://doi.org/10.1080/13575279.2011.621888.

Egan J. Why online surveys top traditional surveys. 2016. https://icitizen.com/blog/why-online-surveys-top-traditional-surveys/.

Fricker S, Galesic M, Tourangeau R, Yan T. An experimental comparison of web and telephone surveys. Publ Opin Q. 2005;69(3):370–92.

Graham AL, Papandonatos GD, Bock BC, Cobb NK, Baskin-Sommers A, Niaura R, Abrams DB. Internet- vs. telephone-administered questionnaires in a randomized trial of smoking cessation. Nicotine Tob Res: Off J Soc Res Nicotine Tob. 2006;8(Suppl 1):S49–57.

Greenlaw C, Brown-Welty S. A comparison of web-based and paper-based survey methods: testing assumptions of survey mode and response cost. Eval Rev. 2009;33(5):464–80. https://doi.org/10.1177/0193841x09340214.

Hohwü L, Lyshol H, Gissler M, Jonsson SH, Petzold M, Obel C. Web-based versus traditional paper questionnaires: a mixed-mode survey with a nordic perspective. J Med Internet Res. 2013;15(8):e173. https://doi.org/10.2196/jmir.2595.

Hoogendorn AW, Daalmans J. Nonresponse in the requirement of an Internet panel based on probability sampling. Surv Res Methods. 2009;3:59–72.

Kellner P. Can online polls produce accurate findings. Int J Mark Res. 2004;46(1):3–22.

Kongsved MS, Basnov M, Holm-Christensen K, Hjollund HN. Response rate and completeness of questionnaires: a randomized study of internet versus paper-and-pencil versions. J Med Internet Res. 2007;9(3):e25. https://doi.org/10.2196/jmir.9.3.e25.

Kramer ADI, Guillory JE, Hancock JT. Experimental evidence of massive-scale emotional contagion through social networks. Proc Natl Acad Sci. 2014;111(24):8788–90. https://doi.org/10.1073/pnas.1320040111.

Liljeberg H, Krambeer S. Bevölkerungs-repräsentative Onlinebefragungen. Die Entdeckung des Scharzen Schimmel? Planung und Analyse, Sonderdruck: online, social, mobile: what's next? 2012.

Marleen MHJ, van Gelder MM, Reini W, Roeleveld N. Web-based questionnaires: the future in epidemiology? Am J Epidemiol. 2010;172(11):1292–8. https://doi.org/10.1093/aje/kwq291.

McColl E, Jacoby A, Thomas L, Soutter J, Bamford C, Steen N,..., Bond J. Design and use of questionnaires: a review of best practice applicable to surveys of health service staff and patients. Health Technol Assess. 2001;5(31):1–256.

McKoy Davis JG, Willie-Tyndale D, Mitchell-Fearon K, Holder-Nevins D, James K, Eldemire-Shearer D. Caregiving among community-dwelling grandparents in Jamaica. GrandFamilies: Contemp J Res Pract Policy. 2017;4(1):7–40

Mhaka-Mutepfa M, Cumming R, Mpofu E. Grandparents fostering orphans: influences of protective factors on their health and well-being. Health Care Women Int. 2014;35(7–9):1022–39. https://doi.org/10.1080/07399332.2014.916294.

Morales-Vives F, Vigil-Colet A, Lorenzo-Seva U, Ruiz-Pamies M. How social desirability and acquiescence affects the age–personality relationship. Personal Individ Differ. 2014;60 (Supplement):S16. https://doi.org/10.1016/j.paid.2013.07.370.

Newman JC, Des Jarlais DC, Turner CF, Gribble J, Cooley P, Paone D. The differential effects of face-to-face and computer interview modes. Am J Public Health. 2002;92:294–7.

Nosek B, Hawkins C, Frazier R. Implicit social cognition: from measures to mechanisms. Trends Cogn Sci. 2011;15(4):152–9.

Pew Research Centre.The demographics of social media users. 2015. http://www.pewinternet.org/2015/2008/2019/the-demographics-of-social-media-users/.

Ponto J. Understanding and evaluating survey research. J Adv Pract Oncol. 2015;6(2):168–71.

Poushter J. Smartphone ownership and internet usage continues to climb in emerging economies. Pew research center. 2017. Pewglobal.org. http://www.pewglobal.org/2016/2002/2022/smartph one-ownership-and-internet-usage-continues-to-climb-in-emerging-economies/.

Rankin KM, Rauscher GH, McCarthy B, Erdal S, Lada P, Il'yasova D, Davis F. Comparing the reliability of responses to telephone-administered vs. self-administered web-based surveys in a case-control study of adult malignant brain cancer. Cancer Epidemiol Biomark Prev. 2008;17(10):2639–46. https://doi.org/10.1158/1055-9965.EPI-08-0304.

Schoen KL, Crilly N. Implicit methods for testing product preference exploratory studies with the affective simon task. 2012. http://www-edc.eng.cam.ac.uk/~nc266/WebFiles/Schoen_&_ Crilly_(2012)_Implicit_methods_(Design_&_Emotion).pdf.

Spek V, Nyklíček I, Cuijpers P. Internet administration of the Edinburgh Depression Scale. J Affect Disord. 2008;106(3):301–5.

Szolnoki G, Hoffmann D. Online, face-to-face and telephone surveys – comparing different sampling methods in wine consumer research. Wine Econ Policy. 2013;2(2):57–66. https://doi.org/10.1016/j.wep.2013.10.001.

Taylor PA, Nelson NM, Grandjean BD, Anatchkova B, Aadland D. Mode effects and other potential biases in panel-based internet surveys final report. 2009. Retrieved from Wyoming Survey & Analysis Centre.

Tourangeau R, Yan T. Sensitive questions in surveys. Psychol Bull. 2007;133(5):859–83. https://doi.org/10.1037/0033-2909.133.5.859.

Touvier M, Méjean C, Kesse-Guyot E. Comparison between web-based and paper versions of a self-administered anthropometric questionnaire. Eur J Epidemiol. 2010;25(5):287–96.

UN News Centre. Billions of people in developing world still without Internet access, new UN report finds. 2015. www.un.org/apps/news/story.asp?NewsID=51924.

van den Berg MH, Overbeek A, van der Pal HJ, Versluys AB, Bresters D, van Leeuwen FE, … van Dulmen-den Broeder E. Using web-based and paper-based questionnaires for collecting data on fertility issues among female childhood cancer survivors: differences in response characteristics. J Med Internet Res. 2011;13(3):e76. https://doi.org/10.2196/jmir.1707.

van Gelder MMHJ, Bretveld RW, Roeleveld N. Web-based questionnaires: the future in epidemiology? Am J Epidemiol. 2010;172(11):1292–8. https://doi.org/10.1093/aje/kwq291.

Varkevisser CM, Pathmanathan I, Brownlee A. Designing and conducting health systems research projects. Proposal Dev Fieldwork. 2003;1:1–357

Walker T. State of the internet 1st quarter 2012. 2012. comScore.

Wyatt JC. When to use web-based surveys. J Am Med Inform Assoc. 2000;7(5):426–9.

Zhang X, Kuchinke L, Woud ML, Velten J, Margraf J. Survey method matters: online/offline questionnaires and face-to-face or telephone interviews differ. Comput Hum Behav. 2017;71:172–80. https://doi.org/10.1016/j.chb.2017.02.006.

Zuidgeest M, Hendriks M, Koopman L, Spreeuwenberg P, Rademakers J. A comparison of a postal survey and mixed-mode survey using a questionnaire on patients' experiences with breast care. J Med Internet Res. 2011;13(3):e68. https://doi.org/10.2196/jmir.1241.

Epidemiology

33

Kate A. McBride, Felix Ogbo, and Andrew Page

Contents

K. A. McBride (✉)
School of Medicine and Translational Health Research Institute, Western Sydney University,
Sydney, NSW, Australia
e-mail: K.Mcbride@westernsydney.edu.au

F. Ogbo · A. Page
School of Medicine and Translational Health Research Institute, Western Sydney University,
Campbelltown, NSW, Australia
e-mail: F.Ogbo@westernsydney.edu.au; A.Page@westernsydney.edu.au

© Springer Nature Singapore Pte Ltd. 2019
P. Liamputtong (ed.), *Handbook of Research Methods in Health Social Sciences*,
https://doi.org/10.1007/978-981-10-5251-4_91

Abstract

Epidemiology is the core discipline underlying health research. Measuring health, identifying causes of ill health, and intervening to improve health are all key tenets of the discipline. Concerned with the "who," "when," "why," and "where" of health, epidemiology is essential in informing clinical decision-making through evidence-based medicine. Traditionally the study of the occurrence and distribution of disease and determinants, epidemiology is a dynamic discipline that is increasingly being applied in differing disciplines, even beyond health. This chapter is designed as a basic introduction to epidemiology, the terminology used, and the principles in epidemiologic practice. Using public health as a framework, this chapter will also give an overview of both the traditional and more contemporary applications of epidemiology.

Keywords

Measures of association · Prevalence · Incidence · Observational study · Experimental study · Cohort · Case control · Cross sectional · Randomized controlled trial · Ecological study · Random error · Systematic error · Confounding · Selection bias · Measurement error

1   Introduction

This chapter outlines what epidemiology is: a historical perspective of the discipline of epidemiology and core concepts of the discipline, which, along with biostatistics and health services, is a core element of public health (Liamputtong 2016). Essentially, public health is concerned with threats to the health of the entire population rather than health of individuals. Epidemiology is often described as the cornerstone of public health research and practice for a number of reasons (Lee 2016; Graham 2017). First, epidemiology is concerned with assessment and monitoring of health problems in specified populations using sound research methodology. Second, epidemiology deals with identifying predisposing factors for health problems based on the development and testing of hypotheses established in other scientific fields such as behavioral sciences, physics, and biology to explain health-related states and events. Last, epidemiology provides the foundation for predicting the impact of specific exposures on health, which are essential in guiding policies and resource allocation for improving the health outcomes and social environments of people (Cates 1982; Friis 2017; Merrill 2015).

2   What Is Epidemiology?

The word "epidemiology" takes its origin from a combination of Greek words "*epi* meaning on or upon," "*demos* meaning the people," and "*logos* meaning study." Epidemiology simply means "the study of what is upon the people" (Merrill 2017, p. 2). Many definitions of epidemiology have been proposed, but the following definition contains the basic concepts and principles of the field of epidemiology: "The study of the distribution and determinants of health-related states or events in specified populations, and the application of this study to the control of health problems" (Last 2007, p. 61).

Primarily, epidemiology is concerned with the distribution of *frequency* and *pattern* of health-related states and events in a target population (Lee 2016; Graham 2017). Frequency is the number of health-related states and events (such as the number of cases of diabetes or stroke in a target population) and their relationship to the size of the population. In epidemiology, the resulting rate is more meaningful when compared to disease occurrence across different people. Pattern is the occurrence of health-related events by a person (Who is ill?), place (Where do they live?), and time (When did they become ill?). Personal characteristics may comprise sociodemographic factors related to the health-related state or event, including sex, age, educational status, and marital status. Place pattern may include urban-rural differences or geographic variation, while time pattern may be daily, weekly, monthly, seasonal, annually, or any other analysis of time that may impact health-related states or events occurrence. This characterization of health-related states or events by *person, place, and time* are within the domain of **descriptive** epidemiology (Cates 1982; Lee 2016; Graham 2017).

Epidemiologic methods are also used to identify **determinants**, which are factors that can bring change to a health condition or other defined characteristics. Examples of determinants include chemical agents (e.g., chemical carcinogens), biological agents (e.g., bacteria), and other less precise factors (e.g., a lack of physical activity) (Cates 1982; Dicker et al. 2006; Friis 2017). Health-related states or events are used in the definition of epidemiology to capture not only diseases (e.g., stroke or cholera) but also study of events such as drug abuse, injuries, behaviors, and conditions that affect the well-being of a population. Lastly, epidemiology involves applying the knowledge gained from epidemiologic investigations to guide policies and evidence-based practice. Results of epidemiologic studies can also assist individuals to make better and informed decisions about their lifestyle choices (Lee 2016).

3   A Historical Perspective of Epidemiology

The discipline of epidemiology evolved from Hippocrates of Cos (460–375 BC), who recognized that environmental factors could impact on health (Hippocrates 400 BC). John Graunt (1620–1674) measured for the first time patterns of birth, death, and disease occurrence by sex and geography and developed life tables and the concept of life expectancy (Graunt 1977). Thomas Sydenham (1624–689) described occurrence of disease from an observational viewpoint (Lilienfeld and Stolley 1994; Friis 2017). Latterly, William Farr (1807–1883) built on Graunt's work by systematically collecting and analyzing mortality data in Britain. Additionally, Ignaz Semmelweis, Louis Pasteur, Robert Koch, Florence Nightingale, and others have made significant contributions to the discipline of epidemiology (Dicker et al. 2006; Merrill 2015; Friis 2017; Webb et al. 2017). It was not until later in the nineteenth century, however, that the distribution of disease was measured in specific populations with this work by John Snow underpinning many contemporary epidemiologic approaches, as well as being one of the fields of epidemiology's notable achievements.

John Snow was an English physician who made several innovative contributions to the field of modern epidemiology in his pioneering work in explaining the mode of transmission of cholera in London. In the mid-1840s, Snow began an epidemiologic investigation of a cholera outbreak in the Soho and Golden Square districts of London by assessing where in these areas people with cholera lived and worked. He was able to identify each residence on a map of the area, known today as a spot map, showing the geographical distribution of cholera cases (Snow 1857).

Snow believed that contaminated water was a source of infection for cholera; he marked the location of water pumps in the area on his spot map and then considered a relationship between the movement of people with cholera and the geographical location of the pumps. He found that more households with cholera were located within a short distance from Pump A (the Broad Street pump) compared to households with cholera around Pump B or C, which were located in different streets. When Snow interviewed households who lived in the Golden Square area, he found that residents avoided Pump B because it was perceived to be grossly contaminated and that Pump C location was too far for many residents. From this evidence, Snow concluded that the most likely source of infection for most people with cholera in the Golden Square area was the Broad Street pump (Pump A), the primary source of water for most residents. However, Snow observed that two separate populations in the Soho district were not profoundly affected by the cholera outbreak. Further investigation by Snow revealed that a brewery located in the area had a deep well on-site, where brewery workers living in the area collected their water – this protected them from the cholera epidemic (Snow 1855a, 1857).

Snow then obtained additional information from people with cholera epidemic on their water sources to confirm that the Broad Street pump was the primary source of the cholera outbreak. He presented his findings to the district officials that the consumption of water from the Broad Street pump (Pump A) was the most common source of infection among people with cholera. As a control measure and to prevent any reoccurrence of infection, the handle of the pump was removed, and the cholera epidemic ceased. This first cholera outbreak investigation by John Snow was an exercise in descriptive epidemiology.

In 1854, Snow examined data from another cholera outbreak in London (which had occurred in 1853), where he observed that the highest death rates were in districts serviced by two water companies: the Lambeth Water Company and the Southwark and Vauxhall Water Company. Both companies drew water from the Thames River in London, an area susceptible to contamination from London sewerage, which flowed directly into the river. In 1852, however, the Lambeth Water Company had moved its water sources upstream of the Thames River, to an area less vulnerable to contamination. The Southwark and Vauxhall Water Company did not relocate the position of its water source. During this time, household residents in the area were free to obtain water from any water company. Based on these observations, Snow developed comparison tables on deaths by water source and subdistricts (Snow 1855b, 1857).

Snow concluded that water obtained solely upstream by the Lambeth Water Company caused limited deaths compared to water drawn from downstream in areas below sewage openings (thus vulnerable to contamination) by the Southwark and Vauxhall Water Company (Snow 1855b, 1857). This second work of Snow involved an analytical aspect of epidemiologic investigation of the cholera epidemic, as he compared death rates from cholera by water sources, either from the Lambeth Water Company or the Southwark and Vauxhall Water Company. Despite no practical knowledge of the occurrence of microorganisms, Snow was able to show through epidemiologic methods that water could serve as a medium for cholera transmission and therefore that findings from epidemiologic studies could be used to guide public health initiative (Snow 1855b, 1857).

More recently, epidemiology has informed prevention and control methods, with Doll and Hill (1950) famously establishing a clear-cut relationship between lung cancer and tobacco use in the 1950s in the British Doctors cohort study. Not all relationships are as well defined; therefore, newer epidemiologic methods that are more robustly able to establish causality are now needed to examine the multifactorial relationships that may exist between multiple exposures and disease outcomes. Major challenges exist in the areas of emerging infectious disease as well as the need to explore and act on determinants of health and disease to inform prevention and control methods.

4 Study Design

Epidemiologic studies may be of **experimental** or **observational** design. Further, epidemiologic studies are categorized as being either **analytic** or **descriptive** (Lee 2016; Graham 2017). Choosing the appropriate study design is essential in any epidemiologic investigation, as each study design has strengths and weaknesses as well as practical considerations. The simplest study designs, e.g., descriptive observational studies, aim to estimate a single risk, whereas more complex studies, e.g., cohort studies, aim to compare disease occurrence, understanding the cause of a disease or evaluation of the impact of a disease (Rothman 2012). The design of a study also guides how strong or weak the evidence from that study is (see Fig. 1), as well as the ability to make causal judgments from the evidence generated.

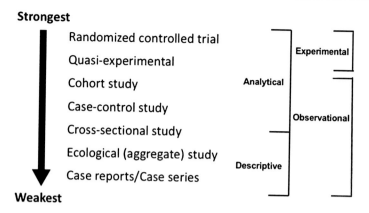

Strongest

Weakest

Fig. 1 Study designs

4.1   Experimental Studies

The aim of an experimental study is to intervene and change a variable in one or more groups of people. This may mean giving a drug to one group and comparing them to a control group who does not receive the drug or restricting a factor, such as dietary fat, in a group and comparing them to the control group who continues with their usual diet. While experimental studies can provide the highest level of evidence in regard to cause and effect, ethical considerations must guide their use as no patient should be denied treatment that is known to beneficial nor should any treatment being tested that is known to be harmful (Bonita et al. 2006). Experimental studies include **randomized controlled trials** and **quasi-experimental studies** (Lee 2016; Webb et al. 2017).

- **Randomized controlled trials** (RCT) are designed to examine the effects of an intervention usually for a specific disease in the form of a clinical trial. In a RCT, participants are randomly allocated to either the intervention or control group. If randomization is conducted correctly, the only difference between the groups should be the intervention; thus this is the best way to minimize confounding (discussed in the next section) as it makes the groups exchangeable. Once allocated to either the intervention or control, individuals are followed to ascertain the effects of an intervention (see also ▶ Chaps. 37, "Randomized Controlled Trials," and ▶ 3, "Quantitative Research").
- **Quasi-experimental studies** are similar to RCTs in that they examine the effects of an intervention. Unlike a true RCT, where treatment assignment is at random, assignment in quasi-experiments may be either by alternate allocation, by self-selection, or by administrator judgment. Because of the nonrandom nature of allocation to either intervention or control group, distinct differences in addition to the intervention itself may exist between the two groups that could influence the outcomes following intervention.

4.2 Observational Studies

Observational studies, as implied by the name, observe nature as it takes its course without any sort of intervention. Observational studies can be both descriptive and analytic. Observational **analytic** studies identify the cause of disease by analyzing the associations between exposure and outcome. An analytic study can answer the why or how questions. Almost all epidemiologic studies are analytic. Observational analytic studies include **cohort, case control,** and **cross-sectional analytic** studies (Lee 2016; Graham 2017; Webb et al. 2017; see also ▶ Chap. 3, "Quantitative Research").

- **Cohort** studies compare the rate of disease in a group of people exposed to a factor with a group that has not been exposed. Participants in a cohort study are, therefore, selected based on their exposure status. One of the key features of a cohort study is the observation of large numbers of people over a long period of time, which can be both time-consuming and expensive, though cost can be reduced with the use of routinely collected data. Cohort studies aim to identify associations between possible risk factors or prognostic factors and the outcome of interest. One advantage of a cohort study is that a number of exposures can be measured at the same time. Cohort studies are usually prospective, where individuals are selected based on their exposure status, then followed through time to see who develops the outcome of interest. This makes measurement of the exposure less susceptible to bias when compared to a case control study. Prospective or historical cohort studies are also possible, where exposure is identified through records of past exposure, e.g., a drug exposure recorded in medical records.
- A **case control** study selects subjects on the basis of disease status with "cases" having the disease (or outcome) of interest and controls *not* having the disease (or outcome) of interest. Cases are then compared with controls with respect to past exposure history. The cases in a case control study should be representative of all cases in a specified population group (study base). The study base should be well defined in terms of time, person, and place. Controls should also be selected from the same study base and be representative of people who could become cases if they had developed the disease. One important part of a case control study is the measurement of exposure, which is done retrospectively after the development of disease. Often, this is done by direct questioning of the case or control; therefore this type of study is prone to bias (discussed in the next section).
- **Analytic cross-sectional** studies examine the association between a risk factor and an outcome. In a cross-sectional study, the exposure (or risk factor) and outcome are measured simultaneously, which can make it difficult to establish if the exposure came before or after the disease.

Descriptive studies describe the health status of a given population by person, time, and place. A descriptive study can answer the who, what, where, and when questions around the relationship between exposure and outcome. Descriptive studies are usually the first foray into an area of inquiry and are a useful way to document the health (or illness) of a population. Generally, they are a prelude to more rigorous studies as they can only provide suggestive findings. Descriptive studies include cross-sectional (prevalence) studies, ecological studies, use of routine data, case reports, and case series (Bonita et al. 2006; Lee 2016; Graham 2017; Webb et al. 2017).

- When a **cross-sectional (prevalence) study** is descriptive, it assesses the burden of a disease in a defined population at a given point in time, like a "snapshot."
- **Ecological studies** are aggregate or correlation studies and related exposure and outcome between populations or groups *not* individuals. Populations can be compared in different geographical areas at the same or different points in time. Many ecological studies use routine published data.
- A **case report** provides a detailed description of an unusual disease or association or a common disease in an unusual person.
- A **case series** simply aggregates individual cases into a report and may also include a hypothesis as to the cause. As with a case report, this type of study is considered to be weak evidence but may prompt further investigation with more rigorous study designs.

Practical considerations play a large part in study selection, including ethical considerations in the case of an experimental study and time/cost considerations for observational studies. Selection of study design, however, is first and foremost related to the one most appropriate for the study question. If the aim is to establish causality, the amount of bias a study is prone to must be taken into consideration.

5   Measures of Disease Frequency

In epidemiology, many different measures are used by researchers and policy decision-makers to describe the health of a specified population over a period of time. These measures allow for comparison and can be divided into two broad categories – prevalence and incidence (see also ▶ Chap. 3, "Quantitative Research").

5.1   Prevalence

Prevalence measures the presence of a disease, condition, or risk factor such as smoking in a particular population at a given time. Prevalence is defined as the proportion of a population who has (or had) a specific attribute in a given time period (Merrill 2015; Lee 2016; Friis 2017; Graham 2017; Webb et al. 2017), that is:

$$\text{Prevalence} = \frac{\text{Number of existing (and new) cases}}{\text{Population at risk}}$$

Prevalence is often reported as a percentage (9% or 9 people out of 100) or as the number of cases per 10,000 or 100,000 people, depending on how common the illness or risk factor is in the population. Common types of prevalence are dependent on the time frame for the estimate:

- **Point prevalence** refers to the proportion of a population that has the disease at a specific time.
- **Period prevalence** refers to the proportion of a population that has the attribute at any point over a period of interest.
- **Lifetime prevalence** refers to the proportion of a population who at some point in life up to the time of assessment ever had the disease or attribute.

5.2   Incidence

Incidence is a measure of the likelihood of new cases of disease or injury in a population within a specified period of time (Merrill 2015; Lee 2016; Friis 2017; Webb et al. 2017), that is:

$$\text{Incidence} = \frac{\text{Number of new cases}}{\text{Time interval}}$$

Incidence is usually expressed as the proportion or rate (with a denominator) of new cases and has two main types:

- Incidence proportion (IP) or cumulative incidence (CI)
  Other synonyms of IR include attack rate, risk, and probability of developing disease.
- Incidence rate (IR) or person-time rate

5.2.1   Incidence Proportion
The incidence proportion (IP) is the proportion of an initially disease-free population that develops the disease within a specified period of time. IP is a proportion because the persons in the numerator, those who develop the disease, are all included in the denominator (the total population).

$$\text{Incidence proportion} = \frac{\text{Number of new cases of disease during specified period}}{\text{Total number of the population at risk}}$$

5.2.2   Incidence Rate (IR)
The incidence rate is a measure of incidence that incorporates time directly into the denominator. IR is usually calculated from a long-term cohort follow-up study, where participants are followed over time and the occurrence of new cases of disease is documented. Like the incidence proportion, the numerator of the incidence rate is

the number of new cases identified within the period of observation. However, the denominator is different. The denominator is the amount of the time each participant was observed, totaled for all persons. This denominator represents the total time the population was at risk of and being observed for disease (Merrill 2015; Friis 2017; Webb et al. 2017). For example, in a long-term cohort follow-up study of a specific disease, each study participant may be followed or observed for several years. One person followed for 5 years without developing disease is said to contribute 5 person-years of follow-up.

$$\text{Incidence proportion} = \frac{\text{Number of new cases of disease during specified period}}{\text{Time each person was observed, totaled for all persons}}$$

5.3   How Does Prevalence Differ from Incidence?

Incidence is a measure of the number of new cases of an attribute (e.g., illness or risk factor) that is present in a population over a specific period (e.g., a month or a year), while prevalence is the proportion of a population who has (or had) a specific attribute in a given time period, irrespective of when they first developed the attribute (Merrill 2015; Friis 2017; Webb et al. 2017). The main difference between prevalence and incidence is in their numerators as the numerators of prevalence are all cases *present* within a given time period, whereas the numerators of incidence are new cases that *occurred* within a given time period.

6   Measures of Disease Association

Observational, analytical studies usually aim not only to report disease frequency but also to examine the associations between exposures (e.g., tobacco smoking) and outcomes (e.g., lung cancer). The key to these epidemiologic analyses is to compare the observed amount of disease in a population with the expected amount of disease. The comparisons can be calculated using measures of association such as risk ratio (relative risk), rate ratio, and odds ratio (Merrill 2015; Lee 2016; Friis 2017; Webb et al. 2017). Broadly, these measures provide evidence about causal relationship between an exposure and disease. However, it is important to note that association does not necessarily imply a causal relationship.

6.1   Risk Ratio

A risk ratio (RR), also called the relative risk, compares the risk of a health states or events (e.g., disease or injury) in one group with the risk in another group. Often, the two groups are differentiated by specific attributes that may include demographic factors as sex (e.g., males vs. females) or by exposure to a potential risk factor (e.g., did or did not smoke cigarettes). The group of primary interest is usually labelled "the exposed group," while the comparison group is labelled the "unexposed group."

$$\text{Risk Ratio} = \frac{\text{Incidence in exposed group}}{\text{Incidence in unexposed (comparison) group}}$$

A risk ratio is interpreted where 1.0 indicates equal risk in both the exposed and unexposed groups (null value). A risk ratio greater than 1.0 shows a positive association, that is, the exposure is associated with an increased risk of the disease. A risk ratio of less than 1.0 means an inverse of negative association, that is, exposure is associated a decreased risk of the disease, indicating that the exposure may be protective against disease occurrence.

6.2 Odds Ratio

An odds ratio (OR) also measures the association between an exposure with two categories and health outcome. The odds ratio is the measure of choice in a case-control study. In most cases, the population size from which the cases are selected is unknown. Therefore, it may be problematic calculating risk ratios from a typical case-control study. However, an odds ratio can be calculated and interpreted as an approximation of the risk ratio, particularly when the outcome is rare in the population of interest.

$$\text{Odds Ratio} = \frac{a \times d}{c \times b}$$

Where:

a = Number of persons exposed and with the disease
b = Number of persons exposed but without the disease
c = Number of persons unexposed but with the disease
d = Number of persons unexposed and without the disease

7 Measures of Public Health Impact

Measures of public health impact are used when a causal relationship has been established between exposure and an outcome. These metrics place the relationship between an exposure and a disease in a public health context and indicate the burden that an exposure contributes to the occurrence of disease in the population (Merrill 2015; Friis 2017; Webb et al. 2017). **Attributable proportion** and **population attributable risk** are two such measures.

7.1   Attributable Proportion

The attributable proportion (also called the attributable risk percent or the attributable fraction) is a measure of the public health impact of a causative factor. The calculation of this measure assumes that the occurrence of disease in the unexposed group is the baseline, that is, it allows you to calculate the *proportion* of disease in the exposed group that can be attributed to the exposure. Alternatively, this measure can be considered as the proportion of disease in the exposed group that could be prevented by eliminating the risk factor (Rockhill et al. 1998). In simple terms, the attributable proportion is the amount of disease in the exposed group attributable to the exposure. The use of attributable proportion in measuring public health impact depends on a single risk factor being responsible for a health state or event. However, when multiple risk factors may be interacting, the attributable proportion may not be appropriate.

$$\text{Attributable Proportion} = \frac{RD}{CI_e} = \frac{CI_e - CI_u}{CI_e}$$

Where:

RD = Risk difference (also called absolute risk reduction)
CI_e = Cumulative incidence in exposed group
CI_u = Cumulative incidence in unexposed group

7.2   Population Attributable Risk

The population attributable risk (PAR) is used to answer the following question: "What is the incidence of disease in a population, associated with the occurrence of the risk factor, or how much does a risk factor contribute to the overall rates of disease in groups of people?" PAR (also sometimes referred to as the population attributable fraction) is the proportional reduction in *population* disease that would occur if exposure to a risk factor were reduced to an alternative ideal exposure scenario (e.g., no physical activity). Alternatively, the attributable proportion for the entire population is the (incidence) risk in the overall population that can be attributed to the exposure.

$$\text{Population attributable risk (PAR)} = (\text{proportion of cases exposed}) \times (\text{attributable proportion in the exposed})$$

8   Validity in Epidemiologic Studies

When conducting epidemiologic research studies, the goal is accuracy in estimation where the results of the study can be generalized to the target population. The results are product of the study design itself, the study conduct, and how the data is

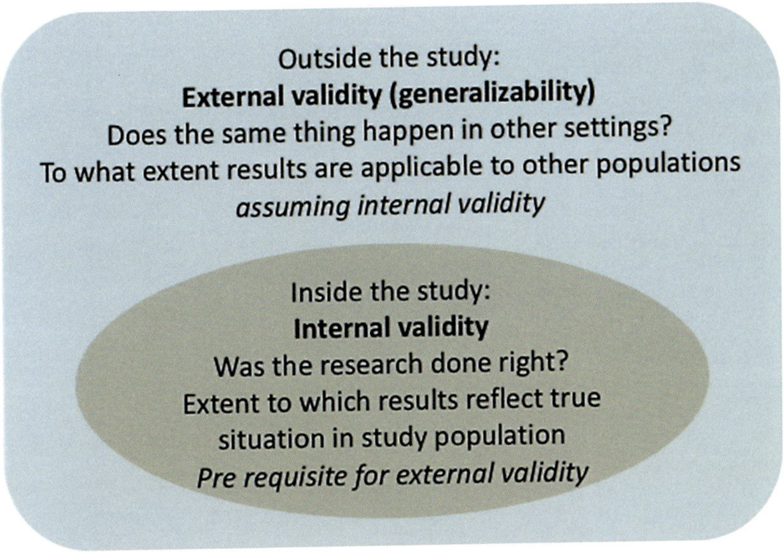

Fig. 2 Study validity

analyzed. Epidemiology is, therefore, an exercise in accurate measurement. Error is common and is classified as random (non-differential) or systematic (differential). Random error refers to a lack of precision, whereas systematic error refers to bias – an estimate that is not biased is valid. Both *internal* validity and precision are components of accuracy in an epidemiologic study. If a study is not internally valid, it cannot have *external* validity (be generalized beyond the study population – see Fig. 2). Internal validity can be affected by error in the selection of study participants, measurement of the exposure, or outcome and confounding.

8.1 Random Error

Any difference found between different groups in a study can occur due to random variation in the study subjects, meaning that the results of the study may have occurred due to chance and, therefore, differ from the true population value. This variance could be due to biological variation of individuals in a study, random sampling error, and random measurement error (Bonita et al. 2006). Random error can be minimized with adequate sample sizes. For example, if a RCT has insufficient participants, one factor may be unevenly distributed between the two groups such as age or gender. The chance of this occurring can be reduced as the number of participants increases as this reduces the amount of unevenness between the groups. Nonetheless, even when everything is constant, there will still be some random variation; therefore, we need tools to assess if any variations observed are real and important as opposed to arising by chance. Determination of random error is a key feature of statistical analyses.

A common method to assess the role of chance is hypothesis testing which produces a p values. The p value assesses whether or not findings are significantly different or not from some reference value, reflecting no effect (e.g., relative or odds ratio of one or a mean difference of zero). P represents probability, which measures the strength of the evidence against the null hypothesis and estimates how likely the difference between study groups is due to chance alone. The p value is a proportion. A smaller p indicates stronger evidence against the null and that the result is less likely due to chance. Conventionally, an arbitrary value of $p < 0.05$ is used to decide whether the results are due to chance, with a p value of <0.05 considered to be statistically significant. It should be noted, however, that the p value is not an estimate of any quantity; rather it is only a measure of the strength of evidence against the null hypothesis. The p value does not tell us anything about the size or direction of any difference. For example, in a very large clinical trial with a very small p value based on a small effect size, may not be important when translated into clinical practice.

Another value which is used to assess random variation is the confidence intervals, which is an estimate of the range of values that the true (population) value lies within. This range is the confidence interval and is commonly set at the same arbitrary level as p values (0.95) and is called the 95% confidence interval. It is important to note, however, that there is only one true value and the confidence interval defines the range where it is most likely to be. The confidence interval is not the variability of the true value or of any other value between subjects. The width of the confidence interval is important in assessing for accuracy or precision. A narrow confidence interval implies high precision, whereas a wide confidence interval implies poor precision (usually due to inadequate sample size).

8.2 Confounding

The presence of confounding can violate interval validity and introduce systematic error (bias) into a study. In epidemiology, confounding is where all or part of an apparent (observed) effect is due to some factors other than the primary exposure of interest. The result of this will mean a lack of exchangeability between exposed and nonexposed groups, which can bias the results of a study. For a factor to be a confounder, it must fulfil three criteria:

1. Be associated with both the exposure and the outcome.
2. Must be distributed unequally among the groups being compared.
3. Must *not* lie on the causal pathway between exposure and outcome.

There are several ways in which confounding can be controlled, either in the study design itself or during the analysis stage. At the study design stage, confounding can be controlled through randomization, restriction, or matching. At study analysis stage, confounding can be controlled through stratification, standardization, or multivariate modelling (Rothman 2012; Webb et al. 2017).

8.3 Selection Bias

Another type of systematic error that can introduce bias into a study is selection bias. There are several types of selection bias including volunteer bias, ascertainment or detection bias, and the healthy worker effect (Webb et al. 2017). Selection bias can also manifest in different ways in different study types. For example, in a RCT or cohort study, selection bias can occur as a result of differential losses to follow up (see ▶ Chap. 37, "Randomized Controlled Trials"). Once selection bias has occurred, it cannot be undone; therefore, the best way to control for selection bias is to have a rigorous study design with clear and appropriate inclusion criteria, while minimizing losses to follow-up and refusals (see also ▶ Chap. 3, "Quantitative Research").

8.4 Measurement Error

The last type of error that can introduce systematic error into a study is measurement error. Also known as observation or information error, this type of bias is any error in measuring or classifying the exposure, outcome, or both. Measurement of known confounders is also prone to this type of error. The measurement error can be either differential (systematic) or non-differential (random). Differential error is when there is a difference in the measurement between the two study groups, e.g., the exposed and unexposed. Differential error can influence the effect estimate either toward or away from the null, depending on the particular error. Non-differential error is measurement error that affects both study groups and will decrease the effect size of the study, that is, bring the value closer to the null (see also ▶ Chap. 3, "Quantitative Research").

9 Evidence Synthesis and Causation

9.1 Some Definitions

The principle aim of analytic (as distinct from descriptive) epidemiologic studies is to determine whether an exposure is a "cause" of an outcome. The rationale for identifying causes in epidemiology is to prevent or modify them and, therefore, prevent disease outcomes in populations. However, identifying whether an exposure is a "cause" of an outcome is often problematic, particularly in observational studies, which are always affected by bias and confounding to a greater or lesser degree (as discussed above).

There are many definitions of a "cause." But a definition that is commonly referred to in a range of epidemiologic textbooks is by Rothman et al. (2008, p. 6): A cause is "an antecedent event, condition, or characteristic that was necessary for the occurrence of the disease at the moment it occurred, given that other conditions are fixed." The concept of conditions that are "fixed" also relates to a more recent interpretation of causation in contemporary epidemiology that applies the concept of the "counterfactual ideal," summarized by Parascandola and Weed (2001, p. 906): A cause is "something that makes a difference in the outcome (or the probability of the outcome) when it is present compared with when it is absent, while all else is held constant." Of course, this condition of ceteris paribus ("all things being equal") is almost always impossible in the real world, and so epidemiologists attempt to approximate the conditions of the counterfactual ideal by using rigorous study design and accounting for bias and confounding. Epidemiologists aim to make comparison groups in a study *exchangeable* (Greenland and Robins 1986), in that the characteristics of an exposed group are the same in all respects as the unexposed group. Thus, any differences that might be observed between the "exposed" and the "unexposed" group – with all things being equal – can be attributed to that exposure factor only.

In epidemiology, "causes" are commonly classified as "sufficient," "necessary," or "component" causes in what has been referred to as the sufficient cause model (Rothman et al. 2008). A "sufficient" cause is a factor (or more usually a particular combination of factors) that will inevitably produce disease. A "component" cause is any component of a sufficient cause that is not an absolute requirement for the development of disease. A "necessary" cause is any sufficient cause, or component of a sufficient cause, required for the development of disease.

A useful way of conceptualizing these types of causes is the causal "pie" diagrams previously proposed by Rothman (1976). For example, Fig. 3 shows two classes of sufficient causes of a disease outcome. Sufficient cause I requires the presence of the component causes of A, B, C, D, and E, whereas sufficient cause II requires the presence of the component causes A, F, G, H, and I. The disease outcome might be ischemic heart disease, and the component causes could represent specific biological, behavioral, or environmental factors, such as "current smoking," "obesity," "genetics," "physical activity," "air pollution," and so forth.

In this example, there are only two sufficient causes of the disease outcome. Thus, the disease will not occur due to sufficient cause I if any of the component causes A,

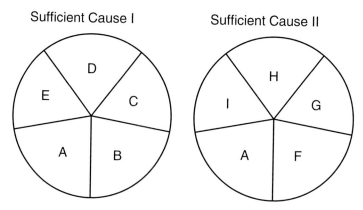

Fig. 3 Two classes of sufficient causes for a disease

B, C, D, or E are not present. Similarly, the disease will not occur due to sufficient cause II if any of the component causes A, F, G, H, or I are not present. In this example, A is a *necessary* cause for the disease, as it is present in both classes of sufficient cause. The sufficient cause model of causation implies that to prevent a given disease outcome, epidemiologists do not need to identify the *all*-component causes of disease; rather disease outcomes can be prevented by identifying single component causes.

9.2 Association Versus Causation

Analytic studies in epidemiology investigate associations between particular risk factors, or interventions, and health outcomes, and the nature of this association can be described using measures of disease frequency (prevalence, incidence) and association (relative and absolute risk), as discussed above. When epidemiologists observe an association between an exposure and an outcome (e.g., higher incidence of ischemic heart disease among current smokers compared to never smokers), the next step is to assess whether the observed association may be causal. Just because a particular exposure is associated with a disease does not automatically mean that the relationship is one of cause and effect. The observed association needs to be critically assessed in the light of how well the study was designed and implemented and how it compares with other evidence.

How might we do this? Epidemiology is an inherently *inductive* (as opposed to *deductive*) discipline – epidemiologists have hypotheses; they collect information and then arrive at conclusions based on observations. The process of induction based on repeated observation can never "prove" a hypothesis to be true. Epidemiology, like any scientific discipline, progresses through a process of elimination of bad hypotheses – a process of "conjecture" and "refutation" of hypotheses. Thus, in practice epidemiologists employ strategies to systematically evaluate whether, on balance, an observed association is more or less likely to be causal. A commonly employed framework for assessing cause and effect was proposed by Sir Austin Bradford Hill (2015). Hill proposed a series of "viewpoints" to consider when assessing an observed exposure-outcome association. Various modifications to this list have been suggested, and many elements remain cornerstones of judgment on whether an exposure is a "cause" of a disease outcome or whether an intervention is effective in treating a disease.

An abridged list of key elements of the Bradford Hill viewpoints is summarized in Table 1. "Temporality" refers to the necessity that the exposure precedes the outcome and is an inarguable criterion in assessing cause and effect. This might seem obvious; however, in practice it may not be straightforward to clearly determine whether an exposure preceded an outcome. For example, it might be observed in a study of stomach cancer that those with a diagnosis of cancer have lower levels of vitamin C. However, can we be sure that lower vitamin C levels preceded the onset of stomach cancer, or might low levels of vitamin C be a result of the disease process (Webb et al. 2017)? Establishing temporality is especially difficult, if not impossible, for

Table 1 The Bradford Hill "viewpoints"

Temporality	Does the exposure precede disease occurrence?
Strength of association	How strong is the association? What is the effect size between the exposure and the outcome?
Consistency	Is the same association evident across a range of studies and in different populations?
Dose-response	Is an increase (or decrease) in the level of exposure associated with an increase (or decrease) the level of the outcome?
Biological plausibility	Is there a plausible biological mechanism by which the exposure might cause the outcome?
Specificity	Is the exposure associated with a specific outcome or multiple outcomes?

cross-sectional study designs where information on both exposure and outcome are collected simultaneously or where exposure information is collected retrospectively.

"Consistency" refers to whether the observed association is similar to the associations observed in other studies or in different populations. If a similar association between exposure and outcome is observed across a number of studies, conducted at different points in time and using different methodologies, then the observed association is less likely to be artefactual. However, a lack of consistency with other studies does not necessarily rule out a causal relationship. Different results across studies may reflect variations in study design, measurement, or population characteristics. Consideration of sources of heterogeneity and reasons for inconsistences across studies can also be important information in any assessing an observed association between an exposure and an outcome.

"Dose-response" refers to whether the amount (or "dose") of the exposure is related to a change in the risk of the occurrence of the outcome. If an increase (or decrease) in the level of an exposure is associated with an increase (or decrease) in the level of the outcome, then this may be evidence for a causal association. However, the lack of a dose-response association does not necessarily preclude cause and effect, as there may be genuine cause-effect relationships where some threshold of exposure is required before it has an effect on the outcome of interest. For example, infectious diseases often have a threshold, below which the number of organisms does not cause the disease (Webb et al. 2017).

"Biological plausibility" refers to whether there is a plausible biological mechanism that can explain an observed association between the exposure and the outcome. If there is a clear biological mechanism, then this can add substantial weight to a causal argument. However, a lack of biological mechanism does not necessarily mean that the observed association is not causal. Given the complexity of human biology, it may be that a plausible biological mechanism for an observed exposure-outcome association is yet to be identified.

"Specificity" relates to the concept that a single "cause" has a single "effect." That is, the cause is specific to that effect and not multiple effects. This does not mean that causation cannot be attributed to instances where there are multiple effects of a single cause. For example, tobacco smoking is associated with a range of poor health outcomes. However, it is not expected that a single exposure variable would be linked to *all* outcomes. Thus, an expectation of a degree of specificity can be used to distinguish some causal hypotheses from noncausal hypotheses and whether the specificity relates to the outcome or to the exposure. For example, bicycle helmet use would be expected to reduce the risk of head injury but not other types of injury (specificity of outcome) (Webb et al. 2017), whereas a study of screening sigmoidoscopy to reduce mortality of colorectal tumors might be expected to reduce mortality-associated tumors within reach of the sigmoidoscope but not mortality from tumors located in other areas of the colon (specificity of exposure) (Rothman et al. 2008).

Importantly, each of these elements needs to be considered together. Integrating each of these perspectives can help to identify those associations that are more or less likely to be causal and the extent to which a hypothesis is sustained or refuted. This is necessarily a somewhat qualitative approach and reflects the inductivist nature of the discipline of epidemiology – ultimately whether an observed association between an exposure and outcome is causal is a matter of judgment and argument and a consideration of findings in the context of all available evidence on the topic (see Sect. 10 below). Bradford Hill (2015, p. 299) summarizes this succinctly: "Is there any other way of explaining the set of facts before us, is there any other answer equally, or more likely than cause and effect?"

10 Evidence Synthesis

While it is important to be able to appropriately interpret findings from a single study, this evidence alone is not going to provide a comprehensive picture of the effect of a given exposure on an outcome. It is important to review the literature more widely in order to identify similar patterns, and also highlight differences, across studies. A systematic review of the literature is a synthesis of *all* relevant primary research studies in response to a *focused* research question (Khan et al. 2003; Wilczynski and McKibbon 2013; see also ▶ Chap. 46, "Conducting a Systematic Review: A Practical Guide").

11 Current Challenges in Epidemiology

There are numerous historical examples of epidemiologic principles being applied to population health problems (perhaps the most prominent example being John Snow's investigation described above). However, epidemiology as a formalized discipline has emerged relatively recently in the last five to six decades. There have been many achievements of epidemiology in improving population health during this period, such as vaccination and control of infectious disease, family planning, fluoridation of drinking water, the recognition of tobacco use as a health hazard, safer work places, healthier mothers and babies, and declines in cardiovascular disease and a range of cancers, to name a few. The challenge for epidemiology is how it can continue to be useful, particularly with the emergence of chronic disease outcomes with complex etiology in both high-income and low- and middle-income populations (Miranda et al. 2008). Causes of disease occur at macro- and microlevels of populations, and the causes of disease are inextricably intertwined with social, economic, and political environments. This is of relevance across the life course (from birth to death), within populations (from genetic to societal influences), and across populations (from one nation or culture to the next).

A "risk factor" approach to understanding disease outcomes that has commonly been employed in the past is limited in the face of this complexity (Galea et al. 2009). Single risk factors from single studies will necessarily overlook the multivariable, and often hard to measure, causes of health outcomes, particularly, chronic diseases. The tools of epidemiology, especially those relating to the design and critique of studies and evidence synthesis, are central to the interpretation of the ever-expanding evidence base in population health.

12   Conclusion and Future Directions

Epidemiology is the scientific study of the distribution (frequency, pattern) and determinants (causes, risk factors) of health-related states and events (diseases, injuries) in specified populations (people in community or health facility), and the application of this study (since epidemiology is a subfield within public health) to the control or prevent threats to population health. There is also a need for epidemiology to adapt due to the now recognized multifactorial nature of most diseases, with the standard set of approaches to be augmented by approaches that better capture the complexity of human populations. This includes the need to begin thinking in interacting *systems* of exposures, outcomes, and interventions and employing systems-level approaches that have a long history in ecology, engineering, and computer science (Galea et al. 2009; El-Sayed and Galea 2017).

References

Bonita R, Beaglehole R, Kjellstrom T, World Health Organization. Basic epidemiology. 2nd ed. Geneva: World Health Organization; 2006.
Cates W. Epidemiology: applying principles to clinical practice. Contemp Obtetrics/Gynecol. 1982;20:147–61.
Dicker RC, Coronado FT, Koo D, Parrish RG. Principles of epidemiology in public health practice; an introduction to applied epidemiology and biostatistics. 3rd ed. Atlanta: Centers for Disease Control and Prevention; 2006.
Doll R, Hill AB. Smoking and carcinoma of the lung; preliminary report. Br Med J. 1950; 2(4682):739–48.
El-Sayed AM, Galea S. Systems science and population health. New York: Oxford University Press; 2017.
Friis RH. Epidemiology 101. 2nd ed. Burlington: Jones & Bartlett Learning; 2017.
Galea S, Riddle M, Kaplan GA. Causal thinking and complex system approaches in epidemiology. Int J Epidemiol. 2009;39(1):97–106.

Graham M. How do we know what we know? Epidemiology in health research. In: Liamputtong P, editor. Research methods in health: foundations for evidence-based practices. 3rd ed. Melbourne: Oxford University Press; 2017. p. 257–73.

Graunt J. Natural and political observations mentioned in a following index, and made upon the bills of mortality. In: Mathematical demography. New York: Springer; 1977. p. 11–20.

Greenland S, Robins JM. Identifiability, exchangeability, and epidemiologic confounding. Int J Epidemiol. 1986;15(3):412–8.

Hill AB. The environment and disease: association or causation? J R Soc Med. 2015;108(1):32–7.

Khan KS, Kunz R, Kleijnen J, Antes G. Five steps to conducting a systematic review. J R Soc Med. 2003;96(3):118–21.

Last JM. A dictionary of public health. New York: Oxford University Press; 2007.

Lee P. Assessing the health of populations: epidemiology in public health. In: Liamputtong P, editor. Public health: local and global perspectives. Port Melbourne: Cambridge University Press; 2016. p. 188–212.

Liamputtong P. Public health: an introduction to local and global contexts. In: Liamputtong P, editor. Public health: local and global perspectives. Port Melbourne: Cambridge University Press; 2016. p. 1–22.

Lilienfeld DE, Stolley PD. Foundations of epidemiology. New York: Oxford University Press; 1994.

Merrill RM. Introduction to epidemiology. Burlington: Jones & Bartlett Publishers; 2015. p. 2.

Miranda JJ, Kinra S, Casas JP, Davey Smith G, Ebrahim S. Non-communicable diseases in low-and middle-income countries: context, determinants and health policy. Trop Med Int Health. 2008;13(10):1225–34.

Parascandola M, Weed DL. Causation in epidemiology. J Epidemiol Community Health. 2001;55:905–12.

Rockhill B, Newman B, Weinberg C. Use and misuse of population attributable fractions. Am J Public Health. 1998;88(1):15–9.

Rothman KJ. Causes. Am J Epidemiol. 1976;104(6):587–92.

Rothman KJ. Epidemiology: an introduction. 2nd ed. New York: Oxford University Press; 2012.

Rothman KJ, Greenland S, Lash TL. Modern epidemiology. 3rd ed. Philadelphia: Lippincott Williams & Wilkins; 2008.

Snow J. On the mode of communication of cholera. London: John Churchill; 1855a.

Snow J. On the mode of communication of cholera. 2nd ed. London: Churchill; 1855b.

Snow J. Cholera, and the water supply in the south districts of London. Br Med J. 1857;1(42):864.

Webb P, Bain C, Page A. Essential epidemiology: an introduction for students and health professionals. 3rd ed. Cambridge, UK: Cambridge University Press; 2017.

Wilczynski N, McKibbon A. Chapter 3: Finding the evidence. In: Hoffman T, Bennett S, Del Mar C, editors. Evidence based practice across the health professions. 2nd ed. Chatswood: Churchill Livingstone; 2013.

Single-Case Designs

34

Breanne Byiers

Contents

B. Byiers (✉)
Department of Educational Psychology, University of Minnesota, Minneapolis, MN, USA
e-mail: byier001@umn.edu

© Springer Nature Singapore Pte Ltd. 2019
P. Liamputtong (ed.), *Handbook of Research Methods in Health Social Sciences*,
https://doi.org/10.1007/978-981-10-5251-4_92

Abstract

Single-case designs (also called single-case experimental designs) are system of research design strategies that can provide strong evidence of intervention effectiveness by using repeated measurement to establish each participant (or case) as his or her own control. The flexibility of the designs, and the focus on the individual as the unit of measurement, has led to an increased interest in the use of single-case design research in many areas of intervention research. The purpose of this chapter is to introduce the reader to the basic logic underlying the conduct and analysis of single-case design research by describing the fundamental features of this type of research, providing examples of several commonly used designs, and reviewing the guidelines for the visual analysis of single-case study data. Additionally, current areas of consensus and disagreement in the field of single-case design research will be discussed.

Keywords

Single-case designs · Single-subject designs · Small-N research · Intervention research · Idiographic research · Operant psychology

1   Introduction

Single-case designs (also called single-case experimental designs) are system of research design strategies that can provide strong evidence of intervention effectiveness by using repeated measurement to establish each participant (or case) as his or her own control. Although the methods were initially developed as tools for studying basic behavioral and physiological principles, the flexibility of the designs, and the focus on the individual as the unit of measurement, has led to an increased interest in the use of single-case design research in many areas of intervention research. Because single-case design studies do not require large numbers of participants, they can be extremely useful when the population of interest is very small or difficult to access, features that would preclude the use of large-scale randomized control trials (Rose 2017; see also ▶ Chap. 37, "Randomized Controlled Trials"). Further, single-case design strategies are better suited than group designs to answering research questions regarding the identification of the most effect procedures for a specific individual (or small group of people), rather than estimating an average effect within a population. Single-case design research can also be useful in the early stages of intervention development, as intervention strategies can be refined during the course of the study without compromising internal validity.

Although the term single-case implies that studies using these methods include only one participant, that is typically not the case. Some single-case designs actually require replication across participants as a form of experimental control. Even when inter-individual replication is not a design requirement, it is often scientifically important for documenting the generality of the effect (i.e., Sidman 1960). Most published studies using single-case design methods include a minimum of three participants for this reason. Further, it should be noted that, though the "case" in single-case often refers to individual participants or organisms, this is not a necessity. It is possible to use the single-case methods to study changes in the behavior of groups of individuals. There are several examples of educational studies that have used single-case designs to evaluate changes in behavior at the classroom- or school-level (e.g., Barrish et al. 1969; Colvin et al. 1997; Putnam et al. 2003). Some researchers have even assessed changes in behavior in larger, less formal groups of individuals using the logic of single-case designs. For example, Brownell et al. (1980) examined the effects of signage on physical activity by monitoring changes in the percent of people taking the stairs in a public subway station on each day of the study. The key factors in using single-case design strategies for studying groups are the identification of a measurement scheme that will result in a single value of the group's behavior on each measurement opportunity and documenting stability in the behavior of the group prior to implementing the intervention (both issues will be described in more detail in sections below).

The development of single-case design logic was pioneered primarily by researchers working in operant psychology in the 1960s (e.g., Sidman 1960; Baer et al. 1968). As such, many of the designs were developed to test the effects of operant behavioral principles, such as reinforcement schedules, and many of the conventions of single-case research can be traced back to that heritage. Since that time, however, single-case designs have been used to generate evidence in a range of fields, including education, social work, communication sciences, and rehabilitation studies (Rose 2017). In 2005, Horner and colleagues estimated that more than 45 scholarly journals had published single-case design studies (Horner et al. 2005).

2 Fundamentals of Single-Case Design Research

There are several key features that differentiate single-case design studies from other research methods, including quasi-experimental pre/post designs, and uncontrolled case studies. These features include: a focus on the individual case as the unit of analysis; repeated measurement of dependent variable(s); and repeated, systematic manipulations of the independent variable(s).

2.1 The Individual Case as the Unit of Analysis

In randomized controlled trials and other group-based research designs, data are most often analyzed and reported at the group level (see ▶ Chap. 37, "Randomized

Controlled Trials"). This means that the results reflect differences in average performance between groups; the degree to which these results represent the performance of any single individual within the group is usually unclear. In contrast, in single-case research, changes in the performance of each individual case is of primary concern. Because of this focus on the individual (as well as the small number of participants), reports of such studies typically include much more detailed descriptions of the individual participants than is possible in group studies. Further, because the research question is being answered at the individual level, single-case designs provide the opportunity to be responsive to individual differences in responding to interventions by modifying the design and intervention to achieve better results without compromising the internal validity of study (Gast and Ledford 2014; Rose 2017). See Morgan and Morgan (2014) for a more extensive comparison of the advantages and disadvantages of group and single-case design methods.

2.2   Repeated Measurement

In single-case design studies, each participant or case serves as its own control. This is achieved through repeated measurement of the dependent variable(s) within and across multiple conditions or phases. Most designs include a pre-intervention (or baseline) phase and at least one intervention phase. A post-intervention, or maintenance phase, is not uncommon in intervention research.

In a group-based intervention research, the performance of each participant is likely to be measured once at each relevant point in time (e.g., pre-intervention, post-intervention, and follow-up). This frequency of measurement is sufficient when the goal of the study is to evaluate differences in the average change in performance between groups. Although, due to measurement error and other factors, the value that each individual contributes to the group is not likely to be a perfect summary of their ability or performance in the domain of interest, if the errors are randomly distributed across members of the group, it is possible to get a reasonable estimate of group performance (and the stability of that estimate) by averaging the values from the individual members. When the distributions of the scores of two groups do not overlap, or overlap very little, there is evidence of a treatment effect. The logic of single-case design studies is similar, except that, rather than multiple measurements from individuals within a group at single points in time, the estimates of performance are based on multiple measurements within a single individual over time. By gathering multiple measurements of the individual within each phase of the study, it is possible to create a picture of how stable the individual's performance is from one measurement opportunity to the next, thereby providing a point of comparison against which to judge changes in performance due to the intervention or treatment. Because the passage of time is a key variable in single-case research, analysis of data needs to involve assessment of data patterns within and across study conditions to rule out the influence of extraneous time-related variables, such as history, repeated testing, and maturation. Ideally, variability between measurement occasions within each phase will be limited, as changes due to treatment effects are easily detected when the distributions of the scores in each phase do not overlap with each other.

2.3 Repeated, Systematic Manipulations of the Independent Variable(S)

The logic of experimental control in single-case research goes as such: If an experimenter repeatedly makes changes to the intervention conditions (e.g., a change from the baseline condition to the intervention condition and then back again), and each change coincides with changes in the participant's performance (especially if changes are large, immediate, and consistent), this provides convincing evidence of a causal relationship between the experimenter's manipulations and the observed changes. This is because it is highly unlikely that the changes in performance were random fluctuations when they consistently occur following experimental manipulations. Many of the potential threats to the internal validity (see Shadish et al. 2002 for a thorough discussion of threats to internal validity) are no longer plausible when such effects are demonstrated multiple times. The conventionally agreed-upon minimum requirement for the number of replications is three separate manipulations, staggered over time (e.g., Kratochwill et al. 2010). As is the case in all experimental research, these manipulations need to be controlled by the experimenter – repeated observations of changes in Y following changes in X do not necessarily demonstrate a causal relationship between the two variables; changes in both might be due to changes in factor Z. By systematically manipulating X at different points in time and continuing to see the systematic variation in Y, however, the experimenter rules out the potential influence of Z (and other, potentially unidentified factors).

3 Baseline Logic

Nearly all single-case designs include a "business as usual" or baseline condition. Baseline conditions allow researchers to document the patterns of responding before any changes to the independent variable, and thereby provide a basis for extrapolation or a prediction of how responding would have continued if study conditions had not been changed. Because the goal of baseline data is to extrapolate, the ideal patterns of responding will be stable over time, with limited variability from one point to the next. Examples of useful and problematic baseline patterns are presented in Fig. 1. In each panel, the predicted patterns of performance during the subsequent intervention phase based on baseline performance are represented by the shaded bars and dashed lines. The top (A) panel shows an ideal stable baseline pattern. In panel B, there is a systematic increasing trend, which is problematic in cases where the intervention is expected to produce increases in responding, as it would make it difficult to disentangle the effects of the existing trend and any true intervention

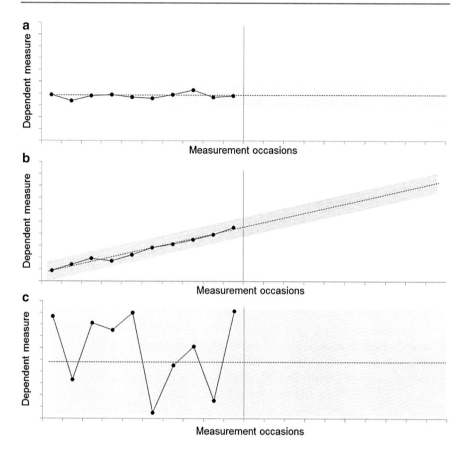

Fig. 1 Hypothetical examples of an ideal baseline data pattern (**a**), a baseline pattern with an increasing trend (**b**), and a variable baseline data pattern (**c**). The dashed lines represent the projected trend of performance during intervention based on the slope of the best-fitting line during the baseline phase. The shaded area represents the projected range of values

effects. If, on the other hand, the intervention was expected to produce decreases in responding, this trend would pose fewer interpretive challenges. In panel C, the wide range of values makes it difficult to predict what the next data point from the ones before, which might obscure any potential treatment effects.

Because baseline conditions are fundamental to the logic of most single-case research designs, it is essential that the researchers identify appropriate baseline conditions that will provide a reasonable test of performance under untreated conditions when designing and conducting a study. Because excessive variability in baseline performance can obscure treatment effects, researchers should consider possible sources of variability that can be controlled during all study sessions, such as changes in the time of day, location, materials present, personnel, etc.

4   Matching Research Questions to Design

In designing single-case research studies, it is essential that researchers select the appropriate research design for the specific research question and dependent variable. The first two factors to consider when making that selection are whether (a) the research question involves demonstrating the effects of a single intervention or comparing the effects of two or more interventions, and (b) the target behavior or process is reversible.

4.1   Demonstration Versus Comparison

There are two primary categories of research questions that can be answered with single-case design methodology: questions regarding whether or not a treatment or intervention condition is effective (compared to no intervention; i.e., demonstration questions), and questions regarding the relative effects of two or more treatment or intervention conditions (i.e., comparison questions). Each of these question types has a corresponding set of design strategies.

4.2   Reversibility of the Target Behavior

An additional factor distinguishing different research designs is whether it assumes the "reversibility" of the dependent variable. The analogy of the intervention or treatment as light switch is often helpful when deciding whether or not a behavior is likely to be reversible. For example, consider two pharmaceutical interventions: insulin treatment for diabetic management and antibiotics for treatment of infection. In the first case, the symptoms would be expected to be reversible: if the patient stops taking the medication, it is likely that there would be measurable changes in her/his blood sugar levels. Therefore, the medication could be used like a light switch to turn on and off the symptoms. In the second case, the symptoms being treated are likely not reversible: once the infection has been adequately suppressed by the medication, it is unlikely to return after the patient stops taking the medication. Therefore, in this case, the light switch analogy does not work, as symptoms can be eliminated but cannot be "turned back on". Examples of reversible and nonreversible dependent variables can likely be identified in all fields of intervention research, and the expectation of reversal should be based on knowledge of the construct being studied and pilot data when available. Even when behaviors or processes should, in theory, be reversible, it may be difficult to achieve reversal in real-world contexts. For example, when interventions are being implemented by stakeholders, such as parents, teachers, or other caregivers in naturalistic environments, those stakeholders may be unable, or unwilling, to return to baseline conditions after seeing the effects of the intervention first-hand. When it is reasonable to expect that the target behavior or process being studied can be reversed, design strategies that incorporate reversals for internal validity evidence can provide the most stringent tests of experimental control in single-case studies. Alternate design strategies for nonreversible behaviors will also be discussed below.

5   Common Design Types

5.1   Demonstration Designs for Reversible Behaviors

The ABAB Design. The ABAB design, sometimes referred to as a reversal or withdrawal design, involves alternating a baseline (A) condition with an intervention (B) condition in sequential phases. Following the initial AB demonstration of the effects of the treatment, the intervention is withdrawn in a second baseline phase, which is expected to result in a decrement in performance, which should then be reversed again in a final intervention phase. Each phase change (A to B, B to A, A to B) represents one opportunity to document a treatment effect, and as such, a successful ABAB study will result in the minimum three demonstrations of effect necessary to establish experimental control.

Figure 2 shows an example of a hypothetical ABAB study. In this example, data in the first baseline phase are relatively stable. There is an immediate change in level following the introduction of the intervention, followed by an increasing trend during the next several sessions. After performance has stabilized, the intervention is withdrawn, and an immediate change in level and trend are observed again. A large and immediate change in level occurs again in the final phase. Taken together, these results would provide strong evidence of a treatment effect as there are three, unequivocal demonstrations of the effects of manipulations of the independent variable.

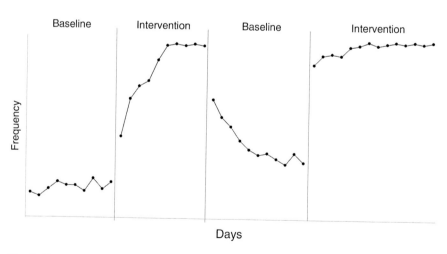

Fig. 2 Hypothetical results of an ABAB study (A = baseline condition, B = intervention condition) in which the intervention is designed to increase the frequency of the target behavior

5.2   Demonstration Designs for Nonreversible Behaviors

Multiple Baseline Design. This design is the most commonly used in contemporary intervention research (Smith 2012). One primary reason for this popularity is that the multiple baseline design does not require reversible behaviors. Instead, the multiple baseline design uses three or more separate series or baselines to provide the necessary independent demonstrations of effect. Each series may represent a different individual or case, or the effects can be replicated within a participant by selecting different target behaviors or contexts/environments. The key factors in identifying an appropriate set of series for a multiple baseline design is that (a) each needs to be expected to respond similarly to the same intervention conditions, and (b) each must be independent from the others (i.e., if the intervention is implemented in the first series, there are no concurrent changes in the others). Independence is relatively simple when each series is a separate participant but can be more challenging when series are defined by different responses, contexts, or tasks within a participant, as the likelihood that the effects of treatment for one response or in one context will affect the other responses or contexts is higher. Nevertheless, many researchers have successfully completed within-participant multiple baseline design studies that have shown strong experimental control. For example, Hersen and Bellack (1976) evaluated the effects of a social skills training package that was sequentially introduced across four target behaviors (i.e., eye contact during speech, frequency of speech disruptions, frequency of smiles, and appropriate affect) with an individual diagnosed with schizophrenia. An example of a multiple baselines across stimulus conditions was reported by Stark et al. (1990) who examined the effects of a treatment package for increasing caloric intake for children with cystic fibrosis and malnourishment that was implemented across four eating opportunities (snack and three meals).

The multiple baseline design can be conceptualized as a set of stacked AB designs, in which a baseline, or "A" phase, is followed by an intervention, or "B" phase, with no reversal replication within each series. Data collection across all of the series is conducted simultaneously, but with the intervention or treatment is introduced at different points in time across the three series to provide independent demonstrations of effect. For all of the series, data collection begins with a baseline condition. When sufficiently stable baseline data have been collected in all of the series, the intervention is implemented in the first series, and baseline data collection continues in the others. Once an effect is detected in the first series, the intervention is introduced in the second, and so on until the intervention has been implemented across all of the series.

A hypothetical multiple baseline across settings study is depicted in Fig. 3. In this case, there is a small but immediate change in level, and a change in trend observed when the intervention is introduced in Setting 1. Looking vertically across the three series, no concurrent changes occur in the other two series, which supports the assumption that the series are independent. A large immediate change is again observed when the intervention is introduced in Setting 2, with no concurrent change in the series for Setting 3. The same type of pattern is observed with the introduction of the intervention in Setting 3. Therefore, this design would provide the three

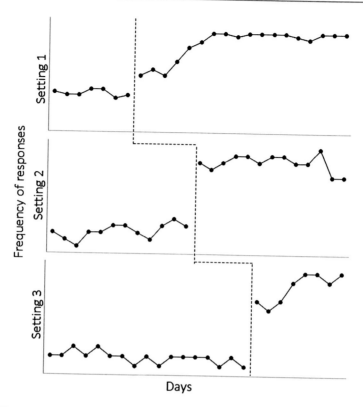

Fig. 3 Hypothetical results of a multiple baselines across settings study in which the intervention is designed to increase the frequency of the target behavior

independent demonstrations of effect across time that is necessary for demonstrating experimental control.

A potential limitation of the multiple baseline design is that the staggered introduction of the intervention results in longer baseline phases for the participants/responses/settings assigned to the lower series. In some cases, this may mean collecting dozens of data points prior to implementing any interventions or procedures, which may be resource intensive and potentially frustrating for participants, clinicians, and researchers. The multiple probe design was introduced to alleviate some of these issues by reducing the number of data collection opportunities required during the baseline phase by collecting data intermittently prior to the introduction of the intervention (see Horner and Baer 1978, for a full discussion).

5.3 Comparison Designs for Reversible Behaviors

Alternating Treatments Design. The most commonly used design for evaluating the relative effects of two or more treatment conditions is the alternating treatments

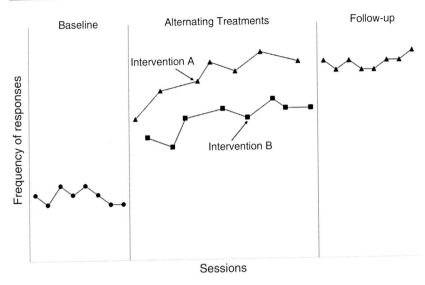

Fig. 4 Hypothetical results of an alternating treatments study comparing the degree to which two interventions increase the frequency of the target behavior relative to baseline levels and each other. This design also includes a follow-up condition evaluating the effects of the "winning" Intervention A to ensure that it remains effective without sessions of Intervention B

design (ATD; Barlow and Hayes 1979). ATD studies involve rapid and repeated manipulation of two or more conditions across observations (i.e., sessions or days), typically in a semi-randomized order. For example, Ahearn and colleagues (1996) compared the effects of two treatment packages (nonremoval of spoon and physical guidance) on food consumption among three children with chronic food refusal. The interventionists alternated between the two conditions across brief (20 spoon presentations) sessions in a semi-randomized order over several days. The results suggested that both treatments were effective in increasing food acceptance, although the physical guidance package was associated with shorter meals and was rated as more acceptable by parents.

Data from a hypothetical ATD study are presented in Fig. 4. In this example, the study begins with a baseline phase, in which responding is relatively low and stable, and continues with an alternating treatment phase, in which the two interventions (A and B) are randomly assigned to each session in the phase. Finally, a follow-up phase is conducted in which the "winning" intervention is tested alone.

The process of visual analysis for ATD studies is slightly different than in sequential phase designs, as the primary comparison of interest is between the data paths representing the two interventions within the same phase. Rather than looking for a level change between phases, visual analysts should look for separation between the data paths. In the example graph, the two data paths never cross, suggesting that performance under the conditions of Intervention A is consistently higher than those of Intervention B.

The follow-up phase, although not necessary, is recommended for this type of design, as the process of alternating between the two intervention conditions may have affect performance in one or both of the conditions, an effect called multiple treatment interference. As it is often difficult to ascertain the effects that are due to multiple treatment interference, including a follow-up phase in which only the most effective intervention is implemented over a longer period of time can be useful in demonstrating that the intervention remains effective on its own (see Higgins Hains and Baer 1989 for a more detailed discussion of this and related issues).

5.4 Comparison Designs for Nonreversible Behaviors

Adapted Alternating Treatments Designs. In the example just described, it is reasonable to assume that the response of food acceptance would be a "reversible" response, as it is likely to vary from session to session based on the intervention conditions in place. For nonreversible behaviors, however, the alternating treatments design is not a feasible design option for comparing the effects of multiple conditions. The adapted alternating treatments design (AATD; Sindelar et al. 1985) addresses this issue by assigning equivalent sets of stimulus items to each intervention or treatment condition. This design is used most frequently in educational settings to evaluate the effects of instructional practices when it is assumed that the responses being learned are not reversible. For example, Schlosser and Blischak (2004) evaluated the effects of three different feedback conditions of the spelling accuracy of four children with autism spectrum disorder. Because it was expected that the children would continue to spell each word correctly once it was learned, the researchers identified sets of target words that were of comparable difficulty to be assigned to each of the conditions for each child. By evaluating the condition associated with the fastest acquisition of the target words across sessions, the researchers were able to identify which type of feedback was most effective for each child. An important challenge for AATD studies, however, is ensuring comparability of the stimulus sets across conditions.

5.5 Combined Designs

The ways in which the more traditional single-case designs can be combined is nearly limitless. For example, in the Ahearn et al. (1996) study described in the previous section, the ATD comparison was embedded within multiple-baselines across participants design, such that the introduction of the intervention comparison phase was staggered across participants. Other studies have embedded reversal/withdrawal elements within multiple-baseline designs to provide further evidence of functional relations. See Kazdin (2011) for an extensive discussion of several additional design options for addressing different research questions.

6   Assessment of Single-Case Design Data

6.1   Graphical Presentation of Raw Data

Historically, single-case design researchers eschewed statistical analysis of data in favor of graphical presentation and visual analysis of the results. Although there is some evidence that opinions have generally shifted in favor of incorporating quantitative analyses of single-case design data in published studies, nearly all groups writing on the subject continue to support the concept that visual inspection of the raw data is an important step in the evaluation of single-case design results (e.g., Smith 2012). The conventional depiction of the data includes time (measures in sessions, days, or weeks, for example) along the horizontal axis, and the dependent measure along the vertical axis. The raw data are then plotted as individual data points for each measurement session (connected with a line within phases), and vertical lines representing changes in study conditions.

By presenting the raw data in this way, all consumers of the study results have access to all of the relevant information about participant performance. It has also been argued that using visual analysis of graphically presented data rather than relying on statistical analyses means that only large, incontrovertible effects are likely to be identified, resulting in more robust clinical interventions (i.e., Parsonson and Baer 1978). There is some evidence, however, that certain qualities of data patterns may affect the sensitivity of visual analysis, resulting in increased type I error rates (e.g., Jones et al. 1978; Matyas and Greenwood 1990; Fisch 2001), although the validity of these types of studies have been questioned by some single-case researchers (e.g., Parsonson and Baer 1992). The issues of the primacy of visual analysis, as well as its reliability, remain somewhat contentious among researchers (Manolov et al. 2014).

Although there may be some disagreement among researchers regarding whether visual analysis should be the only method of analysis for single-case experimental data, most agree that it should play a role in the interpretation of study results. As previously noted, a causal relation can be demonstrated in a single-case design study when there is a minimum of three demonstrations of changes in the dependent variable following manipulations of the independent variable. A demonstration of effect is documented when performance in one condition differs substantially from what would have been expected based on performance in the previous condition. It is important to note that, in order to minimize the potential confounding effects of time-based threats to internal validity, only data patterns from temporally adjacent phases should be compared directly, although the consistency of the results across all phases of the study should be evaluated globally.

6.2   Visual Analysis Guidelines

In his seminal textbook on single-case design methodology, Alan Kazdin (1982, p. 233) states that "in cases where intervention effects are very strong, one need not carefully scrutinize or enumerate the criteria that underlie the judgment that the effects are veridical." When effects are less dramatic, however, several factors need to be considered when making a judgment regarding presence and magnitude of treatment effects. Specifically, researchers and consumers should consider six major features: (1) change in level/mean, (2) change in trend, (3) change in variability, (4) immediacy of changes, (5) degree of overlap between phases, and (6) consistency of the effects across similar phases (Parsonson and Baer 1978; Kratochwill et al. 2010; Kazdin 2011; Horner et al. 2012).

Changes in mean and *changes in level* are both indicators of the magnitude of a treatment effect that refer to the changes in the value of the dependent variable between phases. Changes in mean values are fairly intuitive, as it refers to shifts in the average performance between phases. Figs. 5 and 6 both show changes in means between phases. Changes in level occur when there is a discontinuity in the data paths across phases resulting from an immediate change in the value of the dependent variable that coincides with the phase change, such as in Fig. 5. In contrast, Fig. 6 shows a change in means without a corresponding change in level, as the first data point in the intervention phase does not differ from the last data point in the baseline phase. *Trend* refers to systematic within-phase patterns of increasing or decreasing

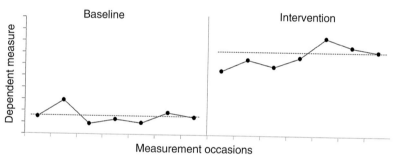

Fig. 5 Hypothetical data showing a change in means (represented by the horizontal dashed lines) and a change in level (the discontinuity identified by the arrow) between the baseline and intervention conditions

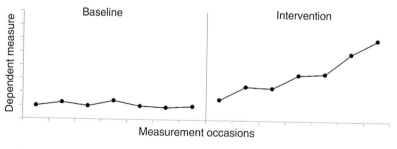

Fig. 6 Hypothetical data showing a change in trend between the baseline phase (no trend) and the intervention phase (increasing trend)

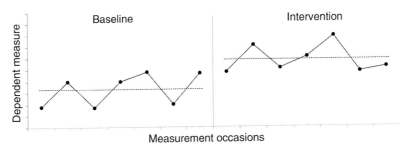

Fig. 7 Hypothetical data showing a change in means (represented by the horizontal dashed lines), but no corresponding change in level, as there is no discontinuity in the data pattern between the baseline and intervention phases

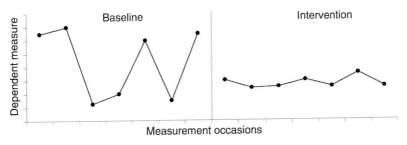

Fig. 8 Hypothetical data showing a change in variability between the baseline phase (variable) and the intervention phase (stable)

values, or the slope of the best-fitting straight line. Fig. 7 shows an example of a between-phase change in trend, as there is no trend during the baseline phase, followed by a systematic increasing trend in the intervention phase. *Variability* refers to the instability of the values of the dependent variable from one measurement point to the next, often summarized by the range or standard deviation of values within a phase. Fig. 8 shows a between-phase change in variability in which a variable pattern of responding during the baseline phase leads into a stable pattern of responding during the intervention phase (without any obvious corresponding changes in level or trend). *Immediacy of change* refers to how long it takes to see an effect following a change in phase conditions. Whereas Figs. 2, 3, 4, 5, 6, and 7 all document changes that occur immediately following a phase change; Fig. 9 shows a delayed change in level. Evidence of a treatment effect is stronger when the effects are immediate, but sometimes delayed effects are anticipated based on the nature of the behavior being evaluated. *Overlap* refers to the degree to which the range of values in one phase overlaps with the range of values from the phase against which it is being compared. For example, in Fig. 5, all of the intervention phase values fall above the range of values observed during the baseline phase. In contrast, in Fig. 8, none of the intervention phase values fall outside of the range of the baseline values, and in Fig. 6, three of the seven (43%) of the intervention phase values fall within the range

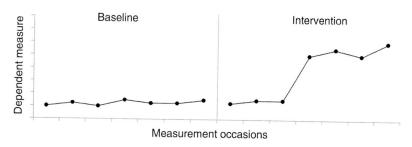

Fig. 9 Hypothetical data showing a delayed change in level during the intervention phase. Delayed effects typically reduce the confidence that any changes can be directly attributed to the effects of the intervention

Table 1 Factors to consider during visual analysis of single-case design data

Factors that increase confidence in experimental control and treatment effectiveness
Assessment of within- and between-condition data patterns
1. A sufficient number of data points in each phase/condition have been collected to allow assessment of data patterns
2. Within-phase data show limited variability
3. Baseline data show little or no trend
4. Changes in level, trend and/or variability occur immediately following manipulation of the independent variable
5. Little to no overlap in the range of values across conditions
6. The data pattern of the intervention phase(s) differ substantially from the path projected based on the baseline data
Overall results
1. There are at least three unambiguous demonstrations of the experimental effect that occur at different points in time
2. The magnitude, direction, and latency of changes are consistent across phase changes of the same type
3. The change in performance represents a clinically-meaningful improvement based on the target behavior and participant characteristics
Procedural considerations
1. Measurement of the dependent variable(s) has been adequately described and reliability data are provided
2. The independent variable (i.e., the intervention) has been adequately described, and the fidelity of implementation has been measured and reported
3. Plausible alternative explanations of the results have been considered and adequately addressed
Design-specific evidence of functional relations and treatment effectiveness
Multiple-baseline designs
1. There is no evidence of treatment diffusion across series: There is no evidence of changes during baseline for series in which the independent has not yet been manipulated (i.e., prior to the implementation of the intervention)
Alternating treatments designs
1. There is limited overlap between conditions, as defined by clear separation between data paths and few instances of data paths crossing

of the baseline values. In general, less overlap between conditions provides stronger evidence of experimental control. Finally, *consistency of the effects across similar phases* refers to the degree to which the same patterns of data are obtained every time the same type of change is made to the independent variable. For example, although the delay observed in Fig. 9 may call into question the functional relationship between the intervention and changes in the dependent variable, if this pattern of delayed effects were replicated within or across participants, this replication would minimize this concern.

All six of these features need to be considered in tandem when evaluating the results of a single-case experimental study, as treatment effects may result in simultaneous changes to multiple features of the data patterns (Table 1) (see Kratochwill et al. 2010; Kazdin 2011; Horner et al. 2012; Parsonson and Baer 1978 for more extensive discussions of the process of visual analysis).

7   Other Considerations for Designing and Conducting Single-Case Research

7.1   Selection of Dependent Measures

Selection of appropriate dependent measures is an essential component of any intervention study. In the context of single-case research, the researcher must select dependent variables that can be measured repeatedly over time, with minimal changes in performance due to repeated testing. Historically, single-case design studies have used directly observed counts of behavior as the primary dependent measures. In observational measurement, one or more members of the research team observes the study participant(s) during each scheduled study session and records the relevant feature(s), such as frequency, duration, latency, or force of the target behavior. Unlike other measurement systems, such as paper- or computer-based testing, or self-report measures, directly observed behaviors may be less prone to biases due to repeated testing when appropriate safeguards are in place. Monitoring of interobserver agreement throughout a study is one essential safeguard. This process involves having two or more trained, independent observers record data on the target behavior during a random sample of sessions during each phase. The scores from the observers are compared to ensure reliability of the measures. Several textbooks provide extensive discussions of interobserver agreement procedures (e.g., Barlow et al. 2009; Kazdin 2011; Gast and Ledford 2014).

Other sources of data can be used in the context of single-case design research, as long as the source can provide a single score for each target behavior or process for each measurement occasion, the scores produced are reliable, valid, and relatively stable over time when measured repeatedly. Wearable and remote sensors that can collect hundreds or even thousands of data points for a given metric in a day may provide an opportunity to expand the reach of single-case design research (Dallery et al. 2013). The historic focus on direct observation of behavior has meant that single-case design studies have been extremely resource-intense to conduct: multiple observers needed extensive training to reliably record occurrence and nonoccurrence of the target behavior, and at least one observer needed to be physically present at each measurement opportunity. With remote sensors, it may be possible to evaluate any number of clinically important variables, such as sleep quality, heart rate, physical activity, and medication adherence, among many others. Prior to implementing such measures in the context of single-case design research, however, researchers should consider the stability of the measures over time, the likelihood of participants' adhering to the use of the devices as designed, and the validity of the measures being collected.

7.2   Procedural Fidelity

Just as the measurement of the dependent variables need to be carefully operationalized and validated, implementation of the independent variable should be diligently described and documented. This includes providing detailed descriptions of the procedures in place during each of the different conditions in the study, including the baseline condition. This practice is important as interpretation of the findings of intervention research is predicated on the assumption that baseline and intervention conditions differ only by the variables relevant to the study. In addition adequate descriptions of the conditions, systematic measurement of the fidelity with which those are conditions are implemented is necessary. As described by Wolery (1994), collection of procedural fidelity data plays at least three roles in intervention research. First, it is used to monitor the occurrence of relevant variables, such that any unplanned changes in the implementation of variables can be detected. This gives researchers the opportunity to correct such shifts and may also be helpful in explaining unexpected variability in the data. Second, it provides documentation that the experimental conditions occurred as planned, which provides credibility for consumers of the research, and also facilitates replication. Finally, it provides a base from which generalizations and recommendations of the findings can be made. Specifically, by documenting procedural fidelity of an intervention, researchers can communicate with others about the conditions that need to be in effect for the intervention to have a high likelihood of success. Ledford and Gast (2014) provides a more thorough discussion of how to design and implement measurement systems for monitoring procedural fidelity.

7.3   Response-Guided Decision-Making Versus Randomization

Single-case researchers have historically used response-guided decision-making when conducting studies. Response-guided decision-making refers to using patterns in the data as they emerge to make decisions about when to make changes to the independent variable during the course of the study. The main advantage of this strategy is that the research team has the power to avoid making phase change decisions when patterns in the data might obscure treatment effects. For example, researchers might opt to extend a baseline phase if the responding is variable or shows a trend in the direction of the expected treatment effect. Although these types of decisions frequently result in a more convincing demonstration of the effects of the intervention, it has been argued that response-guided decisions may compromise the internal validity of single-case design studies by capitalizing on chance fluctuations in the dependent variable. Specifically, it has been demonstrated that response-guided experimentation allows the researchers the opportunity for repeated visual tests of the data as they are collected, which increases the likelihood of Type I error (Allison et al. 1992).

Proponents of response-guided methods argue that the ability to be responsive to patterns in the data means that clinicians and researchers can modify treatments as needed throughout the study in order to individualize treatment components. Further, as visual analysis remains the gold standard method of analysis for single-case experimental data, some degree of responsiveness to data patterns may be necessary to avoid potential ambiguous results that can be introduced when phase change decisions are not response-guided (e.g., Ferron and Ware 1994). Many behavior-analytically oriented researchers would also argue that systematic replication alone of the effect is sufficient for experimental control, as explained by Donald Baer (1977, p. 168): "In the individual-subject paradigm, a judicious defense against chance is available. [...] If behavior repeated under the repeated "A's" is repetitively different from behavior repeated under the repeated "B's", the scientist will conclude that such consistency cannot be a product of chance. After all, it has been repeated quite repetitively".

Others argue that replication is insufficient to guard against the dangers of response-guided decision-making, leading several groups to call for incorporating randomization into single-case design studies (see Dugard et al. 2012; Kratochwill and Levin 2014 for more detailed discussions). As stated by Dugard et al. (2012, p. 137): "The reason why internal validity cannot be established when the timing of the intervention is response-guided is that there is no way of telling what the baseline would have looked like if it has been allowed to continue for a few more observations. We know that it probably would have appeared less stable if we had stopped it a few observations earlier; otherwise, we would have stopped it then. It might also have looked less stable if we had waited for a few more baseline observations before introducing the intervention. After all, we stopped when things seemed to be going particularly well". Further, including randomization elements in a study design means that the data can more reasonably be subjected to inferential hypothesis tests. A full discussion of the current state of research and thinking on hypothesis testing in single-case designs is beyond the scope of the current chapter but is currently an active domain of discussion in the literature. The interested reader is referred to the 2014 special issue on analysis and meta-analysis of single-case designs in the *Journal of School Psychology* for a broad sampling of approaches and opinions (Volume 52, Issue 2).

Ultimately, the decision to use response-guided decision-making versus randomized assignment will likely depend on the goals of the research team, the research question being asked, and the training and philosophical orientation of the research team. One potentially interesting methodological compromise involves "masked" visual analysis, in which an evaluation team who is blind to the current intervention conditions evaluates the data for potential treatment effects (e.g., Byun et al. 2017). Although intriguing, few studies using this strategy have been published to date, and additional work is needed to determine the utility of these methods in real-life intervention contexts.

8 Conclusions and Future Directions

Single-case design research can be used to address a number of different types of research questions regarding treatment effectiveness that are likely to be of interest to researchers working in clinical settings. Important quality control features of well-designed single-case design studies include an appropriate match between the research question and features of the target behavior and the design selected, reliable and valid measurement of the independent and dependent variables, and a minimum of three unequivocal demonstrations of effect that are staggered over time. Ongoing debates regarding single-case design research include whether randomization is a necessary component of all single-case design studies, and identifying the most appropriate quantitative analytic strategies for evaluating single-case design data for inferential testing and meta-analyses.

References

Ahearn WH, Kerwin ML, Eicher PS, Shantz J, Swearingin W. An alternating treatments comparison of two intensive interventions for food refusal. J Appl Behav Anal. 1996;29(3):321–32.

Allison DB, Franklin RD, Heshka S. Reflections on visual inspection, response guided experimentation, and type I error rate in single-case designs. J Exp Educ. 1992;61(1):45–51.

Baer DM. Perhaps it would be better not to know everything. J Appl Behav Anal. 1977;10:167–72.

Baer DM, Wolf MM, Risley TR. Some current dimensions of applied behavior analysis. J Appl Behav Anal. 1968;1:91–7.

Barlow DH, Hayes SC. Alternating treatments design: one strategy for comparing the effects of two treatments in a single subject. J Appl Behav Anal. 1979;12(2):199–210.

Barlow DH, Nock M, Hersen M. Single-case experimental designs. 3rd ed. 2009.

Barrish HH, Saunders M, Wolf MM. Good behavior game: effects of individual contingencies for group consequences on disruptive behavior in a classroom. J Appl Behav Anal. 1969;2:119–24.

Brownell KD, Stunkard AJ, Albaum JM. Evaluation and modification of exercise patterns in the natural environment. Am J Psychiatr. 1980;137:1540–5.

Byun TM, Hitchcock ER, Ferron J. Masked visual analysis: minimizing type I error in visually guided single-case design for communication disorder. J Speech Lang Hear Res. 2017;60: 1455–66.

Colvin G, Sugai G, Good RJ, Lee YY. Using active supervision and pre-correction to improve transition behaviors in an elementary school. Sch Psychol Q. 1997;12:344–63.

Dallery J, Cassidy RN, Raiff BR. Single-case experimental designs to evaluate novel technology-based health interventions. J Med Internet Res. 2013;15:e22.

Dugard P, File P, Todman J. Single-case and small-n experimental designs: a practical guide to randomization tests. New York: Routledge; 2012.

Ferron J, Ware W. Using randomization tests with responsive single-case designs. Behav Res Ther. 1994;32:787–91.

Fisch GS. Evaluating data from behavioral analysis: visual inspection or statistical models? Behav Process. 2001;54:137–54.

Gast DL, Ledford J. Single case research methodology. 2nd ed. New York: Routledge; 2014.

Hersen M, Bellack AS. A multiple-baseline analysis of social-skills training in chronic schizophrenics. J Appl Behav Anal. 1976;9(3):239–45.

Higgins Hains AH, Baer DM. Interaction effects in multielement designs: inevitable, desirable, and ignorable. J Appl Behav Anal. 1989;22:57–69.

Horner RD, Baer DM. Multiple-probe technique: a variation of the multiple baseline. J Appl Behav Anal. 1978;11:189–96.

Horner RH, Carr EG, Halle J, McGee G, Odom S, Wolery M. The use of single subject research to identify evidence-based practice in special education. Except Child. 2005;71:165–79.

Horner RH, Swaminathan H, Sugai G, Smolkowski K. Considerations for the systematic analysis and use of single-case research. Educ Treat Child. 2012;35(2):269–90.

Jones RR, Weinrott MR, Vaught RS. Effects of serial dependency on the agreement between visual and statistical inference. J Appl Behav Anal. 1978;11:277–83.

Kazdin AE. Single-case experimental designs: methods for clinical and applied settings. New York: Oxford University Press; 1982.

Kazdin AE. Single-case research designs: methods for clinical and applied settings. New York: Oxford University Press; 2011.

Kratochwill TR, Hitchcock J, Horner RH, Levin JR, Odom SL, Rindskopf DM, Shadish WR. Single-case designs technical documentation. 2010. Retrieved from What Works Clearinghouse website: http://ies.ed.gov/ncee/wwc/pdf/wwc_scd.pdf.

Kratochwill TR, Levin JR. Enhancing the scientific credibility of single-case intervention research: randomization to the rescue. In: Kratochwill TR, Levin JR, editors. Single-case intervention research: methodological and statistical advances. Washington, DC: American Psychological Association; 2014. p. 53–90.

Ledford JR, Gast DL. Measuring procedural fidelity in behavioural research. Neuropsychol Rehabil. 2014;24:332–48.

Manolov R, Gast DL, Perdices M, Evans JJ. Single-case experimental designs: reflections on conduct and analysis. Neuropsychol Rehabil. 2014;24(3–4):634.

Matyas TA, Greenwood KM. Visual analysis of single-case time series: effects of variability, serial dependence, and magnitude of intervention effects. J Appl Behav Anal. 1990;23:341–51.

Morgan DL, Morgan RK. Comparing group and single-case designs. In: Morgan DL, Morgan RK, editors. Single-case research methods for the behavioral and health sciences. Thousand Oaks: SAGE; 2014.

Parsonson BS, Baer DM. The analysis and presentation of graphic data. In: Kratochwill T, editor. Single subject research. New York: Academic; 1978. p. 101–66.

Parsonson BS, Baer DM. The visual analysis of data, and current research into stimuli controlling it. In: Kratochwill TR, Levin JR, editors. Single-case research design and analysis: new directions for psychology and education. Hillsdale: Lawrence Erlbaum Associates; 1992. p. 15–40.

Putnam RF, Handler MW, Ramirez-Platt CM, Luiselli JK. Improving student bus-riding behavior through a whole-school intervention. J Appl Behav Anal. 2003;36:583–90.

Rose M. Single-subject experimental designs in health research. In: Liamputtong P, editor. Research methods in health: foundations for evidence-based practice. Melbourne: Oxford University Press; 2017. p. 217–34.

Schlosser RW, Blischak DM. Effects of speech and print feedback on spelling by children with autism. J Speech Lang Hear Res. 2004;47(4):848.

Shadish WR, Cook TD, Campbell DT. Experimental and quasi-experimental designs for generalized causal inference. Boston: Houghton Mifflin; 2002.

Sidman M. Tactics of scientific research. Boston: Authors Cooperative, Inc; 1960.

Sindelar P, Rosenberg M, Wilson R. An adapted alternating treatments design for instructional research. Educ Treat Child. 1985;8(1):67–76.

Smith JD. Single-case experimental designs: a systematic review of published research and current standards. Psychol Methods. 2012;17:510–50.

Stark LJ, Bowen AM, Tyc VL, Evans S, Passero MA. A behavioral approach to increasing calorie consumption in children with cystic fibrosis. J Pediatr Psychol. 1990;15:309–26.

Wolery M. Procedural fidelity: a reminder of its functions. J Behav Educ. 1994;4:381–6.

Longitudinal Study Designs

35

Stewart J. Anderson

Contents

Abstract

Longitudinal study designs are implemented when one or more responses are measured repeatedly on the same individual or experimental unit. These designs often seek to characterize time trajectories for cohorts and individuals within cohorts. Three broad categories of longitudinal designs include (1) repeated measures or growth curve designs, where multiple responses for each individual are observed over time or space under the same intervention or other conditions; (2) crossover designs, where individual responses are measured over *sequences* of interventions so that individuals each "cross over" from one intervention to another; and (3) follow-up studies, where individuals in a cohort are followed

S. J. Anderson (✉)
Department of Biostatistics, University of Pittsburgh Graduate School of Public Health,
Pittsburgh, PA, USA
e-mail: sja@pitt.edu

© Springer Nature Singapore Pte Ltd. 2019
P. Liamputtong (ed.), *Handbook of Research Methods in Health Social Sciences*,
https://doi.org/10.1007/978-981-10-5251-4_70

until the time that they either have an "event" (e.g., death, depressive episode) or have not had an event but have no further follow-up information. Longitudinal designs may be either *randomized* where individuals are randomly assigned into different groups or *observational* where individuals from different well-defined groups are observed over time. In this chapter, I briefly discuss the nature of each of the three designs above and more deeply explore visualization and some analysis techniques for repeated measures design studies via examples of the analyses of two datasets. I conclude with discussion of recent topics of interest in the modeling of longitudinal data including models for intensive longitudinal data, latent class models, and joint modeling of survival and repeated measures data.

Keywords

Longitudinal data · Survival analysis · Repeated measures · Crossover designs

1   Introduction

Throughout the history of science, a primary philosophical goal has been to establish the cause and effect of natural phenomena (Pearl 2000). Such cause and effect mechanisms can often be established in the physical sciences where one can tightly control experimental designs and outcomes are precisely defined. However, cause and effect phenomena are not as easily established in the social, biological, and medical sciences due to the difficulty in establishing feasible outcomes, the inability to tightly control experimental conditions, and the large variability observed both across individuals and within each individual's changing biological or clinical characteristics or state of mind (Hedeker and Gibbons 2006; Fitzmaurice et al. 2009).

A statistical study design that attempts to at least partially explain the cause and effect phenomena for the analysis of medical and biological data is called a "longitudinal" design (Hedeker and Gibbons 2006; Fitzmaurice et al. 2009). In the broadest sense, a longitudinal design is one where measurements or observations on each individual or experimental unit in a study are made at more than one (often many!!) point(s) in time. Due to the temporal nature of longitudinal designs, one may establish cause and effect in some cases, for example, how measured phenomena influence a response of interest over time. The evolution of a response may be due to aging, the effect of the intervention, or another factor.

The statistical antithesis of the longitudinal design is the *cross-sectional* design where only one measurement is made on each individual or experimental unit at some given point in time (Rosner 2010). Ideally, such designs seek to characterize a cross-section of information on one or more cohorts at an instant in time. Such designs have the advantage that they are cheaper and easier to implement. Their major disadvantage is that no cause and effect mechanism can be identified whatsoever.

It should be noted that the language used in this chapter reflects that of a broad literature concerning longitudinal statistical designs. For example, the terms outcome, response, or measurement are used in different areas of psychology, biological sciences, or medical sciences to mean the same thing. Likewise, one may use the terms subject, individual, animal, experimental unit, or patient synonymously in a statistical or analytical sense. Furthermore, intervention, therapy, and treatment may also be interchangeable depending on the scientific application. What is important about these entities is their interrelationships. They follow a hierarchy, and hence, longitudinal designs are sometimes known as *multilevel* designs (see Fig. 1). The top level is the cohort or group. That level is the one that we most often wish to infer upon, for example, we may wish to infer about the differences in groups with respect to treatment, therapy, and so on. The level below that, often nested within group, is the individual. At the lowest level in most longitudinal designs is the response or outcome measured within individual. Because of the wide range of longitudinal designs, a few, e.g., crossover designs (discussed in Sect. 3.2.2), do not follow the exact hierarchy outline in Fig. 1, but almost all are associated with some degree of hierarchy (two-level, three-level, or more).

The rest of this chapter is organized as follows. In Sect. 2, some advantages and disadvantages of longitudinal designs are introduced. In Sect. 3, many types of longitudinal designs are described. In Sect. 4, a few analysis techniques are discussed. Two examples of analytical approaches are given for a repeated measures design.

2   Advantages and Problems Associated with Longitudinal Designs

Longitudinal designs have an advantage over cross-sectional designs in that they can facilitate the understanding of how different phenomena change over time. In particular, together with the use of randomization in longitudinal designs, one may detect *moderators* of intervention (therapy) effectiveness utilizing measurements unrelated to the intervention but which *precede* and are related to the response of interest (Baron and Kenny 1986; Kraemer 2013). An example is that of pre-

Fig. 1 The multilevel nature of a typical longitudinal study design

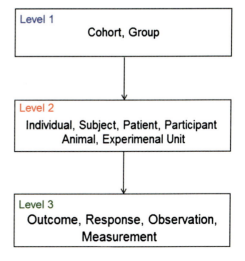

existing comorbid medical and cognitive characteristics that may moderate the effectiveness of depression interventions over time in older adults. Longitudinal designs also enable one to detect *mediators* which are variables measured *after* intervention that modify outcomes and allow insight into understanding how the variables affect or mediate the relationship between the intervention and the outcome. Such a relationship might occur in variables measuring side effects of an active therapy that may attenuate its effectiveness as compared to a placebo because of the higher incidence of the side effects in the active therapy.

From a purely statistical standpoint, longitudinal study designs have an advantage in efficiency over cross-sectional designs due to the fact that fewer subjects are needed because of the correlation of subjects across the population of interest with their own observations. Furthermore, within a cohort measured longitudinally, each subject serves as her/his own control allowing one to characterize the change of condition over time for both the at individual and cohort levels.

3 Types of Longitudinal Design

3.1 Follow-Up Studies

A very common type of design is one where one or more cohorts of individuals are measured over a period of time and the outcome of interest for each individual is the time of an "event" such as death, relapse, or depression. In some cases, individuals drop out of the study before having events, or the analysis of the data is done before all of the individuals have had events. Such occurrences prior to the event of interest are referred to as being "censored."

These types of studies, while technically longitudinal in nature, form their own unique category known as survival or follow-up studies because the measurement of interest for each individual is often a single entity, namely, the time to event or censor. Due to issues of power, these studies are often employed in very large cohorts that are observed over long periods of time. Many involve population studies comparing individuals with different demographic, biological, or clinical characteristics or large clinical studies where individuals are randomized to receive different treatments, interventions, or therapies. I do not elaborate further in this chapter about this type of study, but excellent references are available that detail their design and analysis (see Kalbfleisch and Prentice 2002; Klein and Moeschberger 2003).

3.2 Repeated Measures

3.2.1 Classic Repeated Measures and Growth Curve Designs

One of the most common types of longitudinal designs is one where each member of one or more cohorts is followed over a long period of time under the same conditions (Fleiss 1986; Hedeker and Gibbons 2006). The "same conditions" here means that the members within each cohort have the same treatment or intervention or have the same characteristics (e.g., same gender or same level of depression or some other characteristic to be tested). Consequently, each subject only has one treatment or characteristic of interest. Since subjects within each cohort share the same treatment or characteristic and are associated with only one such treatment or characteristic, their individual effects are said to be "nested within" the treatment or characteristic effect.

If the period of measurement is long enough, measurements within subject tend to have correlation structures that do not widely vary over time. In other cases, where the time frame is short, measurements that are closer together tend to be more highly correlated than those that are farther apart in time.

With the advent of technology to rapidly monitor measurements within individuals, an adaptation of the growth curve design is that of intensively measured longitudinal studies (Wallis and Schafer 2006). The purpose of such studies is to repeatedly measure hundreds or thousands of measurements over either a short or a long period of time to detect both gross and subtle patterns of change. Like the standard growth curve design, each unit or subject is measured under the same condition or treatments. Examples of this type of design are cardiac studies that monitor patient output each second over several hours or days and studies that use electronic diaries to monitor physical and behavioral activity over time.

The ideal design for growth curve studies is to have each subject measured at equally spaced times. Of course, ideally, one would not want to have any missing measurements. Unfortunately, in human or animal studies, individuals do not show up for visits, or they show up several days or weeks after their scheduled visits, or they withdraw or die before the end of the study or a whole host of other things that can happen over time. Moreover, there are certain studies where measurements cannot be observed at regular times due to situations beyond the experimenters' control. Such studies can produce irregularly spaced data.

3.2.2 Crossover Designs

Another type of longitudinal study design is called a "crossover design" (Grizzle 1965; Fleiss 1986; Brown and Prescott 2006). In this type of design, individuals are typically randomized to receive a *sequence* of treatments. Thus, instead of being assigned to a single treatment group, each unit is usually assigned to receive all of the treatments, but different individuals may be assigned to have these treatments in a different sequence. An example of a two-period crossover design is depicted in Fig. 2. As one can see, each subject receives a sequence of treatments A and B. When an individual has completed one treatment, they go through a "washout period" before receiving the next treatment. This guards against what is known as a carryover effect so that the effect of each treatment can be uniquely distinguished. For example, if treatment A is given in the second period, we would want to ensure that effects of treatment B on the outcome of interest had not "carried over." Otherwise, possible differences in the effects would be attributed to the previous treatment and hence, the sequence that individuals were assigned to rather than the treatments themselves. Of course, crossover designs are not limited to two periods and two treatments, and there are many variations where there are many periods and with all treatments being received by each individual or some subset of treatments

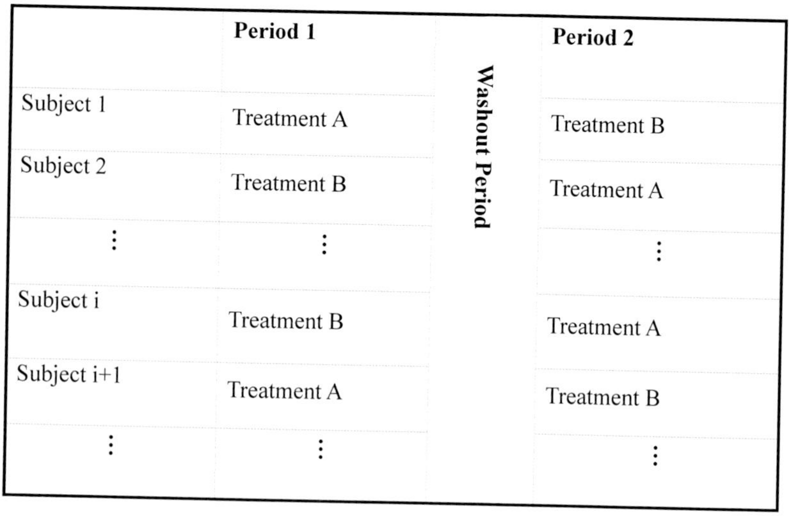

Fig. 2 Schema of a two-period crossover design

being received by each individual. What is common to all crossover designs is that washout periods are necessary between periods and individuals are typically randomized to receive *sequences of treatments* rather than a single treatment. Crossover designs evolved from an earlier incarnation in agriculture studies known as permuted block or Latin square studies (Fleiss 1986).

Crossover designs are appropriate when a disease or condition is *chronic* (e.g., "endogenous" depression, essential hypertension, Fleiss 1986). In this case, the period of administration of the treatment is short as compared to the duration of the condition or disease. Furthermore, crossover designs work best when there is little or no residual or *carryover effect* of a treatment after an appropriate washout period. If, in fact, there are *small* carryover effects of the treatments, then those effects should be close to the same for both treatments. It is not appropriate to use crossover designs when a condition or disease being treated is acute such as with postoperative pain, short-term depression, or a short-term illness like a cold or short-term flu. Crossover designs are also not effective when the residual effects of the treatments are so long that individuals will drop out of the study or if the residual effects of the treatments are either large or vary by treatment group or both.

Crossover studies have a desirable feature that longitudinal growth curve studies do not have in that each participant of the study can receive more than one or possibly all of the treatments being tested (Grizzle 1965; Fleiss 1986). Like the longitudinal growth curve studies, increased precision is obtained by the multiple measurements on an individual being made, and confounding is reduced because each patient serves as his or her own control. However, the potential carryover effect associated with a crossover study is always worrisome, and hence, crossover studies may be difficult to implement because of the complexity of the design. Furthermore, results from crossover studies can be somewhat difficult to interpret even when carryover effect is not present. Finally, like other longitudinal studies, dropouts and missing data are problematic. In crossover studies, a large number of dropouts are particularly troublesome. Sometimes, a complete period of treatment must be dropped due to bias. This can force a longitudinal study to be analyzed as if it were cross-sectional, thus losing the advantages of having repeated observations.

A recent design innovation somewhat related to crossover designs involves *adaptive* designs (Lavori and Dawson 2000; Murphy 2003, 2005). These designs usually involve individuals being randomly assigned to a therapy, being followed (measured) over time, and then, at a particular decision point, being reassigned to a different or the same therapy depending upon whether they were responsive to the original therapy. A study where individuals are randomized to a strategy involving both initial therapies and then re-randomized to receive other therapies depending on their response to the initial therapy is called sequential multiple assignment randomized trial (SMART) (Murphy 2005). What distinguishes a SMART study from a crossover study is that in the former, the sequential strategy is adapted to the response of each individual, whereas in the latter, the sequential treatment pattern is assigned without regard to patient or subject response.

4   Visualizing Repeated Measures Data

One important step that is sometimes forgotten by a busy analyst is to visualize the data. It is useful to do this before formal analyses are performed. Luckily, with modern software, there is an extraordinary amount of routines that allow us to visualize data in many different ways (Anderson 2011). In the sections below, I introduce a couple of simple types of plots that can be quite useful in understanding our data and properly interpreting the formal analyses that we perform on repeated measures data.

4.1   Motivating Example #1

To motivate plotting of repeated measures data, I refer to a famous data set, first presented by Potthoff and Roy (1964) and later by many other authors. The data involve measurements in millimeters (mm) of the distance from the center of the pituitary to the pteryomaxillary fissure in 27 children taken at ages 8, 10, 12, and 14 years by investigators at the University of North Carolina Dental School. Sixteen of the study cohorts were boys and 11 were girls. By the standards of modern investigation, this is a relatively small study. Questions of interest involved characterizing the growth patterns in both cohorts and comparing the growth patterns between genders.

4.2   Spaghetti and Panel Plots

One useful type of plot of these data involves the display of the boys' and girls' trajectories of measurements overlaid over time as interpolated curves (Fig. 3). This

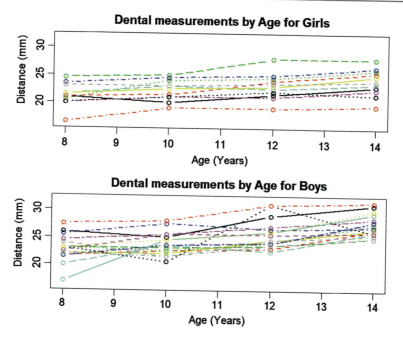

Fig. 3 Spaghetti plots of boys' and girls' dental measurements (Potthoff and Roy 1964)

can enable us to view general patterns in the two cohorts and possibly identify outliers if they exist.

Another useful visualization of repeated measures data is that of *panel plots* where each individual's data is plotted on a graph and then the cohort(s) of individuals are stacked together so that one can view the variability of the trajectories of individuals within the cohort(s) being studied. Panel plots of the dental measurements of each individual boy and girl for the Potthoff and Roy data are presented in Fig. 4.

Both spaghetti and panel plots are very useful for getting a sense of the data in small studies and for identifying potential outliers or influence points in the data. However, one disadvantage of spaghetti and panel plots is that, for larger studies, individual variation leads to morass visualization of the information so that patterns averaged (or summarized) over a cohort are difficult to discern. In such cases, plots of summarized data can be explored (see next section).

4.3   Mean Plots

Another type of useful plot where measurements are on a continuous scale and made over time on each individual within a study is that of means plus or minus either their standard errors (SEM) or means and their pointwise 95% confidence intervals. If data are approximately normally distributed, such plots facilitate the interpretation of statistical inference about one or more cohorts (see Fig. 5 for our example data).

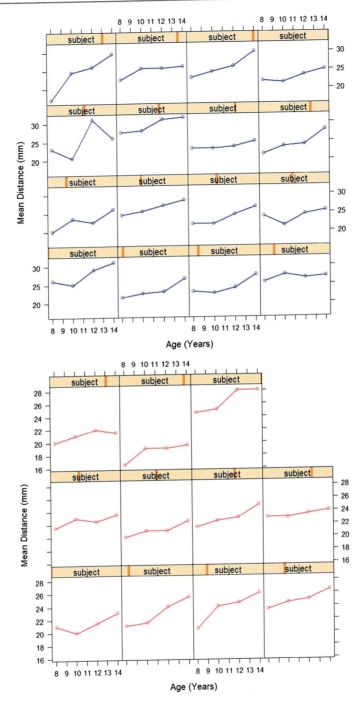

Fig. 4 Panel plots of boys' and girls' dental measurements (Potthoff and Roy 1964)

Fig. 5 Mean plots of dental observations ± standard errors of the mean (SEM) by age and sex

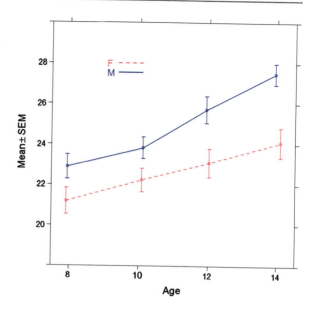

In other cases, where the data are skewed, one can substitute medians for means, and say, the lowest and highest deciles for SEMs, and so on.

5 Fitting Appropriate Models to Repeated Measures Designs

In this section, I will discuss appropriate analyses for some classic repeated measures design.

5.1 Continuous Data

5.1.1 Analysis of Variance Approach

The simplest method of analyzing repeated measures data that is still used in some circles but has very limited utility is the "classical method" which is accomplished using analysis of variance (ANOVA) methods. Suppose that there are I cohorts or treatments in a repeated measures study and for each cohort, $i = 1,\ldots,I$; there are n_i subjects, and there are T equally spaced observations per individual. A typical ANOVA model to analyze such a situation would be written as:

$$y_{ijt} = \mu_i + \gamma_{i(j)} + \tau_t + (\mu\tau)_{ik} + e_{ijt}, i = 1, \ldots, I; j = 1, \ldots, n_i; t = 1, \ldots, T; \quad (1)$$

where y_{ijt} represents a measurement on subject j in cohort i at time t, $\gamma_{i(j)}$ represents a subject effect, τ_t represents a time effect, $(\mu\tau)_{ik}$ represents a potential time × cohort interaction (moderation of cohort effect due to time), and e_{ijt} is the random error associated with measurement t on subject j within cohort i. This is referred to as a

Table 1 ANOVA table for the Potthoff and Roy data

Source	DF	SS	MSE	F-value	Prob > F
Sex	1	140.4649	140.4649	9.29^a	0.0054
Subject (Sex)	25	377.9148	15.1166	7.65	–
Time	3	237.1921	79.0640	40.03^b	<0.0001
Sex*Time	3	13.9925	4.66418	2.36^b	0.0781
Error	75	148.1278	1.97504	–	–
Total	107	917.6921	–	–	–

^{a}Obtained by dividing the sex MSE by the Subject (Sex) MSE
^{b}Obtained by dividing by the error MS

univariate model because the model is based on a single outcome as opposed to a "vector of outcomes" assumed by models developed later.

Here, the cohort and time effects are fixed, i.e., inference is limited to the prespecified levels used in a particular experiment. The "subject effect" is random, so that the subjects (patients, experimental units) in our experiment are assumed to represent a random sample from a large population. Hence, these types of models are known as *mixed effects* models. Furthermore, the subject effect itself can vary and so is associated with a probability distribution which assumed to be normal in the simplest case.

Specific to the ANOVA approach, the observations for each subject are assumed to form a *block* having the property that variability within blocks (i.e.,, within subject) is generally smaller than is the variability across subjects. In studies similar to that given our first example, the subject effect is said to be *nested within* the cohort effect and is denoted as such by the $\gamma_{i(j)}$ term. Hence, each subject is in one and only one cohort (can only be one sex in this case) or has one and only one treatment in treatment studies.

One trick that is used for both univariate and multivariate analyses seeking to fit polynomial trajectories is to transform the (equally spaced) time points into orthogonal coefficients. This is particularly useful as times t, t^2, and t^3 tend to be highly correlated inducing correlation between corresponding coefficients. By orthogonalizing the coefficients, we can get uncorrelated estimates meaning that if, say, one wishes to reduce the degree of polynomial in a model and one or more coefficients are dropped, then the resulting coefficients will have the same values as in the original model, allowing one to make inference about each coefficient independent of all other coefficients. This property is not true for the coefficients of the original data. A more complete discussion of this topic is given in Hedeker and Gibbons (2006).

In Table 1, I fit such a univariate model to the dental measurement data. In this model, the cohort effect is due to sex (boys vs. girls).

Notice that F-test for sex is obtained by dividing the sex MS by the subject within Sex [denoted Subject (Sex)] MSE so that the F-test for comparing sexes has 1 and 25 DF. This is because each child is nested within gender (i.e., can *be* only one gender) so is considered a "block," and hence the variation of interest for the denominator is associated with the subject (child) MS. The Time and Sex by Time effects, assuming the correlations between repeated observations are the same across time, are tested via F-tests, both of which have 3 and 75 DF. Inference for uniformity of correlations across time points in this analysis indicates that the assumption of temporal uniformity of correlation structure for these data is reasonable.

The results indicate that we see highly significant growth in dental measurements in both sexes over time ($p < 0.0001$), but the average values of the dental measurements are significantly greater in boys than in girls ($p = 0.0054$). There is a nearly significant moderation ($p = 0.078$) in growth by sex indicating that the boys' average growth rate is higher but not significantly so than that of girls. An analysis (not shown) using orthogonal polynomials as described above indicate a significant linear growth effect in both boys and girls but the linear growth is significantly higher in boys than in girls. No significant quadratic or cubic growth terms were detected.

5.1.2 Multivariate Approach

Another way to handle nonuniform correlation structure is to model it using multivariate techniques. We can write a general multivariate model as $Y = A B P + E$, where the rows of Y represent the multiple observations of the subjects, the A and P are fixed matrices which allow one to test hypotheses between and within subjects, respectively, and E is a matrix with rows that are each multinormally distributed and have the property that different rows are independent of each other (Rao 1958, 1965; Potthoff and Roy 1964; Grizzle and Allen 1969). To properly develop the covariance structure with these models, one has to *vectorize* the observations and change the other components of the model accordingly. As Potthoff and Roy (1964) pointed out in their paper, with the structure outlined above, one could test differences across sexes with the A matrix and could fit a polynomial model for the subjects with the P matrix.

Multivariate methods provide a big improvement over univariate methods in the modeling of repeated measures data in that they can address a broader set of questions and they allow a general structure to be modeled for the correlation structure in the errors. Hence, in most cases, such models are more efficient. However, they still are limited because:

- They don't allow subject effects to be easily distinguished and predicted.
- They don't handle unequally spaced data very well.
- They don't handle missing data well especially if there a lot of missing observations.

The mixed model regression approach discussed in Sect. 5.1.4 allows the analyst to easily address these problems.

5.1.3 Missing Data

As any investigator overseeing longitudinal studies knows, one of the greatest challenges during a study is to minimize missing information (Hedeker and Gibbons 2006; Fitzmaurice et al. 2009). In particular, one wishes to keep the number of subjects who completely drop out a study to a minimum.

There are three broad categories of missing data (Little and Rubin 2002). The first, called "missing completely at random (MCAR)," can be described as missing measurements that are not related to any other non-missing data nor are they related to latent variables or information that is not collected. This type of missingness is easily accounted for in modern mixed model analyses. The second broad category is called "missing at random" where the missingness can be accounted for by other *observed* measurements such as baseline demographics (age, race, clinical status) or previously observed measurements made longitudinally. As long as the proper observed covariates are accounted for in the analysis, this category poses no serious problem for modeling. The third category, known as missing not at random (MNAR), is when missing measurements are related to information not observed in the longitudinal study (e.g., future mortality status or characteristics not measured in a study) and can cause model inference to be biased.

5.1.4 Mixed Model Regression Approach

Modern methodology has focused on techniques that optimally model repeated measures data even in the presence of missing and/or unequally spaced observations made on each subject. These approaches explicitly model both the fixed and random effects associated with repeated measures data. The seminal papers by Harville (1977) and Laird and Ware (1982) were crucial in the development of these models. Consider the model given by:

$$y_i = X_i\beta + Z_i b_i + e_i, \tag{2}$$

where y_i, $i = 1, \ldots, n_i$, is a *vector of observations* for the ith individual; X_i and Z_i are matrices that allow for cohort level and individual level covariate and time effects, respectively; β and b_i are parameters that relate the cohort and individual effects to the outcome, respectively; and e_i represents the random error *vector* for the ith individual. For continuous data that is approximately normal or transformed to be approximately normal, the e_i are $N(0, R_i)$, the b_i are $N(0, G_i)$, and the correlation between the e_i and the b_i is assumed to be zero. The model given above accommodates both a population parameter vector, β, (denoted without a subscript) which is common to all members of the defined population, and parameter vectors, b_i, which are specific to each individual. The regression approach used in Eq. (2) allows time to be viewed and analyzed on a *contiuum* as it should be rather than being a discrete (fixed) factor as in the ANOVA approach above. Thus, observation times need not be equally spaced to be properly analyzed using this approach. Furthermore, it allows one to model differing numbers of measurements on each individual. Hence, if there are missing values or the measurements being taken are inherently unequally spaced, the above approach can be employed with the caveat that the missingness is MCAR or MAR with appropriate covariates accounted for.

5.1.5 Properly Accounting for Correlation Structure

One challenge of fitting longitudinal models like that displayed in Eq. (2) is the fact that we must try to account for the variances and correlations among two sources of random variation: (1) the coefficients of the random effects, b_i, which allow for individual trajectories to be fitted, and (2) the temporal correlation of the random errors. These estimates usually cannot be calculated in closed form so require optimization programs to maximize the likelihood function of the model. Hence, numerical algorithms must be implemented to approximate the optimal solutions. As the number of nonlinear parameters rises, the probability that solutions converge to a maximum goes down depending on the sample size and the nature of the data itself. For case (1), the number of random coefficients allowed for will determine the number of variance/covariance parameters. If, for example, there is only one random coefficient, then only the variance of that coefficient will require estimation. If, say, a random intercept and slope are to be estimated, then one must maximize the likelihood with respect to three parameters (the variance of each parameter and the covariance between them). For three random effects coefficients, six variance/covariance estimates must be made and any number of random coefficients beyond three usually results in models that do not properly converge.

With regard to properly accounting for the temporal structure in the errors, there are many possibilities, but three structures are the most common: independence in the error structure, uniform correlation over time, or a correlation structure that attenuates as time intervals between measurements increase. If the data are regularly spaced data, autoregressive [usually AR(1)] and Toeplitz structures are common. For irregularly spaced data, extensions known as continuous autoregressive (CAR) structures are often used (Jennrich and Schlucter 1986; Jones and Ackerson 1990).

5.1.6   Motivating Example #2

A recent study published in the Lancet (Lenze et al. 2015) involved a randomized comparison of Aripiprazole, a second-generation antipsychotic drug previously approved by the US Food and Drug Administration for augmentation treatment of major depressive disorder with a placebo. In the study, 468 eligible participants were recruited at three sites; 181 did not remit and were randomly assigned to aripiprazole ($n = 91$) or placebo ($n = 90$). A greater proportion of participants in the aripiprazole group achieved remission than did those in the placebo group participants (44% vs. 29%; odds ratio [OR] 2·0 [95% CI 1·1–3·7], $p = 0·03$). One of the primary endpoints in the study was the Montgomery-Asberg Depression Rating Scale [MADRS]. For the MADRS endpoint, a lower score is better, and a remission was defined as a MADRS score of 10 or less. The data were reasonably normally distributed. Summary statistics are given in Table 2, and mean values plus or minus standard errors (SEM) are plotted by treatment and time point (Fig. 6). Note that there was attrition in both groups as the sample sizes reduced from 91 to 68 in the aripiprazole group and from 90 to 71 in the placebo group. The best fitting trajectories were cubic polynomials, and an overall reduction in the trajectory of the MADRS scores was observed for the aripiprazole group, indicating that the drug was associated with less depressive symptoms (Table 3). Note that, even though some of the differences between the groups in the terms of the cubic polynomials were not significant, the overall difference in the polynomials was significant (overall time × treatment p-value ≈ 0.017). A plot of the trajectories as predicted by the model is displayed in Fig. 7.

Table 2 MADRS data: means and standard deviations by week and treatment group

Trt[a]	Arip	Arip	Arip	Arip	Arip	Arip	Arip	Arip	Arip	Arip
Week	0	1	2	3	4	5	6	8	10	12
Mean	23.47	19.11	16.73	16.06	15.22	15.01	14.18	13.47	14.58	14.37
Std dev	6.43	6.64	8.4	8.2	9.27	8.89	9.03	8.75	9.3	9.34
N	91	89	88	87	86	84	79	75	79	68
Trt[a]	Plac	Plac	Plac	Plac	Plac	Plac	Plac	Plac	Plac	Plac
Week	0	1	2	3	4	5	6	8	10	12
Mean	23.02	20.17	19.26	18.78	18.07	17.18	17.64	18.08	17.77	16.44
Std dev	6.58	7.53	7.79	7.46	8.34	8.41	8.54	9.09	9.65	9.93
N	90	90	90	85	86	82	85	73	74	71

[a]*Arip* Aripiprazole, *Plac* Placebo

Fig. 6 IRL-Grey mean MADRS data standard errors of the mean (SEM) plotted by weeks rounded to the expected visit times by treatment group

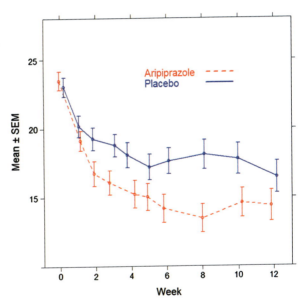

5.2   Models for Binary and Other Discrete Data

In some cases, a variable measured repeatedly on one or more cohorts may be of a binary or, more generally, discrete nature. This is often encountered in the social sciences where simple distinctions, such as one feeling distressed or not or functioning better or worse or cognitive ability decreasing or not, are studied over time (Muenz and Rubinstein 1985; Zeger and Liang 1986). Two primary approaches are used for the analysis of such measurements in a longitudinal study.

Table 3 Type III fixed effects statistics for the aripiprazole study

Source	Num. DF	F-value	Prob > F
Age at entry	1	12.82	0.0004
Site	2	6.59	0.0014
Intercept	1	0.04	0.84
Linear	1	136.63	< 0.0001
Quadratic	1	72.49	< 0.0001
Cubic	1	44.36	< 0.0001
Treatment*TIME	3	–	**0.017**[a]
Treatment*linear	1	4.99	0.026
Treatment*quadratic	1	0.93	0.33
Treatment*cubic	1	0.06	0.80

[a]Calculated using a likelihood ratio test

Fig. 7 IRL-Grey-2 predicted MADRS scores for typical participants

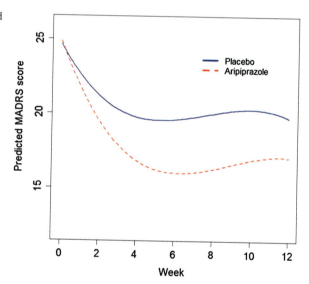

5.2.1 Marginal Models

Here, I present the seminal work by McCulloch and Nelder (1982, 1989) in the development of genera*lized* linear models. Their approach involves "link functions," g, given by:

$$g(\mu) = g(X_i\alpha)$$

where X_i is a matrix with columns containing each individual's (denoted with a subscript i) measurements (covariates) and α is a parameter to be estimated that relates the covariates to the transformed outcomes and hence the mean response, $\mu = X_i\alpha$. Typical link functions include the logit and exponential functions for binary (e.g., yes/no, event/no event) outcomes.

The generalized linear model framework was adapted to fit longitudinal data by the groundbreaking work of Liang and Zeger and others at the Johns Hopkins University (Liang and Zeger 1986; Zeger and Liang 1986). They used "generalized estimating equations" that allowed them to first fit marginal longitudinal data assuming a "working covariance structure" that initially assumed independence and then tried to correctly model the within subject covariance structure.

Using this GEE approach, one can easily handle binary, discrete, or even continuous measurements in a general framework allowing for MCAR or MAR missingness and appropriately accounting for other covariates and repeated outcomes to be modeled. Excellent references for this method are given in Hardin and Hilbe (2003), Brown and Prescott (2006), and Diggle et al. (2002).

5.2.2 Transition Models

Another approach to modeling binary and general discrete outcomes measured repeatedly is done by characterizing the probabilities of transitioning from different binary or discrete "states," for example, the probability of going from "feeling good" to "feeling bad" or vice versa. These models can be adapted to adjust for covariates. Good references for this type of model can be found in Muenz and Rubinstein (1985), Molenberghs and Verbeke (2005), and Diggle et al. (2002).

***

6 Discussion and Future Directions

The purpose of this chapter was to familiarize the reader with the design and analysis of different types of longitudinal studies. Longitudinal study designs allow researchers to investigate trajectories of phenomena of one or more populations over short- or long-term periods of time (or both). Analysis techniques allow one to make inference both at the individual and population levels. Due to the temporal nature of these studies, causal relationships of interventions on outcomes can often be established particularly when the study involves randomization of the participants into different intervention groups.

The field of longitudinal data analysis has exploded in the last 50 years with different methods. As is noted by many authors (Ware 1985; Rizopoulos 2012), most longitudinal studies collect both follow-up (time-to-event or survival) data along with data that is measured repeatedly. Accordingly, in recent years, a common goal of longitudinal studies is to relate a set of repeated observations to a time-to-event endpoint (Henderson et al. 2000; Song et al. 2002; Guo and Carlin 2004; Rizoupoulos 2011, 2012). One example of such a design is in the area of late-life depression research where repeated measurements of cognitive and functional outcomes can contribute to one's ability to predict whether or not an individual will have a relapse of a major depressive episode over a period of time (Reynolds et al. 2011).

Another area of recent research is that of modeling *intensive* longitudinal data (Wallis and Schafer 2006). This field has arisen due to our ability to measure hundreds or thousands of measurements on each subject within a set of cohorts in a longitudinal study and how to account for complex within subject variation while, at the same time, making inference on a cohort level for possibly complicated patterns (e.g., circadian, daily and yearly patterns seen in an overall cohort or set of cohorts). Two examples of such studies are cardiac monitoring and exercise monitoring studies where thousands of typically equally spaced measurements are made per individual. A third is monitoring behavioral or physical activity based on random "prompts" from an electronic diary leading to large numbers of observations that are usually not equally spaced in a single individual and may occur at varying time points over a cohort of individuals (Stone et al. 2007; Shiffman et al. 2015).

Another area of recent research for longitudinal data analysis is that of so-called "latent class" trajectory models (Nagin 1999; Roeder et al. 2011). The primary focus of such studies is to identify classes of trajectories of different subsets of a cohort based on the trajectories themselves and a set of covariate levels that are unique to each subset. These methods are best suited to observational studies or as hypothesis-generating studies in randomized trials.

Because real-time data acquisition continues to get more sophisticated, the future challenges of designing and analyzing longitudinal studies to accommodate such advances will be formidable. For example, data acquisition systems now allow us to measure high-dimensional data such as genomic or imagining data over time. This type of information typically involves data with more variables than there are subjects (Hastie et al. 2009; Zipunnikov et al. 2014; Chen and Lei 2015) and requires appropriate dimension reduction techniques to properly fit models. To design and model studies that allow scientists to accurately model such data, along with other traditional demographic and clinical information measured cross-sectionally and repeatedly, will be a great challenge. Furthermore, relating these measurements to ultimate events such as relapse and mortality will occupy much future research at statistical and, more generally, scientific levels. Lastly, many times, several repeated measurements are made simultaneously, and sometimes it is appropriate to model such processes jointly by themselves or as predictors of time-to-event outcomes. Methods for doing so are another area of ongoing research (Choi et al. 2014).

References

Anderson SJ. Biostatistics: a computing approach. Boca Raton: Taylor & Francis Group, LLC; 2011.

Baron RM, Kenny DA. The moderator-mediator variable distinction in social psychological research: conceptual, strategic and statistical considerations. J Pers Soc Psychol. 1986;51(6):1173–82.

Brown H, Prescott R. Applied mixed models in medicine. 2nd ed. West Sussex: Wiley; 2006.

Chen K, Lei J. Localized functional principal component analysis. J Am Stat Assoc. 2015;110(511):1266–75.

Choi J-I, Anderson SJ, Richards TJ, Thompson WK. Prediction of transplant-free survival in idiopathic pulmonary fibrosis patients using joint models for event times and mixed multivariate longitudinal data. J Appl Stat. 2014;41(10):2192–205.

Diggle PJ, Heagerty P, Liang K-Y and Zeger SL. Analysis of longitudinal data. 2nd ed. Oxford: Oxford University Press; 2002.

Fitzmaurice G, Davidian M, Verbeke G, Molenberghs G, editors. Longitudinal data analysis. Chapman & Hall/Taylor & Francis Group: Boca Raton; 2009.

Fleiss JL. The design and analysis of clinical experiments. New York: Wiley; 1986.

Grizzle JE. The two–period change–over design and its use in clinical trials. Biometrics. 1965;21:467–80.

Grizzle JE, Allen DM. Analysis of growth and dose response curves. Biometrics. 1969;25(2):357–81.

Guo X, Carlin BP. Separate and joint modeling of longitudinal and event time data using standard computer packages. Am Stat. 2004;58:16–24.

Hardin JW, Hilbe JM. Generalized estimating equations. Boca Raton: Chapman & Hall/CRC; 2003.

Harville D. Maximum likelihood estimation of variance components and related problems. J Am Stat Assoc. 1977;72:320–40.

Hastie T, Tibsharani R, Friedman J. The elements of statistical learning. 2nd ed. New York: Springer; 2009.

Hedeker D, Gibbons RD. Longitudinal data analysis. Hoboken: Wiley; 2006.

Henderson R, Diggle P, Dobson A. Joint modelling of longitudinal measurements and event time data. Biostatistics. 2000;1:465–80.

Jennrich RI, Schlucter MD. Unbalanced repeated-measures models with structured covariance matrices. Biometrics. 1986;42:805–20.

Jones RH, Ackerson LM. Unequally spaced longitudinal data with serial correlation. Biometrika. 1990;77:721–31.

Kalbfleisch JD, Prentice RL. The statistical analysis of failure time data. 2nd ed. New York: Wiley; 2002.

Klein JP, Moeschberger ML. Survival analysis: techniques for censored and truncated data. 2nd ed. New York: Springer; 2003.

Kraemer HC. Discovering, comparing, and combining moderators of treatment on outcome after randomized clinical trials: a parametric approach. Stat Med. 2013;32:19.

Laird NM, Ware JH. Random effects models for longitudinal data. Biometrics. 1982;38:963–74.

Lavori PW, Dawson R. A design for testing clinical strategies: biased adaptive within-subject randomization. J R Stat Soc A. 2000;163:29–38.

Lenze EJ, Mulsant BH, Blumberger DM, Karp JF, Newcomer JW, Anderson SJ, Dew MA, Butters M, Stack JA, Begley AE, Reynolds CF. Efficacy, safety, and tolerability of augmentation pharmacotherapy with aripiprazole for treatment-resistant depression in late life: a randomized placebo-controlled trial. Lancet. 2015;386:2404–12.

Liang KY, Zeger SL. Longitudinal data analysis using generalized linear models. Biometrika. 1986;73:13–22.

Little RJA, Rubin DB. Statistical analysis with missing data. 2nd ed. New York: Wiley; 2002.

McCullagh P, Nelder JA. Generalized linear models. London: Chapman and Hall; 1982.

McCullagh P, Nelder JA. Generalized linear models. 2nd ed. London: Chapman and Hall; 1989.

Molengberghs G, Verbeke G. Models for discrete longitudinal data. New York: Springer; 2005.

Muenz LR, Rubinstein LV. Markov models for covariate dependence of binary sequences. Biometrics. 1985;41:91–101.

Murphy SA. Optimal dynamic treatment regimes. J R Stat Soc B. 2003;65(2):331–66.

Murphy SA. An experimental design for the development of adaptive treatment strategies. Stat Med. 2005;24:1455–81.

Nagin DS. Analyzing developmental trajectories: a semiparametric, group-based approach. Psychol Methods. 1999;4(2):139–57.

Pearl J. Causality: models, reasoning and inference. Cambridge: Cambridge University Press; 2000.

Potthoff R, Roy SN. A generalized multivariate analysis of variance model useful especially for growth curve problems. Biometrika. 1964;51(3):313–26.

Rao CR. Some statistical methods for comparison of growth curves. Biometrics. 1958;14(1):17.

Rao CR. The theory of least squares when parameters are stochastic and its application to the analysis of growth curves. Biometrika. 1965;52(3/4):447–58.

Reynolds CF III, Butters MA, Lopez O, Pollock BG, et al. Maintenance treatment of depression in old age: a randomized, double-blind, placebo-controlled evaluation of the efficacy and safety of donepezil combined with antidepressant pharmacotherapy. Arch Gen Psychiatry. 2011;68(1):51–60.

Rizopoulos D. Dynamic predictions and prospective accuracy in joint models for longitudinal and time-to-event data. Biometrics. 2011;67:819–29.

Rizopoulos D. Joint models for longitudinal and time-to-event data, with applications in R. Boca Raton: Chapman and Hall/CRC; 2012.

Roeder, K, Lynch, KG and Nagin, DS. Modeling uncertainty in latent class membership: a case study in criminology. J Am Stat Assoc. 2011;94:766–776.

Rosner B. Fundamentals of biostatistics. 7th ed. Boston: Brooks/Cole; 2010.

Shiffman S, Dunbar MS, Kirchner TR, Li X, Tindle HA, Anderson SJ, Scholl SM, Ferguson SG. Cue reactivity in converted and native intermittent smokers. Nicotine Tob Res. 2015; 17(1):119–23.

Song X, Davidian M, Tsiatis AA. A semiparametric likelihood approach to joint modeling of longitudinal and time-to-event data. Biometrics. 2002;58:742–53.

Stone AA, Shiffman S, Atienza AA, Nebeling L, editors. The science of real-time data capture. New York: Oxford University Press; 2007.

Wallis TA, Schafer J, editors. Models for intensive longitudinal data. New York: Oxford Press; 2006.

Ware JH. Linear models for the analysis of longitudinal studies. Am Stat. 1985;39(2):95–101.

Zeger SL, Liang KY. Longitudinal data analysis for discrete and continuous outcomes. Biometrics. 1986;42:121–30.

Zipunnikov V, Greven S, Shou H, Caffo B, et al. Longitudinal high-dimensional principal components analysis with application to diffusion tensor imaging of multiple sclerosis. Ann Appl Stat. 2014;8(4):2175–202.

Eliciting Preferences from Choices: Discrete Choice Experiments

36

Martin Howell and Kirsten Howard

Contents

Abstract

Discrete choice experiments (DCEs) have been widely used as a research tool to elicit the preferences of patients, clinicians, the community, and policy-makers for a range of health-related questions including complex interventions, treatment options, health programs (e.g., cancer screening) and policies, and health service delivery. In a DCE, treatments or health programs are described by a set of attributes with varying levels, for example, health outcomes (harms and benefits), cost, time, properties of the procedure (e.g., injection or tablet), and so on. The participant is asked to choose their preferred treatment or program. By systematically varying the attribute levels across a range of choices, preferences for health goods and services can be calculated. Unlike other preference elicitation techniques such as ranking or rating, DCEs are underpinned by a well-established and robust theoretical framework that allows estimation of a range of outputs, including the relative importance of individual attributes within a multi-attribute health

M. Howell (✉) · K. Howard
School of Public Health, University of Sydney, Sydney, NSW, Australia
e-mail: martin.howell@sydney.edu.au; Kirsten.howard@sydney.edu.au

© Springer Nature Singapore Pte Ltd. 2019
P. Liamputtong (ed.), *Handbook of Research Methods in Health Social Sciences*,
https://doi.org/10.1007/978-981-10-5251-4_93

623

program (e.g., waiting time, travel time, type of care), the trade-offs individuals may be willing to accept between attributes (e.g., side effects and survival), as well as willingness to pay and uptake of health programs. This chapter provides an overview of the theory and application of DCEs.

Keywords

Discrete choice experiments · Best-worst scaling surveys · Preference elicitation · Preferences and values

1   Introduction

It is increasingly recognized that health interventions/programs, research priorities, policies, and resource allocation should reflect the values, preferences, and priorities of patients, their carers, and the community. Individuals and societies value additional outcomes of health care beyond just clinical outcomes. Indeed, if the long-term benefits and harms are unclear or there are multiple options with differing outcomes including undesirable side effect profiles and serious adverse outcomes, then patient's values, preferences, and priorities for these other outcomes may be as important to decisions as medical factors (Braddock 2013). Issues for patients and their carers extend beyond medical considerations and include a range of social and financial factors all of which may change before, after, and during the course of a health intervention (Martin et al. 2010). Similarly, the relative importance of medical factors may change over the course of treatment, particularly with chronic or incurable conditions, the development of comorbidities, and adverse outcomes. Disparate medical and nonmedical factors may contribute to decisional conflicts arising from the changing need to balance benefits and harms and the potential discord with values, preferences, and priorities (Murray et al. 2009; Mühlbacher and Juhnke 2013). Allocation of finite resources for health-care programs requires a balance between making the best use of a finite resource (efficiency) and being fair (equity). There are four key principles underpinning this balance, which have been reflected in community preferences and priorities, namely, "treating people equally," "favoring the worst off," "maximizing benefits," and "promoting social usefulness" (Persad et al. 2009). Finally, individual preferences are key to assigning a utility weight (Quality Adjusted Life Year weight) to health states used in economic evaluations of health interventions and programs (Richardson and Manca 2004).

Preferences can be measured using a variety of qualitative and quantitative approaches. The most commonly used methods include focus groups, structured interviews, mixed methods such as the nominal group technique, surveys with or without rating and ranking scales, standard gamble and time trade-off (Torrance 1986), and discrete choice experiments (DCEs) (see ▶ Chaps. 23, "Qualitative Interviewing," ▶ 40, "The Use of Mixed Methods in Research," ▶ 42, "Consensus Methods: Nominal Group Technique," and ▶ 32, "Traditional Survey and Questionnaire Platforms").

Discrete choice experiments have been applied to a wide range of health-related questions at a patient, clinical, community, and policy level. These have included eliciting patient, clinician, and community preferences and priorities for outcomes of interventions, treatment options, and a range of health programs such as cancer screening and health service delivery (de Bekker-Grob et al. 2012; Clark et al. 2014; von Arx and Kjær 2014; Wortley et al. 2014) as well as eliciting utility weights for health states (Ratcliffe et al. 2009, 2011; Norman et al. 2013; Viney et al. 2014). One main advantage of DCEs over other approaches such as ratings, rankings, standard gamble, and time trade-off is that they enable assessment of priorities and preferences beyond the simple ordinal scales of these other techniques. Furthermore, trade-offs between desirable and undesirable outcomes, the willingness to pay for services, and estimation of uptake of community wide programs can also be evaluated (Chuck et al. 2009; Regier et al. 2009; Groothuis-Oudshoorn et al. 2014; Howard et al. 2014; Kawata et al. 2014; Laba et al. 2015; Kan et al. 2016).

This chapter provides an overview of discrete choice experiments (DCEs) that have, since the 1990s, become one of the most commonly used quantitative techniques for eliciting preferences and priorities in health-related research questions.

2   What Are Preferences?

Before describing DCEs, it is necessary to understand what is meant by preferences. In health research, despite the terms patient, consumer, community, or individual preferences or values being widely used, there is no clear definition or understanding of either preference or values (Street et al. 2012). Individual "values" have been used interchangeably with preferences, or preferences are seen as a subset of "values" and vice versa. Patient preferences have also been defined as being the final choice or decision in the context of shared decisions (Brennan and Strombom 1998). The Institute of Medicine defines patient values as referring "to the unique preferences, concerns, and expectations that are brought by each patient to a clinical encounter [that] must be integrated into clinical decisions if the patient is to be served" (Committee on Quality of Health Care in America Institute of Medicine 2012, p. 47). Under this definition, "preferences and concerns" is a broad concept and implicitly includes cultural identity, existential and nonexistential beliefs, and personality traits such as aversion to risk (Koltko-Rivera 2004; Daher 2012) as well as treatment-related factors (Blinman et al. 2012). In the context of health, it is more useful to think of values and preferences as separate concepts. Values reflect an individual's identity and world view (Koltko-Rivera 2004), while preferences reflect an individual's evaluation of benefits, harms, costs, and inconveniences of one treatment option compared to another (Blinman et al. 2012) at a particular point in time. This distinction is important. Preferences are underpinned by beliefs about the consequences of outcomes and anticipation of the ability to cope or adjust (Blinman et al. 2012), while "values" largely reflect cultural, social, and other influences not directly related to the intervention (Koltko-Rivera 2004; Daher 2012). For complex questions or decisions, knowledge and experience have a strong influence on an

individual's beliefs of the consequence of an outcome. This could include expectations of improvement or deterioration to quality of life, anticipation of the range and
severity of side effects, and the range and likelihood of adverse outcomes associated
with an intervention. Expectations are in turn strongly influenced by interactions
with health professionals, other patients, family, friends, and the media (Ubel et al.
2005; Epstein and Peters 2009; Hausman 2012; Dirksen et al. 2013).

A prime distinction between "preferences" and "values" is that as expectations
and beliefs change with time, experience, and health (Ubel et al. 2005), "preferences" for treatment options may vary or even reverse (Slovic 1995). This is
particularly so with chronic health conditions (Dipchand 2012; Gordon et al.
2013). Indeed, for complex problems or where there is no clear choice, preferences
may be constructed during the decision process (Slovic 1995). In contrast, individual
values are influenced by cultural and societal norms, religious and nonreligious
beliefs, and close relationships (family and friends). Thus more stable aspects of
an individual's "world view" define values (Koltko-Rivera 2004; Daher 2012).
Furthermore, preferences may be underpinned by beliefs and expectations that are
erroneous or biased (skewed). Moreover, there may be little or no relevant experience to draw on to formulate well-founded beliefs and expectations (Hausman
2012). In contrast to values, erroneous or skewed beliefs of consequences may be
challenged and influenced by health professionals (Légaré et al. 2012).

In summary, patient preferences are best defined as "statements made by individuals regarding the relative desirability of a range of health experiences, treatment
options and health statements" (Brennan and Strombom 1998, p. 259). An individual's preference for a treatment "reflects their evaluation of its relative benefits,
harms, costs, and inconveniences in comparison with a given alternative or alternatives" (Blinman et al. 2012, p. 1104). The goal of patient-centered care or shared
decision-making should be that patient preferences are well informed and lead to
patient decisions that are ultimately aligned with their values.

Broadly speaking, there are two types of preferences, revealed and stated:

- Revealed preferences are inferred by what people actually choose when presented
  with alternatives, for example, choosing whether to purchase private health
  insurance or participate in a cancer screening program.
- Stated preferences are inferred from what an individual says they would do given
  a hypothetical choice. They can be used to predict future choices or identify the
  attributes of an intervention or program such as the risk of adverse outcomes that
  most influence their choices. Stated preferences are also used to assign values to
  health states and health goods and services (Lancsar et al. 2011).

In health-care settings, revealed preferences are difficult to uncover, and the
choices made commonly do not reflect the preferences of the person accessing a
service or choosing an intervention. Available options are commonly constrained by
structural barriers to access, and the preferred alternative may not be available, for
example, rural and remote patients may have limited access compared to urban
settings. Furthermore, as physicians act as the gatekeepers to many if not most health

services, the choice made may reflect the physicians' preferences and not their patients. Also different physicians may offer access to different options even within a single health-care setting. In contrast, the stated preference for an intervention or health state reflects a personal evaluation of what it would be like to live with one set of outcomes or conditions compared to another set. In addition, health research is commonly interested in understanding factors that might influence acceptance (or not) of public health programs or treatment options (Lancsar and Louviere 2008; de Bekker-Grob et al. 2012; Clark et al. 2014) to assist in development of the programs. As a consequence, studies in health mostly address stated rather than revealed preferences.

An important limitation common to all methods of preference elicitation including DCEs is that individual preferences are underpinned by expectations and beliefs that are in turn influenced by the way in which outcomes or services are framed or communicated. In short, stated preferences may be a construct of the elicitation method (Slovic 1995). As such, "data from preference elicitation tasks partly reflect individuals' preferences and partly the manner in which the preferences were elicited" (Lloyd 2003, p. 394).

3 Estimating Preferences from Choices

Discrete choice experiments (DCEs) have become one of the most common techniques to elicit preferences in situations where choosing a commodity (or a health program, intervention, service, and so on) is dependent on multiple aspects or attributes of that commodity. As a simple illustration, when choosing to drive or catch a bus to work, an individual will take into account multiple factors such as the cost, convenience, travel time, availability of parking, and weather. Furthermore, if one or more of these factors change, for example, the weather goes from sunny to rainy, their decision may switch from the bus to the car. A DCE can elicit preferences for the multiple attributes, and these can be used to predict when and how often the bus might be chosen or to develop strategies that encourage greater use of the bus. The advantage of a DCE over simple ranking or rating scales is that behavioral factors underpinning preferences can also be evaluated (Train 2009). There are three broad areas where DCEs have been particularly useful in health-related research:

1. Identifying individual preferences, priorities, and values for health service delivery, health prevention programs, treatment options, etc. and outcomes and preferences for specific health states.
2. Evaluating the trade-offs individuals may be willing to make in balancing benefits and harms. For example, the trade-off between treatment-related side effects and adverse outcomes and disease-free survival.
3. The trade-off individuals may be willing to make to achieve a desirable outcome or avoid an undesirable outcome. This could be in terms of trading the length of life for a shorter but better quality of life or willingness to pay more to attend a clinic with shorter waiting times or to receive an alternate treatment.

3.1   Theoretical Framework for Discrete Choice Experiments

Discrete choice experiments derive preferences by analysis of choice data that has been collected in a systematic way. In a DCE, individuals are asked to make a series of hypothetical choices between two or more options where the options are described by a defined set of attributes. By systematically varying the values of the attributes across multiple choices, the relative preferences for each attribute can be estimated (Bryan and Dolan 2004). Attribute values may be descriptive, ordinal, continuous, or categorical. Figure 1 shows a single choice task for a UK study of preferences for local neighborhood physician practices versus those out of the local area (Lagarde et al. 2015). The attributes characterizing the clinics are flexibility of opening hours; time taken to get an appointment, whether the clinic meets individual health needs; and the experience of the clinic. In this study, participants were shown 16 separate choice tasks across which the values of the attributes have been systematically varied.

The analysis of the choices made in a DCE such as shown in Fig. 1 is underpinned by a theoretical framework of choice behavior and utility maximization proposed and largely developed by econometricians (For an overview of the history of development of choice modeling, refer to Hensher et al. 2016) and applicable to all of the many disciplines in which DCEs have been used (Hensher et al. 2016). This choice theory assumes that when an individual decides between the two clinics, the choice of one over the other is based on a collective assessment of the attributes. The

Question 1 of 16	Practice in your local neighbourhood	Practice outside your local neighbourhood
Open Saturday and Sunday morning	Not open	Open
Open at lunchtime	Always	Sometimes
Extended opening hours	No extended hours	Extended hours
How quickly can usually see a GP	Normally in the next few days	Normally same day
Meets your specific health needs	Yes	Yes
How well the practice knows the health care services	The practice has experience with most providers	The practice has experience with some providers
Which would you choose to register with?	◯	◯

Fig. 1 A single scenario for the choice of a primary care clinic has been adapted from a discrete choice experiment to elicit preferences for GP practices in the UK (Lagarde et al. 2015). The clinics options are described by 6 attributes the values of which are varied across 16 choice scenarios

clinic with the most favorable balance of attributes will be chosen as this maximizes the individual's satisfaction or utility (Hensher et al. 2016). A change in just one attribute, say being open on a Saturday, could result in a switch in choice from one clinic to another. This choice theory is known as random utility maximization or random utility theory (RUT) (McFadden 2001) and has been applied across many economic and noneconomic disciplines including marketing, transport, environmental economics, and health-related research (Clark et al. 2014). Random utility theory underpins both the design and analysis (Hensher et al. 2005; Lancsar and Louviere 2008) of DCEs across all of these fields.

The "random" component in RUT arises from the researcher's inability to identify every single attribute that characterizes the options or to know all of the individual's characteristics that influence their choice. As a consequence, an individual's choice may reflect attributes and characteristics not known to the researcher and, therefore, not included in the DCE. These may be totally unrelated to the included attributes, or an unknown factor may be substituted for one of the defined attributes. In the example shown in Fig. 1, an individual may prefer out of area clinics as they may think local clinics to be more likely staffed by local residents and to present privacy concerns for them. This would be unrelated to the included attributes. In contrast, if an individual had an underlying (latent) preference for smaller clinics, they may consider that extended opening hours are likely to suggest a large busy clinic rather than a small one. In this case, extended opening hours could be substituted for the suspected size of the clinic, and this then influences their choice. The choices made in the DCE are also influenced by errors of judgment arising from the lack of knowledge of the meaning or the consequences of an attribute or option, poorly described or ambiguous attributes, and a range of survey completion errors. Random utility theory recognizes all of these as contributing to a random component of utility. The utility associated with a choice can then be expressed as the sum of the utility associated with known factors and the unexplainable random component (Lancsar and Louviere 2008; Hensher et al. 2016):

$$U_{(ij)} = V_{(ij)} + \varepsilon_{(ij)} \quad, j = 1, \ldots . J$$

Where:

$U_{(ij)}$ is the utility for individual i for choice j
$V_{(ij)}$ is the explainable component as defined by the attributes and characetristics, and
ε_{ij} is the unexplained or random component of utility.

The explainable component of utility is generally considered to be a linear function of the properties of the attributes and known characteristics of the participants such as age, gender, and health according to the following equation (Lancsar and Louviere 2008; Hensher et al. 2016):

$$V_{(ij)} = X_{(ij)}\beta + Z_{(ij)}\gamma$$

Where:

$X_{(ij)}$ is the vector of attributes included in the design,

$Z_{(ij)}$ is the vector of participant characteristics or other covariate, and

β and γ are the vector coefficients to be estimated.

In the example shown in Figure, 1 the utility functions for the local and non-local clinics can be written as:

$$V_{local} = ASC_{local} + \beta_{sat\_sun} * sat\_sun + \beta_{lunchtime} * lunchtime + \beta_{extended} * extended$$
$$+ \beta_{time\_GP} * time\_GP + \beta_{health\_needs} * health\_needs + \beta_{experience}$$
$$* experience$$

$$V_{non-local} = ASC_{non\_local} + \beta_{sat\_sun} * sat\_sun + \beta_{lunchtime} * lunchtime + \beta_{extended}$$
$$* extended + \beta_{time\_GP} * time\_GP + \beta_{health\_needs} * health\_needs$$
$$+ \beta_{experience} * experience$$

Where:

ASC_{local} and ASC_{non_local} are alternative specific constants for the two types of clinics.

β_{sat_sun}, $\beta_{lunchtime}$, etc.are regression coefficients for the attributes

sat_sun, lunchtime, etc. are the attribute values.

Alternative specific constants provide some information on the unobserved effects; in this case, it would indicate if there was an underlying (latent) preference for one clinic over another irrespective of the known attributes. One approach that can be used to minimize the effects of a latent preference and to encourage greater consideration of the individual attributes is not to label the clinics and simply call them A and B. Including a choice option of "I would choose neither clinic" can also encourage greater consideration of the individual attributes.

Estimation of coefficients in the utility function is undertaken using regression models that apply varying assumptions to account for the random error component ε_{ij} (Readers are referred to the following texts for detailed descriptions of the derivation of regression models for choice analysis: Train (2009); Hensher et al. (2016)). The simplest approach assumes that ε_{ij} is independent of and identically distributed across all participants that allows a closed-form computation of a multinomial logit (MNL) regression model. Essentially, this simplification assumes that the unknown portion of utility of one alternative is unrelated and independent of all other alternatives in the DCE. Furthermore, it assumes that each choice is an independent observation, which is not the case as individuals provide multiple choices or observations and these are unlikely to be independent. This greatly simplifies the estimation of the utility function and enables estimation of average preferences across populations. However, it presents limitations in the evaluation of heterogeneity of preferences between individuals and correlation of preferences within individuals (Train 2009). For example, in the UK study shown in Fig. 1, individuals who were 65 years and older and those who look after their family at home (both known characteristics) were more likely to choose the local practice. This preference could reflect distinct and essentially independent unobserved reasons, for example, the lack of access to transport for the elderly and the desire to minimize time away from home for the second group. Alternatively, the reasons could be related or even the same, as the elderly may have to rely on the same family carers for transport. Under the simplifying assumption of an MNL, it is not possible to evaluate these associations. Despite these shortcomings, Train (2009) makes the point that in many situations, the MNL provides a simple and robust approach to the evaluation of preferences and is often sufficient to provide good estimates of average preferences (Train 2009).

A number of approaches have been developed to relax the simplifying assumptions of the MNL. These include forms of MNL models that use simulation methods to allow for heterogeneity within and between individuals with respect to known and unknown factors and to allow for correlations between choices made by individual participants. These mixed or random component MNLs also allow estimation of individual as well as average preferences. Other approaches include latent class and nested MNL models that allow for heterogeneity between the classes or nests while enforcing the simplifying MNL assumptions within the classes/nests (Lancsar and Louviere 2008; Hensher et al. 2016). Table 1, adapted from Hauber et al. (2016), provides a summary of commonly used approaches to DCE estimation methods and advantages and limitations.

3.2   Best–Worst Scaling Surveys

A best-worst scaling (BWS) survey is a specific form of DCE that was initially developed as a means of increasing the information obtained from a DCE by providing partial or complete rankings of choice options rather than a single choice (Flynn et al. 2007). In a BWS, the participant is asked to indicate the most preferred or best option and also the least preferred or worst option. Depending on the number of options, participants could be asked to indicate the best, worst, next best, next worst, and so on. There are three broad types of BWS surveys (Louviere et al. 2015), and these are briefly described below.

Object scaling (Case 1). These are suited to eliciting preferences for a large list of factors or objects. Rather than asking participants to rank all of the factors from least to most important, they are shown a small subset and asked to choose the best and the worst (most important, least important) from that subset. They are shown multiple subsets where the factors presented are varied and the best and the worst is chosen for each subset. Analysis of the choices provides a basis for identifying the relative

Table 1 Advantages and disadvantages of common analysis methods (Adapted from Hauber et al. 2016)

Method	Advantages	Limitations
Multinomial logit (MNL)	Focuses on average preferences Commonly available in software packages	Assumes homogeneity in preferences. Does not account for correlation of choices made by individual participants.
Random-parameters logit or mixed MNL	Models heterogeneity. Accounts for the panel nature of the data. Available in some software packages. Can deal with heterogeneity of preferences and allow for correlation across attributes	More difficult to use than MNL. Requires assumptions about the distribution of parameters across respondents. Requires larger sample sizes than MNL models
Latent-class model	Models latent classes and describes heterogeneity by class. Requires smaller samples than mixed MNL and RPL	Requires specialized software. Judgment required to determine appropriate number of classes. Difficult to interpret results from any given class when the chance of being in all classes is more or less the same across respondents. The required sample size varies with the number of classes in the model

Identify which technology you believe will have the most impact and the least impact in your country over the next 5 to 10 years	Most impact (one only)	Least impact (one only)
Molecular target therapy	☐	☐
Stem cell therapy	☐	☐
Adjuvant/Neo-adjuvant therapies	☐	☐
Transplant technologies	☐	☐
Genetic/Genomic biomarkers	☐	☐

Fig. 2 Example of an object scaling (Case 1) BWS. Adapted from a survey of clinicians regarding their expectations of the impact of 11 emerging technologies for the treatment of hepatocellular carcinoma (Gallego et al. 2012). Participants were shown a series of choice sets each with a different subset of the five technologies

importance of all of the factors on a continuous scale. An example of an object scaling BWS is shown in Fig. 2.

Profile scaling (Case 2). These are applied where a single profile (e.g., a health state, treatment, or service) can be described by a set of attributes that can have varying values or levels. Multiple scenarios of single profiles are shown to participants all of which have the same attributes; however the values of the attributes are varied between scenarios. As for the Case 1, BWS participants are asked to select the best and worst attributes from the profile. The attribute levels may be numeric (the

Consider the following attributes in asthma control. Which do you consider to be the best and which is the worst.		
Best		Worst
☐	**Night symptoms:** 3 days per week	☐
☐	**Wheezing or tightening of the chest:** None	☐
☐	**Changing medication:** To add oral steroids for 5 days	☐
☐	**Emergency visits:** 10 per year	☐
☐	**Limitation to physical activities:** 2 times per month	☐

Fig. 3 Example of a profile scaling BWS (Case 2). Adapted from a study eliciting preferences of parents and adolescents for the control of asthma (Ungar et al. 2014). Each attribute is described by 3 to 4 levels representing the range of possibilities for that attribute (e.g., nighttime symptoms, none, 3 days a week, 5 days a week). Each choice set contains the same five attributes; however the attribute levels are varied across the choice sets

probability that a side effect will occur) or descriptive (the severity of the side effect). It has been suggested that when there are a large number of attributes, the Case 2 BWS survey is less cognitively demanding than a DCE and that it should result in greater consideration of all attributes (Flynn et al. 2007), although this view is not universally accepted (Whitty et al. 2014a, 2014b). An example of a profile scaling BWS is shown in Fig. 3.

Best-worst DCE (Case 3). The best-worst DCE is essentially a standard DCE that includes an additional best-worst task. Each scenario consists of three or more multi-attribute profiles as per a DCE, and participants are asked to show their preferred option and then to select the worst of the remaining profiles (Lancsar et al. 2013). Case 3 aims at augmenting data collected from the DCE by providing partial or complete ranking in addition to the preferred choice. An example of a best-worst DCE is shown in Fig. 4.

Since 2008, BWS surveys have gained increasing use in the evaluation of health-related research questions; however, the absolute number has been relatively small with 53 publications overall and only 15 and 16 published in 2014 and 2015, respectively (Mühlbacher et al. 2016). They can be cognitively less demanding than DCEs, although as noted above, this is not universally accepted (Whitty et al. 2014a, 2014b), and enable assessment of the relative importance of a large number of attributes on a continuous scale avoiding issues associated with ranking and rating techniques. Compared to DCEs, a BWS can provide a more realistic approximation of questions where there is no clear choice to be made. For example, maintenance immunosuppression after transplantation is best described as a series of adjustments with resulting changes to the benefits and harms rather than a discrete choice of one option over another. Another advantage is that attributes are assessed on the same underlying scale, enabling direct comparison of the relative importance of attribute

	Scenario 1	Scenario 2	Scenario 3	Scenario 4
Arm/hand function	Normal	Impaired	Impaired	Impaired
Walking	Wheelchair	Normal	Wheelchair	Wheelchair
Bladder Bowel	Impaired	Impaired	Normal	Impaired
Sexual	Normal	Dysfunction	Dysfunction	Dysfunction
Pain	Occasional	Occasional	Occasional	Persistent
Best	☐	☐	☐	☐
Worst	☐	☐	☐	☐
Second best	☐	☐	☐	☐

Fig. 4 Example of a best-worst DCE task. Adapted from a study of people with a spinal cord injury (Lo et al. 2016). Each attribute is described by two levels representing possible outcomes for that attribute (e.g., walking; wheelchair or normal). In each choice set, the attributes remain the same; however the attribute levels are varied across the choice sets. In this example, participants are asked to select the best (i.e., their choice in the DCE as well as the worst scenario and their second best scenario)

levels both within and between attributes (Marley et al. 2008; Flynn 2010; Louviere et al. 2015).

As BWS surveys are underpinned by the same random utility framework as DCEs, design and analysis follow the same general principles (Marley et al. 2008; Louviere et al. 2015; Mühlbacher et al. 2016). At the simplest level, the best and worst choices can be counted for each attribute level with the best minus the worst score providing a measure of the preference for the attribute levels. While this has been shown to provide a robust estimate of relative importance, it does not allow for estimation of trade-offs between attributes (Louviere et al. 2015).

When using MNL models, there are considerations specific to the completion of a BWS survey that need to be taken into account. The most important are assumptions as to the order in which selections are made, for example, the best first then the worst, or the other way around or both at the same time, and in terms of utility whether a worst selection is a negative mirror of a best. Completion styles may vary between participants, and the magnitude and direction between desirable and undesirable attributes may vary between participants and attributes (Rose 2014). In Louviere and Flynn's view, as the differences are most likely minor, the pragmatic approach is to simplify analysis by assuming the worst is a negative mirror of and selected after the best choice (Louviere et al. 2015).

In summary, a BWS survey is an alternate approach to eliciting preferences under the general framework of random utility theory. They have potential advantages to DCEs including ability to assess attributes and attribute levels on the same underlying scale and to avoid problems associated with rating and ranking techniques. A BWS survey can be used as a data augmentation technique for a DCE or as an alternate solution to elicitation of patient preferences for multiple attributes associated with complex health questions.

4 Steps to Conducting a Discrete Choice Experiment

Comparative studies of preference elicitation techniques suggest that DCEs (and BWS surveys) encourage greater consideration of all attributes and should provide more thorough evaluations of complex questions (Pignone et al. 2013; Wijnen et al. 2015). However, as with all survey techniques (see ▶ Chap. 32, "Traditional Survey and Questionnaire Platforms"), DCEs are subject to a range of biases that need to be addressed in both design and analysis. While a DCE should encourage consideration of all attributes, individuals may still focus on a limited number or just one attribute and ignore the remainder. In addition to ignoring attributes, they may take mental shortcuts (heuristics) in completing the survey and show diminishing attention to attributes as the survey progresses (Cairns et al. 2002; Lloyd 2003; Hensher et al. 2012). Other biases include status quo (Salkeld et al. 2000), framing effects (Howard and Salkeld 2009), affect heuristics (Slovic et al. 2005), and a range of survey completion biases (left to right, top to bottom) (Campbell and Erdem 2015; see also ▶ Chap. 32, "Traditional Survey and Questionnaire Platforms").

Cognitive burden increases with increasing number of attributes and choices, and most DCEs are limited to a relatively small number of attributes anticipated by the researchers as being most important. Over 90% of the health-related studies published from 2001 to 2012 had fewer than ten attributes with the majority having only four or five (Clark et al. 2014). Simplifying a DCE by limiting the number of attributes limits the ability to address complex questions, and participant's choices may be increasingly determined by unknown factors, thus increasing the error component (Hensher et al. 2005; Lancsar et al. 2017). Conversely, completion and nonattendance errors increase with increasing complexity. These issues may be addressed to a varying extent by the use of qualitative studies, pilot testing with and without qualitative assessment tools such as "thinking aloud" during completion (Ryan et al. 2009), and the use of experimental designs aimed at limiting cognitive burden (Lancsar and Louviere 2008; Clark et al. 2014). In short, maximizing the value of DCEs requires careful planning and an iterative approach to design and assessment.

Discrete choice experiments are highly flexible in terms of research questions, design, and analysis. The number of choices, whether they are labeled or unlabeled, whether there is a status quo or opt-out choice, the number and type of attributes (numeric, descriptive, categorical), and the number of levels for each attribute can all be varied to suit the question of interest. The success of a DCE requires a balance between the ease of completion and complexity as being too simple or too complex both introduces errors. The key stages in conducting a DCE in health-related research is outlined in Fig. 5 adapted from Bridges et al. (2011). Each of these stages, described below, is applicable to both DCEs and BWS surveys.

1. **Research Objectives**

   All health-related research questions need to be clearly defined, have a testable hypothesis, and follow a structured approach relevant to the area of research (e.g., clinical research, health economics, health service delivery, and so on). There needs to be a clear rationale for using a DCE over less complex alternate methods.

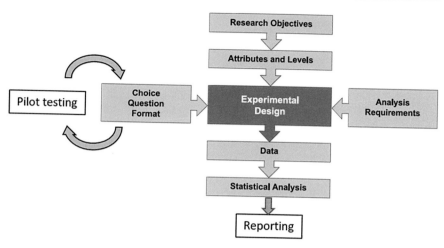

Fig. 5 A checklist for planning and conduct of discrete choice experiments in health-related research adapted from (Bridges et al. 2011)

2. Attributes and Levels

As the objective of a DCE is to provide, as far as possible, an unbiased evaluation of preferences, it is important that the attributes reflect those that are anticipated to be meaningful and likely to be the most important or relevant for the research question and the population. This will generally require a combination of literature reviews, opinions from expert panels, consensus meetings, and/or qualitative research including focus groups, structured interviews, and nominal groups techniques (see ▶ Chaps. 23, "Qualitative Interviewing," and ▶ 42, "Consensus Methods: Nominal Group Technique"). The evidence collected to support attribute selection should be relevant to the population. A DCE that aims to elicit patient preferences, which relies solely on the views of an expert panel, will exclude outcomes valued by patients not identified by the experts and/or place different levels of importance on outcomes. In situations where the number of possible attributes is large, the researcher must identify a subset of attributes for inclusion in the DCE, and these, as far as possible, should be those most relevant to the participants, and not experts or researchers. To minimize the influence of expert opinion or researcher bias in eliminating potential attributes, qualitative research techniques exploring the preferences of relevant stakeholder groups are an important component of the design process. As an example, Fig. 6 shows the findings of a nominal group study undertaken to inform the design of a DCE addressing the trade-offs between outcomes associated with maintenance immunosuppression after a kidney transplant (Howell et al. 2012; Howell et al. 2017). Transplant recipients identified 47 unique outcomes all of which were relevant to the question and in theory could be included in the DCE. Analysis of the priority scores, the reasons underpinning the scores, and a pilot-scale DCE formed the basis for selection of just nine attributes in the final DCE (Howell et al. 2017). Attribute levels must

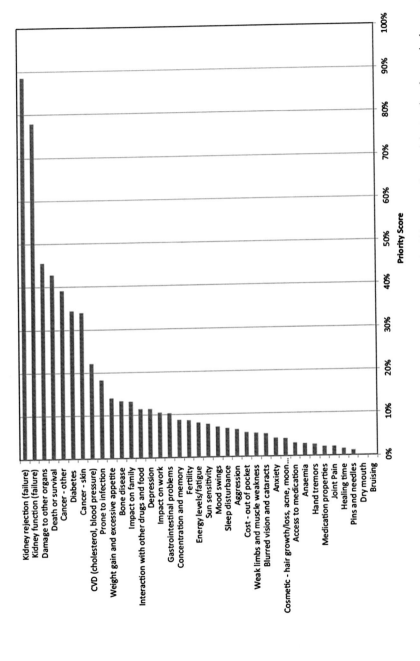

Fig. 6 Mean priority scores for outcomes associated with immunosuppression identified by kidney transplant recipients using the nominal group technique (Howell et al. 2012)

also be realistic and consistent with the research question. As far as possible, clinical outcomes should fall within expected ranges, while descriptive values should not be irrational or present implausible scenarios. However, the levels need to be sufficiently broad so as to influence choices made between profiles (Lancsar and Louviere 2008).

3. **Question Formats**

The DCE relies heavily on the construction of choice tasks that can be understood by the study population and an experimental design that combines attributes in a way that maximizes choice data while minimizing cognitive burden. Constructing questions also needs to consider the way in which profiles are presented, the number of choices within each choice set, whether they should be labeled, whether a status quo or opt-out option should be included, and the way in which attribute levels are described. Numeric levels may be numbers, words, pictograms, or a combination of methods. Figure 7 shows a single choice question from a DCE investigating the heterogeneity of women's preferences for breast screening and demonstrates the use of pictograms, numbers, and alternate ways of describing out-of-pocket costs (Vass et al. 2017).

4. **Experimental Design**

Experimental design refers to the process whereby a subset of all of the possible combinations of attributes and attribute levels is selected and combined into the individual choice sets or tasks (Lancsar and Louviere 2008). Good experimental design is key to achieving a DCE that balances the number of choice tasks and

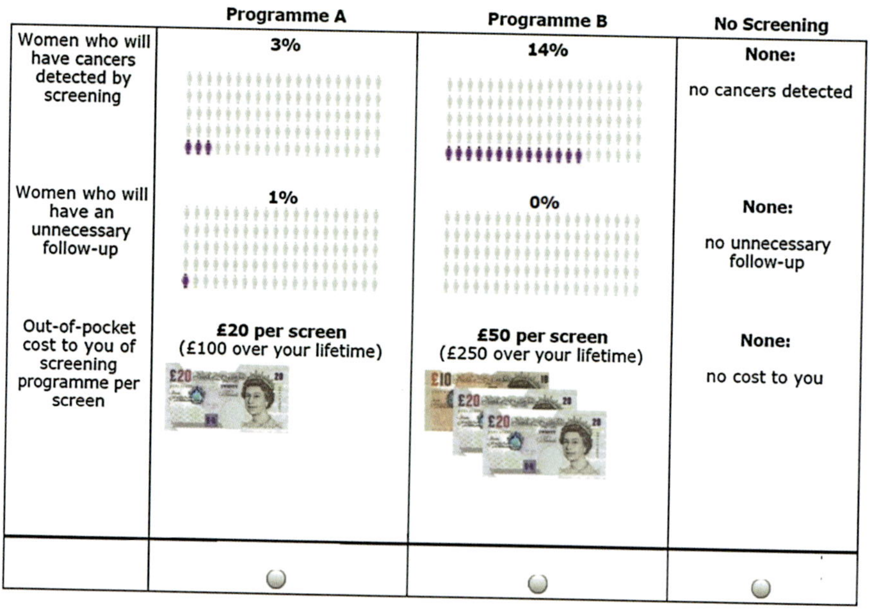

Fig. 7 Single choice task from a DCE investigating heterogeneity in women's preferences for breast screening (Vass et al. 2017)

thus complexity of the survey and the ability to estimate the choice models. There are a number of approaches to experimental design that are beyond the scope of this chapter to detail. Broadly speaking, experimental design is a balance between minimizing errors in responses and maximizing statistical efficiency and is dependent on the analytical requirements for the research question (Johnson et al. 2013). Sample size requirements are also linked to experimental design (Readers are referred to the following for further detail on experimental design: Rose and Bliemer (2013), Rose et al. (2008), Johnson et al. (2013)). It is also important to avoid implausible combinations of attribute levels in profiles, and constraints in the occurrence of certain combinations may need to be applied (Lancsar and Louviere 2008). The final design may also include additional choice sets aimed at identifying internal validity and the quality of the responses. This may take the form of duplicate choice sets for which the choice by a participant should be the same or a dominant choice set where only one response would be considered feasible or rational. While this can be used to exclude participants from the analysis, the size of the survey is increased, and the removal of participant responses assumes that "incorrect" choices are, from a clinical perspective, irrational or implausible which may not be the case for all participants. Consideration of what is implausible should be cognizant of the participant's perspective rather than clinical considerations (Lancsar and Louviere 2008). Lancsar and Louviere (2006) argue against deleting responses that are thought to be "irrational," as this may result in the removal of valid responses and RUT is able to cope with such data. Usual practice is to maximize data collected.

5. **Analysis Requirements**

As detailed in Section 3, there are multiple approaches to the analysis of choice data collected from a DCE. There is no single preferred approach, and the method chosen will reflect the research question and the extent to which simplifying assumptions will affect interpretation. As noted by Train (2009), more complex models that relax the simplifying assumptions of an MNL may not always be necessary. However, as the simplifying assumptions limit the ability of the DCE to evaluate preference heterogeneity and the influence of attribute and participant characteristics, more complex models are more commonly selected. Model selection should identify the simplest (parsimonious) approach to addressing the research question. Model selection influences sample size requirements and experimental design efficiency. Most analyses will also require respondent characteristics, and these should be identified in the planning phase.

6. **Pilot Testing**

Question formatting and experimental design are an iterative process with pilot testing used to refine the final design. The purpose of pilot testing is to test comprehension of the choices, attributes, and levels and the complexity of the task. Piloting may take the form of interviews or "think-aloud" studies (Ryan et al. 2009; Whitty et al. 2014) or having the survey completed by sufficient numbers to allow a preliminary estimation of a choice model (Howell et al. 2016). The regression coefficients from the pilot model can be used to further refine the design.

7. **Data Collection**

Attribute levels may be descriptive, categorical, ordinal, or continuous, and many DCEs will include a mix of formats (see Fig. 7). As such, most if not all attribute levels will need to be dummy coded for analysis and interpretation. The choice of coding is important for both analysis and interpretation (Lancsar and Louviere 2008; Hauber et al. 2016). A single choice from the survey will identify the profile selected and often the attributes and the attribute levels in the profile. The data format required for analysis varies according to experimental design (e.g., DCE vs. BWS), the proposed regression model, and software and may require substantial manipulation. Respondent characteristics may be also collected as part of the survey, and these also need to be coded for inclusion as interaction terms in the utility functions.

8. **Statistical Analysis**

The analysis of choice data can involve complex statistical analysis and choice models that vary according to the objective of the DCE. The requirement for secondary estimates such as trade-off and the willingness to pay introduces another layer of complexity as does the inclusion of respondent characteristics in the utility function. There are a number of general statistical and econometric software packages that provide built-in functions suitable for choice model estimation. Not all models are able to be estimated by all statistical software packages which may present a practical limit to the analysis (Lancsar et al. 2017) (see also ▶ Chap. 54, "Data Analysis in Quantitative Research").

9. **Reporting**

The reporting of the results of the DCE has two prime objectives: firstly, to provide sufficient detail to enable an independent assessment of statistical validity of the model including statistical significance of all parameters and, secondly, to report the findings in the context of the research question. The meaning of attribute coefficients is often not intuitive and can be difficult to explain to the intended audience, and a range of approaches may be required beyond that needed to demonstrate validity. Regression coefficients can be expressed as odds ratios to aid interpretation; however given the mix of numeric and descriptive attribute levels, interpretation may also not be intuitive (see also ▶ Chap. 56, "Writing Quantitative Research Studies").

5 Conclusions and Future Directions

In summary, DCEs are well suited and widely used to elicit the stated preferences of patients, their caregivers, the general public, and health professionals for a range of health-related questions. They are underpinned by a well-established theoretical framework and have the flexibility to address a wide range of questions. The prime limitation in addressing complex health decisions is the cognitive demand of the survey associated with multiple profiles, multiple attributes with multiple levels, and the expectation that individuals will give attention to all scenarios and attributes equally. The more complex the DCE, the greater the errors, and the more biased the estimates. Best-worst scaling surveys have been used as one approach to minimize complexity. Careful planning and implementation are required to ensure the DCE provides reliable and meaningful answers to the research question. Given the flexibility and robust framework for assessment, DCEs are becoming a commonly used tool for preference elicitation across clinical research, health economic evaluation, health service delivery, and patient-centered research. The increasing availability of large data sets from registries of prescription and medical service use and the ability to link these to individual patients raise the potential for DCEs to combine this revealed data with stated preferences using econometric models to explore the extent to which health services align with preferences and values. Given the increasing trend for inclusion of the preferences and values of patients, their carers, and the general public in decisions related to all aspects of health service delivery, DCEs and BWS surveys are likely to become more widely used.

References

Blinman P, King M, Norman R, Viney R, Stockler MR. Preferences for cancer treatments: an overview of methods and applications in oncology. Ann Oncol. 2012;23(5):1104–10.

Braddock CH. Supporting shared decision making when clinical evidence is low. Med Care Res Rev. 2013;70(1 suppl):129S–40S.

Brennan PF, Strombom I. Improving health care by understanding patient preferences. The Role of Computer Technology. J Am Med Inform Assoc. 1998;5(3):257–62.

Bridges JFP, Hauber AB, Marshall DA, lloyd A, Prosser LA, Regier DA, Johnson FR, Mauskopf J. Conjoint analysis applications in health—a checklist: A report of the ISPOR good research practices for conjoint analysis task force. Value Health. 2011;14:403–13.

Bryan S, Dolan P. Discrete choice experiments in health economics. Eur J Health Econ. 2004; 5(3):199–202.

Cairns J, van der Pol M, Lloyd A. Decision making heuristics and the elicitation of preferences: being fast and frugal about the future. Health Econ. 2002;11(7):655–8.

Campbell D, Erdem S. Position bias in best-worst scaling surveys: a case study on Trust in Institutions. Am J Agric Econ. 2015;97(2):526–45.

Chuck A, Adamowicz W, Jacobs P, Ohinmaa A, Dick B, Rashiq S. The willingness to pay for reducing pain and pain-related disability. Value Health. 2009;12(4):498–506.

Clark M, Determann D, Petrou S, Moro D, de Bekker-Grob E. Discrete choice experiments in health economics: a review of the literature. PharmacoEconomics. 2014;32(9):883–902.

Committee on Quality of Health Care in America Institute of Medicine. Crossing the quality chasm: a new health system for the 21st century. Washington: National Academies Press; 2012.

Daher M. Cultural beliefs and values in cancer patients. Ann Oncol. 2012;23(suppl 3):66–9.

de Bekker-Grob EW, Ryan M, Gerard K. Discrete choice experiments in health economics: a review of the literature. Health Econ. 2012;21(2):145–72.

Dipchand AI. Decision-making in the face of end-stage organ failure: high-risk transplantation and end-of-life care. Curr Opin Organ Transplant. 2012;17(5):520–4.

Dirksen C, Utens C, Joore M, van Barneveld T, Boer B, Dreesens D, van Laarhoven H, Smit C, Stiggelbout A, van der Weijden T. Integrating evidence on patient preferences in healthcare policy decisions: protocol of the patient-VIP study. Implement Sci. 2013;8(1):64.

Epstein RM, Peters E. Beyond information: exploring patients' preferences. JAMA. 2009; 302(2):195–7.

Flynn TN. Valuing citizen and patient preferences in health: recent developments in three types of best-worst scaling. Expert Rev Pharmacoecon Outcomes Res. 2010;10(3):259–67.

Flynn TN, Louviere JJ, Peters TJ, Coast J. Best–worst scaling: what it can do for health care research and how to do it. J Health Econ. 2007;26(1):171–89.

Gordon EJ, Butt Z, Jensen SE, Lok-Ming Lehr A, Franklin J, Becker Y, Sherman L, Chon WJ, Beauvais N, Hanneman J, Penrod D, Ison MG, Abecassis MM. Opportunities for shared decision making in kidney transplantation. Am J Transplant. 2013;13(5):1149–58.

Groothuis-Oudshoorn C, Fermont J, van Til J, IJzerman M. Public stated preferences and predicted uptake for genome-based colorectal cancer screening. BMC Med Inform Decis Mak. 2014; 14(1):18.

Hauber AB, González JM, Groothuis-Oudshoorn CGM, Prior T, Marshall DA, Cunningham C, Ijzerman MJ, Bridges JFP. Statistical methods for the analysis of discrete choice experiments: a report of the ISPOR conjoint analysis good research practices task force. Value Health. 2016. https://doi.org/10.1016/j.jval.2016.04.004.

Hausman D. Preferences, value, choice, and welfare. New York: Cambridge University Press; 2012.

Hensher D, Rose J, Greene W. Applied choice analysis: a primer. Cambridge: Cambridge University Press; 2005.

Hensher D, Rose J, Greene W. Inferring attribute non-attendance from stated choice data: implications for willingness to pay estimates and a warning for stated choice experiment design. Transportation. 2012;39(2):235–45.

Hensher D, Rose J, Greene W. Applied choice analysis. Cambridge: Cambridge Books; 2016.

Howard K, Salkeld G. Does attribute framing in discrete choice experiments influence willingness to pay? Results from a discrete choice experiment in screening for colorectal cancer. Value Health. 2009;12(2):354–63.

Howard K, Salkeld GP, Patel MI, Mann GJ, Pignone MP. Men's preferences and trade-offs for prostate cancer screening: a discrete choice experiment. Health Expect. 2014;18(6):3123–35.

Howell M, Tong A, Wong G, Craig JC, Howard K. Important outcomes for kidney transplant recipients: a nominal group and qualitative study. Am J Kidney Dis. 2012;60(2):186–96.

Howell M, Wong G, Rose J, Tong A, Craig JC, Howard K. Eliciting patient preferences, priorities and trade-offs for outcomes following kidney transplantation: a pilot best–worst scaling survey. BMJ Open. 2016;6(1)

Howell M, Wong G, Rose J, Tong A, Craig JC, Howard K. Patient preferences for outcomes after kidney transplantation: a best-worst scaling survey. Transplantation. 2017;101(11):2765–73.

Kan H, de Bekker-Grob E, van Marion E, van Oijen G, van Nieuwenhoven C, Zhou C, Hovius S, Selles R. Patients' preferences for treatment for Dupuytren's disease: a discrete choice experiment. Plast Reconstr Surg. 2016;137(1):165–73.

Kawata A, Kleinman L, Harding G, Ramachandran S. Evaluation of patient preference and willingness to pay for attributes of maintenance medication for chronic obstructive pulmonary disease (COPD). Patient. 2014;7(4):413–26.

Koltko-Rivera ME. The psychology of worldviews. Rev Gen Psychol. 2004;8(1):3–58.

Laba T-L, Howard K, Rose J, Peiris D, Redfern J, Usherwood T, Cass A, Patel A, Jan S. Patient preferences for a Polypill for the prevention of cardiovascular diseases. Ann Pharmacother. 2015. https://doi.org/10.1177/1060028015570468.

Lagarde M, Erens B, Mays N. Determinants of the choice of GP practice registration in England: evidence from a discrete choice experiment. Health Policy. 2015;119(0):427–36.

Lancsar E, Fiebig DG, Hole AR. Discrete choice experiments: a guide to model specification, estimation and software. PharmacoEconomics. 2017; 1–20

Lancsar E, Louviere J. Deleting 'irrational' responses from discrete choice experiments: a case of investigating or imposing preferences? Health Econ. 2006;15(8):797–811.

Lancsar E, Louviere J. Conducting discrete choice experiments to inform healthcare decision making: a user's guide (practical application). PharmacoEconomics. 2008;26(8):661–77.

Lancsar E, Louviere J, Donaldson C, Currie G, Burgess L. Best worst discrete choice experiments in health: methods and an application. Soc Sci Med. 2013;76(0):74–82.

Lancsar E, Wildman J, Donaldson C, Ryan M, Baker R. Deriving distributional weights for QALYs through discrete choice experiments. J Health Econ. 2011;30(2):466–78.

Légaré F, Turcotte S, Stacey D, Ratté S, Kryworuchko J, Graham ID. Patients' perceptions of sharing in decisions. Patient. 2012;5(1):1–19.

Lloyd AJ. Threats to the estimation of benefit: are preference elicitation methods accurate? Health Econ. 2003;12(5):393–402.

Louviere J, Flynn TN, Marley AAJ. Best-worst scaling theory, methods and applications. Cambridge: Cambridge University Press; 2015.

Marley AAJ, Flynn TN, Louviere JJ. Probabilistic models of set-dependent and attribute-level best–worst choice. J Math Psychol. 2008;52(5):281–96.

Martin SC, Stone AM, Scott AM, Brashers DE. Medical, personal, and social forms of uncertainty across the transplantation trajectory. Qual Health Res. 2010;20(2):182–96.

McFadden D. Economic choices. Am Econ Rev. 2001;91(3):351–78.

Mühlbacher A, Juhnke C. Patient preferences versus physicians' judgement: does it make a difference in healthcare decision making? Appl Health Econ Health Policy. 2013;11(3):163–80.

Mühlbacher AC, Kaczynski A, Zweifel P, Johnson FR. Experimental measurement of preferences in health and healthcare using best-worst scaling: an overview. Heal Econ Rev. 2016;6(1):1–14.

Murray MA, Brunier G, Chung JO, Craig LA, Mills C, Thomas A, Stacey D. A systematic review of factors influencing decision-making in adults living with chronic kidney disease. Patient Educ Couns. 2009;76(2):149–58.

Norman R, Viney R, Brazier J, Burgess L, Cronin P, King M, Ratcliffe J, Street D. Valuing SF-6D health states using a discrete choice experiment. Med Decis Mak. 2013;34(6):773–86.

Persad G, Wertheimer A, Emanuel EJ. Principles for allocation of scarce medical interventions. Lancet. 2009;373(9661):423–31.

Pignone M, Howard K, Brenner A. Comparing 3 techniques for eliciting patient values for decision making about prostate-specific antigen screening: a randomized controlled trial. JAMA Intern Med. 2013;173(5):362–8.

Ratcliffe J, Brazier J, Tsuchiya A, Symonds T, Brown M. Using DCE and ranking data to estimate cardinal values for health states for deriving a preference-based single index from the sexual quality of life questionnaire. Health Econ. 2009;18(11):1261–76.

Ratcliffe J, Couzner L, Flynn T, Sawyer M, Stevens K, Brazier J, Burgess L. Valuing child health utility 9D health states with a young adolescent sample: a feasibility study to compare best-worst scaling discrete-choice experiment, standard gamble and time trade-off methods. Appl Health Econ Health Policy. 2011;9(1):15–27.

Johnson FR, Lancsar E, Marshall D, Kilambi V, Mühlbacher A, Regier DA, Bresnahan BW, Kanninen B, Bridges JFP. Constructing experimental designs for discrete-choice experiments: report of the ISPOR conjoint analysis experimental design good research practices task force. Value Health. 2013;16(1):3–13.

Regier DA, Friedman JM, Makela N, Ryan M, Marra CA. Valuing the benefit of diagnostic testing for genetic causes of idiopathic developmental disability: willingness to pay from families of affected children. Clin Genet. 2009;75(6):514–21.

Richardson G, Manca A. Calculation of quality adjusted life years in the published literature: a review of methodology and transparency. Health Econ. 2004;13(12):1203–10.

Rose J. Interpreting discrete choice models based on best-worst data: A matter of framing. Paper No. 12–3103-1. Transportation Research Board 93rd Annual General Meeting. Washington, DC. 2014.

Rose JM, Bliemer MCJ. Sample size requirements for stated choice experiments. Transportation. 2013;40(5):1021–41.

Rose JM, Bliemer MCJ, Hensher DA, Collins AT. Designing efficient stated choice experiments in the presence of reference alternatives. Transp Res B Methodol. 2008;42(4):395–406.

Ryan M, Watson V, Entwistle V. Rationalising the 'irrational': a think aloud study of discrete choice experiment responses. Health Econ. 2009;18(3):321–36.

Salkeld G, Ryan M, Short L. The veil of experience: do consumers prefer what they know best? Health Econ. 2000;9(3):267–70.

Slovic P. The construction of preference. Am Psychol. 1995;50(5):364–71.

Slovic P, Peters E, Finucane ML, Macgregor DG. Affect, risk, and decision making. Health Psychol. 2005;24(4 Suppl):S35–40.

Street RL Jr, Elwyn G, Epstein RM. Patient preferences and healthcare outcomes: an ecological perspective. Expert Rev Pharmacoecon Outcomes Res. 2012;12(2):167–80.

Torrance GW. Measurement of health state utilities for economic appraisal. J Health Econ. 1986; 5(1):1–30.

Train KE. Discrete choice methods with simulation. Cambridge: Cambridge University Press; 2009.

Ubel PA, Loewenstein G, Schwarz N, Smith D. Misimagining the unimaginable: the disability paradox and health care decision making. Health Psychol. 2005;24(4):S57–62.

Vass CM, Rigby D, Payne K. Investigating the heterogeneity in Women's preferences for breast screening: does the communication of risk matter? Value Health. 2017;

Viney R, Norman R, Brazier J, Cronin P, King MT, Ratcliffe J, Street D. An Australian discrete choice experiment to value eq-5d health states. Health Econ. 2014;23(6):729–42.

von Arx L-B, Kjær T. The patient perspective of diabetes care: a systematic review of stated preference research. Patient. 2014;7(3):283–300.

Whitty JA, Ratcliffe J, Chen G, Scuffham PA. Australian public preferences for the funding of new health technologies: a comparison of discrete choice and profile case best-worst scaling methods. Med Decis Mak. 2014a;34(5):638–54.

Whitty JA, Walker R, Golenko X, Ratcliffe J. A think aloud study comparing the validity and acceptability of discrete choice and best worst scaling methods. PLoS One. 2014b;9(4):e90635.

Wijnen B, van der Putten I, Groothuis S, de Kinderen R, Noben C, Paulus A, Ramaekers B, Vogel GC, Hiligsmann M. Discrete-choice experiments versus rating scale exercises to evaluate the importance of attributes. Expert Rev Pharmacoecon Outcomes Res. 2015;15(4):721–8.

Wortley S, Wong G, Kieu A, Howard K. Assessing stated preferences for colorectal cancer screening: a critical systematic review of discrete choice experiments. Patient. 2014;7(3): 271–82.

Randomized Controlled Trials

37

Mike Armour, Carolyn Ee, and Genevieve Z. Steiner

Contents

M. Armour (✉) · C. Ee
NICM, Western Sydney University (Campbelltown Campus), Penrith, NSW, Australia
e-mail: m.armour@westernsydney.edu.au; c.ee@westernsydney.edu.au

G. Z. Steiner
NICM and Translational Health Research Institute (THRI), Western Sydney University, Penrith, NSW, Australia
e-mail: g.steiner@westernsydney.edu.au

© Springer Nature Singapore Pte Ltd. 2019
P. Liamputtong (ed.), *Handbook of Research Methods in Health Social Sciences*,
https://doi.org/10.1007/978-981-10-5251-4_94

Abstract

This chapter covers the current gold standard for evaluating the effectiveness of therapeutic interventions, the randomized controlled trial (RCT). Key features of the RCT, regardless of sub-type, are randomization, allocation concealment, and blinding. These key features help reduce bias and the influence of confounding variables, making the randomized controlled trial eminently suitable to determine cause and effect relationships. Protocol design and registration prior to trial onset are important factors in determining the quality of the trial, and various trial design sub-types, including parallel, factorial, crossover, and cluster, are outlined and the strengths and weakness of each examined. Various checklists such as SPIRIT and CONSORT can be used to ensure proper reporting of both trial protocols and trial findings, to ensure clear, concise reporting. Finally, the shortcomings of RCTs and newer trial designs, such as comparative effectiveness research and pragmatic studies, designed to overcome some of these issues are examined, and ways to make clinical trial results more clinically applicable are discussed.

Keywords

Randomized controlled trials · Blinding · Factorial · Pragmatic · Factorial · Cluster · Crossover · Pragmatic · Comparative effectiveness

1   Introduction

> Not everything that counts can be counted, and not everything that can be counted, counts – Albert Einstein

Humans have continuously attempted to find a cure for disease. In pre- and ancient history (from the dawn of civilization to AD 1000), medicine was a product of cultural beliefs and theories and, later, some clinical experience. In prehistoric times, illness was viewed as a spiritual event and was treated likewise. Later, healers used herbal or surgical treatments that seemed to work (using clinical experience to judge). The Arabs began studying chemistry and advocating the use of chemical medications and defined rules for adding, subtracting, multiplying, and dividing in AD 800 (Mayer 2004).

During the Renaissance and Industrial Revolution, there were revolutionary changes in the understanding of the basic sciences and advances in mathematics and statistics. Vesalius rejected Galen's incorrect anatomical theories (which were based on the dissection of animals), and Paracelsus advocated for the use of chemical instead of vegetable medicines. In 1202, Fibonacci first introduced numbers to the European civilization. Prior to this, the use of Roman numerals had made complex calculations impossible. Probability theories developed, and in 1619, Gataker expounded on the meaning of probability by noting that natural laws, and not divine providence, governed these outcomes (the ancient Greeks had previously believed that the Gods decided all life). The microscope was invented in the sixteenth century, and Harvey put forward the theory of blood circulation in the seventeenth century.

Development of modern medicines in the eighteenth century saw the introduction of digitalis, the use of inoculation against smallpox, and the postulation of the existence of vitamins. Pascal refined the theories of statistics, and actuarial tables began to be used to determine insurance for merchant ships (Mayer 2004).

These changes set the scene for a new approach to defining the best current medical practice in the twentieth century. However, apart from a few pioneers, the art of medicine up to this time remained largely anecdotal and continued to be based on deductions from experiences and induction from physiological mechanisms (Mayer 2004). Sir Richard Doll wrote that in the 1930s, new treatments invariably arose as a result of physicians observing the effects in small numbers of patients (the case series) (Doll 1998). This method of using uncontrolled observations may be reliable when the intervention in question has dramatic effects, such as those of insulin on type I diabetes (Gluud 2006). For the majority of interventions though, especially for those with an outcome that is somewhat subjective, observation alone introduces bias and fails to control for factors that may lead to an apparent improvement that is not related to the intervention in question (confounding factors). Pierre Louis illustrated this in the 1800s when he demonstrated, via a retrospective analysis of a case series, that bloodletting (the most popular therapeutic invention of the day) was of no benefit in pneumonia (Doherty 2005).

During the twentieth century, biomedical research progressed in leaps and bounds, beginning with a rise in the numbers of research studies in physiology and other basic sciences and a move away from empirical observation of cases. Austin Bradford Hill published a series of articles in *The Lancet* in 1937 on the use of statistical methodology in medical research and went on to direct the first true modern randomized controlled trial (RCT) to be published – which examined the efficacy of streptomycin versus standard care for the treatment of pulmonary tuberculosis in 1948. During the 1950s, the RCT became the standard for excellent clinical research. Archie Cochrane later drove the movement to perform systematic reviews of RCTs (Mayer 2004).

2   Randomized Controlled Trials: General Aspects

Even though the first RCT was not published until 1948, controlled trials have been documented since the Old Testament. The Book of Daniel describes an experiment comparing the effects of a vegetarian diet with the standard Royal Babylonian diet in servants. James Lind, a naval surgeon, conducted a small RCT ($n = 12$) while on board the HMS Salisbury in 1747, which compared the effects of citrus fruit (two oranges and a lemon daily) against five other control groups, who had other additives to their diet (vinegar and seawater being among these), and observed that the citrus group recovered from scurvy after 14 days (Doherty 2005).

The RCT is now considered the "gold standard" in evaluating the effectiveness of a therapeutic intervention. It attempts to control or minimize confounding factors that affect outcomes and distort apparent treatment effects (Manchikanti 2008). A well-designed RCT provides the best evidence on the efficacy of healthcare interventions and should identify situations where the difference in outcomes between two or more intervention groups was not due to chance or confounding factors (Moher et al. 2001).

The main features of an RCT that help to minimize bias are *randomization, allocation concealment, blinding*, and the use of an *intention-to-treat (ITT) analysis*. Inadequacies in these key methodological approaches often lead to a distortion of treatment effects, usually by overestimating the effects of the intervention (Schulz et al. 1995). The extent to which bias is controlled, and to which results reflect the true effect of the intervention, is referred to as internal validity (Juni et al. 2001). For academic rigor, RCTs should conform to the CONSORT (Consolidated Standards of Reporting Trials) statement, which was developed to improve the quality of reporting of RCTs (Moher et al. 2012) (see later section Reporting of RCTs).

3 Features of RCTs

3.1 Type of Randomization

Randomization, or random allocation, is the process by which each participant has a known probability of receiving each intervention before one is assigned. However, the actual intervention is determined by a chance process and cannot be predicted (Moher et al. 2001). Using chance to decide treatment allocation eliminates selection bias and can ensure that prognostic factors and unknown confounding variables are similar between comparison groups, as these are expected to balance out on average with randomization (The James Lind Library 2007a). Proper randomization also facilitates blinding (which reduces bias after assignment of interventions) and permits the use of probability theory to express the likelihood that any difference in outcome between two intervention groups is due to chance. Randomization is a crucial component of high-quality RCTs (Moher et al. 2001).

Adequate randomization methods include a computer-generated random sequence, drawing from a table of random numbers, lots or envelopes, tossing a coin, or shuffling cards or dice. Using alternation, date of birth, or case record numbers does not constitute adequate randomization (Juni et al. 2001). Non-randomized studies may produce misleading results even when comparison groups appear similar (Manchikanti 2008) as they tend to overestimate treatment effects by up to 160% or underestimate effects by up to 76% (Gluud 2006).

There are a range of different types of randomization strategies including simple, block, stratified, and adaptive, and selection will vary depending on the study and intervention design. *Simple* randomization involves using a single sequence of random assignments for each participant per treatment group (e.g., a coin flip to determine treatment (heads) or control (tails)) (Altman and Bland 1999b). *Block* randomization allocates participants into equally sized blocks/groups. The number of blocks should be a multiple of the block size, the product of which equals the total sample size (Altman and Bland 1999a). *Stratified* randomization allows for the control of covariates that may influence the trial's results (e.g., age, disease severity).

Once covariates have been identified, participants are then allocated to separate blocks for each combination of covariates, and then simple randomization is conducted for each block (Suresh 2011). *Adaptive* randomization involves sequentially assigning participants to treatment groups based on the previous group assignment of other participants (Kalish and Begg 1985). Adaptive randomization is typically used when controlling for covariates to allow for an even distribution of potentially outcome altering covariates between treatment allocations.

3.2   Allocation Concealment

For the randomization process to be considered adequate, it must both be generated by a chance process and followed by proper allocation concealment (Moher et al. 2001). Allocation concealment is the process used to ensure that the individual deciding to enter a participant into an RCT does not know (and cannot influence) the comparison group into which that individual will be allocated (Higgins et al. 2011). Proper randomization followed by adequate allocation concealment protects the allocation sequence and eliminates selection bias by shielding those who enroll patients from being influenced by knowledge of the next allocation. The use of a third party (e.g., independent pharmacies or a centralized telephone system) is ideal. Sequentially numbered opaque sealed envelopes (SNOSE; as long as they cannot be transilluminated) or numbered containers are also acceptable and can provide a more cost-effective option, providing the procedure is followed diligently (Doig and Simpson 2005). Writing the participant's name and details on envelopes prior to opening them is suggested to ensure adequate allocation concealment (Moher et al. 2001). Allocation concealment methods can be adopted to facilitate a variety of randomization types including those outlined above (simple, block, stratification, and adaptive). Studies which are labeled randomized but fail to report adequate allocation concealment yield larger estimates of treatment effects and may exaggerate them by up to 30% (Schulz et al. 1995).

3.3   Blinding

Blinding refers to the process of keeping participants, healthcare providers, evaluators, and sometimes those who manage and analyze data unaware of treatment allocation so that they will not be unduly influenced by this knowledge. When healthcare providers and participants are adequately blinded, it eliminates performance bias that is associated with patient and investigator expectations (Gluud 2006). Blinding of participants prevents treatment responses being influenced by expectations. That is, participants may have favorable expectations if assigned to the treatment group or disappointment if assigned to a control group, while healthcare providers may treat participants differently if group assignment is known (Moher et al. 2001).

Unlike allocation concealment, blinding may not always be appropriate or possible (Manchikanti 2008). The relevance of using blinding varies according to circumstances. Blinding and the use of placebo controls are not always possible or ethical, for example, in trials involving surgical procedures (Moher et al. 2001). Some interventions may be difficult to blind, for example, drugs that cause significant adverse effects (Gluud 2006).

The use of blinding assists in evaluating the efficacy of an intervention. For example, the use of placebo controls attempts to control bias by controlling for non-specific effects (all potential influences on the apparent course of the disease apart from those arising from the intervention itself) (Manchikanti 2008). The earliest use of a placebo control involved a sham device (made of wood) to test claims that a metal tractor cured through "electrophysical force." A crossover study failed to detect any benefit from the metal tractors (The James Lind Library 2007b). Placebo controls should be identical to the intervention in taste, smell, appearance, and so on (Gluud 2006) but not contain the actual drug or procedure.

However, not all placebo controls are completely inert or inactive, for example, the injection of sodium chloride into a nerve root or joint may exert a significant pain-relieving effect. Placebo-controlled RCTs have the advantage of providing the maximum ability to distinguish adverse effects from a drug or procedure but may create ethical concerns and patient and physician practical concerns and may cause patients to withdraw due to perceived lack of treatment response even if they were assigned to the active group (Manchikanti 2008).

Even in cases where blinding of patients or investigators is not achievable, blinding of outcome assessors is always possible and may theoretically be one of the most important considerations, preventing detection bias (Gluud 2006). It is more important when the outcome measures involve some subjectivity, such as assessment of pain or functional status, but is probably less important in situations such as assessing the effect of an intervention on mortality (Manchikanti 2008).

Blinding may be evaluated; however, if participants do successfully identify their assigned interventions more than would be expected by chance, it may not necessarily indicate that blinding was unsuccessful (Manchikanti 2008). Adverse events may also provide clues, as might treatment response; hence, participants are more likely to assume that a favorable outcome or the experience of side effects means they were allocated to the active intervention (Moher et al. 2001). Trials in which participants and caregivers/outcome assessors are not blinded exaggerate treatment effects by an average of 17% (Schulz et al. 1995). In a placebo-controlled trial, the assessments by unblinded, but not blinded, investigators showed an apparent benefit of the intervention (Moher et al. 2001).

3.4   ITT Analyses

RCTs rarely run smoothly, and deviations from protocol and loss to follow-up are sometimes inevitable. Analyzing results on an "intention-to-treat" (ITT) basis is recommended to reduce attrition bias. An ITT analysis is generally conducted by including all patients, regardless of eligibility, treatment received, withdrawal, or deviation from protocol (Hollis and Campbell 1999), or all available participants according to their original group assignment (Moher et al. 2001). However, there is still debate about the validity of excluding specific cases within each of these categories, and full application of an ITT analysis is only possible when outcome data are available for all randomized participants (Hollis and Campbell 1999) at all prespecified points during the trial (Juni et al. 2001). It is common for some participants to not complete a study, and these participants are not included in the ITT analysis (Moher et al. 2001). It has been suggested that ITT analyses may be more suitable for trials of effectiveness rather than explanatory trials of efficacy (Hollis and Campbell 1999).

There is no consensus on how to deal with missing data (Hollis and Campbell 1999). Suggestions for imputation include carrying forward the last observed response or calculating the most likely outcome based on that of other participants (Gluud 2006), but assumptions about missing data cannot be verified in most clinical trials (Hollis and Campbell 1999). Here, prevention is the key – all attempts should be made to minimize dropouts and missing data (Gluud 2006). ITT analyses may be less important in terms of internal validity than allocation concealment, randomization, and blinding. When these components were adjusted for, trials that excluded participants yielded similar treatment effects compared with trials that did not exclude participants (Schulz et al. 1995).

Presenting the flow of participants through a trial is highly recommended. This should include descriptions of the numbers of participants randomly assigned, receiving intended treatment, and completing the protocol, and analyses for the primary outcome (preferably through the use of a diagram, such as the CONSORT flow diagram; see below) (Moher et al. 2012). The possibility of attrition bias should be considered and discussed in reports on the trial (Juni et al. 2001). Describing those participants who dropped out may be important for external rather than internal validity (Moher et al. 2001).

3.5   Sample Size Calculation

Even after making efforts to minimize bias through randomization, allocation concealment, blinding, and ITT analyses, chance may still produce misleading results. An apparent difference between treatment groups may be purely due to chance alone (The James Lind Library 2007c); alternatively, there may not be enough participants to show differences between groups. There are two major types of statistical errors: type I error, concluding that there are statistically significant differences when actually there are none (false positive), and type II error, concluding that there are no differences between groups (e.g., due to small sample size) when in fact there is a difference (false negative). Sackett (1979, p. 61) says this clearly: "Samples which are too small can prove nothing; samples which are too large can prove anything."

A study is considered adequately powered when it has a high probability, through a minimum sample size, of detecting a statistically *and* clinically significant difference between groups. Larger sample sizes are required to detect more modest differences. Formal power calculations are recommended. Many studies are considered too small to have enough power to detect a difference between groups, therefore increasing the likelihood of type II error (Moher et al. 2001).

3.6 Pilot Studies

Randomized controlled trials are unwieldy tools, with practical problems being common. Some trials may never be completed due to issues relating to recruitment or acceptability of the intervention. Despite the tidy way that RCTs are reported when published, in reality they rarely run smoothly. RCTs are also relatively expensive to run, and the bodies that fund these studies have to consider what the probability is that a study will be completed and yield useful data, in addition to balancing ethical elements such as participant burden, risk of side effects, and reimbursement for participation. Increasingly, pilot studies are required to justify the launch of a large-scale RCT (Arnold et al. 2009).

A pilot study is a smaller-scale preliminary version of a larger RCT (the parent study) with similar methods and procedures (Jairath et al. 2000) that are specifically designed to inform the design and conduct of the parent study (Arnold et al. 2009). The term pilot study does not apply to all small-scale studies; studies with small sample sizes are only considered pilot studies if they directly or indirectly yield data to justify a larger study (Jairath et al. 2000).

Pilot studies must have explicitly defined objectives (Arnold et al. 2009). These usually relate to an assessment of feasibility of the trial and may extend to the feasibility and acceptability of the intervention, the trial design and procedures, data collection, and data analysis (Jairath et al. 2000). Costs and timelines are also assessed. When research funding is sought, an additional objective is to gather sufficient preliminary data to justify a grant award (Jairath et al. 2000). Findings from pilot studies should be used to improve trial design as appropriate.

Because they are usually designed to be a dress rehearsal for the full performance, simulating all the procedures of the parent study (including randomization and allocation concealment), pilot studies offer a unique opportunity to identify and prepare for the challenges of conducting a large RCT (Feeley et al. 2009). They function as a test to ensure future trials are designed optimally and can be implemented in practice (Arnold et al. 2009) and serve to justify planned research (Jairath et al. 2000). As such, they contribute significantly to increasing the methodological rigor of an RCT.

There are important limitations to pilot studies. Sample sizes are small, and overanalysis of outcomes (e.g., sub-analyses) should be avoided and de-emphasized as these are likely to be significantly underpowered. Hence, pilot studies cannot be used to determine treatment effects. They also cannot be relied on to provide valid estimates of event rates and should be used cautiously or not at all to guide power calculations (Arnold et al. 2009).

4   Protocol Design

Writing a clinical trial protocol is important for a number of different reasons. A protocol helps outline key steps in the trial for those involved in coordinating and delivering the trial intervention. When written correctly, it also provides the detail required for external review, whether by ethics committees or institutional review boards, funding agencies, and other researchers to assess the aims, interventions, and outcomes of the work as well as to identify any ethical concerns in the conduct of the trial. The guidelines for Good Clinical Practice (GCP), first published in 1996, are an international ethical and scientific quality standard for designing, conducting, recording, and reporting trials that involve the participation of human participants (ICH 1996). The objective of these guidelines is to provide an international unified standard for ensuring that the rights of participants are protected and that data generated by the trial are credible. However, writing a complete and clear trial protocol that adheres to the GCP standard can be challenging, especially for those new to the field, and missing key components can lead to having to submit multiple amendments to the original protocol. To help guide authors in writing complete and concise trial protocols, there are a number of expert guidelines, the most commonly used being the SPIRIT (Standard Protocol Items: Recommendations for Interventional Trials) statement for clinical trials (Chan et al. 2013). The SPIRIT statement provides guidelines for a minimum set of scientific, ethical, and administrative elements that should be addressed in a clinical trial protocol that adheres to GCP standards and provides a useful checklist to cover off the content of each section of the clinical trial protocol. The entire SPIRIT checklist is outside the scope of this chapter and is being updated on an ongoing basis to best reflect any changes in reporting standards; therefore, checking the latest version will provide the most up-to-date guidance.

5   Protocol Registration

A common practice in recent years is the publication of the clinical trial protocol during the recruitment phase of the clinical trial itself. This is often performed in conjunction with registering the trial itself in a clinical trial registry. There are a number of clinical trial registries which can be location specific, such as the ANZCTR for Australian and New Zealand trials (http://www.anzctr.org.au), while others such as ClinicalTrials.Gov (http://www.clinicaltrials.gov) index a number of international registries. The motivation behind trial registration is multifaceted, primarily driven by the observation that either negative results are not published, their hypotheses are altered after the data has been analyzed or only certain outcome measures are reported, to make the findings more positive (Fanelli 2010). Additionally, registration helps prevent duplication of work that is already ongoing. Preregistration of clinical trials is now mandatory for most academic journals (De Angelis et al. 2004) to try and ensure that all preplanned outcomes are included in the final manuscript, and registration before the enrolment of the first participant is strongly encouraged. However, individual journal editors have been slow to enforce this convention, and the implementation of this policy is uneven, at best (Scott et al. 2015). Publication of clinical trial protocols often occurs in specific journals catering to trial protocols. The benefit for authors is this allows elaboration on the somewhat terse information contained in clinical trial registries and allows reference back to the original protocol in any subsequent publications, providing more detail than might otherwise be available.

6 Categories of RCTs

There are four current major categories of RCT design commonly used in medical research: parallel, factorial, cluster, and crossover. Specific designs provide unique advantages and disadvantages.

6.1 Parallel

The most common type of RCT uses two parallel groups, often considered the "classic" RCT design, where participants are randomized to either an active treatment (A) or control group (B). Each group does not require exactly equal numbers, but in most cases, an almost equivalent number would be expected in each group, unless there had been an unequal randomization ratio specifically chosen (e.g., A to B ratio: 3:1). Each group receives only one type of treatment during the treatment phase (in contrast to crossover designs). The control group is commonly a placebo but increasingly can be either the current "gold standard" treatment or even another dose of the same medication. Parallel studies would be preferred over crossover designs if the disorder is cyclical (e.g., menstrual pain), if the condition is expected to progress in severity over time (e.g., dementia), or if the condition is acute (e.g., musculoskeletal pain).

6.2 Factorial

A factorial design includes two or more "factors" and two or more "levels" for each factor. The most common variant is a *2 × 2 factorial design*, where there are two factors and two levels for each factor. This type of design allows the researcher to examine both the influence of each factor separately (the main effects) and the interaction between any of the factors, on the outcome variable(s). This is more efficient than running a parallel RCT where the influence of a single factor is usually examined. Running multiple parallel RCTs on different factors will also not allow any possible interaction between factors to be observed. An example of a 2 × 2 factorial design would be examining the effect of a restricted versus normal calorie diet and resistance training exercise versus rest on the outcomes of body weight and maximum leg press weight. In this example, calorie intake and exercise are factors, while restricted calories versus a normal diet (without calorie restriction) and rest versus resistance training are levels for each factor, respectively. This would result in four possible comparisons:

1. Calorie restriction and rest
2. Calorie restriction and resistance training
3. Normal diet and rest
4. Normal diet and resistance training

In a parallel group design, this would equate to two separate studies, one examining the effect of calorie restriction and another the effect of resistance training. However, factorial design allows for the evaluation of both factors separately in one trial (the main effects) as well as the possible interaction between the two (i.e., the potential effect of one factor on another on the outcome). If calorie restriction was superior for weight loss compared to no calorie restriction, regardless of exercise, we would say there is a main effect for calorie intake, reporting the factor rather than the level. For examining interactions, we might find that while the group using calorie restriction and resistance training may cause the greatest bodyweight reduction, it may reduce maximum leg press weight due to the significant calorie deficit. This would be an interaction between calorie intake and exercise. Again, we report the *factors* not the levels. This ability to examine multiple factors in a single experiment and to look for interaction effects is a significant strength of the factorial design.

6.3 Cluster

Cluster randomized trials involve randomizing intact groups of participants, rather than individuals, and typically involve randomizing by site or study location (e.g., hospitals, clinics, communities) (Cornfield 1978). The primary advantage of cluster randomization is to avoid contaminating the intervention and control groups between participants. Other advantages include administrative efficiency and enhanced participant compliance. Cluster randomization is an appropriate trial design for educational intervention trials in healthcare settings, for example, a trial that compares changes in health outcomes between patients with end-stage kidney disease trained by nurses who received an educational intervention versus usual care. If nurses within the same renal unit were randomized to either the treatment or usual care group, it is likely that there would be a significant contamination between the groups, with nurses in the usual care group having the potential to be exposed to elements of the training received by the training group (e.g., patient remarks, overhearing conversations about the training, seeing the training materials, and so on). Some RCTs must employ a cluster randomized design by necessity, such as community trials that test the efficacy of public health interventions ("Community Intervention Trial for Smoking Cessation (COMMIT): I. cohort results from a four-year community intervention" 1995).

The primary disadvantage of cluster randomization compared to individual randomization is the "design effect" which involves lower statistical efficiency caused by variance inflation due to clustering (i.e., more participants are required to reach the same statistical power). Other disadvantages include clustering or collinearity between the sampled individuals within a cluster (i.e., high degree of similarity in the outcome measures between individuals within clusters), in addition to an increased level of complexity in the design and analysis of the data compared to RCTs involving individual randomization such as in parallel trial designs (Donner and Klar 2004).

6.4 Crossover

Crossover studies are repeated-measures studies (i.e., within participants), where participants receive different treatments in a different sequence depending on their group allocation. In contrast to parallel group studies, where participants receive treatment A or B, in crossover designs, each participant receives both A and then B or vice versa. Crossover designs offer three primary advantages over parallel designs: statistical efficiency, reduction of confounding variables, and a decreased ethical burden compared to parallel group trials. Increased statistical efficiency in crossover designs (Viboud et al. 2001) means that fewer participants are required in a trial compared to a parallel design to reach the same statistical power because each participant effectively acts as their own control. As participant recruitment is usually the most costly and difficult part of running an RCT, this can be a significant advantage. As each participant is their own control, the problem of imbalance of covariates/confounding variables (such as age or weight), which may occur in parallel RCTs, is solved. Finally, for placebo-controlled trials, there may be ethical concerns about withholding suitable treatment in parallel-design RCTs. This is especially relevant where those delivering the intervention are not blinded and may feel significant conflict around delivering a known placebo intervention (Barr et al. 2016). Crossover trials avoid this concern due to the fact that each participant will receive both active and placebo treatments.

Despite these advantages, there are several significant drawbacks that reduce the real-world usefulness of crossover studies. Carry-over effects, where the effect of the first treatment (treatment A) may still be present when starting the second treatment (treatment B), is of significant concern. This can be avoided by using a "washout" period, where a period of time is added between the two treatments, so that any effects from treatment A have completely resolved prior to starting treatment B (and vice versa). However, this requires detailed knowledge of the pharmacokinetics of the treatment and other parameters which are often not clear. Additionally, if using therapies such as psychotherapy or acupuncture where the effects may be long-lasting, it may be inappropriate to use crossover studies. Order effects, where the order of the treatment can affect the outcome, are also important. If treatment A is time-consuming, such as an exercise program, participants may lose motivation by the time they start treatment B. Alternatively, they may have become better by "practicing," so if treatment A and treatment B are both different exercise programs, participants may be better at whichever they are assigned to as the second treatment. Crossover designs are not suited to conditions where the condition may change significantly (e.g., an acute pain episode may have resolved, or a degenerative disease may have progressed) by the time that the second treatment is given; these are called "period effects." They are, however, suited to chronic conditions that do not change significantly over time (e.g., hypertension). Finally, as each participant is their own control and thus contributes several "sets" of data, each dropout results in a much greater proportion of missing data than would be seen in a parallel trial.

7   Superiority, Equivalence, and Non-inferiority Trials

RCTs, regardless of category, can be used to determine if a treatment is "superior" to a current treatment or placebo (superiority trial), "roughly the same" as a current treatment (equivalence), or "not much worse" than a current treatment (non-inferiority trial) (Lesaffre 2008). In medical research, the superiority trial is the most common (Piaggio et al. 2006), especially when comparing the active intervention to a placebo. Both equivalence and non-inferiority trials compare the intervention to an active comparator and are most commonly used during the development of a new intervention, such as a new version of a medication. The distinctions of "superior," "roughly the same," and "not much worse" are based on both statistical and clinical significance (Ganju and Rom 2017), where trial findings may be statistically significant (due to type II error in a large trial) but not necessarily *clinically* significant. Generally, the margin of clinical significance for equivalence and non-inferiority will be the largest margin possible before a clinically noticeable difference that would impact clinical practice is seen (Committee for Proprietary Medicinal Products 2001). The type of trial must be decided a priori, and only one trial type can be used at any one time. It must not be altered after the data has been examined.

8   Reporting of RCTs

Similar to the SPIRIT guidelines outlined above for publishing trial protocols, the CONSORT statement (Moher et al. 2001, 2012) is an established set of guidelines for the reporting of RCTs. The aim of CONSORT is to provide a clear and standardized set of minimum requirements for the reporting of RCTs in order to facilitate "complete and transparent reporting, and aiding ... critical appraisal and interpretation" (The CONSORT Statement 2017). CONSORT also includes a 25-point checklist and flow diagram that many peer-reviewed journals require RCT authors to complete when making a submission. The CONSORT statement and checklist are subdivided into the major sections related to reporting an RCT (title, abstract, introduction, methods, results, and discussion) and focus on essential aspects of trial design, for example, participants, interventions, outcomes, sample size, randomization, blinding, statistical methods and analyses, participant flow, recruitment, baseline data, numbers analyzed, outcomes including effect sizes, harms, limitations, generalizability, and interpretation of results. As with the SPIRIT guidelines, readers should check for the latest version of the CONSORT statement before use.

9 Shortcomings of RCTs

Although RCTs are a valuable way of evaluating therapeutic interventions, they are not a perfect approach. The disadvantages of RCTs are that they allow for less generalizability, are slow and expensive to conduct, and cannot answer as broad a range of questions as observational studies. RCTs also do not simulate real practice – they attract a particular type of patient, who may be more willing to comply with interventions (which can introduce selection bias), do not have multi-morbidity, and result in more attention and education given to patients than would normally be afforded in usual clinical practice (Katz 2006). In reality, patients are often far more complex than those enrolled in RCTs with narrowly defined eligibility criteria. RCTs also rarely study long-term outcomes and require large sample sizes to achieve statistical significance in order to equalize confounding factors (Manchikanti 2008). However, there are some possible design modifications that can be used to alleviate the shortcoming of RCTs.

9.1 Pragmatic

Every clinical trial is situated somewhere along the pragmatic-explanatory continuum, with few trials being purely pragmatic or purely explanatory (Thorpe et al. 2009). Pragmatic trials "help users choose between options for care" and answer the question "does this intervention work under usual conditions?," while explanatory trials "test causal research hypotheses" and answer the question "can this intervention work under ideal conditions?" (Thorpe et al. 2009, p. 464). Pragmatic trials tend to follow on from explanatory trials in pharmacological research (Thorpe et al. 2009). In other words, pragmatic trials inform practice and policy as to whether an intervention works under usual conditions, compared with usual care, while explanatory studies inform as to whether the observed effect is due to the intervention or due to other non-specific factors, such as regression to the mean, the placebo response, and spontaneous improvement.

Explanatory trials have the advantage of maintaining high internal validity or the ability to reduce bias or systematic error; some of this bias can be prevented by blinding. However, external validity, or the ability to generalize the findings, may be compromised. On the other hand, pragmatic trials may have higher external validity than explanatory trials, due to less stringent eligibility criteria and an intervention that closely reflects clinical practice. This may come at the expense of lower internal validity, and larger sample sizes may be needed (Gartlehner et al. 2006).

Pragmatic studies provide valuable evidence about whether an intervention does what it is purported to do. However, pragmatic trials do not inform why the treatment worked – was it because of the specific (active) effect of the treatment or non-specific effects? As an editor so eloquently wrote in the *British Medical Journal* in the middle of the twentieth century: "In treating patients with unproved remedies we are, whether we like it or not, experimenting on human beings, and a good experiment well reported may be more ethical and entail less shirking of duty than a poor one" (Hill 1952, p. 119).

9.2   Comparative Effectiveness Research (CER)

Comparative effectiveness research (CER) is defined as "the generation and synthesis of evidence that compares the benefits and harms of different treatment options to prevent, diagnose, treat, and monitor a clinical condition or to improve the delivery of care" (Witt et al. 2012, p. 1). Additional CER has been recommended particularly for clinical conditions that are "common and costly to society and that have a great degree of variation in their treatment." Features of CER include generating evidence to inform a specific clinical decision and comparison of at least two interventions, each with the potential to be the "best practice." Additionally, CER is conducted in the routine clinical setting to which the intervention belongs.

Larger sample sizes are needed for CER, which is a practical and financial challenge, as recruitment is a considerable challenge in clinical trials. However, broadening eligibility criteria could assist with feasibility of recruitment. Larger sample sizes would also allow for adequately powered subgroup analyses, assessing the impact of confounding factors such as use of co-interventions, and different clinical styles. Multicenter or multinational trials may be conducted to further expand external validity. Some novel methods have been suggested for CER RCTs such as dynamic allocation to balance treatment arms across prognostic factors, rank minimization, response-adaptive allocation, partially randomized patient preference design, and cluster randomization (Witt et al. 2012).

The disadvantages of CER are that it involves a greater financial cost due to the larger sample sizes needed and considerable time and effort in the protocol design phase if methods such as stakeholder consultation and consensus methods are to be used. CER also does not inform on component efficacy.

10   Conclusion and Future Directions

RCTs are currently the gold standard for evaluating the effectiveness of therapeutic interventions. Careful a priori choices regarding which type of RCT is chosen to answer the specific clinical question are vital to ensure that the strengths of the RCT are maximized while avoiding the limitations. Where suitable, crossover and factorial designs can provide increased statistical power compared to parallel groups, while cluster and crossover designs can increase participant compliance. Both are significant in terms of cost reduction, cost being a major disadvantage to most RCTs, due to smaller sample sizes and reduced dropouts. The standard features of RCTs, random allocation, allocation concealment, and blinding, provide a countermeasure to confounding factors and bias. However, they must be implemented carefully to ensure that they can perform these functions; simply performing an RCT does not equate to good evidence (Landorf 2017).

Good-quality, randomized controlled trials are still lacking for a surprising number of common interventions used in healthcare and are relatively rare in many non-pharmaceutical interventions such as surgery, physical activity, or diet. While placebo-controlled, parallel RCTs are still the most common design, especially in efficacy studies, these do not always provide the most clinically relevant information and can be difficult to design when using more complex interventions such as lifestyle changes. Therefore, if the focus is on changing clinical practice, researchers should consider pragmatic and comparative effectiveness research frameworks to ensure that their results can be translated into changes in practice. Future RCTs need to be undertaken with these research designs in mind, especially when considering healthcare challenges, such as obesity, that require a reflection of the complex nature of the problem when considering the trial design to ensure that relevant, effective interventions are developed and deployed.

References

Altman DG, Bland JM. How to randomise. BMJ. 1999a;319(7211):703–4.

Altman DG, Bland JM. Statistics notes. Treatment allocation in controlled trials: why randomise? BMJ. 1999b;318(7192):1209.

Arnold DM, Burns KEA, Adhikari NKJ, Kho ME, Meade MO, Cook DJ. The design and interpretation of pilot trials in clinical research in critical care. Crit Care Med. 2009;37(1 Suppl):S69–74.

Barr K, Smith CA, de Lacey SL. Participation in a randomised controlled trial of acupuncture as an adjunct to in vitro fertilisation: the views of study patients and acupuncturists. Eur J Integr Med. 2016;8(1):48–54. https://doi.org/10.1016/j.eujim.2015.10.006.

Chan AW, Tetzlaff JM, Altman DG, Laupacis A, Gotzsche PC, Krleza-Jeric K,... Moher D. SPIRIT 2013 statement: defining standard protocol items for clinical trials. Ann Intern Med. 2013;158 (3): 200–7. https://doi.org/10.7326/0003-4819-158-3-201302050-00583.

Committee for Proprietary Medicinal Products. Points to consider on switching between superiority and non-inferiority. Br J Clin Pharmacol. 2001;52(3):223–8.

Community Intervention Trial for Smoking Cessation (COMMIT): I. cohort results from a four-year community intervention. Am J Public Health. 1995;85(2): 183–92.

Cornfield J. Randomization by group: a formal analysis. Am J Epidemiol. 1978;108(2):100–2.

De Angelis C, Drazen JM, Frizelle FA, Haug C, Hoey J, Horton R,... International Committee of Medical Journal, E. Clinical trial registration: a statement from the International Committee of Medical Journal Editors. N Engl J Med. 2004;351(12): 1250–1. https://doi.org/10.1056/NEJMe048225.

Doherty S. History of evidence-based medicine. Oranges, chloride of lime and leeches: barriers to teaching old dogs new tricks. Emerg Med Australas. 2005;17(4):314–21.

Doig GS, Simpson F. Randomization and allocation concealment: a practical guide for researchers. J Crit Care. 2005;20(2):187–91.; discussion 191–83. https://doi.org/10.1016/j.jcrc.2005.04.005.

Doll R. Controlled trials: the 1948 watershed. BMJ. 1998;317(7167): 1217–20.

Donner A, Klar N. Pitfalls of and controversies in cluster randomization trials. Am J Public Health. 2004;94(3):416–22.

Fanelli D. Do pressures to publish increase scientists' bias? An empirical support from US states data. PLoS One. 2010;5(4):e10271. https://doi.org/10.1371/journal.pone.0010271.

Feeley N, Cossette S, Cote J, Heon M, Stremler R, Martorella G, Purden M. The importance of piloting an RCT intervention. Can J Nurs Res. 2009;41(2):85–99.

Ganju J, Rom D. Non-inferiority versus superiority drug claims: the (not so) subtle distinction. Trials. 2017;18(1):278. https://doi.org/10.1186/s13063-017-2024-2.

Gartlehner G, Hansen R, Nissman D, Lodhr K, Carey T. Criteria for distinguishing effectiveness from efficacy in systematic reviews. 2006. Retrieved from.

Gluud LL. Bias in clinical intervention research. Am J Epidemiol. 2006;163(6):493–501. https://doi.org/10.1093/aje/kwj069.

Higgins JP, Altman DG, Gotzsche PC, Juni P, Moher D, Oxman AD,... Sterne JA. The Cochrane Collaboration's tool for assessing risk of bias in randomised trials. BMJ. 2011;343: d5928. https://doi.org/10.1136/bmj.d5928.

Hill AB. The clinical trial. N Engl J Med. 1952;247(4):113–9. https://doi.org/10.1056/nejm195207242470401.

Hollis S, Campbell F. What is meant by intention to treat analysis? Survey of published randomised controlled trials. BMJ. 1999;319(7211):670–4.

ICH. ICH harmonised tripartite guideline. Guideline for Good Clinical Practice E6(R1) 1996.

Jairath N, Hogerney M, Parsons C. The role of the pilot study: a case illustration from cardiac nursing research. Appl Nurs Res. 2000;13(2):92–6.

Juni P, Altman DG, Egger M. Systematic reviews in health care: assessing the quality of controlled clinical trials. BMJ. 2001;323(7303):42–6.

Kalish LA, Begg CB. Treatment allocation methods in clinical trials: a review. Stat Med. 1985;4(2):129–44.

Katz MH. Study design and statistical analysis: a practical guide for clinicians. Cambridge: Cambridge University Press; 2006.

Landorf KB. Clinical trials: the good, the bad and the ugly. In: Liamputtong P, editor. Research methods in health: foundations for evidence-based practice. 3rd ed. Melbourne: Oxford University Press; 2017. p. 275–90.

Lesaffre E. Superiority, equivalence, and non-inferiority trials. Bull NYU Hosp Jt Dis. 2008;66(2):150–4.

Manchikanti L. Evidence-based medicine, systematic reviews, and guidelines in interventional pain management, part I: introduction and general considerations. Pain Physician. 2008;11(2):161–86.

Mayer D. Essential evidence-based medicine. Cambridge: Cambridge University Press; 2004.

Moher D, Schulz KF, Altman DG. The CONSORT statement: revised recommendations for improving the quality of reports of parallel-group randomized trials. Ann Intern Med. 2001;134(8):657–62.

Moher D, Hopewell S, Schulz K.F, Montori V, Gotzsche PC, Devereaux PJ,.... Consort. CONSORT 2010 explanation and elaboration: updated guidelines for reporting parallel group randomised trials. Int J Surg (Lond). 2012;10(1): 28–55.

Piaggio G, Elbourne DR, Altman DG, Pocock SJ, Evans SJ, Group C. Reporting of noninferiority and equivalence randomized trials: an extension of the CONSORT statement. JAMA. 2006;295(10):1152–60. https://doi.org/10.1001/jama.295.10.1152.

Sackett DL. Bias in analytic research. J Chronic Dis. 1979;32(1–2):51–63.

Schulz KF, Chalmers I, Hayes RJ, Altman DG. Empirical evidence of bias. Dimensions of methodological quality associated with estimates of treatment effects in controlled trials. JAMA. 1995;273(5):408–12.

Scott A, Rucklidge JJ, Mulder RT. Is mandatory prospective trial registration working to prevent publication of unregistered trials and selective outcome reporting? An observational study of five psychiatry journals that mandate prospective clinical trial registration. PLoS One. 2015;10(8):e0133718. https://doi.org/10.1371/journal.pone.0133718.

Suresh KP. An overview of randomization techniques: an unbiased assessment of outcome in clinical research. J Hum Reprod Sci. 2011;4(1):8–11. https://doi.org/10.4103/0974-1208.82352.

The CONSORT Statement. The CONSORT statement. 2017. Retrieved from http://www.consort-statement.org/

The James Lind Library. Avoiding biased comparisons. 2007a. Retrieved from http://www.jameslindlibrary.org/essays/bias/avoiding-biased-comparisons.html

The James Lind Library. Differences in the way treatment outcomes are assessed. 2007b. Retrieved from www.jameslind.org

The James Lind Library. Taking account of the play of chance. 2007c. 17 Dec 2009. Retrieved from www.jameslind.org

Thorpe KE, Zwarenstein M, Oxman AD, Treweek S, Furberg CD, Altman DG,... Chalkidou K. A pragmatic-explanatory continuum indicator summary (PRECIS): a tool to help trial designers. J Clin Epidemiol. 2009;62(5): 464–75. https://doi.org/10.1016/j.jclinepi.2008.12.011.

Viboud C, Boelle PY, Kelly J, Auquier A, Schlingmann J, Roujeau JC, Flahault A. Comparison of the statistical efficiency of case-crossover and case-control designs: application to severe cutaneous adverse reactions. J Clin Epidemiol. 2001;54(12):1218–27.

Witt CM, Aickin M, Baca T, Cherkin D, Haan MN, Hammerschlag R.,... Berman BM. Effectiveness Guidance Document (EGD) for acupuncture research – a consensus document for conducting trials. BMC Complement Altern Med. 2012;12:148. https://doi.org/10.1186/1472-6882-12-148.

Measurement Issues in Quantitative Research

38

Dafna Merom and James Rufus John

Contents

Abstract

Measurement is central to empirical research whether observational or experimental. Common to all measurements is the systematic application of numerical value (scale) to a variable or a factor we wish to quantify. Measurement can be applied to

D. Merom (✉)
School of Science and Health, Western Sydney University, Penrith, Sydeny, NSW, Australia

Translational Health Research Institute, School of Medicine, Western Sydney University, Penrith, NSW, Australia
e-mail: d.merom@westernsydney.edu.au

J. R. John
Translational Health Research Institute, School of Medicine, Western Sydney University, Penrith, NSW, Australia

Capital Markets Cooperative Research Centre, Sydney, NSW, Australia
e-mail: jjohn@cmcrc.com

© Springer Nature Singapore Pte Ltd. 2019
P. Liamputtong (ed.), *Handbook of Research Methods in Health Social Sciences*,
https://doi.org/10.1007/978-981-10-5251-4_95

663

physical, biological, or chemical attribute or to more complex factors such as human behaviors, attitudes, physical, social, or psychological characteristics or the combination of several characteristics that denote a concept. There are many reasons for the act of measurement that are relevant to health and social science disciplines: for understanding aetiology of disease or developmental processes, for evaluating programs, for monitoring progress, and for decision-making. Regardless of the specific purpose, we should aspire that our measurement be adequate. In this chapter, we review the properties that determine the adequacy of our measurement (reliability, validity, and sensitivity) and provide examples of statistical methods that are used to quantify these properties. At the concluding section, we provide examples from the physical activity and public health field in the four areas for which precise measurements are necessary illustrating how imprecise or biased scoring procedure can lead to erroneous decisions across the four major purposes of measurement.

Keywords

Measurement · Reliability · Validity · Sensitivity · Bias · Error

1   Introduction

Measurement is central to empirical research whether observational or experimental. A study of a novel, well-defined research question can fall apart due to inappropriate measurement. Measurement is defined in a variety of ways (Last 2001; Thorndike 2007; Manoj and Lingyak 2014), yet common to all definitions is the systematic application of numerical value (scale) to a variable or a factor we wish to quantify. Measurement can be applied to physical, biological, or chemical attribute or to more complex factors such as human behaviors, attitudes, physical, social, or psychological characteristics or the combination of several characteristics that denote a concept, for example, "disability" or "quality of life." Hence, when researchers intend to measure complex or abstract concepts, the act of measurement involves two more steps prior to assigning the numerical value; first, to identify and define the quality or the attribute of the concept they wish to measure and, second, to determine the set of operations by which the attribute may be isolated (Thorndike 2007). For example, if we wish to measure people's diet, we will first need to define "what are the qualities of the diet that we wish to measure?" Let us assume that in diet we wish to measure "eating habits." In this case, the operationalization of this concept can be (a) pattern of eating or (b) quality of the diet. Regardless of how the concept is operationalized (a or b), each decision will be associated with breaking down the broad definition to domains that need further quantifications and a scoring system for these domains. For example, if (a), eating could be quantified by the number of meals per day, time lapse between meals, or snacking frequency, whereas "(b) quality of diet" needs to be broken down to a list of foods within domains such as "healthy food items" and "unhealthy food items." The next decisions will relate to scoring, numerical value, of these domains. What will be the scale metric for pattern of eating? Is it a continuous variable with values range from 1 (e.g., one episode of eating) to 24 episodes of eating (i.e., episode for every hour) whereby the higher the score, the closer to "overeating pattern." We may also wish to give greater weight for one domain such as snacking compared to a usual meal? In short, measurement of complex concepts involves a lot of thinking and decisions to make before the development of the instrument and during the process of assigning numerical values to each item.

There are many reasons for the act of measurement. Here, we suggest four main purposes that are relevant to health and social science disciplines:

1. Aetiological – understanding relationship between attributes and determining causality
2. Evaluation – determining success or failure of programs and interventions
3. Monitoring – identifying secular changes in important factor
4. Decision-making/actions – classifying population according to norms, using score to predict an outcome

Regardless of the specific purpose, we should aspire that our measurement be adequate.

Inadequate measurement can lead to spurious conclusions about the nature of aetiological relationship, undermining or overstating the effectiveness of program/intervention, erroneous conclusions about secular trend, or misclassifying population as having health risk or not or predicting whether a person will be a successful medical school graduate.

In the following sections, we will review the properties that determine the adequacy of our measurement: reliability, validity, and sensitivity. We will use the general term "instrument" when referring to measurement properties, which could be a technological device that were designed to assess specific characteristics or behavior, survey questionnaires, laboratory tests, or human observations. We will then give examples on the effects of measurement errors on the validity of conclusions in aetiology, evaluation, monitoring, or decision-making.

2   Reliability

Reliability of the data and findings is one of the essential components of any research process. Reliability is defined as the extent to which results are consistent over repeated testing periods and an accurate representation of the study population under similar methodology (Golafshani 2003; Griffiths and Rafferty 2014). In principle, reliability deals with repeatability, consistency, reproducibility, and dependability of the result findings which are critical elements of any research (Nunan 1992; Leung 2015). The lack of reliability of an instrument will invariably affect the validity of such instrument (Bolarinwa 2015). Hence, having a reliable instrument is essential for the precision, accuracy, and adequacy of the measurement. For example, a weighing scale measuring a person's weight cannot be considered reliable if it provides different readings of the same person in a short period of time that is inconceivable to gain weight, and, hence, it may indicate a faulty scale.

2.1   Sources of Risk for Reliability

When conducting research, it is imperative to be aware of the different conditions that endanger reliability of the findings. Any measure is subject to some degree of imprecision or error which inversely affects reliability. For example, the greater is the degree of imprecision or error, the less accurate the result findings. The errors can be at random, which is less concerning, or systematic known as biases. Brink (1993) collectively categorized the major sources of risk for reliability (biases) into four groups as seen in Fig. 1.

2.1.1   Researcher

A researcher or observer is frequently used as the instrument for data collection, and therefore, it is important to consider the source of errors associated with observer as an instrument. First, observer can produce systematic error that occurs during querying, recording, and interpreting of the data commonly known as "observer bias" (Pannucci and Wilkins 2010). Therefore, an observer's knowledge of the hypothesis, disease, or exposure status contributes to bias with varying levels of misclassification, which in turn affects the reliability of the study findings (Delgado-Rodríguez and Llorca 2004). This may not always be deliberate but may involve subtle changes in the way the interviewer interacts and chooses to put emphases on selective questions (Davis et al. 2009). For example, if the interviewer is aware that the patient has lung cancer, he/she may put more emphasis on certain questions ("Are you sure you've never smoked? Never? Not even once?"), among cases prompting answers toward one direction (smoking) in cases but not in controls, leading to differential bias.

Fig. 1 Validity source dilemma – a purpose-led approach is here

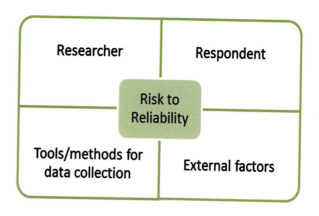

2.1.2   Respondent

Reliability of the source is directly associated with the quality of research. The accuracy of information received from respondent is a key concern when responses are collected via interviews and questionnaires. Bias may be introduced when respondents try to provide information better or worse than they really are (Delgado-Rodríguez and Llorca 2004). Participants might also be less reliable, being hesitant and deliberately withholding some critical information, known as social desirability bias. In other cases, participants may find cognitive difficulties to reconstruct information from the past, what is known as recall bias. For example, patients are more likely to recall their medical and treatment history associated with an exposure compared to people who are free of disease and selected as comparison group.

2.1.3   Tools/Methods for Data Collection

As aforementioned, research instruments could be in the form of interviewer, questionnaires, or measurement devices. Accuracy in measurement is crucial, and, therefore, for an instrument to be reliable, it has to repeatedly measure what it is supposed to measure in a consistent manner (Leung 2015). In addition to researcher and participants' biases, measurement errors due to methods of data collection can affect reliability. For example, faulty devices (e.g., timeworn, broken) can produce inconsistent reading. Tools that are not calibrated can introduce bias to readings, either systematically over- or underestimation of the accurate value. Therefore, it is important to conduct regularly calibration of instruments by pretesting or pilot testing against a standard value in order to identify and minimize such errors (Kimberlin and Winetrstein 2008).

2.1.4   External Factors

External factors, such as environmental and social context in which the research is carried out, also influence the degree of reliability. First, mechanical devices can be sensitive to extreme weather conditions, such as extreme heat of coldness, hence affecting the consistency of the reading in comparing to "normal" conditions. If the measurement is conducted by humans under extreme weather conditions, they may not have the same concentration level or energy to perform the measurement. Participants may also exhibit different behaviors under different social environments. For example, patients may not provide sensitive information when they are in a group, whereas they are likely to reveal this information in a more solitary environment (Brink 1993).

Similarly, gender and cultural differences between researcher and respondents may also affect the reliability of result findings. For example, women would be more likely to withhold information on their sexual activity or breastfeeding activity in the presence of a male interviewer. Ursachi and colleagues (2015) examined the reliability of three instruments used in market research and have shown that the reliability coefficients of each instrument varied by socioeconomic factors, rural or urban residency, and degree of religiosity.

2.2   Types of Reliability and Their Associated Statistical Methods

Although it is not feasible to provide an exact measure of reliability, different measures can be used to achieve an estimate of reliability. There are three general estimators of reliability which are as follows:

1. *Inter-rater or interobserver reliability* – This test is used to establish the level of agreement between different instruments or between two or more observers measuring the same factor (Heale and Twycross 2015). The inter-rater reliability determines the equivalence of the measures. For example, consistent readings obtained from two different sphygmomanometers measuring a person's blood pressure relates to a good level of inter-rater reliability of the instruments. Inter-rater reliability can be assessed by statistical methods such as the Cronbach's α test for quantitative measures and Cohen's kappa statistic for categorical measures (Bowling and Ebrahim 2005).
2. *Test-retest reliability* – This form is used to assess the consistency (AKA stability) of a measure over time (Griffiths and Rafferty 2014). Test-retest reliability requires using the instrument (say a questionnaire on attitudes, knowledge, or behavior) on the same sample population at two different time periods and comparing the scores for consistency. It is important to note that the two time periods should be within the timeframe of the recall period. For example, if a questionnaire queries about past week behavior, the test-retest should be undertaken within the same week that their behavior is recalled. It is important to note that test-retest reliability for activities that change from day to day is not recommended for factors that changes over a short period of time. For example, the frequency and duration of walking to do errands may change from one day to the other. Hence, if you ask to recall walking trips of the past week, the repeatability may be reduced because it is hard to recall this behavior over a week. It is preferable to ask about walking for errands in the past 24 h to increase accuracy, but then the retest should be within the same day. Statistics such as Cohen's kappa coefficient (for dichotomous scale) and Pearson's correlation (for continuous scale) or Spearman rank correlation (for scale that is not normally distributed) can be used to quantify the test-retest reliability (Manoj and Lingyak 2014). In general, a test-retest correlation of 0.70 is acceptable, 0.80 is good, and 0.90 is excellent (Manoj and Lingyak 2014).
3. *Internal consistency reliability* – This test measures the extent to which each item in the instrument is related to other items (Manoj and Lingyak 2014), in simple words, the degree to which items are held together. Internal consistency determines the homogeneity of the measures. Essential requirement is that the scale has more than one item that measures the phenomenon. There are three primary approaches to estimate internal consistency: Cronbach's α, Kuder-Richardson coefficient, and split-half reliability (Manoj and Lingyak 2014). Regardless of methods, all values range from 0 to 1, with values closer to 1 represent strong correlation and, hence, high reliability of the instrument and vice versa. Cronbach's α is the most commonly employed method to evaluate the internal

consistency of an instrument. It takes the mean of the individual item-to-item correlations and adjusting for the total items in the instrument. Some classifications of reliability estimates were offered: coefficient below 0.6 is considered unacceptable, 0.6–0.65 is undesirable, whereas correlations range between 0.7 and 0.9 are acceptable, good, or excellent (0.8–0.9). It is also suggested that if a coefficient is too high (>0.9), there is a possibility that some items may not be necessary and the scale can be shortened (Manoj and Lingyak 2014).

In split-half reliability, the result findings of an instrument or test are split into two equal halves (e.g., either by random selection, odd and even numbered, or first and second half), followed by calculating the correlations using by Spearman-Brown split-half reliability methods and comparing the correlation across both halves and to the entire test (Manoj and Lingyak 2014; Heale and Twycross 2015). A drawback of this method is that the reliability will vary depending on the methods of splitting chosen (Manoj and Lingyak 2014). The Kuder-Richardson test is a variant of split-half method and is appropriate for a scale that consists of dichotomous responses such as true/false or yes/no and alike. Another method, not so much in use, is the item-to-total correlation test, which is performed to find the relationship between each item with the total scale, thus eliminating inconsistent items before determining factors that represent the construct. A correlation value less than 0.2 suggests that the corresponding item does not correlate with the overall scale, whereas high inter-item correlations (>0.8) indicate the presence of repetitions which should be removed (Streiner and Norman 2003).

3   Validity

All measurements require validity evidence. Validity answers the question whether the instrument measures what it requires to do. Unlike reliability, which is an attribute of the instrument, validity is not an attribute but rather the extent of support for interpretation of scores on an instrument when it is being used for its intended purpose. When we carried out a measurement and we ask "how valid is it," we are inquiring whether the instrument measures what we want to measure, all of what we want to measure and nothing but what we want to measure (Thorndike 2007).

Validation research uses theory, data, and logic to argue for or against specific score interpretations: "validity is always approached by hypothesis, such that the desired interpretative meaning associated with assessment data is first hypothesized and then data are collected and assembled to support or refute the validity hypothesis" (Downing 2003, p. 830). It is agreed that assessment data for validity study are more or less valid for some very specific purpose, meaning, or interpretation, at a given point in time and only for some well-defined population (Downing 2003; Schmidt and Bullinger 2003). This may explain why, for example, one questionnaire has been validated repeatedly in different populations and presented a range of coefficients.

Several varieties are distinguished as though they established different "types" of validity, which can be confusing "am I using the correct name for the type of validity study?" For example, accelerometer is used to validate physical activity questionnaires, but some researchers refer to this source as "criterion validity" and others as "construct validity;" which one is the correct term? In general, there are three main types: content validity, construct validity, and criterion validity. Other terms such as "face validity," "convergent validity," "concurrent validity," and "predictive validity" are also in use, but these may be considered as sub-type within the three main types.

3.1 Types of Validity Evidence and Their Associated Statistical Methods

Content validity – this source of evidence answers the question "to what extent the content of the instrument incorporates the phenomenon under study." For example, complex constructs such as "functional health" can be assessed by questionnaire, but the instruments should grasp several domains such as activity of daily living and occupational, family, and social functioning. The developers of the instrument are best placed to provide the rationale of the domains used to measure the phenomenon and why these items were selected to represent each domain. But the judgment about the plausibility of the rationale should be made by others known as "experts." Content –related evidence involves several steps:

(a) The process by which the instrument was developed (e.g., panel of experts, consultation with users). Grant and Davis (1997) present important consideration for how to select experts and how to utilize the information they provide.
(b) Provision of a clear, explicit statement (definition) of what is being measured, referred to by Thorndike (2007) as "blueprint" and table of specification, which is the specification of the content to be covered in the instrument. This needs to include full account for whether the instrument omits important elements or included irrelevant ones.
(c) The plausibility of the explicit or implicit rationale linking the content of the instrument to the definition of what it is to be measured.
(d) Provide evidence for the acceptability of instrument by the target population; it is of no use to have a rationale and good coverage of the construct if the participants are not able to provide the answers.

The methods of quantifying experts' agreement are qualitative in nature, and this is where the term "face validity" takes place. This is the most basic type of validity of the instrument; it means that the experts think that the item appears to measure the construct under consideration; be appropriate for assessing this concept in the target population on the "face" of things. The panel, for example, is sent the "blueprint" of the instrument with instruction to determine the face validity and the content validity of the operational definitions of each variable within each domain that form the whole instrument. This forms the first draft of the instrument. Their comments on the first draft lead to the second draft with the aim to reach a consensus among panel members. The instrument is finalized when consensus is reached.

Construct validity – this source of validity investigates the extent to which the items of the instrument correspond to the phenomenon under the study overall and how well they related to each other. A simple example from health science would be, if on theoretical grounds, the phenomenon should change with age, a measurement with construct validity should reflect such change. Construct validity also answers whether there is a logical structure of the instrument and whether the items measure a unidimensional construct or a multidimensional construct. An example from the field of psychology would be an instrument that was developed to measure a trait such as "motivation to achieve;" this instrument should correlate with college grades and with completion of more items on a speeded test, given the ability being equal. If these predictions are met, we can say the instrument measured meaningfully the "construct motivation to achieve" (Thorndike 2007). A statistical approach to measuring construct validity is confirmatory factor analysis, which is a complex method that is used for construct validation. It tests whether the number of factors/domains and their loading on indicator variables conform to what we expect based on theory or hypothesis. Indicator variables are selected based on pre-established theory. For example, the concept "walkability" refers to features of the environment that encourage/or discourage (if absent) walking by residents. Researchers have used confirmatory factor analysis to develop a questionnaire measuring the perception of walkability, and these factors should correlate with the level of walking. This long instrument was abbreviated, and confirmatory factor analysis was used to test whether the abbreviated version retains similar factors (Cerin et al. 2006). More details on the technicality of factor analysis and sample size can be found in Manoj and Lingyak (2014, pp. 127–141). One sub-type of construct validity is "convergent validity" which is often used to compare two different instruments measuring the same construct. For example, two measures that assess quality of life are compared; theoretically they should be related to each other, but by how much?

Criterion validity – the extent to which the instrument correlate with an external criterion of the phenomenon under study. Criterion is often considered "a gold standard" of the phenomenon. For example, a visual inspection of a wound for evidence of infection can be validated against bacteriological examination of the wound, which is the "gold standard" criterion that is used to validate doctors' judgment. However, in social science, validation against a criterion is not that straightforward. Some behavior has no clear "gold standard." Thorndike (2007) further explains that criterion needs to fulfill four qualities: it has to be relevance, free from bias, reliable, and available. For example, fitness test is often considered as a criterion against scores produced by self-reported physical activity questionnaires. According to theory, a person who is regularly active is expected to achieve a better cardiorespiratory fitness score than a person who less is frequently active, yet genetics can also play a role in explaining high fitness test even if a person does not exercise at all. Therefore, it cannot be accepted as a "gold standard." Fitness test, however, is an objective physiological measure of cardiorespiratory fitness, and, therefore, it is free from common biases of self-report instruments that involve recall or social desirability biases, which reduce their accuracy. Fitness test is also a stable measure, unless a person purposefully engaged in intervention to improve fitness, the results of one test over time are expected to be the same under similar laboratory condition, and the test is accessible and reasonably cheap. This makes fitness test a good criterion for validation of self-reported questionnaire.

There are two broad classes of criterion-related validity (Last 2001):

(a) Concurrent validity – this examines the correlation between two instruments that supposedly measure the same thing and the scores are obtained concurrently (at the same time). Self-report consumption of surgery food in the past 24 h is compared to sugar level in the blood in the same day.

(b) Predictive validity – this is when performance on one instrument is used to forecast the performance in the future, for example, relationship between scores on a college admissions test and subsequent grades in college.

The statistical methods that are used for concurrent validity produce correlation coefficients (r); it can be Pearson coefficient or, if data is not normally distribute, the Spearman rho coefficient. If the predictive validity is sought, a regression equation can be computed, which tells us what our "best guess" of a person's score on the criterion would be, given the person's score on the predictor instrument (Thorndike 2007).

3.2   Which Validation Type to Choose

Current view is that construct validity encompasses all evidence and rationales for supporting trustworthiness of score interpretations. A key feature is that there is some organizing theoretical or conceptual framework to serve as a guide to score interpretation. If the construct validity is "the whole validity," why do we need content and criterion validity? In the context of social science, primarily education, Downing (2003) argues that validity is a unitary concept, with construct validity as the whole of validity, but has five related units: content, response process, internal structure, relationship to other variables, and consequences. This means that all other types of validity sit under construct validity. Thorndike (2007) views content and construct validity as "interpretive inference" and criterion validity as an "action inference" and argues that every instrument requires both. This means that ideally all sort of validity evidence should be provided. Here, we present a more pragmatic view that focuses on the purpose of the study as depicted in Fig. 2.

Ideally, researchers would like to present all sources of validity for a particular instrument, but there are some that will be most important and other less, dependent on the context or purpose of their research. In case researchers wish to develop a new instrument from scratch, they must present face and content validity of their instrument following by construct validity. If researchers wish to adapt existing instrument to other population or to create an abbreviated version of already existing instrument, it is good to present face or content validity of the adapted/abbreviated version (e.g., cultural aspects that were not addressed in the existing instrument), but not less important is to conduct construct validity to justify retention or removal of certain items. Some good example of instruments of quality of life in abbreviated or long

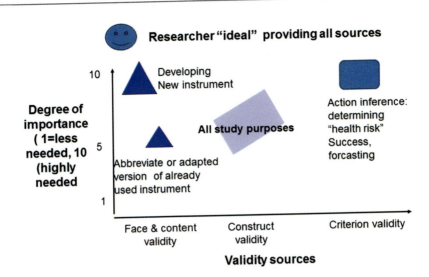

Fig. 2 Different sources of errors that risk the reliability of an instrument is here

form can be found in the paper by Busija and colleagues (2011) with data on validity sources. When the purpose of the study is to action based on the score of an instrument (e.g., classification of "health risk," "likely to succeed in job," or meeting certain benchmark) most important is to demonstrate a criterion validity, although it will add value if the construct validity of the instrument is provided to enhance the theoretical support behind the criterion validity.

4 Sensitivity

Sensitivity of an instrument is of the most essential measurement property to examine in the context of intervention, because the lack of sensitivity of the instrument measuring the main outcomes may result in wrong conclusion about efficacy/effectiveness of the intervention. Although the term "sensitivity to change" is often used interchangeably with "responsiveness," they can be distinguished as two separate entities (Walters 2009). Sensitivity is the ability of an instrument to detect changes between groups, for instance, between two groups in a clinical controlled trial (Fok and Henry 2015). In terms of experimental research, an instrument is reported to be sensitive only if it can identify clinically significant changes between and within patients over time (Gadotti et al. 2006). In addition, the instrument must show changes in the variable being assessed while being unaffected by changes in other variables under study, for example, in the comparison group that received no intervention (Deyo and Centor 1986).

Responsiveness on the other hand consists of two aspects: internal and external. The widely used internal responsiveness refers to the ability to accurately detect small but clinically significant or meaningful changes in a phenomenon or concept being measured over time (De Bruin et al. 1997; Terwee et al. 2003). On the other hand, the external responsiveness refers to the extent to which changes in a measure over time relate to corresponding changes in a reference measure of health status (Husted et al. 2000). For example, a study conducted to measure the responsiveness of the Lachman test to detect knee improvement in patients after ligament reconstruction surgery, test from before and after the surgery showed significant changes (improvement in knee function) that suggests this test is responsive (Gadotti et al. 2006). External responsiveness differs from internal responsiveness in that, measure is not in and of itself a primary interest but a relationship between change in the measure and the change in the external standard used as reference (Husted et al. 2000).

4.1   Sources of Risk for Sensitivity

Similar to reliability, the potential sources of risk for sensitivity of instruments could be in the form of faulty devices which are either timeworn or insensitive to detect changes over time or under different conditions (e.g., temperature, air pressure). For example, if a device is designed to detect improvement in plasma glucose concentration, implementation of the scale must be effective both in the ICU and in a self-management setting. It is considered to be poorly sensitive if it is not responsive to detect changes in both the settings. In addition, wrong use of research instruments which are not specific to measure a particular variable under study could also contribute to failure of responding to changes. Thus, it is imperative to conduct repeated testing of instruments against an evidence-based gold standard device in order to verify its responsiveness (Kimberlin and Winetrstein 2008).

4.2   Statistical Methods for Evaluating Sensitivity

While sensitivity can be determined by cross-sectional studies, responsiveness can only be assessed in longitudinal design, whether experimental design comparing efficacy of a treatment (intervention) to nontreatment group, or in observational studies where medical treatments are carried out on patients over a period of time (Revicki et al. 2008).

There are various statistical methods that can be used to evaluate sensitivity and responsiveness. There are as follows:

1. *Paired t-test*: This test statistic is used to calculate the changes in the scores by an instrument over two time points (Walters 2009). This test exclusively relates to the statistical significance of the observed change in the scores, which in turn is dependent upon sample size and variability of the measure (Husted et al. 2000).
2. *Effect size*: This test has been widely employed as a method for evaluating responsiveness. In contrast to paired t-test, this test measures the magnitude of

change in the measure by the standard deviation of the baseline measure of the instrument (Walters 2009).

3. *Standardized response mean (SRM)*: This test is a measure of ratio of magnitude of changes and the standard deviation reflecting the variability of the change scores (Husted et al. 2000).

In addition to the abovementioned methods, other methods such as repeated measures analysis of variance, responsive coefficient, Guyatt's responsive statistic, receiver operating characteristic curves, and regression models are also used (Deyo et al. 1991; Terwee et al. 2003).

5   Putting It All Together: The Effect of Inadequate Measurement on the Conclusions Drawn from the Study

In this section, we would like to present few examples where study conclusions were erroneous or compromised due to measurement issues. We will review examples from the physical activity and public health field in the four areas for which precise measurement is necessary: aetiology, evaluation of intervention, monitoring, and decision-making.

Aetiology: Epidemiological studies involve thousands of people at the minimum; therefore, self-report questionnaires are the most practical available method to measure complex behaviors such as diet or physical activity, both are established risk factors of many chronic diseases. One interesting question that occupied physical activity epidemiologists since the early 1990s is whether physical activity must be at high intensity to gain health benefits or lower intensities are also health enhancing. An epidemiological study by Yu et al. published in 2003 entitled "What level of physical activity intensity protects against premature cardiovascular death?" was one of several others that assessed this question. The researchers used the Minnesota Physical activity questionnaire which had excellent test-retest repeatability (0.88–0.82) and good concurrent validity (0.33–0.58) or construct validity (coefficient with fitness test 0.48). The investigators found that only vigorous-intensity leisure-time physical activity and not moderate or light intensity protected against premature mortality (Yu et al. 2003). This finding was inconsistent with all other longitudinal studies which consistently found that moderate-intensity leisure-time physical activity protected against coronary heart disease, stroke, and cardiovascular disease mortality. A careful examination of the methods described in this published paper indicated that the operationalization of intensity levels for each leisure-time physical activity deviated from the common acceptable intensity category level: light intensity included activities that raise the body energy expenditure from 2.5 up to 4 times above the resting metabolic rate, whereas moderate-intensity range is classified at 3 times and up to 6 times above rest including walking for exercise. At that study, swimming was classified as vigorous-intensity physical activity, like jogging or running, when it is usually classified at the top range of moderate-intensity level. It is most likely that the operationalization of intensity levels created systematic misclassification of the population according to intensity level; in some participants, efforts were underestimated, and in others (like the swimmers), efforts were overestimated leading to biased conclusion that moderate-intensity exercise has no benefit. This is an example how researchers' view affects their conclusion leading to them being inconsistent with 90% of the epidemiological studies.

Evaluation of interventions: In the past two decades, the increasing trends toward a sedentary lifestyle have brought the specific behavior of walking to public health attention. Walking is characterized by its both voluntary (incidental) and intentional nature and, as such, poses a great measurement challenge (Merom and Korycinski 2017). People may be able to recall well their walking as a form of exercise (planned walks) but less likely to recall incidental walking. This is why the repeatability coefficients of walking questions are varied according to what is asked. For example, a question that assesses all source of walking, for travel and for exercise, such as in the Active Australia questionnaire, a moderate repeatability coefficient (0.53) was noted, compared with a very good (0.78) coefficient for a question that assesses only walking to get fit or exercise, such as is in the Australia National Health Survey (Brown et al. 2004). This lack of stability leads to high variance around the mean estimate and poses a great challenge in the evaluation of the effect of walking promotion. In the evaluation of the step-by-step walking promotion, Merom et al. (2007) used several measures; one was the Active Australia walking question; one was the Harvard Alumni leisure-time physical activity questionnaire which asked about any planned recreational or exercise activities of the past 3 months, including walking; and a pedometer step count. A definitive effect was seen only for the pedometer step counts and the Harvard Alumni questions but none on the Active Australia walking question (Merom et al. 2007). It is now very common to use objective measures of physical activity along with self-report questionnaire in order to prove effectiveness of walking interventions due to the large measurement errors of the self-reported questions (Harris et al. 2016).

Monitoring: Physical activity is an important behavior to monitor due to its health-enhancing potential. The most important requirement of effective monitoring is to keep the instrument exactly the same as the baseline measurement; otherwise the changes over time will be contaminated. People may interpret differently the wording of the new question or will be asked to recall differently the amount of physical activity which will not be comparable with the old version. For example, positive change toward increased physical activity levels was reported in the UK over a 15-year period, but the authors alerted the readers that the walking question within the National Health Survey was changed in the middle of the period and included walking for everyday purposes (Stamatakis et al. 2007). Merom et al. (2007) also reported a biased estimate of walking trend over time which could be due to increased population's awareness of messages about the health-enhancing potential of everyday walking. For example, when Australians were asked to report on all types of exercise, recreation, or sports activities that they did in the past 12 months, the major increase was noted in walking behavior. The question is whether this was a real increase or a biased finding? A content analysis of what Australians consider as recreational, sport, and exercise walking revealed that about 4% of responders considered travel-related walking as part of exercise. These changes in perception may explain the increasing trend – while the behavior of walking for exercise remains the same, it is possible that the responders now perceived travel-related walking as exercise when previously they would not do so (Merom et al. 2009).

Decision-making: Health professionals often make decision by rough classification of individuals into categories that predict risk of disease. For example, obesity is considered at a body mass index equivalent to 30 and overweight at 25. Similarly, population is classified as "physically inactive" when they failed to reach the minimal recommendations for health. However, physical activity screening used by general practitioners (GPs) is a poor proxy of human movement. Primary care physical activity initiatives (e.g., the SNAP, the Lifescript, the Australian Better Health Initiatives) utilize a short self-report questionnaire to identify inactive clients, but individual-level inference from questionnaires is problematic due to large error margins. The predictive validity of the (1) "Lifescript" questionnaire administered by GPs, (2) a GP's own assessment methods, and (3) patients' step count using a pedometer were compared against accelerometer criterion of ≥ 30 min of moderate-to-vigorous physical activity. Pedometer had the highest accuracy, 82% sensitivity and 100% specificity, using 7500 steps per day as a criterion for active individuals. GPs' subjective assessments performed better than the self-reported questionnaire (Winzenberg and Shaw 2011). At the population level, there will be large discrepancy between estimates of the prevalence of physically inactive adults if the classification is based on questionnaire or accelerometers (50% vs. 12%, respectively) or if researchers use the same self-reported questionnaire but the operationalization of the scoring system to derive the cut point for "physical inactivity" differs (Brownson et al. 2000).

6   Conclusion and Future Directions

Good measurement skills are fundamental to research and practice. The quality of the knowledge base and the decisional process that follows the advancement in knowledge are all influenced by the precisions of the instruments that are in use. Therefore, the art of improving the measurement properties is everlasting and under examination for proper reasons. In public health and social science, the information gathered typically rely on large number of people at different points of time; hence, self-reported questionnaires are the most cost-saving and practical solution. Therefore several issues need to be considered: (1) how much information can we include in the questionnaire without posing too much burden that can lead to missing of answers due to the burden and (2) should we include information that is cognitively challenging (i.e., very difficult to remember or quantify) on account of reducing the accuracy of the response, that is, reducing the reliability? The balance between the amount of information sought and the risk to the measurement properties of the questionnaire will always need to be considered. More questions to increase content validity may compromise measurement properties. The rapid advancement in technology and the reduction in cost of the new technologies suggest that new devices will always be introduced to the market. Hence, previous information derived from older versions may not be equivalent to the new information generated from the state-of-the-art devices. Researchers and practitioners must be aware of the model they use, year of production, and how to adjust the new values to the old versions. Finally, in this chapter, we cover only few statistical methods as examples. This may change profoundly according to type of variable scales, whether it is a binary variable, categorical variable with a nominal or ordinal scale, or continuous variable. The way the data is manipulated and its distribution will determine the statistical methods, and this was out of the scope of this chapter.

References

Bolarinwa OA. Principles and methods of validity and reliability testing of questionnaires used in social and health science researches. Niger Postgrad Med J. 2015;22(4):195.

Bowling A, Ebrahim S. Key issues in the statistical analysis of quantitative data in research on health and health services. In: Handbook of health research methods: investigation, measurement and analysis. England: Open University Press McGraw Hill Education Birshire; 2005. p. 497–514.

Brink H. Validity and reliability in qualitative research. Curationis. 1993;16(2):35–8.

Brown WJ, Trost SG, Bauman A, Mummery K, Owen N. Test-retest reliability of four physical activity measures used in population. J Sci Med Sport. 2004;7(2):205–15.

Brownson RC, Jones DA, Pratt M, Blanton C, Heath GW. Measuring physical activity with the behavioral risk factor surveillance system. Med Sci Sports Exerc. 2000;32(11):1913–8.

Busija L, Pausenberger E, Haines TP, Haymes S, Buchbinder R, Osborne RH. Adult measures of general health and health-related quality of life: Medical Outcomes Study Short Form 36-Item (SF-36) and Short Form 12-Item (SF-12) Health Surveys, Nottingham Health Profile (NHP), Sickness Impact Profile (SIP), Medical Outcomes study Short Form 36-Item (SF-36) and Short Form 12-Item (SF-12) Health Surveys, Nottingham Health Profile (NHP), Sickness Impact Profile (SIP), Medical Outcomes Study Short Form 6D (SF-6D), Health Utilities Index Mark 3 (HUI3), Quality of Well-Being Scale (QWB), and Assessment of Quality of Life (AQoL). Arthritis Care and Research. 2011;63(Supll S11):S383–S4121.

Cerin E, Saelens BE, Sallis JF, Frank LD. Neighborhood environment walkability scale: validity and development of a short form. Med Sci Sports Exerc. 2006;38(9):1682–91.

Davis RE, Couper MP, Janz NK, Caldwell CH, Resnicow K. Interviewer effects in public health surveys. Health Educ Res. 2009;25(1):14–26.

De Bruin A, Diederiks J, De Witte L, Stevens F, Philipsen H. Assessing the responsiveness of a functional status measure: the Sickness Impact Profile versus the SIP68. J Clin Epidemiol. 1997;50(5):529–40.

Delgado-Rodríguez M, Llorca J. Bias. J Epidemiol Community Health. 2004;58(8):635–41.

Deyo RA, Centor RM. Assessing the responsiveness of functional scales to clinical change: an analogy to diagnostic test performance. J Chronic Dis. 1986;39(11):897–906.

Deyo RA, Diehr P, Patrick DL. Reproducibility and responsiveness of health status measures statistics and strategies for evaluation. Control Clin Trials. 1991;12((4):S142–58.

Downing SM. Validity: on the meaningful interpretation of assessment data. Med Educ. 2003;37:830–7.

Fok CCT, Henry D. Increasing the sensitivity of measures to change. Prev Sci. 2015;16(7):978–86.

Gadotti I, Vieira E, Magee D. Importance and clarification of measurement properties in rehabilitation. Braz J Phys Ther. 2006;10(2):137–46.

Golafshani N. Understanding reliability and validity in qualitative research. Qual Rep. 2003;8(4):597–606.

Grant JS, Davis LL. Focus on quantitative methods: Selection and use of content experts for instrument development. Research in Nursing and Health. 1997;20:269–74.

Griffiths P, Rafferty AM. Outcome measures (Gerrish K, Lathlean J, Cormack D, editors), 7th ed. West Sussex, UK: Wiley Blackwell; 2014.

Harris T, Kerry SM, Limb ES, Victor CR, Iliffe S, Ussher M, . . . Cook DG. Effect of a primary care walking intervention with and without nurse support on physical activity levels in 45- to 75-year-olds: the Pedometer And Consultation Evaluation (PACE-UP) cluster randomised clinical trial. PLoS Med. 2016;14(1):e1002210. https://doi.org/10.1371/journal.pmed.1002210.

Heale R, Twycross A. Validity and reliability in quantitative studies. Evid Based Nurs. 2015. https://doi.org/10.1136/eb-2015-102129.

Husted JA, Cook RJ, Farewell VT, Gladman DD. Methods for assessing responsiveness: a critical review and recommendations. J Clin Epidemiol. 2000;53(5):459–68.

Kimberlin CL, Winetrstein AG. Validity and reliability of measurement instruments used in research. Am J Health Syst Pharm. 2008;65(23):2276.

Last MJ. A dictionary of epidemiology. 4th ed. New York: Oxford University Press; 2001.

Leung L. Validity, reliability, and generalizability in qualitative research. J Fam Med Prim Care. 2015;4(3):324.

Manoj S, Lingyak P. Measurement and evaluation for health educators. Burlington: Jones & Bartlett Learning; 2014.

Merom D, Korycinski R. Measurement of walking. In: Mulley C, Gebel K, Ding D, editors. Walking, vol. 11–39. West Yorkshire, UK: Emerald Publishing; 2017.

Merom D, Rissel C, Phongsavan P, Smith BJ, van Kemenade C, Brown W, Bauman A. Promoting walking with pedometers in the community. The step-by-step trial. Am J Prev Med. 2007;32(4):290–7.

Merom D, Bowles H, Bauman A. Measuring walking for physical activity surveillance – the effect of prompts and respondents' interpretation of walking in a leisure time survey. J Phys Act Health. 2009;6:S81–8.

Nunan D. Research methods in language learning. Cambridge: Cambridge University Press; 1992.

Pannucci CJ, Wilkins EG. Identifying and avoiding bias in research. Plast Reconstr Surg. 2010;126(2):619.

Revicki D, Hays RD, Cella D, Sloan J. Recommended methods for determining responsiveness and minimally important differences for patient-reported outcomes. J Clin Epidemiol. 2008;61(2):102–9.

Schmidt S, Bullinger M. Current issues in cross-cultural quality of life instrument development. Arch Phys Med Rehabil. 2003;84(Suppl 2):S29–34.

Stamatakis E, Ekelund U, Wareham NJ. Temporal trends in physical activity in England: the Health Survey for England 1991 to 2004. Prev Med. 2007;45:416–23.

Streiner D, Norman G. Health measurement scales: a practical guide to their development and use. Oxford: Oxford University Press; 2003.

Terwee C, Dekker F, Wiersinga W, Prummel M, Bossuyt P. On assessing responsiveness of health-related quality of life instruments: guidelines for instrument evaluation. Qual Life Res. 2003;12(4):349–62.

Thorndike RM. Measurement and evaluation in psychology and education. 7th ed. Upper Saddle River: Pearson Prentice Hall; 2007.

Ursachi G, Horodnic IA, Zait A. How reliable are measurement scales? External factors with indirect influence on reliability estimators. Procedia Economics and Finance. 2015;20:679–86.

Walters SJ. Quality of life outcomes in clinical trials and health-care evaluation: a practical guide to analysis and interpretation, vol. 84. West Yorkshire, UK: Wiley; 2009.

Winzenberg T, Shaw KS. Screening for physical activity in general practice a test of diagnostic criteria. Aust Fam Physician. 2011;40(1):57–61.

Yu S, Yarnell JW, Sweetnam PM, Murray L. What level of physical activity protects against premature cardiovascular death? The Caerphilly study. Heart. 2003;89(5):502–6.

Integrated Methods in Research

39

Graciela Tonon

Contents

Abstract

The process of integration of methods in research is not without its difficulties. In some cases, the literature does not specify the differences between triangulation, mixed methods, and integrated methods. The integration of methods in research springs from triangulation, as far as the validation process of the completed research is concerned, and converges in the use of mixed methods as a strategy to complement and expand the combination of the quantitative and qualitative methods. This chapter aims to promote integration based on the notion that it can be achieved if the researcher thinks in a holistic way from the outset of the research process. An integrated methods study first requires the researcher to consider quantitative and qualitative methods in a way that does not result in contradictions, as both methods should collaborate with and complement each

G. Tonon (✉)
Master Program in Social Sciences and CICS-UP, Universidad de Palermo, Buenos Aires, Argentina

UNICOM- Universidad Nacional de Lomas de Zamora, Buenos Aires, Argentina
e-mail: gtonon1@palermo.edu

© Springer Nature Singapore Pte Ltd. 2019
P. Liamputtong (ed.), *Handbook of Research Methods in Health Social Sciences*,
https://doi.org/10.1007/978-981-10-5251-4_96

other in pursuit of a common aim. This is only possible where the researcher adopts an open and creative stance and avoids extreme positions. Finally, these considerations allow us to conclude that the integration of methods in research is a form of innovation.

Keywords

Research methods · Integration · Innovation · Triangulation · Mixed methods

1   Introduction

This chapter discusses the concepts and use of integrated methods in health social science research starting from the consideration that integration is a systemic process whose complexity lies in the relationships between the dimensions it involves (Grbich 2017). In this sense the combination of different individual elements is what makes up a coherent whole when they are brought together. The lack of integration is problematic in studies where greater understanding or more valid results might have been obtained if all types of available data had been considered together (Bazeley 2009), so we propose the use of integration of methods in health social science research to achieve a holistic and deeper analysis of the data generated in the studies.

For a proper discussion of integrated methods in research, it is necessary to discuss three considerations: the distinction between qualitative and quantitative methods, the review of the definitions of the concepts of triangulation and mixed methods, and the role of the researcher and the possibility of considering integration as a kind of innovation.

First and considering the distinction between qualitative and quantitative methods (see also ▶ Chap. 63, "Mind Maps in Qualitative Research"), we will remember Huston (2005) when he says that this distinction was really between variable-oriented and person-oriented approaches. Maxwell (2010, p. 477) argues that "the distinction between qualitative and quantitative is the distinction between thinking of the world in terms of variables and correlations and in terms of events and processes." Jones and Summer (2007, p. 5) remind us that the terms qualitative and quantitative are used to refer to:

> types of methodology – the overall research strategy used to address the research questions or hypotheses; types of methods of data collection – i.e. the specific methods; types of data collected – i.e. the raw data; types of data analysis – i.e. the techniques of analysis; and types of data output – i.e. the data in the final report or study.

Second it is necessary to review the definitions of the concepts of triangulation and mixed methods, as in some cases the term "triangulation" has been used interchangeably with "mixed methods" or "integration of methods." This has created some problems as it has obscured the difference between "the processes by which methods (or data) are brought into relationships with each other (combined, integrated, mixed) and the claims made for the epistemological status of the resulting knowledge" (Moran-Ellis et al. 2006, p. 2). In addition, it has blurred the distinction between the outcomes and the process.

Third, the chapter goes on in the conceptualization of the term integration by making a reflection on the role of the researcher and proposing recommendations for research projects utilizing integrated methods which lead to the possibility of considering integration as a kind of innovation.

2   Triangulation

A brief historical review of the concept of triangulation shows that in its origins, much emphasis was placed on increasing confidence in the results obtained. This was referred to by various authors as *increasing validity* (Campbell and Fiske 1956; Webb et al. 1966; Fielding and Fielding 1986). Some years later, other authors (Lincoln and Guba 1985; Smith and Hershusius 1986) adopted a different approach according to which "methods can be triangulated to reveal the different dimensions of a phenomenon and to enrich understandings of the multi-faceted complex nature of the social world" (Moran-Ellis et al. 2006, p. 6).

Prior to that, Denzin (1978) had already defined triangulation as the combination of methodologies for the study of the same phenomenon. Jick (1979, quoted in Tonon 2015) outlined some of the advantages of using triangulation, namely, that it allowed researchers to be more confident of their results and it stimulated the development of creative forms of gathering data and facilitated the synthesis of theories as well as the contemplation of contradictions.

Kelle (2001) presents the term "triangulation" taking into account that many authors consider it as a central concept for method integration. The author states that "this notion carries systematic ambiguities, at least when transferred to the integration of qualitative and quantitative methods – triangulation does not represent a single integrated methodological concept but a metaphor with a broad semantic field" (Kelle 2001, p. 1). Finally, Kelle distinguishes between three types of triangulation: as mutual validation, as the integration of different perspectives on the investigated phenomenon, and in its original trigonometrical meaning.

According to Creswell et al. (2004), research studies using triangulation typically organize separate sections for each of the methods (quantitative and qualitative), so that one section deals with data collection and quantitative analysis, while the other contains a discussion of the data collection and a qualitative analysis. In the final section, the researchers discuss the results of both analyses. This leads us to conclude that the term triangulation has been, and is still, used to indicate the use of more than one method in a research process.

3   Mixed Methods

In 2002, Sale, Lohfeld and Brazil published an article on the combination of quantitative and qualitative methods in health studies. The authors, following Guba (1990, p. 18), argue that each of these methods is based on a particular paradigm, a patterned set of assumptions concerning reality (ontology), knowledge of that reality (epistemology), and the particular ways of knowing that reality (methodology) (see also ► Chap. 6, "Ontology and Epistemology"). The authors, thus, state that each of these methods does not study the same phenomenon and propose a new solution for using mixed methods in research. They further argue that the distinction of phenomena in mixed methods research is crucial and can be clarified by labeling the phenomenon examined by each method, rather than merely using the strengths of each method to notice the weaknesses of the other, or capturing various aspects of the same phenomenon. They finally propose carrying out each method simultaneously or sequentially in a single study or series of investigations.

Prior to that study, in the field of nursing research, Shih (1998, p. 633) had pointed out that the "analytic density" rationale does not mix methods to obtain more reliable and valid findings but to get a wider and deeper picture from all angles (quoted in Fielding 2012, p. 127).

Pawson (1995 quoted by Moran-Ellis et al. 2006, p. 13) criticizes multiple methods and data approaches which primarily generate more data about a phenomenon without addressing how the plurality of data will be combined analytically. Sandelowski et al. (2009) discuss the nature of data and quote Wolcott (1994, pp. 3–4) when he argues that "[e]verything has the potential to be data, but nothing *becomes* (italics in the original) data without the intervention of a researcher who takes note – and often makes note – of some things to the exclusion of others." Punch (2005, p. 246) identifies three important factors for the use of mixed methods: whether the methods are taken as equal, whether or not they influence the operationalization of each other, and whether they are conducted simultaneously or analytically.

Fielding (2012) summarizes the work of Ivankova and Kawamura (2010), which consisted in an extensive bibliometric survey of mixed methods practice. On the basis of searches of five databases (PubMed, ERIC, PsychInfo, Academic OneFile, Academic Search Premier) and two journals (Journal of Mixed Methods Research, International Journal of Multiple Research Approaches), the authors found a consistent growth in mixed methods research since 2000, from $N = 10$ (2000) to $N = 243$ (2008).

Johnson and Onwuegbuzie (2004) argue that mixed methods are plural and complementary, allowing the researcher to adopt an eclectic approach. In a subsequent writing, Johnson et al. (2007) define mixed methods as an approach to theoretical and practical knowledge that takes into consideration multiple points of view, perspectives, positions, and outlooks, both at a qualitative and quantitative level. Along these lines, Maxwell (2010, p. 478) warns us that:

> the use of numbers per se, in conjunction with qualitative methods and data, does not make a study mixed-method research.......the systematic (although not necessarily explicit) use of both ways of thinking is what is most distinctive of, and valuable in, mixed-method research.

In addition, according to Greene (2007), the value of mixed methods research is in creating a dialogue between different ways of seeing, interpreting, and knowing, not simply in combining different methods and types of data. This leads us to agree with Creswell (2018) when he points out that a mixed methods design is useful to capture the best of both quantitative and qualitative approaches (see also the ► Chaps. 4, "The Nature of Mixed Methods Research," and ► 40, "The Use of Mixed Methods in Research").

4   Conceptualizing the Term "Integration"

The term "integration," first defined in an article published in the 1981 issue of the ECLAC (UN) Review, arises from the confluence of the fields of economics and politics. The author of the article, Cohen Orantes (1981), points to the need to conceive integration as a process deriving from a set of activities taking place in a continuous manner, aimed at intensifying interdependence between its elements (participants) for the achievement of mutual benefits.

Based on this definition, we start from the premise that integration involves the combination of different individual elements that make up a coherent whole when they are brought together. In the field of research methods, according to Moran-Ellis et al. (2006, p. 3) "integration in multi methods/multi data research must be understood as a particular practical relationship between different methods, sets of data, analytic findings or perspectives."

Integration has been variously described as being undertheorized and understudied (Greene 2007, cited in Bazeley and Kemp 2012, p. 55). The lack of integration is problematic in studies where greater understanding or more valid results might have been obtained if all types of available data had been considered together (Bazeley 2009). Bazeley and Kemp (2012, p. 56) remark that "not only is integration of methods undertheorized and understudied but also the level of integration practiced in many mixed methods studies remains underdeveloped." Other authors have considered several forms of integration, such as separate methods and integrated analysis (Bazeley 2002; Moran-Ellis et al. 2004) and separate methods, separate analysis, and theoretical integration (Green 2003).

Bazeley and Kemp (2012) examine the metaphors used to describe the process of integration of analyses in mixed methods research, discovering different ways in which researchers think and write about integration. The authors conclude their analysis by identifying principles that guide the effective integration of analyses in mixed methods research, among which they highlight the following: there are different ways to integrate data; integration might begin at any stage within a study; integration needs to occur before conclusions, crucially during analysis or during the analytic writing of results; the level of integration must be appropriate to the goals and purposes of the study; the product of the integration must be such that it would not have been available without that integration; finally, an integrated study should not be written up as separated components, as this is antithetical to the concept of interdependence (Bazeley and Kemp 2012).

Pawson (1995, quoted in Moran-Ellis et al. 2006) refers to the work of integration as *synthesis*, which includes preplanning, the maintenance of the modalities of the different types of data while at the same time dissolving barriers among them. The methods thus interface and enmesh with each other.

Bryman (2006) has published an important article in which he studied how quantitative and qualitative research is integrated and combined in practice. The article is based on a content analysis of 232 social science articles. His examination of the research methods and research designs employed suggested that on the quantitative side, structured interview and questionnaire research within a cross-sectional design tends to predominate, while on the qualitative side, the semi-structured interview within a cross-sectional design tends to predominate. His research reported that there is considerable value in examining both the rationales that are given for combining quantitative and qualitative research and the ways in which they are combined in practice but that it is important to recognize that when concrete examples of research are examined, there may be a disjuncture between the two.

In addition, Bryman (2007) refers to the so-called genuine integration by discussing how the components of mixed methods research are effectively integrated. In his article, Bryman reports the findings of a study in which he interviewed 20 UK researchers that worked with mixed methods. He identified different types of barriers to the integration of findings in mixed methods research and the factors that hinder the capacity of researchers to engage in such integration, pointing out that "the key issue is whether in a mixed methods project, the end product is more than the sum of the individual quantitative and qualitative parts" (Bryman 2007, p. 8).

In the field of health services research, Fetters, Curry and Creswell (2013) examine the key integration principles and practices in mixed methods research and point out that the extent to which mixed methods studies implement integration remains limited. Fetters and colleagues identify different levels of integration in mixed methods studies: integration at the design level, at the methods level, at the interpretation level, and at the level of reporting the findings. Integration at the methods level can occur through connecting, building, merging, and embedding. Integration through connecting occurs when one type of data links with the other through the sampling frame. Integration through building occurs when results from one data collection procedure inform the data collection approach of the other procedure, the latter building on the former. Integration through merging of data occurs when researchers bring the two databases together for analysis and for comparison. Integration through embedding occurs when data collection and analysis are being linked at multiple points and is especially important in interventional advanced designs, but it can also occur in other designs. Finally, the authors point to the existence of the fit of data integration, which refers to the coherence of the quantitative and qualitative findings, and state that the assessment of the fit of integration leads to three possible outcomes: confirmation, expansion, and discordance.

The combination of different methodologies and interpretive approaches does not necessarily enhance validity but can extend the scope and depth of understanding (Fielding and Fielding 1986; Fielding and Schreier 2001; Denzin and Lincoln 2005; Fielding 2012). For Fielding (2012), integration is the heart of the whole mixed methods exercise because the purpose of mixing methods is to get information from multiple sources and so the issues in bringing together the information are crucial. It is not so much the stage when integration occurs but additionally what types of data are being integrated and how we integrate them (Fielding 2012).

In a study conducted in 2015 by Guetterman, Fetter, and Creswell, the authors studied exemplar joint displays by analyzing the various types of joint displays being used in published articles on mixed methods research in health. The authors searched for empirical articles that included joint displays in three journals that publish state-of-the-art mixed methods research. They thus identified and analyzed joint displays to extract the type of display, mixed methods design, purpose, rationale, qualitative and quantitative data sources, integration approaches, and analytic strategies. The analysis focused on what each display communicated and its representation of mixed methods analysis. Joint displays appear to provide a structure to discuss the integrated analysis and assist both researchers and readers in understanding how mixed methods provide new insights.

We can, thus, suggest that the concept of integration should be used in those studies in which integration occurs from the moment when the original research problem – that is, the question to be solved – is constructed and then extends over the entire research process. The original definition of the research question is the key to establishing that a study has been carried out through the integration of research methods and that integration should extend over the entire process. In other words, precisely specifying the research question is the key thing, and from this, a sense of the best methodological combination will emerge. Researchers must always be ready to adjust the design in light of what is found. Research design is not a stage; it is a process (Fielding and Fielding 2008).

Integration is a multidimensional and complex process, which is built from a social, historical, and cultural perspective. Far from conceiving it as a photograph – that is, a static image – we can compare integration to a film and view it as a construction in motion and permanent change. Indeed, what is interesting about integration is what stems from it.

5   The Role of the Researcher in the Integration of Research Methods

We consider that researchers are social subjects that are culturally conditioned and guided by their personal experiences, which, according to Cipriani (2013, p.53) "are too significant to be completely cancelled."

At this point, it is worth mentioning the difference between experience and life experience. De Souza Minayo (2010, p. 254), following Husserl (2001), contends that the world of experience is the world of life, the base of all action, as well as all knowledge operation and scientific elaboration. She differentiates the experience from life experience. She says that the latter is constituted as the elaboration of the experiences, and, at the same time, it is related to the historical conditions, making the life experiences an individual experience full of collective sense.

Morse and Field (1995) suggests that the approach chosen by each researcher is his/her own product, given that his/her previous knowledge and the methods he/she will employ define his/her decision when doing research. According to Sandelowski et al. (2009), "data" are not simply "given"; they are constructed by researchers from their perceptions and experiences in interacting with the phenomena studied. Therefore, the approach to be adopted by the researcher affects the research process and its outcomes. The approach is "the fundamental point of view of man and the world that the scientist brings with him or her or adopts in connection with his or her work as a scientist, whether this point of view has become explicit or remains implicit" (Giorgi 1970, p.126, cited by Ray 2003, p. 149).

May (2006), quoting Benner (1984), points out that it is interesting to consider researchers' previous experience and notes that expert researchers usually view situations in a holistic manner, based on data drawn from past experiences. May also highlights the importance of pattern recognition, that is, the so-called ability to know where to look, and remarks that the processes of methodologic and disciplinary socialization contribute to such ability. May finally argues that, in addition to his/her extant knowledge – based both on theory and on his/her lived experiences – the expert researcher also relies on intuition and creative reasoning, while she notes that "moving from intuition to insight, from an interesting but quirky question to an important revelation – these processes are not governed by chance" (May 2006, p. 26).

In relation to lived experiences, Hubbard et al. (2001, p. 119) state that "[u]sing our own personal experiences in the field, we present a range of emotional encounters that qualitative researchers may face." Through an exploration of her own experiences as a researcher of intimate and sensitive topics, Brannen (1988, quoted in Hubbard et al. 2001) argues for the need to protect both the researcher and the subjects of research. The theoretical interpretations made by researchers will also be permeated by those previous experiences. As Morse (2003, p. 7) argues "our perception limits what we consider to be a research problem."

Sotolongo y Codina and Delgado Diaz (2006, p. 53) point out that persons can only be the result of a process of constitution of subjectivities, and each subjectivity, instead of being centered in itself, is concocted from a context that transcends it and articulates it to others. In these sense, all research process is played in a field of the intersubjectivities "paving the way to interactional and reticular facts as constitutive sources of reality" (Sotolongo and Codina and Delgado Diaz 2006, p. 62).

At this point, it is worth taking into account Cipriani's assertions (2013, p. 63) quoting Glaser (1978), when he introduces the concept of *theoretical sensitivity*, that is, the ability to gain insight into the subtleties in the meaning of data, distinguishing between what is relevant and what falls outside of the research problem.

For a proper use of integrated research methods, it is necessary for the researcher to first consider quantitative and qualitative methods in a way that does not result in contradictions, as both methods should collaborate with and complement each other in pursuit of a common aim (Bryman 2016). This is only possible where the researcher adopts an open and creative stance and avoids extreme positions.

6 Innovation and the Use of New Technologies for Integrated Methods in Research

Fielding (2012, p. 134) identifies a new approach to methodological integration and points out that "data integration is always a matter of innovation."

Innovation can be defined as the contribution of something new to scientific knowledge. Innovation cannot be conceived as an individual process as the new combinations of knowledge are based on the interaction and communication of those that possess such knowledge. The contribution that innovation produces must necessarily be examined within its development context (Dogan and Pahre 1993). Dogan and Pahre also identify different types of innovations, including the so-called methodological innovation as well as theoretical and conceptual innovation (see chapters in the ▶ Chap. 61, "Innovative Research Methods in Health Social Sciences: An Introduction" section). The European Commission's Green Book (1995, p. 4) refers to innovation as being synonymous with the successful production, assimilation, and exploitation of novelty that offers new solutions to problems and allows meeting the needs of both individuals and society. Drawing upon these definitions, we argue that the integration of methods in research can be considered a true process of innovation.

Fielding (2012) explains three innovative forms of data integration that rely on computational support: the integration of geo-referencing technologies and methodologies, the integration of multistream visual data, and the integration of qualitative and quantitative data. The convergence of geographical and social science enables data integration to link outcome-based spatially defined inequalities with process-based investigations of their origins (Fielding 2012). Being able to code, annotate, and analytically manipulate visual representations of physical space helps researchers integrate visual images, words, and numbers not just for context but for analytic reasons (Kwan 2002).

In considering the integration of multistream visual data, Fielding (2012) argues that there is a growing social science interest in visual data, and emergent technologies offer significant enhancements to visual resources. A current innovation involves the increasing range of technologies that capture visual data so that recordings of meetings or fieldwork activities at locations remote from the researcher can be integrated with textual and statistical data.

Finally, the integration of qualitative and quantitative data is perhaps the most popular form, as it involves the use of software such as NVivo, which enables the integration of interviews and observation data with rating scales or survey responses (Fielding 2012).

According to Bazeley (2006), integration has greatly benefited from technology, in particular, computer software (see ▶ Chap. 52, "Using Qualitative Data Analysis Software (QDAS) to Assist Data Analyses"). However, this is an emerging process that has developed only in recent times, and, while it has gained prominence in the field of research in social sciences, it still has a long way to go until it is finally accepted by the scientific community.

7 Recommendations for Research Projects Utilizing Integrated Methods

Having pointed out the advantages and challenges of using integrated methods, we now present some recommendations that may prove useful when employing these methods:

(a) As the integration of methods requires knowledge and management of qualitative and quantitative methods, researchers aiming at doing research with integrated methods are advised to be well-trained in each of these methods or to organize a research team that has expertise in using them. The use of integrated research methods is not advised for novice researchers that have no previous experience in the use of these methods. As argued by Bazeley (2006, p. 65):

> Integration of data analysis requires a breadth of skills that has not been commonly available in a single researcher, or alternatively a close-knit multi-skilled team; it requires the capacity to imagine and envision what might be possible – to tread new paths – along with the logic (and skills) required to bring that about.

(b) Before attempting an integrated methods study, researchers should accurately identify their theoretical, philosophical, methodological, and technical position, so as not to impose their point of view and lose theoretical sensitivity. According to Bergman (2008, p. 28), the separation between qualitative and quantitative methods is related to "delineating and preserving identities and ideologies rather than to describe possibilities and limits of a rather heterogeneous group of data collection and analysis techniques."

(c) The increasing use of methodological proposals based on integrated research methods is a major advance in the field of research methodology. However, care should be taken that this is not just a passing fad – which could in the future be replaced by another fad – but that it can truly represent a methodological advance.

8   Conclusion and Future Directions

More than five decades ago, Wright Mills (1961) argued that there are three kinds of interludes in a scientific community: on problems, methods, and theory. In reflecting on the methods, we have discussed in this chapter the origin and characteristics of research processes utilizing integrated methods. First, we agree with Fielding (2012, p. 134) when he argues that "methods have moved from being solely a resource to also being a topic in their own right. Such an approach has a dynamic and demanding view of what makes for an adequate understanding of social phenomena."

In addition, if we analyze the process of integration of methods in research, we find that the mere inclusion of quantitative data in a qualitative study or the inclusion of qualitative data in a quantitative study does not make the research an integrated methods study. For this to happen, the concept of integration needs to be present from the moment the initial research question is constructed and extend over the entire research process (Tonon 2015).

In this respect, it is worth noting that combining or interacting is not the same as integrating, as integration requires the researcher to adopt a holistic view of the use of methods. According to Moran-Ellis et al. (2006, p. 10) "integration is a particular type of relationship among methods, data, analytic methods, or theoretical perspectives which carries significant implications for how that part of the research process functions." The issue of theory and ideology deserves separate discussion, and in this sense, it is worth bearing in mind, as Maxwell (2010, p. 477) points out, that researchers' ontological and epistemological assumptions influence research designs, research questions, conceptual frameworks, methods, and validity concerns.

In this chapter, we also view the integration of methods in research as an innovative process, as "the originality of innovation lies in the process that allows a specific change to take place" (Rodríguez Herrera and Alvarado Uriarte 2008, p. 23). This is an important point to consider for researchers in the twenty-first century because innovation entails the creation of something new which is unknown until that moment.

Dogan and Pahre (1993) point out that innovation is the contribution of something new, but this contribution has to be examined within its development context. At the same time, it is important to consider that the recognition of the value of using integrated methods in research needs to be accompanied by a recognition of the pragmatic and epistemological implications (Moran-Ellis et al. 2006).

Innovative forms of data integration that rely on computational support as the integration of geo-referencing technologies and the integration of multistream visual data are new types of methodological strategies. Today, but much more in the future, the use of software will constitute a common practice, so it is necessary for researchers, particularly the younger ones, to learn how to use it.

Finally it is important to say that the integration of methods in social sciences and health research is a view in the horizon. The possibility of taking this challenge is not only an opportunity but a necessity as well.

References

Bazeley P. The evolution of a project involving an integrated analysis of structured qualitative and quantitative data: from N3 to NVivo. Int J Soc Res Methodol. 2002;5(3):229–43.

Bazeley P. The contribution of computer software to integrating qualitative and quantitative data and analyses. Res Sch. 2006;13(1):64–74. Mid-South Educational Research Association. Retrieved from https://www.researchgate.net/profile/Pat_Bazeley/publication/253326153_The_Contribution_of_Computer_Software_to_Integrating_Qualitative_and_Quantitative_Data_and_Analyses/links/548233cd0cf2f5dd63a89882/The-Contribution-of-Computer-Software-to-Integrating-Qualitative-and-Quantitative-Data-and-Analyses.pdf. 6 Dec 2017

Bazeley P. Analysing mixed methods data. In: Andrew S, Halcomb EJ, editors. Mixed methods research for nursing and the health sciences. Oxford: Blackwell; 2009. p. 84–118.

Bazeley P, Kemp L. Mosaics, triangles, and DNA: metaphors for integrated analysis in mixed methods research. J Mixed Methods Res. 2012;6(1):55–72. https://doi.org/10.1177/1558689811419514. Retrieved from http://jmmr.sagepub.com. 5 Dec 2017

Benner P. From novice to expert. Manlo Park: Addison-Wesley; 1984.

Bergman M. The straw men of the qualitative-quantitative divide and their influence on mixed method research. In: Bergman M, editor. Advances in mixed method research. London: Sage; 2008. p. 11–21.

Brannen J. Research note. The study of sensitive topics. Sociol Rev. 1988;36:552–63.

Bryman A. Integrating quantitative and qualitative research: how is it done? Qual Res. 2006;6(1):97–113. London: Sage

Bryman A. Barriers to integrating quantitative and qualitative research. J Mixed Methods Res. 2007;1(1):8–22. https://doi.org/10.1177/2345678906290531. Sage Publications. Retrieved from http://jmmr.sagepub.com. 5 Dec 2017

Bryman A. Social research methods. 5th ed. Oxford: Oxford University Press; 2016.

Campbell DT, Fiske DW. Convergent and discriminate validation by multitrait multidimensional matrix. Psychol Bull. 1956;56:81–105.

Cipriani R. Sociologia cualitativa. Las historias de vida como metodología científica. Buenos Aires: Editorial Biblos; 2013.

Cohen Orantes I. El concepto de integración, Revista de la CEPAL n° 15. Chile: ONU- Santiago; 1981. p. 149–59. Retrieved from http://repositorio.cepal.org/bitstream/handle/11362/10232/015149159_es.pdf?sequence=1. 7 Nov 2017

Comisión Europea. Libro Verde de la Innovación. Luxembourg; 1995. Retrieved from http://sid.usal.es/idocs/F8/FDO11925/libroverde.pdf. 24 Nov 2017.

Creswell J. Research design: qualitative, quantitative and mixed methods approach. 4th ed. Thousand Oaks: Sage; 2018.

Creswell J, Felters M, Ivankova N. Designing a mix-methods study in primary care. Ann Fam Med. 2004;2(1):7–12.

Denzin N. The research act: a theoretical introduction to sociological methods. New York: Praeger; 1978.

Denzin N, Lincoln YS, editors. Handbook of qualitative research. London: Sage; 2005.

Dogan M, Pahre R. Las nuevas ciencias sociales. La imaginación creadora. México: Grijalbo; 1993.

Fetters M, Curry L, Creswell JW. Achieving integration in mixed methods designs – principles and practices. Health Serv Res. 2013;48(6 Pt 2):2134–56. https://doi.org/10.1111/1475-6773.12117.

Fielding N. Triangulation and mixed methods designs: data integration with new research technologies. J Mixed Methods Res. 2012;6:124–36. https://doi.org/10.1177/1558689812437101. Published online 30 Mar 2012. Retrieved from http://mmr.sagepub.com/content/early/2012/03/28/1558689812437101. 5 Nov 2017

Fielding N, Fielding JL. Linking data: the articulation of qualitative and quantitative methods in social research. Beverly Hills: Sage; 1986.

Fielding J, Fielding N. Synergy and synthesis: integrating qualitative and quantitative data. In: Alasuutari P, Bickma L, Brannen J, editors. The sage handbook of social research methods. London: Sage; 2008. p. 1–30. Retrieved from http://epubs.surrey.ac.uk/231711/3/Brannen%20handbook%20chapter%2033%20for%20OA%20.pdf. 6 Dec 2017.

Fielding N, Schreier M. On the compatibility between qualitative and quantitative research methods. Forum Qual Soc Res. 2001;2(1):1–13. http://www.qualitative-research.net/index.php/fqs/article/%20view/965/2106, Retrieved 5 Feb 2018.

Giorgi A. Psychology as a human science. New York: Harper & Row; 1970.

Glaser BG. Theoretical sensitivity. Advances in the methodology of grounded theory. Mill Valley: Sociology Press; 1978.

Grbich C. Integrated methods in health research. In: Liamputtong P, editor. Research methods in health: foundations for evidence-based practices. Melbourne: Oxford University Press; 2017. p. 361–74.

Green S. What do you mean "What's wrong with her?": stigma and the lives of families of children with disabilities. Soc Sci Med. 2003;57(8):1361–74.

Greene J. Mixed methods in social inquiry. San Francisco: Jossey-Bass; 2007.

Guba EG. The alternative paradigm dialog. In: Guba EG, editor. The paradigm dialog. Newbury Park: Sage; 1990. p. 17–30. Retrieved from https://books.google.com.ar/books?hl=es&lr=&id=n1ypH-OeV94C&oi=fnd&pg=PA9&dq=Guba+EG.+The+alternative+paradigm+dialog.+In:+Guba+EG,+editor.+The+Paradigm+Dialog.+Newbury+Park,+CA:+Sage%3B+1990.+pp.+17%E2%80%9330.&ots=ISo03viJUE&sig=36uGVUTI8g-iZqN873zWSj6jgF0#v=onepage&q&f=false. 5 Nov 2017.

Hubbard G, Backett-Milburn K, Kemmer D. Working with emotion: issues for the researcher in fieldwork and teamwork. Int J Soc Res Methodol. 2001;4(2):119–37. https://doi.org/10.1080/13645570175015886. Retrieved from https://www.researchgate.net/profile/Gill_Hubbard/publication/248988604_Working_with_emotion_Issues_for_the_researcher_in_fieldwork_and_teamwork/links/53ce90b80cf24377a65dcd94.pdf. 6 Dec 2017

Husserl E. Idéia de fenomenologia. Lisboa: Edições; 2001, p. 70.

Huston AC. Mixed methods in studies of social experiments for parents in poverty. In: Weisner TS, editor. Discovering successful pathways in children's development: mixed methods in the study of childhood and family life. Chicago: University of Chicago Press; 2005. p. 305–15.

Ivankova N, Kawamura Y. Emerging trends in the utilization of integration designs in the social, behavioral and health sciences. In: Tashakkori A, Teddlie C, editors. The sage handbook of mixed methods in social and behavioral research. 2nd ed. London: Sage; 2010. p. 581–611.

Jick TD. Mixing qualitative and quantitative methods: triangulation in action. Adm Sci Q. 1979; 24:602–11.

Johnson RB, Onwuegbuzie A. Mixed methods research: a research paradigm whose time has come. Educ Res. 2004;33(7):14–26. Published by: American Educational Research Association. Retrieved from http://mintlinz.pbworks.com/w/file/fetch/83256376/Johnson%20Mixed% 20methods%202004.pdf. 6 Nov 2017

Johnson RB, Onwuegbuzie A, Turner L. Toward a definition of mixed methods research. J Mixed Methods Res. 2007;1(2):112–33. Retrieved from https://www.researchgate.net/profile/R_Johnson3/ publication/235413072_Toward_a_Definition_of_Mixed_Methods_Research_Journal_of_Mixed_ Methods_Research_1_112-133/links/55d0cd5308ae6a881385e669.pdf. 6 Nov 2017

Jones N, Summer A. Does mixed methods research matter to understanding childhood well-being?. WeD Working Paper 40. ESRC. Research Group on Wellbeing in Developing Countries. December. Bath: University of Bath-UK; 2007. p. 1–36. Retrieved from http://www.bath.ac. uk/soc-pol/welldev/research/workingpaperpdf/wed40.pdf. 12 Oct 2017.

Kelle U. Sociological explanations between micro and macro and the integration of qualitative and quantitative methods. Forum Qual Soc Res. 2001;2:1–22. http://www.qualitative-research.net/ index.php/fqs/article/view/966/2109, Retrieved 5 Feb 2018.

Kwan M-P. Is GIS for women? Reflections on the critical discourse in the 1990s. Gend Place Cult. 2002;9:271–9.

Lincoln YS, Guba EG. Naturalistic inquiry. Beverly Hills: Sage; 1985.

Maxwell J. Using numbers in qualitative research. Qual Inq. 2010;16:475–82. Retrieved from http://journals.sagepub.com/doi/pdf/10.1177/1077800410364740. 5 Nov 2017

May K. Conocimiento abstracto: un caso a favor de la magia en el método. In: En Morse J, editor. Asuntos críticos en los métodos de investigación cualitativa. Medellín: Editorial Universidad de Antioquia; 2006. p. 14–27.

Moran-Ellis J, Alexander VD, Cronin A, Dickinson M, Fielding J, Sleney J, Thomas H. Following a thread – an approach to integrating multi-method data sets, paper given at ESRC Research Methods Program, Methods Festival Conference, Oxford. 2004 July.

Moran-Ellis J, Alexander VD, Cronin A, Dickinson M, Fielding J, Sleney J, Thomas H. Triangulation and integration: processes, claims and implications. Qual Res. 2006;6(1):45–59.

Morse J, editor. Asuntos críticos en los métodos de investigación cualitativa. Colombia: Editorial Universidad de Antioquia; 2003.

Morse J, Field P. Qualitative research methods health professionals 2nd ed. Thousand Oaks/ London/New Delhi: Sage; 1995.

Pawson R. Quality and quantity, agency and structure, mechanism and context, dons and cons. BMS Bull Methodologie Sociol. 1995;47:5–48.

Punch K. Introduction to social research quantitative and qualitative approaches. 2nd ed. London: Sage; 2005.

Ray M. La riqueza de la fenomenología: preocupaciones filosóficas, teóricas y metodológicas. In: En Morse J, editor. Asuntos críticos en los métodos de investigación cualitativa. Colombia: Editorial Universidad de Antioquia; 2003. p. 140–57.

Rodríguez Herrera A, Alvarado Uriarte H. Claves de la innovación social en América Latina y el Caribe. Santiago de Chile: CEPAL; 2008. Retrieved from http://repositorio.cepal.org/bitstream/ handle/11362/2536/S0800540.pdf?sequence=1. 24 Nov 2017

Sandelowski M, Voils CI, Knafl G. On quantitizing. J Mixed Method Res. 2009;3:208–22.

Shih F. Triangulation in nursing research. J Adv Nurs. 1998;28:631–41. Retrieved from http:// onlinelibrary.wiley.com/doi/10.1046/j.1365-2648.1998.00716.x/full. 5 Nov 2017

Smith J, Hershusius L. Closing down the conversation: the end of the quantitative–qualitative debate among educational enquirers. Educ Res. 1986;15:4–12.

Sotolongo Codina P, Delgado Díaz C. La revolución contemporánea del saber y a complejidad social. Buenos Aires: CLACSO; 2006.

de Souza Minayo MC. Los conceptos estructurantes de la investigación cualitativa. Salud colectiva. 2010;6 (3, sep./dic.):251–261. Retrieved from http://www.scielo.org.ar/pdf/sc/v6n3/v6n3a02. pdf. 7 Dec 2017.

Tonon G. Qualitative studies in quality of life methodology and practice. Social Indicators Research Series, vol. 55. Dordretch; 2015.

Webb EJ, Campbell DT, Schwartz RD, Secherst L. Unobtrusive measures: non-reactive research in the social sciences. Chicago: Rand McNally; 1966.

Wolcott HF. Transforming qualitative data: description, analysis, and interpretation. Thousand Oaks: Sage; 1994.

Wright Mills C. La imaginación Sociológica. México: Fondo de Cultura Económica; 1961.

The Use of Mixed Methods in Research

40

Kate A. McBride, Freya MacMillan, Emma S. George, and Genevieve Z. Steiner

Contents

K. A. McBride (✉)
School of Medicine and Translational Health Research Institute, Western Sydney University,
Sydney, NSW, Australia
e-mail: K.Mcbride@westernsydney.edu.au

F. MacMillan
School of Science and Health and Translational Health Research Institute (THRI), Western Sydney
University, Penrith, NSW, Australia
e-mail: F.MacMillan@westernsydney.edu.au

E. S. George
School of Science and Health, Western Sydney University, Sydney, NSW, Australia
e-mail: E.George@westernsydney.edu.au

G. Z. Steiner
NICM and Translational Health Research Institute (THRI), Western Sydney University, Penrith,
NSW, Australia
e-mail: G.Steiner@westernsydney.edu.au

© Springer Nature Singapore Pte Ltd. 2019
P. Liamputtong (ed.), *Handbook of Research Methods in Health Social Sciences*,
https://doi.org/10.1007/978-981-10-5251-4_97

Abstract

Mixed methods research is becoming increasingly popular and is widely acknowledged as a means of achieving a more complex understanding of research problems. Combining both the in-depth, contextual views of qualitative research with the broader generalizations of larger population quantitative approaches, mixed methods research can be used to produce a rigorous and credible source of data. Using this methodology, the same core issue is investigated through the collection, analysis, and interpretation of both types of data within one study or a series of studies. Multiple designs are possible and can be guided by philosophical assumptions. Both qualitative and quantitative data can be collected simultaneously or sequentially (in any order) through a multiphase project. Integration of the two data sources then occurs with consideration is given to the weighting of both sources; these can either be equal or one can be prioritized over the other. Designed as a guide for novice mixed methods researchers, this chapter gives an overview of the historical and philosophical roots of mixed methods research. We also provide a practical overview of its application in health research as well as pragmatic considerations for those wishing to undertake mixed methods research.

Keywords

Mixed methods · Concurrent triangulation · Sequential exploratory · Sequential explanatory · Convergent parallel · Embedded design · Transformative design · Multiphase design

1   Introduction

Mixed methods research involves the systematic integration of quantitative and qualitative data within a single study, project, or program. This is based on the principle that integration provides a more complete and synergistic utilization of data, versus discrete quantitative and qualitative data collection and analysis. Mixed methods research methodology has advanced significantly in recent years (Creswell and Plano Clark 2018), with its application now bridging a broad range of research questions. Key characteristics of a mixed methods study include the collection and analysis of both quantitative and qualitative data, rigorous methods to collect and analyze the data, use of techniques to execute both data components (either concurrently or sequentially) among either the same participant sample or different samples, and integration of both types of data during data collection, analysis, or discussion. Mixed methods research has several advantages which are discussed throughout the following chapter, but most importantly, mixed methods provide a more complete picture than either qualitative or quantitative data could provide alone.

2   Historical and Philosophical Roots

In 1962, Thomas Kuhn introduced the concept of a paradigm, which has been described as "an agreed upon theory, worldview, or methodology embodied in the beliefs, practices, and products of a group of scientists" (Johnson and Gray 2010, p. 85). Mixed methods research has been recognized as the third methodological paradigm (in addition to the positivist quantitative and constructivist qualitative dichotomy) with applicability to health and social science research (Doyle et al. 2009). Tashakkori and Creswell (2007, p. 4) define mixed methods as "research in which the investigator collects and analyzes data, integrates the findings, and draws inferences using both qualitative and quantitative approaches or methods in a single study or a program of inquiry." The mixed methods approach combines distinct ideas and practices that distinguishes the approach from quantitative and qualitative paradigms (Denscombe 2008).

While debate exists around the timeframe in which mixed methods research emerged, the origins of mixed methods research can be traced back to the early twentieth century. Research between the 1950s and mid-1970s was primarily quantitative and dominated by the positivist paradigm, in which reality is believed to be measured and observed objectively (Denscombe 2008; Tariq and Woodman 2013). From the mid-1970s to 1990s, the constructivist paradigm, which is associated with qualitative research and the belief that there can be multiple realities and interpretations of the world (Appleton and King 2002), started to emerge as an alternative approach to research. The "paradigm wars" (Bryman 2008; Johnson and Gray 2010) is often recognized as the catalyst for the development of mixed methods research which began to emerge as a distinct paradigm from the 1990s onwards (Denscombe 2008).

Mixed methods have potential to combine the strengths of both quantitative (deductive) research and qualitative (inductive) research, while offsetting some of the weaknesses associated with each individual approach. Pragmatism is considered the "philosophical partner for the mixed methods approach" (Denscombe 2008, p. 273) as it provides a practical method of inquiry and set of assumptions that underpins the mixed methods approach (Johnson and Onwuegbuzie 2004). Pragmatism is outcome-oriented and requires researchers to engage in an iterative process, as they are forced to evaluate their beliefs and approaches in relation to the success and applicability of their practical application.

3   Designing Mixed Methods Research Questions

Plano Clark and Badiee (2010) identify three key elements central to the success of a research study or program – the content area, the purpose of the study, and the research questions. On the surface, developing a research study or program incorporating these three elements seems straightforward. However, the developmental process can be challenging as there are many different approaches for a researcher to consider. The first step in conceptualizing a mixed methods research study is to start with the big picture by identifying the overall content area. As a researcher, you might be interested in cancer screening, migrant health, dementia, or physical activity. While these topics are quite general, in this first step, it is acceptable to identify a broad, overarching topic as you will start to narrow down your research focus when developing your purpose statement and research questions.

As with any other research methodology, the researcher must have a strong understanding of their chosen topic early on, including the current research evidence and the key research problems. To ensure this, as with any other research study, it is important to conduct a robust review of relevant literature in the chosen topic area (for information on conducting database searching and collating research evidence see ▶ Chaps. 45, "Metasynthesis of Qualitative Research," and ▶ 46, "Conducting a Systematic Review: A Practical Guide"). To make a unique contribution to this content area, it is imperative that the research contributes new knowledge to the chosen area; fills a gap in current literature, policy, or practice; and/or identifies an underrepresented population group or issue (Plano Clark and Badiee 2010). It is, therefore, important to conduct your literature review early on, as this will inform development of the purpose statement and research questions.

Once the researcher has developed a strong understanding of the content area and any current deficiencies or opportunities for knowledge advancement in this area, it is time to start narrowing the focus by articulating the purpose of the research. The purpose outlines the study objectives and the primary intent of the research (Plano Clark and Badiee 2010) and sets the direction for the project. Some scholars argue that the need to use mixed methods should be determined by the purpose of the research. That is, if the purpose of the research indicates a need for both quantitative and qualitative approaches, the researcher must then select the most appropriate mixed methods approach (Teddlie and Tashakkori 2009).

Others suggest that it is the research questions, which are more defined and extend from the research purpose, that should dictate the choice in methodology (Johnson and Onwuegbuzie 2004). Bryman (2007), for example, notes that research questions play a pivotal role in the early stages of research, as decisions about the research design and methodology should be made to effectively answer research questions. This is particularly important in mixed methods research, where well-defined research questions guide and keep a study contained by helping to set boundaries and clear directions (Teddlie and Tashakkori 2009). Those who take a pragmatic viewpoint tend to abide by the so-called dictatorship of the research question (Tashakkori and Teddlie 2003, p. 65), considering the research question to be more important than the research method or the underlying paradigm (Tashakkori and Teddlie 2003). Newman et al. (2003), however, propose that researchers follow an iterative, systematic process by considering both the research purpose and research questions when deciding upon an appropriate methodological approach. This approach may also encourage researchers to further refine the focus of their study by developing deeper, more substantial research questions and a "greater awareness of potential multiple purposes" (p. 186).

Regardless of the approach a researcher takes, mixed methods research has been identified as an approach that has great potential to effectively combine insights from both quantitative and qualitative research into a workable solution, "offering the best opportunities for answering important research questions" (Johnson and Onwuegbuzie 2004, p. 16). Creswell and Plano Clark (2018) add that mixed methods research can help answer complex research questions that cannot be answered using quantitative or qualitative methods alone. This is particularly important in complex fields such as health and social science. For example, earlier we mentioned migrant health as a potential content area of interest. If we decided that the purpose of our study was to understand whether migrants in Western Sydney, Australia, are accessing health services, we could consider asking questions such as: *How frequently do migrants access health services in Western Sydney? Are migrant adults less likely than Australian-born adults to access health services?* We could answer these questions using a quantitative approach, by analyzing data on the number of migrants who access health services. From these data, we might be able to determine whether there are particular migrant groups who are more or less likely to access services, how many migrants are accessing these services, or whether services are more commonly accessed in specific areas, but this approach cannot tell us *why* this is the case. If we were interested in exploring migrant adults' perceptions of the Australian health system, the services they use regularly and the reasons for engagement, or the barriers they face in accessing health services, we would need to formulate qualitative research questions and use qualitative methods. Both of these approaches can offer important insights that relate to the overarching purpose of the study. The study's purpose then guides considerations for selecting the most appropriate mixed methods study.

4   Mixed Methods Study Designs

Each mixed methods study design can have different variations, purposes, philosophical assumptions, specific considerations, and strengths and weaknesses (Creswell and Plano Clark 2018). Traditionally, mixed methods study designs have been categorized into two main areas: sequential and concurrent (Castro et al. 2010). Sequential designs are characterized by either the qualitative or quantitative data collection being conducted first, which is then sequentially followed by a second (or further) stage of data collection of the other data type. Concurrent designs, on the other hand, involve the collection of both types of data simultaneously. These data collection approaches have contributed to the development of four major design categories: convergent, explanatory, exploratory, and embedded (Creswell and Plano Clark 2018).

4.1   Sequential Explanatory Design

Sequential *explanatory* design studies have a two-phase design, where quantitative data collection precedes qualitative data collection. The secondary qualitative data collection phase is used to either explain or further explore the quantitative findings in more depth, as well as aid with the interpretation of the quantitative results (Center for Innovation in Teaching in Research 2017). Sequential explanatory studies can be particularly useful when trying to explain relationships in quantitative data and provide more detail on the mechanisms of those relationships, particularly for surprising or unexpected results (Creswell and Plano Clark 2018). Other strengths of the explanatory sequential design include its straightforward nature (conceptually easy to design, logistically simple to implement, and ease of reporting), its emergent approach (the design of the qualitative phase can be based on the findings from the quantitative phase), its appeal to quantitative researchers as findings can be explored in more depth, and that it can be conducted by a sole investigator (Creswell and Plano Clark 2018). Weaknesses of the sequential explanatory design include challenges in gaining ethics approval for the entire project at the beginning (as researchers may not know how the participant selection for the second phase will be conducted until the first phase has been completed), subjectivity in deciding which quantitative findings require further explanation and who to sample, and that it can be time consuming (due to the two phases of data collection) (Ivankova et al. 2006).

Despite being time consuming, sequential explanatory designs are a commonly used design for randomized controlled trial (RCT) feasibility studies, amongst other applications. The EASE Back study – Evaluating Acupuncture and Standard care for pregnant women with BACK pain – utilized a sequential explanatory approach to inform feasibility of a large randomized controlled trial (RCT) (Foster et al. 2016). Quantitative data on acupuncture and standard care among pregnant women with low back pain was first collected via a survey. This data then informed collection of focus group and interview data, which explored the perceptions of midwives and pregnant women on acceptability and feasibility of acupuncture for the treatment of low back pain during pregnancy. Together both types of data provided a robust rationale for the proposed RCT as well as information on feasibility, recruitment, and consent procedures for the trial. A pilot study of a pedometer-based walking intervention for older adults delivered via primary care in the UK (Mutrie et al. 2012) reported a significant increase in objectively measured physical activity. Focus groups with participants and primary care staff following the intervention identified key elements of the intervention essential to its success but also highlighted the issues of moving such an intervention into routine practice. Nurse-led one-to-one physical activity consultations were highlighted by all focus group participants as being essential to the intervention's success, whereas the use of pedometers was important for some but not all. Walking groups initiated as part of the intervention were poorly attended but were reported as a vital motivator for the small number of regular attendees. Thus, use of a sequential explanatory approach here allowed the specific active ingredients of the intervention to be teased out (Mutrie et al. 2012).

4.2   Sequential Exploratory Design

Studies with a sequential *exploratory* design are very similar to explanatory study designs except that qualitative data are collected first, and this is then followed by a phase of quantitative data collection. Sequential exploratory designs are most frequently used to identify variables for further exploration (hence the exploratory design), development of an instrument (e.g., a survey), or a classification or theory for testing (Center for Innovation in Teaching in Research 2017). The purpose of the sequential exploratory design is to facilitate the generalizability of results from a small qualitative phase to a larger sample. The strengths of the sequential exploratory design are similar to sequential explanatory design, for example, appeal to quantitative researchers (as qualitative results can be substantiated in the second phase) and that it is also straightforward to implement. An additional strength is that a new instrument, such as a survey, can be produced based on the results of the research process (Creswell and Plano Clark 2018). Weaknesses of the sequential exploratory design are also similar to that of the explanatory design, that is, ethics approval can be sometimes difficult to obtain, the approach can be time-consuming, and the selection of data for exploration in the second phase can be highly subjective. Furthermore, any instruments developed should be valid and reliable, thus additional validation studies are often required (Center for Innovation in Teaching in Research 2017).

As with sequential explanatory designs, sequential exploratory designs are also frequently used in health research, with a common application being the use of qualitative data to inform development of a survey instrument. Keeney et al. (2010) used this approach to examine knowledge, attitudes, and behaviors around cancer prevention among middle-aged individuals in Northern Ireland. This study conducted an in-depth qualitative investigation (through focus groups) on the attitudes and beliefs around cancer prevention behaviors before using the information collected to inform the development of a large-scale cross-sectional survey that ascertained generic beliefs to cancer prevention. Use of this research design facilitated collection of not only in-depth contextual information but also the development of a content valid survey, and thus collection of robust population-based data.

4.3 Concurrent Triangulation Design (Convergent Parallel Design)

The concurrent triangulation design (also known as convergent parallel design) (Castro et al. 2010) is a popular design type, which involves the implementation of both the quantitative and qualitative phases of the research at the same time. Studies with a concurrent triangulation design prioritize both quantitative and qualitative methods equally; however, results are separated for analyses, but then integrated, or triangulated, during interpretation (Tashakkori and Creswell 2007). The main purpose of the concurrent design is "to obtain different but complementary data on the same topic" (Morse 1991a, p. 122) and to bring together the strengths as well as dilute the weaknesses of separate quantitative and qualitative research approaches. Strengths of the concurrent triangulation design include its efficiency of time (both quantitative and qualitative data are collected simultaneously), data analysis can be conducted independently (unlike sequential designs), and that it is intuitive (Creswell and Plano Clark 2018). Despite its popularity, the concurrent triangulation design also has a number of weaknesses and can be challenging to work with. For example, methodological and interpretation complexities emerge when combining datasets due to differing data, samples, sample sizes, and possibility conflicting data (Tashakkori and Creswell 2007). Logistically, concurrent triangulation studies can be challenging in scope for a research team given the concurrent nature of quantitative and qualitative data collection. A concurrent nested design is similar to the concurrent triangulation design, only instead of weighting both quantitative and qualitative data sources equally, one is given more weight than the other (Castro et al. 2010).

Wellard et al. (2013) used a concurrent triangulation design to identify operational and staffing factors which may influence care quality for individuals living in residential aged care in Victoria, Australia. Cross-sectional survey data on diabetes knowledge of residential staff was collected during the same data collection period as qualitative data was collected via focus groups with a subsample of staff who had also completed the survey. A clinical audit of a subset of case files of residents in the aged care facilities were also examined for diabetes care approaches and incidents related to diabetes. Use of this methodology provided a mechanism that corroborated findings from what was a relatively limited sample size, as well as provides an insight into a range of issues around diabetes care in this setting.

4.4 Embedded Design

The embedded design involves embedding quantitative and qualitative data collection and analysis within a traditional research design (that can be either quantitative or qualitative). For example, researchers conducting an RCT wanting to examine participants' experiences during the trial (e.g., recruitment and retention processes to explore enrolment, compliance, and attrition) could conduct a qualitative exit interview to gather this information. It is important to note that the qualitative study (exit interview) is embedded within the primary quantitative study (RCT) and is not related to answering its aims (i.e., whether the treatment has efficacy), distinguishing the embedded design from the convergent design. Strengths of embedded designs include improvements to (and richer data from) the primary study, being able to incorporate the secondary phase of data collection with relatively little time and resources, and additional research outputs from a single study that may be publishable in higher impact journals (Tashakkori and Creswell 2007). Embedded designs have several limitations including adding complexity and participant burden to the design of the primary study, additional expertise required to analyze the secondary data type (e.g., clinical trialists may need to upskill in qualitative research design, collection, and analysis), and the potential for treatment bias (this can be mitigated by collecting the additional data after the primary endpoint assessment has been completed).

Hoddinott et al. (2009) embedded a qualitative data collection element into their embedded mixed methods cluster RCT, which investigated clinical effectiveness and maternal satisfaction with provision of extra breastfeeding groups for breastfeeding women in Scotland. The primary outcome was the number of babies being breastfed at 5–7 days, 6–8 weeks, and at 8–9 months (ascertained via routinely collected national databases). Qualitative case study data on operational factors that may affect delivery of the extra breastfeeding groups were collected from a small proportion of primary care staff. These data suggested that limited staff resources, organizational change, and style of management and leadership all affected implementation of the cluster RCT. Addition of these qualitative data partly explained why there was no increase in breastfeeding rates in this area of Scotland despite a small increase in the breastfeeding groups.

4.5   Transformative Design (Sequential or Concurrent)

Studies with a transformative design have two phases and can either be sequential or concurrent (Center for Innovation in Teaching in Research 2017). However, the nature and order of data collection is theory-driven, and integration of results occurs during the interpretation phase (Center for Innovation in Teaching in Research 2017). The transformative lens is used to advance the needs of marginalized or underrepresented populations through a theoretical framework (e.g., feminist theory, disability theory, racial theory, sexual orientation theory) (Mertens 2012). In transformative designs, mixed methods are utilized for ideological purposes, rather than for logistical or methodological reasons (Creswell and Plano Clark 2018). Strengths of transformative designs include the action-oriented nature of the research, empowering communities to action change, and the participatory nature of the involvement of participants in the research (Tashakkori and Creswell 2007). There are several weaknesses associated with transformative designs. Relationships need to be built with participants to establish trust, additional justification is required for using transformative methods, and there is not much structure and guidance on the best way to utilize transformative approaches (Tashakkori and Creswell 2007).

Overarching theories and frameworks can also guide mixed methods studies, such as a quasi-experimental study that utilized mixed methods to explore the effectiveness of a walking program in a disadvantaged area of Brazil (Baba et al. 2017). A sequential transformative approach was undertaken by Baba et al. with the Reach, Effectiveness, Adoption, Implementation, and Maintenance [RE-AIM] framework for evaluating health interventions guiding each phase of their research. The intervention consisted of practical physical activity sessions plus education on behavior change strategies. Quantitative data were used to measure the reach of the intervention to the target group (% of people at the first day of the program/those that were potentially aware of the program through the recruitment strategies), the efficacy of the intervention on increasing physical activity (using a subjective physical activity questionnaire and accelerometers as an objective measure), as well as adoption of the intervention by health professionals at health centers (using quantitative questionnaires). The program was found to increase physical activity over 6 months using the quantitative measures. Despite this, the intervention was not sustained by the health centers, due to workload burden, low attendance, and limited infrastructure and human resources, as identified in the qualitative component.

Without the quantitative component, it may have been assumed the program lacked efficacy. Without the qualitative component, it may have been concluded that the program could be effectively translated into practice over the long term without issues in maintenance (Baba et al. 2017).

4.6   Multiphase Design (Sequential or Concurrent)

Projects with a multiphase design involve an iterative sequence of connected quantitative and qualitative studies that can be either sequential or concurrent. Each study builds on the previous, and together they aim to answer an overarching research question. Multiphase designs are particularly useful in large projects. There are a number of strengths associated with the multiphase design including the provision of a large framework that allows the advancement of an entire program of research, the flexibility of multiple elements of mixed methods research design, applicability in the areas of program evaluation and development, and multiple research outputs (Creswell and Plano Clark 2018). Multiphase designs are also challenging to implement and have several limitations including adequate time and resources (e.g., multiple ethics applications) are required given the scope of projects, problems need to be anticipated early to ensure the adequate adjustment of subsequent studies and the integration and connection of the overall package of the research, and the translation of findings needs to be a continuing focus of the project (Tashakkori and Creswell 2007).

5   Choosing a Mixed Methods Study Design and Getting Started with Your Research

Selection of study design will be partly dependent on the time available for data collection. For example, concurrent approaches are the least time consuming as both types of data are collected at the same time but require more resources at a single time point. Consideration should be given to what is the most practical way of beginning the study, your own research skills (whether you are stronger qualitatively or quantitatively), resources available, and the questions that are being asked.

5.1   Procedural Diagrams

Once the study design has been decided on, the next step in any mixed methods study is to draw a procedural diagram which should include information on data collection, data analysis, and data interpretation. This type of diagram can be very useful in visualizing all the components of a mixed methods research study, can assist with planning of a project, and can easily be constructed in PowerPoint or Word. When drawing a diagram, using notation that will be recognized by other mixed methods researchers (particularly if you are planning to publish your work) is important. One of the more commonly used notation systems is the one developed by Morse in 1991 where use of capitalization indicates the prioritization given to each form of data (Morse 1991b). The symbols "+" and "→" also indicate sequence, with "+" meaning concurrent and "→" indicating sequential data collection. So, for example, if you were prioritizing quantitative data over qualitative data but collecting the data concurrently, the notation would be:

$$QUAN + qual$$

If on the other hand, you were planning to give equal weight to both, but collect quantitative data first, your notation would be:

$$QUAN \rightarrow QUAL$$

Other key elements to include in a procedural diagram are boxes to indicate data collection and analysis for both types of data, circles for interpretation phases of the study, arrows to indicate study sequence, and products and procedures listed as bullet points alongside the boxes (Ivankova et al. 2006). If you were planning a convergent parallel study (Creswell and Plano Clark 2018), for example, with equal weight for each type of data and interpretation at the discussion stage, the diagram would be:

Note that in addition to this basic procedural diagram design, it can also be useful to add in the procedures and/or products of the research at each stage (Creswell and Plano Clark 2018).

5.2 Study Design Description

At the planning stage, it is also useful to write a succinct paragraph describing the study design, including identification of the design (and variant design if applicable), explanation of distinctive features of the design including weight given to each data type, timing of data collection and integration plans, and a rationale for study design selection. This paragraph should be included at the start of the methods section of a paper. It is important not only for self-clarification on appropriate study design selection but is also useful for potential reviewers given many are still unfamiliar with mixed methods research designs.

5.3 Mixed Methods Data Collection and Sampling

Once these initial steps have been carried out, the next stage is to start data collection based on the sequence defined in the study design, either concurrent or sequential data collection. Qualitative and quantitative sampling and data collection are fully described elsewhere in this textbook. Quintessentially, mixed methods data collection is informed by these standard data collection and sampling methods. It is important to be familiar with these methods, though the exact procedures are dependent on the type of mixed methods design chosen.

5.3.1 Concurrent Data Collection and Sampling

Concurrent data collection is used for concurrent triangulation designs as well as embedded designs and entails independent collection of both qualitative and quantitative data. Both types of data may be independently collected from the same participants in the same time period. For example, McBride et al. (2017) investigated the psychosocial effects of annual whole body MRI screening among individuals diagnosed with a cancer predisposition syndrome by independently collecting qualitative interview data and quantitative survey data within the same time period from the same purposively sampled group of participants. Collecting data from the same participants allows for the data to be converged more easily (Creswell and Plano Clark 2018).

Both types of data may be collected from different participants at the same time, though there should be consideration of confounding when comparing data due to the introduction of individual differences (Creswell and Plano Clark 2018). However, collecting from different groups of participants can be useful in comparing viewpoints. This approach is often used to compare clinician and patient perspectives. Harding et al. (2013) recently used this approach to examine the clinician and patient perspectives of a new triage model designed to reduce waiting times for community rehabilitation programs. In that study, qualitative data were collected from both clinicians and patients at the same time as program appointment waiting times. During data collection, either equal or unequal weight can be given to each type of data. Concurrent embedded designs, for example, often give lesser weight to qualitative data collection during clinical trials.

5.3.2 Sequential Data Collection and Sampling

Several guidelines inform the collection of data and sampling for sequential study designs (sequential explanatory, sequential exploratory, or sequential embedded designs) as these data are collected in stages. Data collected sequentially is not independently collected but instead is related, with one type of data informing collection of the other (Teddlie and Yu 2007). Qualitative or quantitative data can be collected first, with either the first or second data collection weighted more heavily (Creswell and Plano Clark 2018). Sampling of the same or different participants for each phase of the study is also dependent on the study design, with the approach used informed by the research question and emphasis on each type of data in the study.

As detailed above, an exploratory design collects qualitative data first, an approach used by Haider et al. (2017) to investigate a patient centered approach to collection of sexual orientation data in an emergency department (ED). Qualitative interview data were first collected via purposive sampling of patients with ED experiences and ED clinical staff. Once analyzed, these data were then used to inform the development of a national quantitative survey, with participants randomly sampled using random digit dialing and address-based sampling techniques. As described earlier, an explanatory design collects quantitative data first, with the data used to inform qualitative data collection and sampling. Prades et al. (2017) used an explanatory sequential approach to examine mode of radiotherapy delivery among women with breast cancer by first collecting quantitative data on mode of radiotherapy delivery across several radiotherapy centers, before using these data to inform purposive sampling of department heads for qualitative data collection.

As this brief overview of data collection and sampling techniques indicates, there are a number of approaches to data collection and sampling that can be adopted in mixed methods research. Sampling and data collection can be further complicated with the introduction of more advanced variants of mixed method study design. Selecting a sampling and data collection scheme involves complex decision-making not just on sample size but also on how participants will be selected and the circumstances under which selection will happen. Onwuegbuzie and Collins's paper (Onwuegbuzie and Collins 2007) contains a useful and detailed exposition of these considerations.

5.4   Mixed Methods Data Analysis and Integration

Analysis of mixed methods data initially involves individual analysis of both qualitative and quantitative data components (see also ► Chaps. 47, "Content Analysis: Using Critical Realism to Extend Its Utility," ► 48, "Thematic Analysis," ► 49, "Narrative Analysis," and ► 54, "Data Analysis in Quantitative Research"). It is important to have skills in both types of data analysis in mixed methods research. Analysis timing, order, and integration timing are informed by the rationale for the mixed methods study itself, the overall study design, and the data collection order. Generally speaking, as with data collection and sampling, data is analyzed either sequentially or concurrently (Fig. 1). For example, if a study is conducted sequentially, then data are usually analyzed following each phase with this analysis used to

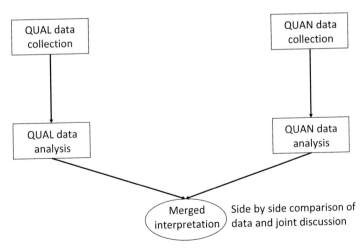

Fig. 1   Procedural diagram convergent parallel study

inform the next data collection and sampling phase. If a study is conducted concurrently, data analysis takes place at the same time with data integration happening immediately following concurrent data collection.

Once the data have been analyzed, they need to be integrated. Integration, when done well, should optimize the strengths of each data collection approach whilst simultaneously minimizing any weaknesses. Effective integration is key to maximizing the mixed methods approach. The concept of data integration can be one of the most challenging concepts in mixed methods research, given that it involves the reconciliation of words and numbers. How data are integrated largely depends on the type of mixed methods research design used. Integration can be done in several ways including merging or consolidation of data, connection of data, and embedding of data (Caracelli and Greene 1993; Creswell and Plano Clark 2018).

1. **Data merging** is where both datasets are jointly reviewed and combined. This form of integration is also sometimes called **data transformation,** which essentially is the conversion of one type of data in to the other so it can be analyzed as one dataset. An example would be transforming qualitative interview data into numeric ratings or converting quantitative data into qualitative narrative (Caracelli and Greene 1993). This integration approach is generally used for concurrent designs as sequential designs are not appropriate for data merging (triangulation), as findings from the first data collection could influence findings from the second data collection, which may result in bias (Onwuegbuzie and Collins 2007).

2. **Connecting the data** is where the analysis of one type of data is used to guide subsequent data collection with no direct comparison of results. This type of integration is used in sequentially designed projects.

3. **Embedding of data** is when more weight is given to one type of data; the other data type is embedded within the first. This approach is commonly used in clinical trials where a small qualitative element is used to supplement data from the trial.

5.5   Reporting of Mixed Methods Data

As well as the commonly expected standards necessary for the write up of any scholarly research, there are additional considerations in the reporting of mixed methods research. First, given that mixed methods research can be complex and often poorly understood, it is usually necessary to define the choice of research design and an emphasis placed on the usefulness of the methodology. Including procedural diagrams (discussed earlier in the chapter) can greatly assist in illustrating research designs. Second, diligent use of subheadings which separate the quantitative and qualitative data collection and analysis can also aid in the understanding of a mixed methods study report. Third, purpose statements for use of each type of data, as well as coherence and cohesiveness of reporting between both data types throughout, will give clarity to the rationale for selecting both data types. Last, any report should be structured in line with each major mixed methods design, with a well-thought-out structure to facilitate the reader's understanding (Creswell and Plano Clark 2018; see also ▶ Chaps. 56, "Writing Quantitative Research Studies," and ▶ 60, "Appraising Mixed Methods Research").

6 Challenges in Using Mixed Methods

Mixed methods studies are challenging to plan and implement, especially when they are used to evaluate complex interventions. The main challenges include (Curry et al. 2013):

1. *Aligning the research aims with methods and the research team's capacity:* A rationale for why a mixed methods approach is appropriate and necessary to answer the research question is not always provided in papers. Multi-expertise teams are necessary to conduct mixed methods research, but having such diverse teams can be challenging in handling data as researchers may be biased towards the use of qualitative or quantitative data, particularly if they do not have much mixed methods experience. However, having an interdisciplinary team involved in analysis can be a strength in facilitating rigorous interpretation of the data from several disciplinary viewpoints.
2. *Following best practice in methodology for each element of the research to ensure rigor throughout*: Rigor is defined differently for quantitative and qualitative research, and where multi-expertise teams are involved, it can be challenging to ensure that guidelines are strictly followed for the respective study type (e.g., sample size requirements are different).
3. *Ensuring appropriate integration of qualitative and quantitative elements:* For integration to be evident, one type of data needs to explain, enhance, confirm, challenge, or quantify the other type of data or one element of the research needs to precede another element, which may lead to subsequent new understanding. Deciding how and when to integrate qualitative and quantitative substudies can also be difficult to decide. This should be driven by the aim and objectives of the study.
4. *Following guidelines for writing up mixed method work:* With journal word count limits, it can be challenging to fully report on mixed method research in a single paper. Specific guidance for those writing up mixed method studies has been published, such as Good Reporting of Mixed Methods Studies, GRAMMS (O'Cathain et al. 2008). Quality and rigor of mixed methods research is highly variable in the current literature. Transparency in reporting is essential but often difficult to juggle with word count limits. Reporting guidelines exist to encourage standard reporting and should be followed to assist in the planning phase of studies as well as when writing up completed work (see also ▶ Chap. 60, "Appraising Mixed Methods Research").

Additional challenges include that mixed methods studies can take longer to conduct than stand-alone qualitative or quantitative studies and often require more funding and resources. Furthermore, limited methodological guidelines on the conduct of mixed method studies exist. Due to the diversity in such research, researchers have acknowledged that such guidelines would be extremely complex to develop (Zhang 2014). Furthermore, as well as journal word count limits, publication is a challenge in mixed methods research, particularly in the health sciences, where few journals publish this type of study design.

7   Handling Conflicting Results from Mixed Method Studies

The combination of qualitative and quantitative data can lead to different conclusions being drawn instead of a reliance on one type of data alone. Using a framework to assess and explain discordance in results is important for methodological rigor. Moffatt et al. (2006) used such an approach in their RCT which explored the impact of a primary care-based welfare rights intervention on health inequality in older adults from the UK, which collected qualitative and quantitative data concurrently. Impact of the intervention was measured quantitatively using a questionnaire which included subscales on health and wellbeing, lifestyle behaviors, psychosocial interaction, as well as socioeconomic status. Qualitative semistructured interviews on a subset of participants were also conducted to examine perceptions of the intervention, intervention outcomes, and acceptability of the research. Findings from the quantitative data revealed no significant differences in the impact of the intervention between the intervention and control groups. The qualitative data, however, found that the intervention was viewed positively by all those interviewed in the intervention group and that the intervention had important, wide-ranging impacts on those receiving it. To address this discord in findings, the results of that study were further analyzed using this six-step framework:

1. *Explore the qualitative and quantitative data separately*: Each element answered distinct, though related, research questions and was therefore expected to return different information.
2. *Examine the methodological rigor of each subcomponent of the study*: This included identifying possible reasons why measureable effects may not have been identified in the quantitative dataset. In the qualitative subcomponent, it may have been possible that the subsample differed in some way to the larger sample participating in the intervention study.
3. *Compare the participants from each subcomponent*: This revealed no differences between the quantitative and qualitative participants based on social and economic factors or on baseline health and wellbeing data.
4. *Collect further data to allow more comparisons*: Longer-term data were collected at 24 months, which supported the initial findings of each substudy.
5. *Examine if the intervention functioned as planned*: Unexpectedly, the qualitative interviews identified that prior to the study, many participants were in receipt of welfare benefit for types of services other than the target of this intervention.

6. *Identify if quantitative and qualitative elements match*: The qualitative interviews identified several areas of health and well-being that were not captured by the questionnaires and some areas of mental health were not being interpreted and captured as expected in the quantitative questions. Also, some of the questionnaires were perhaps not as appropriate for this older age target group who reported in the qualitative interviews that they did not expect to see improvements in their physical health at their age (but rather improvements in coping and managing health issues).

Following this framework, the authors were able to conclude that there was a disjunction between the outcomes being assessed quantitatively in the RCT and the outcomes discussed as being important by participants in the qualitative study. This highlights the importance of including mixed methods in evaluations of complex interventions (Moffatt et al. 2006).

8   Conclusion and Future Directions

Integration of quantitative and qualitative data in mixed methods studies has the potential to add rigor and richness to a research inquiry. However, it can be a complex process from the onset given that to date; there is a lack of consensus on methodological approaches. Nonetheless, if the most appropriate and practicable study design is used for the research question, data integration is rigorous and results well reported a deeper and more meaningful understanding of a diverse range of research objectives can be achieved.

Complex and chronic health problems, influenced by aging populations, income disparities, poor lifestyle behaviors, and rising urbanization are becoming increasingly common at a global level. Mixed methods health research will be of growing importance in the future due to the need for an enriched understanding of these health problems. Only by utilizing a pragmatic, mixed methods approach can the complex contributing factors to chronic health conditions be identified, and the active ingredients of interventions be fully understood through comprehensive evaluation.

References

Appleton JV, King L. Journeying from the philosophical contemplation of constructivism to the methodological pragmatics of health services research. J Adv Nurs. 2002;40(6):641–8.

Baba CT, Oliveira IM, Silva AEF, Vieira LM, Cerri NC, Florindo AA, de Oliveira Gomes GA. Evaluating the impact of a walking program in a disadvantaged area: using the RE-AIM framework by mixed methods. BMC Public Health. 2017;17(1):709. https://doi.org/10.1186/s12889-017-4698-5.

Bryman A. The research question in social research: what is its role? Int J Soc Res Methodol. 2007;10(1):5–20. https://doi.org/10.1080/13645570600655282.

Bryman A. The end of the paradigm wars. In: Alasuutari P, Bickman L, Brannen J, editors. The Sage handbook of social research methods. Thousand Oaks: Sage; 2008. p. 13–25.

K. A. McBride et al.

Caracelli VJ, Greene JC. Data-analysis strategies for mixed-method evaluation designs. Educ Eval Policy Anal. 1993;15(2):195–207. https://doi.org/10.3102/01623737015002195.

Castro FG, Kellison JG, Boyd SJ, Kopak A. A methodology for conducting integrative mixed methods research and data analyses. J Mixed Methods Res. 2010;4(4):342–60. https://doi.org/10.1177/1558689810382916.

Center for Innovation in Teaching in Research. Choosing a mixed methods design. 2017. Retrieved 15 Nov 2017 from https://cirt.gcu.edu/research/developmentresources/research_ready/mixed_methods/choosing_design.

Creswell JW, Plano Clark VL. Designing and conducting mixed methods research. 3rd ed. Thousand Oaks: Sage; 2018.

Curry LA, Krumholz HM, O'Cathain A, Clark VLP, Cherlin E, Bradley EH. Mixed methods in biomedical and health services research. Circ-Cardiovasc Qual Outcomes. 2013;6(1):119–23. https://doi.org/10.1161/circoutcomes.112.967885.

Denscombe M. Communities of practice: a research paradigm for the mixed methods approach. J Mixed Methods Res. 2008;2(3):270–83.

Doyle L, Brady A-M, Byrne G. An overview of mixed methods research. J Res Nurs. 2009;14(2):175–85.

Foster NE, Bishop A, Bartlam B, Ogollah R, Barlas P, Holden M, … Young J. Evaluating acupuncture and standard carE for pregnant women with back pain (EASE back): a feasibility study and pilot randomised trial. Health Technol Assess. 2016;20(33):1–236. https://doi.org/10.3310/hta20330.

Haider AH, Schneider EB, Kodadek LM, Adler RR, Ranjit A, Torain M, … Lau BD. Emergency department query for patient-centered approaches to sexual orientation and gender identity the EQUALITY study. JAMA Intern Med. 2017;177(6):819–828. https://doi.org/10.1001/jamainternmed.2017.0906.

Harding KE, Taylor NF, Bowers B, Stafford M, Leggat SG. Clinician and patient perspectives of a new model of triage in a community rehabilitation program that reduced waiting time: a qualitative analysis. Aust Health Rev. 2013;37(3):324–30. https://doi.org/10.1071/ah13033.

Hoddinott P, Britten J, Prescott GJ, Tappin D, Ludbrook A, Godden DJ. Effectiveness of policy to provide breastfeeding groups (BIG) for pregnant and breastfeeding mothers in primary care: cluster randomised controlled trial. Br Med J. 2009;338:a3026, 10. https://doi.org/10.1136/bmj.a3026.

Ivankova NV, Creswell JW, Stick SL. Using mixed-methods sequential explanatory design: from theory to practice. Field Methods. 2006;18(1):3–20. https://doi.org/10.1177/1525822x05282260.

Johnson B, Gray R. A history of philosophical and theoretical issues for mixed methods research. In: Tashakkori A, Teddlie C, editors. Sage handbook of mixed methods in social and behavioral research. 2nd ed. Thousand Oaks: Sage; 2010. p. 69–94.

Johnson BR, Onwuegbuzie AJ. Mixed methods research: a research paradigm whose time has come. Educ Res. 2004;33(7):14–26. https://doi.org/10.3102/0013189x033007014.

Keeney S, McKenna H, Fleming P, McIlfatrick S. Attitudes to cancer and cancer prevention: what do people aged 35–54 years think? Eur J Cancer Care. 2010;19(6):769–77. https://doi.org/10.1111/j.1365-2354.2009.01137.x.

McBride KA, Ballinger ML, Schlub TE, Young MA, Tattersall MHN, Kirk J, et al. Psychosocial morbidity in TP53 mutation carriers: is whole-body cancer screening beneficial? Familial Cancer. 2017;16(3):423–32. https://doi.org/10.1007/s10689-016-9964-7.

Mertens DM. Transformative mixed methods: addressing inequities. Am Behav Sci. 2012;56(6):802–13. https://doi.org/10.1177/0002764211433797.

Moffatt S, White M, Mackintosh J, Howel D. Using quantitative and qualitative data in health services research – what happens when mixed method findings conflict? BMC Health Serv Res. 2006;6:28. https://doi.org/10.1186/1472-6963-6-28.

Morse JM. Approaches to qualitative-quantitative methodological triangulation. Nurs Res. 1991a;40(2):120–3.

Morse JM. Principles of mixed methods and multimethod research design. In: Tashakkori A, Teddilie C, editors. SAGE handbook of mixed methods in social and behavioral research. Thousand Oaks: Sage; 1991b.

Mutrie N, Doolin O, Fitzsimons CF, Grant PM, Granat M, Grealy M, et al. Increasing older adults' walking through primary care: results of a pilot randomized controlled trial. Fam Pract. 2012; 29(6):633–42. https://doi.org/10.1093/fampra/cms038.

Newman I, Ridenour C, Newman C, De Marco G. A typology of research purposes and its relationship to mixed methods in social and behavioral research. Thousand Oaks: Sage; 2003. p. 167–88.

O'Cathain A, Murphy E, Nicholl J. The quality of mixed methods studies in health services research. J Health Serv Res Policy. 2008;13(2):92–8. https://doi.org/10.1258/jhsrp.2007. 007074.

Onwuegbuzie AJ, Collins KMT. A typology of mixed methods sampling designs in social science research. Qual Rep. 2007;12(2):281–316.

Plano Clark VL, Badiee M. Research questions in mixed methods research. In: Tashakkori A, Teddlie C, editors. SAGE handbook of mixed methods in social and behavioral research. 2nd ed. Thousand Oaks: Sage; 2010. p. 275–304.

Prades, J., Algara, M., Espinas, J. A., Farrus, B., Arenas, M., Reyes, V., . . . Borras, J. M. (2017). Understanding variations in the use of hypofractionated radiotherapy and its specific indications for breast cancer: a mixed-methods study. Radiother Oncol, 123(1), 22–28. doi:https://doi.org/ 10.1016/j.radonc.2017.01.014.

Tariq S, Woodman J. Using mixed methods in health research. JRSM Short Rep. 2013;4(6):1–8. https://doi.org/10.1177/2042533313479197.

Tashakkori A, Creswell JW. Editorial: the new era of mixed methods. J Mixed Methods Res. 2007; 1(1):3–7. https://doi.org/10.1177/2345678906293042.

Tashakkori A, Teddlie C. The past and future of mixed methods research: from data triangulation to mixed model designs. In: Tashakkori A, Teddlie C, editors. Handbook of mixed methods in social & behavioral research. Thousand Oaks: Sage; 2003. p. 671–701.

Teddlie C, Tashakkori A. Foundations of mixed methods research: integrating quantitative and qualitative approaches in the social and behavioral sciences. Thousand Oaks: Sage; 2009.

Teddlie C, Yu F. Mixed methods sampling a typology with examples. J Mixed Methods Res. 2007; 1(1):77–100. https://doi.org/10.1177/2345678906292430.

Wellard SJ, Rasmussen B, Savage S, Dunning T. Exploring staff diabetes medication knowledge and practices in regional residential care: triangulation study. J Clin Nurs. 2013; 22(13–14): 1933–40. https://doi.org/10.1111/jocn.12043.

Zhang W. Mixed methods application in health intervention research: a multiple case study. Int J Mult Res Approaches. 2014;8(1):24–35. https://doi.org/10.5172/mra.2014.8.1.24.